MODERN INORGANIC CHEMISTRY

J. J. LAGOWSKI

Department of Chemistry
The University of Texas at Austin

MARCEL DEKKER, INC.   **New York**

MARCEL DEKKER, INC.
270 Madison Avenue, New York, New York 10016

Library of Congress Catalog Card Number: 72-90374
ISBN: 0-8247-1391-5

Current printing (last digit):
10 9 8 7 6 5 4 3 2

PRINTED IN THE UNITED STATES OF AMERICA

PREFACE

In the recent past the study of inorganic chemistry has reachieved a level of interest comparable to that exhibited in the earlier phases of the development of the subject of chemistry as a whole. This renewed interest has come to pass largely because of the parallel development of numerous—practically useful—theoretical principles and experimental techniques for synthesis, separation, and structure determination. As a result, an enormous amount of information has accrued in the past twenty years in those general areas classically associated with inorganic chemistry. In addition, compounds containing metal-carbon bonds have received considerable attention from inorganic chemists during this time. Thus, inorganic chemistry as it is presently constituted is more properly described as the chemistry of the elements, with little distinction being drawn to the first member of Group IV in the periodic classification. This text is intended to give senior college or first-year graduate students, as defined in American universities, a working knowledge of the essence of modern inorganic chemistry.

The evolution of this volume has been dictated by two factors: the broad scope of inorganic chemistry and the time normally available in the chemistry curriculum for the study of inorganic chemistry. Under these constraints, the inevitable points of decision concerning the relative emphasis to be placed on certain areas are quickly reached. In some instances the decisions to exclude and/or minimize subjects were rather arbitrary; undoubtedly choices other than those presented could be successfully debated. In an attempt to offset partially some of the restrictions imposed by these constraints, a limited set of selected references to the research literature is included. Supplementary references to more extensive expositions—usually in the review literature—are also included. I hope that these devices will alleviate the shortcomings that some users may see in this volume because of the initial constraints placed upon it.

Since experiment and theory are intimately intertwined, several logically defensible strategies are available for a text of this nature. It is my personal preference to separate these two aspects of chemistry because it is possible to know the unity of the experimental results without using extensive theoretical arguments. Put in another way, a great many experimental results can be correlated, and predictions made on such correlations, without confusing the issue with theory. The point is that these facts still will be valid if current theories are unsuitable or, indeed, even do not exist. Hopefully the student will obtain an appreciation of the experimental results which form the basis for the theoretical discussions organized separately. In this respect several older—and presently outmoded—concepts have been introduced to give the student a sense of the evolutionary nature of theoretical arguments and to put the present theories in proper perspective. In a broad sense, it is often more important to understand the basis on which changes in theoretical outlook are made, whether they be drastic or subtle, than it is to know all of the details of the most modern theory.

I should like to acknowledge my indebtedness to Dr. Maurits Dekker for his suggestions at the inception of this volume and his encouragement during its development. Professor Gilbert Gordon read the manuscript in its entirety and offered numerous suggestions for its improvement, for which I express my appreciation. However, errors which may be present are mine alone. The publisher and his staff have provided invaluable assistance in the production of this volume. Finally, but most importantly, I must acknowledge my gratitude to my wife, Jeanne, who not only suffered the usual indirect consequences of living with an author, but also contributed materially by serving as resident professional reader, grammarian, editor, and typist. Without her continual encouragement and assistance, the preparation of this volume would never have been possible.

Austin, Texas   J. J. LAGOWSKI

CONTENTS

CHAPTER 10 / GROUP III REPRESENTATIVE ELEMENTS   255

CHAPTER 11 / GROUP IV ELEMENTS   329

MODERN
INORGANIC
CHEMISTRY

ATOMS
AND THEIR
CHARACTERISTICS

Atoms and their varied characteristics are the basic focus of interest of chemists. Since a vast array of chemical compounds with different properties and characteristics can be prepared from a relatively few atomic types, there is the hope that an understanding of the nature of atoms can lead directly to an elucidation of the chemical unity of the elements, the compounds they form, the structures of such compounds, and the properties, both physical and chemical, of these compounds. Thus it is fitting to begin this discussion with a consideration of the nature and properties of atoms as we presently perceive them.

1 / THE ARCHITECTURE OF ATOMS

Our knowledge of the microstructure of atoms stems from studies on (1) the conduction of electricity through gaseous matter, (2) radioactivity, (3) the

1

production of x rays, and (4) spectroscopy. It is not our purpose here to give a detailed description of all of the pertinent experiments, results, and reasoning in each of these areas which led to establishment of the present model of atomic structure. Rather, we present a synopsis of a few of the key experiments and their contribution to our understanding of atomic structure (*1*).*

1.1 / Electrical Discharge through Gases

Cathode Rays   Studies on the conduction of electricity through matter have provided considerable insight for understanding its constitution. In particular, gases at low pressure ($< 10^{-4}$ torr) conduct an electric current when subjected to a high potential difference; a schematic diagram of the apparatus used in experiments under these conditions is shown in Fig. 1.1. It

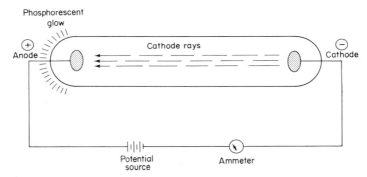

FIG. 1.1 A schematic diagram for an apparatus to create an electrical discharge through a gaseous sample. A phosphorescent glow appears about the anode when current flows through the circuit.

might appear that with a good vacuum in the tube no current would flow through the circuit. However, electricity does flow through the external circuit, as indicated by the ammeter in the diagram, implying that an equivalent charge must pass through the tube. In 1859 Plücker observed that the glass at the anode end of a discharge tube emits a phosphorescent glow which persists as long as the circuit is closed. He suggested that the glow arose from radiation, which he named cathode rays, emanating from the cathode. In addition to glass, substances such as zinc sulfide glow brightly when exposed to cathode rays. Crookes proposed that cathode rays were really particles, and Perrin showed that these particles carry a negative charge.

In a series of brilliantly conceived experiments J. J. Thomson elucidated the fundamental nature of cathode rays; a description of one of these experi-

---

* Collateral readings are indicated by Roman numerals and collected at the end of each chapter.

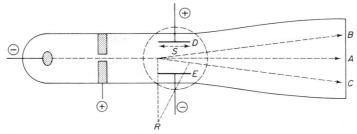

FIG. 1.2 A schematic diagram of the discharge tube used by Thomson to establish the q/m ratio for cathode rays. An electric field could be established perpendicular to the cathode ray beam by charging plates D and E. The dotted circle represents the position of the pole faces of a magnet which is perpendicular to the plane of the paper.

ments is sufficient to show the thrust of his logic (1).* Figure 1.2 is a schematic diagram of the discharge tube Thomson used to determine the ratio of charge to mass (q/m) of cathode rays. Cathode rays are generated in the left side of the tube, and they are accelerated toward the anode, which contains a hole that is collinear with the cathode. Cathode rays pass into the right side of the tube, where they strike the far wall which is coated with zinc sulfide. The tube also contains two parallel plates (D and E in Fig. 1.2) which can be charged by an external potential source. The dotted circle in Fig. 1.2 represents the pole faces of a magnet located above and below the tube. Thus, the cathode rays can be subjected to an electric field acting perpendicular to their path through the discharge tube and a magnetic field perpendicular to both of these directions. The experiment involves balancing the deflection of a magnetic field of known intensity by an electric field. The magnetic field causes the beam to deflect from point A along the line BC, the direction of the deflection depending upon the orientation of the magnetic poles. The beam can be brought back to its original position by applying an electrostatic field generated by charging the plates (D and E) in the appropriate direction. Applying the quantitative relationships which describe the interaction of an electric and a magnetic field with moving charge (a current), Thomson showed that under the conditions of balance the velocity v of the moving charge is related to the magnetic field H, the length S of the condenser plates, and the potential difference V between the plates:

$$v = V/HS. \tag{1}$$

Removal of the electric field while maintaining the magnetic field causes the cathode beam to deviate to point C. Under these conditions q/m is given by

$$\frac{q}{m} = \frac{V}{SH^2R}, \tag{2}$$

---

* References to the research literature are designated by arabic numerals and are collected at the end of each chapter.

where R, the radius of curvature of the deflected beam, can be calculated by simple geometry from the dimensions of the discharge tube and the distance CA. The results obtained for v (at a given potential difference between the cathode and anode) and q/m by this method were in agreement with those obtained from several other types of experiments conducted by Thomson. Of interest here is the comparison of the value obtained for q/m for cathode rays (2×10^{11} C kg^{-1}) with those obtained from the electrolysis of solutions of ionic substances. In 1833 Faraday had observed that the mass of a given substance liberated in electrolysis experiments is directly proportional to the quantity of electricity passed through the circuit. Thus, q/m ratios were known for many chemical species; a selection of values is given in Table 1.1.

TABLE 1.1   The charge-to-mass ratio (q/m) obtained for some ions from electrolysis experiments

Ion	q/m, C kg^{-1}
H^+	9.6×10^7
Cu^{2+}	3.0×10^6
Ag^+	8.9×10^5
Cd^{2+}	1.7×10^6
Fe^{3+}	5.2×10^6

A comparison of the q/m ratios of cathode rays with those for known ionic species strongly indicates that the former are not charged atoms. The ratio of q/m for cathode rays is about 2000 times larger than that of the lightest known ion (H^+), indicating that the particle has either a very small mass or an unusually high charge. Since charges on the majority of ionic species are small multiples of the charge of the hydrogen ion, Thomson reasonably assumed that the difference between the q/m ratios for chemical species and cathode rays primarily arises from a difference in mass. Cathode rays are free negatively charged particles, the discrete unit of negative electricity which Stoney had called an electron many years earlier.

Canal Rays   In 1886, Goldstein detected radiation emitted from a cathode but moving in the direction opposite to that of the cathode rays. A schematic diagram of the Goldstein discharge tube is shown in Fig. 1.3. The discharge tube contains a gas at low pressure, a perforated cathode A, and an anode. When a high potential is impressed across the electrodes, the usual effect of the cathode rays is observed on the glass around the anode. In addition, bundles of light appear in the other compartment, D, of the discharge tube. Goldstein observed that the color of these beams, or canal rays, varied with the gas, and Wein showed that they are deflected by magnetic and

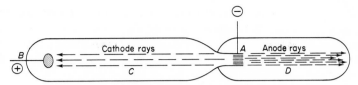

FIG. 1.3 A schematic diagram of a discharge tube designed to study canal, or anode, rays. The left side of the tube is characteristic of the design of a cathode ray tube. Cathode A is perforated. Anode rays which emanate from the cathode are studied in the right side of the tube.

electrostatic fields in opposite directions to those observed for cathode rays. Although Wein showed that the q/m ratios for canal rays depend upon the nature of the gas in the tube and are smaller than that observed for cathode rays, it was Thomson and his students who ultimately described the significance of the canal rays (*II*). Canal rays are positively charged species and are more massive than electrons.

The discharge experiments indicate that neutral matter consists of positively and negatively charged particles, the latter being found in all samples of matter. The nature of the positively charged particles varies with the sample of matter placed in the discharge tube.

Other Evidence for the Existence of Electrons   Two other observations which support Thomson's conclusion that electrons can be dislodged from neutral matter are available. In 1883 Edison discovered that a filament heated to incandescence produces particles which carry a negative charge. A schematic diagram of the apparatus appears in Fig. 1.4. If the filament A which is sealed in an evacuated tube is cold, the ammeter C indicates that no charge flows through the circuit. If the filament is heated to incandescence

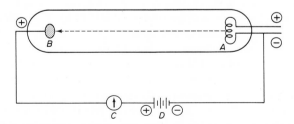

FIG. 1.4 A schematic diagram of the apparatus designed to study the thermionic effect. Heating the filament A gives electrons sufficient energy to escape from the metal surface; the potential difference across the tube established by the battery D accelerates these electrons to the anode. The fact that the internal circuit is completed is indicated by ammeter C.

using an external source of current, a current flows through the circuit if electrode B is connected to the positive pole of the battery. Reversal of the polarity of the battery stops the flow of current even though the filament is incandescent. These results were interpreted by postulating that the hot filament emits electrons which are then accelerated to the anode, thus completing the current.

The photoelectric effect is the second phenomenon which relates to the presence of electrons in matter. Figure 1.5 illustrates an apparatus with

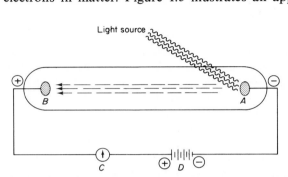

FIG 1.5   A schematic diagram of the apparatus used to study the photoelectric effect. When the photoelectrode A is irradiated by light, a current flows through the circuit as indicated by ammeter C.

which the photoelectric effect may be demonstrated. Electrode A is coated with an active metal such as an alkali or alkaline-earth metal or zinc, and the tube is evacuated. With the negative pole of the battery D connected to the active metal electrode, current does not flow through the circuit as long as electrode A is kept dark. A current flows in the system, however, when the active metal electrode is irradiated with visible light. Reversal of the polarity of electrode A causes the current to cease flowing. The photoelectric effect arises because electrons in the metal receive sufficient energy from the light beam to escape from the surface, after which point they are accelerated toward the positive electrode in the tube.

Quantitatively, the results of irradiating the photoelectrode with monochromatic light are shown in Fig. 1.6. When there is no potential difference impressed across the tube shown in Fig. 1.5 a steady current flows through the circuit (point A on the curve in Fig. 1.6). As the potential difference across the tube is increased, with the photoelectrode negative, a slight increase in current occurs (region B, Fig. 1.6). However, a potential is soon attained after which there is no further increase in current (region C of Fig. 1.6). If the potential difference across the tube is reversed, with the photoelectrode made positive, the photocurrent steadily decreases with increasing potential difference. Ultimately a potential difference is attained (point D of Fig. 1.6),

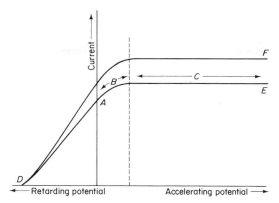

FIG. 1.6 The current–voltage curve observed for the photoelectric experiment. Curve F represents the results obtained with monochromatic light of a greater intensity than that used to obtain curve E.

where current ceases to flow; this is often called the retarding potential and represents the potential required to overcome the kinetic energy of the electrons emitted from the photoelectrode by the monochromatic beam of light. Interestingly, increasing the intensity of light does not increase the kinetic energy of the electrons emitted from the photoelectrode, since the retarding potential remains constant. The magnitude of the photocurrent does, however, increase with increasing intensity of light (F in Fig. 1.6). An increase in the frequency of the irradiating light leads to a linear increase in the retarding potential (Fig. 1.7a), and there is a frequency below which the photoelectrode is inactive; i.e., the radiation is not sufficiently energetic to dislodge electrons from the metal surface. This frequency (v_t in Fig. 1.7b) is called the threshold frequency, and it is characteristic of the photoelectrode material. Examples of threshold frequencies for several metals are listed in Table 1.2.

The fact that the retarding potential for a given frequency of light does not change with the intensity of the radiation is difficult to interpret on the basis of classical electromagnetic theory. Einstein in 1905 suggested a solution to this problem by assuming that the energy associated with electromagnetic radiation comes in discrete packets or quanta which he called photons. Since the kinetic energy of the photoelectron increases with increasing frequency, it was reasonable to suggest that the energy of the exciting photons increases with increasing frequency. The amount of energy per photon was taken as hv, where h is a fundamental constant of nature called Planck's constant of action ($h = 6.6238 \times 10^{-27}$ erg-sec) and v is the frequency of the radiation. Earlier, Planck had shown that the distribution of frequencies found in the electromagnetic radiation emitted from a heated body could be predicted if he assumed a similar relationship to that proposed by Einstein.

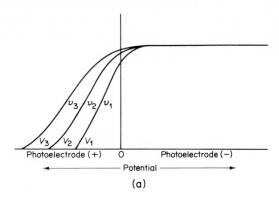

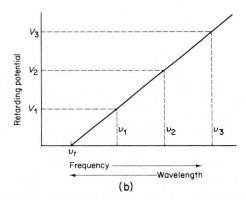

FIG. 1.7   (a) An increase in the frequency (ν_1, ν_2, ν_3) of light falling on the photo-electrode produces an increase in the retarding potential (V_1, V_2, V_3). (b) The relationship between frequency and retarding potential is linear.

TABLE 1.2   Threshold frequencies for some metals

Metal	Threshold frequency, $\times 10^{-13}$ Hz
Na	51.5
Al	63
Mg	98.5
Zn	80
Sn	83
Bi	91
Cu	100
Pt	104

Application of the principle of conservation of energy to the interaction of a photon with an electron in a metal surface leads to the following relationship:

$$KE = P - W, \tag{3}$$

where KE is the kinetic energy of the emitted electron, P is the energy of the photon, and W is the energy required to separate an electron from the metal surface. This last quantity is called the work function, and is characteristic of the particular metal. According to Einstein's postulate the energy of the photon is given by

$$P = h\nu, \tag{4}$$

and the kinetic energy of the electron can be expressed as

$$KE = \tfrac{1}{2}mv^2 = Ve, \tag{5}$$

where m and v represent the mass and velocity, respectively, of the photo-electron, V the retarding potential, and e the electronic charge. Equations (3), (4), and (5) lead to

$$eV = h\nu - W, \tag{6}$$

which can be rearranged as follows:

$$V = \frac{h\nu}{e} - \frac{W}{e}. \tag{7}$$

Thus, according to the Einstein interpretation of the photoelectric effect, the retarding potential should be a linear function of frequency, with slope h/e. Moreover, all metals should exhibit the same slope but different intercepts when the results of photoelectric experiments in which they act as photo-cathodes are plotted according to Eq. (7). These predictions are in agreement with experimental observations; the value of Planck's constant as determined from Eq. (7) for a variety of metals is in good agreement with the values obtained for this constant from a variety of other experimental methods.

1.2 / The Nuclear Atom

The experiments discussed in the previous section indicate that atoms contain positive and negative particles. Thomson suggested that atoms were spheres of "uniform positive electrification" which contained the smaller negatively charged electrons moving about within them. The clue to the distribution of fundamental particles within an atom was provided by Geiger, who reported the scattering of α particles, which had been identified previously as He^{2+}, as they passed through a thin metal foil (2). The apparatus used by Geiger in these experiments (Fig. 1.8) consisted of a glass tube about

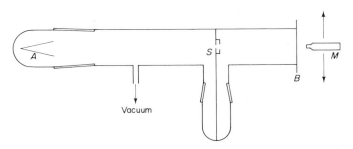

FIG. 1.8   A schematic diagram of the apparatus used by Geiger to study the scattering of α particles as they pass through a metal foil. *A* represents the source of α particles, *S* a collimating slit, and *B* a zinc sulfide screen. Thin metal foils can be placed behind the slit.

two meters long, which could be evacuated, containing a source (*A*) of α particles at one end and a zinc sulfide screen (*B*) at the other end. The beam of α particles from the source passes through a collimating slit (*S*) and strikes the screen where the individual α particles can be detected with the aid of a traveling microscope (*M*). With a good vacuum practically no scintillations are observed on the screen outside the geometrical image of the slit. A thin metal foil is placed over the slit and the number of scintillations at definite intervals from the center is observed. The graph in Fig. 1.9 illustrates the type of data Geiger obtained in this experiment. Curve *A* is the distribution with no foil in the slit; curves *B* and *C* are obtained with one and two thicknesses of gold foil over the slit. These results indicate that α particles are scattered strongly as they pass through dense matter. In other experiments of this type Geiger and Marsden showed that the degree of scattering of α particles depends upon the atomic weight of the metal comprising the foil.

In 1911 Rutherford proposed the presently accepted model of the nuclear atom to explain the results of the Geiger–Marsden experiments (3). The observation that most α particles passed through metal foils suggests that they are small compared to the size of the atom. The fact that some α particles suffer large deflections leads to the hypothesis that the region of the atom which causes the deflection is small in comparison to the size of the atom and carries the same charge as the α particle. The observed experimental results are compatible with a nuclear model which contains a nucleus about 10^{-4} times smaller than atomic dimensions ($\approx 10^{-8}$ cm).

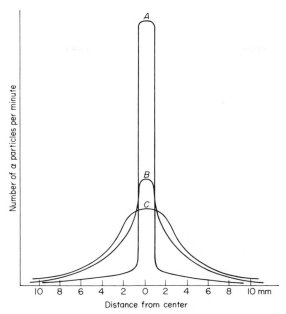

FIG. 1.9   The results of the α-scattering experiment using gold foil. Curve *A* represents the scattering with no foil behind the slit; curve *B*, with one thickness of foil in the beam; and curve *C*, with two thicknesses of foil present.

The Properties of X Rays   The observation that the magnitude of α-particle scattering depends upon the atomic weight of the scattering atom also suggests that the charge on the nucleus increases with atomic weight. A study of the nature of x rays by Moseley provided the basis for determining the nuclear charge (*3–5*). X rays are generated when high velocity electrons strike a target (Fig. 1.10); a narrow beam of x rays can be obtained by passing the initial wide beam through two slits (S_1 and S_2). X rays are electromagnetic radiation; i.e., they are not deflected by magnetic or electric fields, but a beam of x rays can be separated into its component wavelengths with suitable diffracting elements. Moseley employed a crystal of potassium ferrocyanide as a diffracting unit (*A* in Fig. 1.10) and a photographic plate to detect the x-ray spectrum of an element. Qualitatively, Moseley's experimental results indicated that x rays produced from a metal target formed a continuum with discrete lines (peaks) superimposed on it (Fig. 1.11). The breadth of the continuum depends upon the potential difference impressed across the x-ray tube, but the wavelength of the lines depends upon the nature of the target material. The x-ray spectra of the elements exhibited a regularity which is simply described by the following equation:

$$v^{1/2} = a(N - b), \tag{8}$$

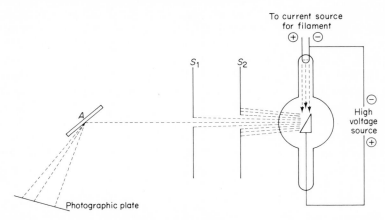

FIG. 1.10 A schematic diagram of the apparatus used to generate x rays. Electrons from a heated filament are accelerated to an anode by a high potential difference, generating x rays when they strike the anode. S_1 and S_2 are slits used to collimate the x-ray beam. A is a diffracting element.

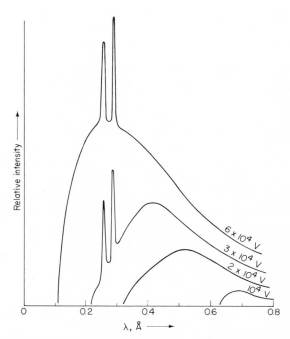

FIG. 1.11 The x-ray spectrum of silver. Below 2×10^4 V the spectrum is featureless, but two intense lines appear at higher potentials.

where v is the frequency of the x-ray line, a and b are constants characteristic of the spectral line used in establishing the relationship, and N is an integer characteristic of the atom. The integer N, starting with a value of 1 for hydrogen, became associated with the number of positive charges, or protons, in the nucleus. Moseley predicted the existence of several elements unknown in his time on the basis of discontinuities observed in the x-ray spectra of the known elements.

The Neutron   The experiments discussed thus far not only yield a gross description of atomic structure but also suggest that an additional fundamental atomic particle exists. For example, several methods assign an atomic weight of ≈ 4 to helium, but its atomic number of 2 indicates that it contains two protons and two electrons and should have an atomic weight of ≈ 2. The discrepancy was accounted for by postulating the existence of a neutron, a neutral nuclear particle with a mass number of approximately 1. Using an argument based on energy conservation, Chadwick (6) identified the neutron in the penetrating radiation emitted when beryllium is irradiated with α particles:

$$^{11}\text{Be} + {}^{4}\text{He} \longrightarrow {}^{14}\text{N} + {}^{1}n. \tag{9}$$

1.3 / The Bohr Atom and Quantum Theory

The Bohr Atom   The Rutherford nuclear atom provides the theoretical basis for understanding much of chemistry. Moseley's atomic number, which is characteristic of atomic species, represents (indirectly) the number of electrons in the extranuclear region of a neutral atom. The number and distribution of extranuclear electrons are of interest because the electron configurations dictate the physical and chemical characteristics of the atoms. Our present description of extranuclear electronic distributions arises from considering the source of the lines observed in the emission spectra of the elements. The original interpretation of emission spectra using quantum theory was given by Bohr (7). Although the Bohr theory of atomic structure is no longer significant in modern chemistry, it does represent an important point in the development of chemical theory. As early as 1811, Helmholtz suggested that electrical forces play an important role in the formation of molecules. The Bohr theory focused attention on the extranuclear region of atoms with far-reaching consequences.

The key experimental result on which the Bohr theory focuses is the fact that the emission spectrum of each element consists of distinctive groups of lines characteristic of the element. The lines within a group usually exhibit a relatively simple relationship. For example, the general regularity observed for the hydrogen spectrum in the visible–ultraviolet region (the Balmer series, shown in Fig. 1.12) is simply expressed by the empirical relationship

$$\bar{v} = R\left(\frac{1}{n'^2} - \frac{1}{n^2}\right),\qquad(10)$$

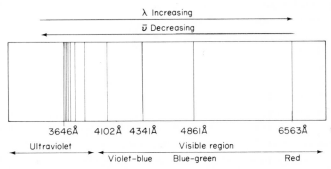

FIG. 1.12 The Balmer series of lines in the spectrum of atomic hydrogen. The series limit occurs at 3646 Å.

where $\bar{v}$ is the position of a line in wave numbers ($\bar{v} = 1/\lambda$), R is an empirical constant determined to be 109,737.309 cm^{-1}, n' is an integer, and n an integer dependent upon n' as follows:

$$n = (n' + 1), (n' + 2), (n' + 3), \ldots .\qquad(11)$$

Four series of lines exist in the hydrogen spectrum (Fig. 1.13), each with characteristic values of n' and n.

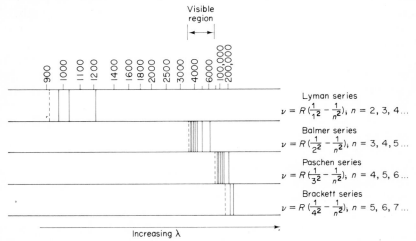

FIG. 1.13 The four series of spectral lines for hydrogen atoms. The series have been separated for clarity, and the dotted lines represent the series limit.

The fact that discrete lines are observed in the emission spectra of the elements indicates that processes occur within the atom which involve discrete amounts of energy. Moreover, these processes involve different amounts of energy for different atoms. It should be clear from the picture of the nuclear atom which is developed in the previous section that the extranuclear electrons (one, in the case of hydrogen) will be attracted towards the positively charged nucleus. If the electrons are stationary the system is unstable, since the electrons will be drawn into the nucleus. This simple argument leads to the suggestion that the extranuclear electrons move in an orbit-like motion which offsets the attractive force of the nucleus. However, a particle moving in a nonlinear orbit must be, according to classical mechanics, accelerating. Under these conditions the fundamentals of classical electrodynamics lead to an impossible situation, since an accelerating electric charge emits electromagnetic radiation. If the motion of the electron about the nucleus is constrained to a circular path, electrodynamic principles dictate that energy must be radiated from the system. Thus, the electron should eventually lose all its energy in the form of radiation and fall into the nucleus. If the constituent particles which comprise atoms obey classical laws, all atoms should be in various states of losing energy and the radiation emitted from a collection of atoms should contain all possible frequencies. Under such conditions atomic spectra should exhibit continua, a conclusion not reflected in experimental fact (Fig. 1.12). Bohr suggested an atomic model which incorporated quantum arguments to account for the line spectra of the elements.

The Bohr model of the hydrogen atom assumes that an electron of mass m and charge $-e$ moves with a velocity v about a nucleus with a charge of $+Ze$ and is constrained by the charge on the nucleus to a circular orbit (Fig. 1.14).

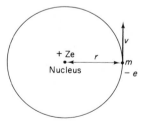

FIG. 1.14   In the Bohr model of the hydrogen atom, an electron with charge $-e$ and mass m is constrained to move about a nucleus with charge $+Ze$. At any moment the electron has a velocity v, the vector of which is tangential to the orbital path.

In the formalism of the Bohr theory the following postulates are made:

1. Atomic systems possess a number of states (called stationary states) for which no emission of radiant energy occurs even if the classical laws of

electrodynamics predict such to be the case because of the motion of charged particles relative to each other.

2. Emission or absorption of radiant energy corresponds to a transition between stationary states.

3. The equilibrium of the system in a stationary state is governed by Newton's laws of motion.

4. The angular momentum of an electron in the lowest stationary state is equal to $h/2\pi$; in any higher state the angular momentum is an integral number of times this basic unit $(nh/2\pi)$.

The first postulate eliminates the embarrassing conclusions which would follow if *some* of the classical laws were obeyed, the second postulate relates the model to spectroscopic observations; the third postulate provides a means of calculating certain parameters of the system; and the fourth postulate explicitly introduces the principal quantum number n which takes on integral values. The direct application (I) of these postulates to the model of the hydrogen atom shown in Fig. 1.14 leads to expressions for the radii of the allowed or Bohr orbits:

$$r = \frac{n^2}{z}\frac{h^2}{4\pi^2 e^2 m} \tag{12}$$

and for the energy of an electron in a stationary state:

$$E_n = -\frac{1}{n^2}\frac{z^2 2\pi^2 e^4 m}{h^2}. \tag{13}$$

Equations (12) and (13) indicate that the size of the Bohr orbits, as well as the energies of the stationary states, are quantized.

In the Bohr view of atomic structure, the spectrum of hydrogen arises because hydrogen atoms are excited to different stationary states. When each atom returns to the ground state a discrete amount of energy is emitted which is dependent upon the difference in energies of the two states. According to Planck the frequency of electromagnetic radiation is related to its energy by the relationship

$$E = h\nu. \tag{14}$$

The energy difference $h\nu$ between two states, which corresponds to a line in the spectrum, is given by

$$h\nu = E_n - E_{n'}. \tag{15}$$

Equations (13) and (15) lead to Eq. (16), which predicts the frequency dependence of the series of lines found in the spectrum of hydrogen shown in Fig. 1.13:

$$v = \frac{z^2 2\pi^2 m e^4}{h^3}\left(\frac{1}{n'^2} - \frac{1}{n^2}\right).$$

(16)

The energies of the stationary states and the assignment of the transitions for some of the lines in the hydrogen spectrum are given in Fig. 1.15. The Lyman, Balmer, Paschen and Brackett series of lines shown in Fig. 1.13 arise from transitions from higher energy levels with principal quantum numbers 1, 2, 3, and 4, respectively.

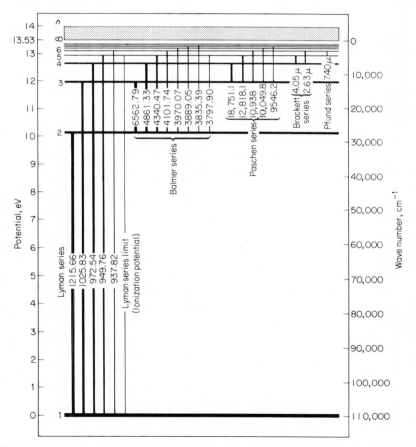

FIG. 1.15   The energy levels of the hydrogen atom as described by the Bohr theory. The wavelengths of the individual lines in each series are given in angstrom units, except in the cases of the Brackett and Pfund series, where they are expressed in microns.

The Bohr theory not only successfully predicts the gross features of the hydrogen spectrum shown in Fig. 1.13, but it also provides the value of the Rydberg constant R in Eq. (10). The relationship between the frequency v and wave number $\bar{v}$ is given by

$$\bar{v} = \frac{1}{\lambda} = \frac{v}{c}, \tag{17}$$

which permits Eq. (16) to be converted into

$$\bar{v} = \frac{z^2 2\pi^2 m e^4}{h^3 c}\left(\frac{1}{n'^2} - \frac{1}{n^2}\right). \tag{18}$$

A comparison of Eq. (18) with Eq. (10) shows that they are of the same form and that the Rydberg constant R can be identified with the factor $(z^2 2\pi^2 m e^4/h^3 c)$; evaluation of this factor using the best values of fundamental constants gives $109{,}700 \text{ cm}^{-1}$, which is in remarkable agreement with the experimental value.

The Bohr theory also successfully predicts other characteristics of the hydrogen atom. For example, the energy required to remove an electron from a hydrogen atom, as in Eq. (19):

$$H \rightarrow H^+ + e^- \tag{19}$$

(i.e., the atom's ionization potential) corresponds in the Bohr theory to the transition of an electron in the ground state ($n = 1$) to a state where it is not under the influence of the nucleus ($n = \infty$). This transition energy can be calculated from Eq. (18); under these conditions the factor $1/n^2$ goes to zero and the transition energy is expressed by

$$\bar{v} = \frac{z^2 2\pi^2 m e^4}{h^3 c} = R. \tag{20}$$

In other words the ionization potential of hydrogen corresponds to the Rydberg constant; the experimental value for the ionization potential as determined from a variety of techniques is 13.6 eV, whereas the value of the Rydberg constant expressed in the same units is 13.54 eV. Finally, the radius of the hydrogen atom in the ground state can be estimated from Eq. (12) to be 0.53 Å, which is of the order expected for atomic species.

The Bohr theory is applicable to any species which has a hydrogen-like structure, i.e., one electron associated with a nucleus. Thus, Eqs. (12), (13), and (16)–(18) are applicable to species such as He^+, Li^{2+}, Be^{3+}, etc., by using the appropriate value for the nuclear charge in these systems.

Modification of the Bohr Theory   The Bohr theory was outstandingly successful in describing hydrogen-like atoms, but it failed to account for the spectra of atomic systems containing more than one electron. In addition, other spectroscopic observations could not be accounted for by the simple Bohr theory. For example, certain lines in the hydrogen spectrum observed under high resolution consist of groups of closely spaced lines, and some lines in the spectrum split to several components when the emitting source is placed in a strong magnetic field. The appearance of several lines under certain conditions implies the existence of several energy levels with very similar energies.

In 1916 Sommerfeld suggested a modification of the Bohr theory to account for the multiplicity of lines observed under high resolution. Sommerfeld's extension provided for elliptical orbits, with the atomic nucleus at a focus, as well as for circular Bohr orbits. The shape of the elliptical orbit is governed by another quantum number, designated by l. The angular momentum of the electron in an elliptical orbit is related to l in much the same way as its angular momentum in a circular orbit is defined by Bohr's fourth postulate:

$$mvr = \frac{h}{2\pi}(l + 1)^{1/2}.$$ (21)

The quantum condition described by Eq. (21) restricts the electron to certain allowed ellipses, and the quantum number l defines the ellipticity of the orbit. In the Sommerfeld modification, the values of l, sometimes called the angular or secondary quantum number, are determined by the value of the principal quantum number n:

$$l = 0, 1, 2, 3, 4, \ldots, (n - 1).$$ (22)

In effect, the number of orbits of nearly equal energy is specified by the value of n; that is, for $n = 1$, one Bohr–Sommerfeld orbit is possible with $l = 0$; for $n = 2$, two Bohr–Sommerfeld orbits are possible ($l = 0, 1$), etc.

The splitting of certain spectral lines by a magnetic field, called the Zeeman effect (after its discoverer), suggests that energy levels are split when the emitting atoms are placed in a magnetic field. This phenomenon was interpreted in terms of a magnetic quantum number m which defined the orientation of the angular-momentum vector of an electron in a Bohr–Sommerfeld orbit with the external magnetic field. The fact that spectral lines are split into a finite number of discrete lines indicates that only a finite number of orientations are possible; i.e., the interaction of the angular momentum with the magnetic field is quantized. The values of the magnetic quantum number are integral and depend upon the value of l:

$$m = l, (l - 1), (l - 2), \ldots, 0, -1, -2, \ldots, -(l - 1), -l.$$ (23)

There are a total of $2l + 1$ possible values of m.

Finally, a fourth quantum number was suggested to account for a doubling of certain spectral lines, a phenomenon which could not be interpreted in terms of the previous quantum numbers. In 1925 Uhlenbeck and Goudsmit proposed that an electron spinning on an axis as it moved in a Bohr–Sommerfeld orbit would have two possible orientations, as shown in Fig. 1.16. The

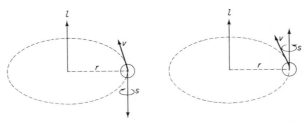

FIG. 1.16 The electron-spin momentum s can be oriented in one of two ways with respect to the orbital momentum l.

electron moving in an orbit exhibits an orbital angular momentum which is represented by a vector perpendicular to the plane of the orbit (l in Fig. 1.16). A spinning electron generates spin angular momentum, and the total energy of the system depends upon the orientation of the electron spin with respect to its orbital motion. There are two possible orientations of the electron spin which lead to two values of the spin quantum number s, *viz.* $+\frac{1}{2}$ or $-\frac{1}{2}$ in terms of $(h/2\pi)$-units of angular momentum.

The Bohr postulates lead to a resolution of the difficulties created by Rutherford's nuclear atom by breaking with classical mechanics and permitting atomic systems to possess only certain allowed—or quantized—energies. The quantum numbers in the Bohr–Sommerfeld model were devised to describe the energy of such systems. Unfortunately several other irreconcilable problems existed with the Bohr–Sommerfeld model for atomic structure.

2 / WAVE MECHANICS AND ATOMIC STRUCTURE (*III, IV*)

The basic difficulties of the Bohr–Sommerfeld theory lay in treating the electron as a discrete particle traveling along a predictable path. It was de Broglie's suggestion that electrons have a dual nature which provided an alternative to the Bohr–Sommerfeld atom.

2.1 / The Dual Nature of the Electron and the Schrödinger Wave Equation

The dual nature of light can be demonstrated by means of certain experiments: diffraction phenomena are best interpreted in terms of electromagnetic waves, while the photoelectric effect is more easily understood on a particulate basis. de Broglie suggested that certain properties of electrons can also be ascribed to a characteristic wave associated with electronic motion. The basic relationships between the wave nature of an electron and its particulate properties are given by

$$\lambda = \frac{h}{mc},\tag{24}$$

where λ is the wavelength of the "electronic wave," m is the mass of the electron, and c is the velocity of light. de Broglie's suggestions concerning the existence of "electronic waves" were verified when Davisson and Germer (8) discovered that electrons could be diffracted in the same manner as electromagnetic waves. Like electromagnetic waves, the waves associated with electronic motion can either reinforce each other to give a standing wave or interact in a destructive manner. For example, assume that the electronic motion is such that the associated waves are superimposed on each other; Fig. 1.17a is a two-dimensional representation of this process. For a

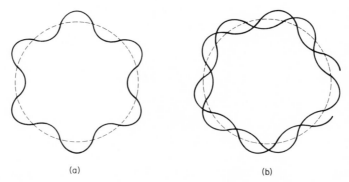

(a)   (b)

FIG. 1.17   The de Broglie waves for the circular orbits for a stationary state (a) and for an excluded state (b) where the waves are destroyed by interference.

given wavelength, there are only certain conditions for which the waves reinforce each other to form a standing wave; in all other cases, such as that shown in Fig. 1.17b, the same waves will interact destructively. It should be recalled from Eq. (14) that the frequency of an electromagnetic wave is related to its energy. Thus, there are only discrete energies associated with the electron which will give a standing wave. Just as the intensity of electromagnetic radiation is related to the number of photons in a light beam, the intensity or amplitude Ψ of the electronic wave is associated with the number of electrons moving along a given path.

The concept of waves associated with electrons, and its consequences, lead to a description of the atom different from that suggested by Bohr. In the wave description the exact positions of electrons involved are not as important as they are in the Bohr theory. The emphasis is placed on describing the waves associated with the motion of an electron as it moves about the nucleus. It is not necessary to consider all possible wave motions of the electron, since the only periodic motions that have a physical meaning are those in which the electronic waves reinforce each other to establish a standing wave (Fig. 1.17). An electron motion that gives a summation of waves that has a zero (or low) amplitude corresponds to an unfavorable electron energy. On the other hand, the motion for which the waves reinforce each other to give a standing wave corresponds to a favorable electron energy. Thus, the establishment of standing waves is equivalent to Bohr's condition that electrons in atoms have discrete energies.

The mathematical description of the waves associated with electronic motion about a nucleus was first given by Erwin Schrödinger in a general form:

$$\frac{\partial^2 \Psi}{\partial x^2} + \frac{\partial^2 \Psi}{\partial y^2} + \frac{\partial^2 \Psi}{\partial z^2} + \frac{8\pi^2 m}{h^2}(E - V)\Psi = 0. \tag{25}$$

At any point in space the three-dimensional wave associated with an electron possessing mass m, total energy E, and potential energy V will have a certain amplitude Ψ which is a function of the coordinates of that point. The solutions to the Schrödinger wave equation, called wave functions, are of the form $\Psi = f(x, y, z)$ and satisfy Eq. (25). Mathematical details for the method of obtaining the solutions of a partial differential equation such as Eq. (25) are not of primary interest here, but the results of this process are important to the modern chemist (V). Only a brief outline of the method of solution will be presented.

First, the Schrödinger equation is transformed into polar coordinates so that $\Psi = f(\theta, \phi, r)$, and an expression for the potential energy of the electron is introduced. If the potential energy of the electron arises from electrostatic interactions, V can be replaced by the quantity $-e^2/r$. Under these conditions the Schrödinger equation takes the form

$$\frac{1}{r^2}\frac{\partial}{\partial r}\left(r^2\frac{\partial \Psi}{\partial r}\right) + \frac{1}{r^2 \sin^2 \theta}\frac{\partial^2 \Psi}{\partial \phi^2} + \frac{1}{r^2 \sin \theta}\frac{\partial}{\partial \theta}\left(\sin \theta \frac{\partial \Psi}{\partial \theta}\right)$$

$$+ \frac{8\pi^2 m}{h^2}\left(E + \frac{e^2}{r}\right)\Psi = 0. \tag{26}$$

A series of wave functions Ψ_i exist that are solutions to Eq. (26). For convenience, the solutions are given in terms of a radial part, $R(r)$, and angular parts, $\Theta(\theta)$ and $\Phi(\phi)$:

$$\Psi_i(r, \theta, \phi) = R(r)\Theta(\theta)\Phi(\phi). \tag{27}$$

The general form of the solutions represented by Eq. (27) contains three quantum numbers, i.e., integers that may be assigned specific values. Since wave-mechanical quantum numbers exhibit the same mathematical interrelationships (Table 1.3) as those developed in the Bohr–Sommerfeld

TABLE 1.3   The relationships among the quantum numbers

Symbol	Name	Allowed values
n	principal	$1, 2, 3, 4, \ldots$
l	secondary	$0, 1, 2, 3, \ldots$
m	magnetic	$0, \pm 1, \pm 2, \ldots, \pm(l-2), \pm(l-1), \pm l$

theory, the same symbols are used. The quantum number n appears in the $R(r)$ portion of the wave function. The secondary quantum number l appears in the $\Theta(\theta)$ part of the wave function, and the magnetic quantum number m in the $\Phi(\phi)$ part. Fourteen solutions to the polar form of the Schrödinger equation for the hydrogen atom are shown in Table 1.4; since these wave functions include the explicit values of the quantum numbers given in Table 1.3, it is apparent that the wave functions can be identified by the values assigned to the quantum numbers n, l, and m. Generally the wave functions are identified by giving the value of the principal quantum number n and the value of the secondary quantum number l; the latter number has been historically designated by a letter according to the following scheme:

value of l	0 1 2 3 4 5 6
letter designation	s p d f g h i

2.2 / The Physical Significance of Ψ

The wave functions Ψ_i that are solutions to the wave equation [Eq. (26)] can, upon substitution into the latter, yield the corresponding values of the energy E_i of the stationary states. The energies that correspond to the solutions of the wave equation are, of course, identical to those determined from the Bohr theory for electrons occupying the corresponding Bohr orbits (Fig. 1.15).

TABLE 1.4 The angular part [$\Psi_e(\Theta, \Phi)$] and radial part [$\Psi_{ne}(R)$] of the hydrogen-like wave functions[a]

Quantum numbers			Angular part, $\Psi_e(\Theta, \Phi)$	Radial part,[b] $\Psi_{ne}(R)$	Orbital designation
n	l	m			
1	0	0	$\Psi(s) = \left(\dfrac{1}{4\pi}\right)^{1/2}$	$\Psi(1s) = 2\left(\dfrac{Z}{a_0}\right)^{3/2} e^{-\sigma/2}$	$1s$
2	0	0	$\Psi(s) = \left(\dfrac{1}{4\pi}\right)^{1/2}$	$\Psi(2s) = \dfrac{1}{2\sqrt{2}}\left(\dfrac{Z}{2a_0}\right)^{3/2}(2-\sigma)e^{-\sigma/2}$	$2s$
2	1	0	$\Psi(p_z) = \left(\dfrac{3}{4\pi}\right)^{1/2}\cos\Theta$	$\Psi(2p) = \dfrac{1}{2\sqrt{6}}\left(\dfrac{Z}{a_0}\right)^{3/2}\sigma e^{-\sigma/2}$	$2p_z$
2	1	1	$\Psi(p_x) = \left(\dfrac{3}{4\pi}\right)^{1/2}\sin\Theta\cos\Phi$	$\Psi(2p) = \dfrac{1}{2\sqrt{6}}\left(\dfrac{Z}{a_0}\right)^{3/2}\sigma e^{-\sigma/2}$	$2p_x$
2	1	-1	$\Psi(p_y) = \left(\dfrac{3}{4\pi}\right)^{1/2}\sin\Theta\sin\Phi$	$\Psi(2p) = \dfrac{1}{2\sqrt{6}}\left(\dfrac{Z}{a_0}\right)^{3/2}\sigma e^{-\sigma/2}$	$2p_y$
3	0	0	$\Psi(s) = \left(\dfrac{1}{4\pi}\right)^{1/2}$	$\Psi(3s) = \dfrac{1}{9\sqrt{3}}\left(\dfrac{Z}{a_0}\right)^{3/2}(6-6\sigma+\sigma^2)e^{-\sigma/2}$	$3s$
3	1	0	$\Psi(p_z) = \left(\dfrac{3}{4\pi}\right)^{1/2}\cos\Theta$	$\Psi(3p) = \dfrac{1}{9\sqrt{6}}\left(\dfrac{Z}{a_0}\right)^{3/2}(4-\sigma)\sigma e^{-\sigma/2}$	$3p_z$
3	1	1	$\Psi(p_x) = \left(\dfrac{3}{4\pi}\right)^{1/2}\sin\Theta\cos\Phi$	$\Psi(3p) = \dfrac{1}{9\sqrt{6}}\left(\dfrac{Z}{a_0}\right)^{3/2}(4-\sigma)\sigma e^{-\sigma/2}$	$3p_x$

TABLE 1.4—*continued*

Quantum numbers			Angular part, $\Psi_e(\Theta, \Phi)$	Radial part,[b] $\Psi_{n\ell}(R)$	Orbital designation
n	l	m			
3	1	-1	$\Psi(p_y) = \left(\dfrac{3}{4\pi}\right)^{1/2} \sin\Theta \sin\Phi$	$\Psi(3p) = \dfrac{1}{9\sqrt{6}}\left(\dfrac{Z}{a_0}\right)^{3/2}(4 - \sigma)\sigma e^{-\sigma/2}$	$3p_y$
3	2	0	$\Psi(d_{z^2}) = \left(\dfrac{5}{16\pi}\right)^{1/2}(3\cos^2\Theta - 1)$	$\Psi(3d) = \dfrac{1}{9\sqrt{30}}\left(\dfrac{Z}{a_0}\right)^{3/2}\sigma^2 e^{-\sigma/2}$	$3d_{z^2}$
3	2	1	$\Psi(d_{xz}) = \left(\dfrac{15}{4\pi}\right)^{1/2} \sin\Theta \cos\Theta \cos\Phi$	$\Psi(3d) = \dfrac{1}{9\sqrt{30}}\left(\dfrac{Z}{a_0}\right)^{3/2}\sigma^2 e^{-\sigma/2}$	$3d_{xz}$
3	2	-1	$\Psi(d_{yz}) = \left(\dfrac{15}{4\pi}\right)^{1/2} \sin\Theta \cos\Theta \sin\Phi$	$\Psi(3d) = \dfrac{1}{9\sqrt{30}}\left(\dfrac{Z}{a_0}\right)^{3/2}\sigma^2 e^{-\sigma/2}$	$3d_{yz}$
3	2	2	$\Psi(d_{x^2-y^2}) = \left(\dfrac{15}{4\pi}\right)^{1/2} \sin^2\Theta \cos 2\Phi$	$\Psi(3d) = \dfrac{1}{9\sqrt{30}}\left(\dfrac{Z}{a_0}\right)^{3/2}\sigma^2 e^{-\sigma/2}$	$3d_{x^2-y^2}$
3	2	-2	$\Psi(d_{xy}) = \left(\dfrac{15}{4\pi}\right)^{1/2} \sin^2\Theta \sin 2\Phi$	$\Psi(3d) = \dfrac{1}{9\sqrt{30}}\left(\dfrac{Z}{a_0}\right)^{3/2}\sigma^2 e^{-\sigma/2}$	$3d_{xy}$

[a] The solutions to the Schrödinger equation have the form $\Psi_i - \Psi_e(\Theta, \Phi)\Psi_{n\ell}(R)$.

[b] $\sigma = \dfrac{2Zr}{na_0},\ a_0 = \dfrac{h^2}{4\pi^2 me^2}.$

The wave functions are also related to the probability of finding an electron at a given point, the probability P being given by

$$P = \Psi(x, y, z)\, \Psi^*(x, y, z), \qquad (28)$$

where Ψ^* is the complex conjugate of Ψ. Since Ψ can have imaginary or negative values, the product of Ψ and Ψ^* gives a probability consisting of real numbers. If Ψ is related to probability, it must have certain mathematical characteristics: it must be single-valued at each point in space; it must not go to infinity at any point; and its absolute value at all points must be such that Eq. (29) is obeyed:

$$\int_{-\infty}^{+\infty} \int_{-\infty}^{+\infty} \int_{-\infty}^{+\infty} \Psi(x, y, z)\, \Psi^*(x, y, z)\, dx\, dy\, dz = 1. \qquad (29)$$

That is, the summation over all points of the probability of finding an electron at each point in space is unity. A wave function that satisfies the conditions expressed by Eq. (29) is said to be normalized.

There are several ways to gain insight into an understanding of the interpretation of the wave function. First, it is possible to consider the relative probabilities of finding an electron at various distances from the nucleus. If we are willing for the moment to ignore the angular dependence $[\Theta(\theta)\Phi(\phi)]$ of the wave function and concentrate only on the radial dependence $R(r)$ of Ψ, we see that Ψ can be either negative or positive (Fig. 1.18). A more meaningful

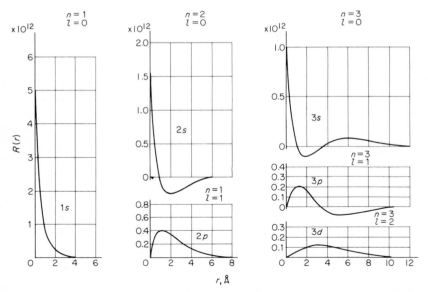

FIG. 1.18   The variation of the radial part $R(r)$ of the wave function can have either positive or negative values.

two-dimensional representation of Ψ comes from a plot of the radial distribution function $r^2 R(r)^2$ against r (Fig. 1.19). In essence the radial distribution function represents the probability Ψ^2 of finding an electron within a

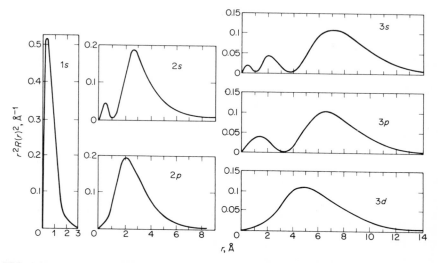

FIG. 1.19   The values of the radial distribution function $r^2 R(r^2)$ for hydrogen-like wave functions plotted against the distance from the nucleus.

spherical shell that has an inner radius r and an outer radius $(r + dr)$. It is interesting to note that for the ground-state wave function Ψ_{1s} the maximum in the radial distribution curve occurs at 0.53 Å, which is the same as the Bohr radius for the ground state of the hydrogen atom.

Although the radial distribution function is useful for many purposes, a representation of the electron probability in three dimensions is equally, if not more, important. Such a representation must, of course, include the angular parts of the wave function as well as the radial portion; since the angular dependence is the same for all values of r, a discussion of the angular parts of the wave function is sufficient to make the necessary points. The angular parts $\Theta(\theta)\Phi(\phi)$ of the wave function for $l = 0$, $l = 1$, and $l = 2$ are shown in Fig. 1.20. The surfaces of the three-dimensional figures depicted enclose a very large portion of the wave function ($\approx 90\%$); it should be noted that the sign of the wave function can vary in different portions of space.

As in the case of the radial functions, the angular dependence for the probability of finding an electron in space is proportional to $\Theta(\theta)^2\Phi(\phi)^2$, the square of the angular part of the function. The three-dimensional representation of these functions for $l = 0$, $l = 1$, and $l = 2$ appears in Fig. 1.21.

The wave-mechanical description of atomic structure has led to a development of simple representations for the solutions to the Schrödinger equation

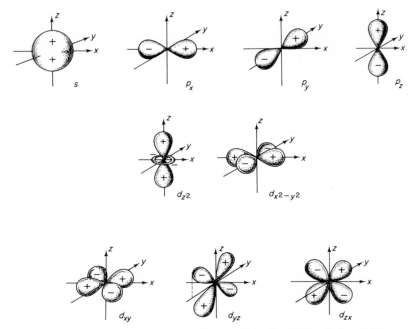

FIG. 1.20   The angular part $\Theta(\theta)\Phi(\phi)$ of the wave function for $l = 0$, 1, and 2 (s, p, and d orbitals, respectively). The surfaces enclose approximately 90% of the values. Note that the wave functions can have positive as well as negative values. The drawings are not to scale.

[Eq. (26)]. As mentioned in the previous section, the mathematical functions which satisfy Eq. (26) are known as wave functions (Table 1.4) and are represented by a unique set of quantum numbers. The set of solutions also leads to characteristic energies or energy levels, as well as to electron probability descriptions, which are called orbitals. In all of these uses of the wave function, the quantum number designation has been quite common. Thus, when we speak of a $1s$ ($n = 1$, $l = 0$) wave function, a $1s$ energy level, or a $1s$ orbital, we mean the wave function which is a solution to the Schrödinger equation that gives rise to a characteristic energy and a specific electron probability description.

The quantum numbers reflect certain geometrical characteristics of the electron probability distribution (Fig. 1.21). The secondary quantum number l is related to the general shape of the distribution, m is related to the orientation of the distribution with respect to an arbitrary axis, and n gives the relative size of the distribution. Although the concept of spin is not explicitly found in the Schrödinger equation, Dirac has demonstrated that the relativistic form of quantum mechanics can yield the property of spin without a separate hypothesis.

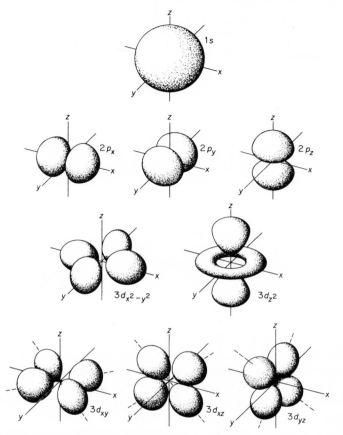

FIG. 1.21 The electronic probability function as calculated from the angular part of the wave function for $l = 0, 1$, and 2. The figures are not drawn to scale.

3 / MULTIELECTRON ATOMS

The relationship among the quantum numbers (Table 1.3) and the fact that the spin quantum number can have the values $\pm\frac{1}{2}$ leads to a description of the orbitals available for occupation by an electron; the orbitals with $n = 1, 2, 3, 4$ are given in Table 1.5. These represent the stationary states available to an electron in a hydrogen atom.

The single electron present in a hydrogen atom can be placed in one of the orbitals available without any particular theoretical difficulty. However, the Schrödinger equation has never been explicitly solved for a system with two or more electrons so that the relative energies of the orbitals in such systems are not known.

TABLE 1.5 The total number of orbitals for $n = 1, 2, 3,$ and 4, and their designations

n	l^a	m	s	Number of designations
1	0 (s)	0	$\pm 1/2$	2
2	0 (s)	0	$\pm 1/2$	2
	1 (p)	$0, \pm 1$	$\pm 1/2$	6
3	0 (s)	0	$\pm 1/2$	2
	1 (p)	$0, \pm 1$	$\pm 1/2$	6
	2 (d)	$0, \pm 1, \pm 2$	$\pm 1/2$	10
4	0 (s)	0	$\pm 1/2$	2
	1 (p)	$0, \pm 1$	$\pm 1/2$	6
	2 (d)	$0, \pm 1, \pm 2$	$\pm 1/2$	10
	3 (f)	$0, \pm 1, \pm 2, \pm 3$	$\pm 1/2$	14

^{a} Quantum number l is indicated either by number or by the letter designation (in parentheses here; see Section 2.1 of this chapter), but not by both together.

In an attempt to overcome these difficulties, a hypothetical process has been evolved to describe the electronic configuration of multielectron atoms in which the structure of each atom is assumed to be derivable from the atom with the next lower atomic number by the addition of one proton to the nucleus and one electron to the extranuclear structure. Thus, the structures of all atoms are ultimately related to that of hydrogen. During this building-up (Aufbau) process two pieces of information are necessary: the order of energies of the atomic orbitals, and the maximum number of electrons that can be accommodated in each orbital. The first problem is usually solved by assuming that the relative order of energies of the orbitals in any atom is the same as that for hydrogen. As we shall see, this is a reasonably good assumption in most cases. The second question is decided by the Pauli exclusion principle, which in its simplest form states that no two electrons in the same atom can have the same energy. Since the solutions to the Schrödinger equation establish the allowed energies for the hydrogen atom, the Pauli principle can be restated in terms of quantum numbers: no two electrons in an atom may be assigned the same combination of quantum numbers. The Pauli exclusion principle immediately provides the answer to the maximum number of electrons possible for each orbital. Table 1.5 indicates that s, p, d, and f orbitals can contain 2, 6, 10, and 14 electrons, respectively. The relative energies of the hydrogen-like orbitals are given in Fig. 1.22.

The following examples illustrate the application of the Aufbau process for multielectron systems. The ground state of hydrogen contains one electron in

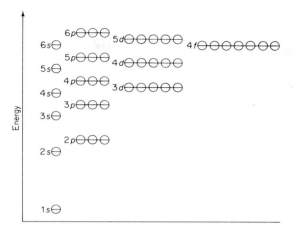

FIG. 1.22   The relative energies of orbitals for hydrogen-like systems.

the 1s level; that is, the wave function for which $n = 1$, $l = 0$, and $m = 0$ is the solution to the Schrödinger equation that gives the minimum energy. The helium $(Z = 2)$ ground state is derived from hydrogen $(Z = 1)$ by the addition of a proton to the nucleus and an electron to the extranuclear region. The Pauli principle indicates that the second or, as it is sometimes called, the differentiating electron for helium has a different set of quantum numbers, i.e., $n = 1$, $l = 0$, $m = 0$, $s = -\frac{1}{2}$, than the first electron (Table 1.5). Correspondingly, a lithium atom $(Z = 3)$ has an electron distribution similar to that of helium, but one more electron must be added. Since the 1s orbital in helium is full, it is apparent from Fig. 1.22 that the next lowest energy level is the 2s orbital, so that the electronic configuration for lithium is $1s^2 2s^1$.* Using similar arguments it is possible to describe the electronic configurations of all elements. A schematic energy-level representation for the electronic configuration of the first ten elements is given in Fig. 1.23, while the simpler line designations for all the elements appear in Table 1.6.

The energy-level representation shown in Fig. 1.22 for the electronic structure of hydrogen and atoms with low atomic numbers indicates that the order of filling of orbitals proceeds first according to the value of the two quantum numbers n and l; that is, the orbital with the lowest value of $n + l$ is filled first. For orbitals that differ only in the value of the magnetic quantum number m, i.e., orbitals that are degenerate, electrons are filled in unpaired

---

* A convenient and often used designation for electron distributions among orbitals is given in symbolic form as nl^x, where n is the principal quantum number, l is the secondary quantum number, and x is the number of electrons present in the orbital; the letter designation is traditionally used in this system.

TABLE 1.6 The electronic configurations of the elements

Z	Element																			
		1s	2s	2p	3s	3p	3d	4s	4p	4d	4f	5s	5p	5d	5f	6s	6p	6d	7s
1	H	1																	
2	He	2																	
3	Li	2	1																
4	Be	2	2																
5	B	2	2	1															
6	C	2	2	2															
7	N	2	2	3															
8	O	2	2	4															
9	F	2	2	5															
10	Ne	2	2	6															
11	Na	2	2	6	1														
12	Mg	2	2	6	2														
13	Al	2	2	6	2	1													
14	Si	2	2	6	2	2													
15	P	2	2	6	2	3													
16	S	2	2	6	2	4													
17	Cl	2	2	6	2	5													
18	Ar	2	2	6	2	6													
19	K	2	2	6	2	6		1											
20	Ca	2	2	6	2	6		2											
21	Sc	2	2	6	2	6	1	2											
22	Ti	2	2	6	2	6	2	2											
23	V	2	2	6	2	6	3	2											
24	Cr	2	2	6	2	6	5	1											
25	Mn	2	2	6	2	6	5	2											
26	Fe	2	2	6	2	6	6	2											
27	Co	2	2	6	2	6	7	2											
28	Ni	2	2	6	2	6	8	2											
29	Cu	2	2	6	2	6	10	1											
30	Zn	2	2	6	2	6	10	2											
31	Ga	2	2	6	2	6	10	2	1										
32	Ge	2	2	6	2	6	10	2	2										
33	As	2	2	6	2	6	10	2	3										
34	Se	2	2	6	2	6	10	2	4										
35	Br	2	2	6	2	6	10	2	5										
36	Kr	2	2	6	2	6	10	2	6										
37	Rb	2	2	6	2	6	10	2	6			1							
38	Sr	2	2	6	2	6	10	2	6			2							
39	Y	2	2	6	2	6	10	2	6	1		2							
40	Zr	2	2	6	2	6	10	2	6	2		2							
41	Nb	2	2	6	2	6	10	2	6	4		1							
42	Mo	2	2	6	2	6	10	2	6	5		1							
43	Tc	2	2	6	2	6	10	2	6	5		2							
44	Ru	2	2	6	2	6	10	2	6	7		1							
45	Rh	2	2	6	2	6	10	2	6	8		1							
46	Pd	2	2	6	2	6	10	2	6	10									
47	Ag	2	2	6	2	6	10	2	6	10		1							

TABLE 1.6—*continued*

Z	Element	1s	2s	2p	3s	3p	3d	4s	4p	4d	4f	5s	5p	5d	5f	6s	6p	6d	7s
48	Cd	2	2	6	2	6	10	2	6	10		2							
49	In	2	2	6	2	6	10	2	6	10		2	1						
50	Sn	2	2	6	2	6	10	2	6	10		2	2						
51	Sb	2	2	6	2	6	10	2	6	10		2	3						
52	Te	2	2	6	2	6	10	2	6	10		2	4						
53	I	2	2	6	2	6	10	2	6	10		2	5						
54	Xe	2	2	6	2	6	10	2	6	10		2	6						
55	Cs	2	2	6	2	6	10	2	6	10		2	6			1			
56	Ba	2	2	6	2	6	10	2	6	10		2	6			2			
57	La	2	2	6	2	6	10	2	6	10		2	6	1		2			
58	Ce	2	2	6	2	6	10	2	6	10	2	2	6			2			
59	Pr	2	2	6	2	6	10	2	6	10	3	2	6			2			
60	Nd	2	2	6	2	6	10	2	6	10	4	2	6			2			
61	Pm	2	2	6	2	6	10	2	6	10	5	2	6			2			
62	Sm	2	2	6	2	6	10	2	6	10	6	2	6			2			
63	Eu	2	2	6	2	6	10	2	6	10	7	2	6			2			
64	Gd	2	2	6	2	6	10	2	6	10	7	2	6	1		2			
65	Tb	2	2	6	2	6	10	2	6	10	9	2	6			2			
66	Dy	2	2	6	2	6	10	2	6	10	10	2	6			2			
67	Ho	2	2	6	2	6	10	2	6	10	11	2	6			2			
68	Er	2	2	6	2	6	10	2	6	10	12	2	6			2			
69	Tm	2	2	6	2	6	10	2	6	10	13	2	6			2			
70	Yb	2	2	6	2	6	10	2	6	10	14	2	6			2			
71	Lu	2	2	6	2	6	10	2	6	10	14	2	6	1		2			
72	Hf	2	2	6	2	6	10	2	6	10	14	2	6	2		2			
73	Ta	2	2	6	2	6	10	2	6	10	14	2	6	3		2			
74	W	2	2	6	2	6	10	2	6	10	14	2	6	4		2			
75	Re	2	2	6	2	6	10	2	6	10	14	2	6	5		2			
76	Os	2	2	6	2	6	10	2	6	10	14	2	6	6		2			
77	Ir	2	2	6	2	6	10	2	6	10	14	2	6	7		2			
78	Pt	2	2	6	2	6	10	2	6	10	14	2	6	9		1			
79	Au	2	2	6	2	6	10	2	6	10	14	2	6	10		1			
80	Hg	2	2	6	2	6	10	2	6	10	14	2	6	10		2			
81	Tl	2	2	6	2	6	10	2	6	10	14	2	6	10		2	1		
82	Pb	2	2	6	2	6	10	2	6	10	14	2	6	10		2	2		
83	Bi	2	2	6	2	6	10	2	6	10	14	2	6	10		2	3		
84	Po	2	2	6	2	6	10	2	6	10	14	2	6	10		2	4		
85	At	2	2	6	2	6	10	2	6	10	14	2	6	10		2	5		
86	Rn	2	2	6	2	6	10	2	6	10	14	2	6	10		2	6		
87	Fr	2	2	6	2	6	10	2	6	10	14	2	6	10		2	6		1
88	Ra	2	2	6	2	6	10	2	6	10	14	2	6	10		2	6		2
89	Ac	2	2	6	2	6	10	2	6	10	14	2	6	10		2	6	1	2
90	Th	2	2	6	2	6	10	2	6	10	14	2	6	10		2	6	2	2
91	Pa	2	2	6	2	6	10	2	6	10	14	2	6	10	2	2	6	1	2
92	U	2	2	6	2	6	10	2	6	10	14	2	6	10	3	2	6	1	2
93	Np	2	2	6	2	6	10	2	6	10	14	2	6	10	4	2	6	1	2
94	Pu	2	2	6	2	6	10	2	6	10	14	2	6	10	5	2	6	1	2

TABLE 1.6—*continued*

Z	Element	1s	2s	2p	3s	3p	3d	4s	4p	4d	4f	5s	5p	5d	5f	6s	6p	6d	7s
															Electron distribution				
95	Am	2	2	6	2	6	10	2	6	10	14	2	6	10	7	2	6		2
96	Cm	2	2	6	2	6	10	2	6	10	14	2	6	10	7	2	6	1	2
97	Bk	2	2	6	2	6	10	2	6	10	14	2	6	10	8	2	6	1	2
98	Cf	2	2	6	2	6	10	2	6	10	14	2	6	10	9	2	6	1	2
99	Es	2	2	6	2	6	10	2	6	10	14	2	6	10	10	2	6	1	2
100	Fm	2	2	6	2	6	10	2	6	10	14	2	6	10	11	2	6	1	2
101	Mv	2	2	6	2	6	10	2	6	10	14	2	6	10	12	2	6	1	2
102	No	2	2	6	2	6	10	2	6	10	14	2	6	10	13	2	6	1	2
103	Lw	2	2	6	2	6	10	2	6	10	14	2	6	10	14	2	6	1	2

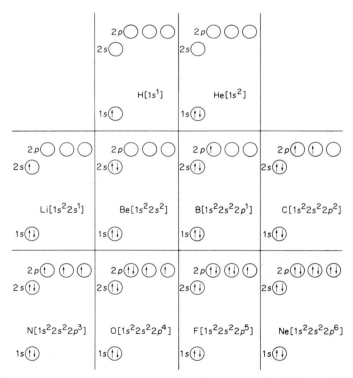

FIG. 1.23 Electronic representations for the first ten atoms in the periodic classification. These electronic distributions involve using the Pauli exclusion principle, the Aufbau principle, and the assumption that the relative energies of the hydrogen-like orbitals are unchanged for the other elements.

before they are paired. This observation was first obtained from an empirical analysis of optical spectra and is known as Hund's rule. Another way of stating Hund's rule is that electrons in degenerate orbitals are at their lowest state when their spins are parallel, a situation which always arises for p, d, and f orbitals. Thus, boron ($Z = 5$), carbon ($Z = 6$), and nitrogen ($Z = 7$) have 1, 2, and 3 unpaired electron spins (Fig. 1.23), and the atoms of these elements are paramagnetic. The paramagnetism of oxygen ($Z = 8$) and fluorine ($Z = 9$) atoms decreases; the neon atom ($Z = 10$) is not paramagnetic.

The electron configurations for the lighter elements (Table 1.6) in general follow the order of filling expected from the order of the energy levels and the application of Hund's rule. However, for heavier elements some apparent anomalies occur between the experimentally observed and the predicted distributions. For example, chromium ($Z = 24$) should have an electron configuration of $1s^2 2s^2 2p^6 3s^2 3p^6 3d^4 4s^2$ on the basis of the arguments developed thus far; the experimentally observed distribution is $1s^2 2s^2 2p^6 3s^2 3p^6 3d^5 4s^1$. This and other minor anomalies occur for the heavier elements because the energy differences between outer orbitals get progressively less. Further discussion of electronic configuration of the heavier atoms appears in Chapter 16.

4 / PERIODICITY AND ELECTRONIC STRUCTURE

Inspection of Table 1.6 leads to the observation that there is a recurring outer electronic structure with increasing atomic number. For example, the elements lithium ($Z = 3$), sodium ($Z = 11$), potassium ($Z = 19$), rubidium ($Z = 37$), cesium ($Z = 55$), and francium ($Z = 87$) all have outer electronic configurations of ns^1. These recurring electronic configurations suggest that the elements can be ordered in a periodic arrangement on this basis. Figure 1.24 shows the elements arranged in increasing order of atomic number; elements with the same or similar electronic configurations are listed in the same columns. Thus, the periodic chart based on electronic structure has many of the same features of the periodic chart originally suggested by Mendeleev on the basis of his study of the chemical and physical properties of the elements. This relationship implies that the properties of the elements are rooted in their electronic structures. We shall develop this point repeatedly in this volume.

Arranging the elements according to their electronic configurations leads to natural groupings in vertical columns (called families or groups) and horizontal sequences (called periods). Some of the families or groups have acquired common names such as the alkali metals (IA), the alkaline-earth elements (IIA), the chalcogens (VIB), the halogens (VIIB), and the rare gases (0). Elements for which the differentiating electron is filling ns and np orbitals

The periodic classification of the elements

Group \ Period	I A	II A	III A	IV A	V A	VI A	VII A	VIII	VIII	VIII	I B	II B	III B	IV B	V B	VI B	VII B	O
1 — 1s	1 H	2 He																2 He
2 — 2s, 2p	3 Li	4 Be											5 B	6 C	7 N	8 O	9 F	10 Ne
3 — 3s, 3p	11 Na	12 Mg											13 Al	14 Si	15 P	16 S	17 Cl	18 A
4 — 4s, 3d, 4p	19 K	20 Ca	21 Sc	22 Ti	23 V	24 Cr	25 Mn	26 Fe	27 Co	28 Ni	29 Cu	30 Zn	31 Ga	32 Ge	33 As	34 Se	35 Br	36 Kr
5 — 5s, 4d, 5p	37 Rb	38 Sr	39 Y	40 Zr	41 Nb	42 Mo	43 Tc	44 Ru	45 Rh	46 Pd	47 Ag	48 Cd	49 In	50 Sn	51 Sb	52 Te	53 I	54 Xe
6 — 6s, 4f, 5d, 6p	55 Cs	56 Ba	71 Lu	72 Hf	73 Ta	74 W	75 Re	76 Os	77 Ir	78 Pt	79 Au	80 Hg	81 Tl	82 Pb	83 Bi	84 Po	85 At	86 Rn
7 — 7s, 5f, 6d	87 Fr	88 Ra	103 Lw	104	105													

57 La	58 Ce	59 Pr	60 Nd	61 Pm	62 Sm	63 Eu	64 Gd	65 Tb	66 Dy	67 Ho	68 Er	69 Tm	70 Yb
89 Ac	90 Th	91 Pa	92 U	93 Np	94 Pu	95 Am	96 Cm	97 Bk	98 Cf	99 Er	100 Fm	101 Md	102 No

s — Representative elements
d — Transition metals
f — Lanthanides and actinides
p — Representative elements

FIG. 1.24 The periodic arrangement of the elements based on their electronic configurations. The differentiating electron for the representative elements enters an s or p orbital; d and f orbitals are being filled for the transition metals and the lanthanides (and actinides), respectively.

are called the representative elements, those for which the differentiating electron goes into a d orbital are the transition elements, and those for which the differentiating electron goes into an f orbital are the lanthanides and actinides (or rare earths).

Many properties of the elements show a periodic variation with increasing atomic number which parallels the periodicity of electronic configurations.

5 / PERIODIC PROPERTIES OF THE ELEMENTS

5.1 / Ionization Potential

The ionization potential of an atom is the work necessary to remove the most loosely held electron from that atom in the gaseous state [cf. Eq. (19)]. Although it is not possible to decide easily which electron is removed when a neutral atom is ionized, it is generally accepted that the outer electrons are removed first. Since more than one electron can be removed from an atom, it is necessary to specify the process for which the ionization energy is measured (e.g., the first, second, or third ionization potential). Figure 1.25 shows the

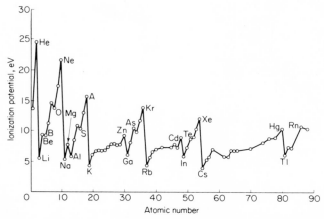

FIG. 1.25 The first ionization potentials, measured in electron volts, of the elements through francium.

values of the first ionization potentials for the elements hydrogen to francium plotted against their atomic numbers; higher ionization potentials are recorded in Table 1.7. Several features are immediately apparent from the plot of the first ionization potentials. First, a periodic variation in the ionization

TABLE 1.7   Ionization potentials of the elements

Z	Element	Con-figuration	I	II	III	IV	V	VI	VII	VIII
							Ionization potential, eV			
1	H	$1s^1$	13.595							
2	He	$1s^2$	24.581	54.403						
3	Li	$2s^1$	5.390	75.619	122.419					
4	Be	$2s^2$	9.320	18.206	153.850	217.657				
5	B	$2s^2 2p^1$	8.296	25.149	37.920	259.298	340.127			
6	C	$2s^2 2p^2$	11.256	24.376	47.871	64.476	391.986	489.84		
7	N	$2s^2 2p^3$	14.53	29.593	47.426	77.450	97.863	551.925	666.83	
8	O	$2s^2 2p^4$	13.614	35.108	54.886	77.394	113.873	138.080	739.114	871.12
9	F	$2s^2 2p^5$	17.418	34.98	62.646	87.14	114.214	157.214	185.139	953.60
10	Ne	$2s^2 2p^6$	21.559	41.07	63.5	97.02	126.3	157.91		
11	Na	$3s^1$	5.138	47.29	71.65	98.88	138.37	172.09	208.444	264.155
12	Mg	$3s^2$	7.644	15.031	80.12	109.29	141.23	186.49	224.90	265.957
13	Al	$3s^2 3p^1$	5.984	18.823	28.44	119.96	153.77	190.42	241.38	284.53
14	Si	$3s^2 3p^2$	8.149	16.34	33.46	45.13	166.73	205.11	246.41	303.07
15	P	$3s^2 3p^3$	10.484	19.72	30.156	51.354	65.007	220.414	263.31	309.26
16	S	$3s^2 3p^4$	10.357	23.4	35.0	47.29	72.5	88.029	280.99	328.80
17	Cl	$3s^2 3p^5$	13.01	23.80	39.90	53.5	67.80	96.7	114.27	348.3
18	Ar	$3s^2 3p^6$	15.755	27.62	40.90	59.79	75.0	91.3	124.0	143.46
19	K	$4s^1$	4.339	31.81	46	60.90	82.6	99.7	118	155
20	Ca	$4s^2$	6.111	11.868	51.21	67	84.39	109	128	143.3
21	Sc	$3d^1 4s^2$	6.54	12.80	24.75	73.9	92	111	139	159
22	Ti	$3d^2 4s^2$	6.82	13.57	27.47	43.24	99.8	120	141	172
23	V	$3d^3 4s^2$	6.74	14.65	29.31	48	65	129	151	174
24	Cr	$3d^5 4s^1$	6.764	16.49	30.95	50	73	91	161	185
25	Mn	$3d^5 4s^2$	7.432	15.636	33.69	52	76	100	119	196
26	Fe	$3d^6 4s^2$	7.87	16.18	30.643	56.8		103	130	151
27	Co	$3d^7 4s^2$	7.86	17.05	33.49			83.1	133	163
28	Ni	$3d^8 4s^2$	7.633	18.15	35.16					168
29	Cu	$3d^{10} 4s^1$	7.724	20.29	36.83					
30	Zn	$3d^{10} 4s^2$	9.391	17.96	39.70					
31	Ga	$4s^2 4p^1$	6.00	20.51	30.70	64.2				
32	Ge	$4s^2 4p^2$	7.88	15.93	34.21	45.7	93.4			
33	As	$4s^2 4p^3$	9.81	18.63	28.34	50.1	62.6	127.5		
34	Se	$4s^2 4p^4$	9.75	21.5	32	43	68	82	155	
35	Br	$4s^2 4p^5$	11.84	21.6	35.9	47.3	59.7	88.6	103	193
36	Kr	$4s^2 4p^6$	13.996	24.56	36.9	52.5	64.7	78.5	110.0	126
37	Rb	$5s^1$	4.176	27.5	40	52.6	71.0	84.4	99.2	136
38	Sr	$5s^2$	5.692	11.027	43.6	57	71.6	90.8	106	122.3
39	Y	$4d^1 5s^2$	6.38	12.23	20.5	61.8	77	93.0	116	129
40	Zr	$4d^2 5s^2$	6.84	13.13	22.98	34.33	82.3	99	116	139
41	Nb	$4d^4 5s^1$	6.88	14.32	25.04	38.3	50	103	125	141
42	Mo	$4d^5 5s^1$	7.10	16.15	27.13	46.4	61.2	68	126	153
43	Tc	$4d^6 5s^1$	7.28	15.26	29.54					
44	Ru	$4d^7 5s^1$	7.364	16.76	28.46					

TABLE 1.7—*continued*

Z	Element	Con-figuration	Ionization potential, eV							
			I	II	III	IV	V	VI	VII	VIII
45	Rh	$4d^8 5s^1$	7.46	18.07	31.05					
46	Pd	$4d^{10}$	8.33	19.42	32.92					
47	Ag	$4d^{10}5s^1$	7.574	21.48	34.82					
48	Cd	$4d^{10}5s^2$	8.991	16.904	37.47					
49	In	$5s^2 5p^1$	5.785	18.86	28.03	54.4				
50	Sn	$5s^2 5p^2$	7.342	14.628	30.49	40.72	72.3			
51	Sb	$5s^2 5p^3$	8.639	16.5	25.3	44.1	56	108	119	
52	Te	$5s^2 5p^4$	9.01	18.6	31	38	60	72	137	
53	I	$5s^2 5p^5$	10.454	19.09	33		71	83	104	170
54	Xe	$5s^2 5p^6$	12.127	21.2	32.1	44	60	83	102	126
55	Cs	$6s^1$	3.893	25.1	35				108	122
56	Ba	$6s^2$	5.210	10.001	35.5					127
57	La	$5d^1 6s^2$	5.61	11.43	19.17					
58	Ce	$4f^2 6s^2$	6.5	12.3	20	33.3				
59	Pr	$4f^3 6s^2$	5.7		23.2					
60	Nd	$4f^4 6s^2$	5.7							
61	Pm	$4f^5 6s^2$								
62	Sm	$4f^6 6s^2$	5.64	11.2						
63	Eu	$4f^7 6s^2$	5.67	11.24						
64	Gd	$4f^7 5d^1 6s^2$	6.16	12						
65	Tb	$4f^9 6s^2$	6.7							
66	Dy	$4f^{10}6s^2$	6.8							
67	Ho	$4f^{11}6s^2$								
68	Er	$4f^{12}6s^2$								
69	Tm	$4f^{13}6s^2$								
70	Yb	$4f^{14}6s^2$	6.23	12.10						
71	Lu	$4f^{14}5d^1 6s^2$	6.15	14.7						
72	Hf	$5d^2 6s^2$	7	14.9						
73	Ta	$5d^3 6s^2$	7.88	16.2						
74	W	$5d^4 6s^2$	7.98	17.7						
75	Re	$5d^5 6s^2$	7.87	16.6						
76	Os	$5d^6 6s^2$	8.7	17						
77	Ir	$5d^9$	9	17						
78	Pt	$5d^9 6s^1$	9.0	18.56						
79	Au	$5d^{10}6s^1$	9.22	20.5						
80	Hg	$5d^{10}6s^2$	10.43	18.751	34.2	72	82			
81	Tl	$6s^2 6p^1$	6.106	20.42	29.8	50.7				
82	Pb	$6s^2 6p^2$	7.415	15.028	31.93	42.31	68.8			
83	Bi	$6s^2 6p^3$	7.287	16.68	25.56	45.3	56.0	88.3		
84	Po	$6s^2 6p^4$	8.43	19.4	27.3					
85	At	$6s^2 6p^5$	9.5	20.1	29.3					
86	Rn	$6s^2 6p^6$	10.746	21.4	29.4					
87	Fr	$7s^1$	3.83	22.5	33.5					
88	Ra	$7s^2$	5.277	10.144						

TABLE 1.7—*continued*

						Ionization potential, eV				
Z	Element	Con-figuration	I	II	III	IV	V	VI	VII	VIII
89	Ac	$6d^1 7s^2$	6.9	12.1	20					
90	Th	$6d^2 7s^2$		11.5	20.0	29.38				
91	Pa	$5f^2 6d^1 7s^2$								
92	U	$5f^3 6d^1 7s^2$	4							
93	Np	$5f^5 7s^2$								
94	Pu	$5f^6 7s^2$	5.1							
95	Am	$5f^7 7s^2$	6.0							
96	Cm	$5f^7 6d^7 7s^2$								
97	Bk	$5f^8 6d^1 7s^2$								
98	Cf	$5f^{10} 7s^2$								

potentials is apparent, the amplitude of which decreases for the heavier elements. Closer inspection of Fig. 1.25 shows that the first member of a given period (an alkali metal) has the lowest ionization potential of the elements in that period, whereas the corresponding rare gas has the highest ionization potential; its configuration is the most stable arrangement of electrons in a period. The more or less regular increase in the ionization potentials for the intermediate atoms in a given period might be expected because these elements correspond to the sequential introduction of a positive charge into the nucleus and the addition of an electron into outer orbitals with the same principal quantum numbers. Under these conditions the differentiating electron is shielded by the same number of inner closed shells, but it feels a progressively larger nuclear charge and hence should be more and more difficult to remove.

There are interesting discontinuities in the ionization potentials of the representative elements which occur at groups IIA (Be, Mg, etc.) and VB (N, P, As, etc.). The outer electronic configuration of the group IIA elements (ns^2) consists of a filled orbital whereas that of the group VB elements $(ns^2 np^3)$ corresponds to a half-filled orbital. Abnormally high ionization potentials or high stabilities for species with filled or half-filled orbitals are often observed. Similar discontinuities appear in the ionization potentials of the elements in the middle of the transition series where the differentiating electron is placed in a d orbital. For example, chromium has a $4s^1 3d^5$ configuration (Table 1.6) rather than the expected $4s^2 3d^4$ configuration; similar variations from the expected configuration to one in which the electron distribution reflects a half-filled or filled orbital configuration are apparent for silver and copper.

Several factors may play an important role in the discontinuities observed at the group IIA elements. Rather than considering that the ionization potential of beryllium is abnormally large, we might assume that the ionization potential of boron is abnormally *small*. This point of view focuses attention on the fact that the ionization of beryllium ($1s^2 2s^2$) and boron ($1s^2 2s^2 2p^1$) involves electrons in different types of orbitals. A consideration of the radial distribution function for s and p electrons (Fig. 1.19) shows that an s electron has a significant electron density nearer the nucleus, or penetrates closer to the nucleus, than the corresponding p electron; accordingly it should be easier to remove a p electron than the corresponding s electron. The shapes of the radial distribution curves for electrons with the same principal quantum number but with different values for l indicate the penetration of these electrons decreases in the order $s > p > d > f$. All other factors being equal, this order also represents the order of decreasing binding of the electron by the nucleus.

A difference in the penetration effects for electrons with different l values cannot be an important consideration in explaining the discontinuities in the trends of ionization potentials for the elements that have half-filled p and d orbitals. However, the possibility that the element which has one electron over that necessary for a half-filled orbital has an electronic configuration that is *less* stable than that corresponding to a half-filled shell can be examined. Thus, in period 2 oxygen has one electron more than is necessary for a half-filled orbital with the electronic configuration $1s^2 2s^2 2p^4$; Hund's rule tells us that two of the four p electrons are unpaired. On this basis it might be expected that the differentiating electron which must be paired for oxygen would be easier to remove than an unpaired electron found in the half-filled shell configuration of the previous element.

The data in Fig. 1.25 show that the ionization potentials of the elements in a given family or group decrease with increasing atomic number. This trend reflects a screening by inner filled orbitals which counteracts the increase in nuclear charge when going from one member of a family to the next. In addition, the differentiating electron for successive members of the family is placed in an orbital that is progressively further removed from the nucleus.

A final feature of interest in Fig. 1.25 is the fact that the ionization potentials of the transition elements are relatively less variable than those of the representative elements. Such trends are undoubtedly related to a combination of the following factors: the location of the differentiating electron in a d orbital, the energetic proximity of the ns and $(n-1)d$ orbitals, and the relative stability of the electron distributions of the atom and the corresponding ion. The ionization potential for a given element is a function of all of the factors described here, and often it is not possible to separate the relative importance of one from another.

5.2 / Atomic Dimensions

If atoms are built up according to the principles discussed in this chapter, it might be expected that the periodic classification should be reflected in atomic dimensions. Well before the details of the electronic configuration of the elements had been elucidated, Lothar Meyer arrived at a classification of the elements from a consideration of their atomic volumes, which is defined as the volume occupied by one gram atom (6.023×10^{23} atoms) of an element. Since this quantity is calculated from the density of an element, it is important that the latter be specified under the same conditions for all elements. The atomic volume should be the volume of the same number of atoms for all elements, but, since the elements vary in their crystal lattices and in their molecularity (e.g., Na, H_2, S_8, P_4) the comparison of atomic volumes calculated from densities does not really reflect the volumes of the atoms. Nevertheless, a plot of atomic volumes against atomic number shows a periodicity (Fig. 1.26) reminiscent of that observed for the ionization potentials.

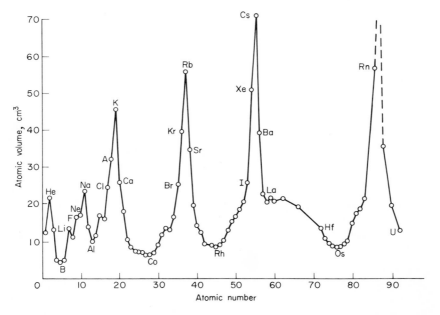

FIG. 1.26   The periodic variation of the atomic volumes of the elements.

A more interesting example of the regularity of atomic dimensions comes from a consideration of atomic radii. X-ray, electron, and neutron diffraction techniques have been used to determine a self-consistent set of atomic radii;

the radii of the metals have been obtained from the interatomic distances in the elements, while the radii of the nonmetals have been obtained from the interatomic distances of substances in which the atoms are held together by covalent bonds. Although metallic and covalent radii are not necessarily related (because of the differences in the types of bonds formed in each case), they often are used interchangeably if qualitative trends are of interest. The metallic and covalent radii of the elements are shown in Fig. 1.27. The

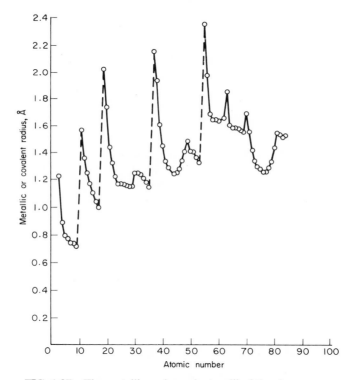

FIG. 1.27   The metallic and covalent radii of the elements.

radial distribution curves suggest that there are limits to the atom as described by quantum mechanics (Fig. 1.19). The regular decrease in atomic radii for the elements in a given period reflects the fact that the differentiating electron is added to the same principal quantum shell for atoms with a regularly increasing positive nuclear charge. Since the increased nuclear charge is not screened during this process, there is a progressive contraction in size of the atoms across a given period. The other obvious feature in Fig. 1.27 is the regular increase in size with increasing atomic number in a given family. This effect arises because the shielding of the outer electron distribution by completed inner orbitals overcomes the increase in nuclear charge which occurs as the atomic number of the elements of a family increases. Although the increase in size for a family of elements with increasing atomic number is a prevalent trend, it is not universally observed. For example, in the family of transition elements composed of Ti (1.324 Å), Zr (1.454 Å), and Hf (1.442 Å), the radius of the third member decreases. In cases like this, the radii in question are usually close in value.

It should be noted that in general the radii of the transition elements or rare earths are relatively invariant compared to the changes observed for the representative elements. This trend can be easily understood if it is recalled that the differentiating electron for the transition elements is placed in an inner orbital. Accordingly, under these conditions we would not expect a large variation in size.

5.3 / Other Physical Properties

Many other physical properties of the elements are periodic functions of the atomic number. For example, both the melting points and the densities of the elements show the typical fluctuation with respect to atomic number (Figs. 1.28 and 1.29). It is often difficult to determine the principal factors which contribute to such a variation of some properties; however these empirical relationships can have very practical consequences in estimating the value of a given element from those known for other elements.

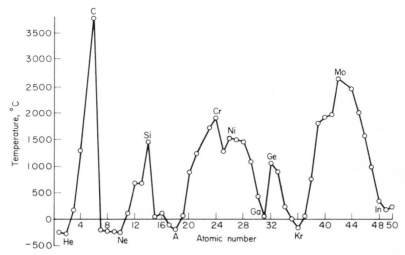

FIG. 1.28   The melting points of the first fifty elements show a periodic variation with atomic number.

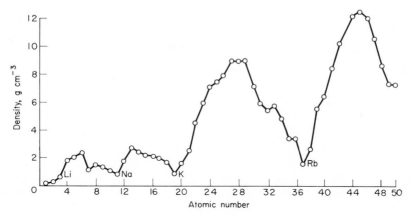

FIG. 1.29 The solid-state densities of the elements show periodic fluctuations.

5.4 / Chemical Properties

The periodic nature of chemical properties has been known since the time of Mendeleev. Since chemical properties such as combining power (valence), formation of covalent compounds, and sizes of the ions formed by the elements are dependent upon their electronic configurations, it should not be surprising that a periodic relationship for such properties exists. Many of these relationships are elucidated and used in subsequent chapters.

Well before the nature of atoms and the significance of their electronic structures were understood, chemists had amassed a vast array of experimental results concerning compounds which also led to a periodic arrangement of the elements that was remarkably similar to the one developed from electronic configurations. Of course, all the elements had not yet been discovered and the form of the chart was slightly different than the modern long form (shown in Fig. 1.24), yet the organization of chemical data on this basis had far-reaching consequences.

Consider, for example, some of the simple halogen-containing compounds of the Period 2 and 3 elements and their properties (Table 1.8). Two interesting observations are immediately apparent from the data in Table 1.8. First, the number of halogen atoms associated with a given atom in the period (i.e., its valence or combining power) rises to a maximum at the center of the period and then decreases. Second, compounds of the first two members of a period possess relatively high melting and boiling points compared with the remaining compounds in the period. The latter observation suggests that the basic structural units in the alkali (Li, Na) and alkaline-earth (Be, Mg) halides are fundamentally different from those found in the remaining

TABLE 1.8   The melting and boiling points of some representative-element halides

Property	Group						
	IA	IIA	IIIB	IVB	VB	VIB	VIIB
	LiF	BeF_2	BF_3	CF_4	NF_3	OF_2	F_2
Mp	842	800^a	-126.7	-183.7	-206.6	-223.8	-223
Bp	1676	—	-99.9	-129	-128.8	-144.8	-188
	NaCl	$MgCl_2$	$AlCl_3$	$SiCl_4$	PCl_3	SCl_2	Cl_2
Mp	801	708	190^b	-70	-112	-78	-103
Bp	1413	1412	182.7	57.57	75.5	59^c	-34.6
	NaF	MgF_2	AlF_3	SiF_4	PF_3	$SF_2{}^d$	ClF
Mp	988	1266	1291^a	-90.2	-151.5	—	-154
Bp	1695	2239	—	-86	-101.5	—	-100.8

^{a} Sublimes.
^{b} Melting point determined at 2.5 atm.
^{c} Decomposes.
^{d} Known only in the gas phase.

compounds of the series. The low melting and boiling points of the latter compounds indicate relatively small forces of attraction between these aggregates compared to compounds with significantly higher melting and boiling points. This argument, together with the observations that the halogen compounds of the elements in groups IIIB, IVB, VB, VIB, and VIIB are virtually non-conductors of electricity in the liquid state while the remaining substances are good conductors, suggests that the low-boiling-point substances consist of individual molecules whereas the high-boiling-point substances contain charged species (or ions).

Thus, any theoretical discussion of compound formation must address itself not only to the pattern of stoichiometry as exemplified by the compounds in Table 1.8, but also must be able to predict the structures and state of aggregation of the compounds formed. G. N. Lewis formulated a simple theory of chemical combination which described the processes involving the outer electrons of an atom by which elements could achieve a stable electronic configuration (*VI*). Briefly, the logic was as follows. Since compounds of the rare gases were unknown to Lewis, he assumed (correctly) that the electronic configuration of a rare-gas atom was the most stable. Accordingly, it was logical to assume that when compounds formed, the elements involved should achieve a rare-gas electronic configuration. According to Lewis, this result could be achieved by either a complete loss or gain of electrons or by sharing of electrons. Elements that had one or two electrons more or less than a rare-gas atom could achieve a stable electronic configuration by either losing or gaining electrons. Thus, atoms such as fluorine $(1s^2 2s^2 2p^5)$ and oxygen $(1s^2 2s^2 2p^4)$ could gain one and two electrons, respectively:

$$F(1s^2 2s^2 2p^5) + 1e^- \longrightarrow [F(1s^2 2s^2 2p^6)]^- \tag{30}$$

$$O(1s^2 2s^2 2p^4) + 2e^- \longrightarrow [O(1s^2 2s^2 2p^6)]^{2-}. \tag{31}$$

In these processes, as in those involving sodium and magnesium,

$$Na(1s^2 2s^2 2p^6 3p^1) \longrightarrow [Na(1s^2 2s^2 2p^6)]^+ + 1e^- \tag{32}$$

$$Mg(1s^2 2s^2 2p^6 3p^2) \longrightarrow [Mg(1s^2 2s^2 2p^6)]^{2+} + 2e^-, \tag{33}$$

each species has achieved the neon configuration $(1s^2 2s^2 2p^6)$. The combining powers of sodium and magnesium are 1 and 2, respectively, because these atoms acquire a $1+$ and $2+$ charge in attaining a rare-gas configuration; this means that it takes one and two counter-ions, respectively, of $1-$ charge to form neutral compounds. Similar arguments obtain for negatively charged ions. Thus, Lewis was able to account for the stoichiometry and the ionicity of the compounds formed by certain groups of elements in the periodic arrangement.

To account for the properties of the nonionic compounds Lewis proposed the concept of the covalent bond. The formation of a covalent bond occurs when two atoms share a pair of electrons. In the formalism of electron counting, the bond is most often formed between two atoms which each contribute one electron to the bond; in some cases one of the atoms supplies *both* electrons to form the bond. These principles are applied to the nonionic compounds of the Period 2 fluorides in Fig. 1.30. When covalent bonds are

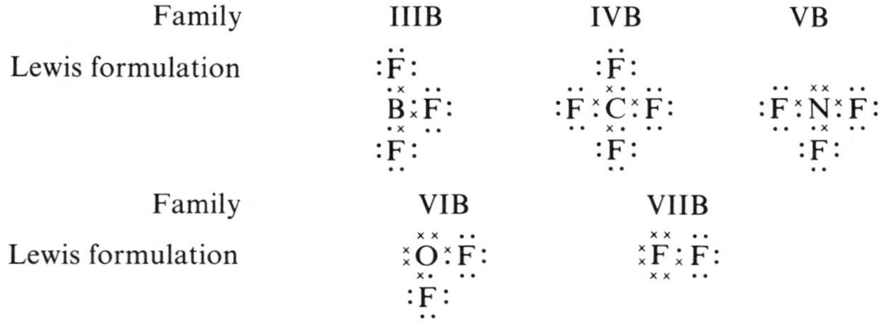

Family	IIIB	IVB	VB
Lewis formulation	$\ddot{:}\ddot{F}:$ $\quad$ $\cdot^\times$ $\ddot{}$ $\quad$ $B\!\cdot^\times\!F:$ $\quad$ $\cdot^\times$ $\ddot{}$ $\quad$ $:\ddot{F}:$	$\ddot{:}\ddot{F}:$ $\quad$ $:\!\overset{\times}{F}\!\overset{\times}{\cdot}\!C\!\overset{\times}{\cdot}\!\overset{\times}{F}\!:$ $\quad$ $:\ddot{F}:$	$:\!\overset{..}{\underset{..}{F}}\!\overset{\times}{\cdot}\!N\!\overset{\times}{\cdot}\!\overset{..}{\underset{..}{F}}\!:$ $\quad$ $:\ddot{F}:$

Family	VIB	VIIB
Lewis formulation	$\overset{\times\times}{\underset{\times\cdot}{\overset{}{O}}}\!\overset{..}{\underset{..}{F}}\!:$ $\quad$ $:\ddot{F}:$	$\overset{\times\times}{\underset{\times\times}{\overset{}{F}}}\!\overset{..}{\underset{..}{F}}\!:$

FIG. 1.30 Lewis dot formulations for the covalent fluorides of the Period 2 elements. The dots and crosses, used for record keeping purposes, represent electrons in the valence shell for the atoms indicated, but must not be construed to imply any actual difference between electrons.

formed each atom in all of the molecules except BF_3 achieves a rare-gas structure; since neutral molecules are formed, there is little association between the molecules except in very special situations. Thus, covalently bound substances exhibit low melting and boiling points.

Although the Lewis theory leads to a simple correlation of many observations, there are several obvious difficulties. For example, the covalent compounds of boron cannot achieve a rare-gas electronic configuration because boron has only three electrons available for covalent bond formation (Fig. 1.30). The electron deficiency of boron compounds is manifested in their ability to react with substances that possess non-bonded electron pairs to form products in which the boron atom has acquired a rare-gas configuration (Fig. 1.31):

$$BF_3 + Na^+F^- \longrightarrow Na^+BF_4^- \tag{34}$$

$$BF_3 + NH_3 \longrightarrow F_3BNH_3. \tag{35}$$

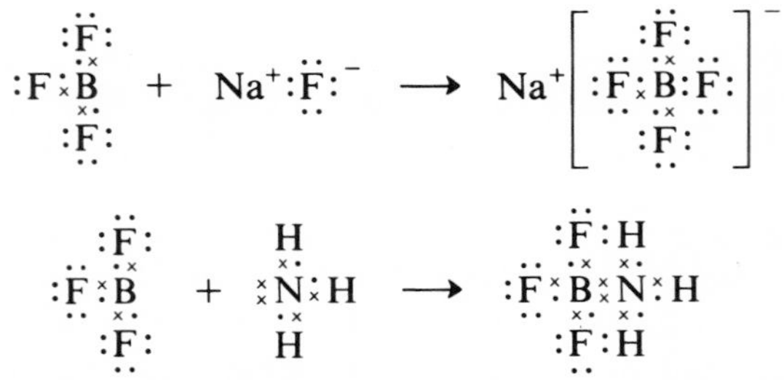

FIG. 1.31 Coordinate-covalent bonds are formed when one of the species involved in a covalent bond contributes the pair of electrons that forms the bond.

The formation of BF_4^- and F_3BNH_3 involves the formation of a covalent bond in which the bonding pair is supplied by one atom, namely, F^- in the case of BF_4^-, and N in the case of F_3BNH_3. Lewis called this type of bond a coordinate-covalent bond. Thus, although some atoms cannot achieve a rare-gas structure in the formation of simple covalent compounds, the chemistry of these substances reflects this electron deficiency. Conversely, some species like NH_3 or Cl^- possess electron pairs that are not involved in covalent bond formation but which *can* be, under certain conditions.

An additional source of difficulty in the Lewis theory arises when the compounds formed by elements in the third and higher periods are considered; several examples of the problem are shown in Table 1.9. Elements in the third and higher periods form covalent compounds exhibiting stoichiometries compatible with the Lewis theory and covalent compounds with higher valences. It is apparent that the electronic configuration of the central atom in the latter compounds cannot be that of the next rare gas.

TABLE 1.9 Some of the halides of the group VB and VIB elements

Period	Group VB		Group VIB	
	Compound	Boiling point, °C	Compound	Boiling point, °C
2	NF_3	− 206.6	OF_2	− 223.8
3	PF_3	− 101.5	SCl_2	—[b]
	PF_5	− 75	SF_4	− 40
			SF_6	63.8
4	AsF_3	− 63	$SeCl_4$	—[c]
	AsF_5	− 53	SeF_6	− 34.5
5	SbF_3	s. 319[a]	$TeCl_2$	327
	SbF_5	149.5	TeF_4	>97
			TeF_6	35.5

[a] Sublimes.
[b] Decomposes at 59°C.
[c] Decomposes at 288°C.

Thus, the Lewis theory accounts for many observations, but it obviously cannot be applied successfully to all systems. Perhaps the most important aspect of the theory is that it focused attention on the existence and nature of the covalent bond. Many of these points are discussed in detail in subsequent chapters.

REFERENCES

1. J. J. Thomson, *Phil. Mag.*, **44**, 294 (1897).
2. H. Geiger, *Proc. Roy. Soc.*, **81**, 177 (1908).
3. E. Rutherford, *Phil. Mag.*, **21**, 669 (1911).
4. H. G. J. Moseley, *Phil. Mag.*, **26**, 1024 (1913).
5. H. G. J. Moseley, *Phil. Mag.*, **27**, 709 (1914).
6. J. Chadwick, *Proc. Roy. Soc.*, *Ser. A*, **136**, 701 (1932).
7. N. Bohr, *Phil. Mag.*, **26**, 2 (1913).
8. C. Davisson and L. H. Germer, *Phys. Rev.*, **30**, 705 (1927).

COLLATERAL READINGS

I. J. J. Lagowski, *The Structure of Atoms*, Houghton-Mifflin, Boston, 1964. A detailed description of the early experiments which led to the modern description of atomic structure.
II. J. J. Thomson, *Rays of Positive Electricity and Their Application to Chemical Analyses*, Longmans, Green and Co., London, 1913. A description of the significance of canal rays, which form the basis of present-day mass spectroscopy.
III. R. S. Berry, *J. Chem. Ed.*, **43**, 283 (1966). A concise summary of the use of quantum mechanics to describe atomic systems; contains numerous references.

IV. I. B. Cohen and T. Bustard, *J. Chem. Ed.*, **43**, 187 (1966). A description of the orbital concept, its usefulness and limitations.

V. J. C. Davis, Jr., *Advanced Physical Chemistry*, The Ronald Press, New York, 1965. Chapter 6 incorporates a detailed description of the method used to obtain the solutions of the Schrödinger equation.

VI. G. N. Lewis, *J. Chem. Phys.*, **1**, 17 (1933). A retrospective description of the evolution of the electron-pair bond concept.

STUDY QUESTIONS

1. Describe the expected results if the Rutherford scattering experiment were carried out with β particles. Include in the discussion (a) the experimental set-up, (b) the expected observations, and (c) the theoretical interpretation. Assume the architecture of the atom is still as we believe it to be.

2. Consider the third and fifth energy levels in the species He^+. (a) Calculate the frequency of the radiation which this species would absorb in going from the lower of these levels to the upper level. (b) Assuming circular orbits of the simple Bohr theory, calculate the ratio of the radii of the orbits associated with these levels. (c) Calculate the first ionization potential of He^+.

3. What is the value of the third ionization potential of lithium?

4. Calculate the cross-sectional area of a hydrogen atom in the $n = 20$ state in terms of its cross-sectional area for the ground state.

5. The Balmer series of lines in the hydrogen spectrum originates from transitions between the $n = 2$ state and higher states. Compare the wavelength for the second line of the Balmer series with the corresponding line expected for the Li^{2+} ion.

6. What is the energy required to excite Li^{2+} (the atomic number of Li is 3) from its ground state to its second excited state?

7. Using the Bohr theory, calculate the radius of the Li^{2+} ion.

8. Although the Bohr theory cannot be applied directly to multielectron systems, it is possible to modify the theory for systems that are almost "hydrogen-like," e.g., the alkali metals. For the alkali-metal atoms the valence electron can be thought of as moving in a Bohr orbit under the influence of a nucleus which is shielded by the closed shells of electrons. Assuming the effect of the shielding of the nucleus is given by the constant σ, give an expression for the frequency of the spectral line which arises from a lithium atom when the valence electron falls from $n = 1$ to $n = 2$.

9. Calculate the fourth ionization potential of beryllium.

10. Without referring to a periodic table, give the expected electronic configurations of the following elements and the number of unpaired electrons in each species: (a) the seventh element in the first transition series; (b) the fourth rare gas; (c) the most stable ion of element 38; (d) a representative element with a half-filled $4p$ energy level.

11. Determine the atomic number of the following species without recourse to a periodic table: (a) the third element of the first transition series; (b) the eleventh representative element; (c) the fourth rare gas; (d) the sixth rare earth (or lanthanide); (e) an element with a half-filled $4p$ orbital.

12. Without referring to a periodic chart, describe the most probable ground-state configuration of elements with the following atomic numbers: (a) 29; (b) 17; (c) 57; (d) 63; (e) 39.

13. Discuss the possible electronic structure of *eka*-Pb, the sixth element expected in Group IV.

14. Consider the elements with atomic numbers 4 and 12. Experimentally the energy required to remove the valence electron from the ground state of element 12 is less than that required to remove the valence electron of element 4 after it has been excited to the $3s$ level. Explain this experimental observation without referring to a periodic chart.

15. Element 46 has a first ionization potential of 8.33 eV, while element 47 has a first ionization potential of 7.57 eV. Discuss these observations without recourse to a periodic table.

16. Arrange each of the following series of elements in order of *increasing* first ionization potential: (a) Fe, Co, Ni, Cu; (b) Sb, Te, I, Xe, Cs; (c) V, Nb, Ta.

17. The elements with atomic numbers 7, 8, and 9 have the following values of the first ionization potential:

Atomic number	First ionization potential, eV
7	14.54
8	13.64
9	17.42

Explain these experimental data without recourse to a periodic table.

18. It is more difficult to remove the valence electron from beryllium after it (the valence electron) has been excited to the 3s energy level than it is to remove the corresponding valence electron from magnesium. Explain this experimental observation.

19. Compare and discuss the first ionization potentials of zirconium and titanium as well as the fourth ionization potentials for these elements.

20. Discuss the basis for the differences in the normal ionization potentials of rubidium and silver and the energy required to remove a 5s electron in the excited hydrogen atom.

21. Without reference to the periodic table, indicate which atom in each of the following pairs (atomic numbers given) would be expected to have the higher ionization potential: (a) 56 and 88; (b) 6 and 7; (c) 8 and 16; (d) 12 and 13. Explain the bases for your choices.

22. In the case of calcium, an electron is more stable in the 4s orbital than in the 3d orbital, but the reverse is true for zinc. Discuss these observations.

23. Using Slater's rules calculate the effective nuclear charge for: (a) 3s electron in chlorine; (b) 3d electron in cobalt; (c) 4s electron in calcium.

24. Which of the following pairs has the greater radius: (a) the element with atomic number 35 or the most probable ion formed by that element; (b) the element with atomic number 18 or the element with atomic number 19; (c) the element with atomic number 22 or the most probable ion of that element? Explain your answers without recourse to a periodic table.

25. Which of the following pairs has the greater radius: (a) the element with atomic number 11 or the most probable ion formed by that element; (b) the element with atomic number 14 or the element with atomic number 32; (c) the element with atomic number 16 or the most probable ion formed by that element; (d) the element with atomic number 18 or the element with atomic number 19? Explain your choices without referring to a periodic table.

26. For the purposes of this question, assume that the following are true: (1) the spin quantum number does not exist; (2) all other quantum numbers and the relationships among them are unchanged; (3) the relative energies of the levels are given first by the values of the principal quantum numbers (the lowest quantum number corresponds with the lowest energy), and within a given principal quantum shell the sublevel with the lowest value of

l has the lowest energy; and (4) the Pauli exclusion principle is valid. (a) Generate a table showing the combination of quantum numbers available for assignment according to the Pauli exclusion principle and the maximum number of electrons that can be placed in each of the first four shells and their corresponding subshells. (b) Sketch the outline of a periodic table for the elements which would correspond to filling the first four shells.

2

IONIC COMPOUNDS

Of the two main classes of compounds recognized by Lewis, ionic compounds are perhaps the most easily understood from a theoretical viewpoint. To a very good first approximation, ionic structures are collections of independent, oppositely charged ions with a discrete size and shape. Thus it might be expected that ionic crystals are formed because the electrostatic interactions (both attractive and repulsive) between ions lead to the most efficiently packed crystals. Our discussion of ionic substances follows two main themes: the geometry of ionic substances, and the energetics of the formation of such structures. For each of these themes, a unified treatment of the experimental results is presented first, followed by the theoretical interpretations.

1 / GEOMETRY OF IONIC CRYSTALS (*I*)

1.1 / Experimental Results

The majority of solids are crystalline in nature; that is, they are composed of atom-sized units that are packed in an orderly way. In ionic substances the building blocks are either simple or complex charged species that are held together by electrostatic forces. Many ionic compounds form crystals with easily discernable plane faces that are arranged in characteristic ways with respect to each other (Fig. 2.1). Indeed, with well formed crystals that

FIG. 2.1 Under appropriate conditions, naturally occurring substances often form large crystals with well defined faces at characteristic angles: (a) sulfur, (b) quartz (SiO_2), (c) calcite ($CaCo_3$), (d) fluorite (CaF_2).

are sufficiently large to permit precise measurements of the angles between face-normals, it is possible to identify substances on the basis of crystal morphology. The process is not often used by chemists because of practical limitations.

Sometimes a substance can crystallize in several different characteristic habits (i.e., in several forms) depending upon the mode of formation. For example, although the common habit of sodium chloride (Fig. 2.2a) is the familiar cubic form it is possible to obtain octahedral crystals (Fig. 2.2d–f) from a solution containing $\approx 10\%$ urea; other possible modifications of the cubic habit are also shown in Fig. 2.2. In such cases it is not unusual to obtain a single crystal that exhibits the characteristics of both habits.

The regular external features of crystals suggest the presence of an ordered microscopic structure. Indeed, Haüy in the last quarter of the 18th century (1784) proposed that a crystal was composed of fundamental units the shape of which reflected the symmetry of the macroscopic crystal. It was not until

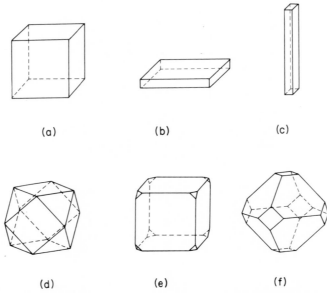

FIG. 2.2   Some variations of the habit of a cubic crystal brought about by external factors. The cubic habit (a) can be elongated parallel to an edge (b, c) or octahedral faces may be more or less apparent (d, e, f).

1913, when Von Laue discovered that x rays could be diffracted by crystalline substances, that a technique was developed to probe the microscopic structure of matter which eventually proved the essential correctness of Haüy's intuitive suggestion.

All crystals consist of a basic pattern of atoms that is repeated a very large number of times in much the same way as is the pattern in a piece of cloth or wallpaper. An example of a repeating array in two dimensions is shown in Fig. 2.3. This pattern can be reproduced by moving a fundamental unit along straight lines in two dimensions. For example, the parallelogram-shaped area designated *ABCD* in Fig. 2.3 can be used to build up the entire pattern. It should be immediately apparent that there are many other units present (such as *EFGH* and *IJKL*) which could serve as the basis for the formation of the overall pattern. Similar relationships occur for three-dimensional crystal structures. The basic repeating unit in a crystal is called a unit cell. As in the case of the two-dimensional example shown in Fig. 2.3, the formulation of *the* unit cell for a crystal is somewhat arbitrary. By convention and as a matter of convenience, unit cells are defined so that they are parallelepipeds with the shortest sides possible and with atoms at the corners. A unit cell with these characteristics is the smallest repeating unit with the full symmetry of the structure of the macroscopic crystal.

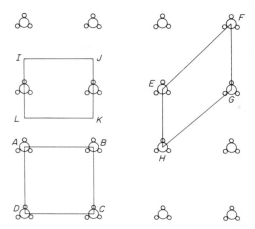

FIG. 2.3 A two-dimensional crystal can be generated by the translation of a fundamental repeating unit. Often the choice of such a unit is arbitrary.

A survey of experimentally observed structures of ionic compounds with the stoichiometries MX and MX_2 shows that only relatively few basic structural types are prevalent. Rather than devise class names for the various structural types, the class is given the name of a common substance which exhibits the structural type in question.

Sodium-Chloride-Type Structure   Sodium chloride and a variety of other ionic uni-univalent ionic substances crystallize in a cubic structure. The sodium-chloride-type structure consists of two interpenetrating face-centered cubic lattices; the unit cell for sodium chloride is shown in Fig. 2.4. An important descriptive feature of crystal structures is the number of nearest neighbors about each ion. In the case of ionic crystals, the nearest neighbors of an ion of a given charge are ions with the opposite charge. Thus, in the case of sodium chloride (Fig. 2.4) each sodium ion is surrounded by six chloride ions which can be imagined as occupying the corners of an octahedron; a similar arrangement of sodium ions exists about each chloride ion. The number of nearest neighbors for a given ion in a crystal is called the coordination number; we can speak of the coordination number of either the cation or the anion. In the case of the sodium-chloride-type structure, both the cation and the anion exhibit octahedral coordination, which is a natural consequence of the lattice type. We shall see that the cation and anion need not have the same coordination number in all structural types.

All of the alkali-metal halides (except CsCl, CsBr, and CsI), the oxides and sulfides of the alkaline earths, and AgF, AgCl, and AgBr possess the sodium-chloride structure. Many other compounds with the general stoichiometry

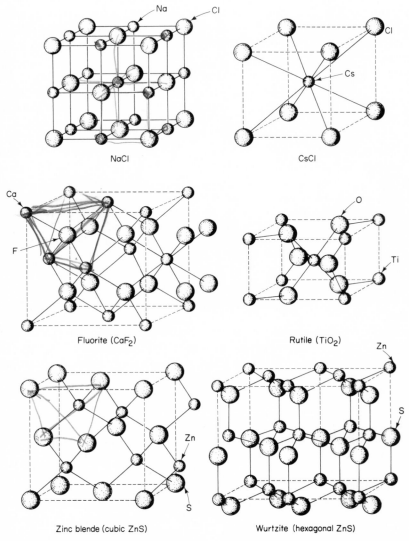

FIG. 2.4   Some of the more common structural types found for ionic substances.

MX in which either the cation, the anion, or both are complex ions rather than simple monoatomic ions also exhibit the sodium-chloride-type structure. Thus, the ionic compounds $[Co(NH_3)_6][TiCl_6]$ and NH_4I crystallize in the sodium-chloride structure; in these cases the complex ions are, in effect, spherically symmetrical. In some instances nonspherical ions can achieve an effective spherical symmetry because sufficient kinetic energy can be attained to permit the ion to rotate freely in the crystal lattice. Common examples of such behavior are the high-temperature forms of KOH, KSH, and KCN, all of which possess the sodium-chloride structure. At lower temperatures, where rotation is restricted, these substances attain different structures.

Many ionic substances of the form MX which contain complex ions that cannot attain spherical symmetry by rotation within the lattice crystallize in a distorted sodium-chloride-type structure. For example, the structure of CaC_2 is essentially that of sodium chloride except that the lattice is distorted in one direction because of the shape of the C_2^{2-} ions which are all arranged so that their long axes are parallel (Fig. 2.5). Similarly, $CaCO_3$

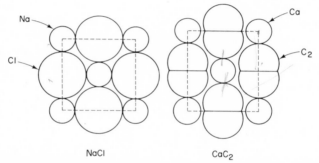

NaCl   CaC₂

FIG. 2.5 The unit cell for CaC_2 is a sodium-chloride-type structure in which the lattice has been distorted to accommodate the linear C_2^{2-} ions. Only one layer of the structure is shown here.

and $NaNO_3$ crystallize in a distorted cubic structure that can be related to the NaCl structure by replacing the spherical anions in the latter by non-planar CO_3^{2-} or NO_3^- ions. Since these complex ions are flat, the basic NaCl-type structure distorts to accommodate them (Fig. 2.6).

Cesium-Chloride-Type Structure   The structure of cesium chloride consists of two interpenetrating cubic lattices of Cs^+ and Cl^- ions; the ions of the one lattice are located at the midpoints of the diagonals of the cubes which are formed by the other ions. The unit cell for this structure, shown in Fig. 2.4, is a cube with ions of one kind at the corners and a single ion of the opposite kind at the center. This basic structure is often called body-centered

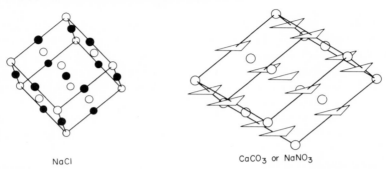

NaCl   CaCO₃ or NaNO₃

FIG. 2.6 Both $CaCO_3$ and $NaNO_3$ crystallize in a sodium-chloride-type structure which is distorted to accommodate the flat anionic units (CO_3^{2-} or NO_3^-). This process, of course, leads to a noncubic lattice.

cubic (bcc). The coordination number of both the cation and anion is 8 (cubic coordination).

The cesium-chloride structure is exhibited by the cesium halides (except CsF), TlCl, the rubidium halides (except RbF) at pressures above 5000 kg cm^{-2}, and RbCl at low temperatures. As is the case in the sodium-chloride-type structure, certain complex anions are nearly spherically symmetrical in the crystal. Thus, $K[SbF_6]$, $Ag[NbF_6]$, $[Be(H_2O)_4]SO_4$, and $[Ni(H_2O)_6][SnCl_6]$ contain nearly spherical complex ions and crystallize in the CsCl-type structure. The anions in CsCN, TlCN, and CsSH attain spherical symmetry by rotation within the crystal lattice, and these substances exhibit the CsCl structure; the anions in NH_4CN and the low-temperature form of $K[HF_2]$ cannot, so these substances crystallize in a distorted CsCl-type structure.

Zinc-Blende-Type Structures   The other common structure observed for binary compounds is that exhibited by ZnS (zinc blende). As shown in Fig. 2.4, the cations are surrounded by a tetrahedral array of anions. The zinc-blende-type structure arises from a cubic closest-packed arrangement of anions in which one-half of the tetrahedral holes are filled with cations (Fig. 2.15a). In addition to ZnS, the sulfur, selenium, and tellurium compounds of zinc, cadmium, and mercury attain the zinc-blende structure.

Many compounds that crystallize in the zinc-blende-type structure also can attain the wurtzite-type structure, shown in Fig. 2.4, which is closely related. Compounds which exhibit two crystalline forms are said to be dimorphic. In the wurtzite structure, as in the zinc-blende-type structure, the anions form a tetrahedral array about the cations; however, the anions occupy a hexagonal closest-packed structure rather than a cubic closest-packed structure.

The Fluorite-Type Structures   A large number of metal difluorides or dioxides crystallize in the fluorite structure. The crystal structure of calcium fluoride (fluorite) is based upon a unit cell with tetrahedral co-ordination of the anions and cubic coordination of the cations (Fig. 2.4). The difluorides of the larger cations (Cd, Ca, Hg, Sr, Pb, and Ba) as well as some lanthanide and actinide dioxides (ThO_2, CeO_2, HfO_2, NpO_2, PuO_2, AmO_2, PoO_2, CmO_2, PrO_2, UO_2, and ZrO_2) crystallize in the fluorite structure. The oxides, sulfides, selenides, and tellurides of lithium, sodium, and potassium crystallize in a structure geometrically identical to the fluorite structure, with the cations and anions interchanged. This structure is known as the antifluorite structure.

The Rutile-Type Structures   The other common structure observed for MX_2 compounds is exemplified by that of TiO_2 (rutile), the unit cell of which is shown in Fig. 2.4. In the crystal of TiO_2 the cations occupy the positions of a body-centered lattice. Each cation possesses a coordination number of 6 and the anions have three coplanar neighbors. The difluorides of magnesium, nickel, cobalt, iron, manganese, and zinc, as well as a large number of transition- and heavy-metal dioxides (GeO_2, SnO_2, MnO_2, RuO_2, OsO_2, IrO_2, CrO_2, PbO_2) crystallize in rutile-type structures.

Defect Structures   Our discussion of ionic crystals has centered about the generation of a perfect structure by a repeating unit cell. In fact, many crystals are imperfect in the sense that they contain defects. Two basic types of defects are known in ionic crystals. A Frenkel defect occurs when an

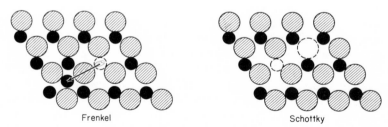

Frenkel   Schottky

FIG. 2.7   Basically two types of defect structures are known. In the Frenkel defect an anion occupies a nonlattice site, whereas a Schottky defect has a cation–anion pair missing from the lattice.

ion occupies a nonlattice site instead of its normal site (Fig. 2.7). Frenkel defects are often observed in compounds such as AgBr, which consist of cations and anions of widely divergent sizes. Silver bromide crystallizes in the sodium-chloride-type structure, which can be imagined as an array of bromide ions with silver ions occupying octahedral holes (Fig. 2.15b). A

Frenkel defect occurs in AgBr when a silver ion vacates the octahedral hole and occupies a tetrahedral hole instead. Schottky defects occur when both a cation and an anion are missing from a structure, leaving vacancies (Fig. 2.7).

Defects are localized irregularities, which can have a marked effect on the properties of a crystalline substance. For example, defects are important in the photochemical decomposition of solids such as silver salts; in reactions involving heterogeneous catalysis, as occur when Cu_2O catalyzes the oxidation of CO to CO_2; and in the luminescence of compounds used for cathode-ray screens. Zinc sulfide fluoresces blue when exposed to ultraviolet radiation or cathode rays after it is heated above 500°C. This phenomenon appears to be associated with a defect structure which possesses vacant sulfur positions. Heating zinc sulfide to 1100°C with a trace of copper, which becomes incorporated within the lattice, gives a defect structure which fluoresces yellow-green when irradiated with cathode rays.

Ionic Radii (*11*)   If, as crystal structures suggest, ionic compounds can be described as collections of independent ions, it might be expected that ions possess characteristic radii. The usual x-ray diffraction data give reasonably precise values of internuclear distances (i.e., the parameters needed to define the unit cell). For example, the internuclear distance, $d(M-X)$, between the cation and anion in binary substances that crystallize in the sodium-chloride-type structure (Fig. 2.4) is related to the length a of the unit cell by the simple relationship $d(M-X) = a/2$. Similar relationships can be derived for any structural type, and a list of internuclear distances could readily be compiled from structural data. But it is not possible to extract individual ionic radii directly from such data. There are, however, indications in the experimental values for internuclear distances that a reasonably consistent set of ionic radii can be obtained. Only one of the arguments is presented here.

If discrete radii can be associated with ions, it would be expected that the differences between radii for different ions should be constant. Although the individual ionic radii cannot be extracted from internuclear distances the differences can. Thus, for example, the difference in radii between Br^- and Cl^-, $r(Br) - r(Cl)$, should be independent of the counter ion:

$$r(Br) - r(Cl) = d(M-Br) - d(M-Cl). \tag{1}$$

In such arguments, of course, it is assumed in order to eliminate possible geometrical effects that ionic compounds of the same structural type are involved. The internuclear distances for several representative pairs of compounds, shown in Table 2.1, exhibit reasonably constant differences.

The problem of dividing internuclear distances into a proper set of self-consistent ionic radii was solved by finding structures in which anion–anion

TABLE 2.1 Differences in the internuclear distances for some alkali-metal halides exhibiting the sodium-chloride-type structure

Ion	Counter ion			
	Li	Na	K	Rb
Br	2.75	2.97	3.29	3.44
Cl	2.57	2.81	3.14	3.29
$r(Br) - r(Cl)$	0.18	0.16	0.15	0.15
	Cl	Br	I	
K	3.14	3.29	3.53	
Na	2.81	2.97	3.23	
$r(K) - r(Na)$	0.33	0.32	0.30	

contact occurred; in such cases half the anion–anion internuclear distance can logically be assigned to the anion radius. Landé, arguing that anion–anion contact should occur in the ionic compound of the smallest cation and the largest anion, suggested that half the I—I distance in LiI should represent the radius of the iodide ion. Lithium iodide crystallizes in the sodium-chloride-type structure so that $d(I—I)$ can be obtained from the internuclear distance $d(Li—I)$ by simple geometry (Fig. 2.8).

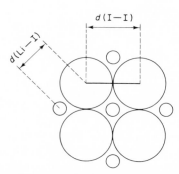

FIG. 2.8 If the anions are in contact in the LiI structure, the anion–anion distance $d(I—I)$ can be obtained from the internuclear distance $d(Li—I)$ by simple geometry. The figure shows a layer of the sodium-chloride-type structure for LiI.

Several other sets of compounds can be used to prove the validity of the Landé-type approach. A consideration of the internuclear distances for magnesium oxide $[d(Mg—O) = 2.10 \text{ Å}]$, sulfide $[d(Mg—S) = 2.60 \text{ Å}]$, and selenide $[d(Mg—Se) = 2.73 \text{ Å}]$, and for manganese oxide $[d(Mn—O) = 2.24 \text{ Å}]$, sulfide $[d(Mn—S) = 2.59 \text{ Å}]$, and selenide $[d(Mn—Se) = 2.73 \text{ Å}]$, which all have the sodium-chloride-type structure, suggests that anion–anion contact occurs in the case of the sulfur and selenium compounds. There is no reason to expect that the radius of Mn^{2+} and Mg^{2+} ions should be the same, and the values for $d(Mg-O)$ and $d(Mn-O)$ support this suggestion. The values of $d(Mn-S)$ and $d(Mg-S)$ [or $d(Mg-Se)$ and $d(Mn-Se)$] represent the distance between the sulfide ion and the center of the hole in which the cation finds itself; since this distance is the same for two sulfides with different-size cations, the only logical conclusion is that anion–anion contact occurs in the sulfides and selenides. Since these compounds exhibit the sodium-chloride-type structure, the radius of the ions [one-half $d(S-S)$ and $d(Se-Se)$] can be calculated according to the method shown in Fig. 2.8, giving $d(S-S)/2 = 1.84\,\text{Å}$ and $d(Se-Se)/2 = 1.93\,\text{Å}$.

Once the radius of one ion is known, it is possible to establish the crystal radii of all other ions from the known internuclear distances. By starting with different reference radii it is possible to establish a reasonably self-consistent list of ion radii such as those listed in Table 2.2.

TABLE 2.2   The empirical radii (in Å) of some common ionic species

									O^{2-}	F^-
Li^+	Be^{2+}								1.45	1.33
0.68	0.30									
Na^+	Mg^{2+}	Al^{3+}							S^{2-}	Cl^-
0.98	0.65	0.45							1.90	1.81
K^+	Cu^{2+}	Ga^+	Ga^{3+}	Ge^{2+}	Ge^{4+}				Se^{2-}	Br^-
1.33	0.94	1.13	0.60	0.93	0.54				2.02	1.96
Rb^+	Sr^{2+}	In^+	In^{3+}	Sn^{2+}	Sn^{4+}				Te^{2-}	I^-
1.48	1.10	1.32	0.81	1.12	0.71				2.22	2.19
Cs^+	Ba^{2+}	Tl^+	Tl^{3+}	Pb^{2+}	Pb^{4+}	Bi^{3+}	Bi^{5+}			
1.67	1.29	1.45	0.91	1.21	0.81	1.16	0.74			

Theoretically, the electron-density maps that can be obtained from a detailed analysis of electron-diffraction data should yield ionic radii (*1–3*). For example, Fig. 2.9 shows a cross section through the nuclei for a pair of lithium and fluoride ions in a crystal. It might be supposed that the point of minimum electron density along the lithium-fluoride internuclear distance divides this distance in proportion to the sizes of the ionic radii. However, it is apparent that the empirical radii (Table 2.2) do not correspond to the radii inferred from the point of minimum density. The empirical radius of the F^- ion (Table 2.2) is about 18% greater than the radius deduced from the minimum (1.13 Å) shown in Fig. 2.9; the empirical radius of the Li^+ ion is correspondingly smaller. A similar analysis of the electron-density map for calcium fluoride shows that the radius of F^- in this compound is

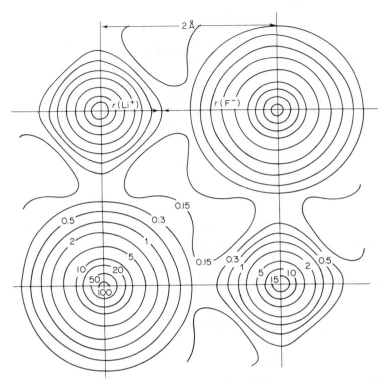

FIG. 2.9 Experimental electron densities for a pair of LiF units in crystalline LiF. Contours are designated in terms of electrons per cubic angstrom unit. The empirical radii of the ions are also indicated.

1.10 Å. In the few cases of ionic crystals for which detailed electron-density maps have been obtained, the density minimum gives a larger value for cation radii and smaller values for anion radii than the corresponding empirical radii. It is interesting to note that for the two substances containing a common ion (LiF and CaF_2) the minimum-electron-density criterion gives reasonably good agreement in spite of the difference in coordination number. If for these structures the larger cation radii are accepted, it is apparent that the concept of anion contact is in question. Unfortunately, detailed electron distribution maps have been obtained for only a small number of ionic structures, so that it is difficult to draw significant conclusions concerning such results. At present the empirically derived ionic radii are the most widely used.

In principle the electron distribution for an ion could be deduced from the type of quantum-mechanical calculations described in Chapter 1. Figure 2.10 shows the theoretical radial electron distribution curve for a

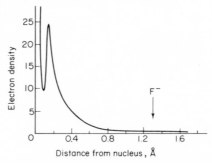

FIG. 2.10   The theoretical radial distribution curve for the fluoride ion. The empirical fluoride radius is indicated.

free fluoride ion. Unfortunately, the asymptotic decrease in the curve makes it difficult to establish an unambiguous "radius". The distance which corresponds to the empirical ionic radius is marked with an arrow in Fig. 2.10; this distance can be altered appreciably without sensibly affecting the total electron density encompassed by the ion. Since quantum-mechanical calculations on a free ion are difficult to interpret because of the ambiguity of defining the boundary of the ion, a better approach to this problem involves similar calculations on ions constrained in a lattice. Unfortunately, such a calculation poses many difficult problems. However, Fig. 2.11 shows the radial distribution curves for Li^+ and H^-, the origins of the curves being placed at the ends of a line representing the experimentally observed internuclear distance in lithium hydride. The minimum electron density occurs at a point corresponding to a radius of 0.70 Å for Li^+ which is in good agreement with the experimental value of 0.68 Å.

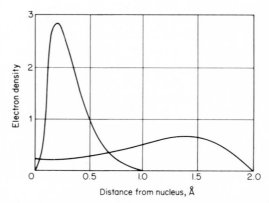

FIG. 2.11   The radial distribution curve for LiH. The electron density is given in arbitrary units. In this calculation the ions have been placed at their relative positions as observed experimentally.

1.2 / Theoretical Considerations of Crystal Structures

If a simple ionic substance such as sodium chloride can be imagined to consist of ions of more or less discrete sizes packed together in the most efficient way because of nondirectional electrostatic forces, it should be possible to predict the most stable three-dimensional arrangement of ions by considering the way in which spheres of different relative sizes can be closely packed. Before discussing this particular point, it may be best to describe the close-packing arrangements of spheres of equal size.

Spheres of equal size can be arranged on a plane surface so that their centers lie at the points of a triangular network (Fig. 2.12a); each sphere touches six others. It is impossible to obtain a greater two-dimensional density of spheres than the arrangement shown in Fig. 2.12a. A three-dimensional close-packed structure can be obtained by placing successive layers of close-packed spheres upon each other. The second layer of spheres can be added to the first in only one way with respect to the first layer to give two close-packed layers, that is by positioning the centers of spheres in the second layer over the holes marked b or c that are formed between three spheres of the first layer (Fig. 2.12b). Assume that the centers of the spheres in the second layer lie over the holes designated as b; because of the size of the spheres, the holes marked c cannot also be covered by the spheres in the second layer. There are now two possible ways in which the third layer of spheres can be stacked upon the second layer. The centers of the spheres in the third layer can be placed either over the centers of the spheres in the first layer (Fig. 2.12c) or they can be displaced so that the centers of spheres in the third layer lie above the holes designated as c in the first layer (Fig. 2.12d). Thus, if A represents the arrangement of spheres in a given layer and the symbols B and C represent the different arrangements possible in the subsequent layers, close-packed layers can be arranged in the sequence $ABAB$, etc., or $ABCABC$, etc. There are, of course, an infinite number of other possible sequences such as $ABABABCABC$, but we shall restrict our discussion to the two simple repeating sequences. In any close-packed structure of equal-sized spheres, each sphere is in contact with twelve other spheres and the spheres occupy 74.02% of the available space within the structure.

The $ABAB$ sequence of close-packed layers has a hexagonal unit cell (Fig. 2.13a), whereas the $ABCABC$ sequence possesses cubic symmetry (Fig. 2.13b). In the latter case the unit cell is a face-centered cube. An alternative description for this structure is the close-packed face-centered structure. The body diagonal of the face-centered cube is perpendicular to the layers used to describe the $ABCABC$ close-packing structure (Fig. 2.13b).

The next most dense structure which equal-sized spheres can attain is the body-centered cubic structure. Each sphere touches only eight neighbors,

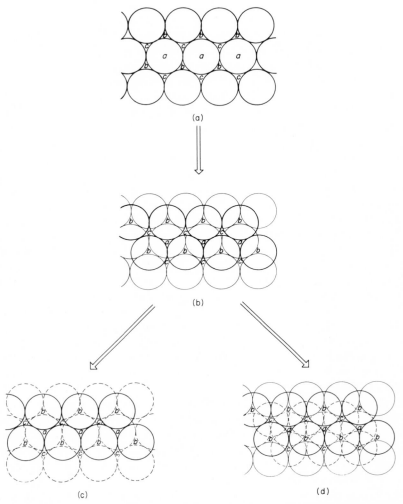

FIG. 2.12   Closest-packing structures can arise in two ways. The first layer (a) contains spheres on a plane surface with their centers at the points of a triangular network. The second layer (b) contains a similar arrangement, but the centers of the spheres are displaced with respect to the first layer. The third layer can be arranged in one of two ways. If the center of the spheres of the third layer occupy positions over those of the first layer (c) the arrangement is *ABAB*, which is hexagonal closest packing. The other possible orientation of the third layer (d) is different from that of the other two layers, leading to an *ABCAB* arrangement which is cubic closest packing.

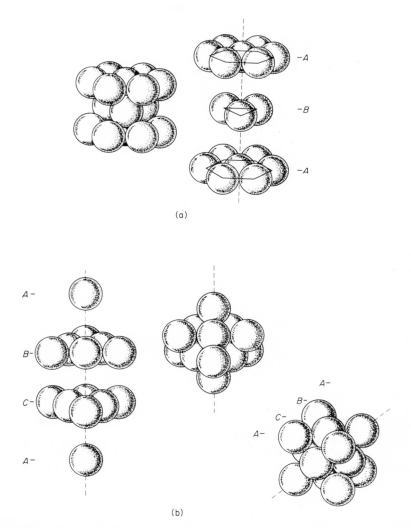

(a)

(b)

FIG. 2.13 The two basic closest-packed structures are (a) hexagonal closest packing where the layers are arranged *ABAB*, and (b) the cubic closest-packed structure with a layer arrangement of *ABCABC*.

and the spheres in this structure occupy 68 % of the available space within the unit cell.

Of all the substances it might be expected that the metals could be composed of equal-sized close-packed spheres. Indeed, many of the metallic elements crystallize in one or both of the closest-packed structures (Fig. 2.14). In addition, the rare gases crystallize in the closest-packing face-centered cubic structure, as do some molecules that have a generally spherical symmetry, such as CH_4, HCl, HBr, and HI.

Li	Be										
B	H										
Na	Mg	Al									
B	H	C									
K	Ca	Sc	Ti	V	Cr	Mn	Fe	Co	Ni	Cu	Zn
C	B, C	H	B	B	B	B	C	C	C	C	H
Rb	Sr	Y	Nb	Nb	Mo	Tc	Ru	Rh	Pd	Ag	Cd
B	C	H	B	B	B	H	H	C	C	C	H
Ca	Ba	La	Hf	Ta	W	Re	Os	Ir	Pt	Au	Hg
B	B	C	B	B	B	H	H	C	C	C	—

FIG. 2.14   The crystal structures of some of the metals. B = body-centered, H = hexagonal closest-packed, C = cubic closest-packed structures.

Ionic Crystals Derived from Closest-Packing Structures   The cubic closest-packing structure can be considered as the parent structure for a variety of ionic crystals. A consideration of this structure (Fig. 2.12d) shows that there are two different types of holes present which are described in terms of the geometry of the spheres that touch to form them. The holes formed by four spheres touching (Fig. 2.15a) are called tetrahedral sites, while the holes formed when six spheres touch are called octahedral sites (Fig. 2.15b). Two layers in the cubic closest-packed structure are redrawn in Fig. 2.16, showing both tetrahedral and octahedral sites. Since each sphere in a cubic closest-packing structure is touching three spheres in the layer above it and three in the layer below it, there are two tetrahedral sites associated with each sphere. A similar consideration shows the presence of one octahedral site for each sphere in the cubic closest-packing structure. Three of the common structures described for ionic substances (Fig. 2.4) can be derived from the cubic closest-packing structure by assuming that the anions occupy the 14 positions of the face-centered cube and that the cations are placed in either the tetrahedral or octahedral sites. For example,

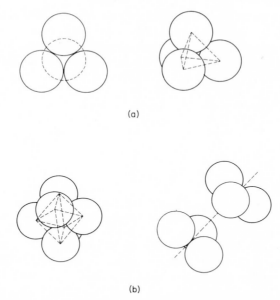

(a)

(b)

FIG. 2.15 Two types of interstices exist in the cubic closest-packing arrangement of equal-sized spheres (cf. Fig. 2.12d). (a) A tetrahedral hole is formed when four spheres touch. (b) An octahedral hole is formed where six spheres touch.

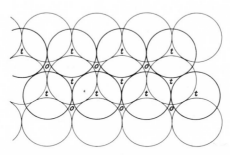

FIG. 2.16 Tetrahedral (t) and octahedral (o) holes in the first two layers of a closest-packed structure.

the sodium-chloride-type structure shown in Fig. 2.4 can be derived from a face-centered-cubic closest-packed array of anions in which all of the octahedral holes are filled with cations; there is no net charge on the structure since the ratio of octahedral sites to spheres is unity. The fluorite (and anti-fluorite) structure (Fig. 2.4) arises if all the tetrahedral holes are filled with cations; if half the tetrahedral holes are filled the zinc-blende structure shown in Fig. 2.4 is formed.

Radius-Ratio Effects   If ions have more or less constant characteristic radii, the most efficient arrangement of anions about cations in a crystal should be governed by the relative sizes of the ions involved. For example, if we limit our discussion to the planar arrangements shown in Fig. 2.17, three

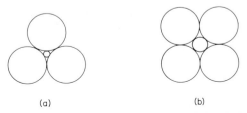

(a)   (b)

FIG. 2.17 The planar-trigonal arrangement of anions (a) has a smaller hole available for a cation than does a square-planar arrangement (b).

anions might easily fit about a cation; however, if the cation size were increased a point would be reached where a fourth anion of the same size could be introduced into the structure. Thus there is a range of cation radii (for a constant anion size) for which the three-coordinate structure would be the most suitable. Cations of a size greater than the values in this range could attain a four-coordinate structure. In other words, cations can achieve either three- or four-coordination for this planar arrangement, depending upon their sizes. Of course, these arguments are valid for three-dimensional structures, although they may be more difficult to visualize.

Since the sizes of *both* the cation and the anion change from one compound to the next it is more logical to discuss the possible modes of coordination with respect to the *ratio* of the radii involved. Assuming that ions are hard spheres, it is possible to determine the radius-ratio (r_+/r_-) limits from a consideration of simple geometry for each coordination number; the results of these calculations are summarized in Table 2.3. The most stable coordination number for a given cation–anion combination is that

TABLE 2.3   Range of radius-ratio values for various cationic coordination numbers

Coordination number	Anion arrangement	Radius ratio, r_+/r_-
2	linear	$\leqslant 0.15$
3	triangular	0.15–0.22
4	tetrahedral	0.22–0.41
4	planar	0.41–0.73
6	octahedral	0.41–0.73
8	cubic	> 0.73

which is dictated by the radius ratio according to the results in Table 2.3. Since each of the commonly observed structure types shown in Fig. 2.4 has characteristic coordination numbers, it is possible to predict the crystal structure for any pair of ions within the limits of these assumptions. For example, from the stoichiometry of the compounds MgF_2 and SrF_2, it is difficult to predict into which of the possible common structures they crystallize, i.e., fluorite or rutile (Fig. 2.4). However, the ionic radii (Table 2.2) lead to radius ratios (MgF_2, 0.60; SrF_2, 0.97) which indicate that the magnesium ion is six-coordinate whereas the strontium ion is eight-coordinate. The common six-coordinate structures are the sodium-chloride-type and the rutile-type structures, but since the former structure contains one anion for each cation, magnesium fluoride should crystallize in the rutile-type structure. Similar arguments lead to the conclusion that strontium fluoride should crystallize in the fluorite-type structure. X-ray diffraction data support these predictions.

Although the validity of these arguments has been verified repeatedly, as shown by data of the type in Table 2.4, there are some instances where an apparent contradiction occurs. For example, zirconium oxide (ZrO_2), on the basis of radius-ratio arguments, should be six- (or four-planar-) coordinate because the radius ratio is 0.62 ($r_{Zr}/r_O = 0.62$). Experiment shows that zirconium oxide crystallizes in the fluorite structure in which the cation is

TABLE 2.4 The radius ratios and structures of ionic compounds

Compound	Radius ratio	Structure type	Cation coordination number
BeO	0.22	wurtzite	4
MgTe	0.29	wurtzite	4
MgSe	0.33	NaCl	6
MgS	0.35	NaCl	6
CaTe	0.45	NaCl	6
MgO	0.46	NaCl	6
SiO_2	0.29	quartz	4
MgF_2	0.48	rutile	6
TiO_2	0.49	rutile	6
SnO_2	0.51	rutile	6
ZnF_2	0.54	rutile	6
PbO_2	0.60	rutile	6
$SrCl_2$	0.62	fluorite	8
CdF_2	0.71	fluorite	8
CeO_2	0.72	fluorite	8
CaF_2	0.73	fluorite	8
HgF_2	0.81	fluorite	8
SrF_2	0.83	fluorite	8
BaF_2	0.99	fluorite	8

eight-coordinate. The fact that the observed coordination number is larger than that calculated from the ion sizes indicates that somehow additional space is made available for the insertion of more anions about the cation. The apparent dilemma is resolved by recognizing the premise upon which the radius-ratio arguments are developed, namely, that ionic substances are composed of discrete units that are held together solely by electrostatic forces. Since the radius ratio for zirconium oxide does not correspond to the structure expected for a collection of Zr^{4+} and O^{2-} ions, the structure must not be ionic. This conclusion is compatible with the increase in coordination, which could be interpreted on the basis of an interpenetration of the ionic electron clouds. Since an interaction of electron clouds corresponds roughly to the Lewis description of a chemical bond (Chapter 1), the interaction between the units in compounds such as zirconium oxide can be described as partially covalent. Thus, ZrO_2 is best described as a polymeric species in which the "molecule" is really the entire crystal.

2 / ENERGETICS OF CRYSTAL FORMATION (*III, IV*)

The stability of an ionic substance arises from the energy gained from the electrostatic interaction of the constituent ions arranged in the manner dictated by their radius ratio. There are, of course, interactions between like-charged ions, leading to repulsive forces, and between unlike-charged ions, leading to attractive forces. In any stable ionic crystalline array the latter must be more important than the former.

2.1 / Experimental Results

The energy associated with the formation of a compound composed of M^+ and X^- ions arises from the formation of a lattice from an equivalent number of anions and cations in the gas phase at infinite separation, as in the following equation:

$$M^+_{(g)} + X^-_{(g)} \rightarrow M^+X^-_{(s)}. \tag{2}$$

The nature of the lattice structural type is, of course, governed by the radius-ratio rules (Table 2.3). The evaluation of the lattice energy, i.e., the energy associated with the process shown in Eq. (2), has been most conveniently accomplished from a thermochemical cycle incorporating Eq. (2), an approach first used by Born and Haber. As an illustration of this argument, consider the alkali halides, for which the necessary thermochemical quantities are readily available. The cycle generally used for these compounds, given in Fig. 2.18, involves the heat of formation ΔH_f of the alkali-metal halide, the sublimation energy ΔH_s and ionization potential I for the alkali metal, the dissociation

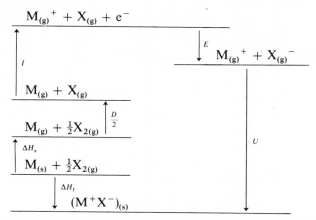

FIG. 2.18 The energy terms associated with the Born–Haber cycle for the alkali–metal halides. The usual thermochemical conventions are given, i.e., the sign of the energy associated with an exothermic reaction is negative.

energy D and electron affinity E for the halogen, and the lattice energy U of the crystalline substance; the lattice energy and the lattice enthalpy ΔH_U are related by

$$\Delta H_U = U + 2RT, \tag{3}$$

if it is assumed that U does not change significantly with temperature; for many calculations the factor RT (≈ 1.2 kcal mole^{-1} at 0°C) is negligible. The cycle in Fig. 2.18 represents two different paths to form the same substance, M^+X^-. Both paths must be of equivalent energy, a constraint which is expressed by

$$U = \Delta H_f - \Delta H_s - I - \tfrac{1}{2}D - E. \tag{4}$$

In the specific cycle described in Fig. 2.18 the reactions representing ΔH_f and D are exothermic, and the numerical values substituted for these symbols, as usually defined, are negative; the reactions represented by the other symbols are endothermic. In the case of the alkali halides the energies for the individual steps have been independently obtained from a variety of experiments, so that the lattice energies for these substances can be determined from Eq. (4). The requisite data and the resultant lattice energies are listed in Table 2.5.

Several interesting features are immediately apparent in the data of Table 2.5. First, the lattice energies of the alkali halides are exothermic, reflecting the stability of these crystalline substances. Indeed, the magnitude of the lattice energy markedly outweighs the other energy terms of the cycle. In addition, the lattice energy varies with ion-size parameters; for example, for a given cation the lattice energy decreases in the order Cl > Br > I, which is

TABLE 2.5   Data used to calculate the lattice energies for the alkali halides[a]

Salt	$-\Delta H_f$	ΔH_s	I	$D/2$	$-E^b$	U
LiF	145.7	38.4	124.4	18.9	79.5	247.9
NaF	136.3	35.9	118.4	18.9	79.5	220.0
KF	134.5	21.5	100.0	18.9	79.5	195.4
RbF	131.8	19.5	96.3	18.9	79.5	187.0
CsF	135.1	18.7	89.7	18.9	79.5	182.9
LiCl	96.0	38.4	124.4	28.9	83.3	204.4
NaCl	98.2	25.9	118.4	28.9	83.3	188.1
KCl	104.2	21.5	100.0	28.9	83.3	171.3
RbCl	103.4	19.5	96.3	28.9	83.3	164.8
CsCl	106.9	18.7	89.7	28.9	83.3	160.9
LiBr	83.7	38.4	124.4	26.8	77.6	195.7
NaBr	86.0	25.9	118.4	26.8	77.6	179.5
KBr	93.7	21.5	100.0	26.8	77.6	164.4
RbBr	93.0	19.5	96.3	26.8	77.6	158.0
CsBr	97.7	18.7	89.7	26.8	77.6	155.3
LiI	64.8	38.4	124.4	25.5	70.6	182.5
NaI	68.8	25.9	118.4	25.5	70.6	168.0
KI	78.3	21.5	100.0	25.5	70.6	154.7
RbI	79.0	19.5	96.3	25.5	70.6	149.7
CsI	83.9	18.7	89.7	25.5	70.6	147.2

[a] All values given in kcal mole^{-1} at 298°K.
[b] From Ref. 14.

the order of increasing anion size (Table 2.2). Similar trends are present with respect to cations. This is precisely the type of variation expected for a system the energy of which is governed primarily by electrostatic interactions.

Thermochemical cycles of the type shown in Fig. 2.18 can be constructed for various types of ionic compounds. Sometimes, however, sufficient independent data are not available to obtain lattice energies; in such cases, it is often possible to calculate the lattice energies theoretically, as is described in the next section.

Finally, it must be stressed that the lattice energies calculated from the Born–Haber cycle are experimental and *not* dependent upon the assumptions made about the nature of the bonding in the crystal.

2.2 / Theory

If an ionic crystal consists of charged particles that are held together by electrostatic forces, it should be possible to calculate the lattice energy from first principles. The general approach used in this type of calculation, first proposed by Born and Landé (4), is illustrated for compounds possessing the sodium-chloride-type structure (Fig. 2.19). In this analysis it is assumed that the ions are hard spheres possessing discrete charge, size, and arrangement, which give the system the lowest possible electrostatic energy. Concentrating attention on one of the cations in the structure (Fig. 2.19) reveals the fact

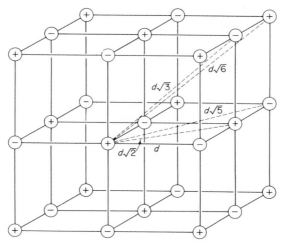

FIG. 2.19   Distances of neighboring ions in a sodium-chloride-type lattice.

that it is surrounded by six anions at a distance d, twelve cations at a distance $d\sqrt{2}$, six cations at $2d$, 24 anions at $d\sqrt{5}$, and so on. In other words, the cation under consideration experiences alternating attractive and repulsive interactions that are related to the crystal structure and the lattice parameters. Assuming that the charged ionic spheres can be replaced by point charges at the same internuclear distance (d), the potential energy released when two ions are brought together is given by

$$E = \frac{(Z_+e)(Z_-e)}{d}, \tag{5}$$

where Z_+ and Z_- represent the charges on the ions. For the collection of ions in the sodium-chloride-type structure the interaction of each ion with the central ion is expressed by Eq. (5), and the total energy for these interactions becomes the summation of all such terms:

$$E = -6\frac{(Z_+e)(Z_-e)}{d} + 12\frac{(Z_+e)(Z_-e)}{d\sqrt{2}} - 8\frac{(Z_+e)(Z_-e)}{d\sqrt{3}} + 6\frac{(Z_+e)(Z_-e)}{2d}$$
$$- 24\frac{(Z_+e)(Z_-e)}{d\sqrt{5}} + \cdots \tag{6}$$

$$E = -\frac{Z_+Z_-e^2}{d}\left(6 - \frac{12}{\sqrt{2}} + \frac{8}{\sqrt{3}} - \frac{6}{2} + \frac{24}{\sqrt{5}} + \cdots\right). \tag{7}$$

The quantity in parentheses in Eq. (7) is an infinite series dependent only on the geometry of the crystal (shown in Fig. 2.19). In the case of sodium-chloride-type structures the series converges toward 1.747558, which is called the Madelung constant for this structure. Other arrangements of ions give rise to different series that have also been evaluated (Table 2.6) (5, 6, V, VI). Thus the total potential energy associated with the sodium-chloride-type structure is given by

$$E = 1.747558 \frac{(Z_+ Z_-)e^2}{d}. \tag{8}$$

TABLE 2.6   Values of the Madelung constant for various structures[a]

Sodium chloride	1.747558
Cesium chloride	1.762670
Zinc blende (ZnS)	1.63806
Wurtzite (ZnS)	1.641
Fluorite (CaF$_2$)	5.03878
Rutile (TiO$_2$)	4.816
Anatase (TiO$_2$)	4.800
Cadmium iodide	4.71
β-Quartz (SiO$_2$)	4.4394
Corundum (Al$_2$O$_3$)	25.0312

[a] Values given are for Z defined as the highest common factor of the ionic charges.

The theoretical model used to derive Eq. (8) is not quite correct, since ions are not hard spheres with the indicated charges. Rather, both the cation and anion possess *negatively* charged electron clouds at their limits, which repel each other strongly as the ions approach (V). It was suggested by Born that the repulsive energy which arises from the interaction of ion electron clouds at short distances takes the following form:

$$E_r = \frac{B}{d^n}, \tag{9}$$

where B and n are constants characteristic of the geometry of the crystal. The total energy of the system is the sum of the attractive [Eq. (8)] and repulsive [Eq. (9)] terms:

$$E_r + E = -1.747558 \frac{(Z_+ Z_-)e^2}{d} + \frac{B}{d^n}. \tag{10}$$

For Avogadro's number N of cations and anions (one mole of substance) the lattice energy is given by

$$U = -1.747558N\frac{(Z_+Z_-)e^2}{d} + \frac{NB}{d^n}. \tag{11}$$

The quantity B may be evaluated by using the fact that the energy of the system is a function of the internuclear distance [Eq. (11)] and must be a minimum when the ions achieve the internuclear distance observed in the crystal ($d = d_0$). Under these conditions

$$\left(\frac{dU}{dd}\right)_{d=d_0} = 0 = \frac{1.747558N(Z_+Z_-)e^2}{d_0^2} - \frac{nNB}{d_0^{n+1}}. \tag{12}$$

Equation (12) can be rearranged to yield an expression for B:

$$B = \frac{-1.747558N(Z_+Z_-)e^2}{n}d_0^{n-1}, \tag{13}$$

which, when substituted into Eq. (11), gives

$$U = -1.747558N(Z_+Z_-)\frac{Ne^2}{d_0}\left(1 - \frac{1}{n}\right). \tag{14}$$

Similar arguments can be made for expressing the lattice energy of any lattice type, the general expression being

$$U = -A(Z_+Z_-)\frac{Ne^2}{d_0}\left(1 - \frac{1}{n}\right), \tag{15}$$

where A represents the value of the Madelung constant as given in Table 2.6.

Values of the Born exponent n have been obtained from the results of compressibility measurements and have been estimated theoretically by Pauling for ions with a rare-gas structure (Table 2.7) (V). The theoretical and

TABLE 2.7   Values of the Born exponent, n, as determined experimentally and theoretically

Experiment		Theory	
Compound	n	Noble-gas configuration	n
LiF	5.9	He	5
LiCl	8.0	Ne	7
LiBr	8.7	Ar	9
NaCl	9.1	Kr	10
NaBr	9.5	Xe	12

experimental values are in reasonably good agreement, especially if it is recognized that an error of unity in the exponent causes an error of less than 2% in the lattice energy [Eq. (15)]. A more refined treatment for calculating lattice energies, which involves a consideration of Van der Waals forces, is available, but Eq. (15) gives results that are sufficiently reliable for most purposes.

Born and Mayer (7) introduced a variation of Eq. (15) which included a quantity (ρ) related to the quantum-mechanical repulsive forces acting between the ionic electronic shells:

$$U = -A(Z_+Z_-)\frac{Ne^2}{d_0}\left(1 - \frac{\rho}{d_0}\right). \tag{16}$$

The quantity ρ is nearly constant at 0.345 Å for most crystals. As in the case of Eq. (15), d_0 represents the internuclear distance between the cation and anion. In many instances, d_0 can be estimated from single-ion radii, but it should be recalled that for some ionic substances where the size of the crystal lattice is governed by anion–anion contact, single-ion radii do not accurately reflect internuclear distances; in these cases the experimentally determined value of d_0 should be used. The Born–Mayer equation [Eq. (16)] has been extended to include dipole-induced dipole forces in the lattice and the zero-point energy of the lattice (III).

Kapustinskii (8, 9, VII) derived a relationship that gives the lattice energies of ionic substances without the knowledge or assumption of crystal structure. The starting point of the Kapustinskii arguments is Eq. (16). If γ is the number of ions in the formula of a substance, the number in a mole is $N\gamma$. Equation (16) can be rewritten as

$$U = \left(\frac{N\gamma}{2}\right)\left(\frac{\alpha Z_+Z_-e^2}{d_0}\right)\left(1 - \frac{\rho}{d_0}\right), \tag{17}$$

where $\alpha = 2A/\gamma$. Although α is not the same for all lattice types, it was observed empirically that changes in α are proportional to changes in interatomic distance. With these results, every crystal may be transformed into a standard lattice without a change in lattice energy if the constants α and d_0 are correspondingly modified to have values equivalent to the ions in that structure; the sodium-chloride-type lattice was chosen as reference. For the sodium-chloride-type lattice $\alpha = 1.745$, and using the average value of $\rho = 0.345$ Å, Eq. (17) becomes

$$U = \frac{287.2\gamma Z_+Z_-}{d_0}\left(1 - \frac{0.345}{d_0}\right) \text{ kcal mole}^{-1}. \tag{18}$$

The value for d_0 can be estimated from the sum of ionic radii determined for octahedral coordination. The results obtained with Eq. (18) are not, of course, as exact as those obtained from the other theoretical expressions (Table 2.8), but relatively little structural information is necessary to use them.

TABLE 2.8   A comparison of the lattice energies of
some alkali-metal halides as determined in different ways

Salt	Experimental value, kcal mole^{-1}	Calculated values, kcal mole^{-1}		
		Eq. (18)	Eq. (16)	Eq. (14)
LiF	241	228	240	239
NaF	216	212	215	214
KF	192	189	190	189
RbF	184	182	182	181
CsF	171	170	174	172
LiI	175	171	174	170
NaI	165	161	164	160
KI	151	147	151	148
RbI	147	141	145	143
CsI	140	135	140	135

As is the case with the other theoretical expressions, the replacement of d_0 by the sum of ionic radii $(r_+ + r_-)$ represents the most serious objection to this approach.

Often the lattice energy for a substance calculated from Eqs. (14), (16), or (18) does not agree with that determined from thermochemical cycles. Since thermochemical cycles reflect the energetics for the interaction of the components of the lattice and are not dependent upon identification of the forces involved, the source of such discrepancies must be in the calculated values. Since the theoretical calculations are based on purely electrostatic models, the discrepancies must reflect the inadequacy of such models. Thus, substances for which the theoretical and thermochemical lattice energies are in disagreement must be appreciably covalent in character; they are, perhaps, best described as giant molecules, and exhibit structures which incorporate infinite chains (e.g., MCl_2, M = Be, Pd, Cu), sheets (e.g., MCl_2, M = Cd, Mn, Fe, Co, Ni), or a continuous three-dimensional network held together by covalent bonds (e.g., ZnS). The data in Table 2.9 illustrate the magnitudes of the discrepancies observed between the experimental and calculated lattice energies. In every case but one, i.e., MnF_2, considerable differences exist between the calculated and experimental lattice energies, and the CdI_2 structure prevails. The CdI_2 structure consists of MX_6 octahedra sharing edges to form continuous sheets. Generally, the experimental and theoretical

TABLE 2.9   **Lattice energies of some metal dihalides**

Compound	Structure	Lattice energy, kcal mole^{-1}	
		Experiment^a	Theory^b
$MgBr_2$	CdI_2	572	511
MgI_2	CdI_2	553	474
MnF_2	rutile	663	660
$MnBr_2$	CdI_2	583	520
MnI_2	CdI_2	569	480

^{a} From Born–Haber cycle.
^{b} Calculated from Eq. (15).

lattice energies of the halides of polyvalent cations are in good agreement for the fluorides, but the correspondence becomes less impressive with increasing size of the halogen.

3 / APPLICATIONS OF LATTICE-ENERGY CALCULATIONS (9, 10, III, IV)

Thermochemical cycles incorporating the lattice energy of an ionic substance together with the methods described to calculate the lattice energy from crystal parameters have been employed to obtain the energies for processes that are often difficult to determine experimentally. For example, the electron affinity of species can be obtained from Eq. (4) if a value for the lattice energy can be determined using the relationships shown in Eqs. (15), (16), or (18).

A slightly different thermochemical cycle can be employed to obtain proton affinities. An example of such a cycle, which can be used to determine the proton affinity of ammonia, is shown in Fig. 2.20. The proton affinity in this case is defined as the energy involved in the following process:

$$NH_{3(g)} + H^+_{(g)} \rightarrow NH^+_{4(g)}. \tag{19}$$

The proton affinity for this process is given by

$$P(NH_3) = -U + I - E + \Delta H_f(NH_{3(g)}) + \tfrac{1}{2}D(H_2) + \tfrac{1}{2}D(Cl_2)$$

$$- \Delta H_f(NH_4Cl). \tag{20}$$

Thus, a knowledge of all the energy parameters shown on the right side of Eq. (20) yields a value for $P(NH_3)$. The average value for $P(NH_3)$ using the

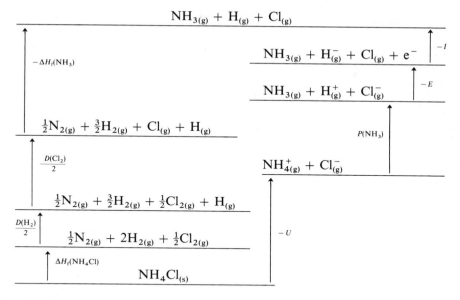

FIG. 2.20   A thermochemical cycle that can be used to estimate the proton affinity P of ammonia.

available data for the four ammonium halides is 214 kcal mole^{-1} (*III*). Using similar methods the following proton affinities (in kcal mole^{-1}) were obtained: NH_2^-, 393; NH^{2-}, 613; H_2O, 182; OH^-, 375; SH^-, 342; O^{2-}, 554; S^{2-}, 550.

Thermochemical cycles incorporating lattice energies have been used to estimate the heats of formation for compounds the existence of which are in doubt. For example, although AuCl and AuBr are known compounds, AuF has never been isolated. Application of equations incorporating the Madelung constant to calculate lattice energies requires the assignment of a structure to the compound in question as well as the estimation of the interionic distance. If the Kapustinskii equation [Eq. (18)] is employed, only estimates of interionic distances are necessary.

The stability of AuF can be estimated from the Born–Haber cycle using Eq. (4), for which all the energy terms are known except ΔH_f and U. The latter can be estimated from the Kapustinskii equation by assuming that the interionic distance can be obtained from the sum of the ionic radii: $r_+ = 1.37$ Å, $r_- = 1.36$ Å, $d_0 = 2.73$ Å. The Kapustinskii equation then yields a value of 186 kcal mole^{-1} for the lattice energy. Thus, the heat of formation can be calculated from the following series of equations, which form the Born–Haber cycle:

$$
\begin{array}{lcr}
 & & \text{Energy,} \\
 & & \text{kcal mole}^{-1}
\end{array}
$$

$$\text{Au}_{(s)} \rightarrow \text{Au}_{(g)} \qquad\qquad +88 \qquad\qquad (21)$$

$$\text{Au}_{(g)} \rightarrow \text{Au}_{(g)}^{+} + e^{-} \qquad\qquad +214 \qquad\qquad (22)$$

$$\tfrac{1}{2}\text{F}_{2(g)} \rightarrow \text{F}_{(g)} \qquad\qquad +19 \qquad\qquad (23)$$

$$\text{F}_{(g)} + e^{-} \rightarrow \text{F}_{(g)}^{-} \qquad\qquad -84 \qquad\qquad (24)$$

$$\text{Au}_{(g)}^{+} + \text{F}_{(g)}^{-} \rightarrow \text{AuF}_{(s)} \qquad\qquad -183 \qquad\qquad (25)$$

$$\text{Au}_{(s)} + \tfrac{1}{2}\text{F}_{2(g)} \rightarrow \text{AuF}_{(s)} \qquad\qquad +54 \qquad\qquad (26)$$

Since the heat of formation [Eq. (26)] is positive, the reaction as written is endothermic. The *free energy* of formation is, of course, the thermodynamic criterion for the stability of compounds, and the heat of formation must be corrected for entropy effects to obtain this quantity. The application of methods for estimating the entropies of formation of substances that are available (*11*) gives a value of about -5 kcal mole^{-1} which makes the free energy of formation of AuF $+59$ kcal mole^{-1}. The large positive value clearly suggests that AuF is thermodynamically unstable with respect to its elements. Many more examples of this type of argument are available in the literature (*12*).

The radii of ions can be estimated using the equations which relate lattice energies to crystallographic parameters by obtaining the lattice energies of the substances in question from thermochemical cycles. The most useful equation for this type of calculation has been the Kapustinskii equation [Eq. (18)]. Radii estimated by this technique have been designated *thermochemical radii*. This method is very useful for compounds which contain polyatomic ions (Table 2.10).

TABLE 2.10   Thermochemical radii of some complex ions

Ion	BrO_4^{3-}	IO_4^{-}	BeF_4^{2-}	MnO_4^{-}	CrO_4^{2-}	PO_4^{3-}	ClO_4^{-}	SO_4^{2-}	BF_4^{-}
Radius, Å	2.68	2.49	2.45	2.40	2.40	2.38	2.36	2.30	2.28

4 / SEMITHEORETICAL VALUES OF CRYSTAL RADII (*3, 13*)

The radii listed in Table 2.2 are derived from a consideration of special crystal structures for which it is possible to decide a priori the proportions into which certain internuclear distances should be divided. Pauling developed a semitheoretical basis for dividing internuclear distances into ionic radii by assuming that the size of an ion is dependent upon the distribution of its outermost electrons. The latter do not feel the effect of the full nuclear charge Z but a somewhat diminished charge because of the intervening filled electron shells. This effective nuclear charge, Z_{eff}, can be calculated from the nuclear charge if the screening effect of the closed electron shells is known:

$$Z_{eff} = Z - S. \qquad (27)$$

Values of the screening constant S have been obtained from theoretical calculations. Slater (*1*) has shown that each electron in a multielectron system can be represented by a wave function for an electron moving about a central charge which is less than that of the true nuclear charge, the difference between the nuclear charge and the charge experienced by the electron being called the screening constant S. The screening constant for a given species can be evaluated from the following rules:

1. An electron is not shielded by other electrons in orbitals of higher energy.
2. The ns and np electrons are considered as a group; the values of the screening constants for these electrons are:
   0.35 for each of the other electrons in the group;
   0.85 for each of the electrons in the next innermost group;
   1.00 for each of the electrons in still lower groups.
3. The nd and nf electrons are considered as a group; the values of the screening constants for electrons in this group are:
   0.35 for each of the other electrons in the group under consideration;
   1.00 for each of the other electrons in groups closer in to the nucleus.
4. A $1s$ electron has a screening constant of 0.30 with respect to the other $1s$ electron.

Pauling suggested that the radius of an ion is inversely related to the effective nuclear charge:

$$r_{ion} = \frac{C_n}{Z_{eff}} = \frac{C_n}{Z - S}, \qquad (28)$$

which can be calculated from the nuclear charge and Slater's rules [Eq. (27)]; C_n is an empirically derived constant dependent upon the quantum number of the outermost electrons.

Assuming that in a given crystal the anion and cation are in contact:

$$d_{MX} = r_M + r_X, \qquad (29)$$

it should be possible to evaluate C_n and determine the ionic radii for an isoelectronic pair of ions from Eqs. (28) and (29) if the internuclear distance is known experimentally. Thus, for the ionic compound KCl which crystallizes in the sodium-chloride-type structure the internuclear distance is 3.14 Å:

$$d = r_{K^+} + r_{Cl^-} = 3.14 \text{ Å}, \tag{30}$$

and the screening constant for the isoelectronic pair of ions K^+ and Cl^- $(1s^2 2s^2 3s^2 3p^6)$ is 11.60:

$$S = 2(1.00) + 8(0.85) + 8(0.35) = 11.60. \tag{31}$$

According to the arguments embodied in Eq. (28), the radii of these ions are given by

$$r_{K^+} = \frac{C_n}{Z_{K^+} - S} = \frac{C_n}{19 - 11.60} \tag{32}$$

$$r_{Cl^-} = \frac{C_n}{Z_{Cl^-} - S} = \frac{C_n}{17 - 11.60}. \tag{33}$$

Equations (30), (32), and (33) can be solved for C_n (1.82 Å) which can then be used to calculate r_{K^+} and r_{Cl^-} from Eqs. (32) and (33). Similar calculations can be made for other isoelectronic pairs of ions in Period 3 using the same value for the screening constant. For example, the radius of the sulfide ion, which is isoelectronic with K^+ and Cl^-, can be calculated from an expression of the type shown in Eq. (32) using the same value of the screening constant; this calculation gives $r_{S^{2-}} = 2.23$ Å. The same method can be used to estimate the value of C_n and the screening constant for ions with a different rare-gas-type electronic structure (for example, LiF and NaCl). Using calculations of this type, Pauling established a table of semitheoretically derived crystal radii (Table 2.11). A comparison of the radii of univalent species in Table 2.11 with the same species in Table 2.2 shows that the correspondence between Pauling's method and those based on structural arguments is very good. There is, however, less agreement when the results for polyvalent ions are compared. The problem, of course, lies in the assumptions made in the Pauling calculation, which treats all the ions as if each were univalent; only in the case of the alkali-metal halides are the ions present uni-univalent. In all other cases the ions that are really multivalent are treated as if they retain their rare-gas-type electronic structures but interact as if they were univalent. As expected, the "uni-univalent radii" (as the radii obtained from the Pauling treatment are called) for the multivalent ions are larger than the experimental crystal radii (cf. Table 2.11). The uni-univalent radii for multivalent ions can be converted into a crystal radii by accounting for the real charge of these ions. The

TABLE 2.11 The uni-univalent radii of some simple ions and the corresponding empirical crystal radii[a]

Li^+	Be^{2+}	O^{2-}	F^-
0.60	0.44	1.76	1.36
(0.68)	(0.30)	(1.45)	(1.33)
Na^+	Mg^{2+}	S^{2-}	Cl^-
0.95	0.82	2.19	1.81
(0.98)	(0.65)	(1.90)	(1.81)
K^+	Cu^{2+}	Se^{2-}	Br^-
1.33	1.18	2.32	1.95
(1.33)	(0.94)	(2.02)	(1.96)
Rb^+	Sr^{2+}	Te^{2-}	I^-
1.48	1.13	2.50	2.16
(1.48)	(1.10)	(2.22)	(2.19)

[a] Empirical crystal radii are given in parentheses.

equilibrium interionic distance in a crystal can be related to constants characteristic of the substance using the general form of Eq. (12), which can be rearranged to give

$$d_0 = \left(\frac{nB}{A(Z_+ Z_-)} \right)^{1/(n-1)}. \tag{34}$$

Assuming that the values of n and B were the same for a multivalent ion and the corresponding uni-univalent ion $(Z_+ = Z_- = 1)$, the crystal radius should be related to the uni-univalent radius by the expression

$$r_{crys} = r_{uni} Z^{-2/(n-1)}, \tag{35}$$

where Z is the charge on the multivalent ion. The crystal radii in Table 2.11 have been calculated from the uni-univalent radii listed in that table and Eq. (35) using the values for n in Table 2.7. A comparison of the semitheoretical crystal radii derived by the Pauling technique (Table 2.11) and the corresponding radii deduced from experimental data (Table 2.2) are in remarkably good agreement.

5 / THE PROPERTIES OF IONIC CRYSTALS

Although ionic substances consist of charged particles, the mobility of these particles is very small in the solid state. Accordingly, ionic crystals are poor conductors of electricity. Since the electrostatic forces which hold ionic crystals together are relatively large (Table 2.5), it is not surprising that the melting and boiling points of these substances are correspondingly high. In fact, the melting points of a series of simple binary substances such as the alkali-metal halides decrease regularly with internuclear distances (Fig. 2.21). Since these distances are related to the lattice energies the trend in melting points is easily understood. As might be expected, an increase in the charge on the ions increases both the lattice energy and the melting point.

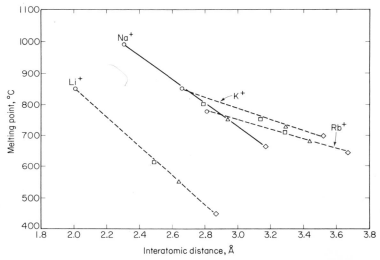

FIG. 2.21 The melting points of the alkali metal halides are linear functions of the internuclear distances in the crystal lattice (circle = F^-, square = Cl^-, triangle = Br^-, diamond = I^-).

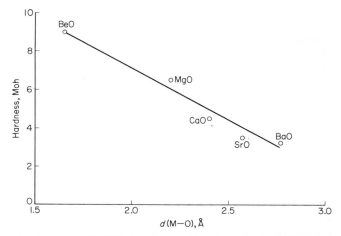

FIG. 2.22 The hardness of an ionic substance is related to the internuclear distance.

The relatively strong electrostatic forces between ions in a crystal lattice should be reflected in the hardness of ionic substances. The data in Fig. 2.22 show a progressive increase in hardness with decreasing internuclear distance; increasing the ionic charge on the ions of a crystal also increases the hardness.

REFERENCES

1. J. C. Slater, *Phys. Rev.*, **36**, 57 (1930).
2. D. R. Hartree, *The Calculation of Atomic Structures*, John Wiley and Sons, Inc., New York, 1957, p. 125.
3. S. Z. Goldberg, *J. Chem. Ed.*, **45**, 638 (1968).
4. M. Born and A. Landé, *Verhandl. Deut. Phys. Ges.*, **20**, 210 (1918).
5. E. Madelung, *Phys. Z.*, **19**, 524 (1918).
6. D. H. Templeton, *J. Chem. Phys.*, **23**, 1826 (1955).
7. M. Born and J. E. Mayer, *Z. Phys.*, **75**, 1 (1932).
8. A. F. Kapustinskii, *Z. Phys. Chem.*, **22B**, 257 (1933).
9. G. J. Moody and J. D. R. Thomas, *J. Chem. Ed.*, **42**, 204 (1965).
10. K. H. Stern, *J. Chem. Ed.*, **46**, 645 (1969).
11. W. E. Dasent, *Nonexistent Compounds*, Marcel Dekker, Inc., New York, 1965.
12. W. M. Latimer, *Oxidation Potentials*, 2nd ed., Prentice-Hall, Inc., New York, 1952, p. 359.
13. L. Pauling, *J. Amer. Chem. Soc.*, **49**, 765 (1927).
14. R. S. Berry and C. W. Reimann, *J. Chem. Phys.*, **38**, 1540 (1963).

COLLATERAL READINGS

I. J. D. Dunitz and L. E. Orgel, *Advan. Inorg. Chem. Radiochem.*, **2**, 1 (1960). A discussion of the stereochemistry of ionic solids.
II. K. H. Stern and E. S. Amis, *Chem. Rev.*, **59**, 1 (1959). A thorough discussion of the methods used to estimate ionic radii.
III. T. C. Waddington, *Advan. Inorg. Chem. Radiochem.*, **1** (1959). A discussion of the calculation of lattice energies using a variety of approaches. Included are the results of calculations on numerous alkali and alkaline-earth metal salts, as well as uses to which such calculations can be put.
IV. M. F. C. Ladd and W. H. Lee, *Progr. Solid State Chem.*, **1**, 37 (1964); *Progr. Solid State Chem.*, **2**, 378 (1965). These two papers contain information of the type described for Collateral Reading *III*; however, it is more extensive in all areas.
V. J. Sherman, *Chem. Rev.*, **11**, 93 (1932). A discussion of lattice energies of ionic compounds and their thermochemical applications.
VI. R. Hoppe, *Angew. Chem.* (Intl. Ed.), **5**, 95 (1966). An analysis of the Madelung constants.
VII. A. F. Kapustinskii, *Quart. Rev.* (London), **10**, 283 (1956). A discussion of the semiempirical Kapustinskii equation and its application to ionic systems.

STUDY QUESTIONS

1. The x-ray diffraction powder pattern for solid krypton shows that this substance exhibits the cubic closest-packed structure. The density of solid krypton is 3.5 g cm^{-3}. (a) How many atoms of krypton are present in a unit cell? (b) What are the dimensions of the unit cell for solid krypton? (c) Estimate the radius of the krypton atom.

2. Estimate the length of the edge of the unit cell of crystalline RbBr.

3. Calculate the theoretical radius ratio for a cubic arrangement of anions about a cation for the case where adjacent anions touch both each other and the cation.

4. Determine the coordination number of magnesium in MgF_2 from the ionic radii. Sketch the unit cell for this coordination number.

5. Discuss the observation that ZrO_2 crystallizes in the fluorite structure.

6. Determine the expected crystal structure for aluminum oxide assuming that it is an ionic substance.

7. Discuss and/or criticize the following statement: Both silicon dioxide and stannic oxide crystallize in the zinc-blende structure.

8. Thallium can exhibit two oxidation states (1+ and 3+). What is the expected coordination number of the cation in the compounds TlCl and $TlCl_3$? Sketch the structure for the most probable unit cell for TlCl.

9. Assuming that cadmium oxide (CdO) is ionic, what is the expected crystal structure for this compound?

10. Using the known interionic distance in NaCl (2.81 Å) estimate the crystal radii of Mg^{2+} and S^{2-}.

11. Estimate the value of the Madelung constant for the CsCl structure to three significant figures.

12. Estimate the lattice energy for TlF_3.

13. Using the Kapustinskii equation as a basis, estimate the effective radius of the hexachloroplatinate anion ($PtCl_6^{2-}$).

14. Estimate the energy involved in the formation of one mole of LiF dimer from LiF ion pairs in the vapor state.

15. Describe a method, based on a thermochemical cycle, for determining the heat of formation of cuprous fluoride (CuF). Indicate which parameters involved might be obtained from the literature.

16. Many of the properties of silver bromide suggest that this substance is appreciably covalent. Knowing that silver bromide crystallizes in the sodium-chloride structure, (a) determine the lattice energy of AgBr, and (b) estimate the extent of the nonionic contribution to the energy of this system.

17. Using thermochemical arguments, discuss (a) the experimental data and (b) the assumptions necessary to estimate the electron affinity of the species NH_2^-.

18. There has been some speculation that an ionic rare-gas compound with the stoichiometry Xe^+F^- can be prepared. Estimate the lattice energy for this hypothetical compound.

19. Assuming lithium hydride to be an ionic substance, derive an expression for the *sum* of the electron affinity of hydrogen and the lattice energy for lithium hydride in terms of a suitable Born–Haber cycle. Incorporate the fact that lithium exists in the form of diatomic molecules in the vapor state.

20. Using calcium oxide as a basis for your calculations, show how you would estimate the total electron affinity of oxygen. In your description identify the quantities involved or give the method of calculation of these quantities.

21. Lithium iodide is soluble in anhydrous ether as well as in water. Show how you would compare the relative enthalpies required to solvate the lithium ion,

$$Li_{(g)}^+ + x\,(solvent) \longrightarrow Li\,(solvent)_x^+,$$

in both solvents, assuming that the energy for the process

$$I_{(g)}^- + (solvent) \longrightarrow I\,(solvent)^-$$

is known.

22. The structures of the aurous halides (AuX) in the solid state are unknown. Using the value 1.37 Å for the ionic radius of Au^+, sketch the unit cell of the structure expected for *each* of these compounds.

23. Using *one* of the compounds in Problem 22, discuss its structure in relation to either the cubic or the hexagonal closest-packed structure.

24. Explain the observation that magnesium oxide and sodium fluoride have the same crystal structure, but magnesium oxide is almost twice as hard as sodium fluoride.

/ **3**

COVALENT
COMPOUNDS

EXPERIMENTAL OBSERVATIONS

As is the case with ionic compounds, any theory which successfully deals
with the formation of covalent compounds must address itself to the observed
geometrical arrangement of atoms in molecules as well as to the energetics
involved in the formation of these structures. This chapter contains a summary of structural data and experimentally observed bond parameters for a
variety of covalently bound species; the theoretical arguments concerning
the constitution of covalently bound species appear in Chapter 4.

1 / THE GEOMETRY OF COVALENTLY BOUND SPECIES

The structures of thousands of compounds containing covalently bound
atoms have been elucidated (I, II) using x-ray- and electron-diffraction
methods as well as a variety of other techniques. We shall concentrate our
discussion on the results obtained with a variety of structural methods and

not attempt to discuss the methods themselves; several good sources exist for this purpose (*III, IV*). Generally these results indicate that some atoms form bonds to as few as one other atom, but in other cases an atom can be bound to ten other atoms. The spatial distribution of the atoms at a given site is generally restricted to only a few geometrical types. The simple Lewis theory of the nature of a covalent bond (Chapter 1) can be used as a basis for correlating the observed structures of covalent species. Initially, Lewis suggested that some atoms form sufficient electron-pair bonds to attain the electronic structure of the next rare gas; in the case of the period 2 elements, the stable electronic structures involve eight electrons in a valence shell. But obviously a group of eight electrons cannot be a requisite for stability if compounds such as PF_5 and SF_6, which have all the characteristics associated with covalently bound compounds, are considered. If electron-pair bonds in the Lewis sense are present in PF_5 and SF_6, a valence shell containing ten or twelve electrons is then possible; indeed, in subsequent chapters we shall have the opportunity to see that species with fourteen to twenty electrons in the valence shell are possible. Thus, the major contribution of the Lewis theory is the formulation of the electron-pair covalent bond, and not the specification of the total number of electrons necessary to form a stable valence shell. In this discussion we shall consider that the electron-pair occupies a volume of space—an orbital—without initially attempting to characterize the orbital, for it is our present purpose to discuss structural generalizations without recourse to extensive theoretical arguments.

If the concept of an electron-pair bond is accepted, it is possible to approximate the known structure of covalent species by considering the distribution of electron pairs about each atom (*V, VI*). Consider, as an example, the compounds CH_4, NH_3, and H_2O, the Lewis formulations of which appear in Fig. 3.1. The central atom in each of these molecules is surrounded by four

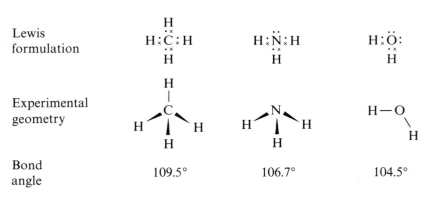

FIG. 3.1   The structures of the isoelectronic compounds CH_4, NH_3, and H_2O can be related if the number of electron pairs on the central atom is considered.

electron pairs; in CH_4 all the pairs are involved in bond formation, whereas one and two electron pairs are not in NH_3 and H_2O, respectively. If we consider the electron pairs in the Lewis formulation, whether they are bonded or not, as stable entities, the spatial distribution of these electron pairs about the central atom can be deduced from simple electrostatic arguments. The minimum electrostatic interaction for four electron pairs would occur if they attained a tetrahedral distribution in space. Thus, the geometrical structures, i.e., the distribution of atoms about the central atom, of CH_4, NH_3, and H_2O should be related because each of the central atoms is surrounded by four electron pairs. In the case of CH_4 all pairs are bonding pairs and the hydrogen atoms should be tetrahedrally disposed about the central carbon atom. Of the four electron pairs about the nitrogen atom in NH_3, three are involved in bonding. Thus, the geometrical arrangement of atoms in NH_3 is pyramidal although the arrangement of electron pairs should be nearly tetrahedral. Finally, the geometrical arrangement of atoms in H_2O should be related to the tetrahedral distribution of the four electron pairs about oxygen. In the case of water only two electron pairs are involved in bond formation; this situation leads to the prediction that water should be a bent molecule with a bond angle that is nearly tetrahedral. The structures predicted for CH_4, NH_3, and H_2O on the basis of electron-pair repulsion arguments are in good agreement with experimental results (Fig. 3.1). The H—C—H bond angles in methane are tetrahedral (109.5°) to within experimental error, whereas the H—N—H and H—O—H angles are close to that value.

These simple electron-pair repulsion arguments can be improved if the relative repulsions of bonding and nonbonding electron pairs are considered. Since a nonbonding pair is constrained by only one atomic nucleus, in contrast to a bonding pair, it should be more diffuse and, hence, closer to other adjacent electron pairs on the central atom. Accordingly, nonbonded pairs should interact electrostatically more strongly than a bonded electron pair. Thus, the electrostatic repulsions between different types of electron pairs should decrease in the following order: lone-pair–lone-pair > lone-pair–bond-pair > bond-pair–bond-pair. These considerations lead to the prediction that the bond angle in NH_3 should be somewhat smaller than the tetrahedral angle because the lone-pair–bond-pair repulsion is greater than the bond-pair–bond-pair repulsion. Since there are two lone pairs on the oxygen atom in H_2O, the bond angle in this molecule should be even smaller than the angle in NH_3. Both predictions are borne out by experimental data (Fig. 3.1). These qualitative suggestions concerning the relative magnitude of electron-pair repulsions are also verified by the fact that the H—N—H bond angles for NH_4^+, in which the lone pair of the NH_3 molecule has now become a bonding pair, is within experimental error the tetrahedral angle. The fact that the bond angles in PH_3 (93.8°), AsH_3 (91°), and SbH_3 (91.5°) are even more distorted from the tetrahedral values than those in ammonia also supports the relative order of repulsions of the different types of electron pairs mentioned earlier; the nonbonding pair is less constrained by the nucleus in these cases than in NH_3, since the valence electrons in PH_3, AsH_3, and SbH_3 are in higher quantum levels. Hence the nonbonded electron pairs in PH_3, AsH_3, and SbH_3 interact more strongly with the bonding pairs in these molecules than in NH_3. As in the case of NH_4^+, the known four-coordinate derivatives of the Group VB elements (PR_4^+, AsR_4^+, SbR_4^+) have a tetrahedral arrangement of atoms around the central atom.

These electron-pair repulsion arguments can be applied to predict the structures for many species (V, VI). We need only to establish the arrangement for any given number of electron pairs for which the electrostatic repulsions are maximized. This process is equivalent to finding the configuration of n points on the surface of a sphere such that the distances between the points are maximal. The mathematical solution to the problem is available (1); the results are summarized in Table 3.1 and shown schematically in Fig. 3.2.

TABLE 3.1   Predicted arrangement
of electron pairs about an atom

n	Geometry
2	linear
3	equilateral triangle
4	tetrahedron
5	trigonal bipyramid or square pyramid
6	octahedron
7	monocapped octahedron
8	square antiprism
9	tricapped trigonal prism
10	bicapped square antiprism
11	icosahedron minus one apex
12	icosahedron

There are two arrangements of five points on a sphere that fit these criteria, a trigonal bipyramid or a square pyramid. As with the isoelectronic compounds CH_4, NH_3, and H_2O, the shape of any molecule is determined by the number of bonded and free electron pairs in the valence shell of the central atom. Thus, for example, three electron pairs occupy the corners of an equilateral triangle. If all electron pairs are bonding, the molecule will be planar with bond angles of 120°, but if only two pairs are bonding the molecule will be bent with bond angles somewhat smaller than 120°, because of the repulsion of the nonbonding electron pair. Similar arguments can be developed for four-, five-, and six-electron-pair systems; some of the molecular shapes are given in Fig. 3.3. Specific examples of species exhibiting such structures are given in Table 3.2.

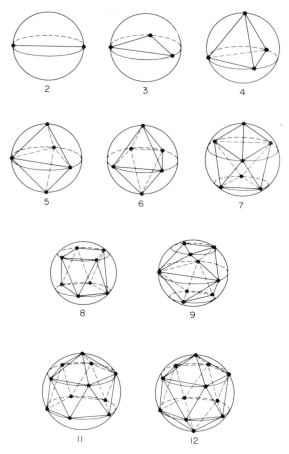

FIG. 3.2 The arrangement of points on the surface of a sphere so that the distance between points is maximized.

A multiple bond contains more than one electron pair and it should occupy a volume of space greater than that of a single bonded electron pair. On this basis we expect a multiple bond to repel a single bonded pair more strongly than the latter repels another single bonded pair. The data in Table 3.3 show that the gross geometries of molecules containing multiple bonds are consistent with this deduction. The experimental bond angles listed in Table 3.3 also reflect the relative magnitudes of the repulsions between multiple and single bonds, i.e., multiple-bonded–single-bonded > single-bonded–single-bonded.

Finally, the small changes in bond angles which occur when the nature of the atom bonded to the central atom changes can also be understood in

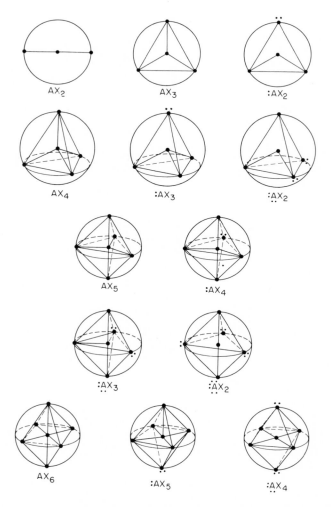

FIG. 3.3 The idealized geometry expected for the arrangement of two to six electron pairs about a central atom. ● Atoms, the bonding electron pair is indicated by lines between atoms; ● : position of nonbonding electron pair.

TABLE 3.2   The arrangement of electron pairs in the valence shell of atoms and their stereochemical consequences

Electron pairs			Arrangement	Molecular type[a]	Molecular shape	Examples
Total	Bonding	Nonbonding				
2	2	0	linear	MX_2	linear	$BeCl_{2(g)}$, HgX_2, ZnX_2
3	3	0	trigonal-plane	MX_3	triangular	BX_3, GaI_3
3	2	1	trigonal-plane	MX_2E	bent	$SnX_{2(g)}$
4	4	0	tetrahedral	MX_4	tetrahedral	CH_4, NH_4^+, GeX_4
4	3	1	tetrahedral	MX_3E	trigonal-pyramid	NX_3, OH_3^+, PX_3
4	2	2	tetrahedral	MX_2E_2	bent	OX_2, SX_2, SeX_2
5	5	0	trigonal	MX_5	trigonal-bipyramid	PX_5, $SbCl_5$, $Sb(CH_3)_3Cl_2$
5	4	1	bipyramid	MX_4E	distorted tetrahedral	SF_4, SeF_4
5	3	2	bipyramid	MX_3E_2	T-shaped	ClF_3, BrF_3
5	2	3	bipyramid	MX_2E_3	linear	ICl_2^-, I_3^-, XeF_2
6	6	0	octahedral	MX_6	octahedral	SF_6, PCl_6^-, AlF_6^{3-}
6	5	1	octahedral	MX_5E	square-pyramid	BrF_5, IF_5
6	4	2	octahedral	MX_4E_2	square	XeF_4, ICl_4^-
7	7	0	pentagonal	MX_7	pentagonal-bipyramid	IF_7
7	6	1	bipyramid	MX_6E	irregular octahedral	$SbBr_6^{3-}$, XeF_6

[a] The symbol E is used to denote a nonbonded electron pair.

TABLE 3.3   The geometries of some molecules containing multiple bonds

Molecule	Lewis formulation	Shape	Angles
CO_2	Ö=C=Ö	linear	$\angle OCO = 180°$
HCN	H—C≡N:	linear	$\angle HCN = 180°$
SO_2	see structure	bent	$\angle OSO = 119.5°$
C_2H_4	see structure	planar	$\angle HCH = 116.8°$ $\angle HCC = 122°$
F_2CO	see structure	planar	$\angle FCF = 108.0°$ $\angle FCO = 126°$
F_2SO	see structure	pyramidal	$\angle FSF = 92.8°$ $\angle FSO = 106.8°$

terms of the electron-pair model. The data in Table 3.4 show that bond angles change when the electronegativity of the substituent changes. Increasing the electronegativity of an atom bound to another atom leads to a decrease in the volume of the bonded electron pair. Accordingly, the interaction between bonded pairs should decrease with an increase in the electronegativity of one of the bound atoms, an effect which is manifested in an increase in the bond angle.

TABLE 3.4   The bond angles in some related compounds

Compound	$\angle XMX$
H_2SO	115.8°
Cl_2O	111.3°
F_2CO	108.0°
$(CH_3)_2SO$	100°
Br_2SO	96°
F_2SO	92.8°
Br_3PO	106°
Cl_3PO	103.6°
F_3PO	102.5°

For the most part, the shapes of the molecules shown in Table 3.2 are almost to be expected on an intuitive basis. However, the stereochemistry associated with 5-coordinate molecules illustrates some interesting problems (*VII*). Neither the trigonal bipyramid, which is the more common structure, nor the square pyramid have equivalent vertices. That is, electron pairs at the axial positions have three nearest neighbors at 90° and are in a different environment than equatorial electron pairs which have only two neighbors at 90°. Generally molecules that exhibit the trigonal bipyramidal structure undergo axial distortion, moving the electron pair with the greatest number of nearest neighbors away from the idealized spherical surface. The magnitude of the effect is seen in Table 3.5. The electron pairs occupying the smallest

TABLE 3.5   Bond distances in some trigonal bipyramidal molecules

	$d(M-X)$, Å	
Molecule	Axial	Equatorial
PCl_5	2.10	2.04
PF_5	1.577	1.534
$P(C_6H_5)_5$	1.987	1.850
$SbCl_5$	2.34	2.29
CH_3PF_4	1.612	1.543

volume should be found at the axial position, where the greatest interaction exists. Thus, when a trigonal bipyramidal structure contains two groups with different electronegativities, the most electronegative group should be in the axial position. This arrangement is found in the compounds $Sb(CH_3)_3X_2$ (X = Cl, Br, I), PF_3Cl_2, $PFCl_4$, R_2PF_3, and R_3PF_2. In each case the most electronegative groups in the compounds occupy the axial positions.

The electron-pair-repulsion model can be applied to aid in predicting the stereochemistry of complex ions containing transition-metal ions. This discussion is delayed until Chapter 17.

In summary, the geometrical arrangement of the atoms in very many complex compounds can be obtained by (a) determining the total number of bonding and nonbonding electron pairs, (b) establishing the spatial distribution of this number of electron pairs using simple electrostatic arguments to obtain the idealized geometry, and (c) taking into account the perturbations arising from the different repulsions of bonding and nonbonding pairs. Subsequent chapters contain many examples of structures that are accurately predicted by this approach.

2 / COVALENT BOND RADII

The suggestion that atoms forming covalent bonds can be assigned a characteristic covalent radius arises from the observed constancy of internuclear distances for compounds containing the same pair of bound atoms. For example, the C—O and C—N bond distances observed for a variety of ethers and amines are remarkably constant at 1.43 Å and 1.47 Å, respectively (Table 3.6). Similar observations can be made for other pairs of covalently bound atoms.

TABLE 3.6   The observed C—O bond distances
in ethers and C—N bond distances in amines

Ether	$d(C-O)$, Å	Amine	$d(C-N)$, Å
$(CH_3)_2O$	1.42 ± 0.03	$(CH_3)_3N$	1.47 ± 0.02
$(CH_2)_2$⟨O,O⟩$(CH_2)_2$	1.44 ± 0.04	CH_3NH_2	1.47 ± 0.02
NH_2OCH_3	1.44 ± 0.02	$(CH_3)_2NCl$	1.47 ± 0.02
$(CH_2=CH)_2O$	1.43 ± 0.03	HC⟨ ⟩NH HC	1.48 ± 0.01
CH_3HC——$CHCH_3$ (O)	1.43 ± 0.03	$(C_2H_5)_2NH$	1.47 ± 0.02

The problem of assigning a set of consistent radii for atoms involved in covalent bonds is essentially the same as that described for the assignment of characteristic ionic radii (Chapter 2). Since only internuclear distances are obtained from x-ray- or electron-diffraction experiments, a method must be devised to divide these experimental distances to give a consistent set of characteristic covalent radii. The problem is further compounded for some covalently bound species because multiple bonds may exist between the atoms; it is not always possible to specify an unambiguous Lewis structure for all molecules, and thus the assignment of a bond distance to a given bond type may be difficult.

There are numerous compounds containing two atoms of the same kind bound to each other for which an unambiguous Lewis formulation containing a single-electron-pair bond can be made. For such substances, it is logical to assume that half the internuclear distance between atoms of the same kind is the single-bond radius of that atom. For example, the Lewis formulation for the diatomic halogen molecules suggests the presence of a single bond in these species; one-half the respective internuclear distance is taken as the radius of the halogen atom. The C—C distance in a variety of alkanes and their derivatives as well as in diamond is 1.54 Å; half this distance is assigned to the single-bond radius of carbon. Single-bond covalent radii for other atoms (M) can be obtained from the known carbon radius and the measured internuclear distance for the methyl derivatives, CH_3—M. The internuclear distances for a few important compounds used to establish single-bond radii are shown in Table 3.7. Using these arguments it has been possible to establish

TABLE 3.7   Bond lengths in selected compounds

Bond	$d(X-Y)$, Å	Compound	Bond	$d(X-Y)$, Å	Compound
C—Br	1.94	CBr_4	N—N	1.47	N_2H_4
C—C	1.54	C_2H_6			
C—C	1.54	diamond	O—H	0.96	H_2O
C=C	1.33	C_2H_4	O—O	1.40	H_2O_2
C≡C	1.20	C_2H_2			
C—Cl	1.77	CCl_4	P—Br	2.28	PBr_3
C—F	1.32	CF_4	P—C	1.84	$P(CH_3)_3$
C—H	1.09	CH_4	P—Cl	2.04	PCl_3
C—N	1.47	CH_3NH_2	P—F	1.54	PF_3
C≡N	1.16	HCN	P—H	1.44	PH_3
C—O	1.42	$(CH_3)_2O$	P—I	2.47	PI_3
C=O	1.22	$(CH_3)_2CO$	P—P	2.21	$P_4{}^a$
C—S	1.82	$(CH_3)_2S$			
			S—Cl	1.99	SCl_2
Ge—Br	2.29	$GeBr_4$	S—F	1.58	SF_6
Ge—C	1.98	$Ge(CH_3)_4$	S—H	1.33	H_2S
Ge—Cl	2.08	$GeCl_4$	S—S	2.08	S_8
Ge—F	1.67	GeF_4			
Ge—H	1.53	GeH_4	Si—Br	2.15	$SiBr_4$
			Si—C	1.89	$Si(CH_3)_4$
N—C	1.47	$N(CH_3)_3$	Si—Cl	2.01	$SiCl_4$
N—F	1.37	NF_3	Si—F	1.55	SiF_4
N—H	1.01	NH_3	Si—H	1.48	SiH_4
			Si—Si	2.35	Si^b

[a] Black phosphorus.
[b] Crystalline.

a reasonably consistent set of single-bond radii (Table 3.8) which can be used to predict internuclear distances. In general, the values predicted for internuclear distances in molecules using the single-bond radii shown in Table 3.8 are in good agreement with experimental values (see Table 3.7). In some cases, however, discrepancies arise which appear to be outside of

TABLE 3.8   Single-bond covalent radii, Å

Be	B	C	N	O	F
0.89	0.80	0.77	0.74	0.74	0.72
	Al	Si	P	S	Cl
	1.25	1.17	1.10	1.04	0.99
	Ga	Ge	As	Se	Br
	1.26	1.22	1.21	1.17	1.14
	In	Sn	Sb	Te	I
	1.44	1.41	1.41	1.37	1.33

experimental error. These apparent discrepancies are discussed in more detail in Chapter 4 where the theoretical aspects of covalent bonding are treated.

Just as it is possible to establish a self-consistent set of radii for singly bound atoms, the radii of atoms involved in multiple bonds can be obtained from appropriate compounds containing the atoms of interest. For example, the double- and triple-bond radii for carbon can be obtained from the C—C distances in ethylene and acetylene, respectively; the Lewis criterion for the number of electrons involved in the C—C bonds in these compounds is unambiguous. Thus, establishment of multiple-bond radii for any atom involves the same types of arguments and experimental data as discussed previously for single-bond radii. Multiple-bond radii for several atoms are shown in Table 3.9.

TABLE 3.9   Multiple-bond covalent radii

	Radius, Å					
Bond order	C	N	O	S	Se	Te
Double bond	0.67	0.60	0.55	0.94	1.07	1.27
Triple bond	0.63	0.55	0.50	0.87	—	—

The most obvious feature in the data of Tables 3.8 and 3.9 is the decrease in covalent radius with an increase in the bond multiplicity (or bond order). Thus, the covalent radii of carbon decrease in the order

$$-\overset{|}{\underset{|}{C}}- > \diagdown_{\diagup}C= > -C\equiv.$$

Similar conclusions are valid for other elements that also form unsaturated compounds. The available data for covalent radii of multiple-bonded atoms in Period 2, i.e., carbon, nitrogen, and oxygen, as well as for sulfur, are plotted in Fig. 3.4. In each case a smooth decrease in covalent radius occurs with increasing bond order.

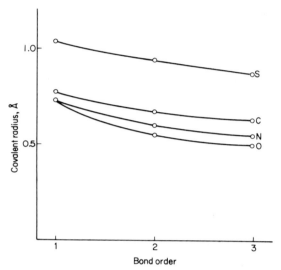

FIG. 3.4 The covalent radii of atoms are smooth functions of the bond multiplicity.

3 / BOND ENERGIES

One of the characteristic properties of a molecule is the energy associated with the formation of the bonds within it. Thus, a system of atoms will form a molecule if the latter is more stable than the former; energy is released in the process. It is usual to assign the energy difference to the formation of covalent bonds. Conversely, energy is necessary to break the bonds in a covalent species to form the corresponding atoms. The energy required for this process is called the dissociation energy. The formation of two atoms from a diatomic molecule represents the simplest example of bond dissociation. For example, the dissociation of molecular hydrogen to give two hydrogen atoms,

$$H_{2(g)} \rightarrow 2H_{(g)}, \tag{1}$$

requires 104.18 kcal mole^{-1} at 298°K.* The bond-dissociation energies $D(X-X)$ of many diatomic molecules have been determined using a variety

---

* Since absolute energies cannot be determined an arbitrary reference is established. A frequently used convention assigns zero enthalpy to each element at 298.15°K (25°C) when it is in the standard state of its most stable form at this temperature.

of experimental techniques. For example, the heat of formation* of a number of compounds has been determined using spectroscopic, photochemical, electron-impact, and thermochemical methods (*VIII, IX*). The results of these measurements for a few diatomic molecules are shown in Table 3.10; it

TABLE 3.10   Bond-dissociation energies[a]

Molecule	$D(X—X)$, kcal mole^{-1}	Molecule	$D(X—X)$, kcal mole^{-1}
CO	256.42	Se_2	64.7
N_2	225.08	ClF	60.6
NO	150.10	Cl_2	58.02
C_2	144	BrF	56.0
HF	134.6	Te_2	53.4
O_2	118.32	BrCl	52.3
P_2	116.9	ICl	50.3
H_2	104.18	Br_2	46.08
HCl	103.16	IBr	42.5
S_2	102	Bi_2	39.7
OH	101.5	F_2	37.8
As_2	90	I_2	36.08
HBr	87.46	Li_2	26.41
HS	85	At_2	22.4
CH	81	Na_2	18.01
HF	71.37	K_2	12.24
Sb_2	69	Rb_2	11.66
B_2	69	Cs_2	10.71

[a] Measurements taken at 298.15°K.

should be noted that some of the molecules listed in Table 3.10 are species not normally encountered under the usual laboratory conditions, for example OH, SH, C_2.

The dissociation energy of more complex molecules is usually a more difficult quantity to determine than the dissociation energy of diatomic molecules (*2*). For example, the dissociation of water can be imagined to occur in two steps,

$$H_2O_{(g)} \rightarrow H_{(g)} + OH_{(g)} \tag{2}$$

$$OH_{(g)} \rightarrow H_{(g)} + O_{(g)}. \tag{3}$$

---

* The heat of formation of a substance is defined as the change in enthalpy ΔH_f accompanying the formation of the substance from its constituent elements in their standard states. For example, the heat of formation of nitric acid corresponds to the enthalpy change for the process

$$\tfrac{1}{2}H_{2(g)} + \tfrac{1}{2}N_{2(g)} + \tfrac{3}{2}O_{2(g)} \rightarrow HNO_{3(l)}.$$

The first step [Eq. (2)] requires 119.7 kcal mole^{-1}, but the dissociation of the hydroxyl radical [Eq. (3)] involves 101.5 kcal mole^{-1}. It is apparent that there is no simple relationship between the respective energies necessary to remove the first and second hydrogen atoms. A detailed analysis of the energy necessary to break a series of bonds sequentially in polyatomic molecules (e.g., CH_4, NH_3, and CO_2) leads to the conclusion that the energies necessary to break bonds between like atoms in a molecule are not generally related. The energies for such processes may be nearly the same, but they are not necessarily equal. Since the energies for the stepwise process of bond dissociation for complex molecules are difficult to determine experimentally, more easily obtained thermochemical data are used to determine that the heat of

$$H_{2(g)} + \tfrac{1}{2}O_{2(g)} \rightarrow H_2O_{(g)} \tag{4}$$

formation of gaseous water is -57.80 kcal mole^{-1}. This datum, together with the dissociation energies for hydrogen and oxygen (Table 3.10), is sufficient to determine the energy involved in the complete dissociation of water into its constituent atoms. The analysis shown in Fig. 3.5 indicates that

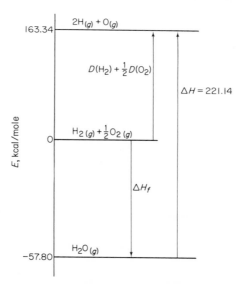

FIG. 3.5 An energy diagram for the dissociation of water.

the total energy for the dissociation of gaseous water is 221.14 kcal mole^{-1}, a value in good agreement with that deduced from the energies of the individual steps [Eqs. (2) and (3)]. If the latter were not known, the average bond-dissociation energy for water would be taken as one-half the total energy or 110.57 kcal mole^{-1}. Similar procedures can be employed to find the average bond energies for polyatomic molecules (e.g., CH_4, C_2H_6, NH_3, CCl_4) from the appropriate thermochemical cycles, and it is possible to establish a consistent set of average bond-dissociation energies. The results for a few systems appear in Table 3.11. Extension of these ideas to complex molecules incorporating multiple bonds has led to the establishment of a set of average multiple-bond energies listed in Table 3.12.

In this chapter we have shown that there are two characteristic properties of covalent species that have received extensive attention: geometry and energetics. A large amount of qualitative structural information can be

TABLE 3.11   Average single-bond energies

Bond	Energy, kcal mole^{-1}	Bond	Energy, kcal mole^{-1}
C—F	116	P—Br	64
As—F	115	S—S	63
H—O	111	As—H	61
C—H	99	Cl—F	61
N—H	93	Se—Cl	59
S—H	88	Te—H	57
C—C	83	C—I	57
C—O	82	As—Br	57
C—Cl	78	S—Br	51
P—H	76	P—I	51
P—Cl	76	Cl—O	50
Sb—Cl	75	N—Cl	48
C—N	70	As—I	42
As—Cl	69	P—P	41
Se—H	67	As—As	40
S—Cl	66	N—N	38
C—Br	66	Sb—Sb	34

TABLE 3.12   Average multiple-bond energies

Bond	Energy, kcal mole^{-1}
N≡N	225
C≡N	212.6
C≡C	194
C=O	179
C=C	148
C=N	147
C=S	114
N=N	100

organized on an extension of the simple Lewis electron-pair model of the covalent bond. In addition, an analysis of intraatomic distances shows that, for the most part, covalently bound atoms exhibit characteristic radii. A consideration of the energy associated with the formation of covalent bonds shows that a characteristic energy can be assigned to each type of bond. Thus, the apparently unrelated profusion of structural and energetic information available exhibits a remarkable unity when expressed in terms of individual bond parameters.

A discussion of the theory of the covalent bond appears in Chapter 4.

REFERENCES

1. H. S. M. Coxeter, *Trans. N.Y. Acad. Sci.*, **24**, 320 (1962).
2. R. A. Howald, *J. Chem. Ed.*, **45**, 463 (1968).

COLLATERAL READINGS

I. *Tables of Interatomic Distances and Configurations in Molecules and Ions*, Special Publications Nos. 11 and 18, The Chemical Society, London, 1958.
II. O. Kennard and D. G. Watson, eds., *Molecular Structures and Dimensions*, Vols. 1, 2, and 3, N.V.A. Oosthoek's Uitgevers, Utrecht, 1970.
III. P. J. Wheatley, *The Determination of Molecular Structure*, The Clarendon Press, Oxford, 1960.
IV. R. S. Drago, *Physical Methods in Inorganic Chemistry*, Reinhold Publishing Corp., New York, 1965.
V. R. J. Gillespie, *J. Chem. Ed.*, **47**, 18 (1970).
VI. R. J. Gillespie, *Angew. Chem.* (Intl. Ed.), **6**, 819 (1967).
VII. E. L. Muetterties and R. A. Schunn, *Quart. Revs.*, **20**, 245 (1966).
VIII. T. L. Cottrell, *The Strengths of Chemical Bonds*, Butterworths, London, 1958.
IX. A. G. Gaydon, *Dissociation Energies and Spectra of Diatomic Molecules*, Chapman and Hall Ltd., London, 1968.

STUDY QUESTIONS

1. Describe the geometry, i.e., the relative positions of all atoms in space, for each of the following species: (a) IBr_4^+; (b) $MgCl_2 \cdot 6H_2O$; (c) XeO_4^{2-}; (d) $OPCl_3$; (e) SO_3^{2-}; (f) H_2O_2.

2. Starting with the values of the heats of combustion for SiH_4 and Si_2H_6, show how the bond energy of an Si—Si bond can be determined. Use symbols for the numerical values of these and any other experimental values you may need.

3. Experimentally, catenated compounds with more than two nitrogen atoms are not known. Estimate the standard heat of formation of tetrazan $(H_2N-NH-NH-NH_2)$ assuming that it is a gaseous compound.

4. Assuming that the $C-H$ bond energy were known, what experimental data would be necessary to determine the $Si-C$ bond energy in the compound $Si(CH_3)_4$? Show in detail how the quantity in question could be obtained from the experimental data.

COVALENT COMPOUNDS

THEORIES OF BONDING

4

1 / INTRODUCTION

Any theory which purports to consider seriously a detailed description of covalent bonding must address itself to the detailed geometry of these substances as well as to the energetics of bond formation, and other properties of molecules such as frequencies and intensities of spectral transitions, dipole moments, and force constants (*I*). The discussion presented in Chapter 3 attempts to illustrate the unity of the various geometrical structures observed experimentally for covalently bound species as well as the relationship among experimental bond energies. Bond energies and geometrical considerations are two important classes of experimental results which must be predicted by successful theories of the nature of the covalent bond.

The electrical nature of the covalent bond was first suggested in 1881 by Helmholtz, and Ramsay, in 1908, attempted to associate bond formation with electrons (*II*). A few years later Lewis and (independently) Kossel described the electron-pair bond. In 1927, a year after the Schrödinger equation had been used as the basis for describing atomic characteristics, Heitler and London applied the same wave-mechanical model to molecular systems.

It will be recalled from Chapter 1 that the Schrödinger equation describes the behavior of electrons based on their wave-mechanical characteristics. The three-dimensional wave equation for a one-electron system is given by

$$\left(\frac{\partial^2 \Psi}{\partial X^2} + \frac{\partial^2 \Psi}{\partial Y^2} + \frac{\partial^2 \Psi}{\partial Z^2}\right) + \frac{8\pi^2 m}{h^2}(E - V) = 0, \tag{1}$$

where E and V are the total and potential energies, respectively, for the system. For the purposes of discussing the strategy used in the wave-mechanical arguments for molecular systems, we shall simplify Eq. (1). The differential

$$\frac{\partial^2}{\partial X^2} + \frac{\partial^2}{\partial Y^2} + \frac{\partial^2}{\partial Z^2}$$

can be conveniently abbreviated as ∇^2 which can be incorporated into Eq. (1) to give*

$$\nabla^2 \Psi + \frac{8\pi^2 m}{h^2}(E - V)\Psi = 0. \tag{2}$$

Equation (2) can be rearranged to give

$$-\frac{h^2}{8\pi^2 m}(\nabla^2 \Psi) + V\Psi = E\Psi, \tag{3}$$

which provides the basis for the introduction of another convenient notation. The symbol $\mathscr{H}$, known as the Hamiltonian operator, is defined as

$$\mathscr{H} = -\frac{h^2}{8\pi^2 m}\nabla^2 + V, \tag{4}$$

which permits us to rewrite Eq. (3) as

$$\mathscr{H}\Psi = E\Psi. \tag{5}$$

At this point it is important to realize that Eq. (5) is merely a shorthand version of the Schrödinger equation [Eq. (1)], but it serves admirably as the basis for further discussion.

If both sides of Eq. (5) are multiplied by Ψ and the resultant equation is rearranged, we obtain

$$E = \frac{\Psi \mathscr{H} \Psi}{\Psi^2}, \tag{6}$$

---

* Mathematically ∇^2 is an operator and does not commute with the function on which it operates, i.e., $\nabla^2 \psi \neq \psi \nabla^2$.

which describes the energy of the system in terms of the wave function; the potential energy V is expressed within the Hamiltonian [Eq. (4)]. Thus, if the potential energy function and the form of the wave function are known for a system, it should be possible to calculate the total energy of the system from Eq. (6). A function expressing the potential energy of a system can usually be obtained, but the larger problem is to obtain functions that approximate the true wave function of the system in question. An approximate energy of the system can be calculated from Eq. (6) if a trial function which approximates the true wave function can be invented. A guide to the correctness of the trial function is the variation principle, which states that the most satisfactory approximation to the true wave function is the trial function which gives the lowest energy for the system as calculated from Eq. (6). Thus, the procedure involves (a) a guess at a trial wave function, (b) a calculation of the energy of the system, (c) adjustment of the trial function, and (d) a recalculation of the energy; the process is continued until a function is obtained that gives the lowest value for the energy of the system. There are elegant mathematical techniques which can be employed to attain the best trial function (i.e., the one which gives an energy minimum), but these will not be discussed in detail here (*I*).

Basically two methods have been used to approximate the solutions of the wave equation for molecular systems. Heitler and London introduced the basis for the valence-bond method. The molecular-orbital method incorporates many of the ideas associated with atomic structure. In their most highly refined forms both methods approach a common point; each has certain advantages as well as disadvantages.

As an example of a simple molecular system, consider the hydrogen molecule, H_2, which consists of two protons and two electrons. Each of the electrons moves under the influence of the two nuclei and the other electron. The potential energy for this system can be expressed as

$$V = -\frac{e^2}{r(A1)} - \frac{e^2}{r(B1)} - \frac{e^2}{r(A2)} - \frac{e^2}{r(B2)} + \frac{e^2}{r(12)} + \frac{e^2}{r(AB)}, \tag{7}$$

where the distances are those specified in Fig. 4.1. Accordingly, the expanded form of the Schrödinger equation [Eq. (2)] can be written as

$$\nabla^2\Psi + \frac{8\pi^2 m}{h^2}\left(E + \frac{e^2}{r(A1)} + \frac{e^2}{r(B1)} + \frac{e^2}{r(A2)} + \frac{e^2}{r(B2)} - \frac{e^2}{r(12)} - \frac{e^2}{r(AB)}\right)\Psi = 0, \tag{8}$$

which can be generalized in the form given in Eq. (5).

For comparative purposes, the approximate solutions to the Schrödinger equation for H_2 [Eq. (8)] will be developed using the valence-bond and the molecular-orbital methods.

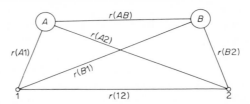

FIG. 4.1 In the hydrogen molecule each of the two electrons (designated 1 and 2) move under the influence of the two nuclei (A and B) and the other electron. At a given instant, the appropriate electrostatic forces act along the lines joining the particles as indicated.

2 / THE VALENCE-BOND METHOD

Imagine that the two hydrogen atoms which will eventually constitute an H_2 molecule are sufficiently separated so that the electrons on each hydrogen atom feel the presence of only one nucleus; i.e., there is no interaction between the atoms. The wave function describing this system is given by

$$\Psi = \psi_A(1)\psi_B(2), \tag{9}$$

where $\psi_A(1)$ and $\psi_B(2)$ are the $1s$ functions for hydrogen, as described in Chapter 1:

$$\psi_{1s} = \frac{1}{\sqrt{\pi}}\left(\frac{1}{a_0}\right)^{3/2} e^{-r/a_0}. \tag{10}$$

If the hydrogen atoms are brought together, as they exist in the hydrogen molecule, it would be expected that the wave function described by Eq. (9) is a poor approximation to the solution of the Schrödinger equation for the system. However, in a hydrogen molecule it is not possible to distinguish electrons as being associated with a specific nucleus. Thus, for a hydrogen molecule there are two acceptable approximate wave functions which arise because of the indistinguishability of electrons:

$$H_A.^{(1)\,(2)}.H_B; \qquad \Psi_I = \psi_A(1)\psi_B(2) \tag{11}$$

$$H_A.^{(2)\,(1)}.H_B; \qquad \Psi_{II} = \psi_A(2)\psi_B(1). \tag{12}$$

Neither Eq. (11) nor Eq. (12) is a good solution to the Schrödinger equation for the system, but a linear combination* of the two approximations is a better solution:

$$\Psi = C_1\Psi_I + C_2\Psi_{II}. \tag{13}$$

---

* The mathematical characteristics of the Schrödinger equation are such that if Ψ_I and Ψ_{II} are solutions, the linear combination $C_1\Psi_I \pm C_2\Psi_{II}$ (where C_1 and C_2 are arbitrary coefficients) is also a solution.

The coefficients C_1 and C_2 reflect the extent to which Ψ_I and Ψ_{II} contribute to the combined function Ψ, which is normalized to ensure that the total probability of finding each electron in all space is unity. Since Ψ_I and Ψ_{II} represent states of equal energy the relationship between C_1 and C_2 is simply

$$C_1 = \pm C_2, \tag{14}$$

giving two possible combinations of Ψ_I and Ψ_{II}:

$$\Psi_+ = C_1(\Psi_I + \Psi_{II}) \tag{15}$$

$$\Psi_- = C_1(\Psi_I - \Psi_{II}). \tag{16}$$

The best value for the constant C_1 is that for which the energy associated with the trial function is a minimum. Practically, C_1 can be evaluated by substituting Eq. (15) into Eq. (6), setting $\partial E/\partial C_1 = 0$, and solving for C_1.

It should be recalled at this point that any trial function for Ψ leads to a value for the energy of the system through Eq. (6) and that the variation principle guides us to the best trial function. Thus, it should be possible to calculate the energy of the system at a given internuclear separation using Ψ_I, Ψ_{II}, Ψ, or indeed any trial function. The energy of the system as a function of internuclear separation for the various trial functions is shown in Fig. 4.2.

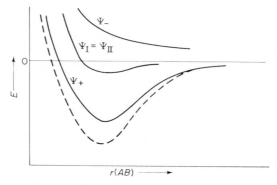

FIG. 4.2 The energy for the H_2 system as a function of internuclear separation for various trial functions. The experimental curve is given as a dotted line.

The numerical results, i.e., the depth of the potential well and the internuclear distance at the energy minimum, are shown in Table 4.1. Taking the reference state for H_2 as the energy associated with two unbound hydrogen atoms, the dissociation energy of H_2

$$H_2 + 104 \text{ kcal mole}^{-1} \rightarrow 2H \tag{17}$$

TABLE 4.1 Calculated bond energy and internuclear separation for the hydrogen molecule using several trial functions[a]

Wave function	E, kcal mole^{-1}	$r(AB)$ at min., Å
Ψ_1 [Eq. (11) or (12)]	6	0.90
Ψ_+ [Eq. (15)]	72.4	0.87
Ψ'_+ [Eq. (20)]	80.3	0.77

[a] The experimental values for these quantities are $E = 108$ kcal mole^{-1} and $r = 0.74$ Å.

becomes one of the fiducial points for how well the trial function approximates the true wave function. The second experimental quantity in this respect is the equilibrium internuclear distance (Chapter 3), which for H_2 is 0.74 Å.

The results shown in Fig. 4.2 and Table 4.1 indicate that the trial function Ψ_1 [Eq. (11)] is a very poor approximation to the true wave function because the binding energy and the internuclear distance calculated with it are considerably different from the experimental values. The linear combination Ψ_- [Eq. (16)] represents a nonbonded state (Fig. 4.2), but Ψ_+ [Eq. (15)] gives values for the binding energy and the internuclear distance considerably closer than Ψ_1. Apparently Ψ_+ is a much better approximation than Ψ_1.

A closer approximation to the true wave function can be obtained by considering the possibility of placing both electrons on one of the hydrogen atoms:

$$H_A{}^{+\,(1)}_{\ (2)}H_B{}^- \, ; \qquad \Psi_{III} = \psi_B(1)\psi_B(2) \tag{18}$$

$$H_A{}^{(1)-}_{\cdot(2)}H_B{}^+ \, ; \qquad \Psi_{IV} = \psi_A(1)\psi_A(2). \tag{19}$$

There are of course two arrangements for this possibility. If hydrogen molecules were to exist with either of these arrangements of electrons we would expect it to exhibit properties characteristic of ionic compounds (Chapter 2). This, of course, is not the case, and it should not be surprising that neither Ψ_{III} nor Ψ_{IV} are good approximations of the true wave function for H_2. However, just as the linear combination of Ψ_1 and Ψ_{II} gave a better approximation than either function by itself, it might be tempting to add Ψ_{III} and Ψ_{IV} to Ψ_+ [Eq. (15)] in the hope of obtaining an even better approximation:

$$\Psi'_+ = C_1(\Psi_1 + \Psi_{II}) + C_3(\Psi_{III} + \Psi_{IV}). \tag{20}$$

Since Ψ_{III} and Ψ_{IV} are also states of equal energy there is only one constant, C_3, to be evaluated [cf. the discussion of Eq. (13)]. The new trial function

Ψ'_+, approximating the solution to the wave equation, gives rise to the energy of the system at any internuclear separation. The best value of C_3 can be obtained by application of the variation principle. The results using Ψ'_+ as a trial solution to the wave equation (Fig. 4.2 and Table 4.1) show that it is indeed a better approximation to the true wave function since both the binding energy (80.3 kcal) and the internuclear distance (0.77 Å) are closer to the experimentally observed values for these quantities.

Further adjustments to the trial function can be made; these are interpreted in terms of the shielding effect of electrons on the nuclei and electron–electron interaction. The effect of nuclear charge screening is usually introduced into this calculation by considering an effective nuclear charge Z^* (cf. Chapter 1) rather than the true nuclear charge. This procedure provides another adjustable parameter in the trial function which can be minimized with respect to the energy of the system. Finally, it has been possible to create a trial function in an essentially empirical manner to obtain virtually complete agreement between the theoretical and the experimental values for the binding energy and the equilibrium internuclear separation. The best results obtained arise from using a 13-term polynomial as the solution of the wave equation; this solution gives a discrepancy of only 0.6% and 0.1% in the calculated and experimental values of the bond energy and bond distance, respectively.

The valence-bond description presented here for the structure of H_2 serves to illustrate explicitly several general points about the method. First, a trial function which is a reasonably good solution to the wave equation for a system can be obtained by the linear combination of a series of approximations which themselves may be much poorer approximations. The latter usually incorporate atomic orbitals and a consideration of various possible distributions of the valence electrons in the system. There is, of course, a very practical reason for using atomic wave functions in this process; explicit mathematical forms for hydrogenic atomic functions are known (Chapter 1). Thus, in the case of H_2 a good approximation to the solution of the wave equation arises from combining the wave functions corresponding to the distribution of electrons shown in Fig. 4.3. The pictorial representations shown in Fig. 4.3 are similar to the structural formulas to which chemists are accustomed, but their meaning is not the

$$H_A \overset{(1)(2)}{:} H_B \leftrightarrow H_A \overset{(2)(1)}{:} H_B \leftrightarrow {}^- H_A \underset{(2)}{\overset{(1)}{:}} H_B^+ \leftrightarrow {}^+ H_A \underset{(2)}{\overset{(1)}{:}} H_B^-$$

$$\underbrace{\Psi_I \qquad\qquad \Psi_{II}}_{\text{Covalent}} \qquad\qquad \underbrace{\Psi_{III} \qquad\qquad \Psi_{IV}}_{\text{Ionic}}$$

FIG. 4.3 Representations of the four principal canonical forms which contribute to the best trial function for H_2.

same. Accordingly, the representations shown in Fig. 4.3 are called valence-bond structures, canonical forms, or resonance forms. They are the individual representations of the wave functions which are linearly combined to approximate the solution to the wave equation for the system. Canonical forms do not exist as such in the chemical sense. The true structure is said to be a resonance hybrid of the canonical forms; i.e., the true wave function can be approximated by a mixture of the wave functions represented by the canonical forms. This mathematical relationship is indicated by the double-headed arrows in Fig. 4.3.

A second point can be made concerning the valence-bond method. The canonical forms are broadly divided into a group which contains forms with no charge distribution, called covalent forms, and a group containing forms with a charge distribution, called ionic forms (Fig. 4.3). The addition of wave functions associated with ionic forms to the final approximation [cf. Eq. (20)] invariably gives a better solution to the wave equation; that is, the calculated binding energy and internuclear distance are closer to the experimental values (Table 4.1). Thus, in general, the valence-bond method gives an approximation [Eq. (21)] to the true wave function for a molecule which contains contributions from covalent and ionic canonical forms:

$$\Psi = \Psi_{covalent} + \Psi_{ionic}. \tag{21}$$

A third point should be stressed in reviewing the strategy of the valence-bond method. The indication whether any trial function is acceptable as a solution to the wave equation for a given system comes from a comparison of experimental parameters such as bond distance and bond energy with the calculated values. Indeed, such experimentally determined parameters can often be used as a guide to the types of canonical forms which might be important contributors to the resonance hybrid. For example, the Lewis formulation for carbon dioxide (**1**) suggests that the molecule contains a pair of carbon–oxygen double bonds. This structure suggests that the C—O bond

$$:\!\ddot{O}::C::\ddot{O}\!:$$

1

distance should be similar to that observed in ketones (R_2CO), which is 1.22 Å. The bond distance experimentally observed for CO_2 is 1.16 Å, considerably shorter than that expected on the basis of (**1**); this distance is not, however, as short as that in carbon monoxide (1.13 Å), the Lewis formulation of which corresponds to a triple bond ($:\!C:::O\!:$). Thus, if Ψ_1 is the wave function for the canonical form which corresponds to the best Lewis structure [i.e., (**1**)], we might expect that Ψ_1 is a completely satisfactory trial function for the solution to the Schrödinger equation for this system.

However, it is possible to obtain a better approximation to the solution by considering canonical forms that incorporate triple bonds; two such forms exist [(2) and (3)] and they happen to be ionic structures. In valence-bond

$$^-:\ddot{O}:C:::O:^+ \qquad\qquad ^+:O:::C:\ddot{O}:^-$$

$$\textbf{2} \qquad\qquad\qquad\qquad \textbf{3}$$

terms, the true structure of CO_2 can be expressed as the resonance hybrid of the canonical forms

$$:\ddot{O}::C::\ddot{O}: \;\leftrightarrow\; ^-:\ddot{O}:C:::O:^+ \;\leftrightarrow\; ^+:O:::C:\ddot{O}:^-,$$

$$\textbf{1} \qquad\qquad \textbf{2} \qquad\qquad \textbf{3}$$

where the best approximation to the wave function is given by

$$\Psi = C_1\Psi_{\text{I}} + C_2\Psi_{\text{II}} + C_3\Psi_{\text{III}}. \tag{22}$$

A comparison of the energy expected for a given canonical form with the experimentally determined energy content of the molecule can also indicate how well that canonical form represents the structure of the molecule. For example, if the Lewis formulation of CO_2 (**1**) is used as the canonical form leading to the wave function, the energy content of the molecule might be expected to be twice the C=O bond energy ($2 \times 179 = 358$ kcal mole^{-1}, cf. Table 3.8). The energy of formation of CO_2 (383 kcal mole^{-1}) is markedly different than the expected value, suggesting that the real molecule is more stable than that represented by the Lewis structure (**1**). In other words the wave function corresponding to this canonical form has to be modified to give a solution yielding a lower energy. As was pointed out earlier in this discussion, the addition of ionic canonical forms to the resonance hybrid leads to lower-energy wave functions. The arguments based on bond distances for CO_2 also result in the same general conclusion.

Finally, it must be emphasized that the process by which acceptable solutions to the wave equation are constructed need not necessarily be related to chemical considerations. Presumably any wave function is acceptable which satisfies the Schrödinger equation using the criteria outlined in this section. The valence-bond method of constructing suitable wave functions reflects an additivity principle which has long been used in other areas of chemistry; that is, the characteristics of a system composed of a number of components can be expressed in terms of the individual characteristics of the components. This type of additivity of characteristics is reflected, for example, in Raoult's law and the establishment of a table of half-cell potentials. The valence-bond method appears to be an extension of these general ideas, in the sense that the electronic distribution of molecules is expressed as a linear combination of the electronic distributions of collections of atomic systems.

2.1 / Hybridization (*III, IV*)

In developing the arguments associated with the valence-bond method we used a simple diatomic molecule (H_2) containing a single bond as an example. Extending these ideas to polyatomic molecules or molecules incorporating multiple bonds leads to the realization that, in general, atomic wave functions do not possess the required spatial orientations. For example, a prediction of the characteristics of the most probable compound formed between carbon and hydrogen using ground-state atomic wave functions leads to a formulation which is not observed experimentally. Carbon has a ground-state configuration of $1s^2 2s^2 2p_x{}^1 2p_y{}^1$, i.e., two of the p orbitals contain unpaired electrons. On this basis it might be expected that two bonds could be formed between carbon in its atomic ground state and hydrogen; a combination of the $1s$ hydrogen orbital with the carbon $2p$ orbital could be used to approximate the wave function for the bond formed between carbon and hydrogen under these conditions:

$$\Psi_{C-H} = C_1[\psi_{1s}(1)\psi_{2p}(2)] + C_2[\psi_{1s}(2)\psi_{2p}(1)]. \tag{23}$$

Since there are two unpaired electrons in the ground state of carbon in orbitals that are perpendicular to each other, the simplest molecule that can be theoretically formed between these two atoms should have the stoichiometry CH_2 with an H—C—H bond angle of 90°. These predictions are, of course, incorrect because the simplest stable compound formed between carbon and hydrogen is methane, CH_4, with an H—C—H bond angle of 109.7° (Chapter 3).

Similar difficulties arise in the application of atomic wave functions to molecules containing double bonds. Thus, there is no way possible of directly incorporating the atomic wave functions suggested by the ground-state configurations of carbon ($1s^2 2s^2 2p_x{}^1 2p_y{}^1$) and oxygen ($1s^2 2s^2 2p_x{}^2 2p_y{}^1 2p_z{}^1$) to approximate the solution to the wave equation for CO_2 which is a linear molecule with multiple bonds.

The problem of associating the directional bonds in polyatomic molecules with the atomic wave function has been solved by introducing the concept of hybridization (*1*). This is a mathematical process of resolving the existing atomic wave functions into components that lie in certain directions, and reconstituting from these components a new set of wave functions called hybrid wave functions (or hybrid orbitals). An atom for which the atomic orbitals have been hybridized is said to be in a valence state. The valence state of an atom is completely imaginary, but it is a useful intermediate in the theoretical description of the electronic distributions in polyatomic molecules.

In the hybridization process, only the angular parts of the wave function are considered; it is assumed that the radial parts for atomic orbitals with the same principal quantum number are sufficiently similar that their differences may be neglected. As an example of the hybridization process, consider one pair of hybrid orbitals that can be derived from s and p atomic orbitals. The angular parts of these wave functions are shown graphically in Fig. 4.4 and are given by

$$\psi_s = 1 \tag{24}$$

$$\psi_{p_x} = \sqrt{3}\cos\theta. \tag{25}$$

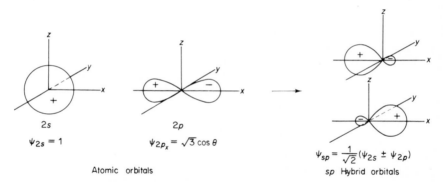

FIG. 4.4 Hybridization involves the mathematical combination of atomic orbitals to give a new set of orbitals with different directional characteristics than the original atomic orbitals.

The relationship between the atomic functions ψ_s and ψ_{p_x} and the hybrid function (ψ_{sp}) is given by

$$\psi_{sp} = a\psi_s + b\psi_{p_x}. \tag{26}$$

Recognizing that ψ_{sp}, ψ_s, and ψ_{p_x} must be normalized and that ψ_s and ψ_{p_x} are orthogonal gives us the basis for evaluating a and b:

$$a = \frac{1}{\sqrt{2}} \tag{27}$$

$$b = \pm\frac{1}{\sqrt{2}}. \tag{28}$$

Thus, there are two hybrid orbitals, called *sp* hybrids, which have the forms

$$\psi_{sp} = \frac{1}{\sqrt{2}}(\psi_s + \psi_{p_x}) \tag{29}$$

$$\psi_{sp} = \frac{1}{\sqrt{2}}(\psi_s - \psi_{p_x}), \tag{30}$$

where ψ_{sp} is a maximum along the Z axis ($\theta = 0°$, $180°$). These results are qualitatively illustrated in Fig. 4.4. In the addition of ψ_{p_x} to ψ_s [Eq. (29)] the larger portion of the hybrid orbital carries a positive sign because the positive values of these two functions coincide in this volume of space. Subtracting ψ_{p_x} from ψ_s [Eq. (30)] is equivalent to inverting the sign on the p_x orbital which still leaves the large part of the hybrid orbital with a positive sign, but this hybrid points in the opposite direction. Thus, the hybridization process yields two equivalent orbitals that have a preferred orientation with respect to each other from two geometrically nonequivalent atomic orbitals. An inspection of the mechanics of hybridization shows striking similarities to the mathematics and strategy of the valence-bond method for describing bond formation.

A consideration of the valence-bond description of BeH_2 (in its monomeric form) serves to illustrate the usefulness of the hybridization concept and to focus attention on the relative energies of the theoretical processes involved in bond formation. Using only the atomic states of Be and H we see an immediate difficulty. Although the valence electron in hydrogen is in a theoretically useful orbital (H, $1s^1$), the valence electrons for beryllium are paired in a $2s$ orbital (Be, $1s^2$, $2s^2$). Thus, from the valence-bond standpoint it is not possible to form the expected linear BeH_2 molecule (cf. Chapter 3) using atomic orbitals. From a theoretical standpoint two objectives must be met. First, the two electrons in the beryllium $2s$ orbital must be unpaired:

$$\text{Be} \quad 1s^2 \, 2s^2 \;\longrightarrow\; \text{Be} \quad 1s^2 \, 2s^1 \, 2p_z{}^1, \tag{31}$$

and then the correct orbitals constructed. Unpairing the $2s$ electrons [Eq. (31)] corresponds to creating a beryllium atom in an excited state, but the valence electrons are now distributed in s and p orbitals which can be hybridized using the process described in Eqs. (26)–(30) to give linear sp hybrid orbitals. Thus, the bond between one of the hydrogen atoms and beryllium can be described by an overlap of a hydrogen $1s$ orbital with a beryllium sp hybrid orbital (Fig. 4.5):

$$\Psi = C_1[\psi_{1s}(1)\psi_{sp}(2)] + C_2[\psi_{1s}(2)\psi_{sp}(1)]. \tag{32}$$

The best values of C_1 and C_2 can be obtained by applying the variation principle, and the energy associated with Ψ calculated in the usual manner.

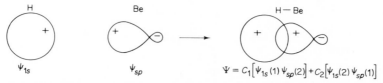

FIG. 4.5 Each Be—H bond in monomeric BeH_2 consists of a linear combination of a hydrogen $1s$ orbital and a beryllium sp hybrid orbital.

A summary of the relative energies corresponding to the valence-bond description of bond formation between beryllium and hydrogen appear in Fig. 4.6. The energy required to take an atom from its ground state to an

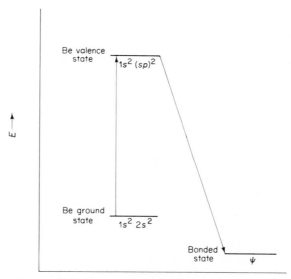

FIG. 4.6 A beryllium atom in an sp hybridized valence state is in an energetically higher state than the ground-state configuration. The bonded state is, however, lower than either of these states.

excited state in which each of the orbitals necessary for the formation of the proper hybrid contain one valence electron is called the promotion energy. Thus, the valence state of an atom is nearly always an excited state. The expenditure of energy to reach an excited-state configuration is more than compensated by the energy released in the formation of a bond.

Since the hybridization is a mathematical process it should not be surprising to learn that any set of atomic orbitals can be combined using these arguments. Some common hybrid orbitals and their characteristic geometries are shown in Table 4.2 (2). Several points should be made on the

TABLE 4.2 The mathematical expressions for and geometrical arrangements of some commonly used hybrid orbitals

Designation	Hybrid orbital	Geometry	Representation
sp	$\dfrac{1}{\sqrt{2}}(\psi_s \pm \psi_{p_z})$	linear	
sp^2	$\dfrac{1}{\sqrt{3}}\psi_s + \sqrt{\dfrac{2}{3}}\psi_{p_x}$ $\dfrac{1}{\sqrt{3}}\psi_s - \dfrac{1}{\sqrt{6}}\psi_{p_x} \pm \dfrac{1}{\sqrt{2}}\psi_{p_y}$	trigonal plane	
sp^3	$\tfrac{1}{2}(\psi_s - \psi_{p_x} - \psi_{p_y} - \psi_{p_z})$ $\tfrac{1}{2}(\psi_s - \psi_{p_x} + \psi_{p_y} + \psi_{p_z})$ $\tfrac{1}{2}(\psi_s + \psi_{p_x} - \psi_{p_y} + \psi_{p_z})$ $\tfrac{1}{2}(\psi_s + \psi_{p_x} + \psi_{p_y} - \psi_{p_z})$	tetrahedral	

TABLE 4.2—*continued*

Designation	Hybrid orbital	Geometry	Representation
dsp^2	$\dfrac{1}{2}\psi_s + \dfrac{1}{2\sqrt{2}}\psi_{d_{x^2-y^2}} - \dfrac{1}{2\sqrt{2}}\psi_{d_{xy}} + \dfrac{1}{\sqrt{3}}\psi_{p_x} - \dfrac{1}{\sqrt{6}}\psi_{p_y}$ $\dfrac{1}{2}\psi_s - \dfrac{1}{2\sqrt{2}}\psi_{d_{x^2-y^2}} + \dfrac{1}{2\sqrt{2}}\psi_{d_{xy}} + \dfrac{1}{\sqrt{6}}\psi_{p_x} + \dfrac{1}{\sqrt{3}}\psi_{p_y}$ $\dfrac{1}{2}\psi_s + \dfrac{1}{2\sqrt{2}}\psi_{d_{x^2-y^2}} - \dfrac{1}{2\sqrt{2}}\psi_{d_{xy}} - \dfrac{1}{\sqrt{3}}\psi_{p_x} + \dfrac{1}{\sqrt{6}}\psi_{p_y}$ $\dfrac{1}{2}\psi_s - \dfrac{1}{2\sqrt{2}}\psi_{d_{x^2-y^2}} + \dfrac{1}{2\sqrt{2}}\psi_{d_{xy}} - \dfrac{1}{\sqrt{6}}\psi_{p_x} - \dfrac{1}{\sqrt{3}}\psi_{p_y}$	planar	
dsp^3	$\dfrac{1}{\sqrt{8}}\psi_s + \dfrac{1}{\sqrt{2}}\psi_{p_z} + \sqrt{\dfrac{3}{8}}\psi_{d_{z^2}}$ $\dfrac{1}{\sqrt{8}}\psi_s - \dfrac{1}{\sqrt{2}}\psi_{p_z} + \sqrt{\dfrac{3}{8}}\psi_{d_{z^2}}$ $\dfrac{1}{2}\psi_s + \sqrt{\dfrac{2}{3}}\psi_{p_x} - \dfrac{1}{\sqrt{12}}\psi_{d_{z^2}}$ $\dfrac{1}{2}\psi_s - \dfrac{1}{\sqrt{6}}\psi_{p_x} + \dfrac{1}{\sqrt{2}}\psi_{p_y} - \dfrac{1}{\sqrt{12}}\psi_{d_{z^2}}$ $\dfrac{1}{2}\psi_s - \dfrac{1}{\sqrt{6}}\psi_{p_x} - \dfrac{1}{\sqrt{2}}\psi_{p_y} - \dfrac{1}{\sqrt{12}}\psi_{d_{z^2}}$	trigonal bipyramid	

TABLE 4.2—*continued*

Designation	Hybrid orbital	Geometry	Representation
d^2sp^3	$\dfrac{1}{\sqrt{8}}\psi_s + \dfrac{1}{\sqrt{2}}\psi_{p_z} + \sqrt{\dfrac{3}{8}}\psi_{d_{z^2}}$ $\dfrac{1}{\sqrt{8}}\psi_s - \dfrac{1}{\sqrt{2}}\psi_{p_z} + \sqrt{\dfrac{3}{8}}\psi_{d_{z^2}}$ $\sqrt{\dfrac{3}{16}}\psi_s + \dfrac{1}{\sqrt{3}}\psi_{p_x} - \dfrac{1}{\sqrt{6}}\psi_{p_y} - \dfrac{1}{2\sqrt{2}}\psi_{d_{xy}} + \dfrac{1}{2\sqrt{2}}\psi_{d_{x^2-y^2}} - \dfrac{1}{4}\psi_{d_{z^2}}$ $\sqrt{\dfrac{3}{16}}\psi_s + \dfrac{1}{\sqrt{6}}\psi_{p_x} - \dfrac{1}{\sqrt{3}}\psi_{p_y} - \dfrac{1}{2\sqrt{2}}\psi_{d_{xy}} - \dfrac{1}{2\sqrt{2}}\psi_{d_{x^2-y^2}} - \dfrac{1}{4}\psi_{d_{z^2}}$ $\sqrt{\dfrac{3}{16}}\psi_s + \dfrac{1}{\sqrt{6}}\psi_{p_x} + \dfrac{1}{\sqrt{3}}\psi_{p_y} - \dfrac{1}{2\sqrt{2}}\psi_{d_{xy}} + \dfrac{1}{2\sqrt{2}}\psi_{d_{x^2-y^2}} - \dfrac{1}{4}\psi_{d_{z^2}}$ $\sqrt{\dfrac{3}{16}}\psi_s - \dfrac{1}{\sqrt{3}}\psi_{p_x} - \dfrac{1}{\sqrt{6}}\psi_{p_y} + \dfrac{1}{2\sqrt{2}}\psi_{d_{xy}} - \dfrac{1}{2\sqrt{2}}\psi_{d_{x^2-y^2}} - \dfrac{1}{4}\psi_{d_{z^2}}$	octahedral	

contents of Table 4.2. Hybrid orbitals incorporating s atomic orbitals will always consist of a large lobe and one or more smaller lobes; the former enclose the positive values of ψ while the latter contain negative values (cf. Fig. 4.4). The small lobes have been omitted for clarity from the sketches of the geometrical distribution of the hybrid orbitals shown in Table 4.2. In addition, it is interesting to note that the hybrid orbitals described in Table 4.2 possess geometrical arrangements which reflect the geometries experimentally observed for a very large number of molecules (Chapter 3). Thus, it is always possible to produce a mixture of atomic orbitals to give any angular disposition of hybrid orbitals. For example, we can consider the possible hybrid orbitals formed from s and p atomic orbitals. The hybrid orbitals sp, sp^2, and sp^3 possess 50%, 33.3%, and 25% s character, respectively, with orbital angles of 180°, 120°, and 109.7°, respectively. The "hybrid orbitals" p^2s^0 (i.e., two pure p atomic orbitals with no s component) have an interorbital angle of 90°. Thus, using only s and p atomic orbitals it should be possible to obtain a set of hybrid orbitals with interorbital angles of any value between 90° and 180°. These arguments can be extended to include the use of d and f atomic orbitals also.

2.2 / Relative Order of Bond Strengths

An integral of the form

$$S = \int \psi_A(1)\psi_B(2)\, d\tau \tag{33}$$

always appears in the calculation of the binding energy for two atoms, A and B, using Eq. (6). The value of this integral, often called the overlap integral, is a measure of the extent to which the wave function of one atom overlaps with that of the second atom. If the pair of atoms were infinitely far apart, the product $\psi_A(1)\psi_B(2)$ would be zero: where $\psi_A(1)$ had a finite value, $\psi_B(2)$ would be zero and vice versa. If the two nuclei merged, i.e., zero internuclear distance, $\psi_A(1)$ and $\psi_B(2)$ would correspond to the same wave function of one of the atoms; since the wave functions in question are normalized, the product $\psi_A(1)\psi_B(2)$ would be unity. At internuclear distances between these two extremes the product is some value intermediate between one and zero and must be evaluated for each case (Fig. 4.7).

In general the extent to which atomic wave functions overlap is a measure of the strength of the bond formed between two atoms. Where the exact form of the wave function is known, the overlap integral can be calculated precisely; this situation does not obtain very often and the overlap integral must be approximated. A useful approximation has been developed by

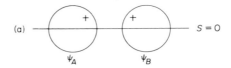

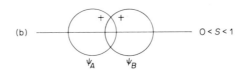

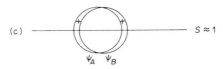

FIG. 4.7 The value of the integral, $S = \int \psi_A(1)\psi_B(2)\,d\tau$, is a measure of the extent to which two wave functions overlap. In the cases shown here, s atomic functions are used.

Pauling. Consider the overlap of an s orbital with s and p orbitals with the same principal quantum number (Fig. 4.8). One point becomes immediately evident: there is an unique orientation to obtain optimum overlap. This is not apparent in the overlap of two s orbitals (Fig. 4.8a), but the importance of the symmetry properties of the two orbitals is evident in the case of s and p interaction (Fig. 4.8b). In the latter instance maximum overlap occurs when the orbitals are concentrated along the internuclear bond distance. If the radial parts of the wave functions for s and p electrons are identical, then the relative magnitude of an orbital in a given direction should be proportional to the angular part of the wave function. The ratio of the angular part of the wave function for s orbitals to that for p orbitals is $1 : \sqrt{3}$ (Chapter 1). Similar arguments can be applied to any orbital whether it is an atomic function (Chapter 1) or a hybrid function (Table 4.2). The relative bond strength obtained by two orbitals is, in this view, taken as the product of radial parts of the wave function, some of which are summarized in Table 4.3. That is, the relative order of bond strengths of an s orbital overlapping with a p and an sp^3 hybrid orbital lies in the ratio of $(1)(\sqrt{3}):(1)(2)$. These arguments are, of course, only crude approximations since they rest on the assumption that the radial parts of the wave functions for the orbitals in question can be compared directly—an assumption that becomes highly questionable when the orbitals are characterized by different values of the principal quantum number.

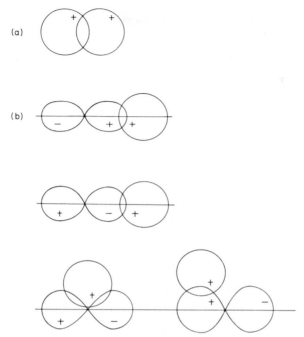

FIG. 4.8 Although symmetry considerations are not important in the criterion of maximum overlap of s orbitals (a), these considerations are important for other cases (b). Thus only one of the types of overlap shown in (b) is acceptable.

TABLE 4.3 The relative strengths of bonds formed by various atomic and hybrid orbitals

Wave function	Relative maximum value of angular part
s	1
p_x, p_y, p_z	$\sqrt{3}$
d_{z^2}	$\sqrt{5}$
d_{xy}, d_{yz}, d_{xz}	$\sqrt{\dfrac{15}{2}}$
$d_{x^2-y^2}$	$\sqrt{\dfrac{15}{2}}$
sp^3	2
dsp^2	2.694
sp^2	1.991
sp	1.932
d^2sp^3	2.923

3   /   THE MOLECULAR-ORBITAL METHOD (*V, VI*)

The electrons present in atomic systems move under the influence of a single nucleus. The molecular-orbital approach considers the movement of electrons under the influence of the collection of nuclei that are found in molecular systems. As in the quantum-mechanical description of atoms, each electron in a molecule possesses a discrete amount of energy that is associated with a molecular wave function with characteristic quantum numbers. The molecular wave functions have the same significance as atomic functions; for example, the value of $\psi^2\, d\tau$ for the molecular wave function is proportional to the probability of finding an electron in the volume element $d\tau$.

As in the case of the valence-bond theory, the problem arises in specifying the exact forms of the molecular wave functions which are solutions to the Schrödinger equation. One of the most useful methods for constructing approximate molecular wave functions involves the assumption that when an electron moves in the vicinity of one of the nuclei in a molecule it will be predominantly under the influence of that nucleus and that part of the molecular wave function will resemble the atomic function of the isolated atom. Thus, to a first approximation the wave function for a molecule is taken as a linear combination of the atomic functions for the atoms that constitute the molecule, due regard being made for geometrical distribution of atomic nuclei and the symmetry of the atomic orbitals within the molecule. As an example of the application of molecular-orbital arguments using a linear combination of atomic orbitals (LCAO), consider again the description of the hydrogen molecule.

3.1   /   The Hydrogen Molecule

The molecular orbital for H_2 is described as a linear combination of the two $1s$ atomic functions $\psi(H_A, 1s)$ and $\psi(H_B, 1s)$:

$$\Psi_{\text{molecule}} = \psi(H_A, 1s) + \lambda\psi(H_B, 1s), \tag{34}$$

where λ is a constant that can be interpreted as indicating the degree to which one orbital is favored over the other in the molecular orbital. The constant λ is essentially a measure of the polarity of the bond and can be evaluated by substituting Eq. (34) into Eq. (6) and applying the variation method. This process gives $\lambda = \pm 1$, a value which might have been expected intuitively for any homonuclear diatomic molecule. The evaluation of λ leads to two molecular wave functions,

$$\Psi_{\text{molecule}} = \psi(H_A, 1s) + \psi(H_B, 1s). \tag{35}$$

$$\Psi^*_{\text{molecule}} = \psi(H_A, 1s) - \psi(H_B, 1s). \tag{36}$$

As in the case of atomic orbitals, molecular wave functions are associated with discrete energies (Fig. 4.9) and give rise to electron densities (Fig. 4.10) (cf. Chapter 1). The linear combination of the two hydrogen 1s orbitals gives rise to two molecular orbitals; one orbital, given by Eq. (35), is lower in energy by an amount E than the original atomic function, and the other

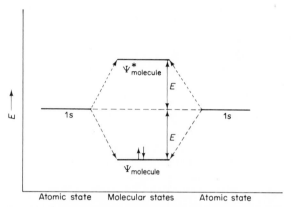

FIG. 4.9   The linear combination of two atomic 1s functions leads to two molecular wave functions. The antibonding level (Ψ^*) is higher in energy than the bonding level (Ψ).

[Eq. (36)] is of higher energy by the same amount E. In general, the LCAO method leads to a low-energy molecular orbital (Ψ), called a bonding orbital, and a higher-energy molecular orbital (Ψ^*), designated an antibonding orbital. As in the case of atomic orbitals, the number of electrons that can occupy each molecular orbital is governed by the Pauli exclusion principle.

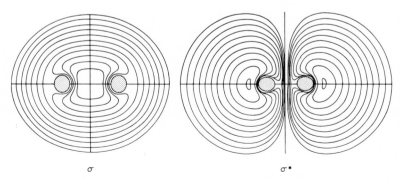

FIG. 4.10   Electron density maps for the bonding and antibonding molecular orbitals have the same significance as those for atomic functions.

The Aufbau principle applied to the set of molecular orbitals available for the hydrogen molecule leads to two electrons with paired spins in the lowest orbital (Fig. 4.9).

The electronic densities corresponding to electrons occupying the bonding and antibonding orbitals are shown in Fig. 4.10. It is apparent that the bonding molecular orbitals correspond to an increase in electronic charge between the nuclei, in contrast to that of the antibonding molecular orbital. The increased electron density between nuclei in the bonding molecular orbital serves to overcome the repulsive interaction of the two nuclei. Thus, the application of the simple LCAO method predicts the existence of an energetically more stable molecular system which reflects an internuclear electron density suggestive of the Lewis description of a covalent bond.

The qualitative usefulness of the LCAO method is apparent if we recognize that the energy-level diagram shown in Fig. 4.9 is essentially unchanged for all molecular systems using $1s$ atomic functions as a basis for bonding. For example, the molecular energy-level diagram for the species H_2^+, HeH, and HeHe should be the same that shown in Fig. 4.9 since the LCAO method applied in each of these cases involves the combination of two $1s$ orbitals. From the distribution of electrons in the available molecular orbitals we might predict that H_2^+ and HeH are stable, and that HeHe is not. The molecular energy-level diagram for H_2^+ in its ground state shows that the single electron occupies the bonding level and is more stable by an amount of energy E than the corresponding atomic states. However, the four electrons in the He–He system must occupy both the bonding molecular orbital and the antibonding orbital. The result is a molecular system with no binding energy, which is another way of saying that the system He–He is less stable than two isolated He atoms. However, the energy-level diagram for the species He_2^+ suggests that it should possess a net binding energy. In the case of the molecule HeH, we would predict that of the three electrons in the system, two would be found paired in the bonding molecular orbital and one would occupy the antibonding orbital. Thus, there should be a net binding energy of E for this molecule relative to the isolated atoms.

These simple arguments lead to predictions that have been verified in fact. The species H_2^+, He_2^+, and HeH have been detected in the vapor phase and the fact that helium is monatomic is well known. The ideas embodied in the molecular energy-level diagram shown in Fig. 4.9 can also be extended even further if it is recognized that the linear combination of any s atomic orbitals with the same principal quantum number should give bonding and antibonding molecular orbitals related in the way shown for the $1s$ combination. Thus, we might expect stable diatomic molecules will be formed by atoms that possess ns^1 atomic configurations. Indeed, the alkali metals are known to exist as dimers (M_2) in the vapor state (Chapter 8).

3.2 / Other Homonuclear Diatomic Molecules (VII)

The simple LCAO method can be used to approximate the molecular orbital formed from the combination of any pair of atomic orbitals. In general, the atomic functions involved should have similar energies, overlap as much as possible, and possess the same symmetry relative to the internuclear axis. The first requirement excludes atomic wave functions associated with inner (closed-shell) electrons; only the outer wave functions are considered. As mentioned earlier, the magnitude of orbital overlap is a measure of bond strength (Fig. 4.7). The importance of the relative symmetries of the combining function has been only casually discussed (cf. Fig. 4.8), and this subject is more fully developed now. The framework for this discussion is the possible types of molecular orbitals in homonuclear diatomic species that can be formed by linear combination of s and p atomic orbitals with the same principal quantum number. As indicated previously, the combination of s orbitals yields bonding and antibonding molecular orbitals, the electron probability distributions for which are shown in Fig. 4.11a. The most efficient linear combination of p-type orbitals is easily understood by recognizing the fact that one p orbital points along the internuclear axis while two are perpendicular to this reference line (Fig. 4.11b

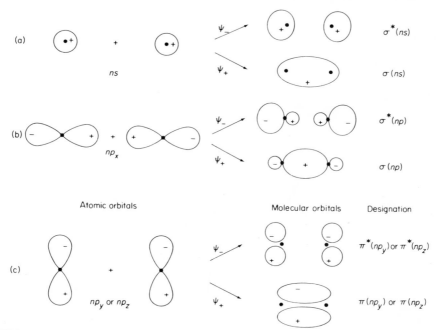

FIG. 4.11   The linear combination of s and p orbitals for homonuclear diatomic molecules.

and c). Taking the x axis as the internuclear axis, the linear combination of p_x orbitals must be made so that the lobes which have the same signs overlap (Fig. 4.11b). This combination also leads to bonding ($\psi_{p_x} + \psi_{p_x}$) and antibonding ($\psi_{p_x} - \psi_{p_x}$) molecular orbitals, the electron probability density plots of which have characteristics similar to those for the linear combination of s atomic orbitals. Molecular orbitals that are cylindrically symmetrical about the internuclear axis are called σ orbitals; the electron density in σ orbitals is concentrated between the nuclei (Fig. 4.10). A convenient method often used to distinguish between σ molecular orbitals arising from different atomic orbitals involves a symbolism incorporating those atomic orbitals. Thus, the $\sigma(1s)$ molecular orbital arises from the linear combination of two $1s$ orbitals (Fig. 4.11a). A $\sigma^*(2p_x)$ molecular orbital is antibonding $[\psi(2p_x) - \psi(2p_x)]$ and is formed from the combination of two $2p_x$ atomic orbitals (Fig. 4.11b).

The combination of either p_y or p_z atomic orbitals leads to molecular orbitals which have different spatial characteristics than the σ orbitals (Fig. 4.11c). These orbitals are perpendicular to the internuclear axis, and the molecular orbitals formed from their combination do not possess simple cylindrical symmetry. The molecular wave function for the bonding combination $[\psi(p_y) + \psi(p_y)]$ or $[\psi(p_z) + \psi(p_z)]$ is of opposite sign on the two sides of the internuclear axis, and the electron density is equally distributed on opposite sides. Molecular wave functions with this kind of symmetry are called π orbitals.

The relative energies of the molecular orbitals that can be formed from ns and np atomic orbitals can, in principle, be obtained from spectroscopic observations on the appropriate molecules. In general, the relative order for the combinations for $2s$ and $2p$ atomic orbitals is given by the sequence

$$\sigma(2s) < \sigma^*(2s) < \sigma(2p_x) < \pi(2p_y) = \pi(2p_z) < \pi^*(2p_y) = \pi^*(2p_z) < \sigma^*(2p_x).$$

A graphical illustration of this order is shown in Fig. 4.12. A similar order of molecular levels appears to be valid for the $3s$ and $3p$ combinations, but these simple results involving higher atomic orbitals must be tempered by the possibility that combinations involving $3d$ orbitals also may occur.

One important point should be made with respect to the generalized molecular energy-level diagram (Fig. 4.12). Just as degenerate atomic levels exist (e.g., $p_x, p_y, p_z; d_{xy}, d_{yz}, d_{x^2-y^2}, d_{z^2}$) it is possible to have degenerate molecular energy levels. The $p\pi$ and $p\pi^*$ molecular levels are degenerate since they arise from a linear combination of the atomic orbitals that are both perpendicular to the x axis and differ only in their relative orientation in space.

The second-period elements possess only s and p outer atomic orbitals. Accordingly, the linear combination of atomic orbitals described in Fig. 4.11 and the corresponding energy-level diagram (Fig. 4.12) should be applicable

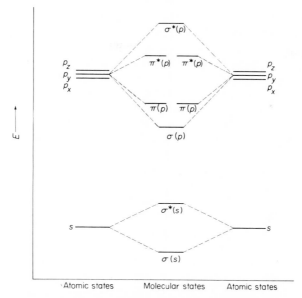

FIG. 4.12   The relative order of molecular energy levels that arise from the linear combination of s and p orbitals.

to the homonuclear diatomic molecules of the elements in this period. The electron distribution within the molecular levels available for the linear combination of $2s$ and $2p$ atomic orbitals in the species Li_2, N_2, O_2 and F_2 is shown in Fig. 4.13. The molecular-orbital description assigns a σ bond to the Li_2 and F_2 molecules. The electrons forming the bond in Li_2 occupy

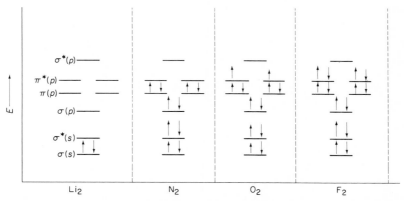

FIG. 4.13   The electronic distribution in the molecular levels of some Period 2 homonuclear diatomic molecules.

the $\sigma(2s)$ bonding level, whereas the bonding electrons in F_2 occupy the $\sigma(2p_x)$ level. In the case of F_2 the bonding and corresponding antibonding levels [$\sigma(2s)$ and $\sigma^*(2s)$; $\pi(2p_y)$ and $\pi^*(2p_y)$; $\pi(2p_z)$ and $\pi^*(2p_z)$] are filled, an arrangement which leads to the conclusion that none of these electrons contribute to the overstability of the F_2 molecule.

Both N_2 and O_2 are characterized by multiple bonds. Of the ten valence electrons in the N_2 molecule, four electrons, that is those in the $\sigma(2s)$ and $\sigma^*(2s)$ levels, do not contribute to bonding the atoms together. In contrast the six electrons distributed in the $\pi(2p_x)$, $\pi(2p_y)$, and $\pi(2p_z)$ orbitals constitute the triple bond formed between the two nitrogen atoms. The O_2 molecule has two more electrons than N_2 which occupy the $\pi^*(2p_y)$ and $\pi^*(2p_z)$ molecular orbitals. In accordance with Hund's rule, these electrons are unpaired in the two degenerate levels. Thus, the molecular orbital description of O_2 suggests that the molecule should be paramagnetic, a result substantiated by experiment (Chapter 13).

Of the remaining diatomic elements in Period 2, the experimental data suggest that the relative energies of the molecular energy levels for Be_2 are given by Fig. 4.12; however, in the case of C_2 and B_2, the $\sigma(2p_x)$ lies above $\pi(2p_z)$ and $\pi(2p_y)$, as shown in Fig. 4.14.

At this point it would be interesting to compare the molecular-orbital descriptions of these molecules with those using the valence-bond theory.

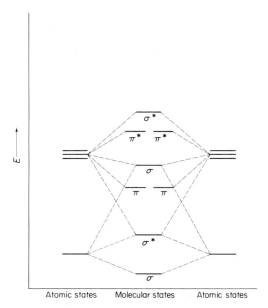

FIG. 4.14 The molecular energy levels for B_2 and C_2.

The principal covalent canonical forms expected for Li_2 and F_2 involve single bonds:

Li:Li $\Psi_{\pm} = C_1\psi_{2s}(1)\psi_{2s}(2) \pm C_2\psi_{2s}(2)\psi_{2s}(1)$

F:F $\Psi_{\pm} = C_1\psi_{2p_x}(1)\psi_{2p_x}(2) \pm C_2\psi_{2p_x}(2)\psi_{2p_x}(1)$.

Of course, the introduction of ionic canonical forms leads to a better approximation. The principal canonical form for N_2 involves a triple bond $(:N:::N:)$, involving either atomic orbitals on each atom (He core $2s^2 2p_x^1 2p_y^1 2p_z^1$) or hybrid orbitals [He core $(sp)^2(sp)^1 2p_x^1 2p_z^1$ or He core $(sp^3)^2(sp^3)^1(sp^3)^1(sp^3)^1$]. With no a priori knowledge of the magnetism of O_2, valence-bond arguments suggest that its primary canonical form is also a multiply bonded structure, $:\ddot{O}::\ddot{O}:$. No other Lewis formulation should be possible for this diatomic molecule formed from an element with an even number of electrons located in Period 2. As indicated previously, O_2 is a paramagnetic species, so that other canonical forms involving either a three-electron bond $(:\ddot{O}:::O:)$ or unpaired electron spins on each atom $(:\dot{O}:\dot{O}:)$ are required to obtain a better approximation of the observed properties. In any event neither of these structures would be seriously considered in the valence-bond formulation of O_2 without the magnetic data.

3.3 / Heteronuclear Diatomic Molecules

The application of molecular-orbital theory to heteronuclear systems follows the general principles outlined above; however, the choice of the atomic orbitals to be used in forming the appropriate linear combinations is not necessarily obvious because the two atoms now have different electronic structures. Indeed, hybrid atomic orbitals may also be employed to construct molecular orbitals (3). Basically, this decision is guided by the requirement that the combining orbitals should be of similar energy, possess maximum overlap, and have the correct symmetry properties with respect to the internuclear axis. Using HF, formed from F $(1s^2 2s^2 2p^5)$ and H $(1s^1)$, as an example, these criteria indicate that the most successful linear combination would involve the hydrogen $1s$ orbital and the fluorine $2p_x$ orbital. That is, the internuclear axis is taken along the axis of the fluorine p_x atomic orbital containing one electron; any other orientation of the $1s$ and $2p$ orbitals leads to poorer overlap than the $1s$–$2p_x$ combination (Fig. 4.8b). Under these restrictions the best molecular orbital can be written as

$$\Psi = \psi(H, 1s) \pm \lambda\psi(F, 2p_x),\qquad (37)$$

where the constant λ obtained by the variation method is no longer expected to be unity. The value of λ reflects the extent to which an atomic orbital contributes to the molecular wave function. The variation method applied to Eq. (37) yields $\lambda > 1$, indicating that the $2p_x$ wave function for fluorine contributes more to the molecular function than does the hydrogen $1s$ function. In other words, the bonding electrons in HF should be more influenced by the fluorine atom than by the hydrogen atom. The value of λ is often used as a measure of the bond polarity.

The valence-bond description of HF would not only incorporate the usual covalent canonical form

$$H \!:\! \ddot{\underset{..}{F}} \!: \qquad \Psi_{covalent} = C_1 \psi_{1s}(1) \psi_{2p_x}(2) + C_2 \psi_{1s}(2) \psi_{2p_x}(1), \qquad (38)$$

but also possible ionic forms

$$H^+ \; : \! \ddot{\underset{..}{F}} \!: ^- \qquad \Psi_{ionic} = \psi_{2p_x}(1) \psi_{2p_x}(2) \qquad (39)$$

$$H \!: ^- \; \ddot{\underset{..}{F}} \!: ^+ \qquad \Psi'_{ionic} = \psi_{1s}(1) \psi_{1s}(2). \qquad (40)$$

The high ionization potential of fluorine and low electron affinity of hydrogen (Chapter 2) suggest that the possible ionic canonical form shown in Eq. (40) does not contribute significantly to the total wave function but that the other ionic form [Eq. (39)] does. Thus, the best valence-bond description for the HF wave function is given by

$$\Psi = C_1 \psi_{1s}(1) \psi_{2p_x}(2) + C_2 \psi_{1s}(2) \psi_{2p_x}(1) + C_3 \psi_{2p_x}(1) \psi_{2p_x}(2). \qquad (41)$$

Since C_1 and C_2 are related, only two constants (C_3 and C_1 or C_2) need be evaluated by the variation procedure.

The relative order of energy levels in more complex heteronuclear systems can be approximated by those for homonuclear systems (Fig. 4.12). On this basis the molecular-orbital description of CO, a system that is iso-electronic with N_2, containing ten valence electrons, indicates that a triple bond is present in the molecule, with all bonding electrons in orbitals arising from the combination of the $2p$ atomic orbitals (Fig. 4.15). Similarly, the NO molecule, which contains eleven valence electrons, should possess something less than triple bond character in addition to exhibiting para-magnetism. The odd electron is not localized on one of the atoms but is delocalized over the whole molecule. This result is consistent with the observed reluctance of NO to undergo dimerization (Chapter 12).

It should be emphasized that the energy-level diagram derived for homonuclear diatomic molecules as shown in Fig. 4.12 is not strictly applicable to heteronuclear systems. Thus, the atomic functions that must be

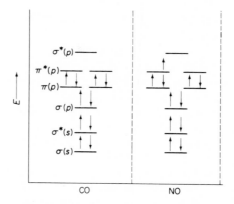

FIG. 4.15 The distribution of electrons in molecular energy levels for CO and NO derived on the assumption that the atomic levels are the same as that for homonuclear diatomic molecules.

combined to give the molecular orbitals in CO and NO are not of the same energy so that the molecular levels are not symmetrically distributed about the atomic levels (Fig. 4.16). Fortunately, the relative order of levels does not change for CO or NO so that the distribution of electrons is the same when using the general scheme shown in Fig. 4.16 as that predicted from Fig. 4.12. This correspondence need not necessarily be always valid.

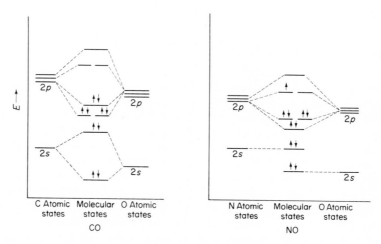

FIG. 4.16 The molecular levels for CO and NO assuming that the original atomic levels do not have the same energy.

4 / MOLECULAR-ORBITAL AND VALENCE-BOND METHODS: A COMPARISON

Both the molecular-orbital and valence-bond methods of describing covalently bound species are attempts to obtain approximate solutions of the equations which govern the characteristic properties of molecules. Thus, in a very real sense, neither is the right or the wrong method. Rather, each method has a different utility under a given set of circumstances. The valence-bond method has found great favor with some chemists since the process of constructing an approximate solution involves many components that are familiar to chemists (e.g., canonical structures than can be conveniently represented by the usual chemical symbolism). Such an easy correspondence of ideas may often lead to misconceptions because these parallelisms are not complete. For example, although several canonical forms can be written as a guide to the form of the approximate solution to the wave equation, it does not follow that these canonical forms have a distinct existence and form a kind of system at equilibrium. In general, the valence-bond method is very useful in describing ground-state phenomena.

On the other hand, the molecular-orbital method gives a useful description of covalently bonded systems if more detailed knowledge of electronic structure is necessary. For example, phenomena involving considerations of the excited states of molecules, as in molecular spectra and photochemistry, are generally more easily understood using molecular-orbital arguments. Further refinements to obtain better wave functions using either method require significantly more extensive calculations than have been indicated here.

5 / IONIC CHARACTER

5.1 / General Considerations

In the general valence-bond treatment the best approximate wave function contains wave functions from covalent ($\Psi_{covalent}$) and ionic (Ψ_{ionic}) canonical structures [Eq. (21)]. If an estimate of the relative contributions of these two classes of canonical structures for a given system can be obtained from experimental data, many tedious calculations could be eliminated. One method (due to Pauling) of estimating the per cent ionic character in a bond is based on the magnitude of electric bond moments. The problem of obtaining individual bond moments from the measured dipole moment of a polyatomic system is dependent upon the geometry of the molecule. To simplify the arguments, let us consider a neutral diatomic molecule, shown in Fig. 4.17, for which the measured dipole moment is the bond moment;

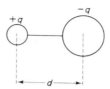

FIG. 4.17 The dipole moment of a diatomic molecule is a measure of the charge separation along the bond axis and the distance by which it is separated, i.e., $\mu = qd$.

in such molecules a charge q^+ is separated by a distance d from a charge of equal magnitude but opposite sign q^-. The dipole moment μ is defined as the product of the charge q and the internuclear separation:

$$\mu = qd. \tag{42}$$

If there is no separation in charge the molecule will not have a dipole moment and the bonding might best be described as purely covalent. In contrast, a full electron transfer corresponds to an ionic structure. The value of the experimental dipole moment will be between these extremes. The per cent ionic character can be determined from a comparison of the experimental dipole moment with the dipole moment expected if a full electron-charge separation occurred:

$$\% \text{ ionic character} = \left(\frac{\mu_{\text{exp.}}}{\mu_{\text{theor.}}}\right) \times 100. \tag{43}$$

The latter quantity can be calculated from the known charge on the electron $(4.8 \times 10^{-10} \text{ esu})$ and the internuclear distance in the molecule under consideration,

$$\mu_{\text{theor.}} = (4.8 \times 10^{-10} \text{ esu})(d). \tag{44}$$

It should be recognized that the calculation of $\mu_{\text{theor.}}$ incorporates the assumption that the internuclear distance does not change with an increase in the magnitude of charge separation.

The per cent ionic character as defined by Eq. (43) for each of the hydrogen halides is listed in Table 4.4. The results show that the per cent ionic character decreases in a regular manner from HF to HI. The contribution of the ionic canonical structures in the valence-bond theory can be easily calculated from Eq. (21) by introducing a weighting factor λ':

$$\Psi = \Psi_{\text{covalent}} + \lambda'\Psi_{\text{ionic}}. \tag{45}$$

When the variation technique is applied to obtain the best wave function, the relative weights the approximate solutions contribute to the best wave

TABLE 4.4 The per cent ionic character of the hydrogen halides

Halide	$\mu_{exp.}$, D^a	d, Å	$\mu_{theor.}$, D^a	%	λ	λ'
HF	1.98	0.92	4.42	45	0.90	0.93
HCl	1.03	1.28	6.07	17	0.45	0.51
HBr	0.79	1.43	6.82	12	0.36	0.41
HI	0.38	1.62	7.74	5	0.23	0.25

^{a} 1 Debye (D) = 4.8×10^{-10} esu Å.

function are in the ratio of the square of the weighting factors. Thus, in the case of Eq. (45) the relative weights of $\Psi_{covalent}$ and Ψ_{ionic} are in the ratio $1 : \lambda'^2$, and the per cent ionic character is given by

$$\% \text{ ionic character} = \frac{(\lambda')^2}{1 + (\lambda')^2} \times 100. \tag{46}$$

Since the per cent ionic character has been estimated from the dipole-moment data (Table 4.4), it should be possible to evaluate λ' and immediately obtain the best wave function without engaging in the details of the variation method. It should be pointed out that Eq. (45) is reminiscent of Eq. (37), which is used in the molecular-orbital method for heteronuclear diatomic molecules. For a given series of compounds the values of λ should show trends similar to those shown by the values of λ' estimated from Eq. (45).

Pauling suggested another method of estimating the ionic character of bonds from experimental data which is more easily obtained than bond moments. The method relies on estimating the ionic contribution to the total bonding energy. Consider the simple diatomic molecule AB. It will be recalled that the wave function $\Psi_{covalent}$ for the purely covalent structure A:B is not a very good approximation for the solution to the Schrödinger equation for the molecule AB. A better solution (i.e., one that predicts a lower energy) is obtained if ionic canonical structures such as $A^+ : B^-$, characterized by Ψ_{ionic}, are introduced. Since each of these wave functions ($\Psi_{molecule}$, $\Psi_{covalent}$, Ψ_{ionic}) has an associated energy, Pauling attempted to estimate the contribution of the ionic canonical structures by considering the corresponding energies. The energy relationship for the molecule AB is given by

$$E_{molecule} = E_{covalent} + E_{ionic}. \tag{47}$$

Recognizing that $E_{molecule}$ is given by the dissociation energy of AB (Chapter 3), which is experimentally determinable, we need only to obtain a value of $E_{covalent}$ to determine the contribution of the ionic canonical forms. Pauling postulated that the bond energy for AB is simply related to the geometric mean* of the bond energies of the molecules AA and BB from which it can be obtained [Eq. (47)]:

$$AA + BB \rightarrow 2AB. \tag{48}$$

Defining the bond energies of the species in Eq. (48) as $D(AA)$, $D(BB)$, and $D(AB)$, the postulate of the geometric mean permits us to estimate $E_{covalent}$ by

$$E_{covalent} = [D(AA) \cdot D(BB)]^{1/2}. \tag{49}$$

Since $E_{molecule}$ is given by $D(AB)$ the ionic contribution can be obtained from Eqs. (47) and (49):

$$E_{ionic} = D(AB) - [D(AA) \cdot D(BB)]^{1/2}. \tag{50}$$

The calculations for the hydrogen halides are summarized in Table 4.5. The ionic contribution to the binding energy in these molecules decreases smoothly from HF to HI, paralleling the change in per cent ionic character estimated from dipole moments (Table 4.4).

TABLE 4.5   The ionic and covalent
contributions in the hydrogen halides

	$D(AB)$	$E_{covalent}$ [a]	E_{ionic}
HF	134.6	61.8	72.8
HCl	103.2	72.7	30.5
HBr	87.5	69.3	18.2
HI	71.4	61.4	10.0

[a] Calculated from Eq. (49) using the following data: $D(HH) = 104$, $D(FF) = 36$, $D(ClCl) = 58$, $D(BrBr) = 46$, $D(II) = 36$.

5.2 / Electronegativity (VIII, IX)

A consideration of the values of the ionic resonance energy E_{ionic} (Table 4.5) for a variety of covalently bound molecules led Pauling to the conclusion that the square roots of these values $(E_{ionic})^{1/2}$ are approximately additive. This behavior suggests that each atom possesses a characteristic property that is reflected in the experimentally determined values of $(E_{ionic})^{1/2}$. Pauling suggested that $(E_{ionic})^{1/2}$ for a covalent bond is related to the

---

* The arithmetic mean is another type of average which might be used. If the bond energies do not differ greatly, the geometric and arithmetic means give essentially the same results.

difference in the electronegativity of the atoms present. Since the electronegativity of an atom is defined as the "power of an atom in a molecule to attract electrons to itself," a reasonably straightforward qualitative argument can be developed to relate $(E_{ionic})^{1/2}$ to electronegativity differences. That is, the greater the electronegativity difference between two bound atoms, the more polar the bond and the greater the ionic contribution to the valence-bond wave function. If the electronegativity of an atom is defined by χ, the electronegativity difference for two atoms A and B can be related to the ionic resonance energy by

$$(E_{ionic})^{1/2} = K|\chi_A - \chi_B|, \tag{51}$$

where K is a proportionality constant. The value of K is empirically determined to be 30 when the energies in Eq. (51) are expressed in kilocalories. Since only differences in the absolute value of electronegativities can be obtained using Eq. (51), it should be apparent that a table of atomic electronegativities cannot be obtained unless the electronegativity of a single atom is established. Pauling created the most widely used electronegativity scale by defining $\chi_F = 4.0$. With this as a definition the electronegativities for other elements can be easily obtained. The Pauling electronegativities for the nonmetallic and metalloid representative elements appear in Table 4.6.

TABLE 4.6   The electronegativities of some representative elements

			H
			2.1
C	N	O	F
2.5	3.0	3.5	4.0
Si	P	S	Cl
1.8	2.1	2.5	3.0
Ge	As	Se	Br
1.8	2.0	2.4	2.8
			I
			2.5

Since ionic character is related to the ionic resonance energy E_{ionic} which in turn is related to electronegativity differences, it should not be surprising to learn that the per cent ionic character can be related to electronegativity differences. This relationship has been established empirically to be

$$\% \text{ ionic character} = 16|\chi_A - \chi_B| + 3.5|\chi_A - \chi_B|^2. \tag{52}$$

Thus, the general scale of characteristic electronegativities serves as a convenient repository of information that can be used to estimate the weighting factor for the ionic contribution of the valence-bond description of a bond formed between any two atoms. In addition, the electronegativity scale has found use in correlating many chemical observations which are difficult to quantify.

A conceptually simpler approach to establishing the electronegativity of an atom is due to Mulliken, who suggested that the ability of an atom in a molecule to compete for an electron in a chemical bond is measured by the average value of its ionization potential I and electron affinity E:

$$\chi_M = \tfrac{1}{2}(I + E). \tag{53}$$

Mulliken electronegativities (χ_M) for the halogens are given in Table 4.7. Although the numerical values of χ_M are different from the Pauling values (Table 4.6) the general variation observed among χ_M values parallels the corresponding χ_P values. The plot of χ_M against χ_P is a straight line with slope 62.5 passing through the origin. Thus the Mulliken and Pauling electronegativities are related to each other by a constant factor (Table 4.7).

TABLE 4.7 **Mulliken electronegativities**[a]

	I	A	χ_M	$\dfrac{\chi_M}{62.5}$
F	401.6	83.5	242.5	3.9
Cl	299.2	88.2	194.0	3.1
Br	272.9	81.6	177.2	2.8
I	241.0	74.6	157.8	2.5

[a] All values in units of kcal mole^{-1}.

Another useful empirical method of estimating electronegativities considers the force exerted by an atomic nucleus in a molecule on a bonding electron (4–6). According to Coulomb's law the magnitude of this force F is given by

$$F = \frac{Z^*e^2}{r}, \tag{54}$$

where Z^* is the effective nuclear charge, and r is the distance between the electron and the nucleus. The effective nuclear charge can be estimated from the known nuclear charge and the shielding parameters assigned to inner electron shells (Chapter 2). The distance r is taken to be the covalent radius of the atom in question. Again, the numerical values obtained from this method do not follow the Pauling values, but similar trends are observed and a linear relationship exists between the Pauling values and those calculated from Eq. (54):

$$\chi = 0.359\frac{Z^*}{r^2} + 0.744. \qquad (55)$$

Since Pauling first attempted to quantify the concept of electronegativity there have been many techniques suggested to estimate the electronegativities of atoms for which the necessary data were not available earlier, as well as recalculations using the original method with more accurately determined thermochemical data (*VIII, IX*). Where such data are still not available, it has been possible to use a method which gives relative electronegativities on a different scale, simply related to the traditional Pauling scale. The current "best" values for the electronegativities of the elements are given in Table 4.8.

TABLE 4.8   The "best" values for the electronegativities of the elements[a]

							Oxidation states										
I	II	III	II	II	II	II	II	II	II	I	II	III	IV	III	II	I
H 2.20																
Li 0.98	Be 1.57											B 2.04	C 2.55	N 3.04	O 3.44	F 3.98
Na 0.93	Mg 1.31											Al 1.61	Si 1.90	P 2.19	S 2.58	Cl 3.16
K 0.82	Ca 1.00	Sc 1.36	Ti 1.54	V 1.63	Cr 1.66	Mn 1.55	Fe 1.83	Co 1.88	Ni 1.91	Cu 1.90	Zn 1.65	Ga 1.81	Ge 2.01	As 2.18	Se 2.55	Br 2.96
Rb 0.82	Sr 0.95	Y 1.22	Zr 1.33	Nb 1.6	Mo 2.16	Tc 1.9	Ru 2.2	Rh 2.28	Pd 2.20	Ag 1.93	Cd 1.69	In 1.78	Sn 1.96	Sb 2.05	Te 2.1	I 2.66
Cs 0.79	Ba 0.89	La 1.10	Hf 1.3	Ta 1.5	W 2.36	Re 1.9	Os 2.2	Ir 2.20	Pt 2.28	Au 2.54	Hg 2.00	Tl 2.04	Pb 2.33	Bi 2.02	Po 2.0	
		Ce 1.12	Pr 1.13	Nd 1.14	Pm —	Sm 1.17	Eu —	Gd 1.20	Tb —	Dy 1.22	Ho 1.23	Er 1.24	Tm 1.25	Yb —	Lu 1.27	
				U 1.38	Np 1.36	Pu 1.28										

The concept of electronegativity has been extended beyond the original Pauling suggestion. Indeed, methods have even been devised to estimate electronegativities of groups of atoms such as CH_3, CF_3, and C_6H_5.

From our discussion of electronegativity thus far it should not be assumed that the electronegativity of an atom is invariable. We should expect that the electronegativity of an atom depends upon its oxidation state and/or the type of orbital used in forming a chemical bond. For example, the electronegativity of tin in a divalent compound is 1.8, but the value increases to 2.0 in tetravalent tin compounds. Intuitively, it might be expected that an atom in a higher oxidation state would have a larger electronegativity than the same atom in a lower oxidation state. For covalent species containing an element with a variable oxidation state there is obviously a relationship between the formal oxidation state, the coordination number, and the nature of the orbitals involved in overlap. Thus, it becomes important in the Mulliken definition of electronegativity to consider the ionization potentials and electron affinities for the correct valence states of the atoms rather than using the ground-state energies. Jaffé and his students (7–10) have applied these ideas to obtain "orbital electronegativities." In a very real sense the electronegativities listed in Table 4.8 must be considered as average values.

REFERENCES

1. J. Waser, *J. Chem. Ed.*, **48**, 603 (1971).
2. H. Eyring, J. Walter, and G. E. Kimball, *Quantum Chemistry*, Wiley, New York, 1960.
3. I. Cohen and J. Del Bene, *J. Chem. Ed.*, **46**, 487 (1969).
4. A. L. Allred, *J. Inorg. Nucl. Chem.*, **17**, 215 (1961).
5. A. L. Allred and A. L. Hensley, *J. Inorg. Nucl. Chem.*, **17**, 43 (1961).
6. A. L. Allred and E. G. Rochow, *J. Inorg. Nucl. Chem.*, **5**, 264, 269 (1965).
7. J. Hinze and H. H. Jaffé, *J. Amer. Chem. Soc.*, **84**, 540 (1962).
8. J. Hinze, M. A. Whitehead, and H. H. Jaffé, *J. Amer. Chem. Soc.*, **85**, 148 (1963).
9. J. Hinze and H. H. Jaffé, *Can. J. Chem.*, **41**, 1315 (1963).
10. J. Hinze and H. H. Jaffé, *J. Phys. Chem.*, **67**, 1501 (1963).

COLLATERAL READINGS

I. W. Kutzelnigg, *Angew. Chem. Intern. Ed.*, **5**, 823 (1966). A survey of the physical and mathematical basis of quantum chemistry stressing methods and applications. Special attention is devoted to the more modern methods of quantum chemistry, the basic assumptions employed, their scope and limitations.
II. J. J. Lagowski, *The Chemical Bond*, Houghton-Mifflin, Boston, 1966. A description of early theories of the chemical bond, including excerpts from pertinent papers.

III. H. A. Bent, *Chem. Revs.*, **61**, 275 (1961). A discussion of the experimental evidence bearing on the relationship of bond properties, such as moments, force constants, dissociation energies, and lengths, to orbital hybridization of the second-period elements.

IV. H. Preuss, *Angew Chem. Intern. Ed.*, **4**, 660 (1965). A critical analysis of the hybridization concept as a basis for predictions about the chemical bond.

V. M. J. S. Dewar and J. Kelemen, *J. Chem. Ed.*, **48**, 494 (1971). An illustration of the LCAO–MO approach for describing chemical bonds using H_2 as an example.

VI. R. M. Gavin, Jr., *J. Chem. Ed.*, **46**, 413 (1969). An application of the modified Hückel molecular-orbital method (HMO) to inorganic systems with particular reference to their geometries.

VII. A. C. Wahl, *Science*, **151**, 961 (1966). A description of the electron densities, as calculated from molecular-orbital theory, for H_2, B_2, C_2, N_2, O_2, and F_2. Contour diagrams for bonding and antibonding orbitals are given.

VIII. H. O. Prichard and H. A. Skinner, *Chem. Revs.*, **55**, 745 (1955). A review of the methods used to estimate the electronegativities of atoms and radicals, applications of the electronegativity concept, and a summary of values.

IX. S. S. Batsanov, *Russ. Chem. Revs.*, **37**, 332 (1968). A more modern version of the information available in Collateral Reading *III*.

STUDY QUESTIONS

1. What structure might you expect for each of the following simple (gaseous) molecules or ions? (a) $SiCl_4$. (b) $TeCl_4$. (c) TeF_6. (d) $CdCl_2$. (e) BrF_3. (f) ZnI_2. (g) $In(CH_3)_3$. (h) NH_3. (i) $NH_4{}^+$. (j) $Fe(CO)_5$. (k) $W(CO)_6$. (l) ICl_2.

2. What canonical forms contribute to the wave function for the substances listed in Problem 1? Defend your inclusion of each canonical form.

3. For each of the following species predict the hybridization of each atom in the structure and indicate the bonds which might be expected to have an order greater than unity, giving the orbitals which could be involved in these bonds: (a) $[(CH_3)_3Si]_2O$; (b) $ClSO_3H$; (c) FN_3; (d) $SO_4{}^{2-}$; (e) H_2NPCl_2; (f) Cl_2O; (g) NO_2.

4. There have been suggestions that the ionic model can be used to calculate the energy of covalent substances. For example, one could consider SiF_4 as being formed in the process

$$Si^{4+} + 4F^- \longrightarrow SiF_4.$$

(a) Calculate the energy for this process using the electrostatic model. (b) Estimate the experimental value for the energy of this process. (c) Compare and discuss the results obtained in parts (a) and (b).

5. Knowing that the bond distances in the species BF_3 and BF_4^- are 1.30 Å and 1.40 Å, respectively, discuss the electronic structures of these species.

6. Give three illustrations, other than those mentioned in this chapter, of species for which (a) $p\pi$–$p\pi$ bonding might be expected, and (b) $d\pi$–$p\pi$ bonding might be expected.

7. Three compounds are known to exist between iodine and fluorine, IF_3, IF_5, and IF_7. The lowest member of the series, IF, has not been characterized, although its existence has been postulated. For this compound, calculate the expected value of (a) the ionic resonance energy, (b) the dipole moment, and (c) the bond energy.

8. The observed bond energy for lithium hydride is $58 \ kcal \ mole^{-1}$. Given the fact that the bond energies of the molecules Li_2 and H_2 are 27 and $104 \ kcal \ mole^{-1}$, respectively, calculate the electronegativity difference between lithium and hydrogen and the per cent ionic character for the Li—H bond.

9. Discuss the observation that the ionization potential of the oxygen atom is less than that of nitrogen, but the electronegativity of nitrogen is less than that of oxygen.

10. Using only a table of electronegativity values, estimate the standard enthalpy of formation of the uncharacterized compound $NI_{3 (g)}$.

11. The heat of formation of gaseous SiF_4 is $360 \ kcal \ mole^{-1}$, whereas the single-bond energy for an Si—F bond has been estimated as $129 \ kcal \ mole^{-1}$. Discuss these data in terms of the electronic structure of SiF_4.

12. On the basis of the arguments developed in Problem 11, give the most probable canonical structures for SiF_4 which would be expected to form the basis of the approximate solution in the valence-bond method to the wave equation for this molecule.

13. Electron diffraction data indicate that the carbon–carbon bond in $(CN)_2$ is about 10% shorter than the carbon–carbon bond in ethane. Write the valence-bond structures which could be used to help explain the electron diffraction data.

14. Consider the bond formed between carbon and fluorine (C—F) in answering the following questions. (a) What is the value of the ionic resonance energy of the C—F bond? (b) What is the value of the bond energy of the purely covalent component in the valence-bond description of the C—F bond? (c) What is the per cent ionic character of the C—F bond?

15. The normal oxide of beryllium, BeO, can be volatilized to give a species in the vapor phase with the molecular formula BeO. (a) Assuming that the atomic orbitals for beryllium and oxygen are of equal energy, discuss the bonding in the species $BeO_{(g)}$ using simple molecular-orbital theory. (b) Discuss the bonding in the species $BeO_{(g)}$ using the correct relative order for the energies of the atomic orbitals. (c) Discuss the bonding in the species $BeO_{(g)}$ using valence-bond arguments.

16. Discuss the most probable reason for the nonexistence of the molecule Ne_2.

17. Assuming that the available atomic orbitals for boron and nitrogen are of equal energy, give the molecular-orbital energy-level diagram for the molecule BN, showing the distribution of electrons in the molecular orbitals available. What would happen to the bond order in this molecule if it formed the species BN^{2-}?

18. The species BF can be made by passing BF_3 over elemental boron which has been heated to 1800°C. Using molecular-orbital arguments, describe the bonding and the magnetic properties of BF for the case where the required atomic orbitals are assumed to be of equal energy.

19. Give a molecular-orbital description of the molecule SN.

20. Using simple molecular-orbital arguments, discuss the structure of the compound $NeO_{(g)}$. Give the equivalent arguments using the valence-bond method.

REPRESENTATIVE ELEMENTS

A BRIEF SURVEY

The representative elements (Fig. 5.1), i.e., those elements that possess an outer electronic configuration of the type $nsnp$, form a group which can be unified from a chemical standpoint by using several simple concepts. These unifying ideas are discussed in this chapter for the representative elements as a group; a detailed examination of the chemistry of these elements is deferred to subsequent chapters.

Consider the stoichiometry of some of the typical compounds formed by the representative elements (Table 5.1). Several striking features are apparent when the compounds are arranged in this manner. First, there is a clear division between the elements in Period 2 and those in higher-numbered periods on the basis of the compounds they form. The valence (i.e., number of single bonds formed) of the elements in the second period rises to a maximum of 4 and thereafter decreases. In contrast to this behavior, there is a general increase in the valence of the elements in the other periods from the first member to the last. The valence of an element in Period 2 never

153

	s^1	s^2	$d^n s^x$ (n = 1 to 10; x = 0, 1 or 2)										$s^2 p^1$	$s^2 p^2$	$s^2 p^3$	$s^2 p^4$	$s^2 p^5$	$s^2 p^6$
$1s$	1 H																	2 He
$2s\,2p$	3 Li	4 Be											5 B	6 C	7 N	8 O	9 F	10 Ne
$3s\,3p$	11 Na	12 Mg											13 Al	14 Si	15 P	16 S	17 Cl	18 Ar
$4s\,3d\,4p$	19 K	20 Ca	21 Sc	22 Ti	23 V	24 Cr	25 Mn	26 Fe	27 Co	28 Ni	29 Cu	30 Zn	31 Ga	32 Ge	33 As	34 Se	35 Br	36 Kr
$5s\,4d\,5p$	37 Rb	38 Sr	39 Y	40 Zr	41 Nb	42 Mo	43 Tc	44 Ru	45 Rh	46 Pd	47 Ag	48 Cd	49 In	50 Sn	51 Sb	52 Te	53 I	54 Xe
$6s\,(4f)\,5d\,6p$	55 Cs	56 Ba	57* La	72 Hf	73 Ta	74 W	75 Re	76 Os	77 Ir	78 Pt	79 Au	80 Hg	81 Tl	82 Pb	83 Bi	84 Po	85 At	86 Rn
$7s\,(5f)\,6d$	87 Fr	88 Ra	89** Ac	104	105													

$f^p d^n s^2$ (p = 1 to 14, n = 0 or 1 (2 for Th))

*Lanthanide series $4f$	58 Ce	59 Pr	60 Nd	61 Pm	62 Sm	63 Eu	64 Gd	65 Tb	66 Dy	67 Ho	68 Er	69 Tm	70 Yb	71 Lu
**Actinide series $5f$	90 Th	91 Pa	92 U	93 Np	94 Pu	95 Am	96 Cm	97 Bk	98 Cf	99 Es	100 Fm	101 Md	102 No	103 Lw

FIG. 5.1   The periodic table of the elements; representative elements are in the clear areas.

TABLE 5.1 Stoichiometries of some compounds of the representative elements[a]

Period	Group							
	I	II	III	IV	V	VI	VII	VIII
2	LiCl	BeCl$_2$[b]	BF$_3$	CH$_4$	NH$_3$	H$_2$O	HF	—
3	NaCl	MgCl$_2$	AlCl$_3$[b]	SiF$_4$	PH$_3$ PF$_5$	H$_2$S SCl$_4$ SF$_6$	HCl ClF$_3$	—
4	KCl	CaCl$_2$	GaCl$_3$[c]	GeF$_2$ GeF$_4$	AsH$_3$ AsF$_5$	H$_2$Se SeF$_4$ SeF$_6$	HBr BrF$_3$ BrF$_5$	KrF$_2$
5	RbCl	SrCl$_2$	InCl$_3$	SnCl$_2$ SnCl$_4$	SbH$_3$ SbCl$_5$	H$_2$Te TeCl$_4$ TeF$_6$	HI IF$_5$ IF$_7$	XeF$_2$ XeF$_4$ XeF$_6$

[a] Compounds in the shaded area are ionic.
[b] At 700°C.
[c] Above 800°C.

exceeds 4, but the valence of elements in higher-numbered periods often exceeds the maximum observed for the first member of a family; for example, compare the compounds formed by nitrogen, oxygen, and fluorine with the higher members of the corresponding family.

1 / THE ELEMENTS OF PERIOD 2

Excluding the elements that form ionic compounds, the geometrical arrangements of atoms in the gaseous molecules shown in Fig. 5.2 for the Period 2 elements agree with structures predicted by the electron-pair-repulsion arguments discussed in Chapter 3. Similarly, BeCl$_2$ and HF are linear molecules, BF$_3$ is trigonal planar, CH$_4$ is tetrahedral, NH$_3$ is pyramidal, and H$_2$O is bent, as indicated by the data in Table 5.2.

The covalently bound atoms of the Period 2 elements (except carbon) can form additional covalent bonds. Those atoms to the left of Group IV (BeX$_2$ and BX$_3$) are unsaturated in the Lewis sense, whereas those to the right (NX$_3$, OX$_2$, FX) have the maximum number of electrons, some of which exist as unshared pairs. Thus, the compounds BX$_3$ and BeX$_2$ are acidic whereas the others are basic; all the Period 2 elements (except carbon)

BH_4^-, CH_4, NH_4^+

NH_3, H_3O^+

H_2O

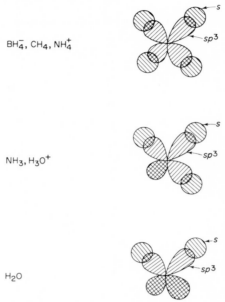

FIG. 5.2 Pictorial representations of the combinations of hybrid orbitals and atomic orbitals in related species. The shading is an attempt to depict electron bookkeeping. Single strokes represent an electronic density equivalent to one electron; cross-hatching corresponds to paired electrons.

TABLE 5.2 Structural parameters of some Period 2 compounds

Compound	XMX angle, degrees	$d(M—X)$, Å
$BeCl_2$	180	1.77
BF_3	120	1.295
CH_4	109.5	1.091
NH_3	107.3	1.008
H_2O	104.52	0.9572
HF	—	0.9170

can expand their coordination numbers* through the formation of co-ordinate-covalent bonds. Inspection of typical compounds of the Period 2 elements containing coordinate-covalent bonds (Table 5.3) indicates that the coordination number never exceeds four. In the Lewis sense, the Period 2

---

* As in the case of ionic compounds, the term coordination number for covalent species refers to the number of nearest neighbors.

TABLE 5.3 Some species formed by Period 2 elements[a]

Bonding type	Group						
	I	II	III	IV	V	VI	VII
Covalent	LiCl	$BeCl_2$	BF_3	CH_4	NH_3	H_2O	HF
Coordinate-covalent	$Li(NH_3)_4{}^+$	$BeCl_2(OH_2)_2$ $BeCl_4{}^{2-}$	$BF_4{}^-$ $BF_3(OH_2)$ H_3NBF_3		$NH_4{}^+$ H_3NBF_3	H_3O^+ (ice)	$(HF)_x$

[a] The compound in the shaded area is ionic.

elements can achieve a rare-gas configuration both by forming compounds expected from the usual rules of stoichiometry and by forming coordinate-covalent bonds. In any event, the number of single bonds formed by an element in this period, irrespective of the source of the bonding electrons, is never greater than four.

2 / THE ELEMENTS IN OTHER PERIODS (*I–IV*)

In contrast with the elements in Period 2, the corresponding elements in higher-numbered periods can form a greater number of single bonds and, hence, achieve a higher coordination number. The stoichiometry of some typical single-bonded compounds of the elements in higher-numbered periods (Table 5.1) indicates that, in some instances, all of the valence electrons of these elements can form single bonds. Thus, some of the elements in Groups V, VI, and VII can form five (PF_5, $AsCl_5$), six (SF_6, TeF_6), and seven (IF_7) single bonds, respectively. This behavior should be compared with the behavior of the corresponding Period 2 elements where the normal valence decreases regularly after carbon (NX_3, OX_2, FX).

Compounds of the elements in Period 3 and in higher-numbered periods which possess an octet of electrons act as Lewis acids and bases just like their counterparts in Period 2; some examples are shown in Table 5.4. That is to say, the elements in Groups II and III are Lewis acids, whereas those in Groups V, VI, and VII are Lewis bases. Compounds for which the non-Period 2 elements have achieved their maximum normal valence can often act as Lewis acids to form higher-coordinate species. Compare, for example, GeF_4 and AsF_5 (Period 4) with the species $GeF_6{}^{2-}$ and $AsF_6{}^-$, respectively. In each of these cases the coordination number of the central metal atom has been expanded by the formation of a coordinate-covalent bond in which the central atom acts as the Lewis acid (Fig. 5.3). In both cases the central atom is surrounded by six electron pairs which, according to electron-pair-repulsion arguments, should lead to an octahedral arrangement of

TABLE 5.4   The formulas of some species formed by Period 4 elements[a]

| Bonding type | \multicolumn{7}{c}{Group} |
	I	II	III	IV	V	VI	VII
Covalent	KCl	$CaCl_2$	$GaCl_3$[b]	GeF_4	AsH_3 AsF_5	H_2Se Cl_2SeO SeF_4 SeF_6	HBr BrF_3 BrF_5
Coordinate-covalent		$Ca(NH_3)_4{}^{2+}$	$H_3NGa(CH_3)_3$		$AsR_4{}^+$		$BrF_4{}^+$ $BrF_4{}^-$
		$Ca(OH_2)_6{}^{2+}$	$GaF_6{}^{3-}$ $GaF_3(NH_3)_3$ $Ga(OH_2)_6{}^{3+}$	$GeF_6{}^{2-}$ $GeCl_6{}^{2-}$	$AsF_6{}^-$		

[a] Compounds in the shaded area are ionic.
[b] Above 800°C.

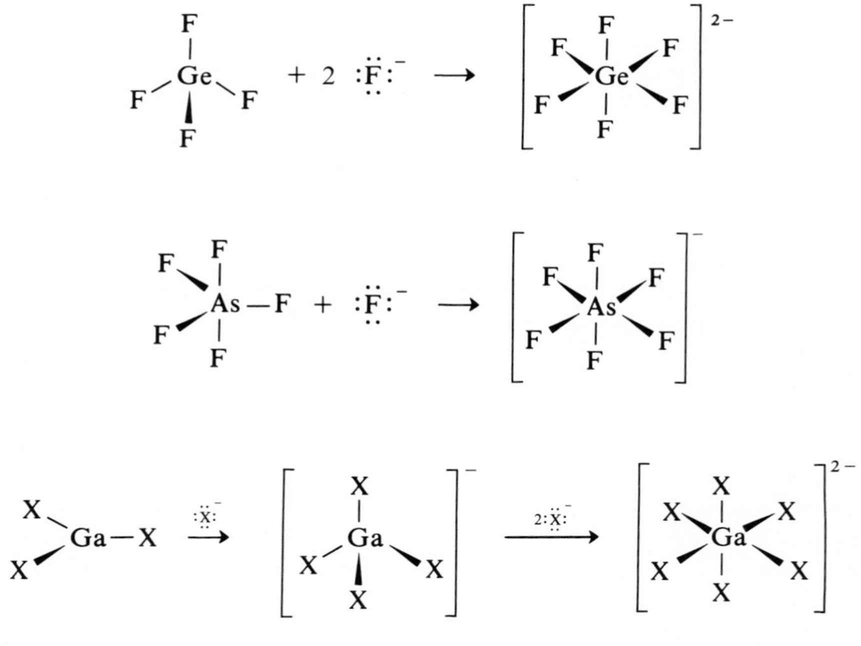

FIG. 5.3   Although germanium, arsenic, and gallium form the expected number of single bonds, they also can act as Lewis acids towards suitable Lewis bases to expand their coordination number.

bonded groups. The additional expansion of coordination number can also occur for the elements in periods higher-numbered than 2, and these elements can also achieve higher valence states. In general, 6 is the maximum coordination number that can be attained for these elements, although occasionally a larger number is observed (e.g., IF_7). Irrespective of whether the bonds are formally normal covalent or coordinate-covalent, the geometry of the covalently bound species of the representative elements can be predicted in nearly every instance from the electron-pair-repulsion arguments.

It might appear that the ability of the heavier elements to form higher-coordinate species is related to their larger size. This, however, cannot be the sole factor, since covalently bound systems containing a small central atom associated with relatively large groups (e.g., BI_4^-) are known. The availability of higher coordination numbers for the elements in Periods 3 and higher fortunately occurs among elements that have d orbitals which are energetically near the s and p orbitals used to rationalize the bonding in compounds of lower coordination. A maximum of four hybrid orbitals can be constructed from s and p orbitals, but the addition of an energetically favorable set of d orbitals provides the theoretical basis for forming five (dsp^3) and six (d^2sp^3) bonding orbitals (Chapter 4) (*III*).

Species which possess a tetrahedral (or nearly tetrahedral) arrangement of valence electrons, whether engaged in bonding or unpaired, can be described theoretically in terms of sp^3 hybrid orbitals. This point of view gives a consistent description of the species BH_4^-, CH_4, NH_3, NH_4^+, H_2O, and the corresponding structures for the elements in Periods 3 and higher. If the valence state of the central atom in such structures is an sp^3 hybrid, the available valence electrons can be distributed in the hybrid orbitals, and electrons that are unpaired will occupy an orbital which can overlap with an orbital on another atom (Fig. 5.2). In the case of CH_4 the central atom has four valence electrons in the four hybrid orbitals and four bonds can be formed. If nitrogen is the central atom, as in NH_3, two of the five valence electrons are paired in one orbital, and the other three orbitals can be combined with hydrogen $1s$ orbitals to form σ bonds; the molecule should be pyramidal with bond angles very near that of the tetrahedral angle. The slightly smaller angle in ammonia (107.5° rather than the theoretical 109.5° expected for a pure sp^3 hybrid) can be accounted for by (a) using a hybrid with less s character, (b) invoking proton–proton repulsions, or (c) considering bonded-pair–nonbonded-pair interactions. All three approaches have been used to account for the discrepancy between the expected bond angle for a pure sp^3 hybrid and the experimentally observed bond angle. The structure of NH_4^+ is exactly like that of NH_3 except that the *vacant* $1s$ orbital of a proton overlaps with the *filled* hybrid orbital on the nitrogen atom. The BH_4^- ion, which is isoelectronic and isostructural with NH_4^+ and CH_4, can be described in exactly the same way, except that the electronic bookkeeping scheme assigns one valence electron each to three of the sp^3 hybrid orbitals on the boron atom; these orbitals are available for combination with half-filled hydrogen $1s$ orbitals. The fourth hybrid orbital on boron, which is empty, combines with a filled hydrogen $1s$ orbital. In all of these descriptions, it must be remembered that the molecular orbitals are constructed from the hybrid wave functions available on the central atom, and the available electrons are placed into the lowest molecular orbitals that emerge from the linear combination.

The theoretical arguments for the bonding of atoms which form species with coordination numbers higher than 4 require more than s and p atomic orbitals to form the required numbers of bonds. An inspection of Fig. 5.4

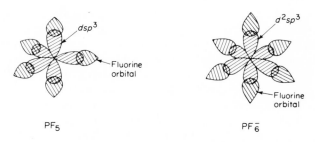

FIG. 5.4   Five- and six-coordinate phosphorus atoms require the use of dsp^3 and d^2sp^3 orbitals, respectively. The conventions in shading are explained in the caption to Fig. 5.2.

shows that the lobes in dsp^3 and d^2sp^3 hybrid orbitals are directed toward the corners of a trigonal bipyramid and an octahedron, respectively. The theoretical description of PF_5 (or similar 5-coordinate species) then requires that the half-filled hybrid orbitals overlap with half-filled orbitals on the fluorine atom. The latter are presumed to be atomic functions p_z since there is no evidence to the contrary for the spatial disposition of the other valence electrons on the fluorine atom. A similar description for the species PF_6^- can be made except that the five valence electrons on phosphorus are unpaired in five of the six hybrid d^2sp^3 orbitals. The vacant sixth orbital is combined with the filled atomic orbital on the fluoride ion. Thus, with relatively few hybrid orbitals the structures of virtually all of the covalently bound species formed by the representative elements can be explained.

In summary, much of the chemistry of the representative elements can be understood and/or correlated with (a) the electron deficiency of the elements in Groups II and III, (b) the presence of unshared electron pairs on the elements in Groups V, VI, and VII, (c) the ability of the elements in Periods 3 and higher to use *all* of their valence electrons to form bonds, and (d) the electron deficiency of these higher-valence structures. The formation of coordinate-covalent bonds can be looked upon as an attempt by atoms that occupy certain positions in the periodic chart to overcome electron deficiencies and achieve a rare-gas structure. In cases where electron deficiency is at a maximum, molecules of one kind can interact with each other to form condensed molecular systems (e.g., Al_2Cl_6). Detailed considerations of these processes appear in the appropriate subsequent chapters.

COLLATERAL READINGS

I. J. I. Musher, *Angew. Chem. Interñ. Ed.*, **8**, 54 (1969). A theoretical description of the compounds of the representative elements which exhibit a coordination number greater than four.

II. R. F. Hudson, *Angew. Chem. Intern. Ed.*, **6**, 749 (1967). A discussion of the compounds of the typical elements in which they exhibit coordination numbers higher than 4.

III. K. A. R. Mitchell, *Chem. Revs.*, **69**, 157 (1969). A consideration of the theoretical aspects involving the use of *d* orbitals in the bonding of the representative elements in Periods 3 and higher.

IV. R. E. Rundle, *Survey of Progress in Chemistry*, **1**, 81 (1963). Section IIIB incorporates a discussion of the use of outer *d* orbitals in bonding of non-transition-metal compounds.

STUDY QUESTIONS

1. Without reference to experimental data, predict the most probable geometry of the following species: (a) IF_3; (b) ClO_3^-; (c) $AsCl_3(CF_3)_2$; (d) $SnCl_2$; (e) $TeCl_4$; (f) GaF_6^{3-}.

2. Give the idealized hybrid orbitals used in bonding for the six species shown in Problem 1.

3. Discuss the Lewis acid–base behavior that might be expected for $SnCl_2$. Include in your discussion the structure of the species involved and the hybridization necessary for tin to form the required bonds.

4. Discuss the structure of the species which might be formed in solution when triethylaluminum is dissolved in ethyl ether.

5. Consider the reaction of each of the following species with each of the other members in the list: $AlCl_3$, $(CH_3)_2O$, $HCl(l)$, $SiCl_4$. Discuss the possible species which may be formed using simple Lewis formulations as the basis for your decision. Give the geometry of the species expected and the hybridization of each atom present.

/6

HYDROGEN

Hydrogen exhibits unique chemical and physical properties because of its ground-state electronic configuration and because it is the lightest and the smallest of the elements. Since the ground-state configuration for the atom is $1s^1$, stable electronic configurations in the Lewis sense can be obtained by either sharing or gaining one electron; in addition, the proton H^+, derived by the removal of one electron from the hydrogen atom, is a well studied species. Thus, simple valence considerations indicate that a hydrogen atom should form compounds containing one covalent bond and compounds containing ionic hydrogen. We shall see that the polarizing effect of covalently bound hydrogen atoms in certain compounds containing very electronegative atoms leads to relatively strong molecular associations. The simplest hydrogen-containing compound is the diatomic molecule H_2, the electronic structure of which is discussed in Chapter 4.

The fact that hydrogen can lose an electron to form a cation, H^+, makes it similar to the alkali-metal atoms; the chemistry of this family of elements

163

is dominated by their tendency to lose electrons. However, the ability of hydrogen to gain an electron to form H^-, which possesses the electronic structure of the next rare gas, is reminiscent of one of the types of reactions the halogens undergo. Thus, on the basis of its chemistry, hydrogen is often placed in both Groups I and VII.

1 / ATOMIC HYDROGEN

Although the hydrogen molecule, H_2, is the simplest stable form of hydrogen under ordinary conditions, atomic hydrogen can be formed from this species if sufficient energy is supplied:

$$H_2 \rightarrow 2H\cdot; \qquad \Delta H_0{}^0 = 103.2\,kcal\,mole^{-1}. \qquad (1)$$

At low pressures the recombination reaction in the gas phase is relatively slow; for example, hydrogen atoms have a half-life of 1.0 sec at a pressure of 0.2 torr. In the presence of an inert gas, at high pressures, or in the presence of certain metals (Pt, Pd, W), recombination occurs very rapidly, since the energy released when H_2 is formed can be removed from the reaction site. Hydrogen molecules can be dissociated thermally (temperatures near 2000°C), by electrical discharge, by irradiation with a mercury arc, or by bombardment with electrons in the 10–20-eV energy range. Hydrogen atoms can be generated in solution using two techniques. When hydrogen-containing solvents are subjected to ionizing radiation, a series of short-lived intermediates is formed, including electrons, ions, excited molecules, and free radicals (I); hydrogen atoms are often among the products. Hydrogen atoms have also been generated in solution by the photolysis of thiols (I):

$$RSH \rightarrow RS\cdot + H\cdot. \qquad (1a)$$

The reactions of hydrogen atoms in solution can be studied by allowing the gas from a hydrogen discharge to flow directly into the solution (2, 3, 4). The reactions of hydrogen atoms with a variety of substances in the gaseous phase and in solution have been extensively studied using kinetic methods because of the transient nature of this species (I).

As might be expected, atomic hydrogen possesses strong reducing properties. The high-valence oxides of the transition metals are reduced to hydroxy compounds, the more noble metal compounds to the metals, and oxyanions (e.g. $CrO_4{}^{2-}$ and $MnO_4{}^-$) to lower-valence ions (Cr^{3+} and Mn^{2+}).

Molecular hydrogen can be activated by salts and complexes of transition elements as well as by compounds of the elements of Groups IB and IIB

(*II*, 5). Three general mechanisms have been identified for this process, which is catalytic in nature. In the presence of other species hydrogen molecules undergo homolytic fission to form hydrido complexes,

$$H_2 + 2M \rightarrow 2HM, \tag{2}$$

which then undergo reaction. An example of this type of reaction involves the complex ion $[Co(CN)_5]^{3-}$ (*6, 7*). A second general process by which molecular hydrogen reacts with species in solution involves heterolytic fission in which a hydride ion and hydrogen ion both add to a species; in a formal sense the latter behaves as both a Lewis acid and base forming a dihydro product:

$$H_2 + M \rightarrow MH_2. \tag{3}$$

The overall process appears to be equivalent to an insertion reaction. An example of this process involves the rhodium complex $RhCl[P(C_6H_5)_3]_3$:

$$\tfrac{1}{2}H_2 + RhCl[P(C_6H_5)_3]_3 \rightarrow RhHCl[P(C_6H_5)_3]_3. \tag{4}$$

The third general process by which hydrogen can be activated involves heterolytic fission with the formation of a monohydrido species:

$$H_2 + M \rightarrow HM^- + H^+. \tag{5}$$

An example of such a reaction occurs in the chemistry of ruthenium:

$$H_2 + RuCl_6^{3-} \rightarrow RuHCl_5^{3-} + H^+ + Cl^-. \tag{6}$$

Many systems containing transition-metal ions exhibit catalytic activity in the reaction of hydrogen with other substances. There is considerable evidence that such species involve the formation of more or less labile complex ions containing hydrogen. A discussion of transition-metal complex ions appears in Chapter 17.

2 / MOLECULAR HYDROGEN

The electronic structure of the hydrogen molecule is discussed in Chapter 4.

All diatomic molecules for which the constituent nuclei have spin exhibit the phenomenon of spin isomerism. That is, the nuclear spins can be parallel (the *ortho* isomer) or opposed (the *para* isomer). It can be shown by quantum mechanics that the *ortho* isomer can occupy only the odd rotational states ($j = 1, 3, 5, \ldots$) whereas the *para* isomer is restricted to the even states ($j = 0, 2, 4, \ldots$). Thus, at any given temperature, an equilibrium should exist between the *ortho* and *para* isomers. For most diatomic molecules which might be expected to exhibit spin isomerism, the energy separation of the rotational states is small compared to kT even at low temperatures. However, in the case of H_2, which has the smallest moment of inertia of any diatomic molecule, the energy difference between states is relatively large and only the lowest states are populated at room temperature and below.

A sample of ordinary hydrogen gas is an equilibrium mixture of *ortho* and *para* hydrogen (Fig. 6.1). The *ortho*-hydrogen portion approaches a

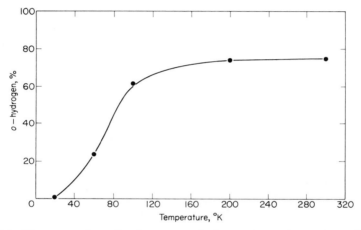

FIG. 6.1   The proportion of *ortho* to *para* hydrogen in a mixture is temperature dependent.

maximum (75%) at about room temperature, which would be statistically expected on the basis of the number of ways in which the nuclear spin of the molecule can combine with its total angular momentum. Interestingly, the conversion of one form into another cannot be accomplished thermally unless complete dissociation and recombination can occur. Thus, the conversion of one form to another can be catalyzed by atomic hydrogen or other paramagnetic substances such as O_2, NO, and transition-metal compounds; the greater the paramagnetism of the catalyst, the greater the rate of conversion. The activation energy for the reaction

$$p\text{-}H_2 + H \rightarrow o\text{-}H_2 + H \tag{7}$$

is $7.25 \text{ kcal mole}^{-1}$.

A sample of ordinary hydrogen containing both *ortho* and *para* isomers can be enriched in the *para* isomer by cooling in the presence of charcoal (cf. Fig. 6.1) which is a good catalyst under these conditions. If the mixture is separated from the catalyst and allowed to warm the adjustment of the relative amounts of the isomers expected on the basis of the data in Fig. 6.1 does not occur unless a paramagnetic substance is added. The half-life for natural conversion of *para* into *ortho* hydrogen is three years.

The two isomers of hydrogen have slightly different properties (Table 6.1), although they are chemically equivalent.

TABLE 6.1 Some physical properties of *ortho*- and *para*-hydrogen

Variety of hydrogen	Mp, °K	Bp, °K	Rotational specific heat at 298°K, cal $mole^{-1}$ deg^{-1}	Heat of fusion at 20°K, cal $mole^{-1}$
ortho	13.93	20.41	1.838	—
para	13.88	20.29	2.186	28.08

3 / THE COMPOUNDS OF HYDROGEN (*III*)

Hydrogen combines with a variety of elements to form compounds with widely differing properties (Fig. 6.2). Elements with very low electronegativities, i.e., about 1.0 or less, form compounds in which the hydrogen appears as an anion, H^-. These compounds have many of the properties associated with ionic substances and are sometimes called *salt-like* or *saline hydrides.*

H																	He
Li	Be											B	C	N	O	F	Ne
Na	Mg											Al	Si	P	S	Cl	Ar
K	Ca	Sc	Ti	V	Cr	Mn	Fe	Co	Hl	Cu	Zn	Ga	Ge	As	Se	Br	Kr
Rb	Sr	Y	Zr	Nb	Mo	Tc	Ru	Rh	Pd	Ag	Cd	In	Sn	Sb	Te	I	Xe
Cs	Ba	La-Lu	Hl	Ta	W	Re	Os	Ir	Pl	Au	Hg	Tl	Pb	Bl	Po	Al	Rn
Fr	Ra	Ac		U	Pu												
Saline hydrides		Metal-like hydrides								Borderline hydrides			Covalent hydrides				

FIG. 6.2 The types of hydrogen-containing compounds formed by the elements.

Elements with electronegativities of about 2.0 or greater form compounds with hydrogen containing covalent bonds. These substances are generally volatile and are often referred to as *covalent hydrides.* The *metal-like hydrides* are formed by the elements with intermediate electronegativities (for example, between 1.0 and 2.0). The metal-like hydrides consist of hydrogen atoms dispersed in the interstices of a close-packed metal lattice. These hydrides sometimes exhibit a stiochiometry expected of true compounds (e.g., TiH_2), and sometimes they do not (e.g., $ZrH_{1.9}$). As in most classification schemes, the division of the hydrogen compounds into three general classes is not perfect. Indeed, there is a group of hydrides which partakes of some of the characteristics of the usual classes.

3.1 / Ionic Hydrides (IV)

The saline hydrides, formed with alkali (MH) and alkaline-earth ($M'H_2$) metals, are colorless crystalline solids which either melt or decompose above 600°. All the hydrides can be formed by direct combination of the elements at elevated temperatures:

$$2M + H_2 \rightarrow 2MH \tag{8}$$

$$M' + H_2 \rightarrow M'H_2, \tag{9}$$

where M represents an alkali metal, M' an alkaline-earth metal. In the case of lithium and sodium, the hydrides can also be prepared by heating the corresponding nitride in a stream of hydrogen:

$$2M_3N + 3H_2 \rightarrow 6MH + N_2. \tag{10}$$

Evidence for the anionic character of the hydrogen in the saline hydrides is obtained from electrolysis experiments. The electrolysis of fused LiH or solutions of hydrides in molten mixtures of alkali-metal halides, such as the LiCl–KCl eutectic which melts at 354°, gives hydrogen at the anode.

The alkali-metal hydrides all exhibit the sodium-chloride-type structure in which the metal–hydrogen distance increases regularly with increasing size of the alkali metal, as would be expected. An attempt to obtain the crystallographic radius of the hydride ion from the observed M—H distances in the alkali-metal hydrides using the arguments developed in Chapter 2 leads to an apparent inconsistency (Table 6.2). The fact that the calculated radii for KH, RbH, and CsH are reasonably constant at 1.53 ± 0.01 Å suggests that the apparent radii calculated from the data for LiH and NaH may reflect incipient covalent bonding in these systems. Since the charge density on the alkali-metal ions is governed by the radii of these ions, Li^+ should possess the greatest polarizing ability in this series. Accordingly the distortion of the anionic electron cloud should be the largest in LiH. Apparently a vestige of this effect is seen in the structure of NaH where the metal–hydrogen distance is still significantly lower than that of the heavier metals.

The metal atoms in the alkaline-earth hydrides occupy an approximately hexagonal closest-packed structure with hydrogen atoms arranged in the largest holes present. Two different metal–hydrogen distances are observed

TABLE 6.2   The crystallographic metal-hydrogen
distances for the alkali-metal hydrides and
the corresponding H^- radius

Hydride	$d(M-H)$, Å	$r(H^-)$, Å[a]
LiH	2.04	1.36
NaH	2.44	1.46
KH	2.85	1.52
RbH	3.02	1.54
CsH	3.19	1.52

[a] The hydride ion radius is calculated from the $M-H$
distance and the ion radius of the corresponding metal
ion (Table 2.2).

in the structures of the alkaline-earth metal hydrides. The dihydrides of
europium and ytterbium possess structures similar to those of the alkaline-
earth metal hydrides.

The reactions of the saline hydrides are governed by the sensitivity of
the hydride ion to protonic species,

$$H^- + H^+ \rightarrow H_2, \tag{11}$$

and the large affinity of hydrogen for many nonmetals (e.g., oxygen and
chlorine). The latter property makes the saline hydrides good reducing
agents in the sense that they can react with, for example, oxygen- or chlorine-
containing compounds to produce free elements:

$$LiH + LiCl \rightarrow 2Li + HCl, \tag{12}$$

or hydrogen-containing compounds:

$$8BF_3 \cdot O(C_2H_5)_2 + 6LiH \rightarrow B_2H_6 + 6LiBF_4 + 8(C_2H_5)_2O \tag{13}$$

$$AlCl_3 + 4LiH \xrightarrow{(C_2H_5)_2O} LiAlH_4 + 3LiCl. \tag{14}$$

Since the hydride ion is a very strong Lewis base the saline hydrides
react readily with substances containing even slightly acidic protons. Thus,
hydrogen is evolved when the alkali-metal hydrides are dissolved in water
or in alcohols:

$$H_2O + MH \rightarrow MOH + H_2 \tag{15}$$

$$ROH + MH \rightarrow MOR + H_2, \tag{16}$$

or when gaseous ammonia or acetylene is passed over the heated hydrides:

$$NH_3 + NaH \longrightarrow NaNH_2 + H_2 \tag{17}$$

$$C_2H_2 + NaH \longrightarrow NaC_2H + H_2. \tag{18}$$

In general the reactivity of the alkali-metal hydrides increases with increasing atomic number of the metal.

The basic character of the hydride ion is illustrated by the formation of compounds of the type $LiAlH_4$ (Eq. 14) and $NaBH_4$ which are constitutionally analogous to the halogen derivatives $LiAlCl_4$ and $NaBF_4$. Both the complex hydrides and halides of the Group III elements can be imagined as arising from the addition of an anion (Lewis base) to the corresponding trivalent derivative (Lewis acid) of the element:

$$MX_3 + :X^- \longrightarrow MX_4^- \tag{19}$$

where M = Group III element, and X = halide or hydride.

3.2 / Metal-like Hydrides

A group of elements, constituting nearly all the transition metals, reacts with hydrogen under a variety of conditions to form substances the structures of which have been termed interstitial. Originally the term interstitial compound implied that the arrangement of metal atoms was essentially the same as that in the pure metal, with hydrogen atoms entering the interstices. Even though this is not generally the case, the term is still used. In contrast with the saline hydrides, which are colorless and occupy less volume than the metal from which they are formed (Table 6.3), the metal-like hydrides

TABLE 6.3 Change in density between some metals and their hydrides (14-16)

Hydride	Density change, %
LiH	+52.8
NaH	+44
CaH_2	+10
BaH_2	+20
$LaH_{2.76}$	−12.8
$CeH_{2.69}$	−17.5
$TaH_{0.76}$	−9.1
$VH_{0.56}$	−6.7

occupy a greater volume than the parent metal, are metallic in appearance, and often possess variable composition.

The metallic hydrides can be prepared by direct combination, usually at elevated temperatures. Thus, platinum, palladium, and iron are permeable to hydrogen at elevated temperatures; tantalum absorbs hydrogen to yield a brittle product. Hydrogen can also be absorbed under more moderate conditions. For example, metallic cathodes used in the electrolysis of aqueous as well as nonaqueous solutions often absorb very large quantities of hydrogen; in some instances (for example, iron) electrolytically deposited metals will also absorb hydrogen.

Although many examples of metallic hydrides are known which possess a stoichiometry suggesting the formation of true compounds (Table 6.4), there is evidence that these substances are not compounds in the usual sense.

TABLE 6.4   A comparison of the structures of some metals and of their hydrides

	Structure[a] of metal atoms in	
Hydride	Hydride	Pure metal[b]
TiH_2	Fluorite[c]	bcc, hcp
ZrH, TiH	Zinc blende[d]	bcc, hcp
Pd_2H	Zinc blende[e]	ccp
Zr_4H	Zinc blende[f]	bcc, hcp
CuH	Wurtzite	ccp
LaH_2, PrH_2, NdH_2	Fluorite[c]	hcp
CeH_2	Fluorite[c]	ccp, hcp
ThH_2	Distorted fluorite[c]	ccp
CrH	Wurtzite	bcc
CrH_2	Fluorite	bcc

[a] See Chapter 2 for sketches of structural types.
[b] Key: bcc, body-centered cubic; hcp, hexagonal closest-packed; ccp, cubic closest-packed
[c] All tetrahedral holes in the cubic-closest-packed structure of metal atoms filled.
[d] One-half of ccp tetrahedral holes filled.
[e] One-fourth of ccp tetrahedral holes filled.
[f] One-eighth of ccp tetrahedral holes filled.

Thus, the hydrogen–zirconium phase diagram shows the existence of α-zirconium (hexagonal closest-packed) which can dissolve several per cent of hydrogen; β-zirconium (body-centered cubic) which is stable at high temperatures and can dissolve up to 50 atom % hydrogen; and the compound ZrH_2, which exists in two modifications (face-centered cubic and tetragonal) (V). As a second example, although well defined stoichiometries for the lanthanum dihydrides and trihydrides have been reported, there are also substances reported with variable compositions in the range $MH_{1.8}$ to $MH_{3.0}$.

The fundamental difference between the ionic hydrides and the metallic hydrides is clearly illustrated by the way the corresponding metals absorb hydrogen (Fig. 6.3). The metals which form the saline hydrides show extended

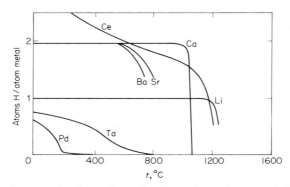

FIG. 6.3   The absorption of hydrogen by various metals (8).

temperature ranges in which hydrogen absorption does not occur until conditions favorable for the formation of the hydride are reached; at this point the expected stoichiometric amount of hydrogen is "absorbed." In contrast, some of the metals that form metallic hydrides absorb hydrogen in amounts that are stoichiometrically correct, whereas for others hydrogen absorption *increases* as the temperature is raised. Little evidence has been found for compound formation in the metallic hydrides.

The nature of the hydrogen atoms bound in the metallic hydrides is not clearly understood. Some of these substances are good reducing agents, suggesting the presence of hydridic or, indeed, atomic hydrogen. The rapid diffusion of hydrogen in metals such as palladium indicates that perhaps unbound hydrogen atoms are present. The constitution of these systems may well vary with composition. For example, up to a composition of $PdH_{0.5}$ the palladium–hydrogen system has the properties associated with a conductor, but at higher concentrations of hydrogen it is a semiconductor. In the same range of stoichiometries the paramagnetism decreases from the value for the pure metal to a minimum at a composition corresponding to $PdH_{0.6}$. It is apparent that a unified description of the metallic hydrides is not presently available.

3.3 / Covalent Hydrides

When hydrogen combines with representative elements in Groups IV, V, VI, and VII, covalent molecules of the general formula MH_{8-n} (where n is the number of the group in the periodic table to which M belongs) are formed (Fig. 6.2); boron also forms a covalent hydride with the expected empirical formula (BH_3), but the simplest boron hydride that has been isolated is the dimer B_2H_6. The covalent hydrides exhibit the stoichiometry expected on the basis of the simple Lewis theory and their molecular geometry follows that predicted by the electron-pair-repulsion arguments outlined in Chapter 3. The nature of the bonding in B_2H_6 and similar compounds is the subject of Chapter 10.

Although it is possible to form compounds of the elements beyond Period 2 with formal valences and/or coordination numbers greater than that predicted by the simple Lewis theory, hydrogen is not sufficiently electronegative to do so. Thus, for example, the Group V hydrides all have the formulation MH_3 even though some of these elements can form more than three single bonds with other atoms (e.g., PF_5, $AsCl_5$, SbF_5). The elements in Groups IV, V, and VI are capable of forming compounds in which several atoms of the same element may be bound to each other by an electron-pair bond; such systems are said to catenated. Catenation becomes less important in the chemistry of an element the further it is removed from carbon (which, of course, is the prime example of an element which forms chains). Thus, more complex covalent hydrides than those predicted on the basis of the Lewis theory are known for the elements in Groups IV, V, and VI (Table 6.5).

TABLE 6.5   Some of the more common covalent hydrides

Group IV	Group V	Group VI
CH_4, C_2H_6, C_3H_8, etc.	NH_3, N_2H_4	H_2O, H_2O_2
SiH_4, Si_2H_6, Si_3H_8, etc.	PH_3, P_2H_4	H_2S, H_2S_2, H_2S_3, H_2S_5
GeH_4, Ge_2H_6, Ge_3H_8	AsH_3	
SnH_4	SbH_3	
PbH_4	BiH_3	

The chemistry of the covalent hydrides of carbon (the hydrocarbons) is not considered here except in certain instances where this information is needed to establish the general unity of the behavior of the elements.

The simple covalent hydrides exist as molecules in all of their states of aggregation and, except for NH_3, H_2O, and HF, relatively weak van der Waals forces exist between these molecules. The relatively larger intermolecular forces which exist for NH_3, H_2O, and HF are revealed in the comparison of, for example, the melting points, boiling points, and heats of vaporization of these compounds shown in Figs. 6.4 and 6.5. The data for

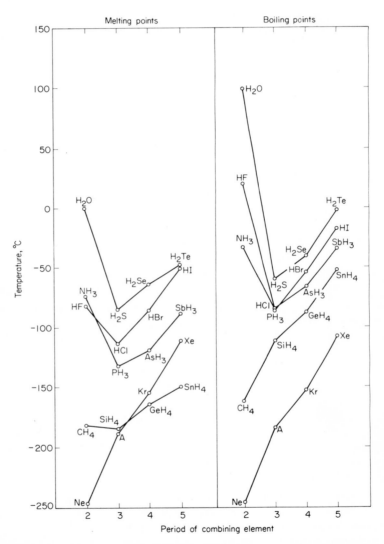

FIG. 6.4 The melting points and boiling points of the simple covalent hydrides and of the rare gases. Compounds of elements in a single group are joined by a line.

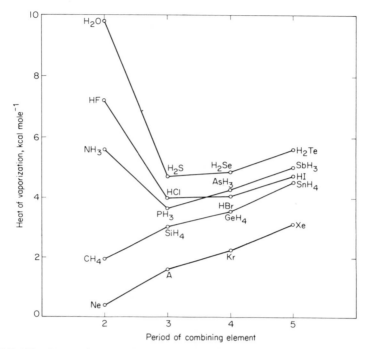

FIG. 6.5   The heats of vaporization of the simple covalent hydrides and of the rare gases. Compounds of elements in a single group are joined by a line.

the rare gases and the Group IV hydrides are included as reference points and show the usual monotonic behavior of these properties. The boiling points, heats of vaporization, and melting points of substances which possess the same structure characteristics (e.g., the rare gases) decrease regularly with decreasing molecular weight. This trend is observed for the heavier hydrides of Groups V, VI, and VII, but the properties of NH_3, H_2O, and HF are anomalous in this respect. Thus, these hydrides have higher boiling points, heats of vaporization, and melting points than predicted on the basis of their respective molecular weights, which suggests a marked change in the nature of the intermolecular forces present. A consideration of the boiling points of some of the derivatives of these elements suggests that the abnormal behavior of the hydrides can be associated with the presence of an X—H bond, where X is a very electronegative element carrying unshared electron pairs (Table 6.6). Using the hydrocarbons as references, the methyl derivatives of ammonia, water, and hydrogen fluoride have higher boiling points than compounds of equivalent molecular weights so long as N—H, O—H, and H—F bonds are present. Upon substitution of all of the X—H bonds in the parent compounds the boiling points approach normal values.

TABLE 6.6   The boiling points and molecular weights of some derivatives of methane, ammonia, water, and hydrogen fluoride

Methane derivative	Mol. wt.	Bp, °C	Ammonia derivative	Mol. wt.	Bp, °C
CH_4	16	−161.4	NH_3	17	−33
CH_3CH_3	30	−88	CH_3NH_2	31	−6.7
$(CH_3)_2CH_2$	44	−42	$(CH_3)_2NH$	45	7.4
$(CH_3)_3CH$	58	−1	$(CH_3)_3N$	59	3.2
$(CH_3)_4C$	72	36			

Water derivative	Mol. wt.	Bp, °C	HF derivative	Mol. wt.	Bp, °C
H_2O	18	100	HF	20	19
CH_3OH	32	65	CH_3F	34	−78.2
$(CH_3)_2O$	46	−24			

The interaction among molecules possessing N—H, O—H, and F—H units is appreciably greater than van der Waals forces, and, as we shall soon see, evidence exists for the directional character in this type of association; this phenomenon is called hydrogen bonding (*VI*).

More complex derivatives of water and ammonia also exhibit properties associated with hydrogen bonding. For example, evidence for both inter- and intramolecular hydrogen bonding can be adduced from the melting and boiling points of the isomeric hydroxybenzaldehydes and nitrophenols (Table 6.7). The *ortho* isomers of certain substituted phenols have markedly

TABLE 6.7   The melting and boiling points of some benzene derivatives

Derivative	Isomer	Melting point, °C	Boiling point, °C
Nitrophenol	*ortho*	45	214
	meta	96	195
	para	114	279
Hydroxybenzaldehyde	*ortho*	−7	197
	meta	106	240
	para	116	sublimes
Chlorophenol	*ortho*	−4.1[a]	176
	meta	29	214
	para	41	217

[a] Three crystal modifications of *o*-chlorophenol exist; the α, β, and γ forms have melting points of 7°C, 0°C, and −4.1°C, respectively.

lower melting and boiling points than the corresponding *meta* and *para* derivatives. In the case of the latter isomers, intermolecular hydrogen bonds can be formed. On the other hand, the *ortho* isomers

have a geometry that is favorable to intramolecular hydrogen bonding which decreases the magnitude of intermolecular forces. As previously mentioned, hydrogen bonds are readily formed in systems which contain the arrangement of atoms $-X-H\cdots X'-$ where X and X' are highly electronegative elements. Although the majority of the work on hydrogen bonding has involved derivatives of oxygen, nitrogen, and fluorine, which are among the most electronegative of the elements, evidence exists for hydrogen bonding in other systems. For example, the isomeric chlorophenols exhibit a pattern of melting and boiling points (Table 6.7) which can be interpreted in terms of hydrogen bonding.

In addition to consideration of the magnitude of boiling and/or melting points, several other methods have been developed for detecting the presence of hydrogen bonding. It might be expected that the position of the vibrational modes associated with the structure $-X-H$ would be altered if this moiety were hydrogen-bonded ($-X-H\cdots X'-$). Thus, compounds containing an OH group exhibit bands in their infrared spectra near 3300 cm^{-1} (broad) and 3600 cm^{-1} (narrow). Only the first of these bands is present in systems containing no hydrogen bonding (e.g., in dilute solutions of alcohols in hydrocarbons) whereas both bands appear in the spectra of hydrogen-bonded systems. *o*-Hydroxybenzaldehyde (**2**) exhibits only the band near 3300 cm^{-1}, but the spectra of pure alcohols contains both bands. Data of this type suggest that the low-energy band is characteristic of hydrogen-bonded structures ($-O-H\cdots X-$) whereas the band at about 3600 cm^{-1} represents the free OH group. Hydrogen-bond formation also has been studied using other spectroscopic methods, such as Raman, nuclear magnetic resonance, electronic, and nuclear quadrupole spectroscopy (*VII*).

Neutron diffraction and proton magnetic resonance studies have been employed to locate the positions of hydrogen atoms in some compounds which possess hydrogen bonds. The available data show that X—X' and X—H distances in hydrogen-bonded systems ($-X-H\cdots X'-$) vary over a reasonably wide range; the data for some representative compounds appear in Tables 6.8 and 6.9. It is interesting to note that some structural evidence

TABLE 6.8   The O—H and —O···O'— distances
in some hydrogen-bonded (—O—H···O'—) compounds

	d(O—H), Å	d(O···O'), Å
$H_2C_2O_4 \cdot 2H_2O$	0.955	2.85
$CaSO_4 \cdot 2H_2O$	0.98	2.70
α-HIO_3	0.99	2.69
$Na_2CO_3 \cdot NaHCO_3 \cdot 2H_2O$	1.01	2.77
H_3BO_3	1.02	2.73
HCO_2H	1.54	2.58

TABLE 6.9   The X—H···X' distance in
some hydrogen-bonded systems

Bond	Compound	d, Å
N—H···O	CH_3CONH_2	2.86
	$CO(NH_2)_2 \cdot H_2O_2$	3.00
N—H···N	NH_3	3.38
	NH_4N_3	2.97
N—H···F	$N_2H_6F_2$	2.62
	NH_4F	2.63
	NH_4HF_2	2.80
F—H···F	NH_4HF_2	2.32
	KHF_2	2.26
	H_2F_2	2.55

exists for the deviation from linearity of the X—H···X bond. The positions
for the hydrogen atoms associated with the water molecules in the compound
$Na_2CO_3 \cdot NaHCO_3 \cdot 2H_2O$ are shown in Fig. 6.6. The H—O—H angle is
107°, which is significantly different from the $O_2CO \cdot\cdot\cdot O \cdot\cdot\cdot OCO_2$ angle

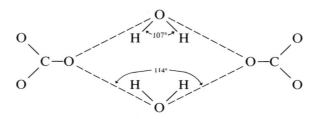

FIG. 6.6 The water molecules in the compound $Na_2CO_3 \cdot NaHCO_3 \cdot 2H_2O$ are
hydrogen bonded to two carbonate moieties. The hydrogen bonds are not on the line
of centers between the oxygen atoms.

(114°), the tetrahedral angle (109.5°), or the H—O—H angle in gaseous water (105.5°). The results of diffraction experiments on boric acid also indicate that the OH···O arrangement in this system is not linear.

A consideration of the available structural data on hydrogen-bonded systems shows that the packing of molecular units is influenced by hydrogen bonding. That is, the structures which are attained are usually not as efficiently packed from the standpoint of density as they could be. In addition, the length of hydrogen bonds between the same functional groups (see Tables 6.8 and 6.9) varies to a considerably greater extent than the inter-atomic distances in molecules or ionic crystals. In spite of this variation, the *average* X···X' separation for chemically similar compounds is less than the corresponding van der Waals distances, which indicates a reasonably strong interaction (Table 6.10).

TABLE 6.10 **A comparison of the X···X' distance in several hydrogen-bonded (—X—H···X'—) systems with the sum of the corresponding van der Waals radii**

X—H···X'	$d(X···X')$, Å	
	Exp.	van der Waals
O—H···O	2.55[a]	2.80
	2.74[b]	2.80
	2.73[c]	2.80
O—H···N	2.80	3.05
N—H···O	2.88[d]	3.05
N—H···N	3.10	3.30
O—H···Cl	3.08	3.20
N—H···F	2.78	3.00
N—H···Cl	3.21	3.45
F—H···F	2.44	2.70

[a] Inorganic acids.
[b] Alcohols.
[c] In salts.
[d] In ammonium salts.

Hydrogen bonds, while stronger than van der Waals forces, are consider-ably weaker than the usual chemical bonds. A variety of methods have been used to measure the enthalpy of association for hydrogen-bonded systems. Representative enthalpy data for different systems appear in Table 6.11. Most of the values lie below $\sim 10\, \text{kcal mole}^{-1}$ although at least one hydrogen-bonded system (KHF_2) is significantly stronger than this.

A satisfactory theoretical description of hydrogen bonding is not yet available. An electrostatic model was suggested by Pauling (9). In this view,

TABLE 6.11   Enthalpy of association in hydrogen-bonded systems

Bond	Compound	Enthalpy of association, kcal mole^{-1}
O—H⋯O	H_2O	4.4–5.0
	CH_3OH	3.2–7.3
N—H⋯N	NH_3	3.7–4.0
	CH_3NH_2	3.4
F—H⋯F	HF	6.7–7.0
	KHF_2	27
C—H⋯N	HCN	3.3

hydrogen possesses only one low-energy orbital and therefore it cannot form more than one covalent bond (see, however, Chapter 10). Thus, the canonical structure 3 cannot contribute significantly to the molecular wave function,

$$X-H-Y \qquad X-H:Y \qquad {}^-X:H-Y^+$$

$$\textbf{3} \qquad\qquad \textbf{4} \qquad\qquad \textbf{5}$$

which must be considered as composed primarily of contributions from structures 4 and 5. In the modern expression of the electrostatic model (*10*), the charge distribution of the nonbonded electrons is an important consideration. A consideration of orbital hybridization (e.g., sp^3 in water, ammonia, or hydrogen fluoride and sp^2 in carbonyl compounds), which accounts for the spatial distribution of electrons of the acceptor in the hydrogen bond, gives a reasonably accurate estimate of the energy of the system and predicts the relative orientations of the other bonds in the system to the hydrogen bond. Several recent attempts at using molecular-orbital methods to describe hydrogen bonding have appeared (*11–13*).

The covalent hydrides of oxygen, nitrogen, and fluorine are of particular interest because their simple compounds form the practical and theoretical basis for elucidating important considerations of solution phenomena. Accordingly, these are treated separately in Chapter 7.

4 / HYDROGEN ISOTOPES

Since hydrogen is the lightest of the elements ($Z = 1$), a change in the number of neutrons in its nucleus leads to a larger per cent change in the weight of the nucleus than for any other element. Thus, considering all the elements and their isotopes, one would expect to find the greatest difference in the chemical and physical properties that are dependent upon mass among the isotopes of hydrogen.

4.1 / Characteristics

Three isotopes of hydrogen are known. The lightest (1.008123 amu), containing one proton in its nucleus, and the most abundant (99.9844 %) is hydrogen (H). The next most abundant isotope (0.0156 %), deuterium (D), contains one neutron and one proton in its nucleus (2.014708 amu). Tritium (T), the heaviest known isotope (3.01707 amu), has a nucleus composed of one proton and two neutrons; this isotope is radioactive, decaying with a half-life of 12.4 years to give an isotope of helium and a β particle:

$$_1T^3 \longrightarrow {_2}He^3 + {_{-1}}e^0. \tag{20}$$

Both H_2 and D_2 crystallize as hexagonal close-packed solids from the corresponding liquids.

The mass differences between deuterium and hydrogen are sufficiently large that a distinct difference exists in the physical properties of the molecules D_2 and H_2 (Table 6.12). As might be expected, the properties of HD lie

TABLE 6.12   A comparison of some properties of H_2, D_2, and HD

Property	H_2	D_2	HD
Atomic weight	1.00813	2.01472	—
Melting point, °K	13.95	18.65	16.60
Boiling point, °K	20.38	23.6	—
Triple point, °K	—	18.72	—
Critical point, °K	33.24	38.35	—
Critical pressure, atm	12.807	16.432	—
Triple-point pressure, atm	53.8	128.5	95
ΔH_f, cal mole^{-1}	28.0	47	37
ΔH_v, cal mole^{-1} at 195 torr	219.7	302.5	263
Zero-point energy, cal	6184	4395	5366

between those of H_2 and D_2. The differences in the physical properties of D_2 and H_2 extend to compounds incorporating these isotopes. Thus, for example, there is a distinguishable difference in the properties of H_2O and D_2O (Table 6.13); a difference in properties is also observed for the more complex hydrides and the corresponding deuterides (Table 6.14).

To a first approximation, all the isotopes of an element are chemically identical; that is to say, they undergo the same reactions and form the same types of compounds. There are, of course, indirect effects related to the mass of an atom forming a bond which affect the rates at which they react as well as the energetics of such reactions. Ordinarily these effects are not detectable for the heavier elements where only a small per cent change in the mass is observed among the isotopes. However, among the lighter atoms

TABLE 6.13   Physical properties of D_2O and H_2O

Property	H_2O	D_2O
Melting point, °C	0	3.82
Boiling point, °C	100	101.42
Molecular volume at 20°C	18.016	18.092
Dielectric constant	81.5	80.7
Viscosity, poise at 20°C	10.09	12.6
Solubility of NaCl, g/100 g	35.9	30.5

TABLE 6.14   Melting points and boiling points
of some hydrides and the corresponding deuterides

Compound	Melting point, °C		Boiling point, °C	
	X = H	X = D	X = H	X = D
CX_4	−182.8	−184.2	—	—
NX_3	−78.0	−73.7	−33.5	−31.2
X_2O	0	3.82	100	101.42
X_2S	−85.7	−86.2	—	—
XCl	−111.1	−115.1	−85.1	−81.6
XF	—	—	19.9	18.6
XCN	−14.2	−12.2	25.1	25.7

the isotope effect is more pronounced and can easily be detected. The lower reactivity of deuterium compounds compared with the corresponding hydrogen compounds arises from the fact that the zero-point energy* for M—D bonds is lower than that for M—H bonds. For example, the zero-point energy for D_2 is about 1.8 kcal mole^{-1} lower than that for H_2 (Table 6.12), which gives D_2 a greater dissociation energy. Similar arguments show that the C--D zero-point energy is about 1.1 kcal mole^{-1} lower than that for C—H. Such differences in the activation energies for reactions of M—D and M—H compounds are reflected in a difference in the relative rates of the reactions of these species (*VII*).

Both deuterium and tritium have found wide use as tracers. In the case of tritium the fate of the labeled compound can be followed by standard radiochemical methods. Deuterium-labeled compounds, on the other hand, require a detection technique such as mass spectroscopy which utilizes the mass difference between hydrogen and deuterium. Deuterium labeling has

---

* The zero-point energy of a molecule is the energy remaining at 0°K, i.e., when the molecule is in its lowest vibrational state.

also found use in infrared spectroscopy. For example, the mass difference between deuterium and hydrogen is sufficiently large that the vibrational bands associated with the M—H and M—D moieties can be distinguished. Thus, it is possible to use deuterium-labeled compounds to aid in the assignment of the absorption bands in the vibrational spectra of hydrogen-containing compounds. In addition, deuterated solvents are very useful in certain aspects of proton magnetic resonance spectroscopy.

4.2 / Preparation

Tritium is made in a nuclear reactor by the reaction of the light isotope of lithium with neutrons:

$$_3Li^6 + {_0}n^1 \longrightarrow {_2}He^4 + {_1}T^3. \tag{21}$$

Tritium can be obtained from the product gases by converting it into a compound such as T_2O which can be easily separated from helium.

Since one atom in about 6000 atoms of hydrogen in nature is deuterium, the most direct way of obtaining this isotope involves the concentration or enrichment of an easily obtained compound of hydrogen. Water is the obvious candidate for this purpose. The thermodynamic and kinetic differences in the reactivity of deuterium compounds compared with the corresponding hydrogen compounds is the basis for the methods used to concentrate D_2O found in natural water. For example, hydrogen is discharged more rapidly than deuterium during the electrolysis of an aqueous solution of sodium hydroxide using nickel electrodes; the separation factor $(H/D)_{(g)}/(H/D)_{(soln)} \approx 6$ under these conditions. Thus, as normal water is electrolyzed the relative concentration of deuterium in the solution is greater than that in the gas being formed. It is possible to achieve high-purity (99%) D_2O using the electrolytic process. The electrolytic process for the concentration of D_2O is attractive in locations where electric power is relatively inexpensive.

There are also a series of isotopic exchange equilibria for which the equilibrium constants favor the enrichment of deuterium in one of the species:

$$HD_{(g)} + H_2O_{(l)} \rightleftharpoons H_{2(g)} + HDO_{(l)} \tag{22}$$

$$K = 3.703 \text{ at } 25°C$$

$$HCl + DI \rightleftharpoons DCl + HI \tag{23}$$

$$K = 1.527 \text{ at } 25°C$$

$$H_2O_{(l)} + HSD \rightleftharpoons HOD_{(l)} + H_2S_{(g)}. \tag{24}$$

$$K = 1.01 \text{ at } 25°C$$

Even though the constants for such equilibria may not be large in the direction of favoring the concentration of deuterium, practical separations can be achieved by physically arranging for the equilibria to be established a large number of times in a flow system.

The most usual primary form of deuterium readily available is either D_2 or D_2O; these species can be converted into other labeled compounds by the appropriate chemical reaction. For example, metal deuterides can be prepared by the processes shown in Eqs. (9) and (10) and deuterosulfuric acid is prepared by dissolving SO_3 in D_2O.

REFERENCES

1. W. A. Pryor, J. P. Stanley, and M. G. Griffith, *Science*, **169**, 181 (1970).
2. T. Henriksen, *J. Chem. Phys.*, **50**, 4653 (1969).
3. H. C. Heller and T. Cole, *Proc. Natl. Acad. Sci. U.S.*, **54**, 1486 (1965).
4. J. N. Herok and W. Gordy, *Proc. Natl. Acad. Sci. U.S.*, **54**, 1287 (1965).
5. F. Nagy and L. Simandi, *Acta Chim. Acad. Sci. Hung.*, **38**, 213 (1963).
6. L. Simandi and F. Nagy, *Acta Chim. Acad. Sci. Hung.*, **46**, 101 (1965).
7. B. de Vries, *J. Catal.*, **1**, 489 (1962).
8. G. Huttig, *Z. Angew. Chem.*, **39**, 67 (1926).
9. L. Pauling, *Proc. Natl. Acad. Sci. U.S.*, **14**, 359 (1928).
10. W. G. Schneider, *J. Chem. Phys.*, **23**, 26 (1955).
11. J. G. C. M. van Duijneveld-van de Rijdt and F. B. van Duijneveldt, *J. Amer. Chem. Soc.*, **93**, 5644 (1971).
12. H. Chojnacki, *Theoret. Chim. Acta. (Berlin)*, **22**, 309 (1971).
13. P. A. Kollman and L. C. Allen, *J. Amer. Chem. Soc.*, **92**, 6101 (1970).
14. A. Sieverts and E. Roell, *Z. Anorg. Allgem. Chem.*, **153**, 289 (1926).
15. A. Sieverts and A. Gotta, *Z. Anorg. Allgem. Chem.*, **187**, 155 (1930).
16. A. Sieverts and A. Gotta, *Z. Elektrochem.*, 105 (1926).

COLLATERAL READINGS

I. P. Ausloos, ed., *Fundamental Processes in Radiation Chemistry*, Interscience, New York, 1968. A collection of reviews dealing with the formation of ions, excited molecules, and radicals in matter when it is irradiated.
II. J. Halpern, *Ann. Rev. Phys. Chem.*, **16**, 103 (1965). A discussion of catalytic processes involving complex compounds.
III. T. R. P. Gibbs, Jr., *Prog. Inorg. Chem.*, **3**, 315 (1962). A detailed review of the properties, structure, and bonding of primary solid hydrides.
IV. C. B. Magee, in *Metal Hydrides* (W. M. Mueller, ed.), Academic Press, New York, 1968, p. 165. A discussion of the chemical and physical properties of the saline hydrides.
V. R. L. Beck, in *Metal Hydrides* (W. M. Mueller, ed.), Academic Press, New York, 1968, p. 241. The chemical and physical properties of zirconium and hafnium are discussed.

VI. M. L. Huggins, *Angew. Chem. Intern. Ed.*, **10**, 147 (1971). A wide-ranging review of the hydrogen bond as a structural concept in a variety of systems. The author originally suggested the existence of hydrogen bonding in 1921.

VII. G. C. Pimentel and A. L. McClellan, *The Hydrogen Bond*, W. H. Freeman and Co., San Francisco, 1960. A definitive work on methods used to study hydrogen-bonded systems.

STUDY QUESTIONS

1. Hydrogen is considered to be the first member of Group VII in some periodic classifications. Discuss the basis for this decision.

2. In the light of the discussion in Problem 1, why is the bond energy for hydrogen (103 kcal mole^{-1}) much greater than that of the halogens (F_2, 37; Cl_2, 58; Br_2, 46; I_2, 36 kcal mole^{-1})?

3. How would you expect atomic hydrogen to react with the following: aqueous copper nitrate, C_2H_4, sulfur, and $SiCl_4$? Write equations to describe the processes.

4. Write the electrode reactions for the electrolysis of a molten mixture of lithium hydride and lithium chloride.

5. Calculate the lattice energy of CsH.

6. Discuss the expected hydrogen-bonded interactions in solid ammonium hydrogen difluoride.

7. Discuss the relative melting points expected for the three isomers of mercaptobenzaldehyde.

8. Sketch the structure of ZrH.

9. Give equations for the preparation of the following compounds from heavy water: (a) DCl; (b) D_2O_2; (c) $Ca(OD)_2$; (d) SiD_4.

10. Assuming lithium hydride to be an ionic substance, derive an expression for the electron affinity of hydrogen using a suitable energy cycle.

/7

SOLVENT PROPERTIES
OF COVALENT
HYDRIDES

The behavior of the covalent hydrides NH_3, H_2O, and HF as solvents is considered here apart from the general discussion of the properties of these compounds (Chapter 6) because their behavior serves as a vehicle for discussing the role of interacting solvents in chemical processes. Most reactions of interest are conducted in a solvent, and very often the presence of the solvent is ignored in considering the detailed chemistry of these processes. Thus, the fact that AgCl precipitates when aqueous $AgNO_3$ and NaCl are mixed does not appear to be compatible with the observation that this reaction does not occur if liquid ammonia is used as a solvent. Faced with such observations, one might well consider whether there are principles which govern, say, the chemistry of silver on a broader basis than those considered for aqueous reaction media. Such principles are described in this chapter using solution phenomena in NH_3, H_2O, and HF for illustrative purposes.

1 / PHYSICAL PROPERTIES

The principle physical properties of NH_3, H_2O, and HF (Table 7.1) show a remarkable similarity. They are all highly associated liquids, the association presumably arising from intermolecular hydrogen bonding (see Chapter 6).

TABLE 7.1   Principal physical constants for ammonia, water, and hydrogen fluoride

Constant	Hydride		
	NH_3	H_2O	HF
Melting point, °C	-77.74	0	-83
Boiling point, °C	-33.35	100	19.4
Heat of fusion, kcal mole^{-1}	1.35	1.44	0.939
Heat of vaporization, kcal mole^{-1}	5.58	9.72	1.6
Trouton's constant	23.28	26.0	24.7
Dieletric constant	$22.7\,(-50°C)$	$87.7\,(0°C)$	$84\,(20°C)$
ΔH formation, kcal mole^{-1}	-11.04 (g)	-57.83 (g)	-64.9 (g)
Critical temperature, °K	133.0	374.0	188
Critical pressure, atm.	112.3	217.7	67.2
Liquid density, g cm^{-2}	$0.6814\,(-33°C)$	$1.00\,(4°C)$	$0.988\,(13.6°C)$
Specific conductance, Ω^{-1} cm^{-1}	2.94×10^{-11} $(-35°C)$	5.95×10^{-8} $(20°C)$	1.4×10^{-5} $(-15°C)$
Autoprotolysis constant	5×10^{-27} $(25°C)$	1×10^{-14} $(25°C)$	2×10^{-12} $(0°C)$
Viscosity, cP	$0.254\,(-33.5°C)$	$1.002\,(20°C)$	$0.207\,(20°C)$
Dipole moment, D	1.49	1.84	1.83
Polarizability $\times 10^{24}$	2.21	1.48	2.46
Ionization potential, eV	10.15	12.6	15.77

Water is, of course, the easiest of the three compounds to manipulate at ambient conditions in the average laboratory. Hydrogen fluoride, with due consideration given to manipulative techniques which must be used because of its corrosive nature, can also be handled at room temperature. Ammonia, on the other hand, must be manipulated in high-pressure apparatus, or at low temperatures if the pressure of the system is to be maintained at one atmosphere or less. In the following discussion, no attempt is made to discuss the technical problems associated with studying these solvent systems (*I*).

2 / SOLVENT PROPERTIES

The solubility of one substance in another involves a consideration of the attractive forces which exist between (a) solvent molecules, (b) solute species, and (c) solute and solvent molecules. High attractive forces existing between solute and solvent molecules, compared with the attraction among molecules of each individual species, favor the solution of a substance in a solvent. The opposite is true if the conditions are reversed. Thus the extent of solution is, as in all chemical processes, dependent upon the magnitude of the free-energy change for the general equation

$$\text{solute}_{(s)} + \text{solvent}_{(l)} \rightleftharpoons \text{solution species}_{(l)}. \tag{1}$$

Several factors are involved in the interaction of covalent species. Polar molecules can interact with other species by electrostatic forces (by virtue of their polarity) and by forces which arise from induction processes (the latter are sometimes called London forces). The interaction between two molecules possessing permanent dipole moments depends upon their relative orientation. The most favored orientation, i.e., dipoles aligned parallel to each other, is opposed by thermal motion. At a given temperature a statistically preferred orientation of the dipoles occurs; the potential energy (E_1) of such a system is given by

$$E_1 = -\frac{2\mu_1{}^2\mu_2{}^2}{3r^6kT}, \tag{2}$$

where μ_1 and μ_2 are the magnitudes of the two dipoles at a distance r from each other.

London (*1*) has shown that two molecules which do not react chemically will attract each other because they possess fluctuating dipole moments, although neither may have a permanent dipole moment. The fluctuating dipole moment, which arises from the circulation of electrons within a molecule, can induce dipoles in neighboring molecules. This process leads to dipole–dipole interactions between induced dipoles. The forces which arise in this process are called dispersion forces; the magnitude of the dispersion energy (E_2) is given by

$$E_2 = -\frac{3}{2}\left(\frac{\alpha_1\alpha_2}{r^6}\right)\left(\frac{I_1I_2}{I_1 + I_2}\right), \tag{3}$$

where α and I represent the polarizability and ionization potentials, respectively, of two molecules at a distance r apart.

The general forms of Eqs. (2) and (3) indicate that, with reference to a given polar solute, substances that have the highest dipole moment, polarizability, and ionization potentials should be the best solvents. This conclusion, of course, is based on the assumption that all other interactions remain constant. As the magnitude of these quantities increases for a series of solvents, however, the interaction among solvent molecules also increases. Thus, it is conceivable that for a given solute the solute–solute interaction can be low and the solute–solvent interaction high, but the solvent–solvent interaction can also be sufficiently high to overcome the apparently favorable relationship of the first two factors. Unfortunately, extensive quantitative solubility data for a series of covalent compounds under the same set of experimental conditions are not available for NH_3, H_2O, and HF as solvents. However, the general validity of the arguments can be seen from qualitative solubility information (Table 7.2).

TABLE 7.2 Qualitative solubilities[a] of carbon compounds in NH_3, H_2O, and HF

		Solubility		
Type	Examples	NH_3	H_2O	HF
Saturated hydrocarbons	C_3H_8, $n\text{-}C_6H_{14}$	insoluble	insoluble	very slightly soluble
Halogen derivatives of saturated hydrocarbons	$CHBr_3$, C_2H_5I	miscible or soluble	insoluble	—
Unsaturated hydrocarbons	C_6H_6, $C_6H_5CH_3$	soluble or moderately soluble	insoluble	slightly soluble to soluble
Halogen derivatives of unsaturated hydrocarbons	$Cl_2C{=}CH_2$	miscible	insoluble	—
Hydrocarbon derivatives containing multiple bonds	CH_3NO_2	miscible	slightly soluble	very soluble
Carbonyl derivatives	$(CH_3)_2CO$, $CH_3(CH_2)_5CHO$	miscible to soluble	miscible to insoluble	very soluble
Alcohols	CH_3OH, $CH_3(CH_2)_6OH$	miscible	miscible to very slightly soluble	very soluble
Amines	CH_3NH_2, $(CH_3)_3N$, C_5H_5N	miscible	miscible to very soluble	very soluble
Ethers	$(C_2H_5)_2O$, $[CH_3(CH_2)_3]_2O$	miscible or soluble	insoluble	very soluble

[a] The solubility observations were made generally at $-33°C$ (NH_3), $20°C$ (H_2O), and $0°C$ (HF).

The forms of Eqs. (2) and (3) suggest that both NH_3 and HF should be capable of generating larger dispersion forces than H_2O, but dipole–dipole interactions for H_2O and HF should be greater than for NH_3. In addition, the ability of these solvents to form hydrogen bonds to unpaired electrons of solute species possessing polar hydrogen atoms accounts for the ready solution of some substances. The importance of hydrogen bonding from the solute to the solvent can be seen from the fact that ethers are insoluble in water but they are markedly more soluble in NH_3 and in HF. The unexpected solubility of ethers in NH_3 and HF can also be interpreted as arising from the large polarizability of these solvent molecules which leads to large dispersion forces, as indicated by Eq. (3).

Two basic types of interaction are possible in the case of ionic species dissolved in covalent polar solvents: ion–dipole interactions and a dispersion-type interaction. Ion–dipole interactions might be expected to be the stronger of the two, although London-type forces may become important for highly polarizable ions. The energy (E_3) arising from the interaction of an ion of charge Ze and a molecule with a dipole moment μ placed at a distance r from the ion is given by

$$E_3 = -\frac{Ze\mu}{r^2}. \tag{4}$$

In this view the larger the dipole moment of a solvent the larger the interaction between its molecules and solute ions. The dipole moments listed in Table 7.1 suggest that H_2O and HF molecules should interact more strongly with ions than does NH_3.

The free energy of solvation of a gaseous ion of charge Ze and radius r in a solvent with a dielectric constant D is given by the Born expression

$$\Delta G = \frac{Z^2 e^2}{2r}\left(1 - \frac{1}{D}\right). \tag{5}$$

The enthalpy change for the process of solvation is given by

$$\Delta H = \frac{Z^2 e^2}{2r}\left[1 - \frac{1}{D} - \left(\frac{T}{D^2}\cdot\frac{\partial D}{\partial T}\right)\right]. \tag{6}$$

The relationship between the bulk dielectric constant, the ionic radius, and the free energy is given graphically by the family of curves in Fig. 7.1. That is, at a given ionic radius, the free energy of solvation increases with increasing dielectric constant, and at a given dielectric constant the smallest ion has the greatest free energy of solvation. Although the Born equation does not give the correct numerical values for the solvation energies of different ions in a given solvent it does give the correct order for these values. The Born equation suggests that in a given solvent the more highly charged ions should be more greatly solvated, while for a given ion compounds with high dielectric constants should be more suitable as solvents than those with lower dielectric constants. Again, extensive solubility data for NH_3, H_2O, and HF are not

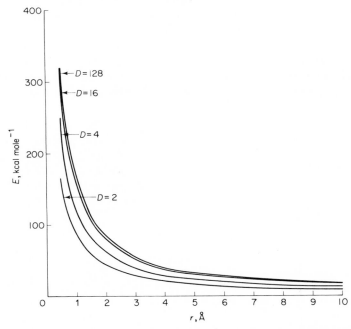

FIG. 7.1   The relationship between the free energy of solvation of an ion, its radius, and the dielectric constant.

available to quantitatively verify these arguments. However, a comparison of the solubilities of a limited series of sodium salts in ammonia and water (Table 7.3) gives insight into the general validity of the previous discussion (under these conditions the sodium salts react with hydrogen fluoride). In each case the solubility of a given salt is greater in water than in ammonia. Within a given solvent, the solubility of the halides falls in the order $Cl^- <$ $Br^- < I^-$, which is the order expected on the basis of the relative lattice energies of these compounds.

TABLE 7.3   Solubilities of some salts in ammonia and water

Salt	Solubility (mole kg^{-1})		Lattice energy, kcal mole^{-1}
	NH_3 ($-0.1°C$)	H_2O ($0°C$)	
NaCl	2.20	6.10	180
NaBr	6.21	7.71	175
NaI	8.80	10.72	161

3 / IONIC EQUILIBRIA IN SOLUTION

Not only are ionic species solvated to varying degrees in different solvents [Eq. (5)], but the electrostatic interaction between ions is dependent upon the properties of the solvent. In general, association would be expected between oppositely charged ions, the equation for the process being given by

$$M^+ + X^- \rightleftharpoons M^+, X^-. \tag{7}$$

The symbol M^+, X^- represents an ion pair. The equilibrium constant for process (7), expressed in the standard manner, is dependent upon the bulk dielectric constant D of the solvent in question:

$$K = \frac{4N}{1000} \left(\frac{|Z_1 Z_2| e^2}{DkT} \right)^3 Q(b), \tag{8}$$

where Z_1 and Z_2 are the charges on the ions in question, k is Boltzmann's constant, T the absolute temperature, and $Q(b)$ is a function of the distance of closest approach of the ions (2). Assuming *all* other factors are equal, the dielectric constants of NH_3, H_2O, and HF (Table 7.1) suggest that ionic species in both HF and H_2O would be less associated than the same species in NH_3. Experimentally this prediction has been verified for these and other solvents; Fig. 7.2 shows the equivalent conductivities of several ionic compounds in liquid NH_3, H_2O, and HF. The fact that the numerical values of the conductivities vary from solvent to solvent can be interpreted in terms of factors, such as viscosity and temperature, which affect ionic mobilities. More important, the general *shapes* of the curves indicate that solutions of potassium salts in HF or H_2O behave as strong electrolytes, whereas solutions of KCl in NH_3 behave as weak electrolytes even in dilute solution. Strong electrolytes in a conductivity sense do not exist in solvents with a dielectric constant of about 36 or lower. Thus, results of conductivity experiments must be used with caution when they are applied to the elucidation of processes which occur in solvents with moderate to low dielectric constants. The pause which occurs at concentrations higher than that shown in Fig. 7.2 in the rapidly decreasing conductivity curve for the KCl–NH_3 system has been interpreted in terms of ion-triplet formation:

$$K^+, Cl^- + K^+ \rightleftharpoons (K^+)_2, Cl^- \tag{9}$$

$$K^+, Cl^- + Cl^- \rightleftharpoons K^+, (Cl^-)_2. \tag{10}$$

Ion pairs are nonconducting, but ion triplets have a net charge and are conducting species.

The complexity of solutions of covalent solutes that undergo partial ionization in solvents with a dielectric constant less than ~ 36 should now be

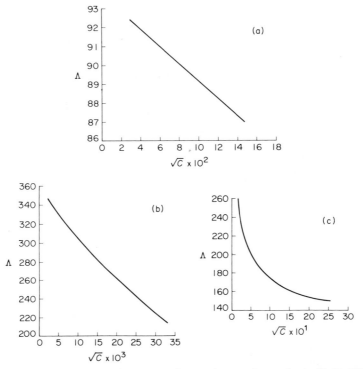

FIG. 7.2 The equivalent conductances of several potassium salts in H_2O, HF, and NH_3. (a) $KCl-H_2O$, $5°C$; (b) $KF-HF$, $-15°C$; (c) $KCl-NH3$, $-34°C$.

apparent. In such solvents a weakly ionized substance (HX) is involved in the usual ionization equilibria

$$HX + S \rightleftharpoons HS^+ + X^-, \tag{11}$$

but the ions so formed associate to an appreciable extent:

$$HS^+ + X^- \rightleftharpoons HS^+,X^-. \tag{12}$$

Both the molecular form HX and the ion pair HS^+,X^- are nonconducting.

4 / SOLVATING POWER OF SOLVENTS

From our previous discussion, it is apparent that the solvating ability of a substance involves an intimate relationship between the various modes of solvent–solute, solute–solute, and solvent–solvent interactions. Specifically such interactions involve hydrogen bonding; ion–dipole, ion–ion, and dipole–

dipole interactions; and dispersion forces. Several attempts have been made to construct generalized scales of the solvating power of substances using relatively simple physical measurements. In such cases no attempts have been made to separate the individual effects; rather they are assimilated in the measurement in some quantitative, but theoretically unspecified, way.

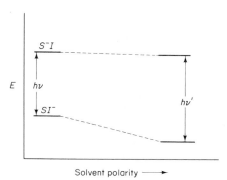

Solvent polarity ⟶

FIG. 7.3 Increasing solvent polarity decreases the energy of the ground state in a charge-transfer-to-solvent spectrum.

A useful measure of solvating power has been developed by observing the wavelength of a band which is sensitive to the nature of the solvent; a variety of ions, such as halide ions, fall into this category. An extensive series of experiments (3) indicate that such solvent effects arise from the fact that the ground state associated with these transitions is ionic (Fig. 7.3). Thus, relatively speaking, the ionic ground state will be more stabilized than the

TABLE 7.4 The energy of the charge-transfer-to-solvent transition for 1-ethyl-4-methoxycarbonylpyridinium iodide in various solvents (4)

Solvent	E, kcal mole^{-1}
H_2O	94.6
CH_3OH	83.6
C_2H_5OH	79.6
$CH_3CONH(CH_3)$	77.9
CH_3CN	71.3
$(CH_3)_2SO$	71.1
$CH_3CON(CH_3)_2$	66.9
$(CH_3)_2CO$	65.7
CH_2Cl_2	64.2
C_5H_5N	64.0
*iso*octane	60.1

excited state; an increase in the polarity of the solvent increases the energy of this transition.

The relative order of the energy of this transition for the 1-ethyl-4-methoxycarbonylpyridinium iodide ion pair has been used as a measure of the polarity of a solvent; typical data appear in Table 7.4 (4).

Another method used to gauge the relative coordinating power of a solvent involves the measurement of the heat of formation of the complex formed between phenol and the solvent:

$$C_6H_5OH + S \rightleftharpoons C_6H_5OH\cdots S. \tag{13}$$

The enthalpy of reaction (13) should be directly related to the ability of the solvent to act as an electron-pair donor in hydrogen-bond formation. The measured heat for the process shown in Eq. (13) using a series of solvents is given in Table 7.5 (5). A comparison of the data in Tables 7.4 and 7.5

TABLE 7.5   The enthalpy for the reaction between phenol and several solvents (5)

Solvent	ΔH, kcal mole^{-1}
CH_3NH_2	-9.27
NH_3	-8.2
C_5H_5N	-8.07
$(CH_3)_2SO$	-6.5
$CH_3CON(CH_3)_2$	-6.4
$(C_2H_5)_2O$	-5.0
$CH_3CONH(CH_3)$	-4.7
$(CH_3)_2CO$	-3.3
CH_3CN	-3.3
C_6H_5CN	-3.3
$C_6H_5NO_2$	-2.04
CH_3NO_2	-1.9

indicates that the two orders of solvating power are not the same. This should not be surprising since the one measurement is related to all factors that can be construed as polar effects whereas the second involves a specific interaction (hydrogen bonding).

5 / ACID–BASE BEHAVIOR (//)

5.1 / The Brønsted–Lowry Approach

Although there are several points of view on acid–base behavior, the Brønsted–Lowry approach to this subject is used in this section, since for the most part we shall be interested in the ionization of covalently bonded hydrogen atoms. Other interpretations are presented in other, more suitable, places in this volume. According to the Brønsted–Lowry theory an acid is a substance that acts as a proton donor, and a base is a proton acceptor:

$$\underset{\text{acid}_1}{\text{HX}} \; + \; \underset{\text{base}_1}{\text{:S}} \; \rightleftharpoons \; \underset{\text{acid}_2}{\text{H:S}^+} \; + \; \underset{\text{base}_2}{\text{X}^-} \; . \qquad (14)$$

The proton-transfer process can also be interpreted in terms of the heterolytic cleavage of one covalent bond (in HX) and the formation of a new covalent bond (in HS^+). In the Lewis sense the new bond formed is a coordinate-covalent bond (Chapter 1) since the basic species carries an unshared electron pair and the proton is deficient in electrons.

The Brønsted–Lowry theory requires a proton acceptor for an acid–base process. In the case of solvent-ionization phenomena, the solvent molecules act as bases when an acid ionizes. In different words, acids ionize in solvents to form solvated protons. Thus in NH_3, H_2O, and HF acids ionize according to the following processes:

$$\text{HX} + :\text{NH}_3 \; \rightleftharpoons \; \text{NH}_4{}^+ + \text{X}^- \qquad (15)$$

$$\text{HX} + \text{H}_2\ddot{\text{O}}: \; \rightleftharpoons \; \text{H}_3\text{O}^+ + \text{X}^- \qquad (16)$$

$$\text{HX} + \text{H}\ddot{\ddot{\text{F}}}: \; \rightleftharpoons \; \text{H}_2\text{F}^+ + \text{X}^-. \qquad (17)$$

Since the molecules NH_3, H_2O, and HF are progressively less basic in the order given, it is not surprising that, all other factors being equal, the same acid, HX, will appear to be progressively less ionized in the same order. The progressively weaker basicity of the molecules NH_3, H_2O, and HF is reflected in the observation that ammonium salts of even weak acids, such as acetic acid, are stable; on the other hand discrete hydronium compounds such as $H_3O^+ClO_4{}^-$ have been isolated only in unusual cases. Hydro-fluoronium compounds ($H_2F^+X^-$) are unknown. The processes in Eqs. (15)–(17) are revealing in several ways. First, the strongest acid in any solvent is the solvated proton. All acids stronger than the solvated proton will react with the solvent to form this species. For example, although there may be an inherent difference in the strength (*II*) of a series of acids such as $HClO_4$, HI, HBr, HCl, and HNO_3, all are stronger proton donors than H_3O^+. Water is said to be a *leveling* solvent for these acids since acids stronger than the solvated proton are converted into this species:

$$\text{HX} + \text{H}_2\text{O} \; \longrightarrow \; \text{H}_3\text{O}^+ + \text{X}^-. \qquad (18)$$

Acids that are weaker proton donors than H_3O^+, such as acetic acid, will react incompletely to form the solvated proton:

$$\text{CH}_3\text{CO}_2\text{H} + \text{H}_2\text{O} \; \rightleftharpoons \; \text{CH}_3\text{CO}_2{}^- + \text{H}_3\text{O}^+. \qquad (19)$$

Because of the way in which the solvent enters into acid–base reactions, the inherent basicity of the solvent molecules becomes an important factor in the apparent acid strength of a substance in that solvent. For example, acetic acid is completely ionized in liquid ammonia:

$$CH_3CO_2H + NH_3 \rightarrow CH_3CO_2^- + NH_4^+, \tag{20}$$

and is as strong an acid as HCl in this solvent:

$$HCl + NH_3 \rightarrow Cl^- + NH_4^+. \tag{21}$$

In water acetic acid behaves as a weak acid [Eq. (19)]. In liquid HF it is a weak base:

$$CH_3CO_2H + HF \rightleftharpoons CH_3CO_2H_2^+ + F^-. \tag{22}$$

Acid–base considerations involve not only the relative ability of a substance to accept or release protons, but also the ease with which ionization processes can occur in the medium under consideration. Since NH_3, H_2O, and HF are reasonably good polar solvents, the general acid–base characteristics of solutes depend upon the relative acidity (or basicity) of the solvent.

The strongest bases in NH_3, H_2O, and HF can be deduced from a consideration of the autoprotolysis reactions which occur in them:

$$2NH_3 \rightleftharpoons NH_4^+ + NH_2^- \tag{23}$$

$$2H_2O \rightleftharpoons H_3O^+ + OH^- \tag{24}$$

$$2HF \rightleftharpoons H_2F^+ + F^-. \tag{25}$$

NH_2^-, OH^-, and F^- represent the basic species in these solvents. As in the case of acids, species inherently more basic than the anion characteristic of the solvent will be converted into that anion. Thus, the very basic hydride ion (as found in NaH) reacts with liquid ammonia to form amide ion,

$$H^- + NH_3 \rightarrow NH_2^- + H_2, \tag{26}$$

and it is not surprising that similar reactions occur in H_2O and HF. On the other hand, alkoxides ($NaOCH_3$) which are sufficiently strong bases to react with water to form the base characteristic of this solvent,

$$OCH_3^- + H_2O \rightarrow HOCH_3 + OH^-, \tag{27}$$

do not form NH_2^- in ammonia. Water is a leveling solvent for alkoxides but ammonia is not.

5.2 / Neutralization

Neutralization reactions in NH_3, H_2O, and HF acting as solvents follow the processes described in Eqs. (23)–(25). That is, the product of the concentrations of solvated protons and the corresponding basic anion must not exceed the ion product for the solvent; the autoprotolysis constants for these solvents are listed in Table 7.1. This is a familiar process in aqueous chemistry; accordingly we shall concentrate on the corresponding reactions in liquid NH_3 and HF.

Since basic solvents tend to level the strengths of acids and acidic solvents have a leveling effect on the strengths of bases, the nature of the solvent is an important consideration in attempting to establish experimentally the relative strengths of acids and bases. For example, the strengths of mineral acids such as $HClO_4$, HBr, HCl, and HNO_3 that are considered strong in water cannot be determined in moderately basic solvents. However, if an acidic solvent such as anhydrous acetic acid is used, the differences in acidity can be detected. The conductivities of solutions of these acids in acetic acid is given in Fig. 7.4. It is apparent from these data that, at any concentration, the conductivities of these acids fall in the order $HClO_4 > HBr > HCl > HNO_3$, which can be taken as the order of decreasing ionization of these substances. This conclusion is based upon the assumption that the extent

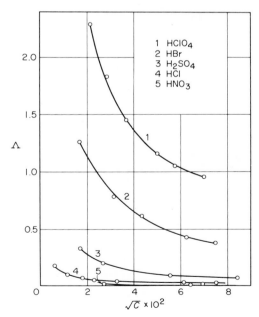

FIG. 7.4 The equivalent conductances of some mineral acids in glacial acetic acid.

of ion-pair formation [cf. Eq. (12)] in acetic acid, which has a dielectric constant of 6.2 at 25°C, is nearly the same for each compound, and that the differences in ionic conductivities of the anion in each system do not contribute significantly to the observed differences shown in Fig. 7.4. Both assumptions have been shown to be experimentally valid.

Substances that are intrinsically very weak acids exhibit marked acidic behavior in liquid ammonia. Thus a variety of compounds which exhibit little or no acidity in water, such as acid amides [$R(CO)NH_2$], sulfonamides (RSO_2NH_2), amines (R_2NH), and hydrocarbons [$(C_6H_5)_3CH$] react with metal amides in liquid ammonia to form salts:

$$R(CO)NH_2 + MNH_2 \rightarrow M[R(CO)NH] + NH_3 \qquad (28)$$

$$M[R(CO)NH] + MNH_2 \rightarrow M_2[R(CO)N] + NH_3 \qquad (29)$$

$$RSO_2NH_2 + MNH_2 \rightarrow M[RSO_2NH] + NH_3 \qquad (30)$$

$$M[RSO_2NH] + MNH_2 \rightarrow M_2[RSO_2N] + NH_3 \qquad (31)$$

$$R_2NH + MNH_2 \rightarrow [R_2N]M + NH_3 \qquad (32)$$

$$(C_6H_5)_3CH + MNH_2 \rightarrow [(C_6H_5)_3C]M + NH_3. \qquad (33)$$

In certain instances substances such as acid amides and sulfonamides can act as dibasic acids [Eqs. (29) and (31)]. As in aqueous solutions, the course of these reactions can be followed using either indicator or potentiometric methods.

At the other end of the scale, substances that are intrinsically very poor bases exhibit a marked basicity in liquid HF solutions. As mentioned previously, substances such as acetic acid which exhibit acidic properties in water [Eq. (22)] act as bases in liquid HF. Indeed even nitric acid can be protonated in liquid HF:

$$HNO_3 + HF \rightleftharpoons H_2NO_3^+ + F^- \qquad (34)$$

Many organic substances with functional groups which possess unshared electron pairs (e.g., alcohols, ketones, and even ethers) dissolve in HF to give very conducting solutions. Like carboxylic acids these substances act as bases toward liquid HF [cf. Eq. (22)].

$$ROH + HF \rightleftharpoons ROH_2^+ + F^- \qquad (35)$$

$$R_2CO + HF \rightleftharpoons R_2COH^+ + F^- \qquad (36)$$

$$R_2O + HF \rightleftharpoons R_2OH^+ + F^-. \qquad (37)$$

Even aromatic hydrocarbons dissolve in liquid HF to give conducting solutions:

$$ArH + HF \rightleftharpoons ArH_2^+ + F^-. \tag{38}$$

The reaction can be interpreted in terms of the protonation of either the π-electron cloud or a specific carbon atom in the ring; in the latter case, rehybridization of a ring carbon atom must occur (6).

5.3 / Solvolysis Reactions

As is the case in aqueous systems, many species which contain no protons undergo reactions with NH_3 and HF to give solvolytic products that can be considered acids or bases. For example, the dissolution of aluminum salts in water leads to solvated aluminum ions, $Al(H_2O)_6^{3+}$, which in turn undergo hydrolysis

$$Al(H_2O)_6^{3+} + H_2O \rightleftharpoons Al(H_2O)_5(OH^-) + H_3O^+ \tag{39}$$

to form solvated protons; amines, on the other hand, dissolve in water to form basic solutions:

$$(CH_3)_3N + H_2O \rightleftharpoons (CH_3)_3NH^+ + OH^-. \tag{40}$$

Thus, highly charged metal ions or anion acceptors react with NH_3 or HF to give acidic solutions:

$$Mg(NH_3)_6^{2+} + NH_3 \rightleftharpoons Mg(NH_3)_5(NH_2)^+ + NH_4^+ \tag{41}$$

$$SbF_5 + 2HF \longrightarrow H_2F^+ + SbF_6^-. \tag{42}$$

In fact many Lewis acids transfer their acid character to these solvents by forming coordinate-covalent bonds with the anion characteristic of the solvent displacing the corresponding autoprotolysis reactions [Eqs. (23) and (25)]. The overall reactions for such processes are (using boron compounds as examples)

$$BR_3 + 2NH_3 \rightleftharpoons BR_3NH_2^- + NH_4^+ \tag{43}$$

and

$$BF_3 + 2HF \rightleftharpoons H_2F^+ + BF_4^-. \tag{44}$$

A more detailed consideration of processes in which metal ions form coordinate-covalent bonds appears in Chapter 17.

6 / STRENGTHS OF BINARY PROTONIC ACIDS

Conventionally a discussion of the acid strength of a substance HX is based upon the extent to which the acid undergoes ionization in a solvent, as given by

$$HX_{(s)} \rightleftharpoons H^+_{(s)} + X^-_{(s)} \qquad (45)$$

In this reaction the products and reactants are depicted as being solvated by an unspecified number of solvent molecules (s). It should be recalled that the equilibrium constant for Eq. (45) is related to the free-energy change ΔG for this process by

$$\Delta G = - RT \ln K, \qquad (46)$$

where R is the gas constant and T the absolute temperature at which the equilibrium is established. Large negative values of the free energy for the process indicate that the reaction proceeds towards the products; this corresponds to large values of the equilibrium constant K. The magnitude of K is, therefore, an indicator of whether or not the reaction is spontaneous (positive or negative value of ΔG) as well as of the extent to which the process proceeds to completion.

It should also be recalled that the free-energy change associated with a process is related to the enthalpy change ΔH by

$$\Delta G = \Delta H - T\Delta S, \qquad (47)$$

where ΔS is the change in entropy during the reaction. Both ΔG and ΔH have often been used in discussing the factors involved in establishing orders of acid strengths of compounds in different solvents. Once again, extensive thermodynamic data for many solvent systems other than water are not available at this time, so that detailed comparisons of ionization processes [Eq. (45)] in several solvents cannot be made. However, some information can be obtained by considering relative acid strengths in water and recognizing that the acidic or basic nature of any other solvent (relative to water) will affect apparent acidities in that solvent.

The factors which affect the ionization of an acid in water can be summarized by the thermochemical cycle shown in Fig. 7.5. The symbols $D(HX)$, $I(H)$, and $E(X)$ represent the bond energy, ionization potential, and electron affinity of the respective species; ΔH represents the enthalpy of solution of the species indicated and ΔH_{ion} is the enthalpy for the ionization process. Summing around the cycle, the enthalpy of ionization (ΔH_{ion}) is given by

$$\Delta H_{ion} = - \Delta H(HX) + \Delta H(H^+) + \Delta H(X^-) + D(HX) + I(H) - E(X). \qquad (48)$$

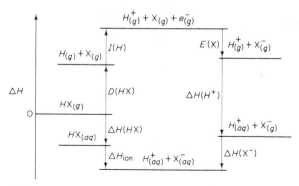

FIG. 7.5 An energy cycle for the ionization of an acid.

There are the usual difficulties in evaluating the solvation enthalpies of individual ionic species and unionized molecules, but suitable approximations to these quantities are available for the hydrogen halides (7). The enthalpies (at 298°K) for the processes which occur in the thermochemical cycle (Fig. 7.5) for the hydrogen halides appear in Table 7.6. In order to

TABLE 7.6 Solvation enthalpies of the hydrogen halides[a]

	Halide			
Quantity	HF	HCl	HBr	HI
$\Delta H(\text{HX})$	-11.5	-4.2	-5.0	5.5
$\Sigma\Delta H(\text{H}^+) + \Delta H(\text{X}^-)$	-281.9	-348.8	-340.7	-330.3
$D(\text{HX})$	134.6	103.2	87.5	71.4
$I(\text{H})$	315	315	315	315
$E(\text{X})$	82.0	87.0	82.0	75.7
$\Delta H_{\text{ion}}{}^{b}$	-3.0	-13.7	-15.2	-14.1
$T\Delta S_{\text{ion}}$	-6	-4	-3	-1
$\Delta G_{\text{ion}}{}^{c}$	3	-10	-12	-13
pK^{d}	2	-7	-9	-10

[a] All entries in units of kcal mole^{-1} at 298°K.
[b] Calculated from Eq. (48).
[c] Calculated from Eq. (47).
[d] Calculated from Eq. (46).

estimate the equilibrium constants for Eq. (45) using Eq. (46), the change in entropy for Eq. (45) is necessary [cf. Eq. (47)]. The change in entropy for Eq. (45) can be obtained from

$$\Delta S = \Delta S_{\text{ion}} + (S_{\text{H}^+} + S_{\text{X}^-}) - S_{\text{HX}}, \qquad (49)$$

where the symbols represent the standard entropies of the species indicated. The entropies of the solvated species can be reasonably approximated by methods which will not be discussed here (8–10); the entropy changes at 298°K for the hydrogen halides which correspond to Eq. (45) also appear in Table 7.6.

At this point we have sufficient data to calculate the free-energy change [from Eq. (47)] and the ionization constant [from Eq. (46)] for the ionization of the hydrogen halides in water [Eq. (45)]. The calculated values, which appear in Table 7.6, indicate that only HF ($pK = 2$) of all of the hydrogen halides should be sufficiently weakly ionized to be detectable by the conventional experimental methods. The other hydrogen halides should be virtually completely ionized. This prediction is experimentally verified, since in aqueous solution HF has a reported pK_a of 3.25, while HCl, HBr, and HI are strong electrolytes in water. An inspection of the data in Table 7.6 shows that the relatively large bond energy for HF compared with the other hydrogen halides is the prime factor in making HF the weakest acid of this series of compounds. The disparity in the entropy term ($T\Delta S_{ion}$) is approximately compensated by the enthalpy of solvation of the ions [$\Sigma\Delta H(H) + \Delta H(X)$].

Similar thermodynamic arguments can be mounted (11) (incorporating a greater number of approximations, however) to estimate the relative acidities of the representative element binary hydrides. The calculated pK values for these substances and the hydrogen halides appear in Table 7.7. In each case

TABLE 7.7   Calculated pK_a values for some hydrides in aqueous solution at 298°K (11)

Group IV		Group V		Group VI		Group VII	
Hydride	pK_a	Hydride	pK_a	Hydride	pK_a	Hydride	pK_a
CH_4	58	NH_3	35	H_2O	16	HF	3.2
		PH_3	27	H_2S	7	HCl	−7
				H_2Se	4	HBr	−9
				H_2Te	3	HI	−10

the process corresponds to the ionization of a covalent hydride forming protons and the corresponding anion

$$M-H_{(aq)} \rightleftharpoons M^-_{(aq)} + H^+_{(aq)}, \tag{50}$$

all species being hydrated. The process in Eq. (50) would never be detected in aqueous solutions for hydrides that are very weak acids (e.g., CH_4, NH_3, PH_3). Correspondingly, hydrides that have very small pK_a values (e.g., HCl, HBr, HI) would be virtually completely ionized. Experimental data indicate that the degree of ionization of the remaining hydrides (H_2O, H_2S, H_2Se,

H_2Te, and HF) decreases in the order indicated by the data in Table 7.7. Indeed, if data obtained in other solvent systems (12) are considered, the relative order of acidities of the very weak and the very strong acids is preserved.

To apply such thermodynamic arguments to the ionization of acids in liquid HF or NH_3 would of course require more extensive entropy and enthalpy data than are presently available. The interplay of factors governing acidities in these solvents should be the same as given in Fig. 7.5.

7 / THE RELATIVE STRENGTHS OF OXYACIDS

A variety of oxyacids containing elements in various oxidation states are known. It might appear that few if any guiding principles would be available to help in our understanding of the relative acid strengths of these substances. However, regularities become immediately apparent if it is recognized that all the oxyacids can be represented by the general formula $(HO)_mXO_n$ where m and n depend upon the oxidation state of the central atom. Thus, the oxyacids of chlorine, i.e., HClO, $HClO_2$, $HClO_3$, and $HClO_4$, can be formulated as $(HO)Cl$, $(HO)ClO$, $(HO)ClO_2$, and $(HO)ClO_3$. In this series m is constant and n increases from 0 to 3, reflecting the increase in oxidation number from $+1$ to $+7$. An inspection of the available experimental data for a variety of oxyacids shown in Table 7.8 indicates that the magnitude of the pK_a values of the oxyacids depends only on the value of n in the general formulation $(HO)_mXO_n$. The oxyacids become progressively stronger as the value of n increases from 0 to 3 with acids with the same n values exhibiting pK's of approximately the same order of magnitude. It is interesting to note that the number of OH groups is apparently also important in considering the acid strength of these substances. For example, the pairs of compounds $(HO)_5IO$ and $(HO)ClO$; $(HO)Cl$ and $(HO)_6Te$; and $(HO)ClO_2$ and $(HO)_2SeO_2$ are classified together on this basis even though in some cases there is a wide divergence in the value of m.

Insight into understanding relationships demonstrated in Table 7.8 comes from a consideration of the relative stability of the anion formed in the ionization of an oxyacid:

$$(HO)_mXO_n + H_2O \rightarrow H_3O^+ + [(HO)_{m-1}XO_{n+1}]^-. \tag{51}$$

The charge on the anion formed can be dissipated onto the nonhydroxyl oxygen atoms in this species. Thus, the greater the number of nonhydroxyl oxygen atoms (larger values of n) the more readily is the charge dissipated. Valence–bond arguments can also lead to a similar conclusion concerning the relative stabilities of the oxyanions. For example, the species involved in the ionization of HOCl and HOClO are depicted in the equations in Fig. 7.6.

TABLE 7.8 The pK_a values of some oxyacids[a] formulated as $[(HO)_mXO_n]$

$n = 0$ $(HO)_mX$		$n = 1$ $(HO)_mXO$		$n = 2$ $(HO)_mXO_2$		$n = 3$ $(HO)_mXO_3$	
Acid	pK_a	Acid	pK_a	Acid	pK_a	Acid	pK_a
$(HO)Cl$	7.2	$(HO)ClO$	2.0	$(HO)ClO_2$	-1	$(HO)ClO_3$	—[b]
$(HO)Br$	8.7						
$(HO)I$	10.0	$(HO)_5IO$	1.6	$(HO)IO_2$	0.8		
		$(HO)_2SO$	1.9	$(HO)_2SO_2$	(-3)		
$(HO)_6Te$	8.8	$(HO)_2TeO$	2.7				
		$(HO)_2SeO$	2.6	$(HO)_2SeO_2$	(-3)		
		$(HO)NO$	3.3	$(HO)NO_2$	-1.4		
		$(HO)_3PO$	2.1				
		$(HO)_2HPO$	1.8				
		$(HO)H_2PO$	2.0				
$(HO)_3As$	9.2	$(HO)_3AsO$	2.3				
$(HO)_3Sb$	11.0						
		$(HO)_2CO$	3.9				
$(HO)_4Ge$	8.6						
$(HO)_3B$	9.2						

[a] Values of pK in parentheses are estimated.
[b] Very large negative value.

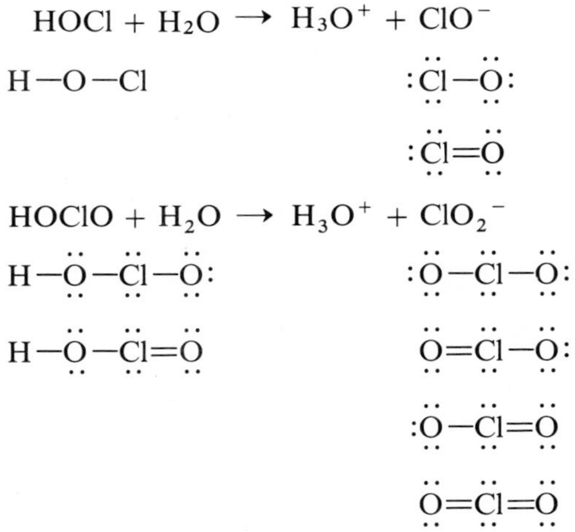

$$HOCl + H_2O \rightarrow H_3O^+ + ClO^-$$

H—O—Cl   :C̈l—Ö:

                           :C̈l=Ö

$$HOClO + H_2O \rightarrow H_3O^+ + ClO_2{}^-$$

H—Ö—C̈l—Ö: :Ö—C̈l—Ö:

H—Ö—C̈l=Ö   Ö=C̈l—Ö:

                           :Ö—C̈l=Ö

                           Ö=C̈l=Ö

FIG. 7.6 The relative stability of the anion formed in the ionization of an oxyacid can be estimated using valence-bond arguments. The greater number of oxygen atoms present in the ClO_2^- anion permits the charge formed upon ionization to be more readily dissipated than in the case of ClO^-.

In such ionization processes the greater the number of oxygen atoms present in the anion, the greater the number of canonical forms that can contribute to the wave function of the anion. It should be recalled from Chapter 4 that a large number of canonical forms is symptomatic of relatively more stable structures. Thus, anions with greater resonance stabilization energy relative to the free acid should have lower energies of formation; this leads to a larger degree of ionization.

Another interesting correlation can be made with the data in Table 7.8. For a given oxidation state of a family of acids (e.g., HOCl, HOBr, HOI), the relative acidity increases with increasing electronegativity of the central atom. An increase in the electronegativity of the central atom should make the adjacent oxygen atom more positive, making loss of a proton an easier process.

Many of the ternary acids are polybasic, e.g., H_3PO_4, H_2SO_3. The pertinent data for the successive ionization constants for several polybasic oxyacids are given in Table 7.9. It is immediately apparent that the differences between

TABLE 7.9 **Successive ionization constants of some polybasic acids**

Acid	pK_1	pK_2	pK_3
$(HO)_4Ge$	8.6	12.7	—
$(HO)_3PO$	2.1	7.2	12.0
$(HO)_3AsO$	2.3	7.0	13.0
$(HO)_2SO$	1.9	7.0	—
$(HO)_2SeO$	2.6	8.3	—
$(HO)_2TeO$	2.7	8.0	—

successive ionization constants for a given acid are four to five orders of magnitude (i.e., 4–5 pK units). The most direct interpretation of this regularity comes from a comparison of the successive ionization processes in terms of simple electrostatics. The loss of one proton from a dibasic acid involves removal of a positively charged species from one that is neutral; the loss of a second proton involves the removal of a positively charged species from one that is negatively charged. Successive removal of protons should be, in an electrostatic sense, successively more difficult. The data in Table 7.9 suggest that the free-energy changes for these processes are not dependent on the nature of the central atom or the number of protons previously removed.

A remarkably simple empirical relationship among the acidity (expressed in pK units) of an oxyacid, the number of nonhydroxyl groups on the acid (m), and the formal charge q on the central atom of the oxyacid has been proposed (13):

$$pK = 8.0 - q(9.0) + m(4.0). \tag{52}$$

The quantities q and m can be determined from an inspection of the formula of the oxyacid as expressed previously, i.e., $(HO)_m XO_n$. The formal charge on the central atom is the difference between the number of valence electrons in the free atom and the number of electrons this atom formally possesses in the covalently bound state. For the purposes of this calculation the second number of this difference is taken as one-half of each electron-pair bond, irrespective of where, in a bookkeeping sense, the electrons in the bond came from. These calculations are summarized for several oxyacids in Table 7.10. It is apparent that hydroxyl groups bonded to the central atoms do not contribute to an increase in the formal charge, but that each non-hydroxyl oxygen atom increases the formal charge by one unit because these are coordinate-covalent bonds in which the central atom has donated two electrons to the bond.

TABLE 7.10   The calculation of the formal charge q in several oxyacids

Formula type	Example	Lewis formulation	Valence electrons	q
HOX	HOCl	H∶O∶Cl∶	7	$7 - 7 = 0$
$(HO)_2 XO$	$(HO)_2 TeO$	H∶O∶Te∶O∶ ∶O∶ H	6	$6 - 5 = 1$
$(HO)_2 XO_2$	$(HO)_2 SO_2$	∶O∶ H∶O∶S∶O∶ ∶O∶ H	6	$6 - 4 = 2$

Equation (52) can be used to approximate the experimental pK values to a reasonable degree of precision. For example, tellurous acid with the formula $(HO)_2 TeO$ should have an estimated pK value of $8.0 - (1)(9.0) + (1)(4.0)$ or 3.0; the experimental value is 2.7 (cf. Table 7.9). For 36 oxyacids, Eq. (52) has been shown to predict the correct pK value of the oxyacid with an average deviation of 0.91 pK units. Equation (52) has found use in verifying the constitution of oxyacids. For example, phosphorous acid, $H_3 PO_3$, can be formulated as $(HO)_3 P$. Equation (52) would predict that the pK value for this formulation should be $8.0 - (0)(9.0) + (0)(4.0)$ or 8 units. The experimental value is 1.8, a very large discrepancy. Certain experimental data (Chapter 12) suggest that phosphorous acid is dibasic and that the correct formulation is $(HO)_2 P(H)O$ in which the phosphorus atom is directly bound to a hydrogen atom. According to this formulation the pK value should be $8.0 - (1)(9) + (1)(4)$ or $3\ pK$ units. The latter value is much closer to the experimental value and strongly suggests that this formulation is correct. Although Eq. (52) gives relatively imprecise results, the estimates are often sufficient to aid in making the correct decision.

Finally, it should be recognized that according to Eq. (52) *all* oxyacids that possess the same formal charge on the central atom and the same number of nonhydroxyl oxygen atoms should have the same pK values. Although this relationship is only expected to be strictly valid for elements in the same family and the same oxidation state, the factors in Eq. (52) which pertain to the formal charge q and the number of nonhydroxyl oxygen atoms could compensate for apparently disparate chemical species and still yield nearly the same pK values.

REFERENCES

1. F. London, *Trans. Faraday Soc.*, **33**, 8 (1937).
2. C. B. Mark, *Electrolytic Dissociation*, Academic Press, New York, 1961, p. 272.
3. M. Smith and M. C. R. Symons, *Trans. Faraday Soc.*, **54**, 346 (1958).
4. E. M. Kosower, *J. Amer. Chem. Soc.*, **80**, 3253 (1958).
5. M. D. Joesten and R. S. Drago, *J. Amer. Chem. Soc.*, **84**, 3817 (1962).
6. M. Kilpatrick and J. Jones, in *The Chemistry of Nonaqueous Solvents* (J. J. Lagowski, ed.), Vol. 2, Academic Press, New York, 1967.
7. J. McCoubrey, *Trans. Faraday Soc.*, **51**, 743 (1955).
8. W. M. Latimer, *Oxidation Potentials*, 2nd ed., Prentice-Hall, Inc., Englewood Cliffs, N.J., 1952.
9. R. E. Powell, *J. Phys. Chem.*, **58**, 528 (1954).
10. D. R. Rosseinsky, *Electrochim. Acta*, **16**, 23 (1971).
11. R. P. Bell, *The Proton in Chemistry*, Cornell University Press, New York, 1959.
12. T. L. Smith and J. H. Elliot, *J. Amer. Chem. Soc.*, **75**, 3565 (1953).
13. J. E. Ricci, *J. Amer. Chem. Soc.*, **70**, 109 (1948).

COLLATERAL READINGS

I. J. Nassler, in *The Chemistry of Nonaqueous Solvents* (J. J. Lagowski, ed.), Vol. 1, Academic Press, New York, N.Y., 1966, p. 214. A detailed description of experimental techniques that have been used to study solutions in liquefied gases.
II. J. J. Lagowski, *Crit. Revs. Anal. Chem.*, **2**, 149 (1971). A discussion of the factors which affect acid strengths in nonaqueous solvents.

STUDY QUESTIONS

1. Using equations illustrate the following phenomena in the H_2O, NH_3, and HF solvent systems (do not use examples given in this chapter): (a) neutralization; (b) displacement; (c) solvation; (d) amphoterism.

2. Given the three solvents H_2O, HF, and NH_3, which would you select for each of the following uses: (a) Use with SO_3 as an oxidizing agent; (b) use with LiH as a reducing agent; (c) for the most complete solvolysis of $NH_4C_2H_3O_2$; (d) for the electrodeposition of an active metal; (e) titration of a very weak acid; (f) titration of a very weak base? If none is suitable, indicate that fact. In each case indicate the reasons for your choice.

3. What species are present in liquid anhydrous hydrogen fluoride containing BF_3?

4. Aluminum trifluoride is insoluble in anhydrous hydrogen fluoride but dissolves readily upon the addition of NaF. Aluminum trifluoride precipitates from the resulting solution when BF_3 is bubbled through it. Write equations for the processes occurring and describe the geometry of all species in solution.

5. Estimate the pK_a values for the following acids: (a) H_2CrO_4; (b) H_5IO_6; (c) H_3AsO_4; (d) $HMnO_4$.

6. Show how you would calculate the proton affinity of the OH^- ion, i.e., the energy for the reaction

$$H_{(g)}^+ + OH_{(g)}^- \rightarrow H_2O_{(g)}.$$

7. Anhydrous hydrogen cyanide, although poisonous in character, has found extensive use as a solvent. Some of the more important physical properties of this substance are listed below.

Property	HCN	H_2O
Freezing point, °C	−13.35	0.00
Boiling point, °C	25.00	100.00
Dielectric constant	123 (15.6°)	81.7 (25°)
Specific conductance, $\Omega^{-1}\,cm^{-1}$	5×10^{-7} (0°)	4×10^{-8} (18°)

What would be the predicted solubility, relative to that in aqueous solution, of ionic substances in anhydrous hydrogen cyanide solution?

8. Using the information given in Question 7, indicate the most probable explanation for the relatively large value of the specific conductance of pure anhydrous hydrogen cyanide.

9. Indicate the acidic and basic species in liquid hydrogen cyanide.

10. Potassium cyanide and organic amines are soluble in anhydrous hydrogen cyanide, and their solutions are much better conductors than is the pure solvent (see data in Question 7). Explain this observation in terms of the species present in solution.

11. It has been observed that acids which are strong in aqueous solution (for example, $HClO_4$, HCl, HNO_3, H_2SO_4) are relatively poorer conductors in anhydrous hydrogen cyanide than are substances like KBr, KCl, NaBr, etc., under equivalent experimental conditions. In view of these observations, would acetic acid be expected to be weaker or stronger in HCN than it is in water? Recall the physical properties of the pure solvent given in Question 7.

12. When a solution containing one mole of potassium cyanide in anhydrous hydrogen cyanide is treated with a solution containing one mole of H_2SO_4 dissolved in anhydrous HCN no reaction apparently occurs. However, upon removing the solvent, $KHSO_4$ can be isolated in quantitative yield. When two moles of KCN are treated with one mole of H_2SO_4, potassium sulfate precipitates from the solution in quantitative yield. Explain these observations in terms of acid–base behavior. Recall the physical properties of the pure solvent given in Question 7.

13. Several metal cyanides can be precipitated from solutions in anhydrous hydrogen cyanide. For example, $FeCl_3$ dissolves in anhydrous HCN to give a clear solution which, when treated with either potassium cyanide or R_3N (R = alkyl), precipitates $Fe(CN)_3$ immediately. However, the precipitate of $Fe(CN)_3$ can be dissolved by adding an excess of either KCN or R_3N. Mercuric perchlorate and silver perchlorate behave in the same manner. Discuss these observations in terms of acid–base behavior, including in your discussion chemical equations which describe the reactions occurring. Recall the physical properties of the pure solvent given in Question 7.

14. Consider the dissolution of the following substances in liquid NH_3, water, and liquid HF: (a) $KOCH_3$; (b) CH_3K; (c) CH_3CO_2H; (d) KF; (e) KOH; and (f) $HClO_4$. What species are present in each solution? Use equations to describe the processes which may occur.

/8

ALKALI METALS

The electronic configurations of the alkali-metal atoms give an insight into their chemical properties as well as into the trends in properties of the compounds these elements form. The ground-state configuration of the alkali metals (Table 8.1) consists of a rare-gas core with a single valence electron in an s orbital. The first ionization potentials of these elements are among the lowest of all the atoms; as would be expected the second ionization potentials are markedly higher than the first. Thus the loss of a single electron is, energetically speaking, a relatively easy process. Since the product ion possesses the much more stable rare-gas structure the chemistry of the alkali metals is dominated by the ease with which they lose electrons and by the nature of the ions formed. Although some systems are known in which the alkali-metal atoms are covalently bonded, the large majority of compounds of these elements are ionic. As such, the stoichiometry of the ionic compounds formed is governed in part by the unit positive charge the ions exhibit. The crystalline structures that are attained can be understood for the most part on the basis of size relationships and electrostatic arguments developed in Chapter 2.

Because the alkali-metal atoms and the corresponding ions possess the simplest of electronic configurations it is not surprising that they form a group for which the effects of size and mass on chemical and physical properties are most clearly exhibited (Table 8.1). Although lithium is clearly an

TABLE 8.1   Some characteristics of the alkali metals

Element (Z)	Electronic configuration	Ionization potentials I	Ionization potentials II	Electro-negativity	Standard electrode potential, (V)	Ionic radius, Å	$\Delta H_{hyd.}$, eV
Li (3)	[He] $2s^1$	5.39	75.6	1.0	3.038	0.68	5.34
Na (11)	[Ne] $3s^1$	5.14	47.3	0.9	2.71	0.97	4.21
K (19)	[Ar] $4s^1$	4.34	31.8	0.8	2.92	1.33	3.34
Rb (37)	[Kr] $5s^1$	4.18	27.4	0.8	2.92	1.47	3.04
Cs (55)	[Xe] $6s^1$	3.89	23.4	0.7	2.93	1.67	2.74
Fr (87)	[Rn] $7s^1$	—	—	—	—	—	—

alkali metal and its general properties are qualitatively those expected for the first member of this series, its properties as well as the properties of its compounds are often markedly different from those expected on the basis of the trends established in the other alkali metals. We shall have occasion to observe a similar behavior of the first members of the other families of elements.

Although francium is properly a member of the alkali-metal series of elements, its chemistry has been investigated in a relatively cursory manner. At least ten isotopes of francium are known, the longest-lived isotope having a half-life of 21 minutes. Thus, the chemistry of francium has been developed largely through radiochemical techniques applied to coprecipitation behavior. We shall not deal any further with the chemistry of francium other than to indicate that its behavior is that expected for the heaviest alkali metal.

1 / THE ELEMENTS

1.1 / Physical Properties

The alkali metals are all silvery white low-melting solids that exhibit the body-centered cubic crystal structure (Table 8.2). Lithium, as mentioned, exhibits exceptional physical properties within this series of elements. The

TABLE 8.2 **Some physical properties of the alkali metals**

Element	Melting point, °C	Boiling point, °C	$\Delta H_{fus.}$, kcal g-atom^{-1}	$\Delta H_{sub.}$, eV	Hardness[a]	Density at 0°C
Li	179.5	1336	0.69	1.61	0.06	0.54
Na	97.8	883	0.63	1.13	0.07	0.972
K	63.5	762	0.57	0.93	0.04	0.859
Rb	38.7	700	0.53	0.89	—	1.525
Cs	39.8	670	0.50	0.82	0.02	1.903

[a] With respect to Moh's scale. The hardest substance on this scale is diamond with a value of 10.

alkali metals sublime to give vapors which are predominantly composed of monatomic species. However, about 1% of the vapor consists of diatomic molecules, M_2 (1). In this respect alkali-metal atoms exhibit behavior similar to that of hydrogen atoms; such similarities might be expected, since both the alkali metals and hydrogen atoms possess an s^1 electronic configuration. The enthalpies of formation of M_2 from the corresponding gaseous atoms (Table 8.3) indicate that the alkali-metal molecules are reasonably stable, with Li_2

TABLE 8.3 **Some properties of diatomic alkali-metal molecules**

Molecule	ΔH_f, kcal mole^{-1}[a]	d_{MM}, Å Experimental	d_{MM}, Å Calculated[b]
Li_2	−27.2	2.67	3.04
Na_2	−18.4	3.08	3.94
K_2	−12.6	3.91	4.76
Rb_2	−11.3	—	5.06
Cs_2	−10.4	4.55	5.44

[a] ΔH of formation given for the process $2M_{(g)} \rightarrow M_{2\,(g)}$.
[b] Using covalent radii for coordination number 12.

being the most stable of the molecules in this series. The internuclear distances observed in these molecules are generally shorter than the expected values for single bonds in these systems. Apparent bond shortening of this type is usually symptomatic of multiple bonding, but a satisfactory solution to the discrepancy in the case of the M_2 molecules is not yet available.

1.2 / Solubility in Liquid Ammonia (1)

Perhaps one of the most interesting properties of the alkali metals is their solubility in liquid ammonia. Starting with pure ammonia at its boiling point ($-33°C$), the addition of a small amount of alkali metal gives a light blue solution which becomes progressively darker as more alkali metal is introduced into the system. After a certain mole ratio of metal to solvent, the addition of more alkali metal causes a phase separation to occur, a bronze-colored liquid phase floating on the very dark blue solution formed previously. As more alkali metal is added the volume of the bronze phase grows at the expense of the dark blue phase until the system contains only the bronze phase. At this mole ratio of constituents, no more alkali metal will dissolve. Figure 8.1 gives the phase diagram for the Na–NH$_3$ system. Region I represents the dilute blue phase, region II (the so-called miscibility gap) represents the composition and temperature combinations which yield both the bronze and the blue phase; the bronze phase alone appears in region III; and excess alkali metal will be present for all systems with parameters that fall within region IV. Although the detailed phase diagrams for all alkali-metal–ammonia solutions are not available, there is reason to believe that their general characteristics are parallel to those depicted in Fig. 8.1. Only in the case of lithium is it possible to isolate a solvate, in which case tetra-hedral Li(NH$_3$)$_4$ units are arranged in a close-packed structure (2).

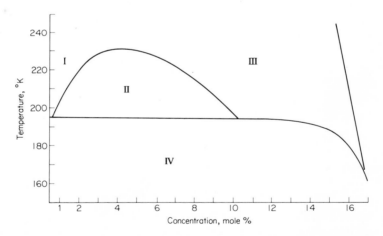

FIG. 8.1   The temperature–composition phase diagram for the sodium–ammonia system.

Both blue and bronze phases are remarkably good electrical conductors; at $-33°C$ the conductivity of a saturated (bronze) solution of sodium in liquid ammonia is about half that of mercury. The mobility of the negative species is reported to be 280 times that of the cation. Clearly the conductivity data available for these solutions cannot be interpreted using conventional ionic species.

The solutions also possess an interesting magnetic and optical behavior. The very dilute solutions are paramagnetic, suggesting that unpaired electrons are present in the system. With increasing concentration the solutions become less paramagnetic, indicating that a spin-pairing process must occur. The dilute solutions formed by each of the alkali metals have exactly the same spectrum for equivalent concentrations, i.e., a broad featureless band centered at about 14,500 Å (3).

The properties of these solutions strongly suggest that alkali-metal atoms ionize in liquid ammonia to form solvated cations and solvated electrons:

$$\text{M}_{(s)} + (x + y)\text{NH}_3 \rightarrow \text{M(NH}_3)_y{}^+ + e(\text{NH}_3)_x{}^-. \tag{1}$$

The detailed description of the species called the solvated electron, $e(\text{NH}_3)_x{}^-$, is not yet available, but the conductivity, magnetic, and optical data on alkali-metal–ammonia solutions point to the existence of such a species. Thus the abnormally high conductivity of these solutions and the high mobility of the negative species in them, the fact that *all* alkali-metal solutions have the same optical properties, and the magnetic behavior of these solutions can be interpreted in terms of the existence of a stable solvated electron. Indeed, the decrease in paramagnetism with increasing analytical concentration of metal–ammonia solutions indicates that solvated electrons undergo a pairing process:

$$2e(\text{NH}_3)_x{}^- \rightleftharpoons e_2(\text{NH}_3)_z{}^{2-}. \tag{2}$$

Considering the moderate dielectric constant of liquid ammonia it is very probable that solvated electrons and solvated metal ions are appreciably associated,

$$\text{M(NH}_3)_y{}^+ + e(\text{NH}_3)_x{}^- \rightleftharpoons \text{M(NH}_3)_y{}^+, e(\text{NH}_3)_x{}^-, \tag{3}$$

as any pair of oppositely charged species in this solvent associate (Chapter 7). The product in Eq. (3) is, like a conventional ion pair, nonconducting, although it is paramagnetic. Presumably, as the analytical concentration of solute increases the association between oppositely charged species increases and the equivalent of ion multiplets form:

$$e_2(\text{NH}_3)_x{}^{2-} + \text{M(NH}_3)_z{}^+ \rightleftharpoons e_2(\text{NH}_3)_x{}^{2-}, \text{M(NH}_3)_z{}^+ \tag{4}$$

$$e(\text{NH}_3)_x{}^- + \text{M(NH}_3)_z{}^+, e(\text{NH}_3)_x{}^- \rightleftharpoons e(\text{NH}_3)_x{}^-, \text{M(NH}_3)_z{}^-, e(\text{NH}_3)_x{}^- \tag{5}$$

$$\text{M(NH}_3)_z{}^+ + \text{M(NH}_3)_z{}^+, e(\text{NH}_3)_x{}^- \rightleftharpoons \text{M(NH}_3)_z{}^+, e(\text{NH}_3)_x{}^-, \text{M(NH}_3)_z{}^+. \tag{6}$$

Some of these higher aggregates are conducting and diamagnetic, some are conducting and paramagnetic, and some nonconducting and paramagnetic.

Thus, as the analytical concentration of alkali metal increases the system contains less and less free solvent and it takes on the general characteristics of a molten metal. Indeed, the properties of the bronze phases have been interpreted as metal-like by some investigators.

Perhaps the best description presently available for the solvated electron in ammonia is the cavity model (4). In this model, the electron is trapped within a cavity surrounded by ammonia molecules. Presumably the solvent dipoles are oriented with the hydrogen atoms pointing inward toward the center of the cavity. In the absence of impurities or metallic surfaces, metal–ammonia solutions are stable for extended periods of time. However, it is possible to cause the reaction

$$e(NH_3)_x^- + NH_3 \rightleftharpoons \tfrac{1}{2}H_2 + NH_2^- + xNH_3 \tag{7}$$

to occur between electrons and the solvent if a catalyst is introduced into the system. The products of the reaction of alkali metals with ammonia are analogous to those formed in their reaction with water. It is interesting that reaction (7) is easily reversible; that is, a solution of alkali-metal amide can be converted into a solution of the corresponding alkali metal by pressurizing the system with hydrogen. The equilibrium constant for the reaction as written in Eq. (7) is 5×10^4 at 25°C (5).

Solutions of solvated electrons can also be prepared by electrolyzing liquid ammonia solutions containing nonreducible species:

$$e_{(cathode)}^- + xNH_3 \rightleftharpoons e(NH_3)_x^-. \tag{8}$$

Solutions so prepared are indistinguishable from the alkali-metal–ammonia solution of equivalent concentration (6).

Solvated electrons have been observed in other solvents. Certain amines dissolve alkali metals to form solutions that are more complex than the metal–ammonia solutions in the sense that they contain diatomic metal molecules and species derived from them

$$M_{2(am)} \rightleftharpoons M_{2(am)}^+ + e_{(am)}^- \tag{9}$$

as well as solvated metal ions and solvated electrons. Certain ethers such as tetrahydrofuran can also form reasonably stable (but very much more dilute) solutions with potassium, cesium, and rubidium. Radiolysis of aqueous solutions gives rise to hydrated electrons, but this species is considerably less stable than the solvated electrons formed in the amines or ethers (7, II, III).

As might be expected, metal–ammonia solutions have found extensive use as chemical reducing agents in a variety of systems (IV). Numerous kinetic studies involving reduction by metal–ammonia solutions have been made in an attempt to formulate the mechanisms by which such processes occur. In general, electron addition to saturated compounds leads to bond cleavage with the formation of either radicals and/or ionic species:

$$X-Y + e^- \rightarrow X\cdot + Y^- \tag{10}$$

$$X-Y + 2e^- \rightarrow X^- + Y^-. \tag{11}$$

The ions formed in such reactions are usually conjugate bases of very weak acids which abstract protons from the solvent or other stronger proton donors in the system:

$$X^- + NH_3 \rightarrow HX + NH_2^- \tag{12}$$

$$X^- + RH \rightarrow HX + R^-. \tag{13}$$

The radicals formed in the original process [Eq. (10)] can take up another electron

$$X\cdot + e^- \rightarrow X^-, \tag{14}$$

undergo dimerization with other radicals

$$2X\cdot \rightarrow X-X, \tag{15}$$

or abstract radicals from molecular species

$$X\cdot + X-Y \rightarrow X-X + Y\cdot. \tag{16}$$

The dimerized species can, of course, arise from the reaction of an anion with the molecular species

$$X^- + X-Y \rightarrow X-X + Y^-. \tag{17}$$

Although there are numerous examples of these pathways available as illustrations, only a representative number appear in Eqs. (18)–(22):

$$(CH_3)_3SnBr + 2Na \rightarrow [(CH_3)_3Sn]Na + NaBr \tag{18}$$

$$ArOAr' + 2Na + NH_3 \rightarrow (ArO)Na + Ar'H + NaNH_2 \tag{19}$$

$$R_4N^+ + 2Na + NH_3 \rightarrow R_3N + RH + NaNH_2 + Na^+ \tag{20}$$

$$2R_2S + 2Na \rightarrow 2(RS)Na + R_2 \tag{21}$$

$$2RCl + 2Na + NH_3 \rightarrow RH + RNH_2 + 2NaCl. \tag{22}$$

The reactions of solvated electrons with unsaturated bonds generally follow similar pathways, except that bond fission need not occur. Either one or two electrons can be added to unsaturated bonds to form a radical anion or a dianion, respectively:

$$X{=}Y + e^- \rightarrow \cdot X{-}Y^- \tag{23}$$

$$\cdot X{-}Y^- + e^- \rightarrow {}^-X{-}Y^- \tag{24}$$

$$X{=}Y + 2e^- \rightarrow {}^-X{-}Y^-. \tag{25}$$

The products from Eqs. (23)–(25) can abstract protons [cf., Eqs. (12) and (13)] and undergo dimerization [cf. Eq. (21)]. Again, numerous examples exist illustrating these possible pathways, but only a few representative reactions are given in Eqs. (26)–(30):

$$RCH{=}CH_2 + 2NH_3 + 2Na \rightarrow RCH_2CH_3 + 2NaNH_2 \tag{26}$$

$$R_2C{=}O + 2Na + 2NH_3 \rightarrow R_2CHOH + 2NaNH_2 \tag{27}$$

$$O_2 + e^- \rightarrow O_2^- \tag{28}$$

$$S_x + 2e^- \rightarrow S_x^{2-} \tag{29}$$

$$9Pb + 4e^- \rightarrow Pb_9^{4-}. \tag{30}$$

Transition metals in high oxidation states can be reduced either to the free metal

$$Cu(NH_3)_4{}^{2+} + 2e^- \rightarrow Cu^0 + 4NH_3 \tag{31}$$

or to species containing the metal in a lower oxidation state

$$Ni(CN)_4{}^{2-} + 2e^- \rightarrow Ni(CN)_4{}^{4-}. \tag{32}$$

Finally, solvated electrons react readily with acidic substances in liquid ammonia to form hydrogen and a compound that can be formally considered as an alkali-metal salt:

$$RH + e^- \rightarrow R^- + \tfrac{1}{2}H_2. \tag{33}$$

This type of reaction is very useful in the synthesis of alkali-metal derivatives of many hydrogen-containing compounds:

$$PH_3 + Na \rightarrow NaPH_2 + \tfrac{1}{2}H_2 \tag{34}$$

$$GeH_4 + Na \rightarrow NaGeH_3 + \tfrac{1}{2}H_2 \tag{35}$$

$$(C_6H_5)_3CH + Na \rightarrow (C_6H_5)_3CNa + \tfrac{1}{2}H_2. \tag{36}$$

1.3 / Preparation

The method of preparation of the alkali metals is governed entirely by the fact that these substances are among the strongest chemical reducing agents known. Correspondingly, the preparation of the metals from these compounds requires a very strong reducing agent. Electrolytic processes are the best known method for the preparation of the free metals. Electrolysis of aqueous solutions of alkali-metal salts produces hydrogen at a solid cathode, but with a mercury cathode, which has a high hydrogen overvoltage, reduction of alkali-metal ions occurs. The free metal forms an alloy with mercury (called an amalgam) in which its activity is markedly reduced:

$$Hg_x(e^-) + M^+ \rightarrow Hg_x(M). \tag{37}$$

Unfortunately it is usually difficult to separate the alkali metals from mercury, so that this method has only limited application. Electrolytic reduction of fused alkali-metal salts or of nonaqueous solutions (e.g., in pyridine) of such salts produces the corresponding pure alkali metals at the cathode (V).

The high volatilities of the alkali metals and the relatively small heats of formation of some alkali-metal compounds permit the preparation of the metals by thermal reduction processes. For example, at high temperatures alkali-metal chlorides can be reduced with calcium,

$$2MCl + Ca \rightarrow 2M_{(g)} + CaCl_2, \tag{38}$$

and alkali-metal carbonates with carbon:

$$M_2(CO_3) + 2C \rightarrow 2M_{(g)} + 3CO. \tag{39}$$

The alkali metals distill from the reaction mixture and can be condensed to give pure products; the remaining reactants and products are either non-volatile or gaseous under the conditions of these reactions. Such chemical reduction methods are useful for relatively small amounts of the alkali metals. Perhaps a better method for the production of alkali metals is the thermal decomposition ($> 300°C$) of the corresponding metal azides:

$$2MN_3 \rightarrow 2M + 3N_2. \tag{40}$$

The starting materials are relatively easy to purify and they are not explosive; the method is thus ideal for the preparation of precise amounts of alkali metals. Lithium reacts readily with nitrogen so this method cannot be used for its preparation.

1.4 / Chemical Properties

As a family the alkali metals are the most reactive of the metals; lithium stands apart from the others of the series in terms of reactivity and the substances with which it reacts.

The alkali metals react to varying degrees with oxygen, hydrogen, and the halogens. In the case of oxygen the nature of the product depends upon the alkali metal. Thus lithium does not react below 100°; however, at 200° Li_2O is formed. Sodium gives a mixture of Na_2O and Na_2O_2. Potassium yields predominantly KO_2 at 200°, but K_2O_2 at higher temperatures. Under certain conditions cesium can form any one of a series of suboxides, Cs_7O, Cs_4O, Cs_7O_2, and Cs_3O. The Rb/Rb_2O phase diagram shows the formation of Rb_3O, which is a copper-colored substance (8). The nature of the oxygen anions is discussed in Chapter 13.

Hydrogen and the halogens yield the corresponding hydrides (Chapter 6) and halides when heated with the alkali metals:

$$2M + H_2 \rightarrow 2MH \tag{41}$$

$$2M + X_2 \rightarrow 2MX. \tag{42}$$

In general, under a given set of conditions the reactivity increases with increasing atomic weight of the alkali metal. For example, lithium reacts only superficially with liquid bromine but potassium explodes. Lithium is unusual in that it reacts readily with gaseous nitrogen to form the nitride

$$6Li + N_2 \rightarrow 2Li_3N, \tag{43}$$

whereas no reaction is observed for the other alkali metals (9). Thus, all the alkali metals react with air, lithium predominantly with the nitrogen present and the remaining members of the series with oxygen.

The alkali metals react with simple molecules containing acidic hydrogen atoms to give the corresponding metal derivatives; hydrogen is usually formed in the process:

$$H_2O_{(l)} + M \rightarrow MOH + \tfrac{1}{2}H_2 \tag{44}$$

$$C_2H_5OH_{(l)} + M \rightarrow C_2H_5OM + \tfrac{1}{2}H_2 \tag{45}$$

$$2H_2S + 2M \rightarrow M_2S + MSH + \tfrac{3}{2}H_2 \tag{46}$$

$$C_2H_2 + M \rightarrow C_2HM + \tfrac{1}{2}H_2 \tag{47}$$

$$NH_{3(g)} + M \rightarrow MNH_2 + \tfrac{1}{2}H_2 \tag{48}$$

$$NH_{3(g)} + 2Li \rightarrow Li_2NH + H_2. \tag{49}$$

The reaction of lithium with gaseous ammonia proceeds at 300–400°C and two products ($LiNH_2$ and Li_2NH) can be formed in the process (*VI*). It should be recalled that alkali-metal amides can also be formed by the catalyzed decomposition of the blue alkali-metal–ammonia solutions. Indeed, all of the reactions given in Eqs. (44)–(49) occur quite readily with solutions of the alkali metals in liquid ammonia.

2 / COMPOUNDS OF THE ELEMENTS

The electronegativities of the alkali metals (Table 8.1) are some of the lowest among the elements, indicating that they form predominantly ionic compounds. Indeed, the structures of many of the alkali-metal compounds can be completely understood in terms of electrostatic interactions of ionic species with characteristic geometries, as described in Chapter 2. Many of the binary ionic compounds can be prepared by direct combination [e.g., Eqs. (41) and (42)]; both binary compounds and compounds containing complex anions can be prepared by the appropriate acid–base reactions:

$$MOH + HX \rightarrow MX + H_2O. \tag{50}$$

As would be expected for ionic compounds, the alkali-metal salts are usually soluble in water and other polar solvents unless the lattice forces are much higher than the solvation energies of the ions involved (Chapter 7). The alkali-metal ions are not strongly hydrated in aqueous solution except for lithium (Table 8.1) which is known to form stable hydrates (e.g., $LiI_3 \cdot 3H_2O$ and $LiClO_4 \cdot 3H_2O$). In fact, lithium salts containing highly polarizable anions, such as iodide ion, have many characteristics reminiscent of covalently bonded species. The lithium ion, the smallest of all the ions except H^+, possesses the highest charge density and should be capable of forming compounds possessing bonds of appreciable covalent character. Thus lithium salts containing large anions, and hence lower lattice energies, are markedly more soluble in organic solvents (such as ethers, alcohols, and acetone) than are the corresponding salts of the other metals in this family.

2.1 / Organometallic Compounds

The marked difference in the nature of the bonds formed by lithium and the other alkali metals is also illustrated by the properties of the organometallic derivatives of the alkali metals. There are a variety of synthetic methods available for the preparation of such compounds, but only a few of the more common ones are presented here (*VII*). Hydrocarbons possessing particularly acidic hydrogen atoms, e.g., acetylene and triphenylmethane, react directly with alkali metals to give the corresponding metal derivatives:

$$RH + M \rightarrow RM + \tfrac{1}{2}H_2 \tag{51}$$

$$RH = C_2H_2, (C_6H_5)_3CH.$$

Such reactions can occur in liquid ammonia solutions or in ethers such as tetrahydrofuran and dimethoxyethane (diglyme) containing a very finely divided suspension of the alkali metal. The reactions occur readily in certain ether solvents, probably because of the ability of these substances to solvate the alkali-metal cations. In special cases crystalline etherates containing alkali-metal ions have been formed, e.g., $Na(diglyme)_2Ta(CO)_6$ and $K(diglyme)_3Mo(CO)_5I$.

Organometallic derivatives of the alkali metals have also been prepared by displacement reactions involving organic halides

$$RX + 2Li \rightarrow RLi + LiX, \tag{52}$$

or other organometallic compounds

$$R_2Hg + M_{(excess)} \rightarrow 2RM + Hg(M). \tag{53}$$

Reaction (52) is most often successful with lithium; the other alkali metals usually yield coupled products

$$2RX + 2Na \rightarrow R-R + 2NaX \tag{54}$$

because the corresponding organometallic compounds are markedly more reactive than the lithium compounds. In preparing organolithium compounds using the process shown in Eq. (52) alkyl chlorides and bromides are preferred; alkyl iodides (except for CH_3I) are more reactive and undergo extensive coupling [Eq. (54)]. In the case of aryl halides, the iodides or bromides are preferred. Butyllithium has found extensive use as an intermediate in the formation of other organolithium compounds by a halogen–metal exchange process:

$$C_4H_9Li + RX \rightarrow C_4H_9X + RLi. \tag{55}$$

The properties of the organometallic derivatives of the alkali metals strongly suggest that, as in the case of the inorganic compounds, organolithium compounds have some characteristics of covalent compounds (*VIII*). The organometallic derivatives of sodium, potassium, rubidium, and cesium, on the other hand, are distinctly ionic. All of the alkyl and simple phenyl derivatives of sodium, potassium, rubidium, and cesium are very reactive, colorless, nonvolatile solids that are insoluble in organic solvents with which they do not react, e.g., hydrocarbons and benzene. Attempts to melt these substances yield complex decomposition products. For example, at 200° methylsodium decomposes according to the following reaction:

$$8CH_3Na \rightarrow 6CH_4 + Na_2C_2 + 6Na. \qquad (56)$$

The product Na_2C_2 is apparently a complex mixture, since a mixture of methane, ethane, acetylene, and hydrogen is obtained when it is hydrolyzed. Thus, these derivatives of the alkali-metal atoms are best formulated as the alkali-metal salts of very weak acids. The crystal structure of KCH_3 shows six discrete CH_3^- units octahedrally arranged about each potassium ion (10); the anions in all but organolithium compounds are most probably carbanions $(R_3C:^-)$. Organoalkali-metal compounds in which the organic group can provide a way to delocalize negative charge are more stable than the simple alkyl derivatives; such derivatives are usually vividly colored. Thus, benzyl sodium $(C_6H_5CH_2^-Na^+)$ and triphenylmethyl sodium $[(C_6H_5)_3C^-Na^+]$ are deep red, crystalline compounds.

In contrast to the properties of the other organoalkali-metal derivatives, the corresponding lithium compounds are soluble in ethers, aliphatic hydrocarbons, and benzene. Except for CH_3Li, the organolithium compounds are volatile or low-melting substances; for example, C_2H_5Li is a colorless solid (mp 95°) and C_3H_7Li is a colorless viscous liquid that can be vaporized at 80–100° under vacuum. Perhaps the most interesting property of organolithium compounds is that they are associated in either benzene or ether solution; typical data are collected in Table 8.4 (11–14). The mass

TABLE 8.4   Association number for alkyllithium compounds [$(LiR)_n$] in hydrocarbons

Compound	Solvent	Degree of association	Method of determination[a]
C_2H_5Li	C_6H_6	6.07 ± 0.35	F
	C_6H_{12}	5.95 ± 0.3	F
n-C_4H_9Li	C_6H_6	6.25 ± 0.06	I
	C_6H_{12}	6.17 ± 0.12	I
$(CH_3)_2Li$	C_6H_6	3.8 ± 0.2	B
	n-C_6H_{14}	4.0 ± 0.2	B

[a] F = freezing-point depression; I = isopiestic method; B = boiling point elevation.

spectra of alkyllithium compounds also indicate that association occurs in the vapor state. We should not expect, on the basis of the number of valence electrons in lithium and the concepts of the covalent bond discussed thus far, to have association between LiR molecules (VIII).

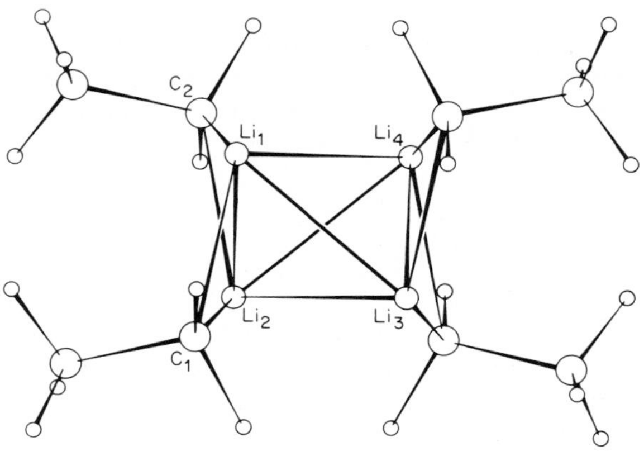

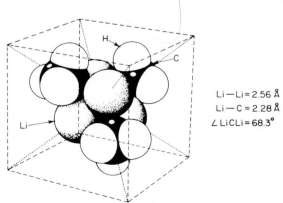

$Li_1 - Li_2 = 2.24\ \text{Å}$
$Li_1 - Li_4 = 2.60\ \text{Å}$
$Li_1 - Li_3 = 2.63\ \text{Å}$
$Li_1 - C_1 = 2.19\ \text{Å}$
$Li_2 - C_1 = 2.52\ \text{Å}$
$Li_3 - C_1 = 2.47\ \text{Å}$

$\angle\ Li_2 C_2 Li_1 = 66.1°$
$\angle\ Li_4 C_2 Li_1 = 67.1°$
$\angle\ Li_4 Li_1 Li_2 = 63.1°$

FIG. 8.2   The structure of $(LiC_2H_5)_4$ in the crystalline state.

X-Ray diffraction data indicate that, in the solid state, ethyllithium possesses the basic tetrameric repeating unit shown in Fig. 8.2. A similar tetrameric structure has been observed for methyllithium (Fig. 8.3). On the basis of infrared data the structure shown in Fig. 8.4 was proposed for the

$Li - Li = 2.56\ \text{Å}$
$Li - C = 2.28\ \text{Å}$
$\angle\ LiCLi = 68.3°$

FIG. 8.3   The structure of $(LiCH_3)_4$ in the crystalline state.

hexamer detected in solutions of alkyllithium compounds (Table 8.4). It is interesting to note that for the polymeric lithium alkyls where internuclear distances have been determined (Figs. 8.2 and 8.3) the Li–Li distances are less than that observed for the Li_2 molecule (Table 8.3), which presumably contains a two-electron bond. In addition the structures contain bridging alkyl groups. The implications of the structures shown in Figs. 8.3 and 8.4

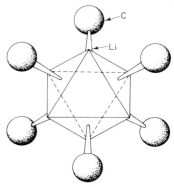

FIG. 8.4 The suggested structure of an alkyllithium hexamer $(LiR)_6$. Only the bridging atoms are shown.

are profound in terms of the theory of the covalent bond. If the shortest distance between two atoms is taken to be the line along which a chemical bond is formed, we observe that the bridging carbon atoms in the tetrameric lithium alkyls form two "bonds" even though the alkyl groups have already formed three bonds; in a chemical sense the alkyl groups are "divalent." Another disturbing feature of this structure is that there are insufficient valence electrons to form the number of "bonds" suggested in Figs. 8.2 and 8.4. We delay the theoretical discussion of the nature of the bonds formed in such electron-deficient structures until Chapter 10, where a more unified approach using a greater variety of compounds can be given.

2.2 / Coordination Behavior

Crystallographically, the coordination numbers of the alkali metals in the predominantly ionic compounds are governed by the radius-ratio rules; the smaller ions (Li^+, Na^+, K^+, Rb^+) give octahedral structures while the larger Cs^+ ion can attain a coordination number of eight in some of its compounds. In general, the alkali-metal ions do not form particularly stable coordination compounds. The structures of some solid hydrates as well as compounds incorporating very easily polarizable ligands such as ammonia give an insight into the coordination behavior of these metals.

Lithium usually exhibits a coordination number of 4, although occasionally under special conditions a coordination number of 6 may be observed. For example, the crystal structure of $LiClO_4 \cdot 3H_2O$ shows that lithium ions are linked in linear chains by water molecules, each lithium being associated with six octahedrally disposed water molecules. The perchlorate ions are interspersed between the chains. A similar arrangement of oxygen atoms about the lithium atom occurs in $LiIO_3$ (*15*). Four-coordinate lithium is undoubtedly present in the very stable ammoniate $LiI \cdot 4NH_3$; the solid-state structure of $LiOH \cdot H_2O$ is polymeric (Fig. 8.5) with each lithium being

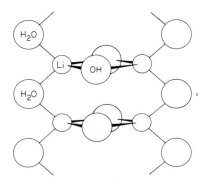

FIG. 8.5 $LiOH \cdot H_2O$ crystallizes in a double chain of lithium atoms with water molecules forming bridges between the lithium atoms in the chain and OH units holding the chains together.

tetrahedrally surrounded by oxygen atoms. Each polymeric chain is attached to an adjacent chain by hydrogen bridges (Chapter 6). Lithium acetylacetonate [the salt of the enol form of acetylacetone, $CH_3-CO-CH=C(OH)CH_3$] forms a dihydrate in which lithium has achieved a coordination number of 4 (Fig. 8.6).

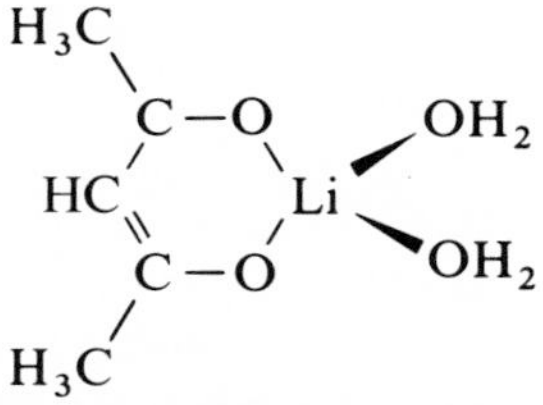

$$
\begin{array}{c}
H_3C \\
\backslash \\
C-O \qquad OH_2 \\
HC \qquad \ \ Li \\
\parallel \qquad \ \ / \quad \\
C-O \qquad OH_2 \\
/ \\
H_3C
\end{array}
$$

FIG. 8.6 Lithium acetylacetonate crystallizes with two water molecules of hydration. The metal ion is tetrahedrally surrounded by oxygen atoms.

The alkali metals larger than lithium usually exhibit higher coordination numbers. For example, sodium, potassium, and rubidium iodides form reasonably stable ammoniates containing six ammonia molecules ($MI \cdot 6NH_3$);

sodium acetylacetonate forms a tetrahydrate in which the sodium ion is octahedrally surrounded by oxygen atoms (Fig. 8.7). Salicylaldehyde, which is acidic, forms compounds with potassium, rubidium, and cesium in which

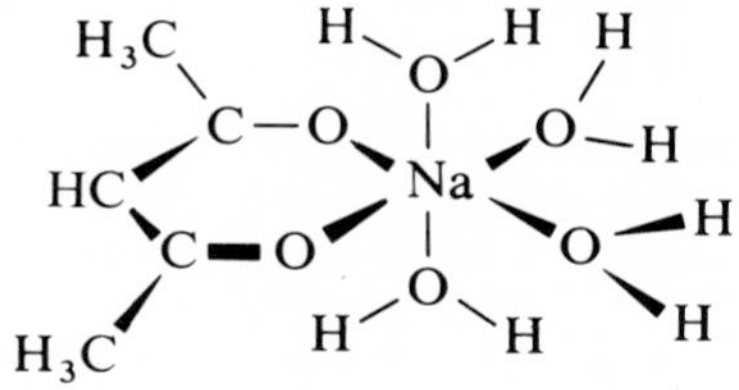

FIG. 8.7 Sodium acetylacetonate forms a tetrahydrate in which the metal atom is octahedrally surrounded by oxygen atoms.

these metals acquire octahedral coordination, as shown in Fig. 8.8. The formulation of these compounds appears to be MOC_6H_4 —$CHO\cdot2HOC_6H_4$ —CHO in which one salicylaldehyde molecule has been converted into an anion and two are present in their molecular forms. The fact that the heavier

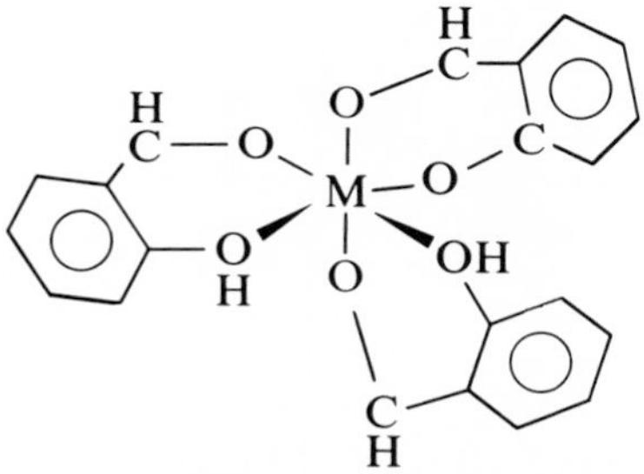

FIG. 8.8 The heavier alkali metals form compounds with salicylaldehyde in which the metal atom is octahedrally surrounded by oxygen atoms. M = K, Rb, or Cs.

alkali metals form 6-coordinate structures does not exclude the formation of structures with a lower coordination number. Thus, NaI also forms a 4-coordinate ammoniate ($NaI\cdot4NH_3$) which has been shown to have a tetrahedral arrangement of ammonia molecules around the sodium ion; salicylaldehyde also forms 4-coordinate structures with all of the alkali-metal ions. In addition, all the alkali metals form acetylacetonates that are dihydrates and presumably contain 4-coordinate cations similar to that shown in Fig. 8.6. The hydrated compounds exhibit solubility properties of covalent compounds; for example, $Na(acac)_2\cdot2H_2O$ is soluble in benzene.

REFERENCES

1. C. T. Ewing, J. P. Stone, J. R. Spann, and R. R. Miller, *J. Phys. Chem.*, **71**, 473 (1967).

2. N. Manmano and M. J. Sienko, *J. Amer. Chem. Soc.*, **90**, 6322 (1968).

3. D. F. Burrow and J. J. Lagowski, *Advan. Chem. Series*, **50**, 125 (1965).

4. J. Jortner, *J. Chem. Phys.*, **30**, 839 (1959).

5. E. J. Kerschke and W. L. Jolly, *Inorg. Chem.*, **6**, 855 (1967).

6. R. K. Quinn and J. J. Lagowski, *J. Phys. Chem.*, **73**, 1374 (1968).

7. F. S. Dainton, D. M. Wiles, A. N. Wright, *J. Chem. Soc.*, 4283 (1960).

8. P. Touzain, *Can. J. Chem.*, **47**, 2639 (1969).

9. C. C. Addison and B. M. Davies, *J. Chem. Soc. A*, 1822, 1827, 1831 (1969).

10. E. Weiss and G. Sauermann, *Angew. Chem. (Intl. Ed.)*, **7**, 133 (1969).

11. T. L. Brown, R. L. Gerteis, D. A. Bafus, and J. A. Ladd, *J. Amer. Chem. Soc.*, **86**, 2134 (1964).

12. T. L. Brown, J. A. Ladd, and G. N. Newman, *J. Organomet. Chem.*, **3**, 1 (1965).

13. D. Margerison and J. P. Newport, *Trans. Faraday Soc.*, **59**, 2058 (1963).

14. M. Weiner, C. Vogel, and R. West, *Inorg. Chem.*, **1**, 654 (1962).

15. A. Rosenzweig and B. Morosin, *Acta Cryst.*, **20**, 758 (1966).

COLLATERAL READINGS

I. W. L. Jolly, *Prog. Inorg. Chem.*, **1**, 235 (1969). An extensive review of the physical properties of metal–ammonia solutions.

II. E. J. Hart, *Science*, **146**, 19 (1964). A discussion of the properties and reactions of the hydrated electron.

III. M. Anbar, *Quart. Revs.*, **22**, 578 (1968). A survey of the reactions of hydrated electrons with inorganic substances.

IV. J. J. Lagowski, *Pure Appl. Chem.*, **25**, 429 (1971). A review of solution phenomena in liquid ammonia. One section of this paper is a discussion of oxidation–reduction processes in this solvent.

V. L. F. Audrieth and H. W. Nelson, *Chem. Revs.*, **8**, 335 (1931). A discussion of the use of nonaqueous solvents as media for electrochemical deposition of metals.

VI. R. Juza, *Angew. Chem. (Intl. Ed.)*, **3**, 471 (1964). A review of the preparation, properties, and structures of the alkali- and alkaline-earth-metal amides.

VII. M. Schlasser, *Angew. Chem. (Intl. Ed.)*, **3**, 287, 362 (1964). A detailed description of the preparation, properties, reactions, and structures of organic derivatives of sodium and potassium.

VIII. T. L. Brown, *Advan. Organomet. Chem.*, **3**, 365 (1965). A discussion of the structure and properties of organolithium compounds.

STUDY QUESTIONS

1. Describe the chemical and physical properties expected for francium.

2. Explain the observation that the heats of formation of the alkali-metal fluorides decrease with increasing atomic weight of the alkali metal, but the reverse is true for the alkali-metal chlorides, bromides, and iodides.

3. Predict the products expected when each of the following substances reacts with a solution of an alkali metal in liquid ammonia: (a) PH_3; (b) NH_4Cl; (c) $AgNO_3$; (d) $(C_6H_5)_3SiH$; (e) H_2O.

4. Using lattice-energy arguments show that the species Na^{2+} would not be expected to exist under normal conditions.

5. The alkali-metal cations have the following equivalent conductivities in aqueous solution at 25°C: Li^+, 39; Na^+, 50; K^+, 73; Rb^+, 78; Cs^+, 77 Ω^{-1}. Discuss these data.

6. Discuss the factors which make lithium fluoride, carbonate, and phosphate markedly less soluble in water than the corresponding compounds formed by the other alkali metals.

7. Give the products expected to be formed in the reaction of methyllithium with the following species: (a) BCl_3; (b) $B(CH_3)_3$; (c) HCl; (d) H_2O; (e) SCl_2.

8. Predict the geometry of the molecules formed in the reactions given in Problem 7.

9

ALKALINE-EARTH METALS

1 / INTRODUCTION

The alkaline-earth metals resemble the alkali metals in many of their properties. There are, of course, differences which arise from the different number of valence electrons in the two groups of elements, but taking these into account there is a remarkable uniformity in the behavior of the two families. The ground-state configuration of the alkaline-earth metals consists of a rare-gas core with two electrons in the next s orbital (Table 9.1). A comparison of the first ionization potentials of the alkaline-earth metals, shown in Table 9.1, and the alkali metals, shown in Table 8.1, suggests that the 1+ oxidation state might be easily attained for the former group. In reality, the 1+ valence is a chemically unimportant state for the alkaline-earth metals; only under unusual conditions have 1+ ions been observed in this family. For example, the electrolysis of conducting aqueous solutions using beryllium electrodes or of pyridine solutions using magnesium electrodes yields highly reducing solutions at the anode which are thought to

TABLE 9.1   Some properties of the alkaline-earth metals and their ions

Element[a]	Electronic configuration	Ionization potentials, eV			Electro-negativity	Standard electrode potential, V	Ionic radius, Å	$\Delta H_{hyd.}$, eV
		I	II	III				
Be (4)	[He] $2s^2$	9.32	18.21	153.85	1.5	1.69	0.31^b	-24.8
Mg (12)	[Ne] $3s^2$	7.64	15.03	80.21	1.2	2.37	0.65	-20.2
Ca (20)	[Ar] $4s^2$	6.11	11.87	51.21	1.0	2.87	0.94	-16.6
Sr (38)	[Kr] $5s^2$	5.69	10.98	—	1.0	2.89	1.10	-15.2
Ba (56)	[Xe] $6s^2$	5.21	9.95	—	0.9	2.90	1.29	-13.7
Ra (88)	[Rn] $7s^2$	5.28	10.10	—	—	2.92	1.50	—

^{a} Nuclear charge Z given in parentheses.
^{b} Estimated.

contain Be^+ and Mg^+, respectively. In addition, phase diagrams for the system $M-MX_2$ (M = Sr or Ba; X = Cl, Br, or I) indicate that the species M_2^{2+} exists in this system (1); the cation Mg_2^{2+} is reported to be the product of the reaction between metallic magnesium and molten $MgCl_2$ (2).

Even though the removal of a second electron from an alkaline-earth metal ion requires approximately twice the energy as is required for the first electron, stable compounds containing 2+ ions are formed. Apparently this increase in energy is more than compensated by lattice formation or by the solvation of these ions. The ions in this family are isoelectronic with the corresponding 1+ ions of the alkali metals. The increase in nuclear charge in going from an alkali metal to the next alkaline-earth metal gives rise to a smaller ion; since the charge on the alkaline-earth ions is also greater their charge densities are greater than those of the alkali-metal ions, as shown in Table 9.2. A consideration of the data in Table 9.2 strongly suggests that the simple compounds of the alkaline-earth metals should be even more ionic than the corresponding alkali-metal compounds. In comparable compounds the alkaline-earth ions would be less polarizable than the alkali-metal ions. However, the data in Table 9.2 indicate that higher electric fields would be generated by the alkaline-earth ions; that is, these ions would be more polarizing in their compounds than the isoelectronic alkali-metal ion. Thus, a clear understanding of the structures and energetics of the ionic compounds of the alkaline-earth metals can be obtained from the radius-ratio rules and the simple electrostatic arguments described in Chapter 2. In considering the structures of the ionic halides we must, of course, recognize that each alkali-metal cation has two counter ions present. This stoichiometry, which arises from the electronic configuration of the metal, requires a more complex structure (e.g., the fluorite or rutile structures) than those observed for the alkali halides. However, many binary compounds

TABLE 9.2 The charge density
on the alkali and alkaline-earth ions

Ion	Charge density, $C \text{ Å}^{-3} \times 10^{-19}$
Li^+	1.22
Na^+	0.42
K^+	0.16
Cs^+	0.12
Rb^+	0.08
Be^{2+}	25.6
Mg^{2+}	2.78
Ca^{2+}	0.92
Sr^{2+}	0.57
Ba^{2+}	0.36

of the alkaline-earth metals containing divalent ions exhibit the common sodium-chloride structure in which the cation exhibits a coordination number of 6 (Table 9.3). Beryllium forms a sufficiently smaller ion so that the lower 4-coordinate wurtzite or zinc-blende structures are preferred.

Although the alkaline-earth metals are more difficult to vaporize (Table 9.4) and ionize than the alkali metals, the higher hydration energies associated

TABLE 9.3 The crystal structures of some of the simple binary compounds of the alkaline-earth elements

Type of compound	Structure				
	Be^a	Mg^a	Ca^a	Sr^a	Ba^a
MO	wurtzite (2530)	NaCl (2802)	NaCl (2587)	NaCl (2430)	NaCl (1923)
MS	zinc blende (—)	NaCl (d.)	NaCl (d.)	NaCl (>2000)	NaCl (1200)
MSe	zinc blende (—)	NaCl (—)	NaCl (—)	NaCl (—)	NaCl (—)
MTe	zinc blende (—)	zinc blende (—)	NaCl (—)	NaCl (—)	NaCl (—)
MF_2	cristobalite (880)	rutile (~ 1350)	fluorite (~ 1350)	fluorite (~ 1300)	fluorite (1285)
MCl_2	chain (430)	$CdCl_2{}^b$ (715)	fluorite (773)	fluorite (870)	fluorite (960)

[a] Melting point in °C given in parentheses.

[b] A layer-type structure containing octahedral $CdCl_6$ groups linked by sharing edges to form infinite layers. Relatively weak forces exist between layers.

with the former ions lead to standard electrode potentials for both groups of elements that are nearly the same (cf. Tables 8.1 and 9.1).

TABLE 9.4   The physical properties of the alkaline-earth elements

Element	Melting point, °C	Boiling point, °C	ΔH_f, kcal(g-atom)$^{-1}$	ΔH_{sub}, kcal(g-atom)$^{-1}$	Hardness[a]	Density, g cm^{-3}
Be	1280	2770	2.34	3.33	—	1.86
Mg	651	1107	2.2	1.56	2.0	1.75
Ca	851	1437	2.2	2.00	1.5	1.55
Sr	800	1366	2.2	1.70	1.8	2.6
Ba	850	1637	1.83	1.82	—	3.6
Ra	960	1140	$(2.3)^b$	—	—	5.0

[a] Moh's scale.
[b] Estimated.

Of the alkaline-earth metals, radium is the least abundant, comprising $\sim 10^{-10}\%$ of the igneous rocks on earth; all the isotopes of radium are radioactive, ^{226}Ra, with a half-life of approximately 1600 years, being the most stable. The known chemical properties of radium are consistent with those of the heavier alkaline-earth metals. We shall address our discussion in this chapter only to the remaining alkaline-earth metals. In this discussion it should be noted that beryllium, like lithium within the alkali-metal family, exhibits properties that are markedly different from the heavier members of its family.

2 / THE ELEMENTS

2.1 / Physical Properties

The alkaline-earth metals are all distinctly metallic in appearance (Table 9.4). Beryllium is dark gray in color and strontium is brass yellow. The other members of the family possess the typical silvery white luster of metals. The four lightest members of the group exhibit close-packed structures. Beryllium and magnesium are hexagonal; calcium is hexagonal and face-centered; strontium is face-centered below 235°C. In contrast, barium crystallizes in the body-centered structure. Beryllium exhibits exceptional physical properties within this series of elements (Table 9.4). Unlike the alkali metals, the alkaline-earth metals vaporize to monomeric species; that is, there is no indication that diatomic or polyatomic species are formed.

2.2 / Solubility in Liquid Ammonia (*l*)

The alkaline-earth metals dissolve in liquid ammonia to form the blue solutions containing solvated electrons. The properties of these solutions, when their concentrations are expressed in terms of equivalents of alkaline-earth metal, parallel those of the alkali metals, described in Chapter 8. For example, the visible spectra of dilute solutions containing equivalent amounts of alkali metal and alkaline-earth metal are indistinguishable. This observation strongly supports the suggestion that the band at $\sim 14,500$ Å arises from transitions associated with the solvated electron and that the alkaline-earth metal atoms dissolve and lose two electrons in liquid ammonia:

$$M + (2x + y)NH_3 \longrightarrow M(NH_3)_y^{2+} + 2e(NH_3)_x^{-}. \qquad (1)$$

In the case of calcium, evaporation of the metal–ammonia solutions yields a substance with the formula $Ca(NH_3)_6$. Attempts to dissolve magnesium in pure ammonia give no indication of forming a blue solution; rather, upon prolonged contact with boiling ammonia, hydrogen is evolved. However, if a strong base such as $NaOCH_3$ or KNH_2 is present, a blue solution readily forms (*3*). This observation can be understood if one recognizes that Mg^{2+} has a relatively high charge density (Table 9.2). Thus, if any magnesium does dissolve according to Eq. (1), the Mg^{2+} formed would be expected to interact sufficiently strongly with the solvent to give ammonolysis products:

$$Mg(NH_3)_x^{2+} + NH_3 \rightleftharpoons Mg(NH_3)_{x-1}^{2+}(NH_2)^- + NH_4^+. \qquad (2)$$

Reactions analogous to Eq. (2) are common in water with small highly charged ions such as Al^{3+}. Thus, the dissolution of magnesium would produce solvated electrons [Eq. (1)] and NH_4^+ [Eq. (2)], which would, of course, react to form hydrogen:

$$NH_4^+ + e(NH_3)_x^- \longrightarrow (x + 1)NH_3 + \tfrac{1}{2}H_2. \qquad (3)$$

In the presence of a strong base the equilibrium in Eq. (2) can be suppressed to the point where the NH_4^+ concentration is negligible. Under these conditions the reaction described by Eq. (3) does not occur and the solution turns blue. There are also reports that under favorable conditions beryllium can be made to dissolve in liquid ammonia. Based upon observations on the dissolution of magnesium, we might expect that similar arguments would be valid for the dissolution of beryllium since the charge density of Be^{2+} is significantly greater than that of Mg^{2+} (Table 9.2).

In general the chemical reactions of liquid-ammonia solutions of alkaline-earth metals follow those described for the alkali metals in Chapter 8, although there have been some reports suggesting that the course of certain reduction processes depends upon the nature of the cation present in solution. Little is known about these apparent differences in mechanism.

2.3 / Preparation

Like the alkali metals, the alkaline-earth metals are prepared by using a very powerful reducing agent with compounds containing these elements in their 2+ oxidation states. Generally, electrolytic reduction processes employing suitable fused salts are the most successful methods known for the preparation of the alkaline-earth elements. Both beryllium and magnesium chlorides are poor conductors of electricity, so NaCl is usually added to the melt to increase the electrical conductivity of the system; electrolysis of mixtures like these under a controlled potential yields the most easily reduced species at the cathode. In both instances the alkaline-earth metal ions are reduced before Na^+. The low conductivity of $BeCl_2$ and $MgCl_2$ in the molten state is not surprising in view of the fact that both metal ions have significantly higher charge densities than the remainder of the alkaline-earth ions (Table 9.2). Apparently even the reasonably small and usually nonpolarizable chloride ion can be distorted sufficiently by these highly charged metal ions to form covalent structures.

The heavier alkaline-earth metals can be produced in chemical reduction processes under certain favorable conditions. Thus, magnesium can be prepared from calcined dolomite

$$MgCO_3 \cdot CaCO_3 \xrightarrow{\Delta} MgO \cdot CaO + 2CO_2, \tag{4}$$

a naturally occurring mineral containing a mixture of $MgCO_3$ and $CaCO_3$, by reduction with silicon in the form of ferrosilicon:

$$2MgO \cdot CaO + Si \rightarrow 2Mg + Ca_2SiO_4. \tag{5}$$

This reduction process can be successfully used to produce magnesium if a low pressure is maintained in the reaction system and if a $MgO \cdot CaO$ mixture is used rather than pure MgO. The mixture yields a silicate as a product [Eq. (5)] rather than SiO_2. Under these conditions the free energy for the process is favorable at $\sim 1100°$. At higher temperatures MgO can be reduced with carbon

$$MgO + C \underset{}{\overset{2000°}{\rightleftharpoons}} Mg + CO \tag{6}$$

if the high-temperature reaction is rapidly quenched. Under the conditions of this reduction, the point of equilibrium lies well toward the right in Eq. (6).

Calcium can be prepared by reducing CaO with metallic aluminum:

$$6CaO + Al \rightarrow 3Ca + Ca_3AlO_6. \tag{7}$$

Again an advantage is gained in this process because all the products and reactants except calcium metal are nonvolatile under the conditions employed.

2.4 / Chemical Properties

The chemical behavior of beryllium, the first member of the alkaline-earth family of elements, is markedly different from that of the other members of this series, although qualitatively each member gives essentially the same type of reaction. This behavior is similar to that observed for the alkali metals (Chapter 8).

The alkaline-earth metals react readily with the oxidizing gaseous elements to form binary compounds containing the metal in the 2+ oxidation state:

$$2M + O_2 \rightarrow 2MO \tag{8}$$

$$M + X_2 \rightarrow MX_2 \tag{9}$$

$$X = F, Cl, Br, I$$

$$M + S \rightarrow MS. \tag{10}$$

In general these reactions occur more readily with the heavier members of the series and the more powerful oxidizing agents. For example, beryllium does not react with cold Cl_2, Br_2, or I_2 but a vigorous reaction occurs if these systems are heated; magnesium reacts very readily with gaseous F_2 and Cl_2 at room temperature, whereas I_2 vapor reacts markedly less vigorously at 600°. Nitrogen reacts with all of the elements to form the corresponding nitrides

$$3M + N_2 \rightarrow M_3N_2, \tag{11}$$

but the reaction with beryllium requires a markedly higher temperature (1000°C) than is needed for the other members of the series. Hydrogen reacts with all of the alkaline-earth metals except beryllium to form the corresponding hydrides

$$M + H_2 \rightarrow MH_2, \tag{12}$$

$$M = Mg, Ca, Sr, Ba,$$

which are salt-like in character (Chapter 6). The alkaline-earth metals, except beryllium, are sufficiently strong as reducing agents to react with many oxygen-containing compounds to give oxides; thus, magnesium reacts with B_2O_3 to give a mixture of MgO and Mg_3B_2.

All of the alkaline-earth metals except beryllium react with water to give hydrogen and the corresponding hydroxide:

$$M + 2H_2O \rightarrow M(OH)_2 + H_2. \qquad (13)$$

The reaction with magnesium proceeds slowly because of the presence of a relatively impervious layer of MgO; amalgamation of the metal prevents the formation of this protective coating and the reaction proceeds vigorously. Alcohols react with the alkaline-earth metals in a similar manner:

$$M + 2ROH \rightarrow M(OR)_2 + H_2. \qquad (14)$$

In the case of magnesium, elevated temperatures or activation by iodine is required to make this reaction occur rapidly. Although water has no action on beryllium, dilute aqueous acids

$$Be + 2HCl \xrightarrow{aq} Be^{2+}_{(aq)} + 2Cl^- + H_2. \qquad (15)$$

or concentrated sodium hydroxide solution will dissolve the metal to give hydrogen.

Aqueous solutions of beryllium salts are extensively hydrolyzed. Since most beryllium salts can be prepared as very stable hydrates, $BeX_2 \cdot 4H_2O$ (Table 9.5), it is reasonable to assume that these substances are best formulated

TABLE 9.5   The numbers of water molecules of crystallization associated with alkaline-earth metal compounds

Type of compound	Molecules of water per formula unit				
	Be	Mg	Ca	Sr	Ba
MCl_2	4	12, 8, 6, 4	6, 4, 2, 1	6, 2, 1	2
MBr_2	4	10, 6, 4	6	6, 2	2, 1
MI_2	—	10, 8, 6	8, 6	6, 2, 1	6, 2, 1
MCO_3	4	5, 3, 1	6, 1	0	0
$M(NO_3)_2$	4	9, 6, 2	4, 3, 2	4	0
MSO_4	4, 2	12, 7, 6, 1	2, 1/2	0	0
$M(ClO_4)_2$	4	6, 4, 2	2	4, 2	3, 1

as $[Be(H_2O)_4]X_2$ on the basis of their properties. For example, $Be(ClO_4)_2 \cdot 4H_2O$ loses no water until the temperature at which it begins to decompose completely. Beryllium chloride dissolves readily in water with the evolution of heat. The compound $BeCl_2 \cdot 4H_2O$ can be readily isolated from this mixture. This compound does not lose water after several months exposure to P_2O_5 under high vacuum. Although a detailed understanding of the nature of aqueous solutions containing divalent beryllium is not available, the properties of these solutions, e.g., high viscosity, low cation mobility, and abnormal freezing-point depression, suggest that these ions are more highly hydrated than other divalent species. Depending upon the counter-ion present, several classes of species are thought to be present in aqueous solutions of beryllium salts. For example, if the counter-ion is chloride or sulfate a series of complex equilibria incorporating $[Be \cdot aq]^{2+}$, $[Be_2(OH) \cdot aq]^{3+}$, $[Be_3(OH)_2 \cdot aq]^{4+}$, $[Be_3(OH)_3 \cdot aq]^{3+}$, $[Be_3(OH)_4 \cdot aq]^{2+}$, $[Be_6(OH)_8 \cdot aq]^{4+}$, and $[Be_6(OH)_9 \cdot aq]^{3+}$ have been described (4); it should be noted that the counter ion does not appear in these species. On the other hand, hydrolysis of Be^{2+} in aqueous fluoride solutions gives the mononuclear species $BeF_n^{(2-n)+}$ ($n = 1, \ldots, 4$) (5).

In general, the alkaline-earth metals heavier than beryllium are not as extensively hydrated in solution. Thus, like beryllium perchlorate, $Mg(ClO_4)_2$ forms a tetrahydrate $[Mg(ClO_4)_2 \cdot 4H_2O]$; indeed, it is possible to prepare several other hydrates of $Mg(ClO_4)_2$. All of these compounds readily lose water at 250° to form the anhydrous salt.

The marked difference in behavior of Be^{2+} when compared with the corresponding ions of the other members of this family can be undoubtedly correlated with the large difference in charge density of the ions. Although there is little direct evidence concerning the nature of the alkaline-earth metal ions in aqueous solution, the trend in ionic mobilities (Be, 30; Mg, 55.5; calcium, 59.8; strontium, 59.8; barium, 64.2 at 25°) indicates that divalent beryllium is the most solvated of these ions.

3 / COMPOUNDS OF THE ELEMENTS

Except for beryllium, many of the simple compounds of the alkaline-earth elements can best be described as ionic. For example, the physical properties and lattice structures of magnesium, calcium, strontium, and barium halides and chalcogenides, shown in Table 9.3, are characteristic of ionic compounds, and in general the structures can be understood in terms of electrostatic interactions between ions. Beryllium compounds, on the other hand, exhibit structures and properties associated with substances which form continuous lattices incorporating covalent bonding such as is found in SiO_2 and diamond. It would be expected that Be^{2+} is the most polarizing of the cations in this family. Accordingly, it is very doubtful if any stable beryllium compound contains the Be^{2+} ion; rather, most compounds containing divalent beryllium have characteristics associated with covalently bonded systems. Thus, not only the organometallic derivatives of beryllium are covalent. Other simple binary compounds containing anions that can be polarized by the Be^{2+} ion, e.g., $BeCl_2$, exhibit a considerable amount of covalent character.

3.1 Organometallic Compounds (II)

The similarity of the electronegativities of calcium, barium, and strontium with the alkali metals suggests that their simple alkyl derivatives (MR_2) would be predominantly ionic in character. Although extensive work has not been done in this area, the data available support this suggestion. For example, the dimethyl derivatives of calcium, strontium, and barium are reported to be white crystalline compounds that decompose under vacuum at 400°, are rapidly hydrolyzed, and burn readily when exposed to air.

In contrast, the markedly greater electronegativities of beryllium and magnesium (Table 9.1) suggest that the organic derivatives of these elements should be more covalently bonded. This prediction correlates with the observed properties of the methyl derivatives of these elements. For example, when heated under vacuum the compounds $Mg(CH_3)_2$ and $Be(CH_3)_2$ sublime before they melt.

The organometallic derivatives of the alkaline-earth elements can, in general, be prepared by the methods described for the alkali organometallic compounds as described in Chapter 8. Undoubtedly the most useful of the organometallic derivatives from the standpoint of synthesis is the so-called Grignard reagent, the mixture formed when magnesium metal is allowed to react with organic halides in ether solution:

$$RX + Mg \xrightarrow{(C_2H_5)_2O} RMgX. \tag{16}$$

Magnesium occupies a rather unique position with respect to the reactivity of the organometallic compounds it forms. Metals that are much more electronegative than magnesium form more reactive derivatives and, of course, are more difficult to handle. Accordingly, the Grignard reagent has achieved a position of great synthetic usefulness because of the relative simplicity of synthesis and the ease with which it can be used. In addition to the reactions that Grignard reagents undergo with organic functional groups, they, together with the organolithium compounds described in Chapter 8, are useful intermediates for the preparation of a variety of other organometallic compounds by halogen-exchange reactions:

$$AsCl_3 + 3CH_3MgI \longrightarrow As(CH_3)_3 + 3MgClI \tag{17}$$

$$SiCl_4 + 4CH_3MgI \longrightarrow Si(CH_3)_4 + 4MgClI. \tag{18}$$

The nature of the Grignard reagent in solution has been the source of considerable controversy (*III*). Ether solutions containing Grignard reagents become very viscous if the solvent is evaporated; however, it is virtually impossible to remove all the ether from these solutions. This behavior suggests that the solvent is chemically combined and that association of the species present occurs as the solvent is removed. Molecular-weight determinations of several Grignard reagents by the boiling-point method verify that association occurs in these solutions (*6*). The results show that the degree of association ranges from about 1 to 4. Since ether is a Lewis base, these results suggest that species such as those shown in Fig. 9.1 may be present.

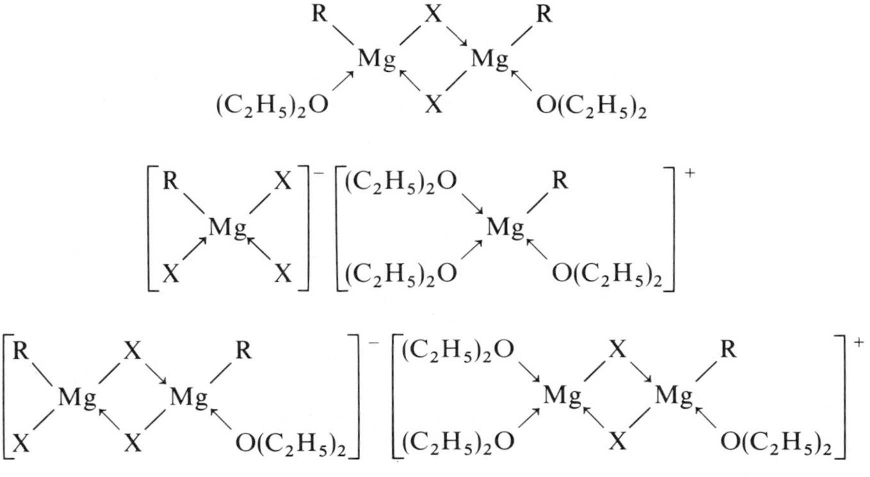

FIG. 9.1 Solutions of the Grignard reagent are associated and conduct an electric current; magnesium is found in both the cationic and anionic species. Shown here are some species which may be present in these solutions.

Addition of dioxane to ether solutions of the Grignard reagent precipitates dioxane complexes of magnesium halides; dialkylmagnesium compounds can be isolated from the remaining solution. Ether solutions containing equivalent amounts of R_2Mg and MgX_2 behave in the same way as solutions prepared from RX and magnesium. With no additional information one might be tempted to suggest that equilibria of the type

$$2RMgX \rightleftharpoons R_2Mg + MgX_2 \tag{19}$$

are present in solution (*7, 8*). However, tracer experiments indicate that practically no exchange occurs between $MgBr_2$ and $(C_2H_5)_2Mg$ in ether solution. Electrical conductivity data show that ions are present in Grignard solutions and that both anions and cations contain magnesium; these observations are consistent with the species shown in Fig. 9.1. Whatever the nature of the species that are present in Grignard solutions, they react as if the carbon atom is negative with respect to magnesium ($^{\delta-}C—Mg^{\delta+}$). The nature of the products formed when Grignard reagents react can be understood on the basis of the attack of the negative carbon atom on the positive part of the other reagent, although this is not necessarily the mechanism of the reaction (IV):

$$\overset{\delta-}{R}—\overset{\delta+}{Mg}X + \overset{\delta-}{Cl}—\overset{\delta+}{Si}Cl_3 \rightarrow RSiCl_3 + MgXCl. \tag{20}$$

3.2 / Coordination Behavior

Crystallographically, the coordination number of the alkaline-earth metals in predominantly ionic compounds is governed by the radius-ratio rules developed in Chapter 2. Ionic compounds containing a divalent anion (MX) possess octahedral coordination; the metal ions in compounds with monovalent anions (MX_2) still exhibit octahedral coordination, but the crystal structures are necessarily more complex than the sodium-chloride structure because of the 1 : 2 ratio of cations to anions. Hence the fluorite structure (Chapter 2) is predominant in these compounds.

Compounds of beryllium which are essentially covalently bound giant molecules (Table 9.3) show predominantly four-coordination; the cristobalite structure of BeF_2 is basically the same as the zinc-blende structure (Chapter 2). Magnesium compounds such as MgF_2 and $MgCl_2$ which crystallize to form giant molecules exhibit the higher octahedral coordination expected for a larger metal ion.

In the more conventional covalently bound species, the coordination number of beryllium is not known to exceed 4. Thus, the hydrates of beryllium salts are four-coordinate (Table 9.6); other neutral Lewis bases form compounds in which the beryllium atom is probably four-coordinate. Anions which can act as Lewis bases, such as OH^-, $OC_2H_5^-$, F^-, and Cl^-, form anionic beryllium-containing compounds in which beryllium is four-coordinate. The known structures of several species (e.g., $Be(H_2O)_4^{2+}$ and BeF_4^{2-}) suggest that all four-coordinate beryllium compounds are tetrahedral.

Although compounds of beryllium with coordination numbers greater than 4 have not been identified, compounds with a lower coordination number are known (Table 9.6). Dialkylberyllium compounds react with trimethylamine to form adducts of the type $R_2Be·N(CH_3)_3$. The compound $(CH_3)_2Be·N(CH_3)_3$ is monomeric in the vapor phase while $[(CH_3)_2CH]_2Be·N(CH_3)_3$ is monomeric in benzene solution. Clearly the beryllium atom in

TABLE 9.6 Some examples of compounds of the alkaline-earth metals exhibiting 3-, 4-, 6-, and 8-coordination

Coordination number	Be	Mg	Ca	Sr	Ba
3	$R_2Be \cdot N(CH_3)_3$ $Be[OCH(CH_3)_2]_2 \cdot N(CH_3)_3$ $(RBeOR')_2$ $[RBeN(CH_3)_2]_3$ $KBeF_3$	$Mg(C_2H_5)_2 \cdot (C_2H_5)_2O$ $Li[Mg(C_6H_5)_3]$			$Ba(acac)_2$
4	$BeCl_2 \cdot 2(C_2H_5)_2O$ $Be(CH_3)_2 \cdot 2N(CH_3)_3$ $BeCl_2 \cdot 2CH_3CN$ $K_2[Be(OH)_4]$ $K_2[Be(OC_2H_5)_4]$ $K_2[BeCl_4]$ $K_2[BeF_4]$ $(BeX_2)_n{}^a$	$Mg(C_2H_5)Br \cdot 2(C_2H_5)_2O$ $MgBr_2 \cdot 2(C_2H_5)_2O$ K_2MgCl_4 K_2MgF_4 $[Mg(H_2O)_4](ClO_4)_2$ $[Mg(NH_3)_4]Cl_2$ $(MgR_2)_n$	$Ca(acac)_2$	$[Sr(H_2O)_4](NO_3)_2$ $[Sr(H_2O)_4](ClO_4)_2$ $Sr(acac)_2$	
6		$[Mg(NH_3)_6](ClO_4)_2$ $[Mg(H_2O)_6](ClO_4)_2$	$Ca(H_2O)_6Cl_2$	$[Sr(H_2O)_6]Cl_2$	$[Ba(H_2O)_6]I_2$
8				$Sr(OH)_2 \cdot H_2O^b$	

^{a} X = Cl, F, CH_3.
^{b} Strontium surrounded by oxygen atoms at the corners of a bicapped trigonal prism.

these species is three-coordinate. A similar amine adduct has been formed with beryllium isopropoxide, $Be[OCH(CH_3)_2]_2 \cdot N(CH_3)_3$. The reaction of dialkylberyllium compounds with alcohols liberates hydrocarbons and compounds which have a stoichiometry that indicates the presence of three-coordinate beryllium:

$$2R_2Be + 2R'OH \rightarrow (RBeOR')_2 + 2RH. \tag{21}$$

The compounds RBeOR′ are dimeric and probably possess the alkoxide bridge structure shown in Fig. 9.2. The reaction of dialkylberyllium compounds with dimethylamine also yields a hydrocarbon and a three-coordinate beryllium compound

$$3R_2Be + 3NH(CH_3)_2 \rightarrow [RBeN(CH_3)_2]_3 + 3RH \tag{22}$$

which is trimeric and probably possesses the cyclic structure shown in Fig. 9.2. In this cyclic structure $RBN(CH_3)_2$ units are formally linked by

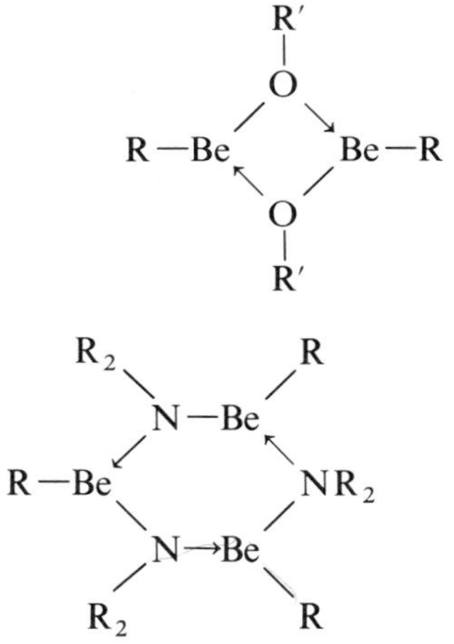

$R = CH_3, CH(CH_3)_2$

$R' = CH_3$

$R = CH_3, CH(CH_3)_2$

FIG. 9.2 Examples of dimeric $[(RBeOR')_2]$ and trimeric $[(RBeNR_2)_3]$ beryllium derivatives.

coordinate-covalent bonds from the lone electron pair on the nitrogen atom, making the latter four-coordinate, as in ammonium salts. In addition to the three-coordinate compounds mentioned here, the phase diagram of $KF-BeF_2$ mixtures contains evidence for the existence of $KBeF_3$ which might be expected to be three-coordinate; however, an x-ray structure of $CsBeF_3$ shows the presence of BeF_4 tetrahedra linked in chains by fluorine bridges (9).

There is no direct evidence for the existence of two-coordinate beryllium-containing compounds, but the physical properties of $[(CH_3)_3C]_2Be$ strongly suggest that this compound is monomeric. For example, di-t-butylberyllium possesses a vapor pressure of 35 torr at 25°, which is comparable to that of the branched-chain C_9 hydrocarbon. Beryllium chloride, which is a polymeric solid at room temperature, gives a monomeric vapor at about 750°. Presumably the species in the vapor state under these conditions are linear $BeCl_2$ molecules. However, the structures of the monomeric alkaline-earth halides show some interesting features that are not totally understood. The infrared spectra of the molecular species BeX_2 (X = F, Cl, Br, I) isolated in an inert matrix indicate that these compounds are linear (*10*). On the other hand, the fluorides of magnesium (158°), calcium (140°), strontium (108°), and barium (100°) exhibit distinctly nonlinear structures using the same experimental technique (*11*).

As is the general observation, the higher coordination numbers are more prevalent for the heavier members of this family (Table 9.6), with octahedral coordination being the most common of these. The crystal structure of

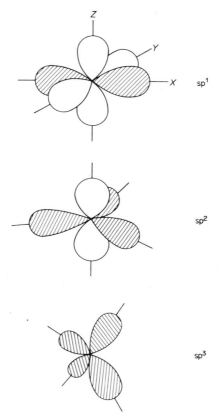

FIG. 9.3 Beryllium should be capable of forming bonds using *sp*, *sp*², and *sp*³ hybrids.

$CH_3MgBr\cdot3THF$ (THF = tetrahydrofuran) shows the presence of five-coordinate magnesium atoms (*12*); sevenfold coordination is known for strontium in SrI_2 and $Sr(OH)_2$ (*13*), while the latter compound forms a monohydrate $Sr(OH_2)\cdot H_2O$ in which the metal is eight-coordinate (*14*). The cation in the structure of $Ba(N_3)_2$ is surrounded by nine azide ions arranged in the form of a right triangular prism (*15*).

3.3 / Covalent Bonding

Theoretically, it should be possible for beryllium atoms to form tetrahedral, trigonal planar, and linear structures by employing sp^3, sp^2, and sp hybrid orbitals, respectively (Fig. 9.3). Thus it should not be surprising that the geometry of covalently bound beryllium species follows closely that observed for carbon compounds since these atoms possess valence orbitals with the same characteristics. Since beryllium has only two valence electrons the array of compounds characteristic of carbon chemistry is not observed.

Although three- and two-coordinate beryllium compounds are known under special conditions, beryllium overwhelmingly prefers four-coordination. Thus, $BeCl_2$, BeF_2, and $Be(CH_3)_2$ are polymeric species. Beryllium dichloride is a colorless crystalline compound which sublimes readily; the vapor is associated and the fused compound is a very poor conductor of electricity. X-Ray diffraction studies of solid $BeCl_2$ show the presence of chains of beryllium atoms with chlorine bridges (Fig. 9.4). In the molten

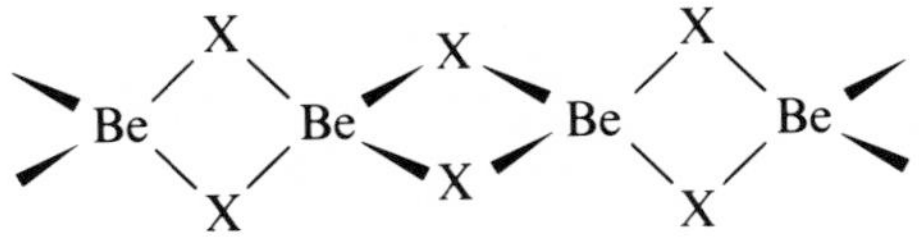

	X = Cl		X = CH_3	
	Exp.	Calc.	Exp.	Calc.
d(Be —X), Å	2.02	1.88	1.93	1.66
d(Be —Be), Å	2.65	1.78	2.09	1.78
$\angle$ XBeX	98°	—	114°	—
$\angle$ BeXBe	82°	—	66°	—

FIG. 9.4 Beryllium compounds are polymeric in the solid state.

state beryllium difluoride is also a poor conductor of electricity; this compound crystallizes in the β-cristolbalite structure similar to that of quartz. The properties of dimethylberyllium are very similar to those of $BeCl_2$. Dimethylberyllium is a volatile solid and the vapor is associated. X-Ray

diffraction studies of the solid indicate a chain structure that is re
similar to that of $BeCl_2$ (Fig. 9.4). The Be—Be distance as well as the distance
between the beryllium atom and the bridging atom are abnormally long for
single covalent bonds (Fig. 9.4). It should be emphasized that the positions
of the hydrogen atoms have not been determined in this structure. Many
beryllium alkyls have characteristics similar to those of dimethylberyllium,
although extensive structural information is not available. The compound
$[(CH_3)_3C]_2Be$ is a monomer in both the liquid and vapor states (16); the
structure in the vapor phase contains a linear C—Be—C skeleton with a
Be—C bond distance of 1.70 Å.

Thus, on the surface, beryllium atoms in the polymeric structures of
BeF_2, $BeCl_2$, and $Be(CH_3)_2$ have all achieved four-coordination. For BeF_2
the environment is tetrahedral and can be easily understood from a theoretical
standpoint by assuming that each beryllium atom is sp^3 hybridized; two
fluorine atoms are formally bonded by shared-electron-pair bonds, and two
are bonded by coordinate-covalent bonds (Fig. 9.5). The four-coordinate

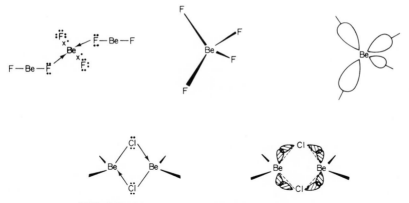

FIG. 9.5 The structures of $(BeF_2)_x$ and $(BeCl_2)_x$.

chain structure is distorted tetrahedral, but the presence of unshared electron
pairs on the chlorine atom still permits a bridging structure involving coordinate-covalent bond formation (Fig. 9.5). Several theoretical arguments
can be made to account for the structure of polymeric $BeCl_2$. Either the
hybridization of beryllium is not sp^3 with the electron density of the bond
situated on the internuclear distance, or the beryllium atom exhibits the
usual sp^3 hybridization but the electron density is off the line of centers
(Fig. 9.5). In the latter interpretation the angles between the nuclei shown
in Fig. 9.5 have no direct relationship to the hybridization scheme of the
atoms that are bonded.

Although the structure of $Be(CH_3)_2$ is consistent with four-coordinate beryllium the theoretical interpretation is more difficult because the bridging group is saturated. That is, once a methyl group is bonded to another atom, the usual bonding arguments do not provide for an additional bond. In other words, methyl groups are usually considered monovalent. A similar problem occurs in the interpretation of the structures of lithium alkyls (Chapter 8). In an attempt to circumvent the theoretical difficulties implied by the structure shown in Fig. 9.4, a new type of bond—the three-centered bond—was developed (*V*). Up to this point in our discussion, covalent bonds were described in terms of the overlap of two orbitals which involve a total electron density equivalent to two electrons. A three-centered bond involves the overlap of three orbitals to form a molecular orbital which can contain a maximum of two electrons (Fig. 9.6). In this view the theoretical interpretation of bonding in polymeric $Be(CH_3)_2$ (Fig. 9.4) consists of three-centered bonds, formed by overlap of two orbitals from adjacent

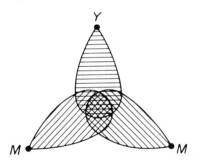

FIG. 9.6 Schematic designation of a three-centered bond.

beryllium atoms and an orbital from the methyl carbon atom (Fig. 9.7). Presumably the orbitals involved from each atom are sp^3 hybrids; the carbon atom still forms σ bonds to hydrogen using three sp^3 orbitals. It must be emphasized again that the position of the hydrogen atoms in $[Be(CH_3)_2]_x$ has not been determined. As is the case in the previous interpretation of the structure of $BeCl_2$, the electron density in the three-centered bond is not along the internuclear distance; accordingly the "bond angle" determined experimentally is not necessarily directly related to the hybridization schemes involved. Presumably a three-centered bond can also be invoked to interpret the structure of $(BeCl_2)_x$. Multicentered bonds have also been used to interpret the structures of the polymeric lithium alkyls described in Chapter 8.

In contrast to the coordination behavior of beryllium, the heavier alkaline-earth metals form six-coordinate as well as four-coordinate compounds (Table 9.6). Only magnesium shows indirect evidence for the formation of three-coordinate compounds in species such as $Mg(C_2H_5)_2 \cdot (C_2H_5)_2O$ and

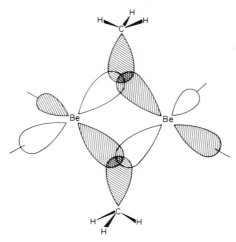

FIG. 9.7 The bridging methyl groups in $[Be(CH_3)_2]_x$ are thought to be involved in the formation of a three-centered bond.

$Li[Mg(C_6H_5)_3]$. If the magnesium atom in these species is surrounded by atoms in a trigonal planar arrangement, the theoretical interpretation for the bonding would involve the sp^2 hybrid orbitals on the metal atom. Four-coordinate species containing magnesium, calcium, strontium, and barium ions are nearly as common as those containing beryllium. As in the beryllium structures, formally neutral compounds of the type MX_2 can be associated with two molecules of neutral Lewis bases such as ether, water, or ammonia (e.g., $Mg(C_2H_5)Br·2(C_2H_5)_2O$, $MgBr_2·2(C_2H_5)_2O$, $[Mg(H_2O)_4](ClO_4)_2$ and $[Mg(NH_3)_4]Cl_2$) or anionic bases such as halide ion (e.g., K_2MgCl_4 and K_2MgF_4). Anions that can form chelate structures, such as acetylacetonate, give four-coordinate neutral species (Fig. 9.8). Magnesium alkyls exhibit

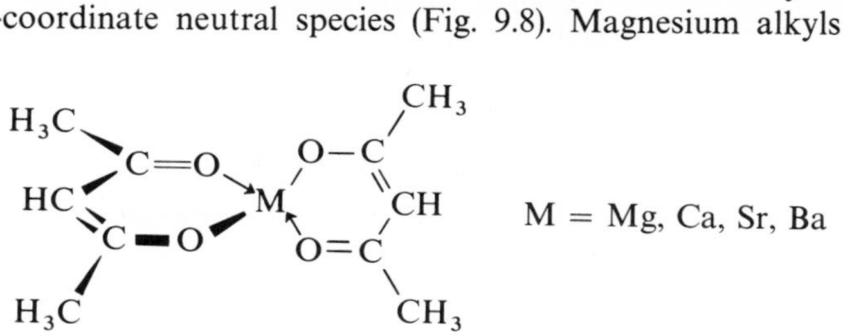

$$M = Mg, Ca, Sr, Ba$$

FIG. 9.8 Acetylacetonates of the alkaline-earth metals contain 4-coordinate metal atoms.

many of those properties of $Be(CH_3)_2$ which led to the conclusion that this latter compound is polymeric. Although the structures of dialkylmagnesium compounds have not been determined, there is good reason to believe that their structures are similar to that of polymeric $Be(CH_3)_2$ (Fig. 9.4). The structure of four-coordinate tetrahedral compounds of the heavier alkaline-earth metals can be interpreted in the same way as the structures of the corresponding beryllium compounds. That is, sp^3 hybrid orbitals on the metal atoms can be used to form either two- or three-centered bonds.

Six-coordinate species (Table 9.6) involve the alkaline-earth metal atoms surrounded by atoms in an octahedral arrangement. In these cases the Lewis base is a neutral molecule (NH_3 or H_2O) and the coordinated metal ion is positively charged. If vacant orbitals on the metal are involved in the bonding in these compounds, they must consist of d^2sp^3 hybrids which possess octahedral symmetry (Chapter 4). It is interesting to note that six-coordinate species have not been observed for beryllium, which does not have d orbitals of sufficiently low energy available for this type of hybridization.

REFERENCES

1. A. S. Dworkin, H. R. Bronstein, and M. A. Bredig, *J. Phys. Chem.*, **72**, 1892 (1968).
2. M. Krumpelt, J. Fisher, and I. Johnson, *J. Phys. Chem.*, **72**, 506 (1968).
3. R. K. Quinn and J. J. Lagowski, *Inorg. Chem.*, **9**, 414 (1970).
4. E. Lanza, *Rev. Chim. Minerale*, **6**, 653 (1969).
5. T. E. Mesmer and C. F. Baes, Jr., *Inorg. Chem.*, **8**, 618 (1969).
6. F. W. Walker and E. C. Ashby, *J. Amer. Chem. Soc.*, **91**, 3845 (1969).
7. M. B. Smith and W. E. Becker, *Tetrahedron*, **23**, 4215 (1967).
8. D. F. Evans and V. Fazakerley, *Chem. Commun.*, 974 (1968).
9. H. Steinfink and G. D. Brunton, *Acta Cryst.*, **B24**, 807 (1968).
10. A. Snelson, *J. Phys. Chem.*, **72**, 250 (1968).
11. V. Calder, D. E. Mann, K. S. Seshadri, M. Allavena, and D. White, *J. Chem. Phys.*, **51**, 2093 (1969).
12. M. Vallino, *J. Organometal. Chem.*, **20**, 1 (1969).
13. C. S. Choi, *Acta Cryst.*, **B25**, 2638 (1968).
14. H. Bärninghausen and J. Weidlein, *Acta Cryst.*, **22**, 252 (1967).
15. C. S. Choi, *Acta Cryst.*, **B25**, 2638 (1968).
16. G. E. Coates, P. D. Roberts, and A. J. Downes, *J. Chem. Soc. A*, 1085 (1967).

COLLATERAL READINGS

I. W. L. Jolly, *Prog. Inorg. Chem.*, **1**, 235 (1959). An extensive review of the physical properties of metal–ammonia solutions.
II. B. J. Wakefield, *Advan. Inorg. Chem. Radiochem.*, **11**, 341 (1968). A review of the properties, preparation, and structure of the alkyl derivatives of the Group II metals.
III. E. C. Ashby, *Quart. Rev.*, **21**, 259 (1967). A review of the composition of Grignard reagents and the mechanisms of their reactions.
IV. R. M. Salinger, *Survey Prog. Chem.*, **1**, 301 (1963). A discussion of the structure of the Grignard reagent and the mechanisms of its reactions.
V. R. E. Rundle, *Rec. Chem. Prog.*, **23**, 195 (1962); *Survey Prog. Chem.*, **1**, 81 (1963). These papers contain sections in which the structural and theoretical aspects of compounds containing three-centered bonds appear.

STUDY QUESTIONS

1. Without reference to experimental data, predict the most probable geometry of the following species: (a) $[Mg(CH_3)_2]_2$; (b) $BeCl_4^{2-}$; (c) $Li[Mg(C_6H_5)_3]$; (d) $[Mg(H_2O)_6]ClO_4$; (e) $[RBeN(CH_3)_2]_3$; (f) $Be(OR)_2 \cdot N(CH_3)_3$.

2. Discuss the observation that BeS crystallizes in the zinc-blende structure.

3. Although the Group II elements commonly exhibit a valence of $2+$, there are a few indications that magnesium can form a $1+$ cation. Predict the stability of MgCl, assuming that it forms a crystal lattice of the same structure as NaCl. Indicate the numerical quantities that might be expected to be found in the literature and those which would have to be estimated. Indicate how you would estimate the latter.

4. Dimethylberyllium reacts with dimethylamine to form a $1:1$ complex which melts at $44°$ with the evolution of one mole of methane. Give the most probable structures of the $1:1$ complex and of the product formed when it melts. What is the hybridization of the beryllium atom in each compound?

5. Molten $BeCl_2$ and $MgCl_2$ are poor conductors of electricity, but the addition of NaCl increases the conductivity markedly. Describe the most probable structures of all the species present in both melts.

6. Sketch the most probable structures of the following species which have been thought to be among the major hydrolysis products of $BeCl_2$: $[Be_2(OH) \cdot aq]^{3+}$, $[Be_3(OH)_2 \cdot aq]^{4+}$, $[Be_3(OH)_3 \cdot aq]^{3+}$.

7. Predict the products formed when dimethylmagnesium reacts with the following substances: (a) H_2O, (b) $AlCl_3$, (c) $LiCH_3$, (d) $(C_2H_5)_2O$.

8. Predict the geometry of the species formed in the reactions given in Problem 7.

9. Give the idealized hybridization for the beryllium atom in the following species: $R_2Be \cdot N(CH_3)_3$, $(RBeOR')_2$, R_4Be^{2-}.

10. Arrange the following compounds in order of increasing ionic character of the bonds formed: $BeCl_2$, NaCl, $MgCl_2$, CsCl, $BaCl_2$.

GROUP III
REPRESENTATIVE
ELEMENTS

1 / INTRODUCTION

The ground-state configurations of the elements in this family are all of the type ns^2np^1. As might be expected from this electronic distribution one electron is relatively more easily removed than the other two (Table 10.1). The ionization potentials for aluminum, gallium, indium, and thallium fall within a relatively narrow range of values that is distinctly smaller than the corresponding ionization potential for boron. In general, boron exhibits the behavior expected for a nonmetal, whereas the other elements in this family are more metallic in their properties. The majority of the compounds of these elements contain the element in an oxidation state of $3+$, which corresponds to the number of valence electrons present. However, there are relatively few compounds in which a $3+$ ion as such is known to be present. In discussing some of the structures of the compounds which are best described as giant molecules (e.g., B_2O_3), it is convenient to assume that such

255

structures consist of an array of ions. Accordingly, ionic radii are assigned to the (hypothetical) $3+$ ions in this family using the methods described in Chapter 2. In addition, covalent radii and metallic radii (summarized in Table 10.1) have been assigned to these elements. Ionic compounds containing the elements of this family are known, but the elements are present as complex cations, as in $[Al(H_2O)_6]Cl_3$, or anions, as in $NaBH_4$. The bonding within these complex species is predominantly covalent.

TABLE 10.1   Some physical properties of the Group III elements

Element[a]	Electronic configuration	Ionization potential, eV			$r_{M^{3+}}$, Å	r_{cov}, Å	r_{met}, Å
		I	II	III			
B (5)	[He] $2s^2 2p^1$	8.30	25.15	37.92	0.20	0.82	0.98
Al(13)	[Ne] $3s^2 3p^1$	5.95	18.82	28.44	0.50	1.18	1.43
Ga (31)	[Ar] $3d^{10} 4s^2 4p^1$	6.0	20.4	30.6	0.62	1.26	1.41
In (49)	[Kr] $4d^{10} 5s^2 5p^1$	5.8	18.8	27.9	0.81	1.44	1.66
Tl (81)	[Xe] $5d^{10} 6s^2 6p^1$	6.1	20.3	29.7	0.95	1.48	1.71

[a] Nuclear charge Z in parentheses.

Aluminum, gallium, and indium form a number of compounds in which these elements exhibit an oxidation state less than $3+$. In most cases the species are formed under high-energy conditions (e.g., AlCl and AlO) but a few compounds, such as B_2Cl_4 and GaI_2, are known with a stoichiometry suggesting that compounds with oxidation numbers less than three are stable under ordinary conditions.

2 / THE ELEMENTS

2.1 / Physical Properties

The physical properties of boron are strikingly different from those of the other members of this family (Table 10.2). Crystalline boron is a black substance with a metallic luster, but it is a poor electrical conductor. Boron is usually produced in the form of a brown finely divided powder. The other elements in this family are typically metallic in their properties. Except for boron the Group III elements are relatively soft; indeed, gallium melts just above normal ambient laboratory temperatures. Since gallium also tends to supercool, it often remains a liquid below its melting point for long periods of time before crystallization sets in.

TABLE 10.2   Some physical properties of the Group III elements

Element	Melting point, °C	Boiling point, °C	Hardness[a]	Density, g cm^{-3}	E^0, V[b]
B	2300	2550	9.3	2.4	—
Al	660	2500	2.9	2.7	1.66
Ga	29.8	2070	1.5	5.93	0.53
In	156	2100	1.2	7.29	0.34
Tl	449	1390	—	11.85	-0.72[c]

[a] Moh's scale.

[b] For the process $M_{(s)} \rightleftharpoons M^{3+}_{(aq)} + 3e^-$.

[c] The 1+ oxidation state is also common for thallium. $E^0 = 0.34$ V for the process $Tl_{(s)} \rightleftharpoons Tl^+_{(aq)} + e^-$.

Elemental boron exists in several crystalline modifications, the most usual being the rhombohedral form. X-Ray studies show that the rhombohedral form contains clusters of boron atoms situated at the corners of a nearly regular icosahedron, as depicted in Fig. 10.1; these clusters are arranged in a slightly deformed close-packed structure. In subsequent discussions it will become apparent that this basic icosahedral structure persists in many boron-containing species. On the other hand, the structures of aluminum, indium, and thallium are all close-packed structures characteristic of metals (Chapter 2). The indium structure is slightly distorted.

Solid gallium exhibits the unusual, complex structure shown in Fig. 10.2. Each gallium atom has only one nearest neighbor at 2.43 Å; six other gallium atoms can be found at distances varying from 2.70 Å to 2.79 Å. X-Ray diffraction data on liquid gallium are different from those obtained from a simple close-packed liquid such as mercury, and it has been suggested that a structure similar to that in solid gallium may persist in the liquid state. Although gallium readily forms alloys with many metals, its solubility in mercury is only 1.3 % at 25°.

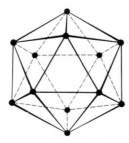

FIG. 10.1   Boron crystallizes as B_{12} units in which the atoms occupy the corners of an icosahedron.

FIG. 10.2 The structure of gallium as seen along the a axis of the orthorhombic crystal.

2.2 / Preparation

The elements of this group can be obtained from suitable compounds using a variety of reduction techniques. For example, an impure amorphous form of boron can be obtained from the reduction of B_2O_3 with magnesium:

$$B_2O_3 + 3Mg \rightarrow 2B + 3MgO. \tag{1}$$

The crude product is washed successively with strong base, HCl, and HF to remove unreacted starting materials and the product MgO. The amorphous brown powder which remains contains microcrystalline boron and metal borides, which are discussed in a subsequent part of this chapter. A crystalline form of boron can be obtained by thermal decomposition of BI_3 at 800–1000°

$$2BI_3 \rightarrow 2B + 3I_2 \tag{2}$$

or by the reduction of BBr_3 with hydrogen at 1300°

$$2BBr_3 + 3H_2 \rightarrow 2B + 6HBr. \tag{3}$$

Aluminum, a major item of commerce, is made by the electrolytic reduction of bauxite, which is a mixture of hydrated aluminum oxides. Bauxite is purified by the following process. It is dissolved in sodium hydroxide, the solution filtered, and the aluminum reprecipitated as $Al_2O_3 \cdot 3H_2O$ with CO_2. The hydrated alumina is heated to form Al_2O_3 which is then dissolved in molten cryolite (Na_3AlF_6) at $\sim 800°C$. This melt is electrolyzed, giving metallic aluminum at the cathode.

Gallium, indium, and thallium are often obtained most directly by electrolysis of aqueous solutions of their salts.

2.3 / Chemical Properties

Aluminum is the most reactive of the Group III elements. All forms of the element are invariably coated with a thin film of aluminum oxide that is impervious to most reagents. For example, metallic aluminum appears to be resistant to attack by dilute mineral acids, solutions containing metal ions which are less electropositive than aluminum, and by atmospheric oxygen. Amalgamation of aluminum destroys the coherence of the oxide film and the reactions expected of a very active metal occur readily. Thus, amalgamated aluminum reacts readily with water to form hydrogen:

$$Al(Hg) + 3H_2O \rightarrow Al(OH)_3(aq) + \tfrac{3}{2}H_2 + Hg. \tag{4}$$

Aqueous solutions of sodium chloride also destroy the coherence of the oxide film.

The fundamental chemical nature of the elements in this family changes with increasing atomic number from essentially nonmetallic through amphoteric to metallic; this change is illustrated in the reactions of these elements with aqueous acids and alkali. Boron does not react with hydrochloric acid, but the other elements in this family react to give hydrogen. On the other hand, boron, aluminum, and gallium react with aqueous alkali to form hydrogen, but indium and thallium do not, as would be expected for more metallic substances. At about 800° aluminum reacts with oxygen or air in spite of the presence of the inevitable oxide film; in air a mixture of oxide and nitride is formed. The other elements in this group are unaffected by the ordinary constituents of the atmosphere at ambient conditions; at elevated temperatures, all react with oxygen to form the expected oxides:

$$4M + 3O_2 \rightarrow 2M_2O_3. \tag{5}$$

Other elemental substances also can act as oxidizing agents with respect to the elements in Group III. Thus, the halogens react to form the corresponding trihalides:

$$2M + 3X_2 \rightarrow 2MX_3. \tag{6}$$

As would be expected, fluorine reacts the most readily of all the halogens; that is, the temperatures at which the boron trihalides are formed according to Eq. (6) increase in the order, room temperature, 410°, 700°, and 1250° for F_2, Cl_2, Br_2, and I_2, respectively. The Group III elements react with sulfur at elevated temperatures to form the corresponding sulfides

$$2M + 3S \rightarrow M_2S_3; \tag{7}$$

B_2Se_3 can be formed in the same manner.

TABLE 10.3   The borides of the elements

I a	I b	II a	III a	III b	IV a	IV b	V a	V b	VI a	VII a	VIII
		Be_4B, Be_2B, BeB_2, BeB_6, BeB_{12}				B_4C					
NaB_6		MgB_2, MgB_4, MgB_6, MgB_{12}		AlB_2, AlB_{10}, AlB_{12}		B_4Si, B_6Si		BP, $B_{13}P_2$			
(KB_6)	CuB_{22}	(CaB_2), CaB_4, CaB_6	ScB_2, ScB_6, ScB_{12}		(TiB), TiB_2		V_3B_2, VB, V_3B_4, VB_2	BAs, $B_{13}As_2$	(Cr_4B), Cr_2B, Cr_5B_3, CrB, Cr_3B_4, CrB_2, (CrB_6)	Mn_4B, Mn_2B, MnB, Mn_3B_4, MnB_2, MnB_4	Fe_2B, FeB; Co_3B, Co_2B, CoB; Ni_3B, Ni_2B, Ni_4B_3, NiB
(RbB_6)	AgB_2	SrB_6	YB_2, YB_4, YB_6, YB_{12}		ZrB_2, ZrB_{12}		Nb_3B_2, NbB, Nb_3B_4, NbB_2		Mo_2B, (Mo_3B_2), MoB, (MoB_2), Mo_2B_5, MoB_4	Tc_3B, Tc_7B_3, TcB_2	Ru_7B_3, $Ru_{11}B_8$, RuB, Ru_2B_5, RuB_2; Rh_7B_3, RhB; Pd_3B, Pd_5B_2

| (CsB₆) | AuB₂ | BaB₆ | LaB₄[a] LaB₆ | HfB₂ | Ta₂B Ta₃B₂ TaB Ta₃B₄ TaB₂ | W₂B WB W₂B₅ WB₄ | Re₃B Re₇B₃ ReB₂ | OsB₂ Os₂B₅ | IrB | PtB |

Actinides:[a]

| | ThB₃ ThB₄ | UB₂ UB₄ UB₁₂ | PuB PuB₂ PuB₄ PuB₆ |

[a]Tetraborides have been made of all of the rare earths with the exception of europium. Lutetium forms a diboride; terbium and lutetium also form dodecaborides.

The nonmetallic nature of boron is manifested in the formation of metallic borides (Table 10.3) (*1*). Virtually all of the metallic elements form compounds with boron in which it formally plays the role of an anionic species; zinc, mercury, thallium, tin, and lead are notable exceptions of metals that do not form borides. Metal borides can be prepared by direct combination; the reactions occur at about 2000°. Reduction of a metal oxide by a mixture of carbon and boron carbide (B_4C) has also been used to form metal borides:

$$2TiO_2 + B_4C + 3C \rightarrow 2TiB_2 + 4CO. \tag{8}$$

The electrolysis of a fused metal borate can also be made to yield the corresponding metal boride. In this process the metal liberated at the cathode reduces borate anions to elemental boron which then combines with excess metal. Metal borides can also be prepared by heating the appropriate amounts of the metal oxide with B_2O_3 in molten sodium or potassium metal at 1600°.

Metal borides with a wide variety of stoichiometric combinations have been observed; the more common types are shown in Table 10.3. The most prevalent formula types are M_2B, MB, M_3B_4, MB_2, and MB_6; some nonstoichiometric compounds also have been reported. The apparent profusion of borides can be understood on the basis of the boron species present (Table 10.4) (*1*). No single metal forms borides which display all of the main structural types shown in Table 10.4. Tantalum borides exhibit five of the six types: Ta_2B (isolated B atoms), Ta_3B_2 (B_2 units), TaB (single B chain), Ta_3B_4 (double B chain), and TaB_2 (layer B network).

The structures which contain individual boron atoms consist of sheets of tetrahedral arrangements of metal atoms with boron atoms in the holes between these sheets. Each boron atom is surrounded by eight nearest neighbors. The distances between boron atoms is much greater than the

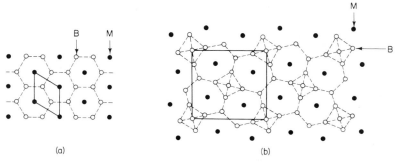

(a)   (b)

FIG. 10.3   (a) Borides with the composition MB_2 crystallize in a simple hexagonal structure; the boron atoms are arranged in sheets. (b) Borides of the type MB_4 crystallize in the tetragonal system. The boron atoms are arranged as octahedra which are linked together by other boron atoms. The projection shown is on the 001 plane.

TABLE 10.4   Classification of metal borides

Boron environment	Atomic ratio	Examples	$d(B-B)$, Å	Schematic representation
Isolated B atoms	M_4B	Mn_4B	>2.10	
	M_3B	Co_3B	>2.10	.....
	M_2B	Be_2B	3.30	
B_2 units	M_3B_2	V_3B_2	1.79	
Single B chains	MB	FeB	1.77	
Double B chains	M_3B_4	Ta_3B_4	1.85 (chain)	
			1.54 (pair)	
		Cr_3B_4	1.74 (chain)	
			1.51 (pair)	
Layers of B networks	MB_2	TiB_2	1.75	
Three-dimensional B networks	MB_4	UB_4	1.7	See Fig. 10.3
	MB_6	CaB_6	1.74	See Fig. 10.4
	MB_{12}	ZrB_{12}	1.75	See Fig. 10.4

distance expected (1.72 Å) for a single bond. It is highly unlikely that significant B—B interactions are present in these structures. In all the other structures internuclear B—B distances strongly suggest that B—B bonds are present. The units can be as small as B_2 (in M_3B_2), continuous single (MB) and double (M_3B_4) chains, or continuous two-dimensional B—B networks (MB_2). If we consider a B—B distance of about 1.72 Å equivalent to a single bond, some of the double chain structures are more weakly held together than others; for example, compare the data for Ta_3B_4 and Cr_3B_4 in Table 10.4.

More complex three-dimensional boron species are observed. The MB_4 species have the rather open three-dimensional network shown in Fig. 10.3 (2). The B_6 unit prevalent in the alkali-metal and alkaline-earth borides is an octahedron of boron atoms with B—B bonds. The B_6 octahedra are arranged as units in the body-centered CsCl-type lattice shown in Fig. 10.4. Concentrating on the metal atoms in these structures we find that they are surrounded by 24 boron atoms which are arranged in a cage-like structure. Compounds with the stoichiometry MB_{12} also possess the open framework shown by the MB_6 compounds (Fig. 10.4), the metal atoms being arranged in a face-centered lattice with each surrounded by a 24-atom boron cage. The

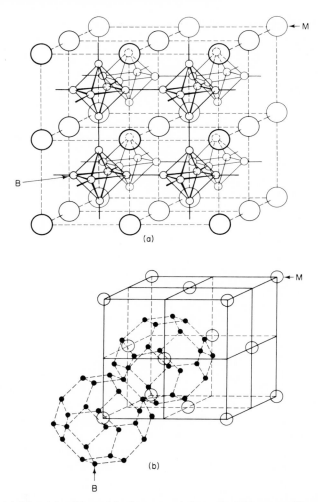

FIG. 10.4 (a) Metal borides with the general formula MB_6 crystallize in a cubic structure which contains boron B_6 octahedra. (b) Metal borides with the formula MB_{12} crystallize in a cubic structure. The structure contains a 24-atom cage of boron atoms arranged at the corners of a cubo-octahedron.

cage of boron atoms is described as a cubo-octahedron, which can be thought of as a cube with the corners cut off so that adjacent cuts intersect each other.

The nature of the bonding in the metal borides is not fully understood at present. It is generally assumed that electric charge is donated from boron to the metal in metal-rich systems such as Fe_2B and FeB (3). The reverse appears to be the case in the boron-rich compounds. In the case of compounds with a layer network of boron atoms (e.g., MB_2) molecular-orbital arguments (4) lead to a graphite-like structure for the boron framework if the metal is divalent (e.g., magnesium). Under such conditions the metal would formally be a closed-shell ion and the B_2-hexagonal network would be isoelectronic with graphite. If the metal in the MB_2 structure has three valence electrons, there should be one extra electron per atom and the B_2^{2-} network should be a semiconductor. The compound YbB_2 has been shown to be a semiconductor (5) and TiB_2 and ZrB_2 which are isomorphous with YbB_2 have ^{11}B nmr spectra which have been interpreted in terms of a graphite-like electronic structure for the boron atoms in these compounds (6).

Similar molecular-orbital interpretations (7) show that the B_6 unit is electron deficient; the B_6 octahedron requires 20 electrons, but six boron atoms provide only 18 electrons. Accordingly, if the B_6 unit is associated with a divalent metal, as in CaB_6, the system should be a nonconductor. But with metals that can attain a higher oxidation state, as in LaB_6, the boron network should be semiconducting (5).

In general the borides are very hard, possess high melting points, and exhibit some useful electrical properties. Table 10.5 gives an indication of

TABLE 10.5 **Some physical properties of typical borides**

Boride	Mp, °C	Resistivity, $\mu\Omega$-cm	Microhardness[a]
CaB_6	2230	0.1×10^6	2750
LaB_6	2230	15	2770
ZrB_2	3040	7–10	2250
ZrB_{12}	2680	60–80	—
VB_2	2400	16–38	2100
W_2B_5	2200	21–56	2650

[a] The Vickers pyramid hardness scale. As a point of reference, the microhardness of diamond is about 8000 on this scale.

the magnitude of some of the physical properties of a few typical borides. (As a point of reference, the resistivities of copper and iron are 7 and 10 $\mu\Omega$-cm, respectively.) NbB, Mo_2B, W_2B, and Ru_7B_3 are superconductors.

The transition-metal borides are generally thermally very stable and do not react with air to about 1200°. These compounds are fairly stable chemically, but they are attacked by strong oxidizing agents, such as peroxides, in a basic environment.

3 / COMPOUNDS OF THE ELEMENTS

Except for the species mentioned previously, the majority of compounds of the Group III elements are covalently bonded. Even in the cases where highly electronegative elements such as oxygen and fluorine are combined with the elements in this group, the compounds formed can be formulated as continuous structures; the question of whether such systems are best described as purely ionic (e.g., Al_2O_3) or as giant molecules (e.g., BN) cannot be determined simply on the basis of their physical properties. The chemical behavior of these compounds and their structures are readily correlated with the fact that the neutral compounds, with the expected valence of 3 (e.g., MX_3), are coordinately unsaturated.

In general, the discussion of the chemistry of these elements is divided into two parts: the chemistry of boron, and the chemistry of the remaining elements in this group. Where important, attempts are made to contrast and correlate the chemistry in these two parts.

All of the elements of Group III form the expected trivalent halogen compounds (Table 10.6) except thallium; the compound with the stoichiometry TlI_3 is isomorphous with RbI_3, which contains the I_3^- ion (Chapter 14). The other properties of TlI_3, such as its deep color, suggest that it should be formulated as $Tl^+I_3^-$.

TABLE 10.6   Melting points of the Group III trihalides[a]

	Trihalide				
Halogen	BX_3	AlX_3	GaX_3	InX_3	TlX_3
Fluorine	-130.7	1290[b]	950[b]	1170	550[c]
Chlorine	-107	193[d]	78	586	$60-70$[e]
Bromine	-46	98	122	436	—
Iodine	43	180	212	210	—

[a] In °C.
[b] Sublimes.
[c] In a fluorine atmosphere.
[d] At 2.24 atm.
[e] In a chlorine atmosphere.

3.1 / Boron Trihalides (*II*)

Physical Properties and Structure   All of the boron trihalides exist as simple molecular species with the expected formula BX_3. The electron-pair repulsion arguments given in Chapter 3 indicate that such molecular species containing trivalent boron should be planar, with XBX angles of

120°. The structural data for the boron halides given in Table 10.7 show that this general expectation is fulfilled. The bond-distance data also suggest that a degree of multiple bonding exists between halogen atoms and the central boron atom. The experimental bond distances shown in Table 10.7 are shorter than expected from the sum of the usual single-bond radii.

TABLE 10.7   Structural parameters for the boron halides

Compound	∠XBX	$d(B-X)$, Å Exp.	Calc.[a]
BF_3	120°	1.30	1.52
BCl_3	120°	1.76	1.87
BBr_3	120 ± 6°	1.87	1.99

[a] Based on the summation of the single-bond radii (see Table 3.8).

The electronic configuration of a Group III element in its ground state (ns^2np^1) would not permit the formation of trivalent compounds with a trigonal arrangement of bonds from the ground-state configuration. However, the use of sp^2 hybrid orbitals with the promotion of one of the paired s electrons (Chapter 4) gives the set of orbitals shown in Fig. 10.5 with the proper geometrical orientation. We might also expect the Group III elements to exhibit tetrahedral structures since the requisite number of valence orbitals are present to form the sp^3 hybrid orbitals shown in Fig. 10.5.

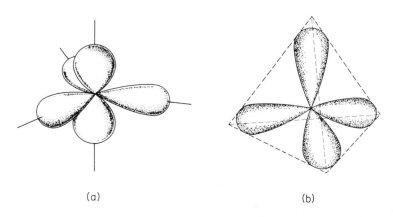

(a)   (b)

FIG. 10.5   Boron uses (a) sp^2 and (b) sp^3 hybrid orbitals for bond formation in many of its compounds.

In the case of the boron trihalides, the bond shortening shown in Table 10.7 is accounted for by introducing multiple bonding, the π component in these bonds corresponding to an internal coordinate-covalent bond; an unshared pair of electrons from the halogen atom is donated to the boron atom. In the valence-bond sense, the appropriate canonical forms are given by structures **1–4**. The most favorable orbital geometry for the halogen atoms for

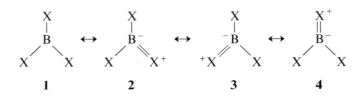

maximum overlap in the π system involves the unhybridized p orbital, which is perpendicular to the plane of the sp^2 hybrid orbitals. A filled p orbital on the halogen atom has a favorable orientation for maximum overlap. Thus the π component in the boron–halogen compound is an internal coordinate-covalent bond, since all of the density in the π system arises from one atom. Application of simple molecular-orbital calculations to the boron halides, using pure p orbitals for the halogens, shows that the π-bond energy decreases regularly with increasing atomic number of the halogen: BF_3, 57 kcal; BCl_3, 34 kcal; BBr_3, 28 kcal $(8, 9)$. Thus the bond shortening in the boron halides parallels the calculated values of the energy of the π component. On a qualitative basis, we might expect bond shortening in any Group III compound attached to an atom which carries unshared electron pairs.

As we shall see, much of the chemistry of boron and the other Group III elements can be correlated with the fact that the trivalent atoms are coordinately unsaturated and are, more or less, involved in π bonding.

Boron halides containing two different halogen atoms have not been isolated, but such compounds are known to exist in equilibrium when two pure halides are mixed:

$$BX_3 + BX_3' \rightleftharpoons BX_2X' + BX_2'X \tag{9}$$

The presence of mixed boron halides such as BCl_2Br and $BClBr_2$ in equilibrium has been established using spectroscopic $(10, 11)$ and mass-spectroscopic methods (12). Although little direct evidence exists on the nature of the processes involved in these exchange reactions, there have been suggestions that the transition state involves a bridged species (**5**). Attempts to isolate the mixed boron halides have always given the pure starting materials.

5

Chemical Properties The boron halides are among the strongest Lewis acids known, and they react with Lewis bases to form four-coordinate species:

$$BX_3 + :L \rightarrow BX_3L \qquad (10)$$

Molecules containing a variety of basic centers can react with the boron halides to form adducts. The Lewis base :L can be either a molecular or an ionic species; several types of addition compounds formed by the boron halides are shown in Table 10.8. Although most of the compounds listed in this table are adducts containing BF_3, similar compounds are known for the other boron halides (*13*).

TABLE 10.8 **Some adducts formed by the boron halides**[a]

$BF_3 \cdot NR_3$	$BF_3 \cdot H_2O$	$BF_3 \cdot R_2O$	$M^+[BF_4]^-$
$BF_3 \cdot PR_3$	$BF_3 \cdot (H_2O)_2$	$BF_3 \cdot RCO_2H$	$M^+[BCl_4]^-$
$BF_3 \cdot AsR_3$	$BF_3 \cdot ROH$	$BF_3 \cdot (RCO_2H)_2$	$M^+[BBr_4]^-$
$BF_3 \cdot RCN$	$BF_3 \cdot (ROH)_2$	$BF_3 \cdot R_2CO$	$M^+[BI_4]^-$
	$BF_3 \cdot H_2S$	$BF_3 \cdot RCHO$	$Na^+[BF_3OH]^-$
	$BF_3 \cdot RSH$	$BF_3 \cdot RCO_2R$	$M^+[BF_3Cl]^-$
	$BF_3 \cdot R_2S$		

[a] R = an organic group.

The structures of a variety of compounds of the type BX_3L show that the bond angles are close to the tetrahedral value. Thus, the planar sp^2 hybridization for boron has been transformed into the familiar sp^3 hybridization shown in Fig. 10.5. Presumably the coordinate bond is formed by overlap of an empty sp^3 orbital on boron with a filled orbital on the ligand.

In some instances where the base contains several sites for coordination, it is difficult to ascertain the location of the coordinate bond without additional information. Boron halides can sometimes form adducts containing two molecules of base. For example, boron trifluoride forms two adducts with water. The monoadduct, $BF_3 \cdot H_2O$, melts at 10.18° and is unionized in the solid state; its structure has not been determined but the compound probably exhibits the expected structure (**6**). The dihydrate, $BF_3 \cdot (H_2O)_2$, melts at

$$F_3B:OH_2 \qquad\qquad H_3O^+(BF_3OH^-)$$

$$\textbf{6} \qquad\qquad\qquad \textbf{7}$$

6.36° and has been formulated as **7**, the hydronium salt of hydroxyfluoroboric acid, $H^+(BF_3OH^-)$ (*III*). The formulation of the dihydrate as **7** is consistent with the monoadduct formed between BF_3 and NaOH; $BF_3 \cdot NaOH$ can be formulated as the sodium salt of **7**, i.e., $Na^+(BF_3OH^-)$. The constitution of the other diadducts of BF_3 which contain two molecules of base (Table 10.8) may be the same as **7**. In each case the Lewis base contains a proton that can be displaced by the boron halide which becomes associated with another molecule of base to form a salt:

$$BF_3 + 2ROH \longrightarrow [ROH_2{}^+][BF_3OR^-] \qquad (11)$$

$$BF_3 + 2RCO_2H \longrightarrow [RCO_2H_2{}^+][BF_3O_2CR^-]. \qquad (12)$$

The cation in such systems could be considered as the solvated proton in the corresponding nonaqueous solvent system, as described in Chapter 7.

Many compounds are known which contain the Group III halide as part of an anion (Table 10.8). The anion in these compounds can be imagined as formed from the neutral trihalide by the addition of an anionic Lewis base:

$$X_3M + :L^- \longrightarrow X_3ML^-. \qquad (13)$$

In some instances the preparative reaction literally corresponds to this process

$$R_4N^+Cl^- + BCl_3 \xrightarrow{CHCl_3} R_4N^+BCl_4{}^-, \qquad (14)$$

whereas other anion adducts are formed by indirect methods.

The relative strengths of the boron halides as Lewis acids has been determined in several ways (*14–17*), giving the following rather surprising order of acid strengths: $BBr_3 > BCl_3 > BF_3$. The reverse order might have been expected on the basis of the relative electronegativities of the halogens. The unexpected order of acid strengths of the boron halides is the reverse of the order of strengths of the π component of the B—X bond. The compound with the largest π component, BF_3, is also the weakest Lewis acid. In the process of adduct formation, the π component of the B—X bond in the planar BX_3 molecule must be broken to form the tetrahedral BX_3L species. The energies required for reorganizing the planar boron trihalide molecules to the corresponding tetrahedral species have been estimated as: BF_3, 48.3 kcal; BCl_3, 30.3 kcal; BBr_3, 26.2 kcal (8, 9).

Some typical reactions of the boron halides are summarized in Fig. 10.6. In many of these reactions an intermediate addition compound is formed which decomposes irreversibly with the loss of a small molecule. The ease

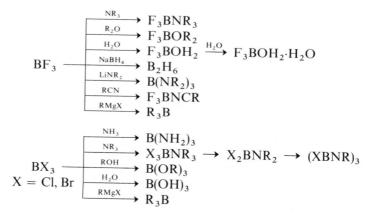

FIG. 10.6   Some reactions of boron halides.

with which these small molecules are eliminated depends to a large extent on the substituents of the Lewis base. If potentially acidic hydrogen atoms are present on the Lewis base, the adduct very often is difficult to isolate, hydrogen halide being readily eliminated. This type of reaction occurs more readily with the heavier boron halides. For example, stepwise replacement of chlorine atoms occurs at $-80°$ when BCl_3 and C_2H_5OH are mixed in the proper proportions

$$BCl_3 + C_2H_5OH \rightarrow BCl_2OC_2H_5 + HCl \tag{15}$$

$$BCl_3 + 2C_2H_5OH \rightarrow BCl(OC_2H_5)_2 + HCl \tag{16}$$

$$BCl_3 + 3C_2H_5OH \rightarrow B(OC_2H_5)_3 + HCl. \tag{17}$$

Another example involves the elimination of hydrogen halide from boron halide adducts of primary and secondary amines. Primary amines can be made to yield mono-, di-, or trisubstituted products

$$(CH_3)_2NH + BX_3 \rightarrow (CH_3)_2NBX_2 + HX \tag{18}$$

$$2(CH_3)_2NH + BX_3 \rightarrow [(CH_3)_2N]_2BX + 2HX \tag{19}$$

$$3(CH_3)_2NH + BX_3 \rightarrow [(CH_3)_2N]_3B + 3HX. \tag{20}$$

Secondary amines yield trimeric compounds which, as a class, are called borazines:

$$3RNH_2 + BX_3 \longrightarrow [RNBX]_3 + 3HX. \tag{21}$$

These boron–nitrogen compounds are sufficiently important that they are considered in a later section.

3.2 / Boron Subhalides

Physical Properties and Structures   Boron subhalides are compounds which contain less than three halogen atoms for every boron atom. Basically, two methods can be used to prepare boron subhalides. One involves the action of an electric discharge on boron trihalides, while the other involves a halogen exchange reaction between boron trihalides and compounds containing a B—B linkage.

When either BCl_3 or BBr_3 are passed through a mercury arc discharge, diboron tetrahalides (B_2X_4) together with boron monohalides (BX) are formed (18, 19); BI_3 can be converted into B_2I_4 by a radio frequency discharge (20). However, diboron tetrafluoride cannot be prepared directly by discharge reactions. Diboron tetrachloride can be made in very good yield by passing BCl_3 over hot boron monoxide (21). This is a particularly useful reaction since boron monoxide can be prepared in rather large yields. The reaction of BCl_3 with $[(CH_3)_2N]_4B_2$ is also reported to yield B_2Cl_4 (22). The latter reactant can be prepared from the reaction of $[(CH_3)_2N]_2BCl$ with finely dispersed sodium (22, 23):

$$[(CH_3)_2N]_2BCl + 2Na \longrightarrow [(CH_3)_2N]_4B_2 + 2NaCl. \tag{22}$$

The compound B_2Cl_4 is important since it is the precursor to the other diboron tetrahalides as well as other diboron derivatives. It reacts with SbF_3 or allyl fluoride to yield B_2F_4 (24, 25), and with BBr_3 to give B_2Br_4 (26):

$$4BBr_3 + 3B_2Cl_4 \longrightarrow 4BCl_3 + 3B_2Br_4. \tag{23}$$

The diboron tetrahalides are volatile compounds (Table 10.9) which decompose readily. The most stable of these halides, B_2F_4, decomposes at about 200°, whereas the others are unstable at room temperature. X-Ray diffraction experiments on solid B_2Cl_4 and B_2F_4 indicate the presence of centrosymmetric planar molecules, as shown in Fig. 10.7a, whereas electron diffraction studies of B_2Cl_4 vapor indicate that the BCl_2 units are at right angles to each other, as shown in Fig. 10.7b. A nonplanar structure is also suggested for B_2Cl_4 on the basis of spectroscopic arguments (27, 28). The

TABLE 10.9 Physical properties of diboron tetrahalides

Compound	Mp, °C	Bp, °C	Comments
B_2F_4	−56.0	−34	40% decomposition at 200°
B_2Cl_4	−92.95	65.5	decomposes at 0°
B_2Br_4	0.5–1.5	—	liquid thermally unstable
B_2I_4	—	sublimes at 60–70° at $1\,\mu$	decomposes at room temperature

molecular parameters for B_2F_4 and B_2Cl_4 are given in Table 10.10. The bond angles suggest that the bonding in these compounds involves sp^2 hybrid orbitals on the boron atom. The B—X bond distances are nearly the same as those of the corresponding boron trihalides (Table 10.7); the abnormally short B—X bonds presumably have essentially the same π character

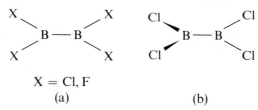

X = Cl, F

(a)   (b)

FIG. 10.7   (a) In the solid state B_2X_4 compounds have a planar structure, but acquire a nonplanar structure in the vapor state (b).

as those bonds do in the boron trihalides. The B—B bond in B_2Cl_4 is virtually the same as is found in the metal borides containing polyatomic boron units. The B—B bond in B_2F_4 is markedly shorter than that in many of the borides, which suggests that a π-electron system might involve the whole molecule. In the case of B_2Cl_4, where the B—Cl π interaction is less, the energy barrier for rotation has been estimated at 1.7 kcal mole^{-1}.

TABLE 10.10   Molecular parameters for B_2F_4 and B_2Cl_4

Compound	∠XBX	$d(B—B)$, Å	$d(B—X)$, Å Exp.	$d(B—X)$, Å Calc.
$B_2F_4{}^a$	120°	1.67 ± 0.05	1.32 ± 0.04	1.52
$B_2Cl_4{}^b$	120°	1.75 ± 0.05	1.73 ± 0.02	1.87

[a] Data from Ref. (77).
[b] Data from Refs. (78) and (79).

The properties of several other boron subhalides appear in Table 10.11. In addition to these compounds, subhalides of the type B_2F_3 are found in the boron–fluorine system when the species BF_2, formed in the high-temperature

TABLE 10.11   **Physical properties of boron subhalides**

Compound	Properties
B_4Cl_4	Pale yellow volatile solid. Stable to 70° in vacuo[a]
B_8Cl_8	Red solid, soluble in organic solvents[b]
B_9Cl_9	Yellow-orange solid[c]
B_4Cl_3Br	Yellow crystalline solid[c]
$(BBr)_n$	Red solid, insoluble in organic solvents[a]
B_7Br_7	Dark-colored crystalline solid[c]
B_8H_8	Red crystalline solid[c]

[a] Ref. (*18*).
[b] Refs. (*80*) and (*81*).
[c] Ref. (*82*).

reduction of BF_3 by boron, is allowed to condense (*29*). The structure of B_4Cl_4, given in Fig. 10.8, shows a tetrahedral arrangement of boron atoms; each chlorine atom is attached to a boron atom. There are several important points to be made here. First, the B—B bond distance in this compound is slightly shorter than those found either in B_2Cl_4 or in the boron polyhedra

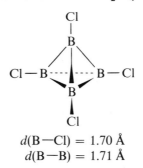

$$d(B-Cl) = 1.70 \text{ Å}$$
$$d(B-B) = 1.71 \text{ Å}$$

FIG. 10.8   Structural parameters of B_4Cl_4. The boron atoms are at the corners of a tetrahedron.

found in the metal borides. Also, if we assume that the chlorine atoms are attached to boron by a normal covalent σ bond, there are insufficient electrons available to make the required number of remaining two-electron bonds. Tetraboron tetrachloride is an electron-deficient molecule in the same sense as is $[Be(CH_3)_2]_x$, which is discussed in Chapter 9.

A similar problem arises for the compounds B_8Cl_8 and B_9Cl_9, which have the cage structures of boron atoms shown in Fig. 10.9. The first of these compounds exhibits a distorted square antiprism of boron atoms, each boron atom attached to a chlorine atom. The compound B_9Cl_9 has a cage structure in the shape of a tri-capped trigonal prism; again, each boron atom has a terminal chlorine atom attached. We shall defer a discussion of the bonding in such boron compounds until a subsequent section.

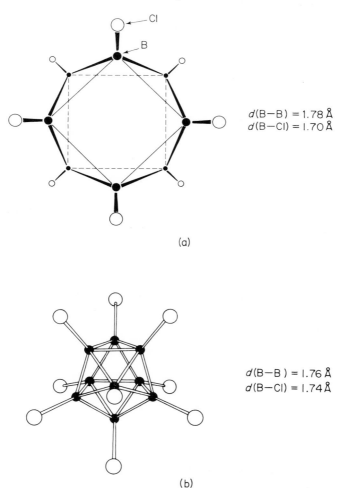

$d(B-B) = 1.78$ Å
$d(B-Cl) = 1.70$ Å

(a)

$d(B-B) = 1.76$ Å
$d(B-Cl) = 1.74$ Å

(b)

FIG. 10.9 Both B_8Cl_8 (a) and B_9Cl_9 (b) contain polyhedral boron cages with each boron atom carrying a terminal chlorine atom.

Chemical Properties   The structures of diboron tetrahalides appear to be similar to that of the trivalent boron halides. With two acidic sites in the molecule we might expect B_2X_4 to form two different adducts,

$$B_2X_4 + L \rightarrow LB_2X_4 \tag{24}$$

$$B_2X_4 + 2L \rightarrow L_2B_2X_4. \tag{25}$$

However, the stability of the diadduct is usually much greater than that of the monoadduct, so that the latter is not often observed:

$$B_2Cl_4 + 2(CH_3)_3N \longrightarrow [(CH_3)_3N]_2B_2Cl_4 \tag{26}$$

$$B_2Cl_4 + 2H_2S \xrightarrow{-78°} (H_2S)_2B_2Cl_4 \tag{27}$$

$$B_2Cl_4 + 2(C_2H_5)_2O \longrightarrow [(C_2H_5)_2O]_2B_2Cl_4. \tag{28}$$

As in the case of the boron trihalide adducts, the presence of acidic hydrogen atoms on the base leads to elimination of hydrogen halide:

$$B_2Cl_4 + 4H_2O \rightarrow B_2(OH)_4 + 4HCl \tag{29}$$

$$B_2Cl_4 + 4ROH \rightarrow B_2(OR)_4 + 4HCl. \tag{30}$$

Adducts of the type $[NR_4]_2[B_2Cl_4X_2](X = OH, Cl, CN)$ containing anionic Lewis bases have been prepared by either indirect methods or by direct combination of reactants,

$$2(CH_3)_4NCl + B_2Cl_4 \rightarrow [(CH_3)_4N]_2B_2Cl_6. \tag{31}$$

In addition, diboron tetrahalides react with olefins. For example, diboron tetrahalides react with ethylene to give a product in which the B—B bond has been cleaved and the parts have been added to the carbon–carbon double bond (8).

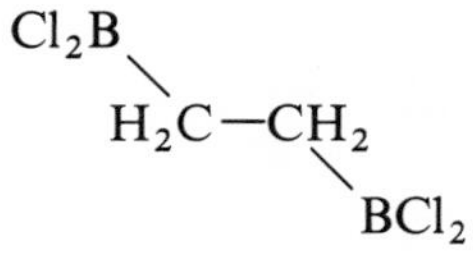

$$\begin{array}{c} Cl_2B \\ \diagdown \\ H_2C-CH_2 \\ \diagdown \\ BCl_2 \end{array}$$

8

3.3 / Aluminum, Gallium, Indium, and Thallium Trihalides

Physical Properties and Structure   The unusually high melting points of AlF_3, GaF_3, and InF_3 (Table 10.6) suggest that these compounds exist as continuous giant-molecule-type structures. The crystal structures of GaF_3 and InF_3 are unknown. However, AlF_3 exhibits the interlocking structure shown in Fig. 10.10 in which each Al^{3+} ion is surrounded by six octahedrally arranged F^- ions; each fluorine atom acts as a bridge between

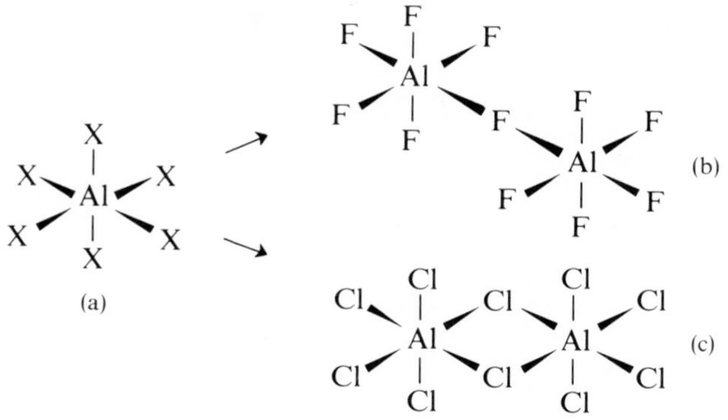

FIG. 10.10   The solid-state structures of AlF_3 and $AlCl_3$ arise from a basic octahedral environment (a) in which aluminum atoms share corners (b) or edges (c).

octahedra. Aluminum has basically the same coordination in solid $AlCl_3$, but two chlorine atoms in each octahedron are shared between adjacent aluminum atoms. A similar structure is observed for $InCl_3$ and $TlCl_3$ in the solid state. The higher octahedral coordination number of these elements in the solid-state structures is consistent with the general observation that the elements in Period 3 and subsequent periods achieve coordination numbers greater than 4.

In general, aluminum, gallium, and indium halides vaporize to give dimeric molecules, M_2X_6, which contain halogen bridges (**9**). Electronically,

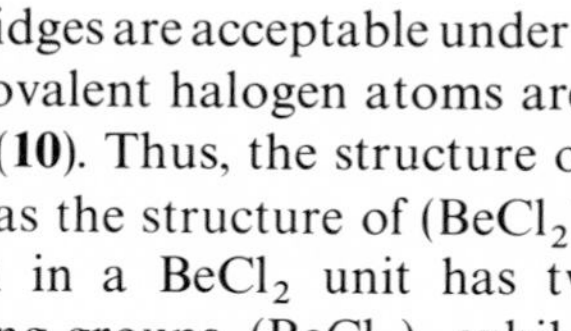

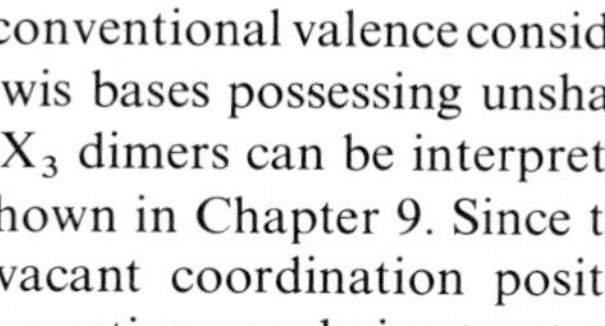

halogen bridges are acceptable under the conventional valence considerations, since monovalent halogen atoms are Lewis bases possessing unshared electron pairs (**10**). Thus, the structure of MX_3 dimers can be interpreted in the same way as the structure of $(BeCl_2)_n$, shown in Chapter 9. Since the beryllium atom in a $BeCl_2$ unit has two vacant coordination positions *and* two bridging groups, $(BeCl_2)_x$ exhibits a continuous chain structure. In the case of the Group III halides, there is only *one* coordination position for three possible bridging groups, and where the chain terminates a dimeric species is formed. It is curious that GaI_3 exists as a monomeric species under the conditions where all the other halides form dimers.

The general geometrical parameters of the dimeric trihalides, shown in Table 10.12, also are reminiscent of those of the continuous-chain beryllium

TABLE 10.2   Vapor-phase structure of some Group III halides

$$X' \diagdown \quad X \diagup \quad X'$$

(structure diagram: X′ and X′ bonded to M, two M centers bridged by two X atoms, terminal X′ groups)

| | $\angle$ X'MX' | $\angle$ XMX | Bond distances, Å | | |
			$d(X'-M)$	$d(M-X)$	Calc.
Al_2Cl_6	$118 \pm 6°$	$79 \pm 10°$	2.06	2.21	2.24
Al_2Br_6	$115 \pm 5°$	$87 \pm 6°$	2.21	2.33	2.39
Al_2I_6	$112 \pm 5°$	$102 \pm 5°$	2.53	2.58	2.58
Ga_2Cl_6	—	—	2.09	2.29	2.25
Ga_2Br_6	—	—	2.25	2.35	2.40
GaI_3	$120°$	—	2.44	—	2.59
In_2Cl_6	—	—	2.46^a	2.46^a	2.43
In_2Br_6	—	—	2.58^a	2.58^a	2.58
In_2I_6	—	—	2.76^a	2.76^a	2.77

^{a} Average value.

compounds. As in the case of the beryllium compounds, the X—M—X bond angles suggest either that the Group III element does not exhibit sp^3 hybridization in the dimers, or that the electron density involving the bridging halogen atoms is off the line of centers. Although electronically there is no need to do so for halogen-bridged compounds, the dimer could be described in terms of the three-centered bond described in Chapter 9. For the lighter halogen-containing dimers, the available data, shown in Table 10.12, indicate a noticeable shortening of the terminal halogen–metal bonds, a symptom which suggests that this bond incorporates a π component. Presumably the source of electrons for this π component is the unused p electrons on the halogen atoms. The three p orbitals of the Group III atom are involved as hybrids forming σ bonds to the four halogen atoms surrounding it. If a π component is present in these species, it must involve the d orbitals of the Group III element (**11** or **12**): $3d$ for aluminum, $4d$ for gallium, $5d$ for indium.

X — M (orbital diagram labeled with + and − signs)

11

$^+X \rightleftharpoons M^-$

12

In a sense, this type of coordinate-covalent π component is similar to the $p\pi$–$p\pi$ bonding invoked previously to account for the bond shortening in the boron halides.

The Group III halides are also dimeric in nonpolar solvents such as benzene. Crystalline aluminum bromide consists of close-packed layers of bromine atoms with aluminum atoms occupying adjacent tetrahedral holes; the structure in the solid state appears to be an orderly arrangement of Al_2Br_6 units.

The dimeric M_2X_6 species can be dissociated thermally or by the action of a Lewis base. The enthalpies of dissociation of the aluminum halide dimers

$$Al_2X_{6(g)} \rightleftharpoons 2AlX_3 \qquad (32)$$

are of the order of 11–15 kcal mole^{-1}.

Chemical Properties   Like the boron halides, the trivalent halides of the other members of this family are strong Lewis acids. Four-coordinate complex compounds of the type $X_3M:L$ are known where $:L$ represents a neutral Lewis base, such as an amine or an oxygen-containing compound. Also, $:L$ can be an anion, such as a halide ion; thus, compounds such as Cl_3AlSH_2, Cl_3AlNR_3, $M^+[AlCl_4^-]$, and $M^+[GaCl_4^-]$ are well known. Although the structures of all the four-coordinate species have not been determined, those that are known show a tetrahedral arrangement of atoms about the central atom. Presumably these four-coordinate species involve the use of sp^3 hybrid orbitals in bonding. In the case of crystalline $Na^+[AlCl_4^-]$, an Al—Cl distance of 2.13 Å has been reported, which is still markedly shorter than the value of 2.24 Å expected for a single bond between these atoms. Multiple bonding of the type described for the terminal aluminum–halogen bonds in the dimer (Al_2X_6) might still be present.

As would be expected for elements in Periods 3 and higher, the trihalides form compounds of the type MX_3L_3 in which a coordination number greater than four has been attained; $:L$ can either be a neutral Lewis base or an anion. Many salts are known in which the anion contains a hexacoordinate Group III element, e.g., M_3AlF_6, M_3GaF_6, M_3InCl_6, M_3InBr_6, M_3TlCl_6, M_3TlBr_6. Where structure determinations have been made, the central Group III atom is surrounded by an octahedral arrangement of halide ions (**13**). Presumably the central atom uses d^2sp^3 hybrid orbitals for bond formation in these species. At first glance, complex fluorides such as Tl_2AlF_5 and NH_4AlF_4

M = Al, Ga, In, Tl

X = halogen

13

might appear to have a lower coordination number than six, but x-ray analysis of the crystalline compounds shows that AlF_6^{3-} octahedra are still present in these compounds. In the case of Tl_2AlF_5 the octahedra are linked by one bridging fluorine atom; the corners of the octahedra are shared. Two corners of the octahedra are shared (two bridging fluorine atoms) in the compound NH_4AlF_4, as shown in Fig. 10.11. An interesting example of a species in which three bridging groups occur is the ion $[Tl_2Cl_9]^{3-}$. In this anion, two distorted octahedra share a common face (Fig. 10.11).

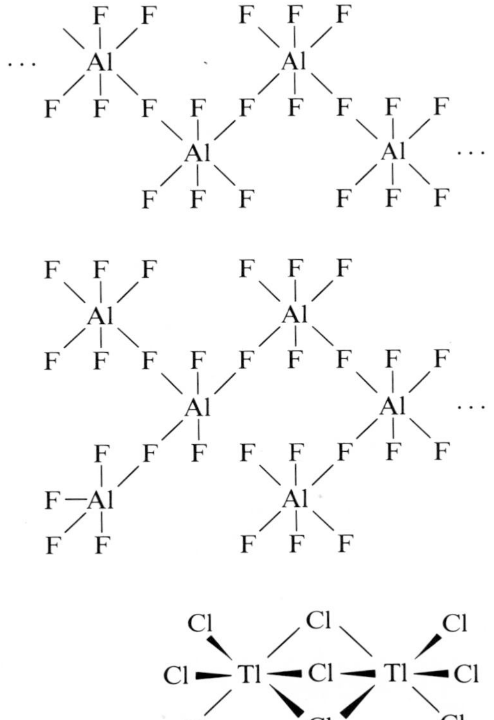

FIG. 10.11 The structures of many of the Group III halides are based on octahedral coordination, even though the stoichiometry suggests otherwise.

Dissolution of the Group III halides in strongly basic solvents often leads to six-coordinate species. For example, dissolution of aluminum chloride in water gives the compound $[Al(H_2O)_6]Cl_3$ upon evaporation. Solvent molecules have displaced halide ions to give complex cations containing the Group III element. Similarly, compounds such as $[Al(NH_3)_6]Cl_3$ and $[Al(OS(CH_3)_2)_6]Cl_3$ are known. Mixed six-coordinate species are known in compounds such as $(NH_4)_2[InCl_5(H_2O)]$ and $Cs_2[TlCl_5(H_2O)]$.

3.4 / Aluminum, Gallium, Indium, and Thallium Subhalides

The chemistry of the Group III elements is predominantly that of the $3+$ oxidation state, although there is a marked increasing tendency of the heavier members of the series to form compounds containing stable $1+$ or $2+$ states, as shown in Table 10.13. Except for species that are stable under high-energy conditions, the $+1$ state for boron and aluminum is unknown.

TABLE 10.13   Halides of aluminum, gallium, indium, and thallium in a $1+$ oxidation state

Element	Compound
Al	AlX^a
Ga	$Ga[GaX_4]^{b,c}$
In	$In[InI_4]^d$
	InX^b
Tl	TlX^e

^{a} X = Cl, Br.
^{b} X = Cl, Br, I.
^{c} "GaX_2".
^{d} "InI_2".
^{e} X = F, Cl, Br, I.

Short-lived boron (BF, BCl, and BBr) and aluminum (AlCl, AlBr) mono-halides have been identified. In the case of aluminum, additional indirect evidence exists for a species of lower valence. For example, if gaseous $AlCl_3$ or $AlBr_3$ is passed over metallic aluminum at 1000°C under reduced pressure, the aluminum rapidly evaporates and condenses in the colder portion of the apparatus. These observations suggest that an equilibrium involving a volatile lower-valent aluminum halide has been established under these conditions,

$$2Al + AlX_3 \rightleftharpoons 3AlX. \tag{33}$$

In contrast to the first two members of this family, the existence of compounds containing Ga^{1+} and In^{1+} ions is well established. The "dihalides" of gallium, which have the empirical stoichiometry GaX_2 and should be paramagnetic if they contain gallium in a $2+$ oxidation state, are really Ga^+ salts containing the GaX_4^- anion ($Ga^+GaX_4^-$) (*IV*). Crystals of "GaX_2" are isomorphic with those of $GaAlCl_4$, which is an ionic compound containing Ga^+ ions. The Raman spectrum of the fused salt "GaX_2" indicates that GaX_4^- ions are also present in the melt. Nuclear magnetic resonance spectroscopy using ^{71}Ga also shows that two types of gallium atoms are present in $GaBr_2$ and $GaCl_2$ (*30*). Although less is known about the corresponding

InX_2 compounds, the x-ray powder pattern of $InCl_2$ strongly suggests that it should be formulated as $In^+InCl_4^-$. Monovalent indium compounds such as $InCl$, $InBr$, and InI have been characterized; $InBr$ and InI crystallize in the cesium-chloride-type structure.

Although the existence of compounds containing the $1+$ oxidation states for Ga and In have been established for the solid state, these ions are unstable in aqueous solution. Dissolution of compounds such as $Ga^+GaX_4^-$, $In^+InX_4^-$, or InX in water leads to disproportionation of the monovalent ion

$$3M^+ \xrightarrow{H_2O} 2M + M^{3+}. \tag{34}$$

The last member of this family of elements forms the most stable $1+$ oxidation state. Ionic compounds containing the Tl^+ ion have been characterized in the solid state as well as in aqueous solution. The thallium halides TlX ($X = F, Cl, Br, I$) crystallize in structures characteristic of the alkali-metal halides; TlF crystallizes in a deformed NaCl-type lattice, whereas $TlCl$, $TlBr$, and TlI crystallize in the cesium-chloride structure. The thallous ion has an ionic radius of 1.40 Å, which lies between that of Rb^+ (1.48 Å) and K^+ (1.33 Å). In addition the ionic mobility of Tl^+ is nearly the same as that for K^+ and Rb^+. Thus, it is not surprising that many of the properties of thallous compounds are similar to those of the corresponding alkali-metal salts. For example, thallous hydroxide (TlOH) is very soluble in water, giving a solution with strongly basic properties; this solution absorbs carbon dioxide readily to give Tl_2CO_3. Many thallium salts, e.g., Tl_2SO_4, Tl_3PO_3, Tl_2HPO_4, TlH_2PO_4, $TlClO_3$, and $TlClO_4$, are isomorphous with the corresponding potassium salts.

3.5 / Boron–Oxygen Compounds

Boron Oxides   Two oxides of boron are known. The normal oxide, B_2O_3, is obtained when amorphous boron is heated in air or when boric acid, H_3BO_3, is dehydrated. Boron trioxide readily forms a glass, but a crystalline form has been obtained by careful dehydration of boric acid. The structure of crystalline B_2O_3 has been described as consisting of BO_4 tetrahedra, but the details are not well worked out. Mass spectral evidence shows that liquid

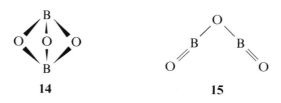

14   15

boron trioxide vaporizes as B_2O_3 molecules. There is considerable question concerning the structure of these molecules, with two different arrangements being postulated [(**14**) and (**15**)] (*31, 32*). Boric oxide dissolves in water to form boric acid and in basic solutions to form borates. These compounds are discussed in the next section.

A lower oxide of boron is known with the formula $(BO)_x$. Although the structure of the solid is unknown, mass spectral studies of the vapor indicate that B_2O_2 species are present. The compound can be prepared by heating B_2O_3 with boron to 1050° or by the dehydration of $B_2(OH)_4$. The solid oxide $(BO)_x$ as well as the molecular species B_2O_2 probably contain B—B bonds, because they can be obtained from compounds known to have B—B bonds by the simple solvolysis or dehydration reactions shown in Fig. 10.12.

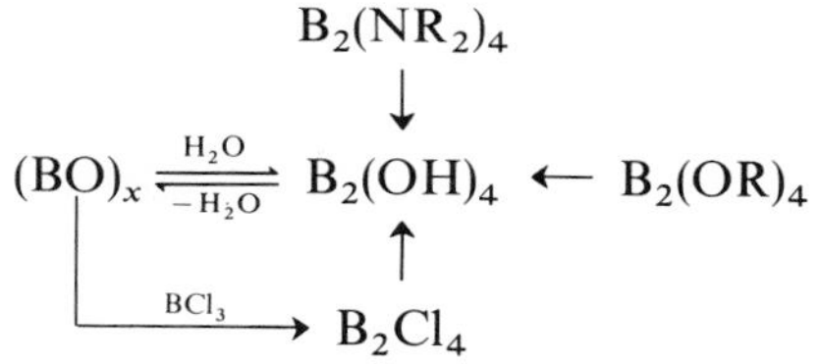

$$B_2(NR_2)_4$$
$$\downarrow$$
$$(BO)_x \underset{-H_2O}{\overset{H_2O}{\rightleftharpoons}} B_2(OH)_4 \leftarrow B_2(OR)_4$$
$$\uparrow$$
$$\xrightarrow{BCl_3} B_2Cl_4$$

FIG. 10.12 The chemical relationships among several diboron compounds.

Boric Acid   Boric acid, H_3BO_3, is formed when boric oxide reacts with water or when any number of trivalent boron compounds, such as boron halides, hydrides, esters, etc., are hydrolyzed. Boric acid is soluble in water and slightly volatile in steam. In aqueous solutions it is a very weak monobasic acid ($pK = 9.00$); in this process it is best described as a Lewis acid, accepting OH^- ions from the solvent:

$$B(OH)_3 + 2H_2O \rightleftharpoons B(OH)_4^- + H_3O^+. \tag{35}$$

Only the species shown in Eq. (35) are present at relatively low concentrations ($<0.025\ M$) of H_3BO_3. At higher concentrations more complex polymerized species are present.

Upon heating H_3BO_3 at 100°, the compound partially dehydrates to give metaboric acid, HBO_2. Further dehydration yields B_2O_3. Both boric acid and metaboric acid exhibit the layer structure shown in Fig. 10.13, which is composed of BO_3 units connected by hydrogen bonds. Each OH group is hydrogen-bonded to an oxygen atom on an adjacent BO_3 unit; in this way each oxygen atom is associated with two hydrogen atoms. Metaboric acid exists in three polymorphs: cubic (stable), monoclinic (metastable), and ortho-rhombic (metastable). The structures of the metastable forms are unknown; the cubic form consists of HBO_2 trimers held together in layers by hydrogen bonds (Fig. 10.13).

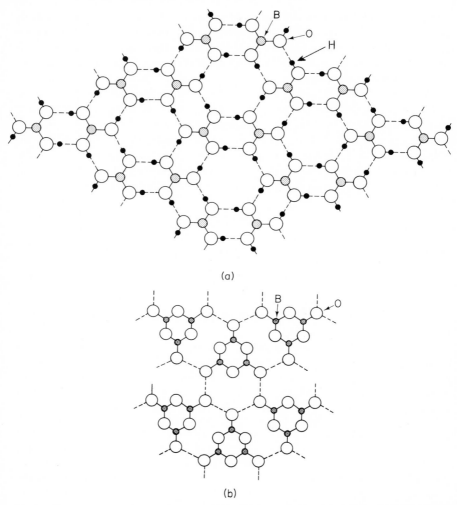

FIG. 10.13   Boric acid (a) and metaboric acid (b) crystallize in a layer structure held together by hydrogen bonds. The hydrogen bonds are indicated by dotted lines; the hydrogen atoms are not shown in (b).

Borate Salts   Boric and metaboric acids form a variety of salts with many metal ions. Anhydrous metal borates can be prepared by fusion of the metal oxide with B_2O_3,

$$Na_2O + B_2O_3 \rightarrow Na_2B_2O_4. \tag{36}$$

Hydrated salts can be obtained by crystallizing the compounds from aqueous solutions. Large quantities of boron are found naturally as hydrated sodium borates [$(Na_2B_4O_7)\cdot10\,H_2O$, borax] or as calcium salts [$(CaB_3O_4(OH)_3\cdot H_2O$, colemanite]. A survey of the stoichiometries of the borates shows that no discernible pattern exists. However, structural studies indicate that many of the simpler borates consist of BO_3 units linked in several different ways, as shown in Fig. 10.14.

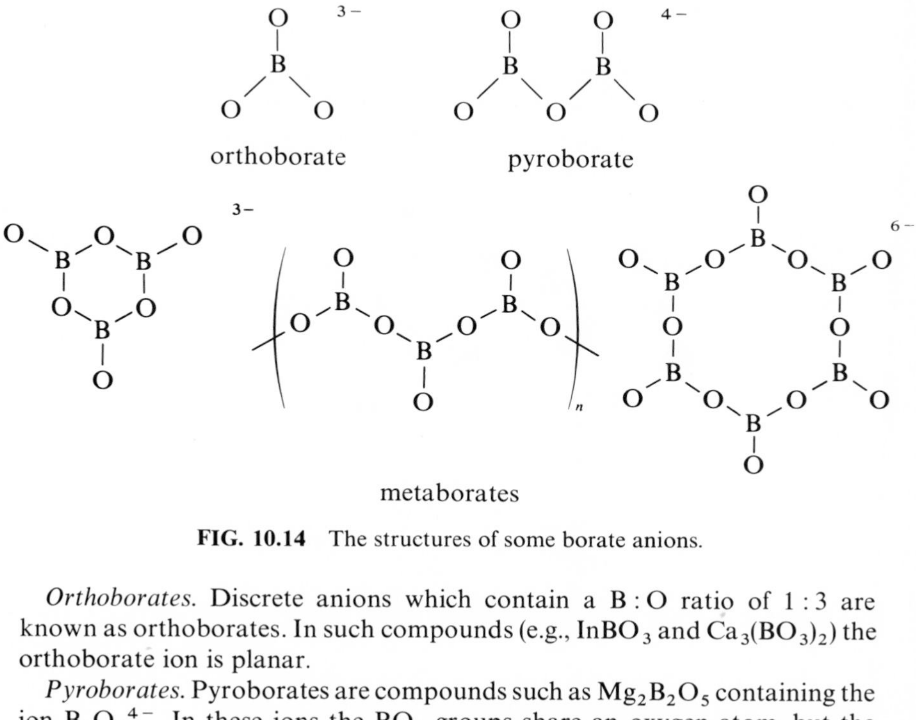

orthoborate   pyroborate

metaborates

FIG. 10.14   The structures of some borate anions.

Orthoborates. Discrete anions which contain a $B:O$ ratio of $1:3$ are known as orthoborates. In such compounds (e.g., $InBO_3$ and $Ca_3(BO_3)_2$) the orthoborate ion is planar.

Pyroborates. Pyroborates are compounds such as $Mg_2B_2O_5$ containing the ion $B_2O_5{}^{4-}$. In these ions the BO_3 groups share an oxygen atom, but the ions, in general, are not perfectly planar.

Metaborates. Metaborates possess structures where each BO_3 unit shares two oxygen atoms. Basically, the ring structures and chain structures shown in Fig. 10.14 are possible. It should be recalled that the basic ring structure exhibited by some metaborates is the same as that found in metaboric acid (Fig. 10.13). In addition, the dehydration of monomethylboric acid, $CH_3B(OH)_2$, gives a cyclic trimer, $(CH_3BO)_3$, which has also this same basic boron–oxygen ring structure (Fig. 10.15). All of the atoms except hydrogen are coplanar, and the B—O bond distance is shorter than the calculated distance expected from single-bond radii (1.54 Å).

The anions containing a $B:O$ ratio of $2:3$ (which corresponds to BO_3 groups sharing three oxygen atoms) are not yet known.

CH3
B
O
O
B
B
O
H3C
CH3

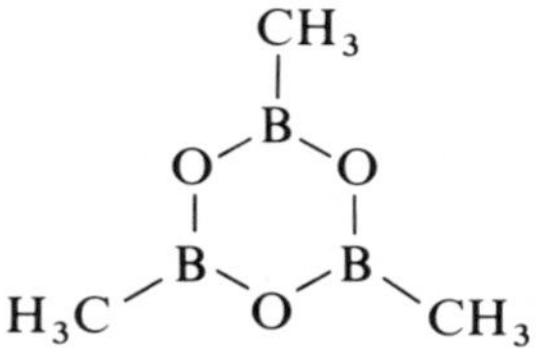

$\angle$ BOB $= 112°$
d(B—O) $= 1.39$ Å

FIG. 10.15 Structural parameters for $(CH_3BO)_3$.

Compounds containing oxyanions in which boron atoms exhibit four-coordination are also well known. As would be expected from previous arguments, oxygen atoms in these compounds surround the boron in a

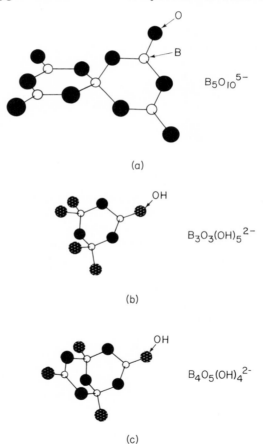

O
B
B5O10 5−
(a)
OH
B3O3(OH)5 2−
(b)
OH
B4O5(OH)4 2-
(c)

FIG. 10.16 The structures of some complex borate ions; the hydrogen atoms are not shown in (b) and (c).

tetrahedral arrangement, discrete tetrahedral BO_4 groups being present in compounds such as BPO_4 and $BAsO_4$. Tetrahedral BO_3OH groups are found in compounds like $CaBSiO_4(OH)$ and $CaB_3O_4(OH)_3 \cdot H_2O$. Tetrahedral $B(OH)_4$ groups are found in minerals such as $Na_2ClB(OH)_4$ and $CaClB(OH)_4$.

Complex Borates. The ortho-, meta-, and pyroborate anions all contain three-coordinate boron atoms. Oxyanions containing only four-coordinate boron atoms are not known. However, a series of oxyanions containing both four- and three-coordinate boron atoms have been found. For example, the ion $B_5O_{10}^{5-}$, found in the compound $KH_4B_5O_{10} \cdot 2H_2O$, exhibits the structure shown in Fig. 10.16a, in which a central four-coordinate boron atom shares oxygen atoms with two three-coordinate boron atoms; the planes of the six-membered B—O rings are perpendicular to each other. The cyclic structure shown in Fig. 10.16b, containing two four-coordinate and one three-coordinate boron atoms, exists in the anion $B_3O_3(OH)_5^{2-}$ which is present in the compound $CaB_3O_3(OH)_5 \cdot 4H_2O$. A related ion is found in borax, which has the empirical formula $Na_2B_4O_7 \cdot 10\,H_2O$. Actually the anion present in this compound is related to the $B_3O_3(OH)_5^{2-}$ anion; it is formed by attaching a bridging planar BO_3 group sharing oxygen atoms with the four-coordinate BO_4 species in the ring to give the structure shown in Fig. 10.16c. Thus, borax should be more properly formulated as $Na_2[B_4O_5(OH)_4] \cdot 8H_2O$.

The average B—O bond length in crystalline borates containing a planar BO_3 group is very close to 1.36 Å. In contrast, the B—O bond distance increases to an average value of 1.48 Å for tetrahedral four-coordinate boron atoms. The change in bond length suggests that there is a change in bond order in going from a planar three-coordinate structure to a four-coordinate structure. Oxygen atoms in these structures possess two lone-pair electrons which can interact with the empty p_z orbital on a planar sp^2 hybridized boron atom (**16**) in much the same way as the bond shortening which occurs for planar boron compounds containing B—F and B—Cl bonds. The

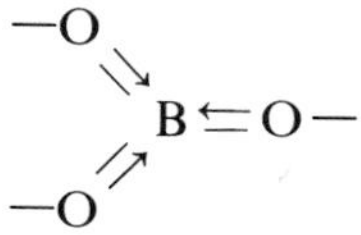

$$
\begin{array}{c}
-O \\
\quad\searrow \\
\qquad B \rightleftharpoons O- \\
\quad\nearrow \\
-O
\end{array}
$$

16

increase in bond length for tetrahedral species correlates with the fact that boron atoms possessing four bonds require the use of sp^3 hybrid orbitals, which does not leave a low-lying orbital available for back-coordination.

Boron Esters   Boric acid or B_2O_3 reacts with alcohols to form esters:

$$B_2O_3 + 6ROH \rightarrow 2B(OR)_3 + 3H_2O \tag{37}$$

$$B(OH)_3 + 3ROH \rightarrow B(OR)_3 + 3H_2O. \tag{38}$$

Esters can also be prepared from the reaction of boron halides with alcohols:

$$BCl_3 + 3ROH \rightarrow B(OR)_3 + 3HCl. \tag{39}$$

In some cases it is possible to prepare boron compounds containing halogen atoms and ester groups:

$$BCl_3 + C_2H_5OH \rightarrow BCl_2(OC_2H_5) + HCl. \tag{40}$$

The lighter trialkoxyboron compounds are all volatile nonassociated liquids which rapidly hydrolyze in water to give boric acid and the corresponding alcohol. Although extensive data on the structures of these compounds are not available, it is very likely that they have a planar structure with an O—B—O angle very close to 120° and a B—O bond distance that is shorter than that calculated from the single-bond radii. For example, electron diffraction experiments on trimethoxyboron indicate that the O—B—O angle is 120° and the B—O bond distance is 1.38 Å. The latter distance is very close to that observed in the metal–borate anions containing planar BO_3 groups (1.36 Å), and certainly shorter than the sum of the single-bond radii for these atoms (1.54 Å).

The strength of boric acid is increased in aqueous solution if polyhydroxy alcohols are present. It is generally agreed that complex bisdialkoxyboron anions which contain four-coordinate boron atoms are formed under these conditions (**17**):

$$B(OH)_3 + 2H_2O \rightleftharpoons B(OH)_4^- + H_3O^+ \tag{41}$$

$$B(OH)_4^- + 2 \begin{matrix} R_2COH \\ | \\ R_2COH \end{matrix} \rightarrow \begin{bmatrix} R_2C-O \quad O-CR_2 \\ \diagdown \diagup \\ B \\ \diagup \diagdown \\ R_2C-O \quad O-CR_2 \end{bmatrix}^- + 4H_2O \tag{42}$$

17

The stability of the four-coordinate anion (**17**) is sufficiently large that the equilibrium shown in Eq. (41) is shifted markedly to the right, increasing the hydrogen-ion concentration in solution. The configuration of the diol is important in the formation of **17**. If there is restricted rotation about the carbon–carbon bond, only diols in the *cis* configuration will be able to enhance the acidity of boric acid. Generally, glycerol and manitol are used to enhance the acidity of boric acid so that it can be analytically determined by titration with a strong base.

When β-diketones such as acetylacetone, which can tautomerize,

$$CH_3-\overset{\overset{\displaystyle O}{\|}}{C}-CH_2-\overset{\overset{\displaystyle O}{\|}}{C}-CH_3 \ \rightleftharpoons \ CH_3-\overset{\overset{\displaystyle O}{\|}}{C}-CH=\overset{\overset{\displaystyle OH}{|}}{C}-CH_3, \qquad (43)$$

react with boron halides, four-coordinate boron-containing cations (18) are formed:

$$3CH_3-\overset{\overset{\displaystyle O}{\|}}{C}-CH=\overset{\overset{\displaystyle OH}{|}}{C}-CH_3 + BX_3 \rightarrow$$

$$\qquad X^- + 2HX. \qquad (44)$$

18

The enol form of acetylacetone reacts as alcohols do, but the carbon backbone has the correct geometry to permit the remaining carbonyl oxygen atom to act as a Lewis base towards the boron atom.

3.6 / Oxygen Derivatives of Aluminum, Gallium, Indium, and Thallium

Oxides   The most common oxide of the heavier elements in Group III is the expected trivalent oxide M_2O_3. Lower-valent oxygen-containing compounds have been reported for these elements. Al_2O is made by heating a mixture of Al_2O_3 and elemental silicon to 1800°, but AlO is known only as a transient species at 3200°, when metallic aluminum is burned in air. Ga_2O and In_2O can be made by reducing the corresponding trioxides. The univalent thallium oxide (Tl_2O) is prepared when metallic thallium is heated in a stream of air. The enthalpies of formation of these oxides, as shown in Table 10.14, decrease steadily with increasing atomic number; conversely, the oxides are more easily reduced in this order. The high heat of formation of Al_2O_3 finds practical use in the thermite process, i.e., the reduction of metal

TABLE 10.14 Enthalpies of formation of the Group III oxides

	B_2O_3	Al_2O_3	Ga_2O_3	In_2O_3
ΔH_f, kcal mole^{-1}	-304.69	-404.08	-258.49	-220.41

oxides with aluminum metal:

$$\frac{2y}{3}Al + M_xO_y \rightarrow xM + \frac{y}{3}Al_2O_3 \qquad (45)$$

Many transition-metal oxides (e.g., MnO_2, Cr_2O_3) are easily reduced in this manner.

The normal aluminum oxide exists in a variety of crystalline modifications between 200° and 1200°. The most familiar of these are α-Al_2O_3—sometimes called corundum—and γ-Al_2O_3. The so-called "β-Al_2O_3" is not pure Al_2O_3, rather it has the composition $Na_2 \cdot 11\,Al_2O_3$. α-Al_2O_3 has a close-packed structure of oxygen atoms with aluminum atoms occupying the octahedral holes; γ-Al_2O_3 exhibits a defect spinel structure. In a perfect spinel structure, the oxygen atoms are cubic closest packed with $\frac{1}{8}$ of the tetrahedral holes and $\frac{1}{2}$ of the octahedral holes filled with cations.

The α-form of Al_2O_3 is the fourth hardest substance known, being exceeded only by diamond, boron nitride, and carborundum. It is virtually insoluble in water, acids, or bases. α-Al_2O_3 forms mixed oxides with other metals. In certain instances, traces of transition-metal oxides give substances that are classified as gem stones. Thus, a ruby is α-Al_2O_3 containing a trace of Cr^{3+}, and blue sapphire contains Fe^{2+}, Fe^{3+}, and Ti^{4+}. Mixtures of Al_2O_3 and certain other metallic oxides on a molar basis also give products that are of gem quality, e.g., alexandrite ($Al_2O_3 \cdot BeO$) and spinel ($Al_2O_3 \cdot MgO$).

γ-Al_2O_3 is metastable and not found in nature, but can be prepared by carefully heating $Al(OH)_3$. It is characterized by small particle size of very high surface area. Because of the latter property, γ-Al_2O_3 finds use as an adsorbant. γ-Al_2O_3 upon prolonged heating yields corundum.

The properties of gallium trioxide follow closely those of Al_2O_3. A stable high-temperature form (α-Ga_2O_3) and a metastable low-temperature form (γ-Ga_2O_3) exist. The crystal structures of these two forms of Ga_2O_3 follow the same pattern as described for the corresponding aluminum oxides. Yellow indium oxide (In_2O_3) and the dark-colored thallium oxide (Tl_2O_3) are known only in one form. Thallium trioxide readily loses oxygen at about 100° to give thallous oxide (Tl_2O).

Hydroxides   When ammonia is added to an aqueous solution of aluminum salts a gelatinous precipitate, best described as a hydrated aluminum hydroxide, is formed. Potentiometric studies on aqueous solutions have been interpreted in terms of a series of complexes of the type $Al[Al_3(OH)_8]_m^{(m+3)+}$, the major constituent (*33*) being $Al_7(OH)_{16}^{5+}$. Aluminum hydroxide is amphoteric, being soluble in acid to form hydrated cations and in base to form aluminate anions:

$$Al(OH)_3 \cdot x H_2O + 3H^+ \longrightarrow Al(H_2O)_6^{3+} + (x-3)H_2O \qquad (46)$$

$$Al(OH)_3 \cdot x H_2O + OH^- \longrightarrow AlO_2^- + (2+x)H_2O. \qquad (47)$$

A nonhydrated, crystalline form of aluminum hydroxide is formed when carbon dioxide is passed into an alkali aluminate solution. Crystalline $Al(OH)_3$ exhibits the layer structure shown in Fig. 10.17; each layer consists

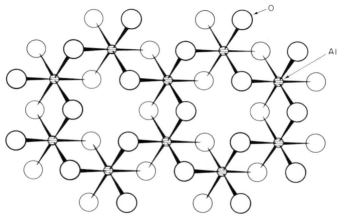

FIG. 10.17   A portion of a layer of $Al(OH)_3$. Only the oxygen and aluminum atoms are shown.

of hexagonally packed OH groups with aluminum atoms in the octahedral holes. Another way of imagining the structure is a series of sheets formed by nearly octahedral $Al(OH)_6$ units sharing edges. Although the details are not shown in Fig. 10.17, hydrogen bonds exist between oxygen atoms within a layer as well as between adjacent layers.

When aluminum hydroxide is heated at 100° partial dehydration occurs to yield boehmite, or γ-AlO(OH). At higher temperatures (150°) complete dehydration occurs to yield γ-Al_2O_3, and ignition above 800° gives α-Al_2O_3. A second modification of AlO(OH) exists called diaspore, or α-AlO(OH), which in turn gives α-Al_2O_3 upon heating.

Both $Ga(OH)_3$ and $In(OH)_3$ are colorless gelatinous substances precipitated from solution when a base is added to aqueous solutions of the corresponding salts. Both substances are amphoteric, dissolving in aqueous acids or bases. The action of strongly concentrated solutions of NaOH ($\sim 10\,N$) on $Ga(OH)_3$ gives the normal gallate species, Na_3GaO_3, but insoluble Na_2HGaO_3 and NaH_2GaO_3 form when less concentrated solutions of base are added.

Addition of base to an aqueous solution containing thallic salts gives Tl_2O_3 rather than the expected hydroxide. Very soluble thallous hydroxide can be prepared by dissolving Tl_2O in water, the reaction being easily reversible. Thallous hydroxide is a strong base, but the aqueous solutions are easily oxidized by air to give the relatively less soluble Tl_2O_3.

Alkoxides   These elements form alkoxides with the expected stoichiometry, $M(OR)_3$ for aluminum, gallium, and indium. The monovalent derivatives are more common for thallium. The compounds are generally insoluble in alcohol and are hydrolyzed by water. However most of these compounds are polymeric in solutions of inert solvents. For example, aluminum t-butoxide is a dimer, $[Al(OR)_3]_2$, and the isopropoxide is a tetramer; the structures of these compounds are given in Fig. 10.18. The

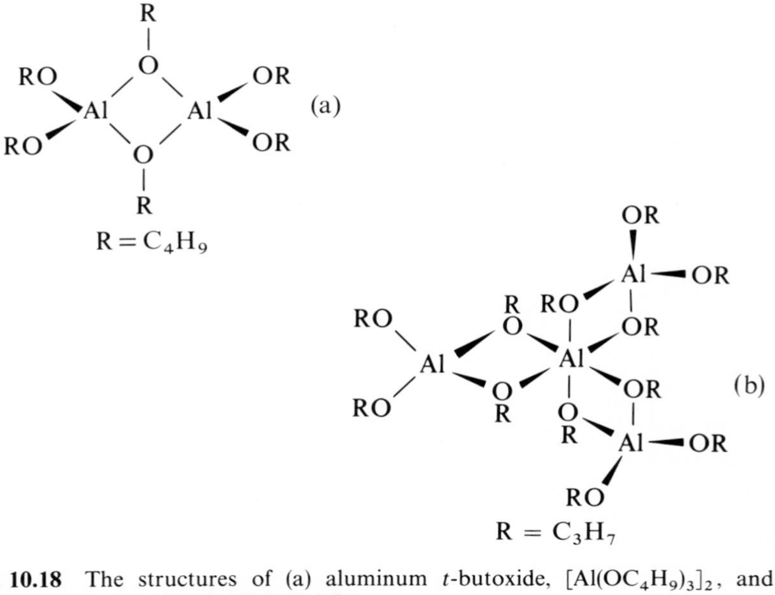

FIG. 10.18   The structures of (a) aluminum t-butoxide, $[Al(OC_4H_9)_3]_2$, and (b) aluminum isopropoxide, $[Al(OC_3H_7)_3]_4$.

dimer has been formulated as an oxygen-bridged species containing four-coordinate aluminum, whereas nmr data (*34*) suggest that the tetramer contains both six- and four-coordinate aluminum atoms. Aluminum alkoxides can be converted into four-coordinate complex anions by alkali-metal alkoxides,

$$Al(OR)_3 + M^+OR^- \rightarrow M^+[Al(OR)_4]^-. \tag{48}$$

Six-coordinate, but neutral, aluminum-containing species are obtained when aluminum ion reacts with the enol form of acetylacetone,

$$Al^{3+} + 3HO-\overset{\overset{\textstyle CH_3}{|}}{C}=CH-\overset{\overset{\textstyle O}{\|}}{C}-CH_3 \rightarrow$$

$$Al(O-\overset{\overset{\textstyle CH_3}{|}}{C}=CH-\overset{\overset{\textstyle O}{\|}}{C}-CH_3)_3 + 3H^+. \tag{49}$$

The carbonyl group present in the anion has the correct geometrical arrangement to act as a Lewis base toward the central atom (**19**). Aluminum acetylacetonate melts at 192°, is virtually insoluble in water, but very easily soluble in many organic solvents. Gallium acetylacetonate (mp 194°) possesses essentially the same properties.

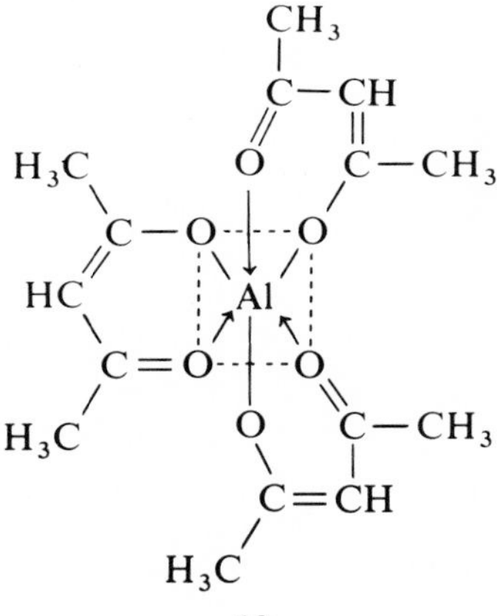

19

Thallous ethoxide, a liquid that freezes at $-30°$, is tetrameric $(TlOC_2H_5)_4$. In general the properties of this substance indicate that it is a covalently bonded species. The tetramer has been assigned a cubic structure incorporating oxygen bridges (**20**).

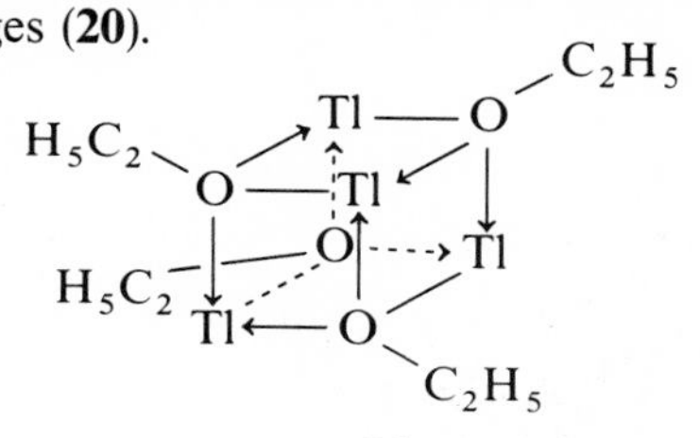

20

3.7 / Nitrogen Derivatives of Boron

The atom pair B—N is isoelectronic with C_2. Therefore, compounds containing boron bonded to nitrogen theoretically can exhibit the range of structure types encountered in carbon chemistry. Thus, compounds containing tetra-, tri-, and dicoordinate boron atoms attached to an aminoid nitrogen atom are isoelectronic with the corresponding C_2 system (Fig. 10.19). Although the BN analogues of the triple-bonded carbon system are not yet known, many examples of the other two possibilities exist.

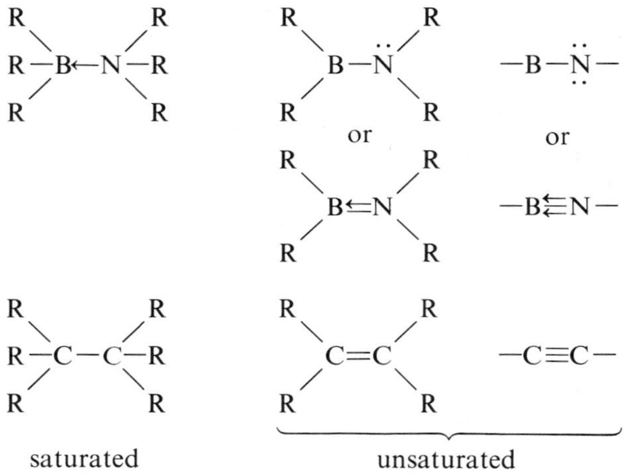

FIG. 10.19 Boron–nitrogen compounds form isoelectronic structures with carbon compounds.

Boron Nitride   Boron nitride, $(BN)_x$, can be prepared by the action of nitrogen or ammonia on elemental boron at 1000° or by heating borax with ammonium chloride:

$$Na_2B_4O_7 \cdot 10H_2O + 2NH_4Cl \rightarrow 2NaCl + 2BN + B_2O_3 + 14H_2O. \quad (50)$$

The compound melts at 3000° under pressure. It is inert to many common reagents, but decomposition occurs when it is heated with acids or with fused alkali.

Boron nitride exists in two modifications. The form normally obtained from the usual preparative procedures exhibits a layer-type structure, shown in Fig. 10.20a, reminiscent of graphite. Each layer consists of a hexagonal arrangement of alternating boron and nitrogen atoms similar to the hexagonal arrangement of carbon atoms in the graphite structure (Fig. 11.2). However, in contrast to the graphite structure, the boron atoms in one layer

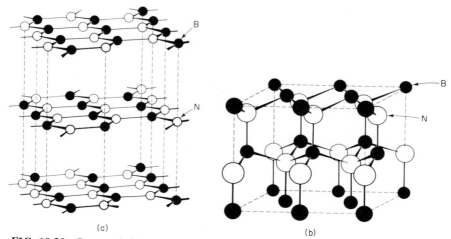

FIG. 10.20   Boron nitride crystallizes in a graphite-like structure (a) as well as in the zinc-blende structure (b). Boron phosphide and arsenide also crystallize in the zinc-blende structure.

of the BN structure are above the nitrogen atoms of the next layer. The B—N distance is 1.446 Å, which is shorter than that expected for a single bond between these atoms (1.54 Å). A comparable decrease in the carbon–carbon distances in the graphite sheets (1.42 Å) is observed over the carbon–carbon single-bond distance (1.54 Å). Again in contrast to the graphite structure, which has a larger interlayer distance (3.35–3.45 Å), the interlayer distance in BN is the same as the distance between boron and nitrogen atoms within a layer. The hexagonal form of boron nitride can intercalate foreign atoms in the same way as graphite (*35*). However, in spite of the structural and chemical resemblance to graphite of the hexagonal form of boron nitride, the physical properties are very different. For example, boron nitride is a white substance and a very good insulator.

When hexagonal boron nitride (density 2.25 g cm^{-3}) is heated to 1400–1800° at 85,000 atm, it is transformed to a more dense form (3.45 g cm^{-3}) which has the zinc-blende structure shown in Fig. 10.20b (*36*). This form, called borazon, is isostructural and isoelectronic with diamond and exhibits many of the properties of diamond. Boron reacts with phosphorus and arsenic—also Group IV elements—to yield a phosphide (BP) and an arsenide (BAs) as well as with Group IV elements to yield binary compounds which also have the zinc-blende structure shown in Fig. 10.20.

Boron–Nitrogen Adducts   Many trivalent boron compounds form adducts with amines and their derivatives

$$R_3B + NR_3 \rightarrow R_3BNR_3, \tag{51}$$

and as mentioned earlier these adducts are isoelectronic and isostructural with the corresponding R_3CCR_3 compounds. Adducts carrying halogen atoms, organo groups, and hydrogen substituents on the boron atom have been prepared with a variety of amines. Boron esters are weak acceptors, undoubtedly because of delocalization of electrons from oxygen to boron (cf., structure **16**); compounds of the type $B(NR_2)_3$ are devoid of acceptor properties probably because of a similar delocalization of electrons from the nitrogen atom to the boron atom.

In a formal sense, the coordinate-covalent bond formed between boron and nitrogen in these adducts is polar (**21**). The planar structure of the three-coordinate, trivalent boron compounds changes to essentially a tetrahedral

$$
\begin{array}{ccc}
R & & R \\
\diagdown & & \diagup \\
R - \overset{-}{B} & \leftarrow & \overset{+}{N} - R \\
\diagup & & \diagdown \\
R & & R
\end{array}
$$

21

structure when coordination occurs (Table 10.15). The known B—N bond distances in the adducts are slightly greater than that calculated (1.54 Å) from the single-bond radii of boron and nitrogen. Thus, the simplest theoretical description of the B—N adducts involves the use of sp^3 hybrid orbitals on both the boron and nitrogen atoms in the bonding scheme.

Aminoboranes (V)   When a trivalent nitrogen atom is substituted for a boron substituent, the resultant compound, R_2BNR_2, is isoelectronic with systems containing a carbon-to-carbon double bond (see Fig. 10.19). Aminoboranes can be made by removing the elements of hydrogen halide from a compound containing a boron–halogen bond and a secondary amine

$$R_2BCl + HNR_2 + (CH_3)_3N \rightarrow R_2BNR_2 + (CH_3)_3NH^+Cl^- ; \quad (52)$$

TABLE 10.15   The structural parameters for some adducts of trivalent boron compounds

$R_3BNR'_3$	$\angle RBR$	$\angle R'NR'$	$d(B-N)$, Å
F_3BNH_3	111°	—	1.60
$F_3BNH_2(CH_3)$	111°	—	1.58
$F_3BN(CH_3)_3$	107°	114°	1.58
$H_3BN(CH_3)_3$	—	—	1.62
H_3NBH_3	—	—	1.56

a base such as trimethylamine often is used to aid the elimination of the hydrogen halide. Although little direct structural evidence exists to support possible double-bond formation in these compounds, their physical properties and infrared spectra suggest that electron delocalization does occur in some derivatives of this type. Depending upon the nature of the substituents on nitrogen or boron, the boron atom in aminoboranes can become electronically saturated by either internal coordination (22) or intermolecular coordination (23) (37). Thus, in the case where the boron substituents are

$$R_2\bar{B} \;{\leftharpoondown}\; \overset{+}{N}R_2'$$

$$\overset{-}{R_2}\overset{}{B}\overset{}{-}\overset{+}{N}R_2'$$
$$\uparrow\quad\downarrow$$
$$\underset{+}{R_2'N}\overset{}{-}\underset{-}{\overset{}{B}R_2}$$

22

23

relatively electronegative, such as in $Cl_2BN(CH_3)_2$, saturation of the boron atom is achieved by dimerization. If the boron substituents are relatively less electronegative [$(CH_3)_2BN(CH_3)_2$], internal coordination yields monomeric species. The monomeric aminoboranes exhibit B—N stretching frequencies that suggest that the bond order has increased over that of the dimeric species (38). In several cases, two forms of the aminoborane can be obtained; one is very reactive and monomeric, whereas the other is usually noticeably less reactive and is dimeric. The reactive form slowly changes to the less reactive; the reverse process can be accomplished by heating. A particularly striking example of this interconversion occurs in the case of $(CH_3)_2$-BNH_2 (39). This substance exists as a liquid (presumably monomeric) which boils at 1° to give a monomeric gas. Upon standing, the liquid deposits colorless crystals (mp 9°). Apparently the equilibrium

$$(CH_3)_2\overset{-}{B}\overset{+}{N}H_2 \;\rightleftharpoons\; (CH_3)_2\overset{-}{B}\underset{\underset{+}{NH_2}}{\overset{\overset{+}{NH_2}}{\diagup\diagdown}}\overset{-}{B}(CH_3)_2 \tag{53}$$

can be easily established in this system. Even aminoboranes containing chloro substituents can be isolated as monomeric species, but they revert to dimers. In such cases, the monomers are very sensitive to moisture and are violently hydrolyzed by water. The dimers, on the other hand, are stable to hot aqueous acid solution.

The properties of the aminoborane monomers can be understood theoretically by using sp^2 hybrid orbitals to form σ bonds to the boron substituents. The unused p orbital then becomes available for back-donation of the unshared electron pair on the nitrogen atom. The dimer presumably then requires sp^3 hybrid orbitals on the boron atom to accommodate the four-coordination at that site.

As might be expected from their position in the periodic chart, both phosphorus (*40, 41*) and arsenic (*42*) form compounds analogous to the aminoboranes. Compounds of the type R_2BPR_2 are known only as polymers, the trimers and tetramers being exceptionally stable. Linear polymers, such as $[(CH_3)_2PBH_2]_n$ with $n \approx 80$, are also known. In one case, $[(CH_3)_2PBH_2]_3$, x-ray diffraction data indicate that the trimer is cyclic, with a B_3P_3 ring system in a chair configuration (**24**); the positions of the boron substituents

24

were not reported. In general the arsenic analogues (R_2BAsR_2) are less stable than the phosphinoboranes. Only in the case of aminoboranes is there evidence for multiple-bond formation in this type of compound.

Borazines (*VI*)   Boron–nitrogen species carrying only one substituent on each atom (see Fig. 10.19) are known, but they exist as trimers, $(XBNR)_3$, and are called borazines. The most satisfactory method for the preparation of borazines involves the condensation of amines with trivalent boron compounds. The condensation reactions can be imagined as occurring via the stepwise elimination of small molecules from the boron–nitrogen adduct, although the mechanism for this reaction is unknown:

$$X_3B + NR_3 \longrightarrow X_3BNR_3 \tag{54}$$

$$X_3BNR_3 \longrightarrow X_2BNR_2 + RX \tag{55}$$

$$3X_2BNR_2 \longrightarrow (XBNR)_3 + 3RX. \tag{56}$$

The condensation reaction [Eqs. (54)–(56)] invariably gives a symmetrically substituted borazine, and in some cases it is possible to obtain unsymmetrical derivatives. It is not necessary to isolate the intermediates in the preparation of borazines, but in several instances the boron–nitrogen adduct X_3BNR_3, as well as the aminoborane (X_2BNR_2), have been isolated before their conversion to the corresponding borazine. The addition compound can be formed by direct combination [Eq. (54)], the displacement of hydrogen halide from an ammonium salt,

$$[R_3'NH]^+X^- + R_3B \longrightarrow R_3BNR_3' + HX, \tag{57}$$

or by the displacement of hydride ion from a borohydride with an ammonium salt,

$$[R_3'NH]^+X^- + MBH_4 \longrightarrow H_3BNR_3' + H_2 + MX. \qquad (58)$$

Conversion of the adduct to the borazine [Eqs. (55) and (56)] occurs most readily when RX is a hydrogen halide, which can be eliminated by pyrolysis of the pure adduct, by refluxing the adduct in a solvent, or by treatment with a strong base like triethylamine. Vigorous pyrolytic conditions ($\sim 450°$) must be used if RX is an alcohol or a hydrocarbon,

$$3(n\text{-BuO})_2BNHR \longrightarrow (n\text{-BuOBNR})_3 + 3n\text{-BuOH} \qquad (59)$$

$$3(CH_3)_3B + 3NH_3 \longrightarrow (CH_3BNH)_3 + 6CH_4. \qquad (60)$$

The molecular structures for a number of borazines have been determined by diffraction methods. The data show that these compounds possess a cyclic structure of alternating boron and nitrogen atoms. All of the atoms in the ring lie in a plane, within experimental error. The boron–nitrogen bond distances are about 1.42 Å, which is considerably shorter than the B—N distance in the boron–nitrogen adducts R_3BNR_3. Moreover, all the bond distances in the ring are equal. These results strongly suggest that the borazine cyclic trimer is best described as containing a B—N multiple bond (**25**) rather than a single bond (**26**). The single-bonded structure (**26**) implies that borazine is

| 25a | 25b | 26 |

a cyclic triamine in which the nitrogen atoms exhibit a tetrahedral arrangement of electron pairs; under these circumstances the ring should be puckered. The two possible valence-bond structures that exist with multiple boron–nitrogen bonds (**25**) are reminiscent of the resonance structures for benzene. The electronic formulations shown in **25** are consistent with the known structural data for the borazine nucleus. Presumably all the ring atoms use sp^2 hybridized orbitals to form the three σ bonds necessary. The unhybridized p orbitals, which are normal to the plane of the ring, form a π system containing three electron pairs which were originally associated with the nitrogen atoms (**27**).

Although the borazine ring has many similarities to the isoelectronic benzene ring, the reactions of the two systems are markedly different. For example, the borazine nucleus readily undergoes addition reactions with molecules containing potentially acidic hydrogen atoms such as hydrogen

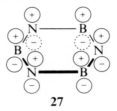

27

halides, water, methanol, and carboxylic acids to give borazanes:

$$(RBNR')_3 + 3HX \rightarrow \left(\begin{array}{c} X \qquad\; H \\ \diagdown\quad\diagup \\ BN \\ \diagup\quad\diagdown \\ R \qquad\; R' \end{array} \right)_3 . \qquad (61)$$

Benzene and its derivatives on the other hand do not undergo such addition reactions readily. The products of Eq. (61) are the cyclohexane analogues of the borazine ring. Indeed, x-ray diffraction studies of the compound $(HBCH_3)_3(HNCH_3)_3$ show it to be a cyclic structure with alternating boron and nitrogen atoms in a chair configuration (**28**) (*43*).

28

Borazines undergo substitution reactions primarily at the boron sites. Borazines carrying a hydrogen atom at the boron site are reducing agents; that is, such hydrogen atoms are hydridic in nature. *B*-Hydrido- and *B*-halogeno-borazines react with Grignard reagents, yielding *B*-substituted derivatives. Such reactions have been used to prepare unsymmetrical *B*-substituted borazines. Substitution reactions rarely occur at the nitrogen sites, and unsymmetrically *N*-substituted borazines are relatively rare.

3.8 / Nitrogen Derivatives of Aluminum, Gallium, Indium, and Thallium

Although the atom pair MN (where M is Al, Ga, In, or Tl) is isoelectronic with C_2, the striking similarity in structures observed for the boron–nitrogen compounds does not occur. The nitrides AlN, GaN, and InN, which have an appreciable amount of ionic character, exhibit the wurtzite structure rather than the zinc-blende structure (Fig. 10.20) of boron nitride. However, the phosphides, arsenides, and antimonides of aluminum, gallium, and indium, which are essentially ionic, crystallize in the zinc-blende structure.

Amine Adducts   In contrast to boron, nitrogen derivatives of aluminum, gallium, indium, and thallium have not been extensively investigated; the greatest amount of information on such systems has been gathered for aluminum–nitrogen compounds. Trivalent compounds readily form 1 : 1 adducts with the general formula $R_3NMR'_3$ (M = Al, Ga, In, Tl) with amines. Thus, compounds such as $(CH_3)_3NAlCl_3$, $(CH_3)_3NAlH_3$, R_3NAlR_3, $H_3NGa(C_2H_5)_3$, $(CH_3)_3NGa(CH_3)_2Cl$, $(CH_3)_3NIn(CH_3)_3$, and $(CH_3)_3$-$NTl(CH_3)_3$ have been prepared and characterized.

Nitrogen derivatives of aluminum, gallium, indium, and thallium that are analogous to aminoboranes have been prepared, but the analogs of borazine are not yet known. The best known analogs of aminoboranes have been prepared by the pyrolysis of ammonia or secondary amine adducts of the alkyl derivatives of these elements:

$$H_3NGa(CH_3)_3 \rightarrow H_2NGa(CH_3)_2 + CH_4 \qquad (62)$$

$$(CH_3)_2NHAl(CH_3)_3 \rightarrow (CH_3)_2NAl(CH_3)_2 + CH_4. \qquad (63)$$

These substances are dimeric and have been formulated as cyclic adducts (**29**).

29

3.9   /   Hydrogen Compounds of Boron (*VII*)

Physical Properties and Structures   Of the Group III elements, the hydrides of boron have been the most extensively investigated. From its position in the periodic table, boron would be expected to form the simplest hydride, with a stoichiometry corresponding to BH_3. In fact, the simplest known hydride of boron possesses the expected stoichiometry, but its molecular weight corresponds to the formula B_2H_6. Indeed, a wide variety of boron–hydrogen compounds, called boranes, have been isolated and characterized (Table 10.16); most of the boranes isolated have molecular formulas corresponding to B_nH_{n+4} or B_nH_{n+6}. The lighter boranes (through B_5) spontaneously react with air, whereas the remaining compounds are air-stable.

TABLE 10.16   The physical properties of boranes

Formula	Name	Mp, °C	Bp, °C
B_2H_6	diborane	−165.6	−92.5
B_4H_{10}	tetraborane	−120	18
B_5H_9 (stable)	pentaborane-9	−46.6	48
B_5H_{11} (unstable)	pentaborane-11	−123	63
B_6H_{10}	hexaborane-10	−62.3	110
B_6H_{12}	hexaborane-12	−82.3	80–90
B_8H_{12}	octaborane-12	−20	—
B_8H_{14}	octaborane-14	—[a]	—
B_9H_{15}	enneaborane	2.6	—
$B_{10}H_{14}$	decaborane	99.7	213
$B_{20}H_{16}$	icosaborane-16	196–199	—

[a] Unstable at room temperature.

The boranes were originally prepared from the hydrolysis of magnesium boride by dilute hydrochloric acid. Diborane is most readily prepared by the reaction of $BF_3 \cdot (C_2H_5)_2O$ complex with lithium hydride:

$$6LiH + 8BF_3 \cdot (C_2H_5)_2O \rightarrow 6LiBF_4 + B_2H_6 + 8(C_2H_5)_2O. \quad (64)$$

It can also be prepared by the reaction of sulfuric acid with sodium borohydride:

$$2NaBH_4 + H_2SO_4 \rightarrow B_2H_6 + 2H_2 + Na_2SO_4. \quad (65)$$

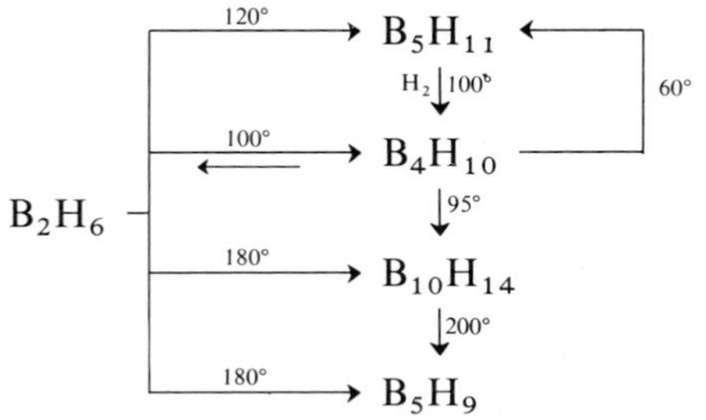

FIG. 10.21   Conditions for the formation of the higher boron hydrides from diborane.

The higher boranes can be made from diborane by heating this substance either alone or with hydrogen (Fig. 10.21) (44).

Extensive data, using a variety of techniques, are available on the structures of the boron hydrides (VIII). Broadly speaking, the higher boron hydrides and their derivatives can be described as cage structures. The structure of diborane, shown in Fig. 10.22, indicates that two types of hydrogen atoms are

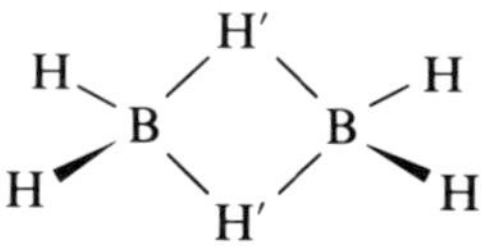

$$d(B-H) = 1.19 \text{ Å}$$
$$d(B-H') = 1.33 \text{ Å}$$
$$d(B-B) = 1.77 \text{ Å}$$
$$\angle HBH = 121.5°$$
$$\angle H'BH' = 97°$$

FIG. 10.22 Structural parameters for diborane.

present. A longer B—H distance is observed for the bridging hydrogen atoms than for the terminal hydrogen atoms. The framework of boron atoms in the lighter boron hydrides (B_4H_{10}, B_5H_9, B_5H_{11}), shown in Figs. 10.23 and 10.24, is related to the octahedral arrangement for the B_6^{2-} ion in CaB_6. In addition, the structure of the heavier boron hydrides (B_9H_{15} and $B_{10}H_{14}$) appears to be related to that of crystalline boron, given in Fig. 10.1.

The boron hydrides form interesting examples of electron-deficient compounds, discussed previously in Chapters 8 and 9. For example, in the case of the simplest borane, B_2H_6, assignment of electrons to form chemical bonds in the conventional way leads to a situation where there are insufficient electrons available to hold the two BH_3 moieties together (30). The structural evidence gathered in Fig. 10.22 indicates the presence of a B—B distance close to that observed in B_2Cl_4 (Table 10.10); the terminal B—H distance is shorter than that in the BH_4^- ion (1.25 Å), whereas the bridging distance is longer.

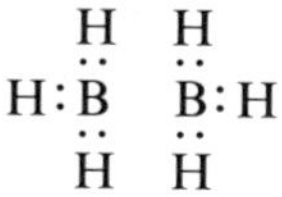

$$\begin{array}{cc} \text{H} & \text{H} \\ \text{H:B} & \text{B:H} \\ \text{H} & \text{H} \end{array}$$

30

The most successful theoretical interpretation of the structure of B_2H_6 involves the formation of a three-centered bond, that is a bond formed from the overlap of three orbitals (IX), although a molecular-orbital description leading to essentially the same result is available (45). Most covalent bonds between atoms are described in terms of the overlapping of only two orbitals (Chapter 4). A pictorial representation of the formation of a three-centered bond is given in Fig. 10.25. The three-centered orbital is lower in energy

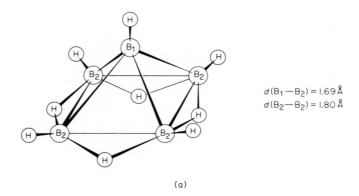

$d(B_1-B_2) = 1.69\,\text{Å}$
$d(B_2-B_2) = 1.80\,\text{Å}$

(a)

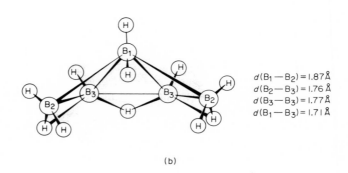

$d(B_1-B_2) = 1.87\,\text{Å}$
$d(B_2-B_3) = 1.76\,\text{Å}$
$d(B_3-B_3) = 1.77\,\text{Å}$
$d(B_1-B_3) = 1.71\,\text{Å}$

(b)

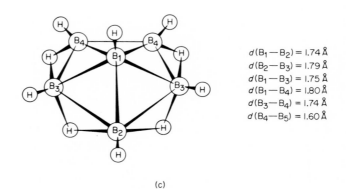

$d(B_1-B_2) = 1.74\,\text{Å}$
$d(B_2-B_3) = 1.79\,\text{Å}$
$d(B_1-B_3) = 1.75\,\text{Å}$
$d(B_1-B_4) = 1.80\,\text{Å}$
$d(B_3-B_4) = 1.74\,\text{Å}$
$d(B_4-B_5) = 1.60\,\text{Å}$

(c)

FIG. 10.23 Structural parameters for (a) B_5H_9, (b) B_5H_{11}, and (c) B_6H_{10}.

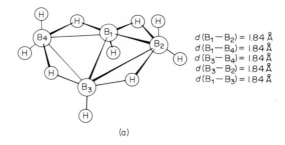

$d(B_1-B_2) = 1.84 \text{ Å}$
$d(B_1-B_4) = 1.84 \text{ Å}$
$d(B_3-B_4) = 1.84 \text{ Å}$
$d(B_3-B_2) = 1.84 \text{ Å}$
$d(B_1-B_3) = 1.84 \text{ Å}$

(a)

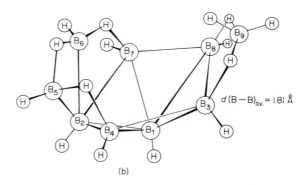

$d(B-B)_{av.} = 1.81 \text{ Å}$

(b)

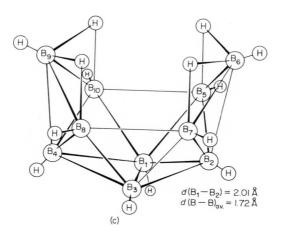

$d(B_1-B_2) = 2.01 \text{ Å}$
$d(B-B)_{av.} = 1.72 \text{ Å}$

(c)

FIG. 10.24 Structural parameters for (a) B_4H_{10}, (b) B_9H_{15}, and (c) $B_{10}H_{14}$.

FIG. 10.25   A three-centered bond forms when three atomic orbitals mutually interact.

than any of the original orbitals. Accordingly, when such orbitals are occupied, a stable bond is formed. In this view, one electron pair can bind three nuclei. Thus, in the case of B_2H_6 (Fig. 10.26) two boron atoms and the two bridging hydrogen atoms form two three-centered bonds while the terminal B—H bonds are conventional two-centered bonds. The boron

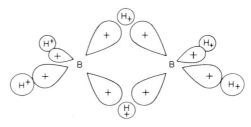

FIG. 10.26   One possible way to form diborane using three-centered bonds for the bridging hydrogen atoms.

orbitals used to form the three-centered bond are obviously *sp*-type hybrid orbitals, the extent of hybridization being estimated from the terminal H—B—H bond angle. It should be recalled from Chapter 4 that sp^2 hybrid orbitals make an angle of 120°, whereas the angle between *sp* hybrid orbitals is

TABLE 10.17   Bonding in the boron hydrides

Symbol	Meaning	Number of electrons
B—H	Terminal hydrogen, two-centered	2
B $\overset{H}{\frown}$ B	Hydrogen bridge, three-centered	2
B—B	Boron–boron, two-centered	2
B $\overset{B}{\frown}$ B	Boron bridge, open three-centered	2
B $\overset{B}{\wedge}$ B	Boron bridge, closed three-centered	2

180°. Apparently an intermediate hybridization is required for B_2H_6, since the terminal H—B—H angle is 121.5°.

The bonding schemes in the higher boron hydrides are more complex than in diborane. Not only are bridging and terminal hydrogen atoms present, but there are also normal B—B bonds as well as three-centered bonds involving three boron atoms. For example, in the case of tetraborane (Fig. 10.24) normal two-centered bonds exist between boron atoms as well as between boron and hydrogen atoms; three-centered bonds involving two boron atoms and a hydrogen atom are also present. In an attempt to formulate the bonding in the boron hydrides simply, the set of symbols shown in Table 10.17, analogous to those used for conventional bonding representation, were developed (*VIII*). In terms of the symbolism shown in Table 10.17, the electronic structures of B_2H_6, B_4H_{10}, and $B_{10}H_{14}$ are shown in Fig. 10.27.

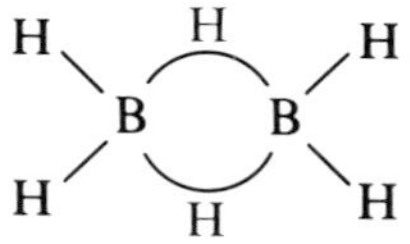

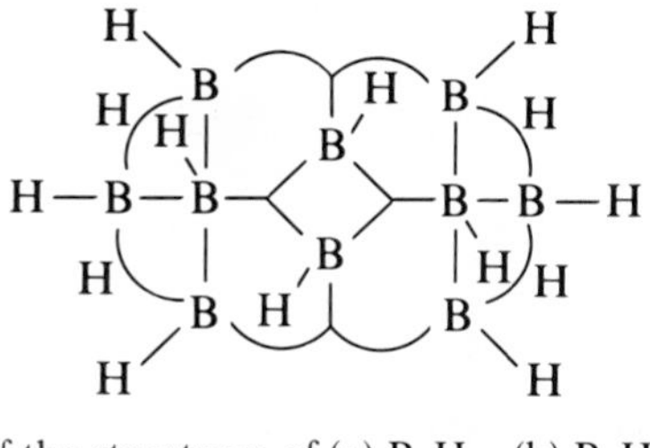

FIG. 10.27 Descriptions of the structures of (a) B_2H_6, (b) B_4H_{10}, and (c) $B_{10}H_{14}$ in terms of direct and bridging B—B bonds and B—H bonds.

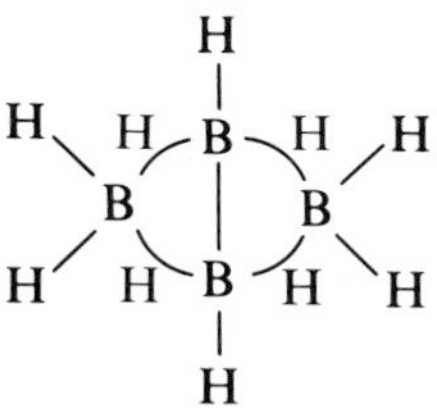

Chemical Properties   The most widely investigated of the boranes is B_2H_6. The dissociation energy of B_2H_6 has been estimated by various methods to be between 28 and 59 kcal mole^{-1} (*46*). Borine, BH_3, is not known to exist for appreciable periods of time (*47*), although it is recognized as an intermediate species in several reactions.

Diborane has been observed to react in several ways (Fig. 10.28). In one type of reaction, the basic BH_2B framework incorporating the three-centered bonds remains intact. Thus, in the presence of $B(CH_3)_3$, boron trihalides, or HCl, diborane yields derivatives in which terminal hydrogen atoms have been replaced (Fig. 10.28). Structural data are available which support this conclusion for several compounds.

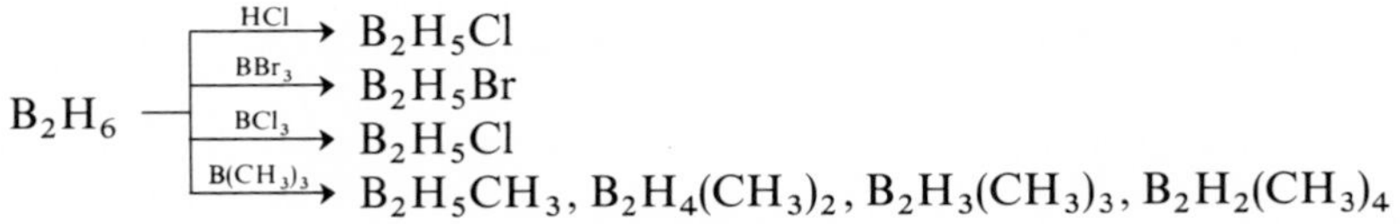

$$B_2H_6 \begin{array}{l} \xrightarrow{\text{HCl}} B_2H_5Cl \\ \xrightarrow{\text{BBr}_3} B_2H_5Br \\ \xrightarrow{\text{BCl}_3} B_2H_5Cl \\ \xrightarrow{\text{B(CH}_3)_3} B_2H_5CH_3, B_2H_4(CH_3)_2, B_2H_3(CH_3)_3, B_2H_2(CH_3)_4 \end{array}$$

FIG. 10.28   Some reactions of diborane.

Diborane is cleaved symmetrically into two BH_3 units by a variety of Lewis bases to form compounds of the type $L:BH_3$ (Fig. 10.29). If the Lewis bases are neutral, the adducts with BH_3 are characteristically covalent, with a tetrahedral arrangement of atoms about the central boron. The electron

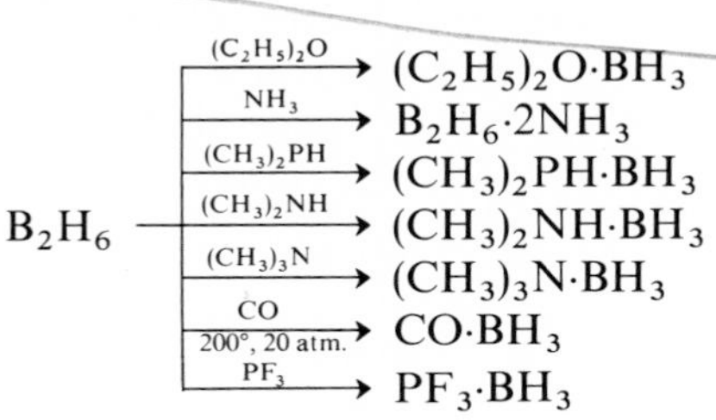

$$B_2H_6 \begin{array}{l} \xrightarrow{(C_2H_5)_2O} (C_2H_5)_2O\cdot BH_3 \\ \xrightarrow{NH_3} B_2H_6\cdot 2NH_3 \\ \xrightarrow{(CH_3)_2PH} (CH_3)_2PH\cdot BH_3 \\ \xrightarrow{(CH_3)_2NH} (CH_3)_2NH\cdot BH_3 \\ \xrightarrow{(CH_3)_3N} (CH_3)_3N\cdot BH_3 \\ \xrightarrow[200°, 20\ \text{atm.}]{CO} CO\cdot BH_3 \\ \xrightarrow{PF_3} PF_3\cdot BH_3 \end{array}$$

FIG. 10.29   Some adducts formed by diborane.

structures of such species are best described in terms of sp^3 hybrid orbitals on the central boron atom. If ammonia is used as a Lewis base, a diammoniate is formed, $B_2H_6\cdot 2NH_3$. This compound has been shown to be a borohydride of the boron-containing cation (31) which arises from the unsymmetrical cleavage of diborane:

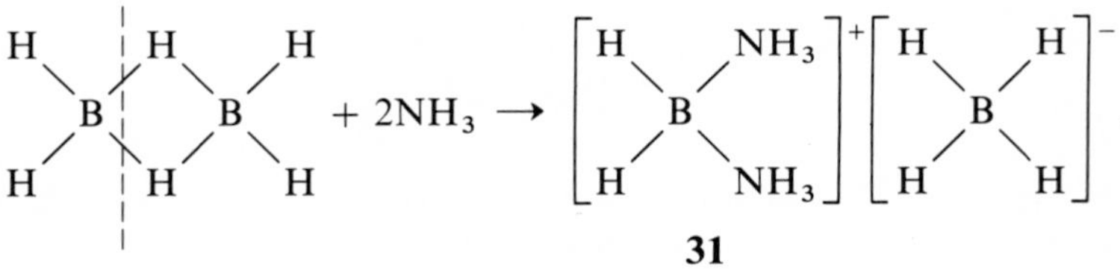

$$(66)$$

31

The first step in all cleavage reactions involves displacement of one of the bridging hydrogen atoms to form a species with a single hydrogen bridge:

$$B_2H_6 + L \rightarrow H_2\underset{\underset{L}{|}}{B}-H-BH_3. \qquad (67)$$

The second step involves displacement of the single-bridged species and leads to two different products,

$$H_2\underset{\underset{L}{|}}{B}-H-BH_3 + L \rightarrow 2H_3BL \qquad (68)$$

$$L + H_2\underset{\underset{L}{|}}{B}-H-BH_3 \rightarrow [H_2BL_2]^+[BH_4]^-. \tag{69}$$

The first case [Eq. (68)] corresponds to symmetrical cleavage of B_2H_6, whereas the second reaction [Eq. (69)] leads to unsymmetrical cleavage products. Three principal factors govern the course of the second reaction (48), namely, the inductive effect of the ligand L after it is coordinated in the first step, the donor character of the ligand, and steric effects.

Unsymmetrical cleavage of B_2H_6 yields species containing boronium ions (XI). Less direct methods are available for the preparation of these species from anionic or neutral borine complexes (49–53):

$$NaBH_4 + 4L + 2I_2 \rightarrow [H_2BL_2]^+I^- + 2LH^+I^- + NaI \tag{70}$$

$$LBH_3 + 2L + I_2 \rightarrow [H_2BL_2]^+I^- + LH^+I^- \tag{71}$$

$$LBH_2R + 2L + I_2 \rightarrow [HRBL_2]^+I^- + LH^+I^- \tag{72}$$

$$L = N(CH_3)_3 \text{ and pyridine}; R = C_6H_5$$

A less basic ligand can be displaced from a boronium ion by a more basic ligand

$$H_2BL_2^+ + 2L' \rightarrow H_2BL_2'^+ + 2L, \tag{73}$$

the ability to displace being given by the order

$$\text{diamines} > \text{amines} > \text{phosphines} \sim \text{arsines} > \text{sulfides.}$$

A large number of borane adducts are known; these can generally be prepared by the reaction of a Lewis base with diborane or the displacement of a hydride ion from BH_4^-. Using the suggestion that BH_3 is isoelectronic with oxygen (54), a variety of adducts have been prepared. For example, this analogy led to the isolation and characterization of borane carbonyl (BH_3CO), and it also can be used to rationalize the formulation of the diamine adduct, $BH_3CO\cdot2NH_3$ (32), in terms of the analogous reaction of CO_2 with ammonia to form the isoelectronic compound ammonium carbamate (33) (55, 56).

$$[H_2N-\overset{\overset{\displaystyle O}{\|}}{C}-BH_3]^-NH_4^+ \qquad [H_2N-\overset{\overset{\displaystyle O}{\|}}{C}-O]^-NH_4^+$$

$$\textbf{32} \qquad\qquad\qquad \textbf{33}$$

The hydrogen atoms in diborane are hydridic and, as such, react with reagents containing acidic hydrogen atoms to yield elemental hydrogen:

$$B_2H_6 + 6H_2O \rightarrow 2B(OH)_3 + 6H_2 \tag{74}$$

$$B_2H_6 + 6ROH \rightarrow 2B(OR)_3 + 6H_2. \tag{75}$$

In the course of these reactions conventional trivalent boron derivatives are formed.

Compounds Related to Boranes   A class of species closely related to the boron hydrides are the boron hydride anions (Table 10.18) (X). Some of these anions can be imagined to arise (not necessarily by chemical reaction) from the coordination of the hydride ion as a Lewis base to a neutral boron hydride; in some instances the corresponding adducts of neutral Lewis bases are known (Table 10.18). The simplest example is the borohydride

TABLE 10.18   The boron hydride anions

Anion	Coordination compound
BH_4^-	$BH_3 \cdot NH_3$
$B_3H_8^-$	$B_3H_7 \cdot NH_3$
$B_{10}H_{14}^{2-}$	$B_{10}H_{12} \cdot 2(CH_3)_2S$
$B_{12}H_{12}^{2-}$	—
$B_{11}H_{14}^-$	—
$B_{20}H_{18}^{2-}$	$B_{20}H_{16} \cdot 2(CH_3)_2S$
$B_{20}H_{19}^{3-}$	—

anion, BH_4^-, in which the borine moiety (BH_3) is coordinately covalently bonded to a hydride ion. Indeed, the BH_4^- anion can be prepared by the direct reaction of a metal hydride with diborane,

$$2LiH + B_2H_6 \rightarrow 2Li(BH_4). \tag{76}$$

A wide range of metal borohydrides are known, including substituted borohydride anions such as $BH(OCH_3)_3^-$. The alkali-metal borohydrides are generally nonvolatile and stable in dry air. Sodium borohydride dissolves in water, tetrahydrofuran, and ethylene glycol dimethyl ether (diglyme) to give solutions that are powerful reducing agents.

The borohydride anions can be prepared from the boranes or their derivatives:

$$NaBH_4 + B_2H_6 \rightarrow NaB_3H_8 + H_2 \tag{77}$$

$$2NaBH_4 + 5B_2H_6 \rightarrow Na_2B_{12}H_{12} + 13H_2. \tag{78}$$

As is the case with the boranes, the known structures of the heavier boro-hydride anions, shown in Fig. 10.30, incorporate boron cage structures which contain B—B bonds. The $(B_3H_8)^-$ anion, on the other hand, contains BHB bridging bonds as well as B—B bonds.

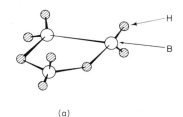

(a)

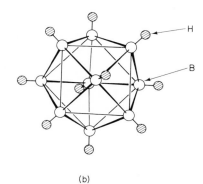

(b)

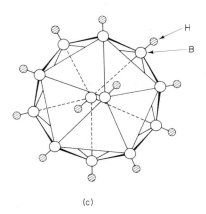

(c)

FIG. 10.30 The structure of some boron hydride anions: (a) $B_3H_8^-$, (b) $B_{10}H_{10}^{2-}$, (c) $B_{12}H_{12}^{2-}$.

The ions $(B_{10}H_{10})^{2-}$ and $(B_{12}H_{12})^{2-}$ are remarkably stable, as might be anticipated from their structures. They are kinetically inert toward acids, bases, and oxidizing agents; aqueous solutions of these ions are thermally stable (58–62). The hydrogen atom in both ions can be replaced by halogens to give ionic species of the type $[B_{10}H_{(10-n)}X_n]^{2-}$ and $[B_{12}H_{(12-n)}X_n]^{2-}$; the fully substituted anions $[B_{10}X_{10}]^{2-}$ and $[B_{12}X_{12}]^{2-}$ have been prepared for all halogens except fluorine. Little is known of the isomerism which theoretically should be possible in the substitution derivatives of $(B_{10}H_{10})^{2-}$ and $(B_{12}H_{12})^{2-}$ if these species maintain the boron framework of the parent ions.

The ion $(B_{10}H_{10})^{2-}$ can undergo an oxidative coupling reaction in aqueous solution, giving rise to two new borohydride anions,

$$2(B_{10}H_{10})^{2-} + 4Fe^{3+} \longrightarrow (B_{20}H_{18})^{2-} + 2H^+ + 4Fe^{2+} \tag{79}$$

$$4(B_{10}H_{10})^{2-} + Ti^{4+} \longrightarrow (B_{20}H_{18})^{2-} + (B_{20}H_{19})^{3-} + Ti^{3+} + 3H^+. \tag{80}$$

It appears that the $(B_{20}H_{18})^{2-}$ ion contains two B_{10} units joined by two B—H—B three-center bonds (34) while the $(B_{20}H_{19})^{3-}$ ion involves only one such bond (35).

34   35

The carboranes are borane structures in which a BH_2 unit (or a BH^- in a boron hydride anion) is replaced, in a figurative sense, by the isoelectronic CH group (Table 10.19) (XII). Carboranes are prepared by the reaction of

TABLE 10.19   Carboranes and their relationship to boron hydrides

Carborane		Parent borane or borohydride anion
$B_3C_2H_5$		—[a]
$B_4C_2H_6$		—[a]
$B_4C_2H_8$	$\xleftarrow[+2CH]{-2BH_2}$	B_6H_{10}
$B_5C_2H_5$		—[a]
$B_{10}C_2H_{12}$	$\xleftarrow[+2CH]{-2BH^-}$	$[B_{12}H_{12}]^{2-}$

[a] Parent boron hydride or borohydride anion is unknown.

acetylene or its derivatives with boron hydrides,

$$B_5H_9 + C_2R_2 \rightarrow B_4H_6C_2R_2 + BH_3 \tag{81}$$

$$B_{10}H_{12} \cdot 2(C_2H_5)_2S + C_2H_2 \rightarrow B_{10}C_2H_{12} + H_2 + 2(C_2H_5)_2S. \tag{82}$$

Probably the most extensively studied carborane is $B_{10}C_2H_{12}$ and its derivatives (XIII). The product of reaction (82) is the so-called *ortho*-isomer, the structure of which is the twelve-atom cage shown in Fig. 10.31. Prolonged

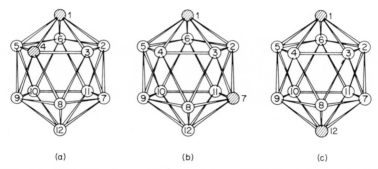

(a)   (b)   (c)

FIG. 10.31   The basic unit of carboranes with the $B_{10}C_2$ grouping is an icosahedron. The three isomers that have been formed with this combination of atoms are called (a) *ortho*, (b) *meta*, and (c) *para* in analogy with benzene.

heating of the *ortho*-isomer in an inert atmosphere gives the *meta*-isomer (*61*) where the carbon atoms have been separated by a BH group. The *para*-isomer has also been prepared in small quantities.

Like the isoelectronic borohydride anion $(B_{12}H_{12})^{2-}$, $B_{10}C_2H_{12}$ is an extremely stable species. That is, the basic framework of atoms is maintained in the presence of acids, bases, and reasonably strong oxidizing agents. The action of chlorine yields chlorine-substituted species. The B—H groups undergo substitution before CH groups in the process.

3.10 / Hydrides of the Other Elements

In addition to the extensive work on boron hydrides, some information is available on aluminum hydride. This compound has the expected stoichiometry AlH_3, but it appears to be a polymeric substance best formulated as $(AlH_3)_x$. There are several methods for the preparation of $(AlH_3)_x$, but perhaps the simplest involves the reaction of $AlCl_3$ with LiH in ether solution to form a solution of AlH_3,

$$AlCl_3 + 3LiH \rightarrow AlH_3 + 3LiCl, \tag{83}$$

which undoubtedly exists as a complex. If this solution is allowed to stand, a solid etherate of variable composition is formed, $(AlH_3)_x[(C_2H_5)_2O]_y$. If excess LiH is used in the preparation, a compound containing the aluminohydride anion (XIV) is formed,

$$LiH + AlH_3 \rightarrow LiAlH_4. \tag{84}$$

The extensive series of aluminum–hydrogen compounds analogous to the boron–hydrogen system is not yet known.

At low temperature, AlH_3 reacts with a variety of Lewis bases to form the expected 1 : 1 adducts (XV),

$$(AlH_3)_x + xL \rightarrow xAlH_3 \cdot L \tag{85}$$

$$L = NH_3, NH_2CH_3, NH(CH_3)_2, N(CH_3)_3, P(C_2H_5)_3.$$

Amine adducts of aluminum hydride can be prepared from the direct reaction of aluminum and hydrogen in the presence of the amine (*64*),

$$Al + \tfrac{3}{2}H_2 + R_2NH \rightarrow AlH_3NR_2H. \tag{86}$$

The secondary amine adducts are unstable and lose hydrogen below room temperature

$$AlH_3NR_2H \rightarrow H_2AlNR_2 + H_2 \tag{87}$$

to form aminoalanes.

Of the remaining elements in this family, only gallium has been reported to give a hydride, Ga_2H_6, which is dimeric and presumably possesses a structure similar to that of diborane (*63*). Gallium hydride is less stable than diborane, decomposing rapidly at room temperature to the elements. Tetramethyl digallane, $Ga_2H_2(CH_3)_4$, has also been reported. Structural information is not available for either of these compounds.

3.11 / Organic Derivatives of Boron

A variety of synthetic methods are available for the preparation of compounds containing a B—C bond, and they generally fall into one of three general types of reactions. The most useful method of preparation on a laboratory scale involves the reaction of a boron trihalide or trialkoxide with an active organometallic compound such as the Grignard reagent,

$$BX_3 + 3RMgX \rightarrow R_3B + 3MgX_2 \tag{88}$$

$$B(OR')_3 + 3RMgX \rightarrow R_3B + 3MgX(OR'). \tag{89}$$

Since such reactions are done in ether solution the boron trihalide is present as an adduct, $(C_2H_5)_2O \cdot BX_3$. Boron trifluoride is usually the preferred reagent since the other boron trihalides cleave ethers readily. In addition to the Grignard reagent, organic-derivatives of lithium, zinc, mercury, and aluminum have been used to alkylate boron compounds.

The second general method for the preparation of B—C bonds involves the addition of B—H bonds to olefins, a reaction described as hydroboration. Thus, diborane adds to olefins under mild conditions to form trialkyl derivatives,

$$B_2H_6 + 6RCH{=}CH_2 \rightarrow 2B(CH_2CH_2R)_3. \tag{90}$$

Boron hydrides also react with acetylenes by *cis* addition,

$$6RC{\equiv}CR + B_2H_6 \rightarrow 2\left(\begin{array}{c} H \\ \diagdown \\ \diagup \quad C{=}C{-\!\!-}B. \\ R \qquad \diagdown \\ \qquad\qquad R \end{array} \right)_3 \tag{91}$$

This reaction is a good route to *cis*-olefins because the B—C bond in such compounds is easily cleaved by acetic acid,

$$\left(\begin{array}{c} H \\ \diagdown \\ \diagup \quad C{=}C{-\!\!-}B \\ R \qquad \diagdown \\ \qquad\qquad R \end{array} \right)_3 + 3HOAc \rightarrow 3 \begin{array}{c} H \qquad H \\ \diagdown \quad\; \diagup \\ C{=}C \\ \diagup \quad\; \diagdown \\ R \qquad R \end{array} + B(OAc)_3. \tag{92}$$

Hydroboration of cyclic olefins also gives a *cis* product.

Hydroboration reactions do not require the handling of diborane since this compound can be prepared in situ from the reaction of sodium borohydride with $AlCl_3$ or BF_3 in ether. Hydroboration reactions have found wide use in organic chemistry as a synthetic tool as well as in structure elucidation (XVI). Hydroboration occurs at terminal carbon–carbon double bonds more rapidly than at nonterminal bonds. Boron–carbon bonds are readily cleaved by H_2O_2 to yield boric acid and the corresponding alcohol.

The third general method of preparing compounds containing a B—C bond involves the reaction between hydrocarbons and boron halides. The reaction is generally limited to unsaturated hydrocarbons,

$$2C_6H_6 + BCl_3 \xrightarrow{\text{Al}} (C_6H_5)_2BCl + 2HCl. \tag{93}$$

This type of reaction has not been exploited as much as the previous reactions.

The organoboron compounds have the expected planar arrangement of three B—C bonds about the central boron atom. These compounds are Lewis acids, forming derivatives with strong Lewis bases that are neutral

$[(CH_3)_3BNR_3]$ or anionic $[Li^+B(CH_3)_4{}^-]$. In addition, triarylboron compounds, which are generally more stable than boron trialkyls, form adducts with OH^-, CN^-, and OR^- that can be isolated as the alkali-metal derivatives $[M^+(C_6H_5)_3BX^-; X = OH^-, CN^-, OR^-]$. The aryl derivatives of boron are sufficiently strong Lewis acids that they react with alkali metals to form radical anions,

$$(C_6H_5)_3B + Na \rightarrow Na^+B(C_6H_5)_3{}^{\cdot-}. \tag{94}$$

Tri-α-naphthylborane adds successively one atom of sodium to give the yellow-brown compound $(C_{10}B_7)_3B^{\cdot-}Na^+$ containing a radical anion and a second atom of sodium to give the deep-violet diamagnetic compound $(C_{10}H_7)_3BNa_2$. There is evidence that the monoadducts dimerize in solution. The stability of these substances is undoubtedly related to their ability to delocalize charge into the aromatic rings.

The boron atoms in these adducts are surrounded by four atoms or groups in a tetrahedral arrangement. Thus, the three-coordinate organoboron compounds, which involve sp^2 hybrid orbitals for bond formation, are converted into four-coordinate compounds, in which sp^3 hybrids are employed.

In general the trialkylboranes are resistant to hydrolysis under normal conditions. These compounds are, however, very sensitive to oxygen; more volatile alkyls are spontaneously inflammable. Controlled oxidation gives alkoxide derivatives,

$$2R_3B + O_2 \rightarrow 2R_2BOR. \tag{95}$$

In one instance, an unstable peroxide $[(CH_3)_2BOOCH_3]$ has been isolated. As mentioned earlier, alkylboron compounds react smoothly with alkaline H_2O_2 cleaving $B{-}C$ bonds. Halogens and hydrogen halides give products arising from $B{-}C$ bond cleavage. The lower boron trialkyls inflame in chlorine or bromine; iodine reacts less readily,

$$(C_3H_7)_3B + I_2 \xrightarrow{150°} (C_3H_7)_2BI + C_3H_7I. \tag{96}$$

Mixtures of alkylboron halide arise when a trialkylboron compound is mixed with a boron trihalide,

$$R_3B + BX_3 \rightleftharpoons R_2BX + RBX_2. \tag{97}$$

The reaction probably proceeds via a bridged intermediate of the type postulated for the exchange of halogen atoms between two different boron trihalides (5).

3.12 / Organic Derivatives of Aluminum (*XVII*), Gallium (*IV*), and Thallium

Preparation and Reactions Of the remaining elements in Group III, the organo-derivatives of aluminum have been most extensively studied. Interest in these compounds is high because they can be used as catalysts in the preparation of polymers from olefins; in combination with certain transition-metal compounds isotactic—or sterically regular—polymers can be formed.

Organoaluminum compounds can be prepared by several routes. Convenient syntheses for laboratory scale involve the reaction of metallic aluminum with organomercury compounds

$$2Al + 3R_2Hg \longrightarrow 2R_3Al + 3Hg, \tag{98}$$

or the reaction of aluminum halides with either organolithium compounds

$$AlX_3 + 3LiR \longrightarrow AlR_3 + 3LiX \tag{99}$$

or Grignard reagents.

A more generally useful method of synthesis involves the reaction of aluminum with alkyl halides

$$2Al + 3RX \longrightarrow R_3Al_2X_3, \tag{100}$$

which yields the mixed alkylaluminum halides. These compounds can be converted into the trialkylaluminum derivatives in any one of a number of ways. Treatment with alkali metals gives the trialkylaluminum derivatives,

$$R_3AlX_3 + 3Na \longrightarrow R_3Al + Al + 3NaX. \tag{101}$$

Magnesium can be used as a reducing agent if it is present in the original reaction mixture:

$$2Al + 3Mg + 6RCl \longrightarrow 2AlR_3 + 3MgCl_2. \tag{102}$$

Aluminum alkyls are spontaneously inflammable, but in contrast with the corresponding boron derivatives they are rapidly hydrolyzed to give aluminum hydroxide and alkanes. Aluminum alkyls act as reducing agents with the formation of an olefin

$$(RCH_2CH_2)_3Al + R_2'C{=}O \longrightarrow$$

$$(R_2'CHO)Al(CH_2CH_2R)_2 + CH_2{=}CHR, \tag{103}$$

and are very useful alkylating agents.

Gallium and indium alkyls can be prepared from the reaction of organomercury compounds and the metal [cf. Eq. (98)], but this reaction is unsuccessful in the preparation of the corresponding trialkylthallium derivatives. Compounds of the latter type can be prepared from the reaction of organolithium derivatives with thallium trihalides:

$$TlX_3 + 3LiR \rightarrow TlR_3 + 3LiX. \tag{104}$$

The use of the less reactive Grignard reagent yields only dialkylthallium halides:

$$TlX_3 + 2RMgX \rightarrow R_2TlX + 2MgX_2. \tag{105}$$

The reactions of gallium, indium, and thallium alkyls generally follow those of the corresponding aluminum compounds. In contrast to the boron trialkyls, the organic derivatives of the other elements in this family form stable adducts with ethers [$(CH_3)_3MOR_2$; M = Al, Ga, In] and other less basic Lewis bases. [$(CH_3)_3AlL; L = P(CH_3)_3, S(CH_3)_2, Se(CH_3)_2, Te(CH_3)_2$. $(CH_3)_3ML$; L = $S(CH_3)_2$, $P(CH_3)_3$, $As(CH_3)_3$, M = Ga, In. $(CH_3)_3$-$TlP(CH_3)_3$.] In addition, a number of derivatives are known where an anionic Lewis base has reacted with aluminum alkyls (*XVIII*):

$$R_3Al + M^+L^- \rightarrow M[R_3AlL]^-. \tag{106}$$

Examples of such compounds include $Na^+Al(C_2H_5)_4^-$ and $M^+Al(C_2H_5)_3F^-$.

Structure   The structures of the alkyls of the heavier members of this series exhibit some interesting features with respect to bonding theory. Many of the trialkylaluminum (Al_2R_6) compounds and all of the alkylaluminum halides ($R_4Al_2Cl_2, R_3Al_2X_3$) are dimeric in solutions of noninteracting solvents or in the vapor phase. Proton magnetic resonance studies show that the two types of methyl groups present in $Al_2(CH_3)_6$ exchange in hydrocarbon solvents by a dissociation mechanism rather than by an intramolecular process (*65*). The gallium trialkyls are monomeric in the vapor phase but cryoscopic measurements suggest that they are associated in solution. Likewise, $(CH_3)_3In$ is monomeric in the vapor phase, but associated (as a tetramer) in solution; the other indium alkyls as well as the thallium alkyls exhibit no tendency toward association.

Early structural studies (*66*) of $Al_2(CH_3)_6$ indicated a bridged structure (Fig. 10.32) similar to that in Al_2Cl_6. However, an electron diffraction study of the monomer $Al(CH_3)_3$ indicates a planar AlC_3 framework with freely rotating methyl groups (*67*); the Al—C bond distances (1.96 Å) are similar to that found in the terminal methyl groups of the dimer. More recent work (*68*) suggests that the units of the dimer are held together by an Al—H—C

(a)

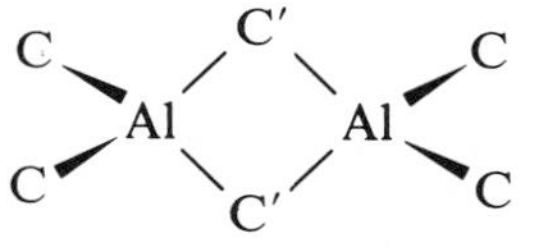

$d(\text{Al}-\text{C}) = 1.99 \text{ Å}$
$d(\text{Al}-\text{C}') = 2.23 \text{ Å}$
$\angle \text{CAlC} = 124°$
$\angle \text{C}'\text{AlC}' = 110°$

(b)

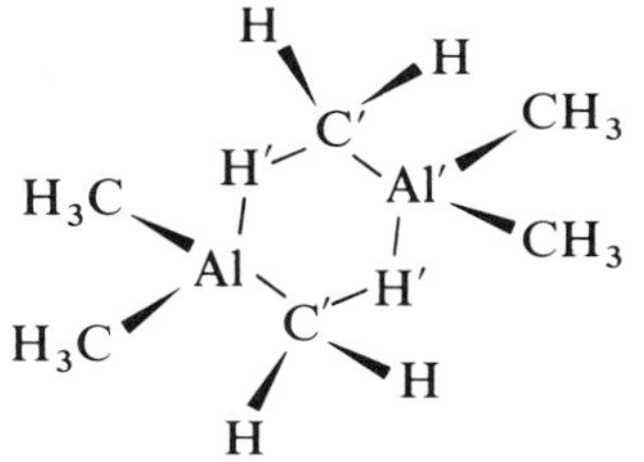

$d(\text{Al}-\text{H}') = 1.78 \text{ Å}$
$d(\text{C}'-\text{H}') = 1.08 \text{ Å}$
$d(\text{C}'-\text{H}) = 1.15 \text{ Å}$
$d(\text{C}'-\text{Al}') = 2.143 \text{ Å}$
$d(\text{Al}'-\text{C}) = 1.985 \text{ Å}$

FIG. 10.32 The earlier structure of $Al_2(CH_3)_6$ (a) did not indicate the position of the hydrogen atoms. A more recent structure (b) suggests that an Al—H—C bridge occurs in this compound.

bridge, as shown in Fig. 10.32. In either event, the theoretical description of the aluminum alkyl dimers must involve a three-centered bond, as postulated earlier in this chapter for B_2H_6. If the bridging atom is a carbon (Fig. 10.32a), the three-centered bond involves sp hybrids from the two aluminum atoms as well as from the bridging carbon atom. In the case of the second structure, the three-centered bond involves sp hybrids on carbon and aluminum and the s orbital of the bridging hydrogen atom. A molecular-orbital calculation indicates that the $3d$ aluminum orbitals are important in stabilizing the structure of $Al_2(CH_3)_6$ (69).

Structural studies on $(CH_3)_4Al_2X_2$ (X = halogen, H) dimers indicate that both halogen and hydrogen atoms act as bridging species; the data are collected in Fig. 10.33. In these cases, the bonding between the two units

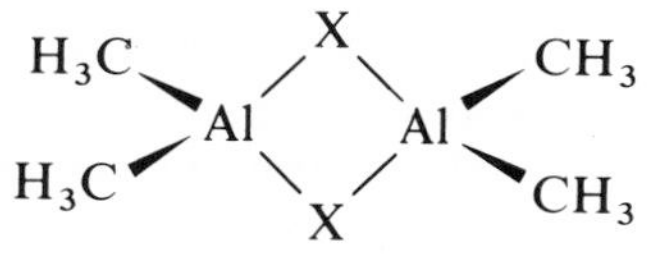

	Angles, degrees		d, Å	
X	XAlX	CAlC	C—Al	X—Al
Cl	89	127.5	1.93	2.31
Br	90	122.5	1.98	2.42
H	77	119	1.95	1.67

FIG. 10.33 Structural parameters for substituted dimethylaluminum compounds.

involves the more conventional coordinate-covalent bond arising from the overlap of the appropriate filled orbital on the halogen atom with the empty hybrid orbital on aluminum. As is the case with B_2H_6, the value of the terminal bond angle in $Al_2(CH_3)_6$ and $Al_2(CH_3)_4X_2$ suggests the nature of the sp hybrid required for the aluminum orbitals.

Indium trimethyl is monomeric and planar in the vapor phase with an In—C distance of 2.16 Å. However, in the solid state the molecules appear to be associated in tetrameric units (Fig. 10.34) (70). Each indium atom has three nearest neighbors with short In—C distances, and two neighbors at markedly larger distances. The bridging methyl groups are unsymmetrically arranged along the In—In internuclear distances. Thus, each indium atom is surrounded by five neighbors arranged in the form of a distorted trigonal bipyramid.

4 / LEWIS ACIDITY OF THE GROUP III ELEMENTS (XIX)

The chemistry of the Group III elements shows that their trivalent compounds are coordinately unsaturated. Because of this, they can act as acids in the Lewis sense. Intuitively there are three factors which play a role in the acidity of these compounds, R_3M: (1) the electronegativity of the central atom M, (2) the inductive effect of the substituents R, and (3) the steric requirements of the substituent when the adduct is formed. As an amplification of the last point, it should be recognized that for the planar R_3M compounds the substituents are further apart than in the tetrahedral R_3ML compounds. Steric requirements for R could be sufficiently large to overcome the formation of a stable four-coordinate adduct. Obviously a similar set of factors can be enumerated for the Lewis base L, with which the compound R_3M reacts. For the purposes at hand we shall assume that the nature of the bond is constant in this discussion. A variety of methods have been used to estimate the strength of the M—L bond in complexes of the type R_3ML. Studies of the degree of dissociation of such complexes in the gas phase (71–73) and nmr methods (74) have been employed.

The electronegativities of the Group III elements (Chapter 4) do not decrease smoothly down the series as might be expected: B, 2.0; Al, 1.45; Ga, 1.8; In, 1.5; Tl, 1.45. Boron and gallium have markedly higher electronegativities than the other three members of the group. All other factors being equal, we might expect that the acidity for a given type of compound towards a given Lewis base would be in the order B > Ga > Al ~ In ~ Tl. However, a study (75) of the stability of the complexes of $(CH_3)_3M$ with $(CH_3)_3N$ gives the order to be B < Al ~ Ga ~ In > Tl.

In addition to the factors mentioned previously, the strength of the interaction between a Lewis acid and base depends upon the nature of the atoms interacting. Two general classifications of Lewis acids and bases

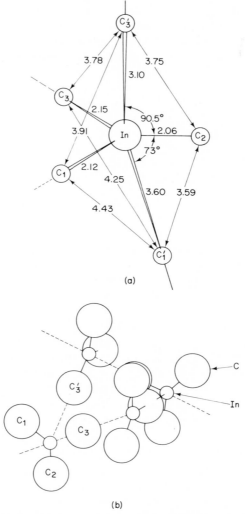

FIG. 10.34 (a) Structural parameters for the indium–carbon skeleton in $[In(CH_3)_3]_4$. Basically, each indium atom has three nearest neighbors in a plane and two further away which belong to two other $In(CH_3)_3$ units. (b) A schematic drawing of the arrangement of the $In(CH_3)_3$ units in a tetramer.

have been established (76). Lewis acids that incorporate small atoms with a high positive charge and that are not easily polarizable are called hard acids; soft acids have acceptor atoms that are large with a small (or zero) charge and are polarizable (XX). A soft base has a donor atom that has easily distorted (or removable) electrons; a hard base, on the other hand, has the opposite characteristics, i.e., it holds its electrons tightly and is not easily polarized. The point to this classification is that much evidence is available to indicate that hard Lewis acids interact strongly with hard Lewis bases and that hard acids are weakly attracted to soft bases. In other words, interactions within one class are strong but they are weak between classes:

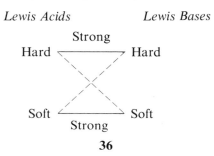

36

Generally, compounds of the lightest element in a family of elements are hard acids or bases; those of the remaining elements are soft (64).

With these ideas in mind, the boron compounds should be the hard acids in the series of compounds R_3M; oxygen compounds and amines should be the corresponding hard bases. Thus, we would expect that addition compounds of the type R_3BOR_2 and R_3BNR_3 should be more stable than, for example, R_3BSR_2 and R_3BPR_3. This expectation is, in general, true although the "hardness" of boron can be modified by the inductive effect of the substituents. For example, boron alkyls are less acidic than boron trihalides towards ethers because of the inductive effect of the alkyl group.

On the other hand, many stable complexes containing soft bases such as R_2S and R_3P with soft acids such as R_3Ga and R_3In are known. It must be emphasized that superimposed upon the general hard–soft concept are the steric and inductive factors described previously.

REFERENCES

1. R. Kiessling, *Acta Chem. Scand.*, **4**, 209 (1950).
2. A. Zalkin and D. H. Templeton, *Acta Cryst.*, **6**, 269 (1953).
3. J. D. Cooper, T. C. Gibb, N. N. Greenwood, and R. V. Parrish, *Trans. Faraday Soc.*, **60**, 2097 (1964).
4. N. Lundqvist, H. P. Myers, and R. Westin, *Phil. Mag.*, **7**, 1187 (1962).
5. R. W. Johnson and A. H. Daane, *J. Chem. Phys.*, **38**, 425 (1963).
6. A. H. Silver and P. J. Bray, *J. Chem. Phys.*, **32**, 288 (1960).
7. H. C. Longuet-Higgins and M. de V. Roberts, *Proc. Roy. Soc., Ser. A.*, **230**, 110 (1955).
8. F. A. Cotton and J. R. Leto, *J. Chem. Phys.*, **30**, 993 (1959).
9. D. R. Armstrong and P. G. Perkins, *Chem. Commun.*, 856 (1969).

10. L. H. Long and D. Dolimore, *J. Chem. Soc.*, 4457 (1954).
11. R. E. Nightingale and B. Crawford, Jr., *J. Chem. Phys.*, **22**, 1468 (1954).
12. R. F. Porter, D. R. Bidinosti, and K. F. Watterson, *J. Chem. Phys.*, **36**, 2104 (1962).
13. D. R. Martin and J. M. Canon, in *Friedel–Crafts and Related Reactions*, Vol. 1 (G. A. Olah, ed.), Wiley–Interscience, New York, 1963, p. 399.
14. J. M. Miller and M. Onyszczuk, *Can. J. Chem.*, **41**, 2898 (1963).
15. M. F. Lappert, *J. Chem. Soc.*, 542 (1962).
16. H. C. Brown and R. R. Holm, *J. Amer. Chem. Soc.*, **78**, 2173 (1956).
17. C. M. Bax, A. R. Katritzky, and L. E. Sutton, *J. Chem. Soc.*, 1258 (1958).
18. G. Urry, T. Wartik, and H. I. Schlesinger, *J. Amer. Chem. Soc.*, **74**, 5809 (1952).
19. A. K. Holliday and A. G. Massey, *J. Chem. Soc.*, 2075 (1960).
20. W. C. Schumb, E. L. Gamble, and M. D. Banus, *J. Amer. Chem. Soc.*, **71**, 3225 (1949).
21. A. L. McCloskey, J. L. Boone, R. J. Botherton, *J. Amer. Chem. Soc.*, **83**, 1766 (1961).
22. H. Nöth and W. Meister, *Chem. Ber.*, **94**, 509 (1961).
23. R. J. Botherton, A. L. McCloskey, L. L. Peterson, and H. Steinberg, *J. Amer. Chem. Soc.*, **82**, 6242 (1960).
24. A. Finch and H. I. Schlesinger, *J. Amer. Chem. Soc.*, **80**, 3573 (1958).
25. P. Ceron, A. Finch, J. Frey, J. Kerrigan, T. Parsons, G. Urry, and H. I. Schlesinger, *J. Amer. Chem. Soc.*, **81**, 6368 (1959).
26. G. Urry, T. Wartik, R. E. Moore, and H. I. Schlesinger, *J. Amer. Chem. Soc.*, **76**, 5293 (1954).
27. M. J. Linevsky, E. R. Shull, D. E. Mann, and T. Wartik, *J. Amer. Chem. Soc.*, **75**, 3287 (1953).
28. D. E. Mann and L. Rano, *J. Chem. Phys.*, **26**, 1665 (1957).
29. P. Timms, *J. Amer. Chem. Soc.*, **89**, 1629 (1967).
30. J. W. Akitt, N. N. Greenwood, and A. Storr, *J. Chem. Soc.*, 4410 (1965).
31. D. A. Daws and R. F. Porter, *J. Amer. Chem. Soc.*, **78**, 5165 (1956).
32. D. White and P. N. Walsh, *J. Chem. Phys.*, **28**, 508 (1958).
33. F. H. Van Cauwelaert and H. J. Bosmans, *Rev. Chim. Min.*, **6**, 611 (1969).
34. V. J. Shriner, Jr., D. Whittaker, and V. P. Fernandez, *J. Amer. Chem. Soc.*, **85**, 2318 (1963).
35. A. G. Freeman and J. P. Larkindale, *Inorg. Nucl. Chem. Letters*, **5**, 937 (1969).
36. R. H. Wenhorf, *J. Chem. Phys.*, **26**, 956 (1957).
37. K. Niedenzu and J. W. Dawson, *J. Amer. Chem. Soc.*, **82**, 4223 (1960).
38. H. Becker and J. Goubeau, *Z. Anorg. Chem.*, **268**, 133 (1952).
39. J. Goubeau and R. Link, *Z. Anorg. Chem.*, **267**, 27 (1951).
40. A. B. Burg and R. I. Wagner, *J. Amer. Chem. Soc.*, **75**, 3872 (1953).
41. R. I. Wagner and F. F. Caserro, Jr., *J. Inorg. Nucl. Chem.*, **11**, 259 (1959).
42. F. G. A. Stone and A. B. Burg, *J. Amer. Chem. Soc.*, **76**, 386 (1954).
43. L. M. Trefones and W. N. Lipscomb, *J. Amer. Chem. Soc.*, **81**, 4435 (1959).
44. A. B. Bayles, G. A. Pressley, Jr., E. J. Sinke, and F. E. Stafford, *J. Amer. Chem. Soc.*, **86**, 5358 (1964).
45. S. F. A. Kettle and V. Tomlinson, *J. Chem. Soc.*, A, 2002, 2007 (1969).
46. H. D. Johnson and S. G. Shore, *Topics Current Chem.*, **15**, 87 (1970).
47. G. W. Mappes, S. A. Fridmann, and T. P. Fehlner, *J. Phys. Chem.*, **74**, 3307 (1970).
48. D. E. Young and S. G. Shore, *J. Amer. Chem. Soc.*, **91**, 3497 (1969).
49. G. Kodama and R. W. Parry, *J. Amer. Chem. Soc.*, **82**, 6250 (1960).
50. W. V. Hough and L. J. Edwards, *Advan. Chem.*, **32**, 184 (1961).

51. R. W. Parry, R. W. Rudolph, and D. F. Shriver, *Inorg. Chem.*, **3**, 1479 (1964).
52. R. Schaeffer, F. Tebbe, and C. Phillips, *Inorg. Chem.*, **3**, 1475 (1964).
53. K. C. Nainan and G. E. Ryschkewitsch, *J. Amer. Chem. Soc.*, **91**, 330 (1969).
54. R. S. Mullikan, *J. Chem. Phys.*, **3**, 635 (1935).
55. J. C. Carter and R. W. Parry, *J. Amer. Chem. Soc.*, **87**, 2354 (1965).
56. R. W. Parry, C. E. Nordman, J. C. Carter, and G. Terharr, *Advan. Chem.*, **42**, 302 (1964).
57. W. H. Knoth, H. C. Miller, J. C. Sauer, J. H. Balthis, Y. T. Chia, and E. L. Muetterties, *Inorg. Chem.*, **3**, 159 (1964).
58. E. L. Muetterties, J. H. Balthis, Y. T. Chia, W. H. Knoth, and H. C. Miller, *Inorg. Chem.*, **3**, 444 (1964).
59. W. H. Knoth, J. C. Sauer, D. C. England, W. R. Hertler, and E. L. Muetterties, *J. Amer. Chem. Soc.*, **86**, 3973 (1964).
60. M. F. Hawthorne and F. P. Olsen, *J. Amer. Chem. Soc.*, **86**, 4219 (1964).
61. H. Schroeder and G. D. Vickers, *Inorg. Chem.*, **2**, 1317 (1963).
62. W. R. Hertler and M. S. Raasch, *J. Amer. Chem. Soc.*, **86**, 3661 (1964).
63. D. F. Shriver, R. W. Parry, N. N. Greenwood, A. Storr, and M. G. Wallbridge, *Inorg. Chem.*, **2**, 867 (1963).
64. E. C. Ashby and R. Kovar, *J. Organomet. Chem.*, **22**, C34 (1970).
65. T. L. Brown, *Acc. Chem. Res.*, **1**, 23 (1968).
66. P. H. Lewis and R. E. Rundle, *J. Chem. Phys.*, **21**, 986 (1953).
67. A. Almenningen, S. Halvorsen, and A. Haaland, *Chem. Commun.*, 644 (1969).
68. S. K. Byram, J. K. Fawcett, S. C. Nyburg, and R. J. O'Brien, *Chem. Commun.*, 16 (1970).
69. K. A. Levison and P. G. Perkins, *Disc. Faraday Soc.*, **47**, 183 (1969).
70. E. L. Amma and R. E. Rundle, *J. Amer. Chem. Soc.*, **80**, 4141 (1958).
71. H. C. Brown and S. Sujishi, *J. Amer. Chem. Soc.*, **70**, 2878 (1948), and references therein.
72. N. N. Greenwood and P. G. Perkins, *J. Chem. Soc.*, 1145 (1960).
73. M. L. Lappert, *J. Chem. Soc.*, 542 (1962).
74. J. F. Deters, P. A. McCusker, and R. C. Pilger, *J. Amer. Chem. Soc.*, **90**, 4583 (1968).
75. G. E. Coates and R. A. Whitcombe, *J. Chem. Soc.*, 3351 (1956).
76. R. G. Pearson, *J. Amer. Chem. Soc.*, **85**, 3533 (1963).
77. L. M. Trefones and W. N. Lipscomb, *J. Chem. Phys.*, **28**, 54 (1958).
78. M. Atoji, P. J. Wheatley, and W. N. Lipscomb, *J. Chem. Phys.*, **27**, 196 (1957).
79. K. Hedberg and R. Ryan, Jr., *J. Chem. Phys.*, **50**, 4986 (1969).
80. A. G. Massey, Ph.D. Thesis, University of Liverpool (1959).
81. M. Atoji and W. N. Lipscomb, *J. Chem. Phys.*, **31**, 601 (1959).
82. G. F. Lanthier and A. G. Massey, *J. Inorg. Nucl. Chem.*, **32**, 1807 (1970).

COLLATERAL READINGS

I. N. N. Greenword, R. V. Parish, and P. Thronton, *Quart. Rev.*, **20**, 44 (1966). A survey of the properties and structures of the metal borides.
II. A. G. Massey, *Advan. Inorg. Chem. Radiochem.*, **10**, 1 (1967). A review of the chemistry, properties, and structures of the normal boron halides (BX_3) and the lower-valent boron halides (B_2X_4, B_4X_4, B_8X_8).
III. D. W. A. Sharp, *Advan. Fluorine Chem.*, **1**, 68 (1960). A review of the chemistry of the fluoroboric acids and their derivatives.

IV. N. N. Greenwood, *Advan. Inorg. Chem. Radiochem.*, **5**, 91 (1963). A review of the chemistry of gallium including compounds containing the element in the $1+$ and $2+$ oxidation states, gallium hydrides, and organogallium compounds.

V. K. Niedenzu, *Angew. Chem. (Intl. Ed.)*, **3**, 86 (1964). A discussion of the chemistry of aminoboranes.

VI. E. K. Mellon, Jr., and J. J. Lagowski, *Advan. Inorg. Chem. Radiochem.*, **5**, 259 (1963). A survey of the preparation, properties, and structural information available for the borazines.

VII. H. D. Johnson, II and S. G. Shore, *Topics Chem.*, **15**, 87 (1970). A review of the recent developments in the chemistry of the lower boron hydrides (B_2H_6, B_4H_{10}, B_5H_9, B_5H_{11}, B_6H_{10}, B_6H_{12}, B_8H_{12}, and B_8H_{14}).

VIII. W. N. Lipscomb, *Boron Hydrides*, W. A. Benjamin, New York, 1963. An extended discussion of the structures, molecular-orbital descriptions, and reactions of the boron hydrides. This work also includes the application of topological theory to the structure of boron hydrides.

IX. R. E. Rundle, *Rec. Chem. Prog.*, **23**, 195 (1962); *Survey Prog. Chem.*, **1**, 81 (1963). These papers contain sections in which the structural and theoretical aspects of compounds containing three-centered bonds appear.

X. B. D. James and M. G. H. Wallbridge. *Prog. Inorg. Chem.*, **11**, 99 (1970). A detailed discussion of the preparation, properties, and structures of the metal tetrahydroborates.

XI. O. P. Shitov, S. L. Ioffe, V. A. Tartakovskii, and S. S. Novikov, *Russ. Chem. Revs.*, **39**, 905 (1970). A review of the preparation, chemical properties, and structures of cationic boron complexes.

XII. R. Koster, *Angew. Chem. (Intl. Ed.)*, **6**, 218 (1967). A discussion of the structure and synthesis of carboranes.

XIII. A series of papers in *Inorg. Chem.*, **2**, 1087–1133 (1963), by various authors, containing descriptions of the preparation and properties of $B_{10}C_2H_{12}$.

XIV. E. C. Ashby, *Advan. Inorg. Chem. Radiochem.*, **8**, 283 (1966). A discussion of the chemistry of the Group I and Group II tetrahydroaluminates.

XV. K. N. Semenko, B. M. Bulychev, and E. A. Shevlyagina, *Russ. Chem. Revs.*, **35**, 649 (1966). A discussion of the methods used to prepare aluminum hydride and its amine complexes; included is a discussion of the chemistry of several types of amine complexes.

XVI. H. C. Brown, *Hydroboration*, W. A. Benjamin, New York 1962. An extensive description of the hydroboration reactions as a synthetic tool.

XVII. R. Koster and P. Binger, *Advan. Inorg. Chem. Radiochem*, **7**, 263 (1965). A discussion of the preparation and properties of organoaluminum compounds, their reactions, and their ability to form complex compounds.

XVIII. H. Lehmkuhl, *Angew. Chem. (Intl. Ed.)*, **3**, 107 (1964). A description of the complex compounds formed by organic derivatives of aluminum.

XIX. D. P. N. Satchell and R. S. Satchell, *Chem. Revs.*, **69**, 251 (1969). A discussion of the quantitative aspects of the Lewis acidity of covalent halides and their organo-derivatives.

XX. R. G. Pearson, *Science*, **151**, 172 (1966). A summary of the concept of hard and soft acids and bases.

STUDY QUESTIONS

1. Predict the geometry of the covalently bound species in each of the following substances : (a) $InCl_3 \cdot 2NH_4Cl \cdot H_2O$; (b) $B(OCH_3)_3$; (c) $[R_2BNR_2']_2$; (d) $AlCl_3 \cdot 6H_2O$; (e) $B_2Cl_2(H_2S)_2$; (f) KBH_4 ; (g) $B[N(CH_3)_2]_3$; (h) $[R_2BPR_2']_3$.

2. Give the idealized hybridization for the central atoms in the species listed in Problem 1.

3. Give the major canonical structures in the valence-bond description of the bonding in the species listed in Problem 1.

4. Draw structural formulas for all the isomers expected for $(CH_3)_2B_2H_4$.

5. Draw the structural formula for the compound with the formulation $B_2H_5NH_2$. Give a description of the bonding in this compound.

6. Explain the acidity of $B_{10}H_{14}$.

7. Draw structural formulas for the phosphorus–boron analogs of benzene, ethane, ethylene, and acetylene, indicating the electronic distribution in the molecules.

8. Indicate which of each of the following pairs of complex compounds would be more stable:

(a) $(CH_3)_3B:NH_3$ or $(CH_3)_3B:PH_3$

(b) $(CH_3)_3Al:NH_3$ or $(CH_3)_3Ga:NH_3$

(c) $Cl_3B:O(CH_3)_2$ or $(CH_3)_3B:O(CH_3)_2$

(d) $(CH_3)_3B:PCl_3$ or $(CH_3)_3B:P(CH_3)_3$

9. Give equations for the reaction of boric acid with (a) alcohols, (b) glycerol, and (c) acid anhydrides.

10. Account for the observation that the following adducts decrease in stability in the order indicated : $BF_3 \cdot (CH_3)_3N > BF_3 \cdot (CH_3)_2O > BF_3 \cdot CH_3F$.

11. The experimental value of the dipole moment of the compound $(CH_3)_2NB(CH_3)_2$ implies that the bond moment of the central $N-B$ bond is essentially zero. Discuss this fact in the light of the large electronegativity difference between nitrogen and boron.

12. The experimentally determined values for bond distances in *B*-trichloroborazine are as follows: B—N, 1.41 Å; B—Cl, 1.75 Å. Sketch the major canonical forms contributing to the wave function for this molecule.

13. Discuss the observation that thallous salts are very often isomorphous with the corresponding salts of potassium and ammonium, but thallous salts are generally less soluble.

14. Discuss the acid–base properties of the following compounds: (a) $(R_2N)_3B$; (b) AlH_3; (c) $GaBr_2$.

GROUP IV
ELEMENTS

1 / INTRODUCTION

The electronic configurations of the elements in Group IV shown in Table 11.1 suggest that these substances readily form tetravalent derivatives. If all four electrons are involved in bond formation the electron-pair repulsion arguments described in Chapter 3 indicate that the four-valent compounds should be tetrahedral. However, as is the case with the other representative elements, the heavier elements of the family also form stable lower-valent compounds, the most common lower oxidation state being $2+$. The compounds of these elements are best described as predominantly covalent, forming either discrete molecular species or continuous structures involving covalent bonds. The $2+$ ions that can be formed by the heavier atoms in this group are sufficiently polarizing that their compounds possess a considerable covalent character; in some cases compounds containing the M^{2+} ion have been described (1). Transient covalent species of the type MX_2 for carbon

329

TABLE 11.1   Some properties of the Group IV atoms

Element[a]	Electronic configuration	Ionization potential, eV				χ	r_{cov}, Å	$r_{M^{2+}}$, Å	E^0, V[b]
		I	II	III	IV				
C (6)	[He] $2s^2 2p^2$	11.26	24.38	47.84	64.48	2.50	0.77	—	—
Si (14)	[Ne] $3s^2 3p^2$	8.15	16.34	33.46	45.13	1.74	1.17	—	—
Ge (32)	[Ar] $3d^{10} 4s^2 4p^2$	7.88	15.93	34.23	45.7	2.02	1.22	—	—
Sn (50)	[Kr] $4d^{10} 5s^2 4p^2$	7.33	14.63	30.6	39.6	1.72	1.40	1.63	0.136
Pb (82)	[Xe] $4f^{14} 5d^{10} 6s^2 6p^2$	7.42	15.03	32.0	42.3	1.55	1.54	1.21	0.122

[a] Nuclear charge Z is indicated in parentheses.
[b] For the process $M \rightarrow M^{2+}_{(aq)} + 2e^-$.

and silicon also have been described. The elements of this family exhibit the entire range of properties expected for metals and nonmetals. For example, all of the elements in this family, except carbon, have the physical appearance of metals. However, silicon and germanium are semiconductors, whereas tin, lead, and one form of carbon—graphite—are good electrical conductors. Generally speaking, the chemical properties of tin and lead indicate that these elements are metallic, whereas carbon and silicon are nonmetallic. Germanium has been described as a metalloid since it partakes of some of the properties of both metals and nonmetals.

Of the elements in Group IV, only carbon possesses an extensive chemistry involving compounds in which two or more atoms of the same element are bound to each other. The ability to catenate—form bonds to elements of the same kind—appears to be related to the strength of the bonds involved (*II*). From the available data on the energies of the bonds formed among the elements in this family which appear in Table 11.2, it is apparent that the carbon–carbon bond is significantly stronger than any of the other bonds formed between elements of the same kind in this period. However, recent mass spectral data suggest that the ease of cleaving M—M bonds in this family is Si—Si > Si—Ge > Ge—Ge, which is contrary to the order of bond energies (*1*). Silicon and germanium form a family of saturated hydrides with the general formula $M_n H_{2n+2}$ containing Si—Si bonds and Ge—Ge bonds. In addition, halogen derivatives containing Si—Si bonds are known. Although the corresponding halides and hydrides of tin are unknown, polymeric organotin derivatives of the type $(R_2 Sn)_n$ have been reported. Catenated lead compounds are unknown, but species containing Pb—Pb bonds are probably present in the liquid solutions which are formed when lead dissolves in metal–ammonia solutions. Compounds such as $Na_4 Pb_9$ and $Na_4 Pb_4$ have been obtained from these solutions; it has been suggested that polyatomic anions are present in these systems.

As shown by the data in Table 11.2, the strengths of covalent bonds formed between a Group IV element and an atom or group with no lone pairs (such as hydrogen) decrease with increasing atomic number. However, stronger bonds are formed between the elements and an atom which possesses lone pairs (i.e., oxygen or chlorine). This phenomenon, which is discussed in

TABLE 11.2   Average values of some single-bond energies in kcal mole^{-1}

	C	Si	Ge	Sn	Pb	H	Cl	O
C	83	77	71	53	35	99	78	84
Si		42	—	—	—	70	86	111
Ge			38	—	—	74	97	86
Sn				34	—	62	76	—

detail in a subsequent part of this chapter, is undoubtedly associated with the ability of silicon and the heavier elements to form $d\pi$-$p\pi$ bonds (*III*). The relative strengths of the bonds between elements of this group is not the sole factor which determines the accessibility of catenated derivatives. For example, silanes and germanes, in contrast to hydrocarbons, rapidly react with water containing a trace of base; Si—Si bonds are cleaved giving the corresponding oxides. The difference between carbon and the other elements in this family is that the latter, as a group, possess low-lying unfilled d orbitals which can act as sites for the formation of five- or six-coordinate intermediates with water or other Lewis bases. For example, $SiCl_4$ rapidly hydrolyzes under conditions where CCl_4 is unreactive, in spite of the fact that the Si—Cl bond energy is greater than that for the bond C—Cl, as shown in Table 11.2. The difference in the behavior of carbon with respect to the remaining elements of this family is, thus, a continuation of the same phenomenon observed for the elements in the previous families.

2   /   THE ELEMENTS

2.1   /   Physical Properties

The elements in this family are notable for the number of allotropic forms they exhibit; a summary of the properties of the known allotropic forms of these elements appears in Table 11.3. Diamond and graphite are the two common forms of carbon. At high pressures (150 kbar) single crystals of graphite have been partially transformed to a cubic form (*2*). Although diamond is more dense than graphite, the latter is the more stable form by 0.69 kcal mole^{-1} at 300°K and 1 atm pressure. Graphite can be converted

TABLE 11.3   Allotropic forms of the Group IV elements

Element	Allotropic modification	Properties
C	diamond	$D = 3.51 \text{ g cm}^{-3}$ $d(C\!-\!C) = 1.54 \text{ Å}$
	graphite	$D = 2.22 \text{ g cm}^{-3}$ $d(C\!-\!C) = 1.415 \text{ Å}$ $\Delta H_s^{300°K} = 171.7 \text{ kcal mole}^{-1}$
	high-pressure graphite	$D = 2.8 \text{ g cm}^{-3}$
	white	hexagonal symmetry $a_0 = 8.945 \text{ Å}, c_0 = 14.071 \text{ Å}$
Si	diamond	$D = 2.32 \text{ g cm}^{-3}$ $d(Si\!-\!Si) = 2.34 \text{ Å}$
Ge	diamond	$D = 5.38 \text{ g cm}^{-3}$ $d(Ge\!-\!Ge) = 2.44 \text{ Å}$
Sn	α-tin (gray), diamond	$D = 5.75 \text{ g cm}^{-3}$ $d(Sn\!-\!Sn) = 2.81 \text{ Å}$
	β-tin (white), deformed octahedral	$D = 7.31 \text{ g cm}^{-3}$ $d(Sn\!-\!Sn) = 3.022 \text{ Å}, 3.181 \text{ Å}$
	γ-tin, distorted close-packed	$D = 6.55 \text{ g cm}^{-3}$
Pb	cubic closest-packed	$D = 11.48 \text{ g cm}^{-3}$ $d(Pb\!-\!Pb) = 3.50 \text{ Å}$

into diamond at 3000°K at pressures in excess of 125 kbar. Normally this conversion is slow, but reasonable amounts of industrial quality diamonds, weighing up to about 0.1 carat, can be prepared by this process if transition metals are present as catalysts. The mechanism for the catalysis is unknown, but it appears that the relative solubility of graphite and diamond in the

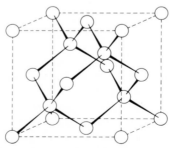

FIG. 11.1   The diamond structure is exhibited by carbon, silicon, germanium, and gray tin.

molten catalyst is an important factor. The so-called amorphous forms of carbon obtained as soot, lampblack, or charcoal in a variety of ways are actually composed of microcrystals of carbon in the graphite structure. These forms of carbon have a very high surface-to-mass ratio, making them excellent adsorbents of gases or of solutes from solution. The reaction of acetylene with $CuCl_2$ in a mixture of methanol and pyridine gives carbon as one of the products. Two crystalline modifications of this product having hexagonal unit cells containing three linear carbon chains have been isolated from this reaction (3). One form, thought to be $(-C\equiv C-)_n$, gives only oxalic acid upon ozonation; the other form gives only CO_2, suggesting the formulation $(=C=)_n$.

The diamond structure, shown in Fig. 11.1, consists of carbon atoms surrounded by a tetrahedral arrangement of other carbon atoms at a distance of 1.54 Å. Recently, diamond has been synthesized with a hexagonal structure (4). Since all of the valence electrons in the normal diamond structure are necessary to form the carbon–carbon bonds, diamond is a nonconductor. On the other hand, the structure of graphite, as shown in Fig. 11.2, consists

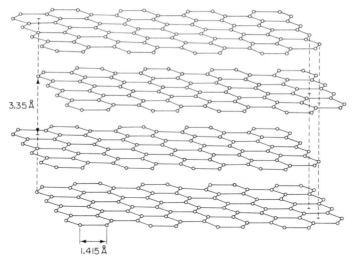

3.35 Å

1.415 Å

FIG. 11.2 The layer structure of graphite.

of carbon atoms arranged in a planar hexagonal array; the carbon atoms in a given plane are equidistant at 1.415 Å but the planes are markedly further apart at 3.35 Å. Two modifications of the graphite structure exist. In the most common modification, the carbon atoms in every other layer lie above each other. In the less common modification of graphite—a rhombic form—the atoms in every third layer are superimposed. Each carbon atom in any layer has three nearest neighbors, so that the best theoretical description of graphite involves the formation of a σ-bonded framework. Adjacent carbon atoms then possess odd electrons which can be paired to form a π system. In terms of valence-bond arguments, the electronic structure within such a layer is derived from a large number of equivalent canonical forms of the type shown in **1**. The high electrical conductivity of graphite has been

1

attributed to the presence of the π system. The C—C distance in graphite can be used as a measure of the bond order for the atoms that lie in the plane. The value of this parameter is 1.3, which should be compared with a bond order of 1.5 for benzene. It is interesting to note that the graphite structure is limited to carbon in this family of elements. A white allotrope of carbon has been produced at the high temperatures ($\sim 2500°$K) and low pressures necessary to sublime graphite.

In contrast to carbon and tin, silicon, germanium, and lead exhibit only one allotropic form (Table 11.3). Silicon and germanium crystallize in the diamond structure shown in Fig. 11.1, whereas metallic lead exhibits the cubic closest-packed structure (Fig. 2.13). Three modifications of tin exist below its melting point (232°C) at one atmosphere,

$$\alpha\text{-Sn} \rightleftharpoons \beta\text{-Sn} \rightleftharpoons \gamma\text{-Sn}. \tag{1}$$

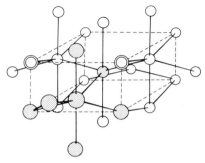

FIG. 11.3   Each atom in white tin is surrounded by six other atoms arranged in a distorted octahedral structure.

Below 13.2°C α-tin, or gray tin, is stable. Above this temperature, the α-form is converted into β-tin, which is sometimes called white tin. γ-Tin is stable above 161°C. α-Tin, which possesses the lowest density of these allotropes, has the diamond structure, but the structure of β-tin is essentially that of a distorted octahedron (Fig. 11.3) with four nearest neighbors. Some have suggested (5) that the electronic structures of tin in the α- and β-forms are distinctly different, using as support for this position the fact that the dissolution of α- and β-tin in aqueous HCl gives two different products. Stannous chloride ($SnCl_2 \cdot 2H_2O$) is isolated from the reaction of β-tin, whereas α-tin gives $SnCl_4 \cdot 5H_2O$.

2.2 / Chemical Properties

Reactions   Excluding graphite for the moment, the elements in this family become generally more reactive with increasing atomic number. The diamond form of carbon is inert to all reagents at ordinary conditions. It burns at about 700°C in both oxygen and fluorine to form CO_2 and CF_4, respectively.

Of the remaining elements, all undergo reaction when heated in oxygen or air to give the corresponding dioxides

$$M + O_2 \rightarrow MO_2 \tag{2}$$

$$M = Si, Ge, Sn$$

except lead, which yields a monoxide

$$2Pb + O_2 \rightarrow 2PbO. \tag{3}$$

Strong oxidizing agents such as the halogens react readily with these elements —elevated temperatures sometimes being required—to form the corresponding tetrahalides, which are covalent in nature

$$M + 2X_2 \rightarrow MX_4 \tag{4}$$

$$M = Si, Ge, Sn; X = halogen.$$

Again, lead is oxidized only to the divalent state with oxidizing agents as strong as chlorine

$$Pb + Cl_2 \rightarrow PbCl_2. \tag{5}$$

Silicon and germanium react at high temperatures with gaseous HCl to form chlorohydrides, $MHCl_3$ (M = Si, Ge), but cold aqueous solutions of HCl have little effect on any of the elements in this family. Concentrated aqueous solutions of oxidizing acids such as H_2SO_4 and HNO_3 dissolve germanium, tin, and lead. Aqueous NaOH dissolves silicon to give hydrogen and a solution of sodium silicate, in agreement with the predominantly non-metallic nature of this element. In contrast, germanium, tin, and lead do not react with the aqueous base.

Graphite Intercalation Compounds (*IV*)   In contrast to the general inertness of diamond, graphite is capable of forming an interesting series of compounds with different reagents. The structure of graphite, shown in Fig. 11.2, suggests that a layerlike structure could still persist if the π system were destroyed. Under such conditions, the carbon atoms would not, of course, be coplanar, because such a process requires hybridization change from sp^2 to sp^3. Two compounds of graphite are known with this arrangement of carbon atoms. When graphite is treated with strong oxidizing agents such as a mixture of H_2SO_4, HNO_3, and $KClO_3$ for several days, a product known as graphite oxide is formed (*V*). The oxygen : carbon ratio varies with the conditions of preparation, but it never exceeds 1 : 2. There is evidence suggesting that the oxygen atoms are present as ethers (C—O—C), ketones (C=O), and hydroxyl groups (C—OH). Graphite oxide has acidic properties, and the dehydrated product can be either methylated by diazomethane or acetylated. The interlayer spacing in a thoroughly dried sample of graphite oxide is 2.5–3.0 Å greater than that in graphite. The spacing increases markedly when water or alcohols are absorbed; for example, when graphite oxide absorbs water the interlayer spacing increases to 11.0 Å. The best available interpretation of the structure of graphite oxide involves oxidation of the double bonds of the original graphite structure, leading to a nonplanar—but continuous—arrangement of carbon atoms. This suggestion is supported by the markedly lower conductivity of graphite oxide compared with graphite. In a sense, graphite oxide is a collection of giant aliphatic cyclic molecules.

A more clear-cut case in which the double bonds in graphite have been saturated by chemical reaction is the product formed when fluorine reacts with graphite. This reaction gives a white product with the formula $(CF)_n$. Gray-colored products which have a lower fluorine content have been prepared also, but these are believed to be completely fluorinated substances. Interlayer spacings in $(CF)_n$ have been reported in the range of 6.9 to 8.2 Å; the C—C bond distance is 1.54 Å. The specific resistance of $(CF)_n$ is about 10^5 times that of graphite. These results suggest that $(CF)_n$ consists of continuous layers of carbon rings in the chair configuration with fluorine atoms in *trans–para* positions, as shown in Fig. 11.4. Graphite also reacts with a variety of other substances to form compounds which have electrical conductivities similar to that of the parent substance. Such compounds—called intercalation compounds—are formed because the reacting species enter the graphite structure between the layers. These reactions can be interpreted in

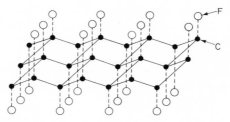

FIG. 11.4 One layer of the $(CF)_n$ structure. The carbon rings are in the chair configuration.

terms of an electron-transfer process, the graphite acting as an electron acceptor in certain cases (e.g., the alkali metals) or donor in others (e.g., the halogens). The ability of graphite to act in this dual role is undoubtedly linked to the existence of closely spaced π-type orbitals which contain loosely held electrons (1). The number of electrons accepted or donated is small compared to the total π electron concentration in each plane.

Graphite reacts readily with molten potassium, rubidium, and cesium or their vapors to form a series of unusual compounds with formulas C_8M, $C_{24}M$, $C_{36}M$, $C_{48}M$, and $C_{60}M$. The colors of the first two compounds are copper-red and steel-blue, respectively. Structural studies show that when any member of this series of compounds is formed the interlayer distance increases and the carbon layers acquire a different orientation with respect to each other than they have in the graphite structure; a graphical summary of the layer structure of these compounds appears in Fig. 11.5. In these intercalation compounds, the layers which are separated by the metal atoms are arranged so that the carbon atoms in one layer are directly above

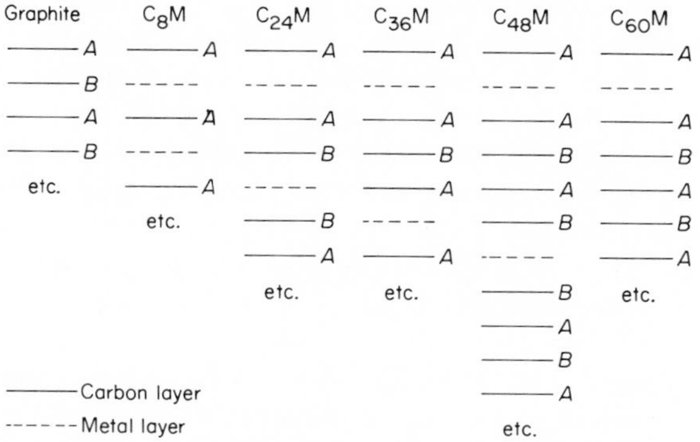

Graphite	C_8M	$C_{24}M$	$C_{36}M$	$C_{48}M$	$C_{60}M$
——A	——A	——A	——A	——A	——A
——B	- - - -	- - - -	- - - -	- - - -	- - - -
——A	——A	——A	——A	——A	——A
——B	- - - -	——B	——B	——B	——B
etc.	——A	- - - -	——A	——A	——A
	etc.	——B	- - - -	——B	——B
		——A	——A	- - - -	——A
		etc.	etc.	——B	etc.
				——A	
				——B	
———Carbon layer				——A	
- - - - - Metal layer				etc.	

FIG. 11.5 The alkali-metal–graphite intercalation compounds possess an expanded graphite structure. The various observed stoichiometries for these compounds correspond to metal atoms occupying different layers of the original graphite structure.

those in the other. Thus, the orientation of the graphite layers changes from ABAB... to AAAA.... In the latter orientation, the metal ions achieve a coordination number of 12 (six carbon atoms in each layer). The metal atoms in the compounds C_8M form a triangular network (Fig. 11.6) whereas those in all of the remaining compounds are arranged in a hexagonal pattern (Fig. 11.6). The relationships among the structures of these compounds is shown in Fig. 11.5.

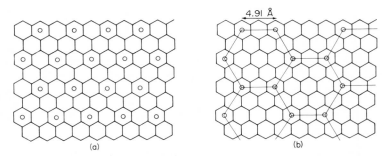

(a)   (b)

FIG. 11.6   The arrangement of alkali-metal atoms in the hexagonal graphite structure for the alkali–metal intercalation compounds. The triangular net of metal atoms (a) is found in the C_8M compounds, whereas the $C_{24}M$ and $C_{36}M$ compounds form a hexagonal net (b).

The volume increment which occurs when these compounds are formed is more nearly that expected for the introduction of ions into the structure rather than the corresponding atoms. This observation is in agreement with the suggestion that these compounds involve the donation of electrons to the graphite π system, $C_x^-M^+$.

Although intercalation compounds containing lithium, sodium, or the alkaline-earth metals are not known, compounds containing the alkali and alkaline-earth metals and ammonia are obtained from the electrolysis of liquid ammonia solutions of the corresponding salts with a graphite cathode. These compounds can be made also by treating graphite with the blue metal–ammonia solutions described in Chapter 8. The compounds exhibit a variable stoichiometry, as shown in Table 11.4, but they appear to form two general classes. One group has a low C : M ratio (10–13 : 1) with an interlayer distance of about 6.6 Å; the other group (C : M = 26–29 : 1) exhibits an interlayer distance of about 9.9 Å.

Bromine, chlorine, and ICl also form intercalation compounds with graphite. The formula C_8X has been established for the bromine and chlorine compounds. The conductivity of these compounds is greater than that of graphite, suggesting that graphite has contributed electrons to the intercalated halogen atoms forming the corresponding halide. The volume increase

TABLE 11.4   Active metal–ammonia intercalation compounds

Composition	d, Å[a]	Composition	d, Å[a]
$C_{10.6}Li(NH_3)_{1.6}$	6.62	$C_{28.8}Li(NH_3)_{1.7}$	9.93
$C_{13.4}Na(NH_3)_{2.0}$	6.63	$C_{26.7}Na(NH_3)_{2.3}$	9.97
$C_{12.5}K(NH_3)_{2.1}$	6.56	$C_{28.7}K(NH_3)_{2.8}$	9.94
$C_{11.9}Rb(NH_3)_{2.0}$	6.58	—	—
$C_{12.8}Cs(NH_3)_{2.2}$	6.58	—	—
$C_{12.1}Ca(NH_3)_{2.2}$	6.62	$C_{26.5}Ca(NH_3)_{4.1}$	9.95
$C_{11.3}Sr(NH_3)_{2.4}$	6.36	$C_{29.5}Sr(NH_3)_{3.4}$	9.87
$C_{10.9}Ba(NH_3)_{2.5}$	6.36	$C_{28.3}Ba(NH_3)_{3.9}$	9.79

[a] Distance between carbon layers.

when these compounds are formed is more nearly that expected for the intercalation of a halide ion in agreement with electron transfer from the graphite to the halogen, $C_8{}^+X^-$.

In addition to the alkali metals and the halogens, a variety of anhydrous metal chlorides (*VI*) form intercalation compounds with graphite (*6*). Only the structure of the C_6FeCl_3 compound is known; this corresponds to layers of $FeCl_3$ molecules between successive layers of graphite. The interlayer spacing is essentially the same at about 9.5 Å for all of these compounds.

3 / COMPOUNDS OF THE ELEMENTS

3.1 / Carbon

The chemistry of carbon is extensive, incorporating the reactions and properties of the hydrides and their derivatives. We shall not consider the chemistry and properties of the carbon hydrides in this discussion, deferring to the more extensive works on the branch of chemistry called organic chemistry. This is not to say that organic derivatives are not treated here when their properties and structures bear on the chemistry of the elements under consideration. Except for the rare gases, virtually all of the elements form compounds in which an organic group is bonded directly through a metal–carbon bond. With the very electropositive elements such as the alkali and alkaline-earth metals (Chapters 8 and 9), the compounds formed are essentially ionic (M^+R^-), exhibiting (a) low solubility in hydrocarbons and (b) decomposition upon heating. Only lithium and beryllium, which are the smallest and most polarizing of the elements in these families, display properties characteristic of covalent compounds. As the electronegativity difference between carbon and the element to which it is bound decreases, the compounds become distinctly covalent, as exemplified by the organic derivatives of the Group III elements (see Chapter 10) and of the transition elements, which are discussed in Chapter 19. Two general types of bonds have been distinguished among the covalent organometallic derivatives of the elements. In one case, the M—R bond involves two electrons in a σ orbital; a variation on the normal σ bond is the three-centered bond found in electron-deficient molecules such as $Al_2(CH_3)_6$ (Chapter 10). The other type of metal–carbon interaction involves the d orbitals of a metal atom and the π electron density of an unsaturated hydrocarbon. This type of interaction is found in ferrocene $[Fe(C_5H_5)_2]$, dibenzenechromium $[Cr(C_6H_6)_2]$, and metal–olefin compounds, and is discussed in Chapter 19 where this subject is more appropriate.

Carbides (*VII*)   Several classes of binary compounds containing carbon are known. Generally speaking, the carbides are compounds formed by the more electropositive elements; compounds of carbon with nonmetals such as the halogens and the first two members of the Groups V and VI elements and of hydrogen are generally not considered in discussions of the carbides.

Preparation. Binary compounds containing carbon can be prepared by three general methods. Direct reaction often occurs when a mixture of carbon and an element is heated to temperatures in excess of about 2200°C. The reduction of metal oxides, or other compounds, with carbon at high temperatures also yields carbides. The third general method involves heating a sample of the metal (often electrically, as in the form of a wire) with a hydrocarbon such as CH_4 at low pressures.

In addition to these reactions, binary carbon compounds of copper, silver, gold, magnesium, zinc, and cadmium can be prepared by passing acetylene into solutions containing compounds of these metals. The alkyl derivatives of the last two named elements react with acetylene in organic solvents

$$R_2M + HC\equiv CH \rightarrow 2RH + MC_2 \qquad (6)$$

$$M = Cd, Zn, Mg.$$

Ammoniacal solutions of the common metal salts of the remaining elements also react with acetylene

$$2M^+ + HC\equiv CH \rightarrow M_2C_2 + 2H^+. \qquad (7)$$

In both cases the reactions reflect the acid character of acetylene. Equation (7) does not accurately represent the reacting species in solution, since undoubtedly the metal is present as a complexed cation and the concentration of protons (and perhaps free acetylene) is very small in the basic media in which such reactions occur. The binary compounds formed in the reactions sometimes are called acetylides; they are extremely sensitive to heat and mechanical shock.

Structure and Reactions. It has been difficult to establish a consistent basis for the classification of carbides. As we shall see, a structural basis for classification does not always reflect consistent chemical properties. Generally, the carbides as a group are divided into salt-like (or ionic) carbides, interstitial carbides, and covalent carbides, a classification reminiscent of that of the hydrides described in Chapter 6.

The Salt-Like Carbides are formed with the more electropositive elements and have properties associated with ionic substances; i.e., they form colorless, transparent crystals that are electrical nonconductors. These compounds contain either discrete carbon atoms, as in Be_2C and Al_4C_3, or C_2^{2-} ions as in K_2C_2 and MC_2 ($M = Ca$, Sr, Ba). The crystal structures of both Be_2C and Al_4C_3 indicate the existence of individual carbon atoms, whereas pairs of carbon atoms exist in the crystals of the other salt-like carbides. For example, CaC_2 has the tetragonally distorted NaCl structure shown in Fig. 2.5, the distortion arising because the linear C_2 units are aligned parallel to the cell axis. The shortest C—C bond distance in CaC_2 is 1.19 Å, which should be compared to 1.20 Å in acetylene. The compounds containing individual carbon atoms hydrolyze readily to give methane as the major product

$$Be_2C + 4H_2O \rightarrow 2Be(OH)_2 + CH_4 \tag{8}$$

$$Al_4C_3 + 12H_2O \rightarrow 4Al(OH)_3 + 3CH_4, \tag{9}$$

suggesting that these compounds are best formulated as containing predominantly C^{4-} ions. On the other hand, hydrolysis of CaC_2 gives acetylene

$$CaC_2 + 2H_2O \rightarrow Ca(OH)_2 + H_2C_2. \tag{10}$$

There is no direct structural evidence for the existence of carbides containing anions more complex than C_2^{2-}; hydrolysis of Mg_2C_3, which is formed in the pyrolysis of MgC_2,

$$2MgC_2 \rightarrow Mg_2C_3 + C, \tag{11}$$

gives allene ($CH_3C\equiv CH$) as the principal product.

The lanthanide and actinide elements form carbides which show some of the properties associated with salt-like carbides. Three stoichiometric ratios have been observed: M_3C, M_2C_3, and MC_2 (7, 8). Carbides with the general formula M_3C can be described as a cubic closest-packed structure of metal atoms, shown in Fig. 2.13, with carbon atoms randomly occupying one-third of the octahedral holes. In the M_2C_3 structure, the carbon atoms are present as C_2 units with the C—C distances in the range 1.30–1.34 Å. The MC_2 compounds, which crystallize in the same way as CaC_2, contain C_2 units with a C—C distance of 1.28–1.34 Å (9). The compounds containing C_2 units (M_2C_3 and MC_2) cannot be simply identified as acetylides for several reasons. First, the bond distances are significantly longer than that in the alkaline-earth acetylides and in acetylene. Some of these carbides, e.g. LaC_2 and ThC_2, are good metal-like conductors, in contrast to CaC_2 which is an insulator. Finally, hydrolysis of these compounds yields mixtures of hydrocarbons (H_2C_2, H_4C_2, CH_4) as well as elemental hydrogen.

The MC_2 compounds are perhaps best described as divalent metal acetylides ($M^{2+}C_2^{2-}$) in which some or all of the metal ions lose an additional electron to the lowest antibonding orbital of the C_2^{2-} system to form $M^{3+}C_2^{3-}$. Under these circumstances we would expect an increase in the C—C distance for the species C_2^{3-}. In addition, there must be extensive delocalized interactions among the ions in the MC_2 compounds because they are good electrical conductors. The compounds M_2C_3 apparently contain trivalent cations and exhibit metal–metal interactions (10).

The Interstitial Carbides, formed by the transition elements, are high-melting and very hard. Basically, there are two groups of compounds which fall into this classification. The truly interstitial compounds are formed by metals which have atomic radii greater than ~ 1.3 Å. Atoms of this size can form close-packed structures with interstices sufficiently large to accept carbon atoms. The elements titanium (1.47 Å), zirconium (1.60 Å), hafnium (1.59 Å), vanadium (1.35 Å), niobium (1.47 Å), tantalum (1.47 Å), molybdenum (1.40 Å), and tungsten (1.41 Å) form carbides with the general formulas MC and M_2C possessing interstitial structures. On the other hand, there is a group of carbides formed by transition elements with radii smaller than ~ 1.3 Å which have been classified historically as interstitial, but which do not exhibit the basic close-packed structure of metal atoms. The carbides of chromium (1.29 Å), manganese (1.37 Å), iron (1.26 Å), cobalt (1.25 Å), and nickel (1.25 Å) fall into this category. The formulas for these compounds (e.g., Cr_7C_3, Cr_3C_2, Mn_4C, Mn_3C, Mn_5C_2, Mn_7C_3, Fe_3C, Co_2C) cannot be explained in terms of the atoms existing in their normal valence states. Indeed, the structural data available suggest that carbon–carbon interactions occur in these systems. For example, the structure of Cr_3C_2 shows a zig-zag carbon chain running through the crystal with a C—C distance of 1.66 Å. only slightly greater than that observed in aliphatic carbon compounds (1.54 Å).

The truly interstitial carbides are chemically inert. For example, TiC is impervious to attack by water or aqueous HCl even at 600°C. The borderline interstitial carbides are markedly more reactive. Thus, Mn_3C yields mixtures of hydrocarbons (CH_4 and C_2H_6) as well as hydrogen when it reacts with water.

The Covalent Carbides are formed when carbon is bonded to elements with nearly the same electronegativity. The most common examples of this type of compound are SiC and B_4C. Silicon carbide—commonly called carborundum —crystallizes in a large number of polymorphs, all of which are modifications of the zinc-blende or wurtzite structure shown in Fig. 2.4.

The structure of B_4C contains icosahedral B_{12} groups (Fig. 10.1) connected to each other through $B\!-\!B$ bonds as well as by C_3 chains; the structure is unusual because the central carbon atom in the chains has only two nearest neighbors. Boron carbide, which forms black, shining, conducting crystals, possesses properties characteristic of metallic systems. The covalent carbides are very hard infusible substances which show little chemical reactivity.

Oxygen and Sulfur Derivatives Four binary oxygen derivatives of carbon are known and characterized. Carbon monoxide (CO) is formed when elemental carbon burns in a deficiency of oxygen

$$2C_{(s)} + O_2 \rightarrow 2CO. \tag{12}$$

With an excess of oxygen, carbon dioxide (CO_2) is the product

$$C_{(s)} + O_2 \rightarrow CO_2. \tag{13}$$

The dehydration of malonic acid with P_2O_5

$$H_2C \overset{\displaystyle CO_2H}{\underset{\displaystyle CO_2H}{\Big\langle}} \xrightarrow{P_2O_5} C_3O_2 + 2H_2O \tag{14}$$

yields carbon suboxide (C_3O_2), and dehydration of benzenehexacarboxylic acid (**2**)—sometimes called mellitic acid—yields a carbon oxide ($C_{12}O_9$) which is really an anhydride containing a benzene ring (**3**).

$$C_6(CO_2H)_6 \rightarrow$$

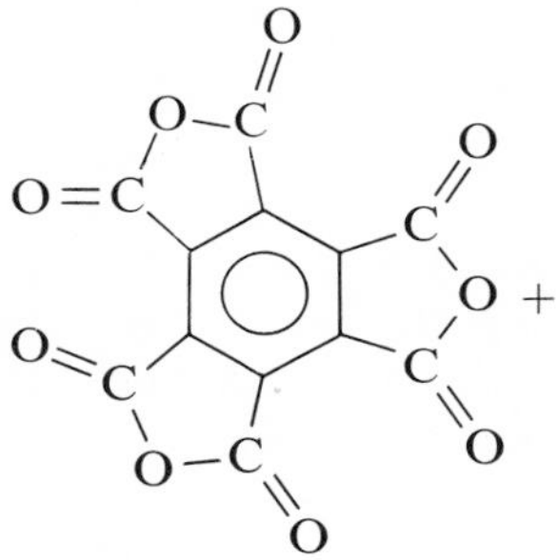

2

3

$$+ \, 3H_2O \tag{15}$$

Carbon Monoxide is a colorless and odorless gas (mp $-205.1°$, bp $-190°$) which burns to give CO_2

$$2CO_{(g)} + O_2 \rightarrow CO_2. \tag{16}$$

Although it is the anhydride of formic acid, carbon monoxide is sparingly soluble in water. However, it reacts with hot basic aqueous solutions, giving formate ions. Formic acid is readily dehydrated to give CO, a very convenient method for synthesizing relatively small amounts of this compound

$$\overset{\overset{\displaystyle O}{\|}}{H-C}-OH \xrightarrow{H_2SO_4} CO + H_2O. \tag{17}$$

The carbon-oxygen framework in formic acid is planar (Fig. 11.7), suggesting that the carbon atom employs essentially sp^2 hybrid orbitals in bonding. At room temperature, 90% of the gaseous formic acid molecules are dimeric. Electron diffraction data, collected in Fig. 11.7, clearly indicate that unsymmetrical hydrogen bonds are involved in the dimer.

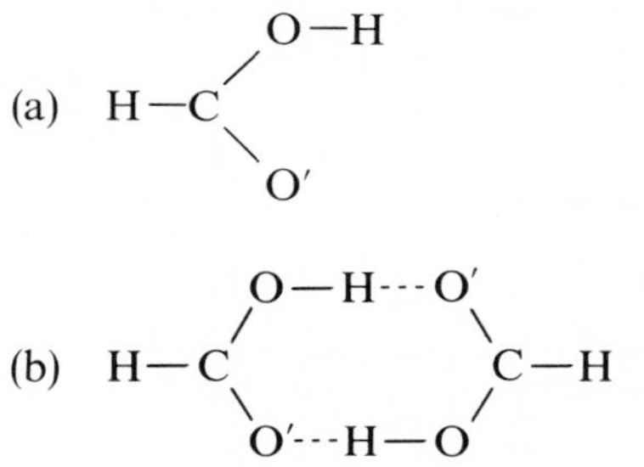

(a)

$d(O-H) = 0.96$ Å
$d(C-O) = 1.36$ Å
$d(C-H) = 1.08$ Å
$d(C-O') = 1.23$ Å
$\angle OCO = 123°$

(b)

$d(O-H\cdots O') = 2.73$ Å
$d(C-O) = 1.36$ Å
$d(C-O') = 1.24$ Å
$\angle OCO' = 121°$

FIG. 11.7   The molecular parameters for formic acid (a) and its dimer (b).

Carbon monoxide is a weak Lewis base toward σ acceptors, one of the few examples of this behavior being the formation of H_3BCO

$$B_2H_6 + 2CO \xrightarrow{700°} 2H_3BCO. \tag{18}$$

However, carbon monoxide forms exceedingly stable compounds with transition-metal atoms which possess filled d orbitals. The strength of the metal–carbon bond in such systems has been traced to the formation of a multiple bond in which the carbon monoxide acts as a σ donor and a π acceptor. A detailed account of the properties, structure, and constitution of these compounds appears in Chapter 19. The poisonous nature of CO arises from the formation of a very stable compound with hemoglobin, carboxyhemoglobin, which cannot act as an oxygen carrier. A victim of carbon monoxide poisoning literally suffocates to death. Carbon monoxide reacts with alkali-metal–ammonia solutions to form compounds with the formulation MCO which are known to contain the anions $[C_2O_2]^{2-}$ (*11*).

The analogous sulfur compound, CS, can be formed by passing an electric discharge through CS_2 vapor, or through a gaseous mixture of hydrocarbon and sulfur. It is a poorly characterized gaseous species which decomposes rapidly at room temperature.

Carbon Dioxide is also a gaseous compound (sublimation temperature is $-78°$ at one atm pressure). The electron-pair-repulsion arguments lead to the correct conclusion that CO_2 should be a linear molecule, but the expected valence-bond structure (**4**) cannot be the only canonical form contributing

$$O=C=O \qquad O-\overset{-}{C}\equiv\overset{+}{O} \qquad \overset{+}{O}\equiv\overset{-}{C}-O$$

$$\textbf{4} \qquad\qquad\quad \textbf{5} \qquad\qquad\quad \textbf{6}$$

to the wave function for the molecule because the experimental C—O bond distance (1.15 Å) is considerably shorter than the corresponding distance in ketones (1.22 Å). Accordingly, canonical forms **5** and **6** must also be important.

Carbon dioxide is the anhydride of carbonic acid, H_2CO_3 (*VIII*). An aqueous solution of CO_2 contains three fundamental equilibria:

$$CO_{2\,(aq)} + H_2O \rightleftharpoons H_2CO_3 \tag{19}$$

$$H_2CO_3 + H_2O \rightleftharpoons HCO_3^- + H_3O^+ \tag{20}$$

$$HCO_3^- + H_2O \rightleftharpoons CO_3^{2-} + H_3O^+. \tag{21}$$

Most of the dissolved CO_2 is in the form of a loosely aquated species $[CO_{2\,(aq)}]$ which is distinctly different from the species called "carbonic acid" (H_2CO_3). It is the latter species which undergoes ionization, as indicated by Eqs. (20) and (21). The conversion of $CO_{2\,(aq)}$ to carbonic acid is very slow, the rate constant for the pseudo-first-order reaction [Eq. (19) as written] being $0.03\ sec^{-1}\,m^{-1}$. Carbonic acid is a weak dibasic acid. The usual values reported for the ionization constants of this acid are $K_1 = 4.16 \times 10^{-7}$ and $K_2 = 4.84 \times 10^{-11}$; the value of K_1 corresponds to the process

$$\text{"carbonic acid"} + H_2O \rightleftharpoons HCO_3^- + H_3O^+ \tag{22}$$

in which no differentiation is made between the species $CO_{2\,(aq)}$ and H_2CO_3 when the first ionization step is written. The true equilibrium constant for Eq. (20) has been estimated as 2×10^{-4}, a value more in keeping with the proposed structure $(HO)_2CO$.

Carbonic acid, as such, has never been isolated, but a white crystalline etherate, $OC(OH)_2 \cdot O(CH_3)_2$ (mp $-47°$), is formed when HCl reacts with Na_2CO_3 in dimethyl ether (12). This substance decomposes at about 5°. Orthocarbonic acid [$C(OH)_4$] is not known either, but the esters, such as $C(OC_2H_5)_4$ (bp 158°), are well known.

Sulfur and selenium analogues of CO_2 are known (VIII). They are generally less volatile [CS_2, mp $-109°$, bp 46.3°; CSe_2, mp $-42°$, bp 125°], less stable, and more reactive than CO_2. Carbon disulfide can be made by direct reaction of the elements; CSe_2 can be obtained in good yield from the action of molten selenium on CH_2Cl_2. Carbon disulfide is slightly soluble in water, but it dissolves readily in concentrated sodium hydroxide solution yielding a mixture of Na_2CO_3 and $NaCS_3$. Free thiocarbonic acid (H_2CS_3) is an unstable yellow oil.

Mixed-chalcogen analogues of CO_2 such as COS (mp $-138.8°$, bp $-50.2°$), COSe (mp $-124.4°$, bp $-21.7°$), CSSe (mp $-85°$, bp 84.5°), and CSTe (mp $-54°$, decomp.) have also been prepared and characterized.

Carbon Suboxide, the anhydride of malonic acid [Eq. (14)], is a foul-smelling gaseous substance (bp $-6.8°$). Like the other carbon-acid anhydrides, it reacts with amines to form amides

$$C_3O_2 + 2R_2NH \rightarrow H_2C\begin{matrix} \diagup CONR_2 \\ \diagdown CONR_2 \end{matrix} \tag{23}$$

Structural studies show that the molecule is linear, with oxygen atoms at the ends of a three-carbon chain. The C—O distance is reported as 1.20 Å, which is only slightly shorter than the normal C—O distance in ketones (1.22 Å). There is only one C—C distance (1.30 Å), which is also shorter than expected for a normal carbon–carbon double bond (1.33 Å). These data suggest that the normal valence-bond structure **7** as well as the canonical forms **8** and **9** must contribute to the best wave function for the molecule.

$$O{=}C{=}C{=}C{=}O \qquad {}^+O{\equiv}C{-}C{\equiv}C{-}O^- \qquad {}^-O{-}C{\equiv}C{-}C{\equiv}O^+$$

$$\textbf{7} \qquad\qquad\qquad \textbf{8} \qquad\qquad\qquad \textbf{9}$$

Nitrogen Derivatives   There are a variety of species containing nitrogen–carbon bonds which are normally included in discussions of the inorganic chemistry of carbon. Many can be related, conceptually if not synthetically, to the general formulation XCN where X is CN, halogen, or hydrogen.

Cyanogen, $(CN)_2$, is a very poisonous colorless gas (mp $-28°$, bp $-21°$) with an odor reminiscent of bitter almonds. It can be prepared by heating silver, mercury, or gold cyanide

$$Hg(CN)_2 \rightarrow Hg + (CN)_2; \tag{24}$$

by oxidizing cyanide ion in aqueous solution

$$2CN^- \rightarrow (CN)_2 + 2e^- \tag{25}$$

using oxidizing agents such as Cu^{2+} or Fe^{3+}; or by oxidizing HCN in the gas phase with air, Cl_2, or NO_2. The last-named method requires the use of solid catalysts such as silver, carbon, or silica for efficient conversion.

Pure $(CN)_2$ is a remarkably stable substance (13), decomposing to CN radicals at about 1000°. The radicals themselves are exceedingly stable. Impure cyanogen polymerizes to $(CN)_x$, called paracyanogen, at about 400°; the solid polymer reverts to cyanogen at higher temperatures. The infrared spectrum of paracyanogen suggests that this substance possesses the unsaturated cyclic structure 10.

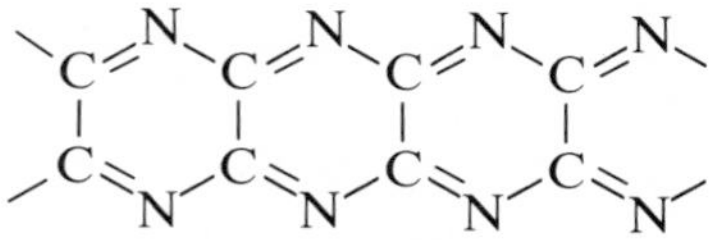

10

Cyanogen can be hydrolyzed in acid solution to give oxamide (11) and

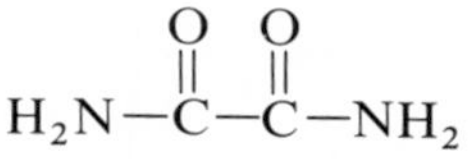

11   **12**

reduced with hydrogen to ethylenediamine (12), products which indicate the presence of a C—C bond in the original compound. Direct structural information shows that the molecule is, indeed, symmetrical and linear, with a C—C bond. The C—C distance is 1.36 Å, which is slightly shorter than that expected for the internuclear distance for two triply bonded carbon atoms. Thus, the normal Lewis configuration (13) probably corresponds to the best canonical structure for for this compound.

$$:N\equiv C-C\equiv N:$$

13

As we shall see, many of the reactions of compounds containing the CN radical are reminiscent of the reactions of the halogens. For example, $(CN)_2$

reacts with aqueous base to give disproportion products in the same way as does chlorine

$$(CN)_2 + 2OH^- \rightarrow CN^- + OCN^- + H_2O. \tag{26}$$

Hydrogen Cyanide (HCN) is a colorless and very volatile liquid (mp $-13.4°$, bp $25.6°$) with a high dielectric constant (107 at $25°$). It may be prepared by the dehydration of ammonium formate

$$NH_4{}^+HCO_2{}^- \xrightarrow{P_2O_5} HCN + 2H_2O; \tag{27}$$

by the reaction of gaseous H_2S with $Hg(CN)_2$

$$H_2S + Hg(CN)_2 \rightarrow 2HCN + HgS; \tag{28}$$

or by the reaction of a strong acid with a cyanide

$$2KCN + H_2SO_4 \rightarrow K_2SO_4 + 2HCN. \tag{29}$$

Hydrogen cyanide is readily soluble in water to give weakly acidic solutions ($K = 4.9 \times 10^{-10}$). Aqueous solutions slowly react to yield ammonium formate [the reverse of Eq. (27)]. The low acidity of the HCN indicates that its ionic metal salts will be extensively hydrolyzed in aqueous systems. Although there is the possibility that hydrogen cyanide could exist in two tautomeric forms

$$HCN \rightleftharpoons CNH \tag{30}$$

no evidence is available to suggest that the *N*-bonded species, hydrogen iso-cyanide, exists.

The cyanide ion, CN^-, is isoelectronic with CO. Being negatively charged, cyanide ion is a better Lewis σ donor than is carbon monoxide. Thus, many complex transition-metal ions form complexes with CN^-; the cyanide ion often forms compounds that are isostructural with the corresponding halide ion.

Sodium cyanide, an important article of commerce, can be prepared by fusing calcium cyanamide with carbon and sodium carbonate

$$CaCN_2 + C + Na_2CO_3 \rightarrow CaCO_3 + 2NaCN. \tag{31}$$

Calcium cyanamide in turn is prepared by heating calcium carbide with nitrogen at about $1100°C$

$$CaC_2 + N_2 \rightarrow CaCN_2 + C. \tag{32}$$

Thus, the commercial preparation of sodium cyanide requires limestone, carbon, and nitrogen.

Cyanogen Halides are compounds containing a halogen atom attached to a cyanogen radical: XCN. They are moderately volatile covalent compounds (FCN, bp $-46°$; ClCN, mp $-6.9°$, bp $13°$; BrCN, mp $51.3°$, bp $61.3°$; ICN, mp $146°$) with a moderate solubility in water. All of the halides except FCN are prepared from the reaction of the halogen with alkali-metal cyanides

$$X_2 + MCN \longrightarrow MX + XCN. \tag{33}$$

Cyanogen fluoride is obtained by pyrolyzing cyanic fluoride at about 1300°

$$(FCN)_3 \longrightarrow 3FCN. \tag{34}$$

The cyanogen halides are linear molecules which polymerize to form trimeric structures containing heteroatomic rings (**14**).

14

Isocyanic Acid. If the trimer of cyanogen chloride (**14**, X = Cl) is hydrolyzed, cyanuric acid is formed

$$(ClCN)_3 + H_2O \longrightarrow (HOCN)_3 + HCl, \tag{35}$$

which could exist either as **15** or in its tautomeric form **16**. An x-ray analysis of cyanuric acid shows that it, indeed, exists in the carbonyl form (**16**), but esters of both forms are known.

15   **16**

Isocyanic acid (mp $-86.8°$, bp $23.5°$) is formed when cyanuric acid is passed through a hot tube; the product reverts to cyanuric acid spontaneously. The substance is a weak acid ($K = 1.2 \times 10^{-4}$); its aqueous solutions rapidly hydrolyze to give ammonia and carbon dioxide. Theoretically isocyanic acid might exist in two tautomeric forms

$$HNCO \rightleftharpoons NCOH, \tag{36}$$

but there is little evidence that the species NCOH or its derivatives exist. The metallic isocyanates are indistinguishable from the cyanates since they both contain the ion CNO^-. Ionic cyanates can be prepared from the corresponding cyanides by oxidation with PbO_2 at elevated temperatures.

Isothiocyanic Acid, HNCS, may be prepared by the action of $KHSO_4$ on KSCN. The substance, which is a gas at room temperature, polymerizes in the liquid state. In aqueous solution, isothiocyanic acid is a strong acid but is readily hydrolyzed by 70 % H_2SO_4

$$HNCS + H_2O \rightarrow NH_3 + OCS. \tag{37}$$

In contrast with isocyanic acid, derivatives of isothiocyanic acid that are *S*-bonded and *N*-bonded are known, such as CH_3SCN (bp 132°) and CH_3NCS (bp 118°). The physical and chemical properties of such isomers, i.e., the nature of hydrogenation products

$$NCSCH_3 + H_2 \rightleftharpoons HCN + HSCH_3 \tag{38}$$

$$CH_3NCS + 3H_2 \rightleftharpoons CH_3NH_2 + CH_3SH, \tag{39}$$

indicate a difference in the position of attachment of the substituent.

Since the ions formed from the two possible tautomeric forms of isothiocyanic acid are indistinguishable, ionic derivatives all possess the same properties associated with the anionic part of the compound. It is of interest, however, that *S*-bonded and *N*-bonded metal complexes containing SCN^- have been prepared.

Halogen Derivatives   The halogen derivatives of carbon are generally considered in detail in organic chemistry textbooks. There are, however, several aspects of the chemistry of such compounds which are relevant to our discussion and will be treated here.

Carbon Tetrahalides are covalent compounds with the expected properties typical of such substances. Carbon tetrafluoride (bp $-185°$, mp $-128°$) is a remarkably stable substance, but CBr_4 (mp 93°, bp 190°) and CI_4 (mp 171°) decompose when heated; CCl_4 (mp $-23°$, bp 76°) is moderately stable. The general decrease in stability of the carbon tetrahalides with increasing molecular weight corresponds well with the decrease in the carbon–halogen bond energy. Carbon tetrachloride can act as a chlorinating agent, as demonstrated by its use to convert metal oxides to the corresponding chlorides. Although CCl_4 is thermodynamically unstable with respect to hydrolysis, the rate of this reaction is sufficiently slow to make it a generally unimportant factor in the chemistry of this substance.

Carbon halides undergo an unusual displacement reaction with certain dibasic Lewis bases (**17**) such as *o*-phenylenebisdimethylarsine (**17**) (which is sometimes abbreviated as diars) to form white crystalline compounds with

17

the stoichiometry $[C(diars)_2]X_4$ and $[C(diars)H_2]X_2$. These substances appear to be best formulated as ionic derivatives of carbon in which halide ions have been displaced by a neutral base. Presumably the carbon atoms are still in a tetrahedral environment, the formal positive charge being distributed over the benzene rings in the Lewis base. Halomethanes are reduced by Cr^{2+} salts in aqueous acetone solutions, forming air-stable compounds which contain cations of the type $(CH_nX_{3-n})Cr^{2+}$ $(X = Cl, Br)$ (*15, 16*).

The Carbonyl Halides, X_2CO $(X = F, Cl, Br, I)$, are formally the acid halides of carbonic acid. Mixed halogen derivatives such as $ClFCO$ are also known. All of the compounds are planar, as expected from the electron-pair repulsion arguments; the C—O distance in the carbonyl halides is shorter (1.13–1.17 Å) than that in acetone (1.24 Å), suggesting that the C—O bond order in these compounds is greater than 2.

As is the case with most acid chlorides, the carbonyl halides are unstable to water and react with substances containing potentially acidic protons, such as alcohols and amines

$$X_2CO + 2H—Y \rightarrow Y_2CO + 2HX \tag{40}$$

$$Y = OH, OR, NR_2, NHR, NH_2.$$

3.2 / Silicon, Germanium, Tin, and Lead

Silicides   Like carbon, silicon forms a large number of binary compounds with the metallic elements; the empirical formulas for these substances are often difficult to understand simply in terms of the usual valence considerations. Some silicides exhibit a typical alloy-type structure in which individual silicon atoms are present. In metal-rich silicides of this type, discrete silicon atoms are located at sites of high coordination; for example, each silicon atom in the compound Mo_3Si has twelve molybdenum atoms as nearest neighbors at 2.73 Å. In silicon-rich silicides, such as $TiSi_2$, the metal atom is entirely surrounded by silicon atoms.

A second general class of silicides exists in which the Si—Si distances are close to the value in elemental silicon (2.35 Å) suggesting that silicon–silicon interactions are important in these compounds. Examples are known in which the silicon groups exist as Si_2 units (U_3Si_2), chains (USi, CaSi), planar hexagonal networks (β-USi_2), packed layers ($CaSi_2$), and three-dimensional frameworks (α-USi_2).

Oxygen and Sulfur Compounds   The oxides of these elements are distinctly different in character and in chemical properties from the oxides of carbon. The latter exist as individual molecule species, whereas all of the oxides of silicon, germanium, tin, and lead either form giant molecules with predominantly covalent bonds or ionic compounds.

Divalent Oxides   Binary oxygen compounds of the type MO (M = Si, Ge, Sn, Pb) are known. SnO and GeO can be made by reducing the dioxide with the elements at elevated temperatures (*17*)

$$MO_2 + M \rightarrow 2MO \tag{41}$$

$$M = Si\,(1450°),\ Ge\,(800°).$$

Tin(II) oxide can be prepared by dehydrating tin(II) hydroxide, and PbO arises when the metal is heated in air. PbO exists in two crystalline forms; the yellow or β-form (massicot) can be converted into the red or α-form (litharge) at $\sim 500°$.

The monoxides of the lighter elements in this family are good reducing agents, as might be expected. For example, at elevated temperatures SiO reduces steam (410°), CO_2 (500°), SO_2 (800°), and $MgCO_3$ (1350°), as well as Ta_2O_5 and Nb_2O_5. The reducing power of these substances decreases with increasing molecular weight.

The infrared spectra of the monoxides, MO (M = Si, Ge, Sn), trapped in a low-temperature matrix indicate that dimeric species with a rhombic structure are present. The structures of SiO and GeO in the solid state are at present unknown. Both SnO and α-PbO possess the unusual layer structure shown in Fig. 11.8 in which the metal atom is associated with four oxygen atoms to form a tetragonal pyramid. These pyramids form an interlocking network through oxygen bridges. Since these compounds contain metal atoms that are formally divalent, each metal atom also possesses an unused pair of

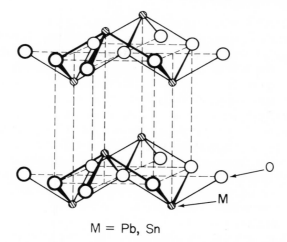

M = Pb, Sn

FIG. 11.8   Both α-PbO and SnO crystallize in the structure shown. Sheets of square-pyramidal MO_4 units sharing oxygen atoms are weakly held together.

valence electrons. It has been suggested that the unshared electron pair occupies an orbital directed away from the oxygen atoms.

Although SiS is not known, the divalent sulfides of germanium, tin, and lead (MS) have been prepared and characterized. These substances can be prepared by precipitation of the divalent cation with H_2S from aqueous solution

$$M^{2+}_{(aq)} + S^{2-} \longrightarrow MS \qquad (42)$$

$$M = Pb, Sn,$$

the reduction of the higher sulfide

$$GeS_2 + Ge \longrightarrow 2GeS, \qquad (43)$$

or from the elements

$$M + 2S \longrightarrow MS_2 \qquad (44)$$

$$M = Sn, Pb.$$

The monosulfides are highly colored solids which crystallize in the sodium-chloride structure or a slightly distorted version of it.

Tetravalent Oxygen Derivatives

Oxides. The tetravalent oxides of these elements have distinctly different chemical and physical properties in comparison with those of CO_2. The latter compound is gaseous under ordinary conditions, and is the anhydride of only one acid. In contrast, other elements in this family form tetravelent oxides that exhibit continuous three-dimensional structures and, at least in the case of silicon, yield an extensive series of acid derivatives. Application of the radius-ratio rules discussed in Chapter 2 to this series of oxides indicates that SiO_2 should crystallize in a four-coordinate structure, whereas SnO_2 and PbO_2 should form six-coordinate (rutile) structures. The radius ratio for GeO_2 is sufficiently close to 0.41 to indicate that either four- or six-co-ordinate structures are possible. As we shall see, these predictions are remarkably good. Under forcing conditions (i.e., high pressure) SiO_2 will crystallize in the six-coordinate structure. Although the radius-ratio arguments presuppose that the structure in question consists of ionic species (M^{4+} and O^{2-}), this does not mean that the bonding in these oxides is wholly ionic. Indeed, the bonds present are best described in terms of the usual covalent links with appreciable ionic character.

At atmospheric pressure silicon dioxide, generally called silica, exists in three crystalline forms which are stable within characteristic temperature ranges; the relationships among these polymorphs are shown in Fig. 11.9.

$$\alpha\text{-Quartz} \qquad \alpha\text{-Tridymite} \qquad \alpha\text{-Crystobalite}$$
$$\downarrow 570° \qquad\qquad \downarrow 120-160° \qquad\qquad \downarrow 200-275°$$
$$\beta\text{-Quartz} \xrightarrow{87°} \beta\text{-Tridymite} \xrightarrow{1470°} \beta\text{-Crystobalite}$$

FIG. 11.9   The temperatures at which the six forms of SiO_2 are stable.

Each of the forms itself can exist in a low-temperature (α) or higher-temperature (β) modification. The structures of quartz, tridymite, and crystobalite consist of SiO_4 tetrahedra linked by bridging oxygen atoms, but the arrangement of tetrahedra is different for each form. In the low-temperature form of SiO_2, quartz, the tetrahedra are arranged so as to form helical chains running through the crystal, as shown in Fig. 11.10a. In a given crystal these chains can either be left- or right-handed, so that it is possible to obtain quartz in two optically active forms. Such isomeric crystalline forms are easily identified and can be mechanically separated. The high-temperature forms of SiO_2 are less dense than quartz and have the more open structures shown in Fig. 11.10b and c. The cubic crystobalite structure is essentially that of zinc blende. It should be recalled that in the zinc-blende structure (Fig. 2.4) the sulfur atoms are cubic closest-packed with zinc atoms occupying the tetrahedral holes. In the crystobalite structure, which is not a close-packed structure, the silicon atoms occupy sites which correspond to the cubic closest-packed positions and the oxygen atoms are at the points associated with the tetrahedral holes. The arrangement of atoms in the tridymite structure is similar to that in the wurtzite structure.

The difference between the α- and β-forms of any of the SiO_2 structures is a slight rotation of the tetrahedra relative to each other without a change

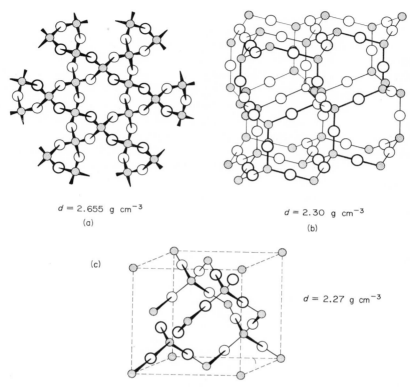

$d = 2.655$ g cm^{-3}

(a)

$d = 2.30$ g cm^{-3}

(b)

(c)

$d = 2.27$ g cm^{-3}

FIG. 11.10 The idealized crystal structures of (a) quartz, (b) β-tridymite, and (c) β-crystobalite.

in the overall arrangement of the tetrahedra. Thus, we can understand why the α–β transition, which does not require bond breaking, occurs at relatively lower temperatures than the conversion of one polymorphic form to another. Quartz is the thermodynamically preferred form at ordinary temperatures, but all forms of SiO_2 are found in nature. The conversion of crystobalite and tridymite into quartz requires breaking and reforming all the bonds in the crystalline array, a process which must have a very high activation energy. Accordingly, these transformations are exceedingly slow. Molten silica, when cooled slowly, gives an amorphous substance which exhibits the characteristics of a glass. There is little long-range structure present in this substance, which can be imagined as consisting of a jumbled array of chains, sheets, and three-dimensional units.

Two other, less widely found, forms of SiO_2 are known. Both are more dense than quartz. Coesite, formed at 35 kbar and 250°C, has a density of 3.01 g cm^{-3} (18, 19). Stishovite (20), formed at 120 kbar and 1300°C, is 60%

more dense than quartz; this form crystallizes in the rutile structure (Fig. 2.4) in which the silicon is six-coordinate.

Many naturally occurring silicon-containing minerals possess structures that can be derived from the forms of silica by a regular replacement of silicon by aluminum with an alkali-metal ion maintaining electrical neutrality. For example, the mineral eucryptite (LiAlSiO$_4$) is a distorted quartz-like structure (*21*) in which aluminum has replaced silicon and Li$^+$ ions occupy the interstices. The holes in tridymite and cristobalite are sufficiently large to accommodate ions of the larger alkali-metal ions. Thus, kalsite (KAlSiO$_4$) exhibits a structure (*22*) closely related to β-tridymite, and the structure of carnegeite (*23*) (NaAlSiO$_4$) is related to that of cristobalite.

Germanium dioxide occurs in two forms. The form stable below 1033° (mp 1086°) crystallizes in the rutile structure. Above 1033° a four-coordinate quartz structure is stable (mp 1116°). Both SnO$_2$ and PbO$_2$ crystallize only in the rutile structure. Lead apparently forms an oxide with an intermediate oxidation number, Pb$_3$O$_4$ (red lead), but its properties and structure, shown in Fig. 11.11, suggest that the correct formulation is PbO$_2$·2PbO. The

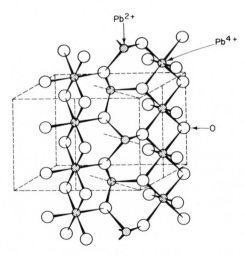

FIG. 11.11   The crystal structure of Pb$_3$O$_4$.

crystal structure of this compound shows chains of PbO$_4$ octahedra sharing opposite corners; the parallel chains are linked by lead atoms surrounded by a pyramidal arrangement of oxygen atoms. Thus, the six-coordinate lead atoms have an environment similar to that of PbO$_2$ while that of the three-coordinate lead atoms resembles the environment in PbO shown in Fig. 11.8.

Hydroxy Compounds. There is little evidence for the existence of discrete compounds with the formulation $M(OH)_4$. Most of the substances reported as hydroxides are probably best described as hydrous oxides $(MO_2 \cdot nH_2O)$. The acidities of the dioxides of this family decrease with increasing molecular weight. Thus, SiO_2 possesses distinctly acidic properties; an aqueous solution of GeO_2 is very weakly acidic $(K = 2 \times 10^{-9})$; SnO_2 is amphoteric; and PbO_2 shows some amphoteric properties.

Orthosilic acid, H_4SiO_4, has not been isolated or characterized, although a variety of derivatives exist. Many silicates which are formally metal derivatives of silic acid are known; these compounds are described more fully in a subsequent section. In addition, esters of the type $Si(OR)_4$ have been characterized.

Two forms of hydrous SnO_2 exist. Low-temperature hydrolysis of Sn(IV) compounds or the reaction of such compounds with base gives the α-form of hydrous SnO_2; the distinctly different β-form results if the hydrolytic reaction is conducted at high temperature. The α-form exhibits amphoteric properties, being soluble in either acid or base. The β-form is markedly less soluble. Both hydrous oxides exhibit the rutile structure of SnO_2, the primary differences being the distribution of water in this structure and the particle size (*24–26*).

Oxyanions   Silicates, the metal derivatives of the hypothetical silicic acid H_4SiO_4, can be classified roughly into water-soluble or water-insoluble substances. Water-soluble silicates have the general composition $aM_2O \cdot bSiO_2 \cdot cH_2O$, where M is an alkali metal. The insoluble silicates are naturally occurring minerals. Water-soluble silicates are prepared by fusing varying proportions of silica with sodium carbonate at about 1300°. Among the crystalline compounds of this type that have been characterized are Na_2SiO_3 (sodium metasilicate), $Na_2Si_2O_5$ (sodium disilicate), Na_4SiO_4 (sodium orthosilicate), $Na_6Si_2O_7$ (sodium pyrosilicate), as well as the hydrates $Na_2SiO_3 \cdot nH_2O$ ($n = 5, 6, 8, 9$). Naturally occurring silicates consist of a wide array of compounds incorporating a variety of linkages for the SiO_4 units present.

A consideration of the structural relationships is perhaps the best basis for discussing the oxyanions of silicon. As is the case with the modifications of silica, the fundamental structural unit in a silicate is the tetrahedral SiO_4 moiety. All silicates can be formally derived from this unit by sharing one or more corners. No silicates are known where two silicon atoms are joined by more than one oxygen atom; that is, only corners of SiO_4 tetrahedra are shared, never edges or faces.

Orthosilicates. Discrete SiO_4 units are found in compounds such as M_2SiO_4 where M is a divalent metal ion such as magnesium, manganese, or iron. These compounds are, in a formal sense, the salts of orthosilicic acid. The SiO_4 units are clustered about the cation, making it six-coordinate; in the case of Be_2SiO_4, where the cation is smaller, the clustering of SiO_4 units leads to four-coordination of the cation. Another description of the structures of the orthosilicates considers that the oxygen atoms are arranged in an approximate hexagonal close-packed structure with silicon atoms in half of the tetrahedral holes (forming the SiO_4 units) and divalent metal ions such as magnesium, manganese, or iron in the octahedral holes; if the metal ion is sufficiently small (e.g., Be^{2+}), it can occupy the other half of the tetrahedral holes.

Many minerals have the basic orthosilicate structure; for example, olivine, which has the general formula $9Mg_2SiO_4 \cdot Fe_2SiO_4$, exhibits the Mg_2SiO_4 structure, but one in ten of the divalent metal atoms occupying the octahedral holes is iron. Garnets are a group of orthosilicates with the

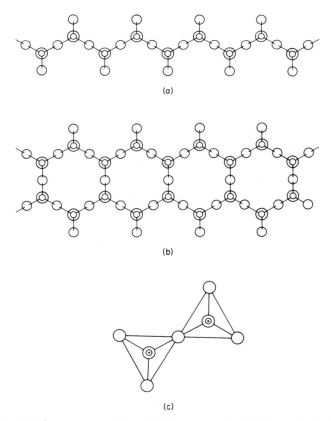

(a)

(b)

(c)

FIG. 11.12   Silicate anions can be formed from tetrahedral SiO_4 units sharing (a) one corner to form a chain or (b) two corners to form a double chain. The $Si_2O_7^{6-}$ ion (c) is the simplest condensed silicate.

general formula $M_3M'_2(SiO_4)_3$, where $M = Ca^{2+}$, Mg^{2+}, or Fe^{2+} and $M' = Al^{3+}$, Cr^{3+}, or Fe^{3+}. The SiO_4 units are packed about M to give eight-coordination and about M' to give six-coordination.

Chain Silicates. The two types of silicate chain structure shown in Fig. 11.12 are known. Simple chains with the composition $(SiO_3)^{2-}$ consist of tetrahedral SiO_4 units joined to form infinite chains. The synthetic metasilicates such as Na_2SiO_3 and Li_2SiO_3 possess an infinite chain structure for the anions, as do the minerals classified as pyroxenes [$MgSiO_3$, enstatite; $CaMg(SiO_3)_2$, diopside; $NaAl(SiO_3)_2$, jadeite; $LiAl(SiO_3)_2$, spodumene]. The chains in such structures are held together by the cations lying between them so that they achieve an appropriate coordination number. For example, the chains in enstatite are arranged to give Mg^{2+} a coordination number of 6, but when Mg^{2+} is replaced by the larger Ca^{2+} ion, as in diopside, the structure accommodates to give the larger Ca^{2+} ion a coordination number of 8.

The other common, infinite-chain structure shown in Fig. 11.12 involves a double chain of SiO_4 units crosslinked so that every other SiO_4 unit shares two of the four oxygen atoms. Such structures are known for that class of minerals called amphiboles, which contain anions of the composition $(Si_4O_{11})^{6-}$. The cations present in such minerals link the adjacent double chains to each other by achieving the appropriate coordination number using the nonbridging oxygen atoms in the chains. The various asbestos minerals are amphiboles; the fibrous nature of such minerals reflects the relatively weak electrostatic forces between the chains compared with the stronger forces holding the SiO_4 units together.

A few silicates are known with short discrete chain lengths. For example, two-unit chains containing discrete $Si_2O_7^{6-}$ ions exist in the structures of $Sc_2Si_2O_7$ and hemimorphite, $Zn_3(Si_2O_7)\cdot Zn(OH)_2\cdot H_2O$. This anion consists of the two SiO_4 tetrahedra sharing a corner (Fig. 11.12).

Cyclic Silicates. Theoretically there could be a very large number of cyclic silicates in which the first SiO_4 unit in a chain shares an oxygen atom with the last SiO_4 unit. However, only two cyclic anions, shown in Fig. 11.13, are known to exist. $Si_3O_9^{6-}$ occurs in $BaTiSi_3O_9$ (benitoite), $Na_2ZrSi_3O_9\cdot 2H_2O$ (catapleite), and $Cu_6Si_6O_{18}\cdot 6H_2O$ (dioptase). The cyclic structure $(Si_6O_{18})^{12-}$ exists in $Be_2Al_2Si_6O_{18}$ (beryl, emerald). The ring anions are arranged in sheets; the metal atoms lie between the sheets holding the rings in different sheets together. When large cations are present, they are six-coordinate as in benitoite. The smaller cations such as Be^{2+} are four-coordinate.

Layer Silicates. When each SiO_4 tetrahedron shares three oxygen atoms with adjacent tetrahedra the infinite sheet structure shown in Fig. 11.14 is formed, with the composition $(Si_2O_5)^{2-}$. The sheet arrangement for

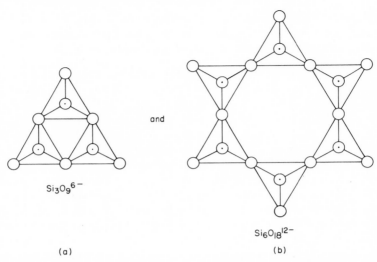

$Si_3O_9^{6-}$

and

$Si_6O_{18}^{12-}$

(a)   (b)

FIG. 11.13   Cyclic silicates can be formed if SiO_4 tetrahedra share two oxygen atoms. (a) The smallest cyclic silicate is $Si_3O_9^{6-}$. (b) $Si_6O_{18}^{12-}$ is a common cyclic silicate ion.

compounds containing the empirical unit $Si_2O_5^{2-}$ is the most common, although recently a discrete ion with this stoichiometry has been isolated in the compound $[Ni(C_2H_4NH_2)_3]Si_2O_5 \cdot 8.7H_2O$ (27). The anion in this substance is $Si_6O_{15}^{6-}$ and contains a prismatic arrangement of silicon atoms, each having three bridging and one terminal oxygen atom.

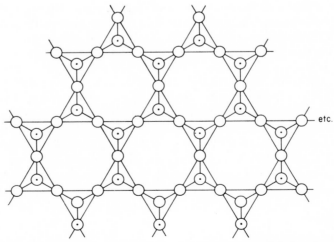

etc.

FIG. 11.14   When SiO_4 units share three oxygen atoms, an infinite sheet can be formed.

There are two possible arrangements of oxygen atoms of the SiO_4 units in a sheet. All of the oxygen atoms can be directed towards one side of the sheet or half can be on one side and half on the other. Many silicates and compounds containing aluminosilicates, in which an appropriate cation has been substituted for silicon, exhibit this layer structure. The counter cations lie between the infinite sheets, holding them together by relatively weak electrostatic forces. Minerals which are found in such layered structures are the kaolins $[Al_2(OH)_4Si_2O_5]$, talc $[Mg_2(Si_2O_5)_2 \cdot Mg(OH)_2]$, and the micas. Kaolins are part of the general class of minerals known as clays; these minerals are soft, easily hydrated, and exhibit cation-exchange properties. Talc is soft, cleaves very easily, and can be used as a dry lubricant. Micas have very nearly perfect cleavage parallel to the layer framework.

Three-Dimensional Silicates. If each SiO_4 tetrahedron shares all of its corners, a three-dimensional network is formed. The possible structures, which are, of course, neutral are described in the section on silica. Isomorphous replacement of aluminum for silicon yields an anion because the atomic number of aluminum is one less than that of silicon. The general formula of such structures is $[AlSi_nO_{2(1+n)}]^{-1}$. The major constituents of igneous rocks are the feldspars which are three-dimensional aluminosilicates. It should be apparent that an almost infinite array of compositions is possible for aluminosilicates. Typical feldspars are $KAlSiO_3$ (orthoclase), $BaAl_2Si_2O_8$ (celsian), $NaAlSi_3O_8$ (albite), and $CaAl_2Si_2O_8$ (anorthite).

A more interesting class of three-dimensional aluminosilicates is the zeolites. Unlike feldspars, these substances have a more open structure, which permits them to absorb liquids and gases rather easily. Generally speaking, three families of zeolites are known. In the first type of zeolite the AlO_4 and SiO_4 tetrahedra are linked into a three-dimensional open framework in which four- and six-membered rings predominate, as shown in Fig. 11.15; chabazite $(Na_2CaAl_2Si_4O_{12} \cdot 6H_2O)$, gemelinite $(Na_2CaAl_2Si_4O_{12} \cdot 6H_2O)$, and erionite $(CaMgNa_2K_2Al_2Si_6O_{16} \cdot 6H_2O)$ are examples of this class.

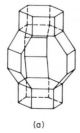
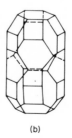

(a)   (b)

FIG. 11.15 Depending upon their composition, zeolites can be produced with different sized cavities: (a) erionite $(CaMgNa_2K_2Al_2Si_6O_{16} \cdot 6H_2O)$ and (b) gemelinite $(Na_2CaAl_2Si_4O_{12} \cdot 6H_2O)$.

Zeolites of this type contain cavities of various sizes which normally contain water molecules. If such a zeolite is strongly heated to about 350° in vacuum, virtually all of the water can be removed. The resulting anhydrous material will absorb and retain molecules with dimensions which permit them to enter the cavities. Zeolites that are highly selective with respect to the geometry of the molecules which can enter the cavities are available. For example, in some cases a straight-chain hydrocarbon can be absorbed whereas a branched-chain isomer can not. Zeolites of this type are truly "molecular sieves."

The second and third groups of zeolites are those which have either a lamellar or a fibrous structure. The detailed structures for such compounds are not well understood, although they must also contain loosely held cations which are easily exchanged.

Oxyanions of Germanium, Tin, and Lead. The information presently available indicates that the oxyanions of these elements do not incorporate the wide structural types described for the silicates. In spite of the fact that the acidity of the dioxides decreases with increasing atomic number, germanates, stannates, and plumbates are known. Crystalline metagermanates (M_2GeO_3) and orthogermanates (M_2GeO_4) have been isolated and characterized; in general these compounds have structures analogous to the corresponding silicon compounds. The cyclic $Ge_3O_9^{6-}$ ion (cf. Fig. 11.13) is present in $SiGeO_3$.

Compounds such as $K_2SnO_3 \cdot 3H_2O$ would appear from the formulation to be alkali-metal derivatives of "metastannic acid," H_2SnO_3. However, structural studies indicate that such compounds contain what is best

TABLE 11.5   Melting and boiling points of the four-valent Group IV halogen compounds

		F	Cl	Br	I
Si	mp	−77	−70	5	120.5
	bp	−65[a]	57.6	153	290
Ge	mp	—	−49.5	26.1	144
	bp	—[b]	86.5	186.5	—
Sn	mp	—	−33	30	143.5
	bp	705	114.1	293	343
Pb	mp	—	−15	—	—
	bp	—	105[c]	—	—

[a] 1810 torr.
[b] Sublimes.
[c] Explodes.

described as six-coordinate tin, and are best formulated $K_2[Sn(OH)_6]$; the water of hydration in these compounds is nonexistent. Plumbates with similar formulations ($K_2PbO_3 \cdot 3H_2O$ or $K_2[Pb(OH)_6]$) are known.

Halogen Derivatives   Except for PbI_4, $PbBr_4$, and PbF_4, the tetrahalides of all these elements have been characterized (Table 11.5) as are some mixed halides of silicon (SiF_3I, $SiCl_2Br_2$, $SiFCl_2Br$). Where such data are available, the structures indicate the expected tetrahedral arrangement about the central atom. This suggests that the central atom utilizes sp^3 hybrid orbitals to form σ bonds. However, bond shortening in the case of fluorine and chlorine derivatives, as shown by the data in Table 11.6, indicates

TABLE 11.6   A comparison of the calculated and experimental bond distances for some Group IV halides (in Å)

	SiF_4	GeF_4	$SiCl_4$	$GeCl_4$	$SnCl_4$
Calculated[a]	1.89	1.94	2.16	2.21	2.40
Experimental	1.54	1.67	2.01	2.10	2.33

[a] From the single-bond radii in Table 3.8.

that some form of π bonding must also be present. Presumably the filled p (or hybrid) orbital of the halogen atom interacts with the empty d orbital of the Group IV element giving a $p\pi$-$d\pi$ component to these bonds. The observed bond distances for the heavier halogen derivatives are more nearly those expected for single bonds.

In addition to the simple tetrahalogen derivatives, silicon forms binuclear compounds (Si_2X_6, X = F, Cl, Br, I) and catenated chloro derivatives up to Si_6Cl_4 (*28–32*). The heavier members of the chlorine series appear to be highly branched structures. SiF_4 reacts with elemental silicon at 1150° to produce SiF_2 (*IX*), which condenses to form a polymer (SiF_2)$_n$ (*33*). Heating the latter compound generates all the saturated catenated fluoro derivatives to $Si_{14}F_{30}$. Of the remaining elements in this family only germanium is known to form a halide with a Ge–Ge bond (Ge_2Cl_6).

The halogen derivatives all more or less readily undergo hydrolysis. The ease of hydrolysis of these compounds is in marked contrast to the stability of corresponding carbon halides. This undoubtedly arises because of the absence of the low-lying d orbitals on carbon which serve as points of attack for the heavier elements. Silicon and germanium tetrafluorides react with water to form a mixture of the corresponding hydrous oxide and hexafluorosilic or -germanic acid

$$MF_4 + (2 + n)H_2O \rightarrow 4HF + MO_2 \cdot nH_2O \qquad (45)$$

$$MF_4 + 2HF \rightarrow H_2MF_6. \tag{46}$$

The other silicon tetrahalides undergo only the hydrolysis reaction [Eq. (45)]; apparently the halide ion formed is not a sufficiently strong base to form hexahalosilicate ions [cf., Eq. (46)]. $GeCl_4$ and $GeBr_4$ are also hydrolyzed in dilute aqueous solution, but in the presence of concentrated hydrohalic acids six-coordinate halogen-containing species such as $[Ge(OH)_nX_{6-n}]^{2-}$ (X = Cl, Br) are formed. $SnCl_4$ and $PbCl_4$ undergo a similar reaction.

The formation of higher-coordinate species of the type MX_6^{2-} (X = halogen) is expected for elements in the Periods 3 and higher. The octahedral arrangement of groups around the central atoms indicates that d^2sp^3 hybrid orbitals are involved in bond formation. Although the bond distances in the six-coordinate fluorides are longer than those of the corresponding four-coordinate compounds (cf. Table 11.6), the bonds are still shorter than expected on the basis of single-bond radii. Compare, for example, the reported bond distances in SiF_6^{2-} (1.71 Å) and GeF_6^{2-} (1.77 Å) with the corresponding data in Table 11.6.

Hydrogen Derivatives   The metal–hydrogen bonds involving silicon, germanium, tin, and lead become less stable with increasing atomic number of the metal; in addition, the ability to catenate also decreases in this order. Thus the hydrogen derivatives of these elements are most numerous for silicon (X); PbH_4 has been prepared in only trace amounts. Silicon and germanium hydrides with the general formula M_nH_{2n+2}, analogous to the alkanes, are known up to $n = 6$ for silicon and $n = 9$ for germanium.

Silanes. A mixture of silanes can be prepared from the reaction of magnesium silicide, Mg_2Si, with hydrochloric acid or with ammonium salts in liquid ammonia (34–36). The hydrolysis products obtained when Mg_2Si is treated with phosphoric acid have been separated into 21 components using vapor-phase chromatography (37). Silicon–halogen compounds are readily reduced with lithium aluminum hydride. Thus $SiCl_4$, Si_2Cl_4, and Si_3Cl_8 yield the corresponding silanes when treated with this reagent. In addition, the silicon–halogen bond in organosilicon halides such as R_3SiCl, R_2SiCl_2, and $RSiCl_3$ can be reduced to the corresponding hydride. Lithium aluminum hydride has also been known to reduce silicon–oxygen compounds to give silanes; for example, $Si(OC_2H_5)_4$ yields SiH_4.

A variety of halosilanes are known. Halogen derivatives of monosilane with the general formula SiH_3X, SiH_2X_2, and $SiHX_3$ (X = F, Cl, Br, I) as well as the mixed compounds $SiHClF_2$ and $SiHClF$ have been characterized. Halosilanes can be obtained from the reaction of hydrogen halides with a silane in the presence of aluminum halide as a catalyst

$$SiH_4 + HX \xrightarrow{Al_2X_6} SiH_3X + H_2 \tag{47}$$

$$SiH_3X + HX \xrightarrow{Al_2X_6} SiH_2X_2 + H_2. \qquad (48)$$

Without the catalyst no reaction occurs, even at elevated temperatures. A very convenient method for the preparation of a halosilane involves the cleavage of a Si—C bond using hydrogen halide. Phenylsilanes have been most often used in this reaction, which occurs smoothly at room temperature

$$C_6H_5SiH_3 + HX \rightarrow C_6H_6 + SiH_3X \qquad (49)$$

$$X = F, Cl, Br, I$$

$$C_6H_5SiH_2X + HY \rightarrow C_6H_6 + SiH_2XY \qquad (50)$$

$$X,Y = Cl, Br, I.$$

Higher halosilanes can be made by the reaction of the silane with chloroform

$$Si_3H_8 + 4CHCl_3 \xrightarrow{Al_2Cl_6} Si_3H_4Cl + 4CH_2Cl_2. \qquad (51)$$

The silanes are all colorless volatile compounds (Table 11.7) with properties characteristic of covalent substances. As far as the known data are concerned, these silanes incorporate tetrahedrally bound silicon atoms.

TABLE 11.7 The melting and boiling points of some silicon and germanium hydrides

Silicon hydride	mp, °C	bp, °C	Germanium hydride	mp, °C	bp, °C
SiH_4	-185	-111.9	GeH_4	-165	-90
Si_2H_6	-132.5	-14.5	Ge_2H_6	-109	$+29$
Si_3H_8	-117.4	$+52.9$	Ge_3H_8	-106	$+110$
Si_4H_{10}	-84.3	$+107.4$			

Presumably the heavier silanes contain silicon chains; both Si_4H_{10} and Si_5H_{12} exist in two isomeric forms, which are presumably straight- and branched-chain forms. The thermal stability of the silanes is less than that of the corresponding carbon compounds. All of the silanes decompose above 500° to hydrogen and silicon. Below this temperature short pyrolysis times yield products reminiscent of cracking reactions in the hydrocarbon series

$$Si_5H_{12} \rightarrow 2(SiH)_x + SiH_2 + SiH_4. \qquad (52)$$

The nonvolatile silicon hydrides $(SiH)_x$ are light brown solids that appear to be polymeric. The silanes react vigorously with oxygen

$$Si_3H_8 + 5O_2 \rightarrow 3SiO_2 + 4H_2O \qquad (53)$$

but do not react with pure water. In the presence of base, however, Si—H bonds are rapidly solvolyzed

$$SiH_4 + (n + 1)H_2O \xrightarrow{OH^-} SiO_2 \cdot nH_2O + 2H_2 \qquad (54)$$

$$SiH_4 + 4ROH \xrightarrow{OR^-} Si(OR)_4 + 2H_2. \qquad (55)$$

Although the hydrogen atoms in the silanes exhibit hydridic properties, it has been possible to prepare silyl potassium, $KSiH_3$,

$$K + SiH_4 \xrightarrow{diglyme} KSiH_3 + \tfrac{1}{2}H_2 \qquad (56)$$

which undergoes reactions characteristic of the SiH_3^- ion (*38*)

$$KSiH_3 + CH_3Cl \rightarrow KCl + CH_3SiH_3; \qquad (57)$$

similar salts of the other alkali metals have also been prepared (*39, XI*). As might be expected, silanes are powerful reducing agents, reducing $KMnO_4$ to MnO_2, Fe^{3+} to Fe^{2+}, Cu^{2+} and Ag^+ to the elements, and Hg^{2+} to Hg_2^{2+}.

The halosilanes are also volatile compounds; all of the monohalosilanes disproportionate

$$2SiH_3X \rightarrow SiH_4 + SiH_2X_2, \qquad (58)$$

the tendency to do so decreasing with increasing atomic weight of the halogen. Nuclear magnetic resonance studies of mixtures of SiH_2X_2 and SiH_2Y_2 (X,Y = F, Cl, Br, or I) indicate that compounds of the type SiH_2YX are formed (*40*). These halogen-transfer reactions probably proceed by way of a four-center transition state, which is favored because silicone can assume coordination numbers greater than four. It will be recalled from Chapter 10 that a similar transition state was postulated for exchange reactions in the boron halides.

All of the halosilanes react rapidly with pure water. Monohalosilanes give disiloxane, the silicon analog of dimethyl ether

$$SiH_3X + H_2O \rightarrow (SiH_3)_2O + 2HX; \qquad (59)$$

di- and tri-halosilanes give polymeric H—Si—O compounds. Thus, trichlorosilane gives a solid with the composition $HSiO_{1.5}$. Aqueous base leads to cleavage of Si—X as well as Si—H bonds. If silyl chloride reacts with excess ammonia, trisilylamine, $(SiH_3)_3N$, is formed. Trisilylamine, the analog of trimethylamine, has virtually no basic properties (*41*). Structural studies show that the three silicon atoms are coplanar with the central nitrogen atom and the Si—N bond distance is significantly shorter (1.738 Å) than expected for a single bond (1.87 Å). A similar planar structure is observed for gaseous $[(CH_3)_3Si]_2NH$, $(SiH_3)_2NH$, and $CH_3N(SiH_3)_2$ (*42–44*), but trimethyl-silylanilines appear to be pyramidal (*45*). These observations lead to the suggestion that in some cases the nitrogen atom uses sp^2 hybrid orbitals to form σ bonds to the silicon atoms. The free electron pair on the nitrogen atom can then be delocalized in three $d\pi$-$p\pi$ bonds using the unhybridized nitrogen p orbital and vacant $3d$ orbitals on the silicon atoms. The $p\pi$-$d\pi$ bonding energy in $(SiH_3)_3N$ has been estimated theoretically at 16 kcal/bond (46). In contrast to the nitrogen derivatives $(SiH_3)_3M$ (M = P, As) is pyramidal.

Germanes. Germanium hydrides can be prepared in much the same way as silanes. Magnesium germanide, Mg_2Ge, reacts with aqueous acid or acid solutions of liquid ammonia to form a mixture of germanes. Although only the lightest members of this series have been well characterized (Table 11.7), there are indications that hydrides as high as octagermanes have been prepared. Monogermane can be prepared from the reduction of $GeCl_4$ with $LiAlH_4$ or $Li(t\text{-}BuO)_3AlH$ in tetrahydrofuran; $NaBH_4$ has also been used as a reducing agent in aqueous solutions. The circulation of monogermane at 0.5 atm through a silent electrical discharge gives a mixture of germanes; a typical reaction gives 20% digermane, 30% trigermane, 6% tetragermane, 0.4% pentagermane, 0.12% hexagermane, 0.1% heptagermane, and 0.04% octagermane. Isomers of Ge_4H_{10} and Ge_5H_{12} have been separated and characterized (47). Monogermane also can be prepared electrolytically by the cathodic reduction of GeO_2 in concentrated H_2SO_4 using lead electrodes. Methylated polygermanes $[(CH_3)_{2n+2}Ge_n]$ have been prepared by the reaction of $Al_2(CH_3)_6$ with GeI_2; homologs of this type up to $(CH_3)_{22}Ge_{10}$ have been identified in this reaction (48).

Considerably fewer halogermanes exist than do halosilanes; monohalo $(GeH_3X, X = F, Cl, I)$, dihalo $(GeH_2X_2, X = Cl, Br, I)$, and trihalo $(GeHCl_3)$ derivatives are known. These compounds can be prepared from reactions that are similar to those used in the preparation of the halosilanes. Halogermanes react with ammonia to form the corresponding substituted amines (49, 50)

$$3H_3GeCl + 4NH_3 \rightarrow 3NH_4Cl + (H_3Ge)_3N \tag{60}$$

$$2(C_6H_5)_3GeBr + 3NH_3 \rightarrow 2NH_4Br + [(C_6H_5)_3Ge]_2NH. \tag{61}$$

Trigermylamine, like the silicon analog, contains a planar Ge–N skeleton with a Ge—N—Ge angle of 120° and $d(\text{Ge}-\text{N}) = 1.836$ Å; the latter value is shorter than the expected single-bond distance (1.88 Å) for these atoms (51).

As indicated by the data in Table 11.7, the germanes as well as halogermanes are volatile compounds that are generally less stable than the corresponding silicon derivatives. In spite of their lower stability, the germanes are less easily oxidized than the corresponding silanes. In addition the Ge—H bond is more stable to hydrolysis, even in the presence of base, than is the Si—H bond. Monogermane reacts with metal–ammonia solutions to form alkali-metal germanyl compounds ($MGeH_3$), which have chemical properties analogous to those of $KSiH_3$ (*38*).

Stannane and Plumbane. At present two binary tin–hydrogen compounds are known, SnH_4 (bp $-52°$, mp $-150°$) and the unstable, poorly characterized Sn_2H_6. Originally SnH_4 was prepared in very low yield by dissolving Sn–Mg alloy in dilute acids or by electrolysis of tin-containing solutions of H_2SO_4 with lead electrodes. A more useful process involves the reduction of $SnCl_4$ with $LiAlH_4$ or aqueous $SnCl_2$ with $NaBH_4$. The latter reaction also yields small amounts of Sn_2H_6. Plumbane has been prepared in only trace amounts from the reactions which yield the hydrides of the other elements in this family.

Stannane decomposes to the elements readily at room temperature; Sn_2H_6 and PbH_4 are also unstable under ordinary conditions. Monochlorostannane, which can be prepared from the reaction of SnH_4 and HCl [cf. Eq. (47)], is noticeably unstable at $-70°$. Stannane is stable to hydrolysis in the presence of either dilute acid or base, which gives it a stability in such reactions intermediate between those of SiH_4 and GeH_4. In concentrated acid or base, the compound undergoes rapid decomposition. As would be expected, SnH_4 is a strong reducing agent. Stannyl derivatives of the alkali metals, $MSnH_3$, can be prepared from the reaction of SnH_4 with metal–ammonia solutions; these compounds undergo reactions similar to those of the corresponding silicon and germanium derivatives (*38, 39*).

Organometallic Derivatives   All of the Group IV elements form a large number of organic derivatives, R_4M (M = Si, Ge, Sn, Pb), which are relatively stable; halogen- (R_3MX, R_2MX_2, RMX_3) and hydrogen-substituted compounds (R_3MH, R_2MH_2, RMH_3) are also known. The stability of M—C and M—M bonds decreases with increasing atomic weight of the Group IV element. For example, chlorine reacts with $Si(C_2H_5)_4$ and $C(C_2H_5)_4$ to give products containing chloroethyl groups; the M—C bonds remain intact. On the other hand, $Ge(C_2H_5)_4$ gives $ClGe(C_2H_5)_3$ and C_2H_5Cl; the corresponding tin compound reacts very rapidly but $C_2H_5SnCl_3$ can be isolated; and tetraethyl lead is rapidly and completely converted into an equivalent amount of $C_2H_5PbCl_3$.

Three general types of organometallic derivatives are discussed here: the tetraalkyl compounds (R_4M), the halogeno alkyls (R_3MX, R_2MX_2, RMX_3), and the alkyl hydrides (R_3MH, R_2MH_2, RMH_3).

Tetraalkyl and tetraaryl derivatives of silicon, germanium, tin, and lead can be readily prepared by the reaction of the Grignard reagent or an organometallic lithium compound with an appropriate halide, MCl_4

$$nRMgX + MCl_4 \rightarrow MR_nCl_{(4-n)} + nMgXCl \tag{62}$$

$$M = Si, Ge, Sn, Pb.$$

Using a mole ratio of reactants less than 4 : 1 it is possible to prepare organo-silicon compounds with one, two, or three halogen atoms per molecule. In the case of lead, the Grignard reaction is conducted with $PbCl_2$ and an excess of the organic halide if R_4Pb compounds are to be prepared. Tetraethyl lead derivatives can also be prepared by the reaction of the alkyl halide with a Na–Pb alloy

$$4RCl + Na_4Pb \rightarrow PbR_4 + 4NaCl, \tag{63}$$

a process which is the basis for the commercial production of tetraethyl lead.

Catenated organometallic derivatives of silicon of the types $(CH_3)_{2n}Si_n$ ($n = 5, 6, 7$) have been prepared by the reaction of $(CH_3)_2SiX_2$ with alkali metals

$$n(CH_3)_2SiX_2 + 2nM \rightarrow (CH_3)_{2n}Si_n + 2nMX, \tag{64}$$

the pyrolysis of linear polymethylsilanes, or a redistribution reaction (52, 53). These substances are not divalent silicon compounds but have a cyclic structure (XII). Open-chain compounds such as $[(CH_3)_3Si]_4Si$ and $[(CH_3)_3Si]_3SiH$ are also known (54). An interesting reaction occurs between triaryl- and trialkylsilicon halides in the presence of magnesium (55)

$$Ar_3SiCl + R_3SiCl + Mg \rightarrow Ar_3SiSiR_3 + MgCl_2. \tag{65}$$

The available data suggest that a Grignard-type intermediate containing a Si—Mg bond is present in this reaction mixture.

In general the tetraalkyl derivatives are stable to oxygen and water, in contrast to the corresponding derivatives of the Group III elements. The lighter alkyls are easily distilled liquids and the heavier derivatives are volatile solids. The known structural parameters for the tetraalkyls indicate that they all exhibit tetrahedral four-coordination of the carbon atoms about the central atom. The metal–carbon bond distances in $(CH_3)_4M$ are those expected from the single-bond radii for silicon, germanium, tin, and lead.

Monohalogeno Derivatives. Several methods are available to prepare mono-halogenated organoderivatives of the Group IV elements in addition to using a deficiency of the Grignard reagent [Eq. (62)]. Mixtures of mono- and dihalogenomethyl derivatives can be prepared from the high temperature reaction of methyl chloride with the metal in the presence of copper

$$CH_3Cl + M \xrightarrow{Cu} (CH_3)_nMCl_{4-n} \tag{66}$$

$$M = Si, Ge, Sn.$$

This reaction has been observed with silicon, germanium, and tin, and is believed to occur by the transient formation of methyl copper (56). Mixtures of mono- and dihalides of tin result when an alkyl halide reacts with tin or a tin–sodium alloy. The monohalides are best obtained by the action of the halogen on the tetraalkyl compound

$$MR_4 + X_2 \rightarrow R_3GeX + RX. \qquad (67)$$

As mentioned previously, the Si—C bond is not readily cleaved by halogens, but $(CH_3)_3GeBr$ and $(CH_3)_3SnCl$ can be prepared by this reaction. The conditions for halogenation must strike a balance between the reactivity of the halogen and the strength of the metal–carbon bond to be broken. The Pb—C bond is much too reactive under the normal conditions for direct halogenation, but hydrogen halides react smoothly

$$Pb(C_2H_5)_4 + HCl \rightarrow (C_2H_5)_3PbCl + C_2H_6. \qquad (68)$$

Tin tetraalkyls and tin tetrahalides undergo redistribution reactions to yield mixed derivatives

$$R_4Sn + SnX_4 \rightarrow SnR_nX_{(4-n)}. \qquad (69)$$

This reaction has been used in several instances for preparative purposes.

The monohalides are generally less volatile than the tetraalkyl derivatives, and are soluble in organic solvents; the known structural parameters indicate that the bonding is best described in terms of the formation of σ bonds using sp^3 hybrid orbitals on the central metal atom. The metal–carbon bonds are in the range expected on the basis of single-bond radii; for example, the observed Si—C and Sn—C distances are in the ranges 1.89–1.86 Å (calculated, 1.84 Å) and 2.19–2.17 Å (calculated, 2.18 Å), respectively. On the other hand, the metal–halogen distances are generally noticeably shorter than expected. The largest discrepancy occurs in the case of $(CH_3)_3SiF$, where the silicon–halogen bond is 17% shorter than expected. The data for other monohalides indicate significantly less shortening in these bonds $[(CH_3)_3SiCl, 5\%;$ $(CH_3)_3SiBr, 4\%]$ or virtually none at all $[(CH_3)_3SiI, 1\%; (CH_3)_3SnCl, 1\%$ $(CH_3)_3SnBr, 2\%]$. Thus, it appears that $d\pi$-$p\pi$ bonds are formed between the lighter halogens and the Group IV elements, a conclusion that is consistent with other considerations.

The monohalogen derivatives are hydrolyzed in neutral or basic aqueous solution to give the corresponding hydroxyl compounds

$$R_3MX + H_2O \rightarrow R_3MOH + HX \qquad (70)$$

which can be easily dehydrated to the corresponding oxides

$$2R_3MOH \rightarrow (R_3M)_2O + H_2O. \qquad (71)$$

Of the silicon and germanium monohalides, the monofluorides are the most resistant to hydrolysis. However, the tin and lead analogues have many properties associated with ionic compounds (e.g., $(CH_3)_3SnF$ and $(CH_3)_3PbF$ decompose without melting). Aqueous solutions of $(CH_3)_3SnF$ are extensively hydrolyzed

$$(CH_3)_3SnF + H_2O \rightleftharpoons (CH_3)_3SnOH_2^+ + F^-; \tag{72}$$

$$\mathbf{18}$$

the hydrated organotin cation (**18**) is a weak acid

$$(CH_3)_3SnOH_2^+ + H_2O \rightarrow (CH_3)_3SnOH + H_3O^+. \tag{73}$$

Trialkylmetal hydroxides (or the corresponding oxides) serve as useful intermediates for the formation of compounds of the type R_3MA, where A is the radical of a weak acid

$$R_3MOH + HA \rightarrow R_3MA + H_2O. \tag{74}$$

Sulfides analogous to the oxides are known for germanium and tin. The sulfur derivatives can be prepared from the monohalides or the oxides

$$2R_3GeBr + Na_2S \rightarrow (R_3Ge)_2S + 2NaBr \tag{75}$$

$$(R_3Sn)_2O + H_2S \rightarrow (R_3Sn)_2S + H_2O \tag{76}$$

$$(R_3Sn)_2O + R'SH \rightarrow R_3SnSR' + R_3SnOH \tag{77}$$

Silicon and germanium monohalides undergo ammonolysis in liquid ammonia to form compounds analogous to the oxides

$$2R_3MX + NH_3 \rightarrow (R_3M)_2NH + 2HX. \tag{78}$$

Compounds containing metal–metal bonds can be prepared from the monohalides using alkali metals (Na, K) to bring about a coupling reaction similar to the Wurtz reaction

$$2R_3MX + 2Na \rightarrow R_3MMR_3 + 2NaX. \tag{79}$$

This reaction has been carried out with alkali metals in the refluxing monohalide, a refluxing inert solvent such as xylene, and with metal–ammonia solutions. These compounds are generally quite stable and react predominantly by cleavage of the metal–metal bond

$$R_3MMR_3 + Cl_2 \rightarrow 2R_3MCl \tag{80}$$

$$R_3MMR_3 + 2Na \longrightarrow 2Na^+MR_3^-. \tag{81}$$

$$\mathbf{19}$$

The alkali-metal salts (19) can be prepared with sodium–ammonia solutions or in ethers; they react as nucleophilic reagents and find use synthetically:

$$R_3M^-Na^+ + R'Cl \longrightarrow R_3MR' + NaCl. \tag{82}$$

In contrast to the behavior of hexaphenylethane, the compounds $(C_6H_5)_3MM(C_6H_5)_3$ have little tendency to form free radicals, although many of their reactions [cf. Eqs (80) and (81)] are the same.

Polyhalogeno Derivatives. The dihalogenodialkyl compounds of silicon, germanium, tin, and lead undergo many of the same types of reactions described for the monohalides; the added factor that two points for reaction exist gives rise to some interesting and important types of compounds. For example, the diols that can be imagined to form upon the hydrolysis of the dihalides

$$R_2MX_2 + 2H_2O \longrightarrow R_2M(OH)_2 + 2HX \tag{83}$$

exist as monomeric substances with the expected formula, as in $(C_2H_5)_2Pb(OH)_2 \cdot 6H_2O$, or as polymeric substances. The tendency to form polymers by loss of water is more pronounced than the formation of diols. Hydrolysis of dihalogeno organosilicon compounds leads to a linear condensation polymer with the structure shown in Fig. 11.16, in which a water molecule has been eliminated from two silicon diol molecules. Such polymers as a class are called silicones. The isolation of $R_2Si(OH)_2$ cannot always be achieved; however, in some cases, such as $(C_6H_5)_2Si(OH)_2$ (mp 139°), the compound has been characterized and yields the silicone upon heating. Dihalides of organogermanium compounds form the low-molecular-weight cyclic polymers $[(CH_3)_2GeO]_4$ and $[(n-C_3H_7)_2GeO]_3$ with the structure shown in Fig. 11.16, or chain polymers $[(C_2H_5)_2Ge]_n$. Except in the case of the t-butyl and aryl derivatives, all the dialkyltin dihydroxides are polymeric, $(R_2SnO)_n$; presumably the presence of bulky groups inhibits condensation so that these compounds exist in the hydroxide form, $R_2Sn(OH)_2$.

Trihalides of these organometallic compounds also undergo rapid reaction with water, followed by dehydration, to yield three-dimensional polymers with the general structure shown in Fig. 11.16 and the general formula $(RMO_3)_n$. Condensation polymers formed from the diols and triols of silicon have found widespread use because they are oils or rubbery solids. These silicones have a low volatility, and are good electrical insulators, thermally stable, and water-repellent. The three-dimensional silicones are

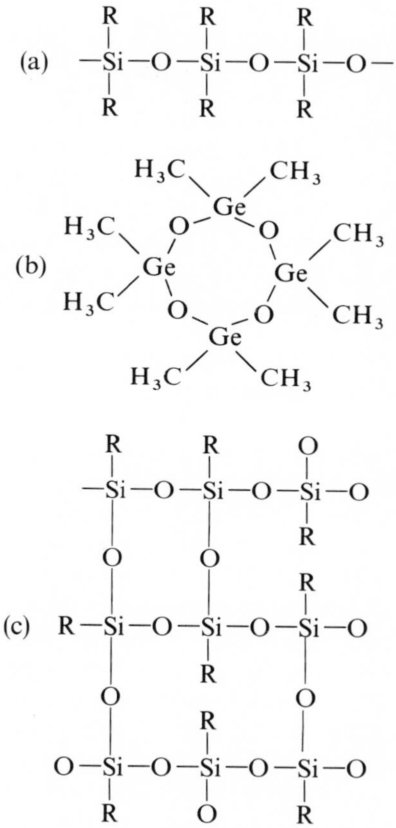

$$
\text{(a)} \quad
\begin{array}{ccccccc}
& R & & R & & R & \\
& | & & | & & | & \\
-Si & -O- & Si & -O- & Si & -O- \\
& | & & | & & | & \\
& R & & R & & R &
\end{array}
$$

$$
\text{(b)} \quad
\begin{array}{c}
H_3C \diagdown \diagup CH_3 \\
Ge \\
H_3C \diagdown \quad O \diagup \diagdown O \quad \diagup CH_3 \\
Ge \qquad\qquad Ge \\
H_3C \diagup \quad O \diagdown \diagup O \quad \diagdown CH_3 \\
Ge \\
H_3C \diagup \diagdown CH_3
\end{array}
$$

$$
\text{(c)} \quad
\begin{array}{ccccccc}
& R & & R & & O & \\
& | & & | & & | & \\
-Si & -O- & Si & -O- & Si & -O \\
& | & & | & & | & \\
& O & & O & & R & \\
& | & & | & & R & \\
R-Si & -O- & Si & -O- & Si & -O \\
& | & & | & & | & \\
& O & & R & & O & \\
& | & & | & & | & \\
O-Si & -O- & Si & -O- & Si & -O \\
& | & & | & & | & \\
& R & & O & & R &
\end{array}
$$

FIG. 11.16 The Group IV compounds can form (a) linear, (b) cyclic, and (c) three-dimensional polymers containing oxygen bridges between metal atoms.

flexible at low temperatures, whereas the viscosity of the two-dimensional polymer containing germanium, tin, or lead is not as interesting from a practical standpoint.

Divalent Organic Derivatives. Organic compounds containing divalent metal atoms are known only for tin and lead, and, at that, the R_2Pb derivatives are poorly characterized. Diaryltin compounds can be prepared by the reaction of the Grignard reagent on stannous halides

$$2RMgX + SnX_2 \rightarrow R_2Sn + 2MgX_2. \tag{84}$$

Dialkyl derivatives can be obtained by reducing tin dialkyldihalides with sodium

$$R_2SnX_2 + 2Na \rightarrow R_2Sn + 2NaX. \tag{85}$$

These compounds are highly reactive and air-sensitive. The structures are not known but they appear to be polymeric $(R_2Sn)_x$. For example, dimethyltin can exist as a linear polymer with a chain length in the region of 12–20 units as well as a cyclic structure $[Sn(CH_3)_2]_6$ (57). The cyclic polymers $[(t\text{-}C_4H_9)_2Sn]_4$, $(R_2Sn)_6$ ($R = C_2H_5, C_6H_5$), and $[(C_2H_5)_2Sn]_9$ have also been reported (58). The structure of $[(C_6H_5)_2Sn]_6$, which is similar to that of cyclohexane, consists of a ring of tin atoms in the chair conformation: the Sn—Sn distances are the same as those observed in gray tin (Table 11.3).

4 / COORDINATION BEHAVIOR

Carbon, unlike boron in Group III, cannot act as a Lewis acid when it has acquired the maximum number of bonds permitted because of its position in the periodic chart. When carbon atoms form four bonds, there is no mechanism available to increase the number of electrons in the valence shell above eight. Thus carbon can exhibit a maximum coordination number of four. Compounds which apparently contain more than four groups associated with carbon, such as $CH_2Cl_2\cdot$diars, are best formulated as ionic species, $[CH_2diars]Cl_2$, on the basis of their properties and chemical behavior. Thus, in such compounds carbon is still maximally four-coordinate.

In contrast to this behavior, the tetracoordinate derivatives of the other members of this family can act as Lewis acids toward suitable Lewis bases, forming five- and six-coordinate derivatives, with the former being less common than the latter. Salts containing the SiF_5^- ion are known (59); the infrared spectra of mixtures of $(C_2H_5)_4N^+Cl^-$ and $SiCl_4$ in nitromethane have been interpreted in terms of the presence of $SiCl_5^-$ ions (60); and evidence exists for GeF_5^- (61–63). With a Lewis base such as acetylacetone all the chloride ions can be displaced from $SiCl_4$

$$\tag{86}$$

to form an ionic compound which has an octahedrally coordinated silicon cation (**20**). Six-coordinate species with anionic Lewis bases are also known for silicon, e.g., K_2SiF_6 which can be imagined as SiF_4 to which two fluoride ions are bound by coordinate-covalent bonds. Germanium, tin, and lead also can form six-coordinate complexes that are ionic $[K_2GeF_6, K_2Sn(OR)_6]$

or neutral ($SnCl_4 \cdot L_2$, L = ROH, R_2O, $R-\overset{\overset{O}{\|}}{C}-R$, $R-\overset{\overset{O}{\|}}{C}-H$, $R-\overset{\overset{O}{\|}}{C}-OR$, RNH_2). In all cases where structural evidence is available, these complex compounds are octahedral. The relative Lewis acidity of the four-covalent compounds depends upon the nature of the groups attached to the central metal atom. For example, $SnCl_4$, $RSnCl_3$, and R_2SnCl_2 form *bis*-pyridine adducts ($SnX_4 \cdot 2C_5H_5N$), R_3SnCl forms a monoadduct ($SnX_4 \cdot C_5H_5N$), but R_4Sn exhibits no Lewis acidity toward pyridine. The *bis*-pyridine adducts are probably octahedral complexes whereas the monoadducts are best described as ionic compounds containing four-coordinate tin atoms, $[R_3SnNC_5H_5]^+Cl^-$. Thus, it would appear that the electron-releasing tendency of alkyl groups is sufficiently large to make trialkyl- and tetraalkyltin compounds noticeably less acidic than the corresponding mono- and dialkyl derivatives. Six-coordinate anionic complexes of alkyl compounds are also known, e.g., $K_2[R_2SnF_4]$, $K_2[RSnCl_5]$, and $K[RPbX_5]$. Lead has been also observed as eight-coordinate in α-$Pb(N_3)_2$ (*64*).

Divalent compounds in this family also exhibit the four- and six-coordination shown by the tetravalent derivatives. Although extensive data are not available on this valence state, compounds such as $(NH_4)_2SnCl_4$ and $SnI_2 \cdot 2H_2O$ have been reported; in addition, three-coordinate compounds containing the SnX_3^- (X = Cl, I) ion, which can be thought of as the anion characteristic of the poorly characterized acids $HSnX_3$, are known. Although the structures of these species are unknown at present, the electron-pair-repulsion arguments suggest that SnX_3^- should be pyramidal with an unshared pair of electrons; on the same basis $SnCl_4^{2-}$ should be isoelectronic with ICl_3, which is a $\perp$-shaped molecule. Six-coordinate divalent lead is presumably present in the compound $K_4[PbF_6]$.

REFERENCES

1. K. M. Mackay, S. T. Hosfield, and S. R. Stobart, *J. Chem. Soc. A*, 2937 (1969).
2. R. B. Aust and H. G. Drickamer, *Science*, **140**, 817 (1963).
3. V. I. Kasatochkin, A. M. Sladkov, Y. P. Kudryavtsev, N. Popov, and V. V. Korshak, *Doklady Chem. (Engl. Transl.)*, **177**, 1031 (1967).
4. F. P. Bundy and J. S. Kasper, *J. Chem. Phys.*, **46**, 3437 (1967).
5. T. Gela, *J. Chem. Phys.*, **24**, 1009 (1956).
6. R. C. Croft, *Australian J. Chem.*, **9**, 184 (1956).

7. W. H. C. Rueggeberg, *J. Amer. Chem. Soc.*, **65**, 602 (1943).
8. M. A. Bredig, *J. Amer. Chem. Soc.*, **65**, 1482 (1943).
9. M. Atoji, *J. Chem. Phys.*, **35**, 1950 (1961).
10. M. Atoji and D. E. Williams, *J. Chem. Phys.*, **35**, 1960 (1961).
11. E. Weiss and W. Büchner, *Helv. Chim. Acta*, **46**, 1121 (1963).
12. G. Gatton and U. Genvath, *Angew. Chem. (Intl. Ed.)*, **4**, 149 (1965).
13. C. F. Cullis and J. G. Yates, *J. Chem. Soc.*, 2833 (1964).
14. R. M. Collinge, R. S. Nyholm, and M. L. Tobe, *Nature*, **201**, 1322 (1964).
15. F. A. L. Anet, *Can. J. Chem.*, **37**, 58 (1959).
16. D. Dodd and M. D. Johnson, *J. Chem. Soc. A*, 34 (1968).
17. J. S. Anderson, J. S. Ogden, and M. J. Ricks, *Chem. Commun.*, 1585 (1968).
18. L. Coes, Jr., *Science*, **118**, 131 (1953).
19. M. J. Buerger and T. Zoltai, *Z. Krist.*, **111**, 129 (1959).
20. S. M. Stishov and S. V. Popova, *Geokhimiya*, **10**, 837 (1961).
21. H. G. F. Winkler, *Acta Cryst.*, **1**, 27 (1948).
22. G. F. Claringbull and F. A. Bannister, *Acta Cryst.*, **1**, 42 (1948).
23. M. J. Buerger, G. E. Klein, and G. Donnay, *Amer. Mineral.*, **39**, 805 (1954).
24. E. Posnjak, *J. Phys. Chem.*, **30**, 1073 (1926).
25. R. Förster, *Phys. Z.*, **28**, 151 (1927).
26. G. F. Hüttig and H. Döbling, *Chem. Ber.*, **60B**, 1029 (1927).
27. Yu. I. Smolin, *Chem. Commun.*, 395 (1969).
28. G. Urry, *J. Inorg. Nucl. Chem.*, **26**, 409 (1964).
29. A. Kaczmarczyk and G. Urry, *J. Inorg. Nucl. Chem.*, **26**, 415 (1964).
30. A. Kaczmarczyk, M. Millard, J. W. Nuss, and G. Urry, *J. Inorg. Nucl. Chem.*, **26**, 421 (1964).
31. A. Kaczmarczyk, J. W. Nuss, and G. W. Urry, *J. Inorg. Nucl. Chem.*, **26**, 427 (1964).
32. J. W. Nuss and G. Urry, *J. Inorg. Nucl. Chem.*, **26**, 435 (1964).
33. P. L. Timms, R. A. Kent, T. C. Ehlert, and J. L. Margrave, *J. Amer. Chem. Soc.*, **87**, 2824 (1965).
34. W. C. Johnson and T. R. Hogness, *J. Amer. Chem. Soc.*, **56**, 1252 (1934).
35. W. C. Johnson and S. Isenberg, *J. Amer. Chem. Soc.*, **57**, 1349 (1935).
36. H. Clasen, *Angew. Chem.*, **70**, 179 (1958).
37. K. Borer and C. S. G. Phillips, *Proc. Chem. Soc. (London)*, 189 (1959).
38. E. Amberger and E. Mühlhofer, *J. Organometal. Chem.*, **12**, 55 (1968).
39. E. Amberger, R. Römer, and A. Layer, *J. Organometal. Chem.*, **12**, 417 (1968).
40. E. A. V. Ebsworth, A. G. Lee, and G. M. Sheldrick, *J. Chem. Soc. A*, 2294 (1968).
41. E. W. Abel, D. A. Armitage, and S. P. Tyfield, *J. Chem. Soc. A*, 554 (1967).
42. C. Glidewell, D. W. H. Rankin, A. G. Robiette, and G. M. Sheldrick, *J. Mol. Structure*, **4**, 215 (1969).
43. L. V. Vilkov and N. A. Tarasenko, *J. Chem. Soc. A*, 1176 (1969).
44. A. G. Robiette, G. M. Sheldrick, W. S. Sheldrick, B. Reagley, D. W. J. Cruickshank, J. J. Monaghan, B. J. Aylett, and I. A. Ellis, *Chem. Commun.*, 909 (1968).
45. E. W. Randall and J. J. Zuckerman, *J. Amer. Chem. Soc.*, **90**, 3167 (1968).
46. P. G. Perkins, *Chem. Commun.*, 268 (1967).
47. K. M. Mackay and K. G. Sutton, *J. Chem. Soc. A*, 2312 (1968).
48. F. Glockling, J. R. C. Light, and J. Walker, *Chem. Commun.*, 1052 (1968).
49. D. W. H. Rankin, *J. Chem. Soc. A*, 1926 (1969).
50. R. E. Highsmith and H. H. Sisler, *Inorg. Chem.*, **8**, 996 (1969).
51. C. Glidewell, D. W. H. Rankin, and A. G. Robiette, *J. Chem. Soc. A*, 2935 (1970).
52. E. Carberry and R. West, *J. Amer. Chem. Soc.*, **91**, 5440 (1969).

53. M. Ishikawa and M. Kumada, *Chem. Commun.*, 567 (1969).

54. H. Bürger and W. Kilian, *J. Organometal. Chem.*, **18**, 299 (1969).

55. N. Duffant, J. Dunogues, and R. Calas, *Compt. Rend.*, **268C**, 967 (1969).

56. A. C. Smith, Jr. and E. G. Rochow, *J. Amer. Chem. Soc.*, **75**, 4103, 4105 (1953).

57. T. L. Brown and G. L. Morgan, *Inorg. Chem.*, **2**, 736 (1963).

58. W. V. Farrar and H. A. Skinner, *J. Organometal. Chem.*, **1**, 434 (1964).

59. H. C. Clark, K. R. Dixon, and J. C. Nicolson, *Inorg. Chem.*, **8**, 450 (1969).

60. G. A. Ozin, *Chem. Commun.*, 104 (1969).

61. K. Behrends and G. Kiel, *Naturwiss.*, **54**, 537 (1967).

62. H. C. Clark and K. R. Dixon, *Chem. Commun.*, 717 (1967).

63. H. C. Clark, P. W. R. Corfield, K. R. Dixon, and J. A. Ibers, *J. Amer. Chem. Soc.*, **89**, 3360 (1967).

64. C. S. Choi and H. P. Boutin, *Acta Cryst.*, **B25**, 982 (1969).

COLLATERAL READINGS

I. J. D. Donaldson, *Prog. Inorg. Chem.*, **8**, 287 (1967). A detailed discussion of the preparation, properties, structure, and theoretical considerations of divalent tin compounds.

II. H. Gilman, W. H. Atwell, and F. K. Cartledge, *Advan. Organometal. Chem.*, **4**, 1 (1966). A review in which the chemistry of the catenated organic derivatives of silicon, germanium, tin, and lead is discussed.

III. H. Schumann, *Angew. Chem.* (*Intl. Ed.*), **8**, 937 (1969). The properties of covalent compounds of the Group IV elements containing phosphorous, arsenic, antimony, and bismuth are reviewed. A part of this review deals with the possible participation of $p\pi$–$d\pi$ components in the bonds formed between the Group IV and Group V elements.

IV. W. Rüdorff, *Advan. Inorg. Chem. Radiochem.*, **1**, 223 (1959); G. R. Hennig, *Prog. Inorg. Chem.*, **1**, 125 (1959). These two review articles contain a wealth of information on the intercalation compounds formed between graphite and a variety of substances.

V. H. P. Boehm, E. Diehl, W. Heck, and R. Sappok, *Angew. Chem.* (*Intl. Ed.*), **3**, 669 (1964). A discussion of the surface oxides formed by carbon.

VI. W. Rüdorff, E. Stumpp, W. Spriessler, and F. W. Siecke, *Angew. Chem.* (*Intl. Ed.*), **2**, 67 (1963). A summary of the reactions of graphite with metal chlorides and the properties of the substances formed.

VII. W. A. Frad, *Advan. Inorg. Chem. Radiochem.*, **11**, 153 (1968). The general methods for the preparation of binary carbides, acetylides, and ternary carbides are discussed, the properties of the products surveyed, and the structure and bonding in these systems described.

VIII. M. Dräger and G. Gattow, *Angew. Chem.* (*Intl. Ed.*), **7**, 868 (1968). A review of the chemistry and properties of the known chalcogencarbonic acids and their derivatives.

IX. J. L. Margrave and P. W. Wilson, *Acc. Chem. Res.*, **4**, 145 (1971). A short summary of the preparation and properties of SiF_2.

X. B. J. Aylett, *Advan. Inorg. Chem. Radiochem.*, **7**, 249 (1965); A. G. MacDiarmid, *Advan. Inorg. Chem. Radiochem.*, **3**, 207 (1961). These two articles contain much useful information on the properties and structures of silicon hydrides and their derivatives.

XI. E. Wiberg, O. Stecher, H. J. Andrascheck, L. Kreuzbichler, and E. Staude, *Angew. Chem. (Intl. Ed.)*, **2**, 507 (1963). A discussion of the recent developments in the chemistry of the metal silyls of the type $M(SiR_3)_n$.

XII. H. Gilman and G. L. Schweke, *Advan. Organometal. Chem.*, **1**, 90 (1964). A discussion of the formation, structure, and properties of cyclosilanes.

STUDY QUESTIONS

1. Germanium oxide is dimorphous; that is, a crystal of the substance undergoes a change in crystal structure when pressure is applied to it. What are the most probable crystal structures for this substance?

2. Give the geometry of the covalently bound species in the following compounds without recourse to structural data. (a) $SiCl_4 \cdot 2KOCH_3$; (b) $(CN)_2$; (c) $Si(OR)_4$; (d) $K_2SnO_3 \cdot 3H_2O$; (e) $(R_3Sn)_2O$; (f) Si_3Cl_8; (g) $(R_3Ge)_2NH$; (h) $[(CH_3)_2SiO]_4$.

3. Which of the species in Question 2 would be expected to exhibit multiple bonds incorporating *d* components?

4. Discuss the Lewis acid–base chemistry expected for the following species: (a) $PbCl_2$; (b) $KSiH_3$; (c) $(R_3Ge)_2NH$; (d) $(R_3Sn)_2O$.

5. Discuss the observation that the adduct $(CH_3)_3NB(CH_3)_3$ is markedly more stable than $(SiH_3)(CH_3)_2NB(CH_3)_3$.

6. Discuss the observation that the Si—N—C group of atoms in H_3SiNCS is linear, but the C—N—C group in H_3CNCS is bent.

7. Comment on the fact that the dipole moments of methyl chloride and silyl chloride are 1.28 and 1.87 D, respectively.

8. Consider the species SiF_2, SiF_4, SiF_5^-, and SiF_6^{2-}. (a) Give the major contributing canonical forms of the wave function for these species using valence-bond arguments. (b) Give a simple molecular-orbital description of SiF_2. (c) Discuss the electronic basis for the expected chemical differences between SiF_4 and CF_4.

9. Give the molecular-orbital description for the bonding in SiO.

10. Discuss four reactions, using equations where necessary, which might be expected for carbon suboxide.

11. Using chemical reactions, discuss the possibility of distinguishing among the three classes of carbides.

12. Discuss the observation that the dipole moments of the following compounds fall in the order indicated: $CH_3Cl > SiH_3Cl < GeH_3Cl$.

GROUP V
ELEMENTS

1 / INTRODUCTION

The ground-state electronic configuration of the Group V elements, as shown in Table 12.1, contains five valence electrons, two paired in an ns orbital and three distributed unpaired in the corresponding np orbitals. Theoretically these elements could achieve a stable electronic configuration by gaining three electrons to form the ionic species M^{3-} or by losing five electrons to form M^{5+}. Compounds containing M^{3-} are known (e.g., Na_3N), but the species M^{5+} has never been observed as such. The heavier elements in this family (antimony and bismuth) tend to form compounds containing cations, but the highest charge observed corresponds to M^{3+}. In such instances, the three p electrons presumably have been lost, the two s electrons becoming inert because of their greater ability to penetrate towards the nuclear charge (Chapter 1).

These elements could also achieve a stable configuration by forming three covalent bonds (e.g., R_3M); in this case the element would have an

TABLE 12.1 Some characteristics of the Group V elements

Characteristic	N	P	As	Sb	Bi
Ground-state configuration	$[He]\,2s^2 2p^3$	$[Ne]\,3s^2 3p^3$	$[Ar]\,3d^{10}4s^2 4p^3$	$[Kr]\,4d^{10}5s^2 5p^3$	$[Xe]\,4f^{14}5d^{10}6s^2 6p^3$
Ionization potential, eV					
I	14.5	10.9	10.5	8.5	8.0
II	29.5	19.6	20.1	18	16.6
III	47.4	30.0	28.0	24.7	25.4
IV	77.0	51.0	49.9	44.0	45.1
V	97.4	65.0	62.5	55.5	55.7
Electronegativity	3.07	2.06	2.20	1.82	1.67
Radii, Å					
ionic (M^{3-})	1.71	2.12	2.22	2.45	—
ionic (M^{3+})	—	—	—	0.92	1.08
covalent	0.70	1.10	1.21	1.41	1.52
Mp, °C	−210	590 (red)	817 (gray)[a]	630	271
Bp, °C	−196	280 (white)	615	1380	1560

[a] Determined at 36 atm.

unshared pair of electrons and should be isoelectronic with the corresponding element in Group IV. Multiple covalent bonds can be formed by the elements in this family in satisfying valence requirements. Thus, nitrogen, like carbon, can form multiple bonds to other atoms in Period 2 (e.g., N_2, HCN, NO_3^-) to achieve a stable configuration. The remaining elements in this family tend to form multiple bonds using d orbitals (_1_).

The Group V elements form compounds in which they exhibit the formal oxidation states from $3-$ to $5+$. It should be emphasized, however, that the oxidation state of an element, while useful for some purposes such as balancing redox equations, has no necessary relationship to structural considerations. For example, the nitrogen atoms in sodium nitride (Na_3N) and ammonia (NH_3) both have oxidation numbers of $3-$, even though one is an ionic substance and the other covalent. A more striking example is the series of compounds NH_3, N_2H_4, and NH_2OH where the formal oxidation states are $3-$, $2-$, and $1-$, respectively. These are all covalent compounds, and have the same pyramidal arrangement of atoms about the nitrogen atom (**1–3**). These structures are consistent with the view that hydrazine (**2**) and hydroxylamine (**3**) are derivatives of ammonia (**1**).

The ligher elements in this family exhibit the greatest ability to catenate, although the number of atoms that can link to each other to form stable species is significantly smaller than for the Group IV elements. Thus, nitrogen is known to form compounds containing a maximum of eight nitrogen atoms in a chain; many of these compounds incorporate multiple-bonded systems such as $-N=N-$. These compounds carry hydrogen or organic substituents, the longer-chain systems being exceedingly unstable. The ability of the heavier members of this family to catenate decreases relative to the ligher elements, paralleling the M—M bond energies in this family, as shown in Table 12.2. The M—M bond energy for carbon is markedly larger than that for any of the elements in Groups IV and V. Although the N—N bond energy is significantly smaller than that for C—C, the data in Table 12.2 indicate that the other elements in Group V have virtually the same energies as their neighbors in Group IV.

The data in Table 12.1 indicate that nitrogen is the most electronegative member of this family. Accordingly it is not surprising that ammonia and its derivatives can act as acceptors in hydrogen-bond formation ($R_3N\cdots H-$) and, if an N—H moiety is present, can form hydrogen bonds to electronegative

**TABLE 12.2   Strengths of M—M single bonds
of the Group IV and Group V elements**

M—M, Group IV	Bond energy, kcal mole^{-1}	M—M, Group V	Bond energy, kcal mole^{-1}
C—C	83	N—N	38
Si—Si	42	P—P	50
Ge—Ge	38	As—As	43
Sn—Sn	34	Sb—Sb	34
Pb—Pb	—	Bi—Bi	—

elements $(R_2N—H \cdots X—)$ (Chapter 7). In fact, trivalent, triply bonded nitrogen atoms such as in HCN also can enter into hydrogen bond formation; the unusual physical properties of HCN have been explained in terms of intermolecular hydrogen bonding $(HCN \cdots HCN)$.

2 / THE ELEMENTS

2.1 / Physical Properties

The low melting and boiling points of nitrogen contrast strikingly with those of the other elements in this series (Table 12.1). Nitrogen, a gas which constitutes 78% by volume of the atmosphere, condenses at $-210°$ to a close-packed solid consisting of N_2 molecules (1). Naturally occurring nitrogen contains two isotopes, ^{14}N and ^{15}N, with a relative abundance of 99.62% and 0.38%, respectively. The heavier isotope, ^{15}N, finds use as a tracer. It is possible to obtain compounds enriched in ^{15}N using chemical methods because certain equilibria are slightly biased in one direction, concentrating a given isotope in a given chemical species. For example, ^{15}N is preferentially concentrated in HNO_3 in the reaction

$$^{15}NO_{(g)} + H^{14}NO_{3(aq)} \rightleftharpoons {}^{14}NO_{(g)} + H^{15}NO_{(aq)} \tag{1}$$

and in NH_4^+ in the process

$$^{15}NH_{3(g)} + {}^{14}NH_{4(aq)}^+ \rightleftharpoons {}^{14}NH_{3(g)} + {}^{15}NH_{4(aq)}^+ \tag{2}$$

because of slight differences in the free energies arising from mass effects. Thus, by physically arranging to have the exchange equilibrium occur a large number of times, it is possible to obtain isotopically enriched ($>95\%$) compounds. One useful method to accomplish this objective involves passing an aqueous solution down a packed column against a countercurrent of the gaseous compound. In both instances described by Eqs. (1) and (2) the nitrogen-containing species dissolved in water becomes enriched in ^{15}N. It then becomes a simple matter of converting the enriched compound (either $^{15}NH_4X$ or $H^{15}NO_3$) into the desired product.

The electronic structure of the nitrogen molecule involves the formation of a triple bond (**4**). This valence-bond description is consistent with the

$$:N\equiv N:$$

4

molecular-orbital description shown in Fig. 4.13, viz.,

$$\sigma(2s)^2\sigma^*(2s)^2\sigma(2p_x)^2\pi(2p_y)^2\pi(2p_z)^2.$$

In this respect the N_2 molecule is isoelectronic with CO and CN^- and has the same distribution of electrons in the bonding orbitals as does the triple bond in acetylene. The $N-N$ bond distance in N_2 (1.095 Å) increases when the short-lived species N_2^+ (1.116 Å) is formed. This observation is consistent with a decrease in bond order predicted from the molecular orbital description.

Electronically, N_2 (**4**) should be capable of acting as a Lewis base, and there is some evidence on this point. The change in the infrared spectrum of BF_3 upon the addition of N_2 has been interpreted as indicating the formation of the 1:1 adduct $N_2:BF_3$ (2).

In contrast to the properties of nitrogen, the other elements of this family are solids at room temperature and they exhibit a number of allotropic forms, the properties of which are shown in Table 12.3. Phosphorus exists in six forms while arsenic and antimony each exhibit three solid polymorphic forms. The molecular weight of phosphorus vapor below 700° corresponds to the species P_4, in which the atoms are situated at the corners of a tetrahedron, as shown in Fig. 12.1. Each atom has three nearest neighbors at 2.21 Å with a

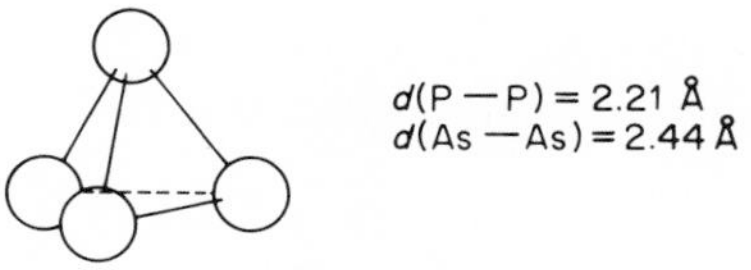

$d(P-P) = 2.21$ Å
$d(As-As) = 2.44$ Å

FIG. 12.1 Both phosphorus and arsenic form tetrahedral M_4 units in the vapor phase.

PPP angle of 60°. Above 700° the P_4 molecules dissociate to P_2 molecules, the dissociation being about 50% complete at 1700°. At even higher temperatures monatomic phosphorus is obtained. When phosphorus vapor condenses under ordinary conditions a yellow-white waxy solid (mp 44.1°)

TABLE 12.3   Properties of the allotropes of the Group V elements

Allotrope	Crystal form	Mp, °C	Specific gravity	Solubility
Phosphorus				
white	cubic (α)	44.1	1.81	C_6H_6, CS_2
	hexagonal (β)	—	2.70	—
violet	rhombehedral	592.5	2.33	no known solvent
scarlet	amorphous	—	1.87	—
black	rhombehedral	—	2.70	—
red	—	592.5	2.20	insoluble
Arsenic				
yellow	cubic	—	1.97	CS_2
black	—	—	4.73	insoluble in CS_2
metallic	rhombehedral	816[a]	5.73	insoluble in CS_2
Antimony				
yellow	cubic	—	—	slightly soluble in CS_2
black	—	—	5.3	insoluble in CS_2
metallic	—	630	6.67	—

[a] Determined at 36 atm.

is formed. White phosphorus has a cubic structure in which P_4 units apparently persist. White phosphorus is soluble in benzene and CS_2, in which it exhibits a molecular weight corresponding to P_4. When white phosphorus is heated in a sealed tube at 530°, the vapor that condenses at 444° gives brilliant, opaque, violet, rhombohedral crystals (mp 592.5°). There are no known solvents for this allotrope and little structural work has been reported on it. If white phosphorus is heated at 200° at 12,000 atm, a black flaky solid is formed. The phosphorus atoms are arranged in the giant molecular structure illustrated in Fig. 12.2, in which each phosphorus atom has three nearest neighbors and three neighbors at a greater distance. This allotrope, unlike the others, is a conductor of electricity. When white phosphorus is heated in a nitrogen atmosphere at 260° it is converted into a red-violet crystalline solid. The structure of this allotrope is shown in Fig. 12.3

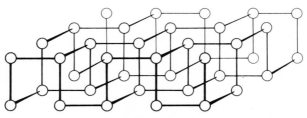

FIG. 12.2   Black phosphorus crystallizes in a hexagonal arrangement of atoms.

and consists of P_8 and P_9 units linked by pairs of phosphorus atoms (3). The entire structure has a pentagonal cross section. Red phosphorus sublimes at $416°$ to P_2 molecules, which quickly dimerize to P_4 molecules since the dimerization equilibrium greatly favors the latter form at these temperatures. If phosphorus vapor at $1000°$ is rapidly cooled, a brown product is formed which slowly converts into a mixture of the red and white modifications.

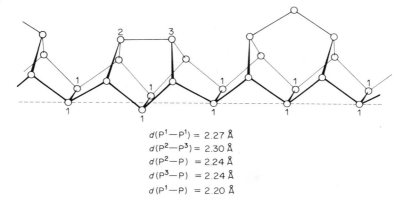

$$d(P^1-P^1) = 2.27 \text{ Å}$$
$$d(P^2-P^3) = 2.30 \text{ Å}$$
$$d(P^2-P) = 2.24 \text{ Å}$$
$$d(P^3-P) = 2.24 \text{ Å}$$
$$d(P^1-P) = 2.20 \text{ Å}$$

FIG. 12.3 Red phosphorus consists of P_8 and P_9 units connected by P_2 groups. These linear tubular arrangements have a pentagonal cross section.

Arsenic exists as tetrahedral As_4 molecules (Fig. 12.1) in the vapor state. The yellow, nonmetallic, metastable allotropes of arsenic and antimony revert to the metallic forms of these elements, which exhibit a layer-like structure. The atoms of a given puckered layer lie in two parallel planes, each atom forming bonds to three atoms in the other layer. In this structure there are three nearest neighbors in an adjacent layer. Black amorphous forms of arsenic and antimony are also known.

2.2 / Chemical Properties

Nitrogen is obtained in large quantities from the liquefaction and fractionation of air. If the commercial product is to be used in the laboratory, the trace quantities of oxygen usually present can be removed by passing the gas through aqueous solutions containing active reducing agents such as V^{2+} or Cr^{2+} or over hot, finely divided copper. A convenient preparation for small amounts of very pure nitrogen involves the thermal decomposition of alkali-metal (except lithium) azides

$$2MN_3 \rightarrow 2M + 3N_2. \tag{3}$$

Nitrogen is relatively inert to most reagents at ordinary conditions, a property undoubtedly related to the dissociation energy of N_2 (225.8 kcal mole^{-1}), which corresponds to breaking a triple bond (**4**). Elemental nitrogen reacts rapidly with lithium at room temperature to give the corresponding nitride

$$6Li + N_2 \rightarrow 2Li_3N. \tag{4}$$

Elemental nitrogen can be fixed in biological systems at room temperature, but the mechanism of such processes is at present unknown. In addition, certain transition-metal complexes have been shown recently to react rapidly with atmospheric nitrogen (*II, III*).

Nitrogen becomes more reactive at elevated temperatures and combines with hydrogen, oxygen, and some metals

$$3H_{2\,(g)} + N_{2\,(g)} \rightarrow 2NH_{3\,(g)} \tag{5}$$

$$O_{2\,(g)} + N_{2\,(g)} \rightarrow 2NO_{(g)} \tag{6}$$

$$3M_{(s)} + N_{2\,(g)} \rightarrow M_3N_2 \tag{7}$$

$$M = Ca, Mg, Ba.$$

A very reactive form of nitrogen, active nitrogen, is formed when gaseous nitrogen is passed through a glow discharge at low pressures. The discharge possesses a characteristic yellow color which can persist for several hours after the current ceases. Active nitrogen consists of nitrogen atoms in their ground state, which recombine only slowly to produce N_2 in an excited, high-spin state. The latter species emits a yellow-colored radiation as it relaxes to the ground state. Active nitrogen reacts readily with many metals (Hg, As, Zn, Cd, Na) and nonmetals (P, S) to yield nitrides. Some metals such as copper and gold catalyze the conversion of active nitrogen to molecular nitrogen without the formation of nitrides.

Elemental phosphorus is prepared by reduction of calcium phosphate (phosphate rock) with coke in an electric furnace

$$2Ca_3(PO_4)_2 + 6SiO_2 + 10C \rightarrow P_4 + 6CaSiO_3 + 10CO; \tag{8}$$

silica is added as a flux and the reaction is conducted at a sufficiently high temperature to vaporize the phoshorus that is formed. The product is condensed and stored under water. The allotropic forms of phosphorus, shown in Table 12.3, have markedly different chemical properties. White phosphorus is considerably more reactive than red phosphorus. It reacts with oxygen spontaneously at room temperature, the reaction being accompanied by a green phosphorescence. A mixture of phosphorus acids is formed when white phosphorus reacts with moist air. On the other hand, red phosphorus does not react with air below about 240°. Phosphorus halides are formed with almost explosive rapidity when white phosphorus is exposed to the halogens; red phosphorus requires the application of heat before reaction occurs. A similar order of reactivity is observed in the reaction of sulfur with white and red phosphorus.

All of the polymorphic forms of arsenic and antimony, described in Table 12.3, are unstable with respect to the metallic allotropes of these elements. Dry air has no effect upon arsenic and antimony at room temperature, but at elevated temperatures oxides with the empirical formulation M_2O_3 are formed. The halogens react readily with these elements to give the corresponding halides. Fluorine is, of course, the most reactive of the halogens, giving a mixture of the tri- and pentafluorides. The remaining halogens yield the trihalides predominantly. Aqueous solutions of non-oxidizing acids do not react with arsenic and antimony, whereas oxidizing acids dissolve these elements.

3 / COMPOUNDS OF THE ELEMENTS

3.1 / Binary Compounds

The simple binary compounds formed by the Group V elements exhibit many of the properties of the binary hydrides (Chapter 6), borides (Chapter 10), and carbides (Chapter 11). The range of properties of the nitrides extends from the ionic derivatives of the most electropositive elements to the compounds formed with hydrogen, carbon, oxygen, sulfur, and the halogens, which are covalent in nature. In addition, there exists a class of compounds formed with some transition metals (*IV*) which appears to be interstitial in nature. The ionic nitrides have formulas corresponding to the normal valence of the metal associated with an N^{3-} ion (*4*). These compounds form clear colorless crystals with high melting points, and they hydrolyze rapidly to give the corresponding hydroxide and ammonia

$$Ca_3N_2 + 6H_2O \rightarrow 3Ca(OH)_2 + 2NH_3. \tag{9}$$

The ionic nitrides can be prepared by direct combination of the metal with nitrogen at elevated temperatures or, in some cases, by heating the metal amide

$$3Ba(NH_2)_2 \rightarrow Ba_3N_2 + 4NH_3. \tag{10}$$

The latter reaction is analogous to that which occurs when some metal hydroxides are heated

$$Mg(OH)_2 \rightarrow MgO + H_2O. \tag{11}$$

Of the alkali-metal nitrides, only Li_3N has been well characterized. The phosphides, arsenides, antimonides, and bismuthides of the alkali metals and the alkaline-earth metals are well known; they are best regarded as ionic crystals containing $3-$ ions. The formulas of some derivatives, such as LiAs, NaSb, and CdP_4, do not conform to the usual valence arguments. These substances have a metallic luster and contain extended chains of atoms of the same kind bound to each other.

The properties of compounds formed between the Group III and Group V elements, as discussed in Chapter 10, indicate that BN and BP are covalent, giant molecular structures, whereas the antimonides and bismuthides have metallic properties. Intermediate compounds appear to be best described in terms of covalent structures with varying amounts of ionic character. All of these compounds crystallize in either the zinc-blende or the wurtzite structure.

Interstitial nitrides are formed by transition metals which crystallize in the cubic close-packed structure with octahedral holes sufficiently large to accommodate a nitrogen atom. In some cases all of the octahedral holes are filled, giving species with the stoichiometry MN (M = Sc, La, Ce, Pr, Ti, Zr, V, Nb, Ta), but nitrides are also known for which half (Mo_2N, W_2N) and one quarter (Mn_4N, Fe_4N) of such holes are occupied. The interstitial nitrides are generally prepared by heating the metal at about 1200° in a stream of ammonia or nitrogen. The interstitial nitrides have many of the properties of the corresponding borides and hydrides; that is, they are hard substances which have melting points that are generally higher than the parent metal. The larger sizes of the remaining elements in this group preclude the formation of interstitial compounds.

3.2 / Oxygen Derivatives of Nitrogen

The oxygen derivatives of the Group V elements are sharply divided into two groups, based upon a consideration of chemical and physical properties, as well as structure. The nitrogen oxides form a group unto themselves, whereas the oxides of the heavier elements in this family have distinctly different characteristics.

Nitrogen Oxides   Of the eight oxides of nitrogen that are known, six are well characterized, as indicated in Table 12.4. In addition to the oxides listed in Table 12.4, nitrogen trioxide (NO_3) and its dimer (N_2O_6) have been reported but little is known of their properties. Two of the common oxides (NO and NO_2) contain an odd number of electrons and many of the properties of these substances derive directly from this fact.

TABLE 12.4 **Some properties of the nitrogen oxides**

Formula	Name	Appearance	Mp, °C	Bp, °C	Trouton's constant
N_2O	nitrous oxide	colorless gas	−90.8	−88.5	21.3
NO	nitric oxide	colorless gas, deep blue liquid and solid	−164	−152	27.1
N_2O_3	dinitrogen trioxide	blue solid	−102	3.5^a	—
NO_2	nitrogen dioxide	red-brown gas	−11.2	21.2	30.9
N_2O_4	dinitrogen tetroxide	colorless solid	$—^b$	$—^c$	—
N_2O_5	dinitrogen pentoxide	colorless solid	41	32.5^c	—

^{a} Extensively dissociated in the gas phase.
^{b} Dissociated in the liquid phase.
^{c} Sublimes.

Nitrous Oxide. Nitrous oxide (N_2O) can be prepared by the controlled reduction of nitrates or the decomposition of hyponitrous acid

$$2HNO \rightarrow H_2O + N_2O. \tag{12}$$

Ammonia or its derivatives are convenient reducing agents for the conversion of nitrates or nitrites into N_2O. Thus the thermal decomposition of ammonium nitrate and the action of hydroxylamine sulfate upon sodium nitrite

$$NH_4NO_3 \rightarrow N_2O + 2H_2O \tag{13}$$

$$(NH_3OH)_2SO_4 + 2NaNO_2 \rightarrow 2N_2O + 4H_2O + Na_2SO_4 \tag{14}$$

give reasonable yields of N_2O. The main impurity in the process described by Eq. (14) is NO, which can be removed by washing the gaseous product with aqueous ferrous sulfate. If NH_4NO_3 contains chloride ion, N_2 is a major product formed in the thermal decomposition reaction.

Dinitrogen oxide is formally the anhydride of hyponitrous acid, HNO (V). Although the gas is reasonably soluble in water, the aqueous solution is neutral and has a conductivity equal to that of pure water. A crystalline hydrate ($N_2O \cdot 6H_2O$) can be obtained from these solutions at low temperatures.

Hyponitrous acid can be prepared by the action of nitrous acid on hydroxyl-amine.

$$HONO + NH_2OH \rightarrow 2HNO + H_2O. \tag{15}$$

Sodium hyponitrite is made by the reduction of $NaNO_2$ or $NaNO_3$ with sodium amalgam.

At room temperature, N_2O is a remarkably stable gas, being inert to the action of O_3, the halogens, or the alkali metals. At elevated temperatures N_2O supports combustion of organic compounds; the alkali metals react at their melting points to form mixtures of the corresponding nitrates, nitrites, and nitrogen. At about 600° N_2O dissociates to the elements. N_2O has a slightly sweet taste and is used as a mild anesthetic.

X-Ray diffraction and spectroscopic data indicate that N_2O is a linear but unsymmetrical molecule (Fig. 12.4) with $d(N-N) = 1.126$ Å and

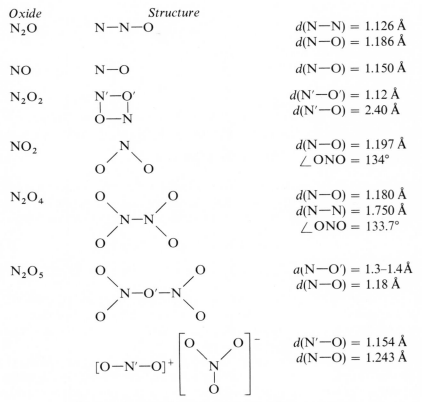

Oxide	Structure	
N_2O	N—N—O	$d(N-N) = 1.126$ Å
		$d(N-O) = 1.186$ Å
NO	N—O	$d(N-O) = 1.150$ Å
N_2O_2		$d(N'-O') = 1.12$ Å
		$d(N'-O) = 2.40$ Å
NO_2		$d(N-O) = 1.197$ Å
		$\angle ONO = 134°$
N_2O_4		$d(N-O) = 1.180$ Å
		$d(N-N) = 1.750$ Å
		$\angle ONO = 133.7°$
N_2O_5		$a(N-O') = 1.3-1.4$ Å
		$d(N-O) = 1.18$ Å
		$d(N'-O) = 1.154$ Å
		$d(N-O) = 1.243$ Å

FIG. 12.4 Structural parameters for the nitrogen oxides.

$d(N—O) = 1.186$ Å. Accordingly, the principal canonical forms contributing to the wave function of the molecule are **5** and **6**, since the observed bond

$$:\overset{-}{\underset{..}{N}}=\overset{+}{N}=\overset{..}{\underset{..}{O}}: \qquad\qquad :N\equiv\overset{+}{N}-\overset{..}{\underset{..}{O}}:$$

5   **6**

distances do not correspond to those expected for $N\equiv N$ (1.10 Å), $N=N$ (1.25 Å), $N=O$ (1.14 Å), or $N—O$ (1.36 Å) (*VI*). The N_2O molecule is iso-electronic and isostructural with CO_2; the two oxides crystallize in the same lattice and their mixtures form a series of solid solutions.

Nitric Oxide. Nitric oxide (NO) is formed when acid solutions of nitrates or nitrites are reduced by metals

$$8HNO_3 + 3Cu \longrightarrow 3Cu(NO_3)_2 + 4H_2O + 2NO \qquad (16)$$

$$2KNO_3 + 6Hg + 4H_2SO_4 \longrightarrow 2NO + 3Hg_2SO_4 + K_2SO_4 + 4H_2O \qquad (17)$$

or other reducing agents

$$2KNO_3 + 4H_2SO_4 + 6FeSO_4 \longrightarrow 3Fe_2(SO_4)_3 + 2NO + K_2SO_4 + 4H_2O \qquad (18)$$

$$2HI + 2HNO_2 \longrightarrow I_2 + 2NO + 2H_2O. \qquad (19)$$

The large-scale preparation of NO involves the oxidation of ammonia above 500° in the presence of platinum

$$4NH_3 + 5O_2 \longrightarrow 4NO + 6H_2O. \qquad (20)$$

Direct combination of the elements occurs under energy-rich conditions

$$N_2 + O_2 \rightleftharpoons 2NO, \qquad (21)$$

but the equilibrium is unfavorable even at high temperatures, about 5 volume per cent of NO being formed at 3200°C.

Nitric oxide is thermodynamically unstable at room temperature, the liquid or compressed gas slowly undergoing disproportionation,

$$4NO \longrightarrow N_2O_3 + N_2O. \qquad (22)$$

Since NO possesses an odd electron it might be expected to undergo radical-type reactions. Indeed, small quantities of NO inhibit chain reactions. Although it generally behaves as expected for a radical species, the degree of reactivity is markedly depressed. Nitric oxide can act as an oxidizing agent;

it supports combustion of carbon compounds as well as elemental carbon and phosphorus at elevated temperatures. In addition it can be reduced by $LiAlH_4$ (to hyponitrous acid), SO_2 (to N_2O), and Cr^{2+} (to NH_2OH). Nitric oxide is also reactive towards oxidizing agents. It reacts immediately with oxygen to form nitrogen dioxide

$$2NO + O_2 \rightarrow 2NO_2, \tag{23}$$

and with the halogens to give nitrosyl halides

$$2NO + X_2 \rightarrow 2XNO \tag{24}$$
$$X = F, Cl, Br.$$

In the reaction with fluorine, an additional product can be formed, F_3NO (5). Aqueous permanganate ion converts NO quantitatively into NO_3^-.

The physical properties of NO, shown in Table 12.4, indicate a rather unusual behavior for a radical species. Contrary to the behavior of most radicals, it is not colored in the gaseous phase, although the pure liquid and solid are reported to be blue. The relatively high value of the entropy of vaporization (Trouton's constant = 27.1) suggests that association occurs in the liquid phase. Nitric oxide in the liquid state is paramagnetic but solid NO is diamagnetic at low temperatures suggesting that intermolecular coupling occurs under these conditions. However an unusual combination of factors found in the NO molecules confuses the obvious interpretation of the temperature dependence of the apparent magnetic moment. The odd electron has an orbital moment about the molecular axis which may be coupled with the spin momentum to give a diamagnetic ground state of the molecule. The paramagnetic excited state is about 350 cal higher in energy than the ground state. The thermal energy available at room temperature ($600\ cal\ mole^{-1}$) is sufficient to keep most of the molecules in the excited spin state, which is paramagnetic. As the sample is cooled, the distribution of molecules shifts toward the diamagnetic state. Thus the temperature dependence of the magnetic susceptibility is not necessarily totally attributable to the dimerization process

$$2NO \rightleftharpoons N_2O_2. \tag{25}$$

In spite of the difficulty in interpreting the magnetic data, evidence for the polymerization of NO in the liquid and solid states comes from a study of infrared-spectral and x-ray diffraction data. Three different dimeric species have been described (6). Two involve an N—N bond in which the planar molecule has a *cis* (**7**) or *trans* (**8**) arrangement of oxygen atoms. The third dimer

has a planar arrangement of atoms in the form of a rectangle (Fig. 12.4).

Valence-bond arguments indicate that the canonical forms **9** and **10** are

$$:N{=}O: \qquad\qquad {}^{-}:N{=}O:{}^{+}$$

9   **10**

the most important contributions to the wave function of the NO molecule. As in the case of N_2O, the observed NO bond distance (1.150 Å) is intermediate between that of a triple and that of a double bond. Although structure **10** possesses an intuitively unfavorable charge distribution, it does account for the rather low (0.16 D) dipole moment of the compound. A more instructive description of the electronic configuration in the NO molecule can be obtained from molecular-orbital theory (7). The approximate arrangement of molecular orbitals for NO appears in Fig. 4.15. In this view, the electronic structure of NO is the same as that of N_2 with an added electron in an antibonding π level. The bond order is 2.5, in agreement with the observation that the value is intermediate between that of a double and that of a triple bond. More important, the molecular-orbital diagram indicates the species NO^- and NO^+ should be easily attained. Many compounds containing the nitrosonium ion (NO^+) are known, e.g., $(NO)ClO_4$ and $(NO)AlCl_4$, and are discussed in a later section. The NO^+ ion, which is isoelectronic with CO, should be diamagnetic with a bond order of 2.0 (*VI*). The observed internuclear distances for NO^+ as it appears in $(NO)ClO_4$ (1.11 Å) and $(NO)AlCl_4$ (1.12 Å) are nearly the same as that for NO. However, a comparison of the N—O stretching frequencies for NO^+ in a variety of compounds (2150–2400 cm^{-1}) with that for NO (1840 cm^{-1}) suggests the presence of a stronger bond in NO^+. The existence of NO^- has not been established. Compounds with a stoichiometry suggesting that NO^- might be present (NaNO) are formed in the reaction of NO with alkali-metal–ammonia solutions, but these substances are diamagnetic. When NaNO is treated with water it yields an aqueous solution of the true sodium salt of hyponitrous acid, $Na_2N_2O_2$; nitrous oxide is also obtained in varying amounts.

Nitrosonium compounds can be prepared from the nitrosyl halides, the chloride being the most useful precursor. Liquid nitrosyl chloride (mp

$-64.5°$; bp $-6.4°$; dielectric constant 225 at $-27°$; specific conductivity $2.7 \times 10^{-6}\,\Omega^{-1}$ at $10°$) forms the basis for an interesting nonaqueous solvent system in which NO^+ and Cl^- are the acidic and basic species, respectively

$$NOCl \rightleftharpoons NO^+ + Cl^-. \tag{26}$$

In the presence of strong Lewis acids, NOCl acts as a Cl^- donor, forming nitrosonium salts

$$NOCl + MCl_3 \rightarrow NOMCl_4 \tag{27}$$

$$M = B, Al, Ga, In, Tl, Bi, Fe, Au$$

$$2NOCl + MCl_4 \rightarrow (NO)_2MCl_6 \tag{28}$$

$$M = Sn, Ti, Pb, Pt.$$

Nitrosonium salts of hexafluoro-anions ($NOMF_6$, M = P, As, Sb, V) have also been prepared. A variety of other nitrosonium salts are available through double decomposition reactions in other nonaqueous solvents. Nitrosonium salts are rapidly hydrolyzed, giving solutions of nitrous acid, which can be regarded as a very weak base

$$NO^+ + 2H_2O \rightarrow H_3O^+ + HNO_2 \tag{29}$$

$$HONO \rightleftharpoons HO^- + NO^+. \tag{30}$$

This point of view is reinforced by the fact that nitrosonium hydrogen sulfate, $NOHSO_4$ (mp $73.5°$), which could be considered as the salt of the weak base HONO, is well characterized.

The nitrosyl halides are bent molecules, as shown by the data in Table 12.5, with halogen–nitrogen distances markedly greater than those expected for

TABLE 12.5 Structural parameters for the nitrosyl halides

		Internuclear distances, Å	
	$\angle XNO$	$d(XN)^a$	$d(NO)$
FNO	$110°$	1.52	1.13
ClNO	$116°$	1.95	1.14
BrNO	$117°$	2.14	1.15

^{a} The expected distances for X—N single bonds are: F, $1.42\,Å$; Cl, $1.69\,Å$; and Br, $1.84\,Å$.

single bonds (*VII*). In addition, the $N-O$ distance is virtually the same as that observed for NO (1.150 Å). These observations suggest that the canonical forms **11** and **12** are important contributors to the wave function for the

11   **12**

molecule. The observed XNO bond angle indicates that, to a first approximation, sp^2 hybrid orbitals on the nitrogen atom are needed to construct the appropriate bonds. The use of an extreme ionic form (**12**) in the wave function is consistent with the ability of these halides to undergo ionization.

The species NO^+ can act as a Lewis acid by the addition of an unshared electron pair to the antibonding π orbitals which exist in this species. Such an interaction is presumably present in the compound formed when nitrosyl chloride reacts with trimethylamine at low temperature, $(CH_3)_3N \cdot NOCl$, the compound being formulated as $[(CH_3)_3NNO]^+Cl^-$. The ability of NO or NO^+ to act as a Lewis base toward transition metal ions is discussed in Chapter 19.

Nitrogen Dioxide. Nitrogen dioxide, NO_2, is formed when concentrated nitric acid reacts with copper

$$Cu + 4HNO_3 \rightarrow Cu(NO_3)_2 + 2NO_2 + 2H_2O, \tag{31}$$

or when NO is treated with oxygen [Eq. (23)]. This substance is a red-brown, paramagnetic gas at room temperature, the latter property being expected because it is a radical. Nitrogen dioxide easily condenses to a brown liquid but freezes to a colorless solid. In the process the system loses its paramagnetism to the point where the solid becomes diamagnetic. This unusual behavior has been interpreted in terms of an equilibrium between monomeric NO_2 and dimeric N_2O_4

$$2NO_2 \rightleftharpoons N_2O_4. \tag{32}$$

The solid is wholly N_2O_4, the liquid contains less than 1% NO_2, and the vapor contains about 90% NO_2 at 100°C. The enthalpy for the dimerization reaction is -12 kcal mole^{-1}. Above 140° the dissociation into NO_2 is complete. Thus, the reported properties of the liquid and gas (below 140°) are those of a mixture of NO_2 and N_2O_4. Attempts have been made to describe the chemical reactivity of NO_2 and N_2O_4 using a variety of methods. Nitrogen dioxide in the vapor state behaves primarily as an oxidizing agent towards metals and nonmetals, as well as towards compounds containing nonmetals in a lower oxidation state:

$$2Cu + NO_2 \rightarrow Cu_2O + NO \qquad (33)$$

$$C + NO_2 \rightarrow CO_2 + \tfrac{1}{2}N_2 \qquad (34)$$

$$7H_2 + 2NO_2 \rightarrow 2NH_3 + 4H_2O \qquad (35)$$

$$X_2 + 2NO_2 \rightarrow 2XNO_2 \qquad (36)$$

$$X = Cl, Br$$

$$CO + NO_2 \rightarrow CO_2 + NO \qquad (37)$$

$$SO_2 + NO_2 \rightarrow SO_3 + NO \qquad (38)$$

$$ClNO + NO_2 \rightarrow ClNO_2 + NO. \qquad (39)$$

With powerful oxidizing agents, higher oxides of nitrogen are formed:

$$O_3 + 2NO_2 \rightarrow N_2O_5 + O_2 \qquad (40)$$

$$NO_3 + NO_2 \rightarrow N_2O_5 \qquad (41)$$

$$H_2O_2 + 2NO_2 \rightarrow 2HNO_3 \qquad (42)$$

$$ClNO_2 + 2NO_2 \rightarrow ClNO + N_2O_5. \qquad (43)$$

In addition, NO_2 has been shown to nitrate aliphatic hydrocarbons in the vapor phase and to add to unsaturated compounds:

$$2RH + 4NO_2 \rightarrow 2RNO_2 + H_2O + N_2O_3 \qquad (44)$$

$$R_2C{=}CR_2 + 2NO_2 \rightarrow R_2C{=}CR_2 \cdot 2NO_2. \qquad (45)$$

On the other hand, liquid dinitrogen tetroxide, which still is an oxidizing agent, undergoes reactions in which the products have a different chemical character. For example, metals react to form nitrates, as do salts which contain oxidizing anions

$$M + N_2O_4 \rightarrow MNO_3 + NO \qquad (46)$$

$$M = Na, Ag, Cu$$

$$KN_3 + N_2O_4 \rightarrow KNO_3 + N_2 + N_2O \qquad (47)$$

$$NaClO_3 + N_2O_4 \rightarrow NaNO_3 + NO_2 + ClO_2. \qquad (48)$$

A consideration of the use of liquid N_2O_4 as a nonaqueous solvent appears in a subsequent section of this chapter.

The structure of NO_2 in the gas phase, given in Fig. 12.4, shows it to be a bent but symmetrical molecule with a bond angle of 134°. The presence of only one N—O internuclear distance (1.197 Å) indicates that the principal canonical forms contributing to the wave function consist of two covalent structures in which the odd electron is localized on oxygen (**13**) and two ionic forms where the electron is localized on the nitrogen atom (**14**). The bond

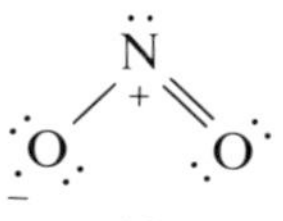

13 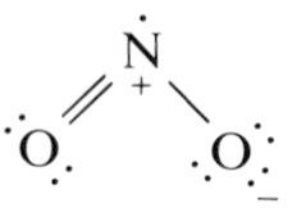   **14**

angle suggests that hybrid sp^2 oribtals are used by nitrogen in bond formation. Formally we might expect NO_2 to be able to acquire an electron as well as to lose an electron (Fig. 12.5). The species NO_2^+ and NO_2^- have been well characterized and are discussed in detail in subsequent sections of this chapter. Electron-pair-repulsion arguments predict that the loss of an electron from NO_2 should give a linear species, and the gain of an electron should decrease the ONO bond angle, a prediction borne out by the data which appear in Fig. 12.5.

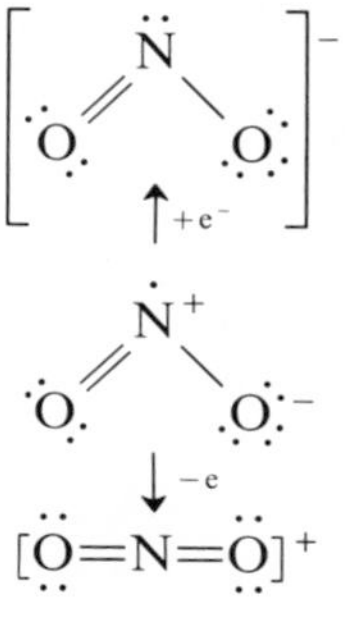

	$d(N-O)$, Å	$\angle$ ONO
	1.236	115.4°
	1.197	134°
	1.10	180°

FIG. 12.5 The relationships between NO_2, NO_2^-, and NO_2^+.

X-Ray and electron diffraction results, given in Fig. 12.4, indicate that planar N_2O_4 is the most stable conformation of nitrogen dioxide. The N—O bond distance in the dimer (1.180 Å) is essentially the same as that in the monomer (1.197 Å), but the N—N distance is considerably longer (1.750 Å) than the normal N—N bond distance (1.47 Å). Two other conformations of the dimer have been reported on the basis of infrared studies at low temperatures in an inert matrix (8, 9). A twisted nonplanar form was found at liquid-nitrogen temperatures, and an oxygen-bonded species, $ONONO_2$, at 4°K.

Liquid dinitrogen tetroxide has a low specific conductivity ($10^{-12} \Omega^{-1}$) and a low dielectric constant (2.42), and ordinarily it would not be considered a very good solvent. However, it dissolves a large number of substances by reacting with them. Thus, the liquid (which, it should be recalled, contains less than 1% NO_2) dissolves halogens, saturated and unsaturated hydrocarbons, alcohols, ethers, amines, and other organic compounds containing oxygen and nitrogen; the organic compounds form adducts with N_2O_4. In addition, a variety of inorganic substances react with liquid N_2O_4 to form useful products.

In most of the reactions of liquid N_2O_4 or its solutions in ether or ethyl acetate, the substance behaves as if it were ionized:

$$N_2O_4 \rightleftharpoons NO^+ + NO_3^-. \tag{49}$$

As was pointed out earlier, the planar configuration of the dimer is apparently the most stable. The reactions may proceed through a small amount of the bridged $ONONO_2$ species that might be present. This could logically ionize according to Eq. (49) (i.e., as nitrosonium nitrate), but this possibility is not supported by experimental evidence.

In the presence of strong proton donors N_2O_4 gives nitric acid and a solution of nitronium ions

$$N_2O_4 + H^+ \longrightarrow HNO_3 + NO^+, \tag{50}$$

a process which can be imagined as occurring because the equilibrium shown in Eq. (49) has been displaced to the right by the removal of NO_3^-. Thus a solution of N_2O_4 in sulfuric acid is essentially a solution of nitrosonium hydrogen sulfate and nitric acid. Nitric acid, of course, can act also as a proton acceptor in concentrated sulfuric acid. Strong acids such as $H_2S_2O_7$, H_3PO_4, and $HClO_4$ also give solutions containing NO^+ when mixed with N_2O_4. Zinc nitrate forms an adduct, $Zn(NO_3)_2 \cdot 2N_2O_4$, which has been formulated as a nitrosonium nitrate complex $[(NO)_2Zn(NO_3)_4]$.

The ionization of N_2O_4 [Eq. (49)] can be assisted if the species NO^+ can be removed from the system. Lewis bases such as amines, nitriles, or ethers form adducts, $L \cdot N_2O_4$, which have been formulated as $(LNO^+)NO_3^-$, containing the ion NO^+ acting as a Lewis acid. Many metals react with liquid N_2O_4 to give the metal nitrates and a reduction product of NO^+ (*VII*):

$$M + NO^+NO_3^- \longrightarrow MNO_3 + NO \tag{51}$$

$$M = \text{alkali metals, Ag}, \tfrac{1}{2}\text{Pb}, \tfrac{1}{2}\text{Zn}, \tfrac{1}{2}\text{Cu}.$$

In some instances such reactions are slow, but they can be hastened if ether or ethyl acetate is added. It has been suggested that these substances increase the degree of ionization of N_2O_4 through the formation of species such as $LNO^+NO_3^-$; presumably electron transfer between the cations present and the metal is enhanced by this process. The metal nitrates formed in this reaction are often solvated with N_2O_4, which can be removed easily in vacuo. Thus reaction (51) becomes a convenient route to the preparation of anhydrous metal nitrates. The reaction of alkali-metal halides with N_2O_4

$$MX + NO^+NO_3^- \rightarrow MNO_3 + XNO \tag{52}$$

$$MX = KCl, KBr$$

also appears to involve a metathetical reaction using the hypothetical nitrosonium nitrate shown in Eq. (49). When potassium azide is dissolved in N_2O_4, the intermediate nitrosonium azide is unstable and decomposes to nitrogen and N_2O

$$N_3NO \rightarrow N_2 + N_2O. \tag{53}$$

Metal oxides and carbonates apparently also form unstable nitrosonium compounds which decompose to N_2O_3

$$CaO + 2NO^+NO_3^- \rightarrow Ca(NO_3)_2 + N_2O \tag{54}$$

$$Na_2CO_3 + 2NO^+NO_3^- \rightarrow 2NaNO_3 + N_2O_3 + CO_2. \tag{55}$$

Remembering that nitrosonium compounds can be considered as salts of the very weak base nitrous acid, HONO [Eq. (30)], the addition of any substance which can supply OH^- to N_2O_4, such as water or NaOH, yields nitrous acid:

$$HOH + NO^+NO_3^- \rightarrow HONO + HNO_3 \tag{56}$$

$$NaOH + NO^+NO_3^- \rightarrow NaNO_3 + HONO. \tag{57}$$

Dinitrogen Trioxide (VIII). Dinitrogen trioxide, N_2O_3, a blue solid which melts at $-102°$, can be prepared by mixing NO and NO_2 below $-100°$. The equilibrium

$$N_2O_3 \rightleftharpoons NO + NO_2 \tag{58}$$

is achieved at $-20°$ with about $10\% \ N_2O_3$ being present at $25°$. Little is known about the constitution of this oxide. It dissolves in aqueous base to produce NO_2^-, but in water it gives a blue solution which decolorizes rapidly, yielding HNO_3 and NO. The rapid exchange between isotopically labeled $^{15}NO_2$ and ^{14}NO under equilibrium conditions [Eq. (58)], as well as ^{14}N nmr experiments on liquid N_2O_3 (*10*), suggest an oxygen-bridged structure, ONONO, for N_2O_3. Direct structural information on N_2O_3 is not available at this time.

Dinitrogen Pentoxide. Dinitrogen pentoxide (N_2O_5) can be prepared in several ways. Ozone will oxidize the lower nitrogen oxides to N_2O_5, and the action of either Cl_2 or $POCl_3$ on silver nitrate will also produce this oxide

$$2AgNO_3 + Cl_2 \rightarrow 2AgCl + N_2O_5 + \tfrac{1}{2}O_2 \tag{59}$$

$$6AgNO_3 + POCl_3 \rightarrow 3AgCl + Ag_3PO_4 + 3N_2O_5. \tag{60}$$

In addition, dehydration of nitric acid by P_2O_5 yields N_2O_5

$$2HNO_3 + P_2O_5 \rightarrow 2HPO_3 + N_2O_5. \tag{61}$$

Dinitrogen pentoxide is a colorless, low-melting ($41°$) solid which slowly decomposes to N_2O_4 and oxygen. As might be expected, it is a good oxidizing agent. It dissolves in concentrated HNO_3 or H_2SO_4 to give nitronium and nitrate ions, such mixtures being good nitrating reagents

$$N_2O_5 \rightleftharpoons NO_2^+ + NO_3^-. \tag{62}$$

This oxide apparently exists in a molecular form in the vapor phase and in CCl_4 solution. An electron diffraction study of the vapor suggests that the molecule contains an oxygen bridge, $O_2N-O-NO_2$, but, as the data in Fig. 12.4 show, accurate values of the bond distances could not be obtained. The substance possesses a dipole moment in CCl_4 solution (1.39 D) suggesting that the $N-O-N$ bridge is not linear. In the solid state N_2O_5 exists as nitronium nitrate, $NO_2^+NO_3^-$. The $N-O$ distances in the ions are close to those reported for other nitronium salts [e.g., 1.10 Å in $(NO_2)^+(ClO_4)^-$] and other nitrates (1.22 Å). This formulation for solid N_2O_5 is consistent with its mode of ionization in strong acid media [Eq. (62)].

Nitryl halides (XNO_2) also contain the NO_2 moiety, but these are typically covalent substances. Nitryl fluoride (mp $-166°$, bp $-72.4°$) can be prepared by the action of fluorine on either nitric oxide at low temperature or sodium nitrite (*IX*)

$$4NO + F_2 \rightarrow 2FNO_2 + N_2 \tag{63}$$

$$NaNO_2 + F_2 \rightarrow FNO_2 + NaF. \tag{64}$$

Nitryl chloride (mp $-147°$, bp $-15°$) is formed when nitrosyl chloride is treated with ozone

$$ClNO + O_3 \rightarrow ClNO_2 + O_2. \tag{65}$$

Both nitryl chloride and nitrosyl chloride are best considered as acid halides of nitric acid. The structural data shown in Fig. 12.6 are consistent with the formulation since they contain nitrogen–halogen bonds. The molecules are planar; the nitrogen–halogen distance in FNO_2 is slightly shorter than

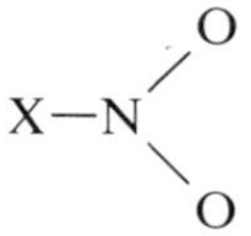

$$X-N\overset{\displaystyle O}{\underset{\displaystyle O}{\diagup\diagdown}}$$

X	d(N—O), Å	d(N—X), Å	$\angle$ONO
Cl	1.840	1.202	130.25°
F	1.40	1.21	130°

FIG. 12.6   Structural parameters for XNO_2 molecules.

expected for a single bond (N—F, 1.46 Å) but in $ClNO_2$ the distance is considerably longer than expected (N—Cl, 1.73 Å). The N—O distances appear to be essentially those observed in NO_3^-.

Nitrogen Oxyacids   Three nitrogen-containing acids have been well characterized. Dinitrogen oxide is formally the anhydride of hyponitrous acid, which has the empirical formula HON. On the same basis, nitrous acid (HNO_2) and nitric acid (HNO_3) can be hydrolytically related to N_2O_3 and N_2O_5, respectively. There are no acids known which correspond to the other common oxides of nitrogen, NO and NO_2. These substances undergo disproportionation when they react with water to form compounds containing nitrogen in different oxidation states

$$4NO + H_2O \xrightarrow{\text{OH}^-} 2HNO_2 + N_2O \tag{66}$$

$$2NO_2 + H_2O \longrightarrow HNO_3 + HNO_2. \tag{67}$$

Hyponitrous Acid (V). Hyponitrous acid ($H_2N_2O_2$) cannot be prepared by the reaction of N_2O with water, although a formal relationship between the acid and its anhydride exists. The compound is best prepared by the oxidation of hydroxylamine with nitrous acid, organic nitrates, or metal oxides such as CuO, Ag_2O, or HgO. In general the yields are small in these processes. However, a better yield of acid can be obtained by treating dry silver hyponitrite with HCl gas in ether. Filtration and evaporation of the solution yields a white, crystalline, explosive compound. Cryoscopic measurements indicate that the acid and its esters are dimeric, $R_2N_2O_2$, and that the diethyl ester has a dipole moment near zero. These results suggest that the acid is best formulated with a —N=N— bond, the nitrogen substituents being in

$$\underset{\displaystyle NOR}{\overset{\displaystyle RON}{\|}}\qquad R = alkyl, H$$

15

the *trans* orientation (**15**). Solutions of hyponitrous acid evolve N_2O

$$H_2N_2O_2 \rightarrow H_2O + N_2O. \tag{68}$$

Hyponitrous acid is a weak dibasic acid in water ($K_1 = 9 \times 10^{-18}$, $K_2 = 1 \times 10^{-11}$ at 25°) but alkali and alkaline-earth metal salts can be prepared. The salts, like the free acids, are explosive

$$3Na_2N_2O_2 \rightarrow 2NaNO_2 + 2Na_2O + 2N_2. \tag{69}$$

Infrared and Raman studies indicate that the hyponitrite anion also has a *trans* configuration (cf. structure **15**), but the position of the N—N frequency suggests a bond order less than 2 (*11*). The dimeric complex ion $[Co(NH_3)_5NO]_2{}^{4+}$ contains a bridging hyponitrite group in a *cis*-planar configuration (**16**) (*12*). Hyponitrous acid and its salts are reducing agents,

$$[(H_3N)_5Co-N\overset{\displaystyle N}{\underset{\displaystyle O}{\diagup}}\diagdown O-Co(NH_3)_5]^{4+}$$

16

but surprisingly these substances are remarkably resistant to reduction.

Nitrous Acid (HNO_2) (X). Aqueous solutions of nitrous acid can be prepared by acidifying nitrites; a solution free of salts is obtained by treating $Ba(NO_2)_2$ with the appropriate amount of H_2SO_4, after which the precipitated $BaSO_4$ can be removed. Pure nitrous acid is unknown, except in the vapor phase where it readily dissociates into the same mixture of nitrogen oxides obtained from the dissociation of N_2O_3 [cf. Eq. (58)]

$$2HNO_2 \rightleftharpoons H_2O + NO + NO_2. \tag{70}$$

Nitrous acid is a weak acid ($K = 4.5 \times 10^{-4}$) in aqueous solution and is an oxidizing as well as a reducing agent. It can be oxidized to nitric acid

$$AgBrO_3 + 3HNO_2 \rightarrow 3HNO_3 + AgBr \tag{71}$$

and reduced to nitrous oxide

$$2Sn^{2+} + 2HNO_2 + 4H^+ \rightarrow 2Sn^{4+} + N_2O + 3H_2O \tag{72}$$

or to nitrogen

$$RNH_2 + HNO_2 \rightarrow ROH + N_2 + H_2O. \tag{73}$$

At elevated temperatures, aqueous solutions of nitrous acid decompose rapidly to give nitric acid

$$3HNO_2 \rightarrow HNO_3 + H_2O + 2NO. \tag{74}$$

Two isomeric organic derivatives are possible. The nitrites are oxygen-bonded (**17**), whereas the nitro compounds are nitrogen-bonded (**18**). The

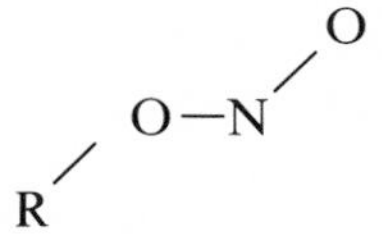

17 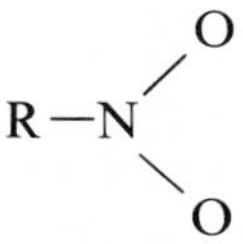   **18**

structural data available for methyl nitrite and nitromethane, which are summarized in Fig. 12.7, indicate that the CON framework for both molecules is planar, with bond distances considerably shorter than those expected for

(a)

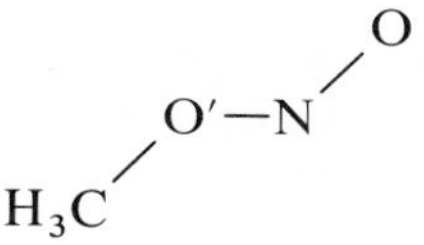

$d(C{-}O') = 1.44$ Å
$d(O'{-}N) = 1.22$ Å
$d(N{-}O) = 1.37$ Å
$\angle(CO'N) = 109.5°$
$\angle(O'NO) = 109.5°$

(b)

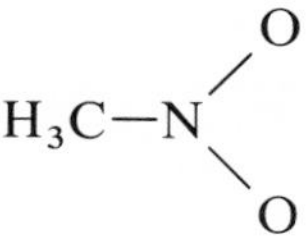

$d(C{-}N) = 1.47$ Å
$d(N{-}O) = 1.22$ Å
$\angle(ONO) = 135 \pm 5°$

FIG. 12.7   Structural parameters for the isomers of methyl nitrite.

single bonds. On the basis of the existence of two isomeric RNO_2 derivatives it might be expected that nitrous acid could exist in two corresponding forms, HNO_2 and $HONO$. Such forms would be tautomers related in the same way as the keto and enol forms of a ketone. The enol form ($HONO$) could theoretically exist in either *cis* or *trans* configurations. The infrared spectrum of gaseous nitrous acid (*13*) shows only the enol form, with the *trans* isomer (cf. **17**) being more stable than the *cis* by about 500 cal mole^{-1}.

The nitrite ion has the bent configuration shown in Fig. 12.5, with bond distances suggesting that structures **19** and **20** are the primary contributing

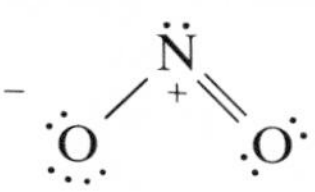

19 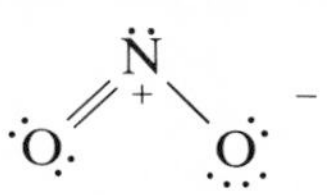   **20**

canonical forms to the wave function of this species. The bent configuration indicates that the nitrogen atom must use sp^2 hybrid orbitals (or some close approximation) in forming the bonds.

Nitric Acid. Nitric acid (HNO_3) can be prepared by distilling a mixture of an alkali-metal nitrate and sulfuric acid. It is prepared commercially from the oxidation of ammonia, which itself is obtained from atmospheric nitrogen, to NO [Eq. (20)]. The nitric oxide thus formed is allowed to react with more air [Eq. (23)] and the product dissolved in water [Eq. (67)]. Eventually a product containing 50–60% nitric acid can be prepared by this procedure. Distillation of aqueous HNO_3 eventually yields a constant-boiling mixture containing 68% acid; distillation from concentrated H_2SO_4 gives a product containing 98% HNO_3. The pure acid, which can be obtained by crystallizing the 98% mixture at $-42°$, undergoes dissociation at room temperature

$$2HNO_3 \rightarrow H_2O + N_2O_5 \tag{75}$$

$$2N_2O_5 \rightarrow 2N_2O_4 + O_2. \tag{76}$$

Heating this mixture essentially leads to a dilution of the acid since N_2O_4 and O_2 are removed in this process. Pure nitric acid in the liquid state undergoes autoionization according to the reaction (XI)

$$2HNO_3 \rightleftharpoons NO_2{}^+ + NO_3{}^- + H_2O. \tag{77}$$

Apparently the protonated acid, $H_2NO_3{}^+$, is unstable with respect to dissociation into nitronium ion and water

$$H_2NO_3{}^+ \rightarrow H_2O + NO_2{}^+. \tag{78}$$

In solvents that are stronger proton donors than nitric acid, the latter acts as a base and these solutions contain nitronium ions (XII). For example, studies on HNO_3 dissolved in sulfuric acid show that the predominating species are nitronium bisulfate and water, which is also protonated in such a strongly acid medium

$$2H_2SO_4 + HNO_3 \rightarrow NO_2{}^+ + 2HSO_4{}^- + H_3O^+. \tag{79}$$

Presumably the first step of the process involves the protonation of HNO_3 molecules, which decompose according to Eq. (78). A similar process occurs in anhydrous perchloric acid. Solutions of HNO_3 in H_2SO_4 have been extensively studied as nitrating agents, and it is well established that the nitronium ion is the active species in nitration reactions in such media.

If nitric acid is dissolved in a substance that is a weaker proton donor (e.g., water), it undergoes the normal ionization process expected of strong acids

$$HNO_3 + H_2O \rightarrow H_3O^+ + NO_3^-. \tag{80}$$

In aqueous solutions less concentrated than $\sim 2M$, nitric acid undergoes reactions typical of strong acids, i.e., it reacts with active metals to form hydrogen. However, in more concentrated solutions the oxidizing power increases because the nitrate ion becomes involved in the chemical processes. Depending upon conditions, nitric acid reacts with every metal except rhodium, tantalum, iridium, platinum, and gold. The latter two elements can be dissolved in aqua regia, which is a mixture of three parts concentrated HCl and one part concentrated HNO_3; dissolution of these metals is aided by the presence of free chlorine, which is strongly oxidizing, and of Cl^-, which forms complex ions with the metals. Chromium and iron are rendered passive by concentrated nitric acid; that is, the metals become inert to attack by most substances which would have normally reacted with them. The nature of the reduction products formed when metals dissolve in nitric acid depends upon the conditions of the reaction as well as the metal involved. Compounds containing nitrogen in every possible oxidation state have been found among the reduction products: HNO_2, NO_2, NO, N_2O, N_2, NH_2OH, and NH_3. In general, metals with standard potentials greater than that of hydrogen give products containing nitrogen in its more reduced states (NH_3, NH_2OH, N_2, N_2O) whereas the other metals yield the more highly oxidized products.

The nitrate salts of all the metals are known, either as anhydrous compounds (VII) or hydrates. Nitrates can be made by neutralizing nitric acid with the appropriate base or by reacting the metal with N_2O_4 [cf. Eq (46)]. Most of the anhydrous metal nitrates sublime without decomposition below $\sim 500°$ under vacuum; at higher temperatures nitrites are formed.

Nitric acid is a planar molecule with two N—O bond distances, as shown in Fig. 12.8. Removal of the proton produces the NO_3^- ion, which is also

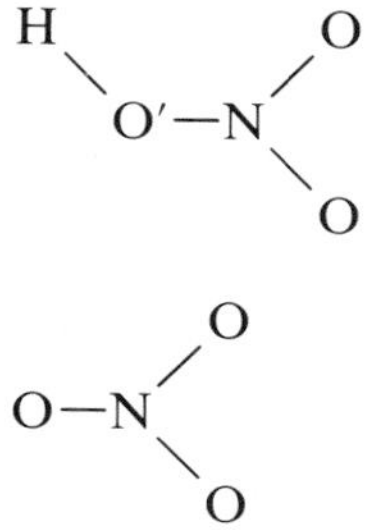

$$d(H-O') = 0.96 \text{ Å}$$
$$d(O'-N) = 1.405 \text{ Å}$$
$$d(N-O) = 1.206 \text{ Å}$$
$$\angle(ONO) = 130°$$

$$d(N-O) = 1.218 \text{ Å}$$
$$\angle(ONO) = 120°$$

FIG. 12.8 Structural parameters for nitric acid and NO_3^-.

planar, with all N—O distances equal (Fig. 12.8). As in the previous nitrogen–oxygen compounds, the theoretical interpretation in terms of the valence-bond theory involves a consideration of several resonance forms for both HNO_3 (**21**) and NO_3^- (**22**) using sp^2 hybrid orbitals on the nitrogen atom.

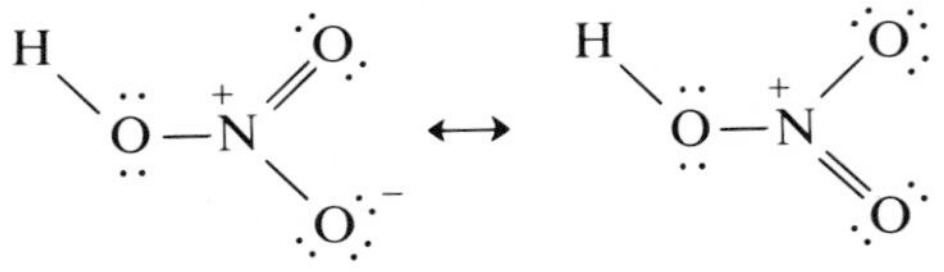

21

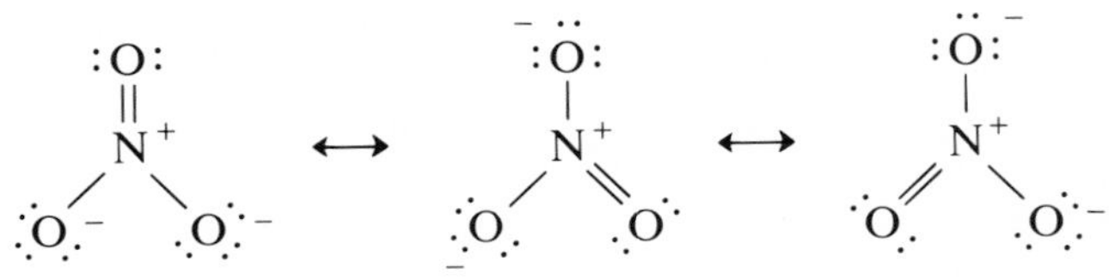

22

The equivalent molecular-orbital description indicates a π system encompassing an NO_2 framework in nitric acid and all four atoms of the nitrate ions.

Nitrates are starting materials for the preparation of another nitrogen-containing anion, viz., N_3^-. The reaction of an alkali-metal nitrate and an amide forms the corresponding alkali-metal azide

$$KNO_3 + 3KNH_2 \rightarrow KN_3 + 3KOH + NH_3. \tag{81}$$

If the reaction is carried out in the molten state ($\sim 175°$) it occurs with explosive violence. The reaction can be performed in liquid ammonia solution under pressure at 90–100°. Azides can also be prepared by reacting nitrous oxide with metal amides

$$N_2O + NH_2^- \rightarrow N_3^- + H_2O, \tag{82}$$

or by the reaction of active nitrogen with metal films. The alkali-metal azides are reasonably stable, but the heavy-metal azides detonate when struck. Hydrazoic acid (mp $-80°$, bp 37°) can be prepared by oxidizing hydrazine with nitrous acid in acid solution

$$N_2H_5^+ + NO_2^- \rightarrow HN_3 + 2H_2O. \tag{83}$$

The pure acid is highly explosive, a weak acid in water ($K = 2.8 \times 10^{-5}$), and decomposes in the presence of finely divided platinum

$$HN_3 + H_2O \rightarrow H_2NOH + N_2. \tag{84}$$

Hydrazoic acid is both an oxidizing and a reducing agent. For example, with hydrogen it forms ammonia, but with zinc and sulfuric acid it forms a mixture of NH_3, N_2H_4, and N_2.

The N_3 unit in metal azides and hydrazoic acid is linear, as shown by the data in Fig. 12.9. As would be expected, the bond distances in the azide ion

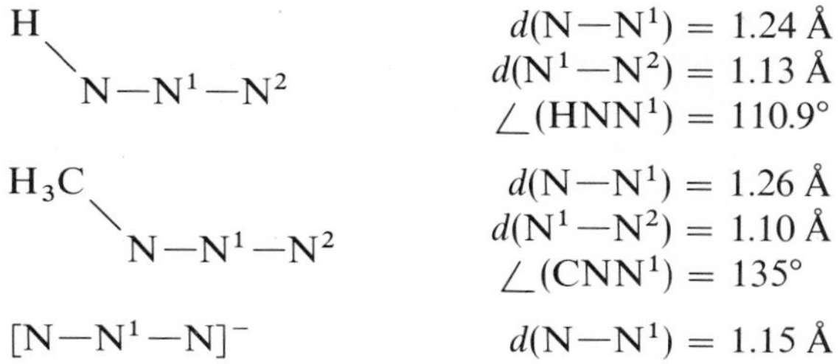

$$
\begin{array}{ll}
\text{H} & d(\text{N}-\text{N}^1) = 1.24 \text{ Å} \\
\quad\diagdown & d(\text{N}^1-\text{N}^2) = 1.13 \text{ Å} \\
\quad\text{N}-\text{N}^1-\text{N}^2 & \angle(\text{HNN}^1) = 110.9° \\
& \\
\text{H}_3\text{C} & d(\text{N}-\text{N}^1) = 1.26 \text{ Å} \\
\quad\diagdown & d(\text{N}^1-\text{N}^2) = 1.10 \text{ Å} \\
\quad\text{N}-\text{N}^1-\text{N}^2 & \angle(\text{CNN}^1) = 135° \\
& \\
[\text{N}-\text{N}^1-\text{N}]^- & d(\text{N}-\text{N}^1) = 1.15 \text{ Å}
\end{array}
$$

FIG. 12.9 Structural parameters for hydrazoic acid and $N_3{}^-$.

are equal (1.15 Å) and not quite as short as that found in molecular nitrogen (1.10 Å). This suggests that the wave function of the molecule is best approximated by the three canonical forms **23–25**. In the case of hydrazoic acid and

$$
:\overset{-}{\underset{..}{\text{N}}}=\overset{+}{\text{N}}=\overset{-}{\underset{..}{\text{N}}}: \qquad \text{N}\equiv\overset{+}{\text{N}}:\overset{..}{\underset{..}{\text{N}}}:^{2-} \qquad {}^{2-}:\overset{..}{\underset{..}{\text{N}}}:\overset{+}{\text{N}}\equiv\text{N}
$$

23 **24** **25**

the covalent azides, the inner N—N distance has increased to near that expected for a N=N bond (1.20 Å), with a corresponding decrease in the terminal N—N distance. This leads to two predominant canonical forms, **26** and **27**, for covalently bound azides. The observed RNN bond angles

$$
\text{R}-\overset{..}{\text{N}}=\overset{+}{\text{N}}=\overset{..}{\underset{}{\text{N}}}:^- \qquad \text{R}-\overset{..}{\underset{..}{\text{N}}}:\overset{+}{\text{N}}\equiv\text{N}:
$$

26 **27**

suggest that this nitrogen atom again employs nearly sp^2 hybrid orbitals for bonding.

3.3 / Sulfur Derivatives of Nitrogen (*XIII*)

Sulfur forms several compounds with nitrogen, but these substances appear to have little in common with the oxygen derivatives. There are at least three compounds with the empirical formula SN, the most thoroughly investigated being S_4N_4. Tetrasulfur tetranitride is a yellow-orange crystalline substance (mp 178°) which detonates when struck or heated rapidly. It can be prepared by the action of silver nitrate on a solution of sulphur dissolved in liquid ammonia. On this basis it has been suggested that the equilibrium

$$10S + 4NH_3 \rightleftharpoons S_4N_4 + 6H_2S \tag{85}$$

exists in liquid ammonia and the addition of silver nitrate displaces the equilibrium toward the products by precipitation of insoluble Ag_2S. However a spectroscopic study of ammonia solutions of sulfur, S_4N_4, and H_2S suggests that the reaction is probably more complex than indicated by Eq. (85) (*14*). Tetrasulfur tetranitride can also be prepared by passing dry ammonia gas into a benzene solution of S_2Cl_2. The compound is very soluble in most organic solvents, in which its molecular weight has been determined.

Heating a stream of S_4N_4 to 300° at low pressure yields the white volatile compound S_2N_2, which can be isolated by quenching the gaseous product. Like S_4N_4, disulfur dinitride is sensitive to shock. Its molecular weight has been established cryoscopically in organic solvents and the planar structure shown in Fig. 12.10 assigned (*15*). S_2N_2 reacts with gaseous or liquid ammonia to form $S_2N_2 \cdot NH_3$. At room temperature dry S_2N_2 polymerizes to $(SN)_x$ which exists as fiber-like brass-colored crystals; in thin layers $(SN)_x$ is dark blue. If a trace of base is present S_2N_2 polymerizes to S_4N_4, but in the presence of moisture both S_4N_4 and $(SN)_x$ are formed. The structure of $(SN)_x$ is unknown, but its color and the fact that it is a semiconductor suggest that the substance consists of long molecules of alternating sulfur and nitrogen atoms with a delocalized electronic structure.

The compound S_2N_2 acts as a difunctional Lewis base forming $1:1$ adducts, such as $S_2N_2 \cdot A$ ($A = BCl_3$ and $SbCl_5$), and $1:2$ adducts, $S_2N_2 \cdot 2A$ (*16*). The S_2N_2 ring is intact in these compounds, and direct evidence is

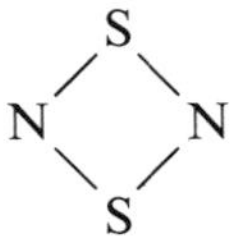

$$d(S-N) = 1.61 \text{ Å}$$
$$\angle(NSN) = 85°$$
$$\angle(SNS) = 95°$$

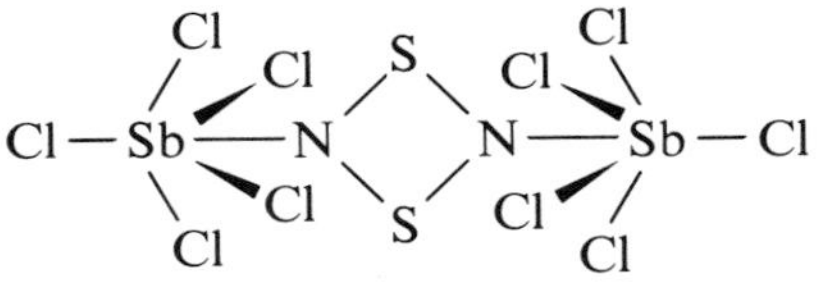

$$d(S-N) = 1.616 \text{ Å}$$
$$d(Sb-Cl) = 2.305 \text{ Å}$$
$$d(Sb-N) = 2.283 \text{ Å}$$
$$\angle(NSN) = 85°$$
$$\angle(SNS) = 95°$$
$$\angle(NSbCl) = 83°$$

FIG. 12.10   Structural parameters for S_2N_2 and $S_2N_2 \cdot 2SbCl_5$.

available for coordinate-covalent bond formation between the acid and the base as shown in Fig. 12.10 (*17*). The dimensions of the ring are little changed from those of the parent molecule. The S—N bond order in the S_2N_2 compounds is approximately 1.3, a value based upon an empirical relationship established between bond order and bond length for S—N compounds (*18*). In the presence of excess S_2N_2, the 1 : 1 complexes are converted into complexes of $S_4N_4 \cdot A$.

The two structures shown in Fig. 12.11 have been proposed for S_4N_4 based upon electron and x-ray diffraction experiments. Both have a tetrahedral arrangement of one kind of atom with four atoms of the other kind

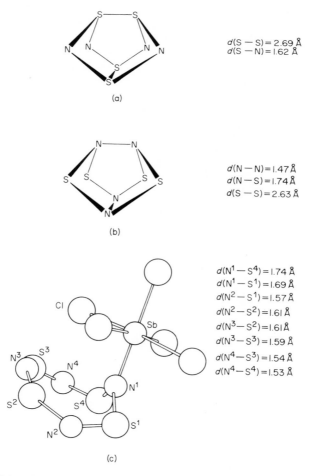

$d(S-S) = 2.69 \text{ Å}$
$d(S-N) = 1.62 \text{ Å}$

(a)

$d(N-N) = 1.47 \text{ Å}$
$d(N-S) = 1.74 \text{ Å}$
$d(S-S) = 2.63 \text{ Å}$

(b)

$d(N^1 - S^4) = 1.74 \text{ Å}$
$d(N^1 - S^1) = 1.69 \text{ Å}$
$d(N^2 - S^1) = 1.57 \text{ Å}$
$d(N^2 - S^2) = 1.61 \text{ Å}$
$d(N^3 - S^2) = 1.61 \text{ Å}$
$d(N^3 - S^3) = 1.59 \text{ Å}$
$d(N^4 - S^3) = 1.54 \text{ Å}$
$d(N^4 - S^4) = 1.53 \text{ Å}$

(c)

FIG. 12.11 Structural parameters of the two proposed structures of S_4N_4 [(a) and (b)], and of $S_4N_4 \cdot SbCl_5$ (c).

forming bridges to give a symmetrical structure. In one structure the nitrogen atoms can be imagined as bridging, while in the other structure the sulfur atoms are bridging. The suggested structures are essentially the same except that the nitrogen and sulfur atoms have switched positions. Both are related to the known structure of the mineral realgar, As_4S_4, shown in Fig. 12.18. The bond distances in both structures indicate that the theoretical description of the bonding in the molecule must include a consideration of the geometry of the molecule as a whole. The S—N distances in both structures are shorter than the sum of the single bond radii. The reported N—N distance in the structure in 12.11b is near the single bond distance, and, even though the S—S distances are longer than the corresponding single-bond distances (2.08 Å), they are markedly shorter than the van der Waals contact radii (3.7 Å). The complex $S_4N_4 \cdot SbCl_5$ mentioned earlier exhibits the structure shown in Fig. 12.11c (19), which is apparently more open than that of either of those reported for S_4N_4.

When S_4N_4 is treated with the reagents shown in Fig. 12.12, a variety of compounds are formed which appear to be sulfur-containing ring structures. Although direct structural evidence does not exist for each of the species, such data are available for tetrasulfur tetraimide ($S_4N_4H_4$) and $S_4N_4F_4$.

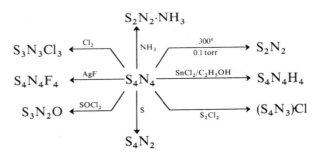

FIG. 12.12 Some reactions of S_4N_4.

Both compounds contain the same basic eight-membered ring; however, the fluorine is substituted on sulfur whereas the hydrogen atoms are on the nitrogen sites (20). Two different alternating S—N bond distances are reported for the fluoride, one being equal to that expected for a double bond (1.54 Å) and the other longer. This implies that a π system of electrons exists over some of the atoms of the ring, but not over the entire ring system. In the imide all the ring bond distances are equal.

Tetrasulfur tetranitride reacts with a number of metal compounds to give derivatives which have been formulated as thionitrosyl (NS) compounds. Thus, solutions of S_4N_4 in ammonia, pyridine, or organic solvents react with metal compounds to give metal thionitrosylates of the type MNS

$(M = Cu, Ag, Hg_2^{2+}, Na), M(NS)_2 (M = Pb, Ag, Cu), $ and $M(NS)_4 (M = Fe, Co, Ni, Pb, Pt)$. These compounds are generally highly colored; some are paramagnetic and some are sensitive to shock. There is little evidence concerning the constitution of these compounds, although some react as if the S_4N_4 ring were intact. For example, AgNS reacts with ethyl iodide to yield $S_4N_4(C_2H_5)_4$. The structure of $Pt(NS)_4$ indicates that the complex **28** is present and that the best formulation is $Pt(N_2S_2)_2$. The available

$$N\overset{S}{\diagdown}\underset{Pt}{}\overset{S}{\diagup}N$$

28

evidence indicates that the metal nitrosylates do not all have the same constitution.

3.4 / Chalcogen Derivatives of Phosphorus, Arsenic, Antimony, and Bismuth

Unlike the oxygen compounds of nitrogen, the oxygen derivatives of the other members of this family are markedly less numerous. In each case a trioxide (M_2O_3) and a pentoxide (M_2O_5) are known. Compounds containing different proportions of oxygen have also been reported for these elements.

The Phosphorus Chalcogenides

Phosphorous Oxide. Phosphorous oxide, P_4O_6 (mp 23.8°, bp 174°) is formed when the element is burned in a deficiency of oxygen. This substance dissolves in ether, CS_2, chloroform, and benzene; cryoscopic measurements on the latter solutions indicate that the molecular weight is twice the empirical formula. The vapor density at low temperatures gives the same results. Phosphorous oxide reacts with air to give P_4O_{10}, and if slightly warmed it inflames. With cold water this oxide slowly hydrolyzes to phosphorous acid

$$P_4O_6 + 6H_2O \rightarrow 4H_3PO_3 \tag{86}$$

a product that is also formed when P_4O_6 is treated with gaseous HCl

$$P_4O_6 + 6HCl \rightarrow 2H_3PO_3 + 2PCl_3. \tag{87}$$

The structure of P_4O_6 as determined by electron-diffraction methods is shown in Fig. 12.13; it is related to the tetrahedral structure of the P_4 molecule, shown in Fig. 12.1. The phosphorus atoms in P_4O_6 are still tetra-hedrally disposed, but six oxygen bridges are also present. There seems to be a

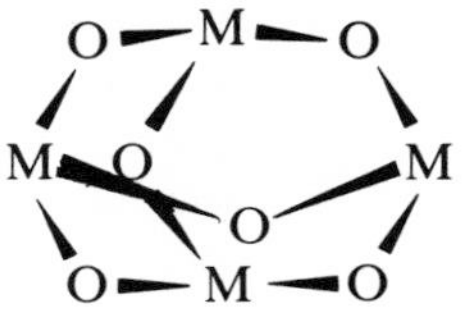

M	$d(\text{M}-\text{O})$, Å	$\angle(\text{MOM})$	$\angle(\text{OMO})$
P	1.638	126.4°	99.8°
As	1.78	99°	128°
Sb	2.0	97°	130°

M_4O_6

P_4O_8

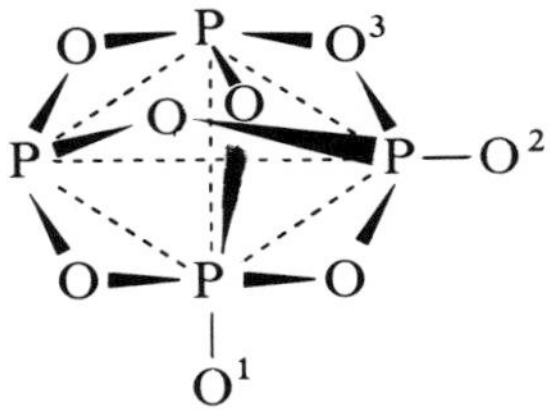

$d(\text{P}-\text{O}^2) = 1.141\ \text{Å}$
$d(\text{P}-\text{O}^2) = 1.556\ \text{Å}$
$d(\text{P}-\text{O}) = 1.664\ \text{Å}$

P_4O_{10}

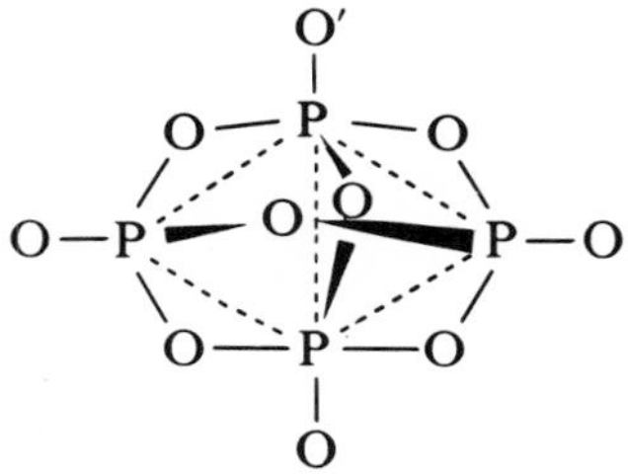

$d(\text{P}-\text{O}') = 1.39\ \text{Å}$
$d(\text{P}-\text{O}) = 1.62\ \text{Å}$
$\angle(\text{POP}) = 123.5°$
$\angle(\text{OPO}) = 101.6°$
$\angle(\text{O}'\text{PO}) = 116.5°$

FIG. 12.13   Structural parameters for the phosphorus, arsenic, and antimony oxides.

considerable amount of double-bond character in the P—O bonds, since the bond distance is markedly shorter than that expected for a single bond (1.84 Å). Since both phosphorus and oxygen have extra electron pairs, it might appear that there are two possible ways to construct a π bond between these atoms. However, oxygen, in contrast to phosphorus, has no low-lying orbitals to accept electron density from the phosphorus atom. Thus, the bond shortening in P_4O_6 can be accounted for by the formation of a $p\pi \rightarrow d\pi$ dative bond where oxygen acts as the donor.

Phosphorous oxide decomposes at about 440° to yield an oxide $(PO_2)_n$, of unknown constitution

$$nP_4O_6 \rightarrow 3(PO_2)_n + nP. \tag{88}$$

The vapor density of this substance at 1400° corresponds to the composition P_8O_{16}. $(PO_2)_n$ dissolves in water, giving an equimolar mixture of H_3PO_3 and H_3PO_4.

Phosphoric Oxide. Phosphoric oxide is the white solid formed when phosphorus is burned with an excess of oxygen. This substance exists in at least four modifications below its melting point (580°). Phosphoric oxide is an excellent dessicant, so much so that it can be used to dehydrate acids to form the corresponding anhydrides

$$2HClO_4 + P_4O_{10} \rightarrow Cl_2O_7 + P_4O_{10} \cdot H_2O. \tag{89}$$

With water P_4O_{10} forms a series of acids which are discussed in a subsequent section.

Phosphoric oxide is a very stable substance which can be sublimed at 360° and 1 atm pressure. The molecular weight of the vapor between 670° and 1100° shows the formula to be a dimer of the empirical formula. Electron-diffraction data obtained from the vapor, and summarized in Fig. 12.13, indicate that the basic P_4 unit found in elemental phosphorus and P_4O_6 is intact. Each phosphorus atom in the P_4O_6 structure carries an additional oxygen atom in P_4O_{10}. A comparison of those two structures indicates that the addition of an oxygen atom to each phosphorus atom in P_4O_6 leaves the basic framework of the latter unchanged. The bridging oxygen distances and bond angles are essentially the same, indicating that the P—O bond order is still greater than unity. The terminal oxygen atoms in P_4O_{10} must use the phosphorus lone pair for bonding, which supports the previous argument that the π-component in the bridging P—O oxygen bond involves donation of electron density from oxygen to phosphorus. The very short terminal P—O distance in P_4O_{10} also suggests a large $p\pi \rightarrow d\pi$ component.

The structure of P_4O_{10}, shown in Fig. 12.13, can be imagined as four PO_4 units, each sharing three oxygen atoms. This basic PO_4 unit appears in the structure of two of the solid modifications of P_4O_{10}. The metastable rhombohedral form contains discrete P_4O_{10} molecules, but the stable orthorhombic form consists of PO_4 units sharing three oxygen atoms to form an infinite array, rather than discrete P_4O_{10} molecules. This three-dimensional network consists of interlocking rings of PO_4 tetrahedra. The third crystalline modification, shown in Fig. 12.14, consists of rings of six PO_4 tetrahedra, giving a layer structure.

When P_4O_{10} vapor and oxygen are subjected to an electrical discharge a deep violet solid consisting of about 5% P_2O_6 is formed. The solid is a powerful oxidizing agent, which loses its color and oxidizing ability when heated to 130°. It has been suggested that P_2O_6 is a peroxy compound of the general constitution O_2POOPO_2.

Finally, an oxide intermediate between P_4O_6 and P_4O_{10} is known. The structure of P_4O_8, shown in Fig. 12.13, contains a basic P_4O_6 unit with only two terminal oxygen atoms (21). Electronically, two phosphorus atoms are pentavalent and two are divalent.

Sulfur Derivatives. Phosphorus forms several binary sulfur compounds which apparently also contain a P_4 unit. P_4S_3 (mp 172°, bp 408°), P_4S_{10}

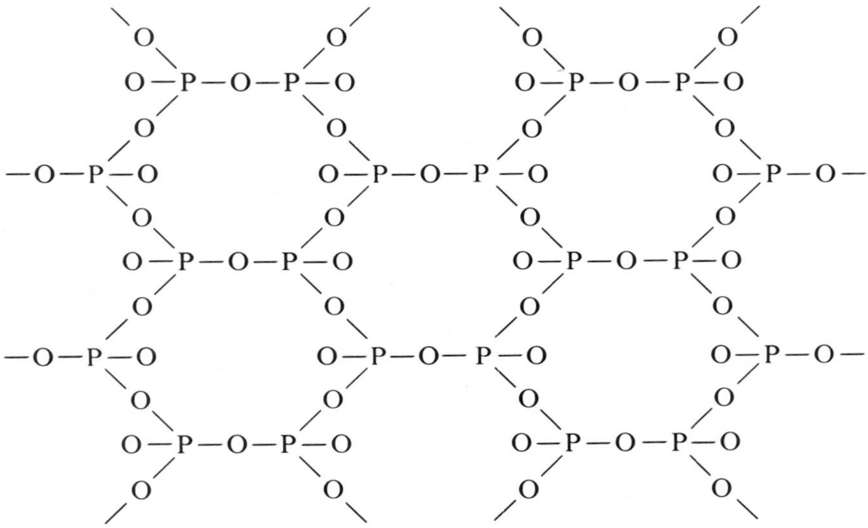

FIG. 12.14 The structure of one of the polymorphs of P_2O_5. Rings of six PO_4 tetrahedra are arranged to give a layer structure.

(mp 290°, bp 530°), P_4S_5, P_4S_7 (mp 30.5°, bp 523°), P_4S_8, and P_4S_9 can be prepared by heating the elements in the proper proportions, alone or in solution (CS_2 or xylene). The molecular weights, as determined cryoscopically, correspond to the formulas given. These compounds are hydrolyzed more or less rapidly to yield phosphorus–oxygen compounds and H_2S.

The data obtained from direct structural determinations are given in Fig. 12.15; all the phosphorous sulfides can be considered as derivatives of the P_4 structure (Fig. 12.1) in much the same way as are the oxides (Fig. 12.13). In the case of P_4S_3, only three sulfur bridges occur; P_4S_5 has four bridges, P_4S_7 five, and P_4S_8 six; the nonbridged sulfur atoms in the compounds are in terminal positions. The structure of P_4S_{10} is exactly analogous to that of the corresponding oxygen compound. It is curious that the P—P bonds in the original P_4 unit are maintained in P_4S_5 and P_4S_7 even though there are sufficient sulfur atoms to form six bridges. In the case of P_4S_3 (2.235 Å) and P_4S_5 (2.21 Å), the P—P bonds are as short as those in the P_4 unit (2.21 Å) of the element; the expected value for a P—P single bond is 2.20 Å. On the other hand, the P—P distance in P_4S_7 is considerably longer (2.35 Å), indicating a bond order less than unity (*XIV*). The selenium analogue of P_4S_3 is also known and has the same structure. When P_4O_{10} is heated with sulfur a violent reaction occurs, yielding the mixed phosphorus oxysulfide $P_4O_6S_4$. This compound contains the P_4O_6 structure with sulfur atoms in the terminal positions. The bridging P—O distances (1.61 ± 0.02 Å) are essentially the same as those of the parent structure.

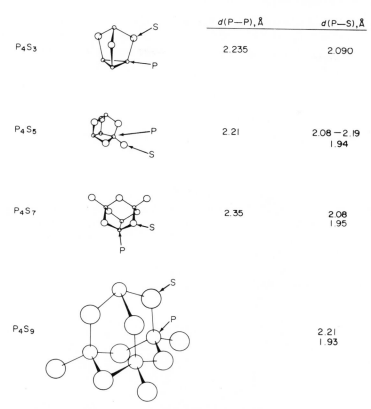

	$d(P-P)$, Å	$d(P-S)$, Å
P_4S_3	2.235	2.090
P_4S_5	2.21	2.08 – 2.19 1.94
P_4S_7	2.35	2.08 1.95
P_4S_9		2.21 1.93

FIG. 12.15 Structural parameters of phosphorus sulfides.

The Phosphorus Oxyacids   From its position in the periodic table we would expect phosphorus to form two acids, H_3PO_3 [$(HO)_3P$] and H_3PO_4 [$(HO)_3PO$]. Indeed, phosphite [$(RO)_3P$] and phosphate [$(RO)_3PO$] esters are known with the correct molecular formulation and structure. The chemistry of these phosphorus acids is exceedingly rich, being considerably more complex than the corresponding nitrogen acids. In fact, the general behavior parallels that of silicon in Group IV.

Phosphorous Acid (H_3PO_3), a colorless, deliquescent compound (mp 73.6°), is prepared by the action of water on PCl_3. H_3PO_3 is a moderate reducing agent, precipitating heavy metals from their salts. When heated to 200°, phosphorous acid gives phosphine and phosphoric acid

$$4H_3PO_3 \rightarrow 3H_3PO_4 + PH_3. \qquad (90)$$

Phosphorous acid is dibasic ($K_1 \approx 0.026$, $K_2 = 1.5 \times 10^{-7}$); two types of salts have been isolated. The normal salt ($CaHPO_3$ or Na_2HPO_3) contains one hydrogen atom which is still acidic. The reducing properties and acid behavior of H_3PO_3 strongly suggest that the correct formulation involves a hydrogen atom directly attached to phosphorus (**30**); structure **30** is the

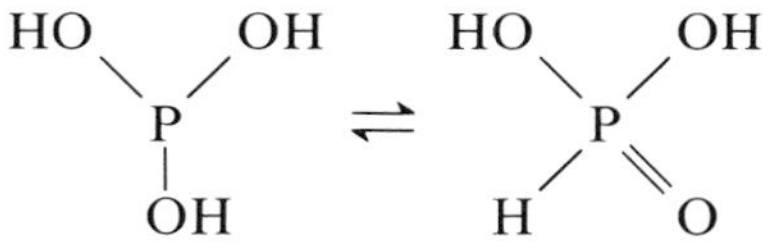

tautomer of the expected structure of phosphorous acid (**29**). Although phosphorous acid appears to exist entirely as the keto form (**30**), organic derivatives of both forms have been prepared by independent routes. Thus, the triethyl derivative of the keto form $[P(OC_2H_5)_3]$ boils at 156° and the isomer $(C_2H_5)(O)P(OC_2H_5)_2$ boils at 198°.

The structure of crystalline H_3PO_3, given in Fig. 12.16, shows a pyramidal arrangement of atoms with phosphorus at the apex. Although the hydrogen

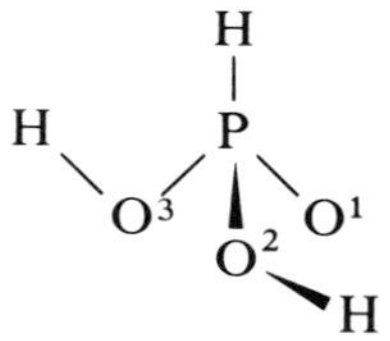

$$d(P{-}O^1) = 1.485 \text{ Å}$$
$$d(P{-}O^2) = 1.544 \text{ Å}$$
$$d(P{-}O^3) = 1.552 \text{ Å}$$
$$\angle(O^1PO^3) = 116°$$
$$\angle(O^1PO^2) = 113°$$
$$\angle(O^2PO^3) = 102°$$

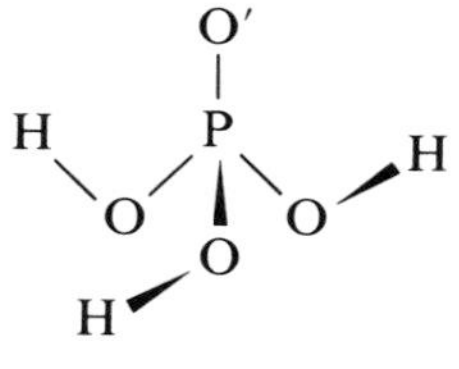

$$d(P{-}O') = 1.52 \text{ Å}$$
$$d(P{-}O) = 1.57 \text{ Å}$$
$$\angle(OPO) = 109.8°$$

FIG. 12.16 Structural parameters for phosphorous and phosphoric acids.

atoms were not located, evidence for structure **30** exists, in that one P—O bond (1.47 Å) is shorter than the other two (~ 1.54 Å). Presumably the fourth position on the phosphorus atom is occupied by the tautomeric hydrogen atom.

Hypophosphorous Acid (H_3PO_2) is formed when phosphine is oxidized with a mixture of iodine and water

$$PH_3 + 2I_2 + 2H_2O \rightleftharpoons H_3PO_2 + 4HI. \tag{91}$$

Salts of hypophosphorous acid can be prepared by boiling white phosphorus with aqueous alkali-metal hydroxides

$$P_4 + 4OH^- + 4H_2O \rightarrow 4H_2PO_2^- + 2H_2. \tag{92}$$

Hypophosphorus acid is a colorless solid (mp 27°) which disproportionates at 130°

$$3H_3PO_2 \rightarrow 2H_3PO_3 + PH_3. \tag{93}$$

It is a good reducing agent and can itself be reduced to phosphine with zinc and sulfuric acid. Although the composition indicates that three hydrogen atoms are present, the acid is monobasic ($pK = 1.2$), forming salts such as NaH_2PO_2. These properties are best understood if, as in the case of phosphorous acid, a tautomeric form $(HO)PO(H_2)$ exists. The crystal structure of the nickel salt $NiH_2PO_2 \cdot 2H_2O$ shows the anion to be tetrahedral.

A variety of acids are known which contain phosphorus in its higher oxidation state. They are all structurally related in that PO_4 tetrahedra are linked through oxygen bridges to form linear chains or cyclic structures.

Orthophosphoric Acid (H_3PO_4) is conceptually the parent of all phosphoric acids. It can be prepared by the hydration of P_4O_{10} or the oxidation of red phosphorus with nitric acid

$$P_4 + 10HNO_3 + H_2O \rightarrow 4H_3PO_4 + 5NO + 5NO_2. \tag{94}$$

Commerically, large quantities of sirupy phosphoric acid (85%) are prepared by reacting phosphate rock with sulfuric acid. Orthophosphoric acid (mp 42.35°) is not an oxidizing agent below $\sim 350°$. This acid is very soluble in water, where it acts as a tribasic acid ($K_1 = 9 \times 10^{-3}$, $K_2 = 6 \times 10^{-8}$, $K_3 = 1 \times 10^{-12}$). The pure acid has an appreciable conductivity, suggesting that it undergoes autoionization

$$H_3PO_4 \rightleftharpoons H_4PO_4^+ + H_2PO_4^-. \tag{95}$$

The protonated acid, $(H_4PO_4)^+$, also occurs when orthophosphoric acid is dissolved in anhydrous perchloric acid.

Many salts of orthophosphoric acid are known. The primary salts (MH_2PO_4) are acidic, the secondary salts (M_2HPO_4) essentially neutral, and the tertiary salts (M_3PO_4) basic in aqueous solutions.

The structure of orthophosphoric acid, as shown in Fig. 12.16, contains an essentially tetrahedrally surrounded phosphorus atom with two different P—O distances. The P—O distances are shorter than that expected for a P—O single bond (1.84 Å). The remaining distance is still shorter, indicating that a considerable amount of π bonding is present in these bonds. Tetrahedral PO_4 units containing essentially equivalent P—O bonds are observed in phosphates (1.56 Å), hydrogen phosphates (1.54 Å), and dihydrogen phosphates (1.54 Å).

Linear Phosphoric Acids (*XV*). Polyphosphoric acids occur when phosphoric acid is dehydrated. Generally, the unbranched linear phosphoric acids are polybasic and possess condensed structures in which PO_4 groups share oxygen atoms. The general formulation for this series of acids is $H_{n+4}P_{n+2}O_{3n+7}$ where n is an integer starting with zero for disphosphoric acid $H_4P_2O_7$. Diphosphoric acid (mp 61°), the first member of the series, can be prepared by heating phosphoric acid at 220°.

$$2H_3PO_4 \rightarrow H_4P_2O_7 + H_2O \tag{96}$$

or by heating a mixture of phosphoryl chloride and phosphoric acid

$$5H_3PO_4 + POCl_3 \rightarrow 3H_4P_2O_7 + 3HCl. \tag{97}$$

The acid takes up water slowly when it is boiled with water. Diphosphoric acid is tetrabasic in aqueous solution ($K_1 = 1.4 \times 10^{-1}$, $K_2 = 1.1 \times 10^{-2}$, $K_3 = 2.9 \times 10^{-7}$, $K_4 = 3.6 \times 10^{-9}$), forming acid salts as well as the tetrasodium salt ($Na_4P_2O_7$) and a tetraethyl derivative [$(C_2H_5)_4P_2O_7$]. The structure of the acid is unknown, but various salts containing the $P_2O_7^{4-}$ ion have been investigated by x-ray methods. The diphosphate ion consists of two PO_4 tetrahedra sharing an oxygen bridge with a POP bond angle of 134°. A similar structure has been observed for the anion of triphosphoric acid as its sodium salt.

Condensed Cyclic Phosphoric Acids (*XV*). The metaphosphates as a class have the empirical formula MPO_3. Metaphosphoric acid, HPO_3, may be obtained as a glassy solid or as a heavy viscous syrup. The structures of several water-soluble sodium salts are known. Among these, the trimetaphosphate $Na_3P_3O_9$ and the tetrametaphosphate $Na_4P_4O_{12}$ contain cyclic anions. Insoluble sodium metaphosphates, $(NaPO_3)_x$, contain very long chains of PO_4 units. The PO_4 units in the rubidium salt $(RbPO_3)_n$ are in the form of continuous spiral chains with a repeat pattern every two PO_4 units.

Phosphoryl Halides (*XVI*). Orthophosphoric acid forms a series of acid halides with the general formula X_3PO, where X can be either a halogen or a mixture of halogens. The phosphorus oxyhalides can be prepared by the partial hydrolysis of the corresponding phosphorus pentahalide

$$PX_5 + H_2O \rightarrow X_3PO + 2HX \tag{98}$$

$$X = F, Cl, Br$$

or by heating a mixture of the pentahalide with P_4O_{10}

$$6PX_5 + P_4O_{10} \rightarrow 10POX_3 \tag{99}$$

$$X = Cl, Br.$$

The properties of these compounds, as given in Table 12.6, indicate that they are typically covalent in nature; the phosphorus atom is tetrahedrally surrounded by the oxygen and three halogen atoms. In general the P—O

TABLE 12.6   The properties of some phosphoryl derivatives

	Bp, °C	Mp, °C	$\angle$(OPX), degrees	Internuclear distances, Å d(P—O)	d(P—X)
POF_3	−40	−39.5	109.5	1.55	1.51
$POCl_3$	107	1.25	106	1.45	2.02
$POBr_3$	189.5	56	108	1.41	2.06
POF_2Cl	3.1	−96.4	106	1.55	1.51 (P—F)
					2.02 (P—Cl)
$POFCl_2$	52.9	−80.1	106	1.55	1.51 (P—F)
					2.02 (P—Cl)
PSF_3	−52.9	−148.8	100	1.85	1.53
$PSCl_3$	125	−36.2	101	1.94	2.02
$PSBr_3$	175 dec.	39	106	1.86	2.13
PSF_2Br	35.5	−136.9	106	1.87	1.45 (P—F)
					2.14 (P—Br)
$PSFBr_2$	125.3	−75.2	100	1.87	1.50 (P—F)
					2.18 (P—Br)

distance is considerably shorter than that expected for a single bond, and the P—X distances also suggest that multiple-bond formation occurs between these atoms.

The phosphorus oxyhalides react readily with reagents containing replaceable hydrogen atoms to give derivatives of phosphoric acid

$$POX_3 + 3MH \rightarrow OP(M)_3 + 3HX \tag{100}$$

$$M = OH, OR, NR_2.$$

In addition, they can act as Lewis bases, the PO group being the basic site. For example, very strong complexes can be formed with transition-metal halides such as $ZrCl_4$, $HfCl_4$, Al_2Cl_6, and $FeCl_3$, and with phosphoryl compounds.

Phosphorus thiohalides are also known (Table 12.6), and can be prepared by the reaction of sulfur with the trihalide

$$PCl_3 + S \rightarrow SPCl_3. \tag{101}$$

These compounds are structurally similar to the oxyhalides, the P—S distances being shorter than expected for a single bond.

The Oxides of Arsenic   Three oxides of arsenic are known: As_4O_6, As_4O_{10}, and As_2O_4. Arsenious oxide (As_2O_6), formed when arsenic burns in air, exists in three crystalline modifications (XVII). The cubic form, which sublimes easily at 135° (mp 275°, bp 465°), is metastable at ambient conditions and slowly changes to a monoclinic form (mp 315°). Arsenious oxide is sparingly soluble in water ($\sim 2\%$ at 25°) forming weakly acidic solutions; in basic solution arsenites, AsO_3^{3-}, are formed. It reacts with sulfuric acid to form arsenyl sulfate, $(AsO)_2SO_4$, and with sulfur trioxide to give $As_2(SO_4)_3$. Arsenious oxide can be reduced to the element by heating with carbon or by treating an HCl solution with $SnCl_2$ or metallic copper

$$As_4O_6 + 12HCl + 12Cu \rightarrow 4As + 12CuCl + 6H_2O. \tag{102}$$

It can also act as a reducing agent, being oxidized to arsenate (AsO_4^{3-}) by ozone, H_2O_2, halogens, $FeCl_3$, dichromate ion, and nitric acid. Arsenious oxide is toxic, creating a special hazard because of its volatility and its lack of odor or taste.

In the vapor state arsenious oxide has a molecular weight corresponding to the dimer As_4O_6 at 800°. The structure of the dimer is analogous to that

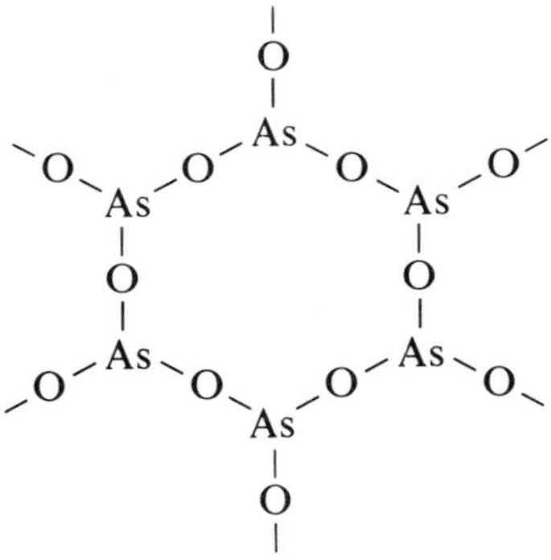

$$\angle(OAsO) = 100°$$
$$\angle(AsOAs) = 123°$$
$$d(As—O) = 1.80 \text{ Å}$$

FIG. 12.17   Structural parameters for As_2O_3.

of P_4O_6 shown in Fig. 12.13; the cubic modification also contains these structural units. The monoclinic form exhibits a puckered layer structure consisting of a hexagonal network of arsenic atoms joined by oxygen bridges (shown in Fig. 12.17).

Arsenic oxide (As_4O_{10}) is a white deliquescent solid, prepared most readily by heating crystalline arsenic acid above 200°

$$4H_3AsO_4 \rightarrow As_4O_{10} + 6H_2O. \tag{103}$$

This oxide decomposes to arsenious oxide and oxygen at red heat, a behavior which is in marked contrast with the stability of P_4O_{10}. Unlike P_4O_{10}, arsenic oxide is an oxidizing agent, for example liberating chlorine from HCl.

There is little structural information available on As_4O_{10} or As_2O_4 at present.

The Acids of Arsenic

Arsenious Acid is unknown except in aqueous solutions. An aqueous solution of arsenious oxide is weakly acidic ($K_i = 8 \times 10^{-10}$). Salts of arsenious acid can be obtained from basic solutions of the oxide. Thus, the normal orthoarsenite, Na_3AsO_3, as well as a metaarsenite, $NaAsO_2$, have been prepared. In contrast to the behavior of phosphorus acid, arsenious acid has no reducing properties, but the arsenites are reducing agents, being oxidized to arsenates

$$AsO_3^{3-} + I_2 + H_2O \rightarrow AsO_4^{3-} + 2HI. \tag{104}$$

Normal esters of arsenious acid [$As(OR)_3$] can be prepared indirectly by reacting arsenic trichloride or arsenious oxide with the appropriate alcohol.

Arsenic Acid is formed when arsenious oxide is treated with concentrated nitric acid. Crystallization of the product gives the hydrate $H_3AsO_4 \cdot \frac{1}{3}H_2O$. Although arsenic oxide is very soluble in water, the products obtained from such solutions are often colloidal in nature. Arsenic acid is tribasic in aqueous solution ($K_1 = 5.6 \times 10^{-3}$, $K_2 = 1.7 \times 10^{-7}$, $K_3 = 3.0 \times 10^{-12}$). When the solid hydrated acid is heated it loses water to yield H_3AsO_4, $H_4As_2O_7$, and $HAsO_3$, acids analogous to those observed in the phosphorus series.

Little information is available on the structures of the arsenic acids. The arsenates appear to be isomorphous with the corresponding phosphates. The salts Ag_3AsO_4 and KH_2AsO_4 contain essentially tetrahedral AsO_4 units [$d(As—O) = 1.75$ Å].

The Sulfur Derivatives of Arsenic   Arsenic forms three well characterized sulfides: As_4S_6, As_4S_4, and As_4S_{10}. Arsenious sulfide (As_4S_6),

sometimes called orpiment, is an easily sublimable yellow solid (mp 230°, bp 707°) which can be prepared by heating As_4O_6 with sulfur, or by passing H_2S into an HCl solution of arsenious acid. Arsenious sulfide burns in air

$$As_4S_6 + 9O_2 \rightarrow As_4O_6 + 6SO_2 \qquad (105)$$

and is insoluble in water or dilute aqueous HCl. However, it dissolves in basic solution to give a mixture of metaarsenites and metathioarsenites

$$As_4S_6 + 4NaOH \rightarrow NaAsO_2 + 3NaAsS_2 + 2H_2O; \qquad (106)$$

with alkali sulfides only metathioarsenites are formed

$$As_4S_6 + 2Na_2S \rightarrow 4NaAsS_2. \qquad (107)$$

Acidification of these solutions reprecipitates As_4S_6.

Arsenious sulfide has the same structure as that of As_4O_6, shown in Fig. 12.13; $d(S-As) = 2.25$ Å and $\angle (SAsS) = 114°$ in the vapor phase. Crystalline arsenious sulfide has the layer structure of the monoclinic modification of arsenious oxide shown in Fig. 12.17.

Realgar, As_4S_4, is a red-orange solid (mp 307°, bp 565°) which changes to a black modification at 267°. On exposure to light in air, As_4S_4 is converted into a mixture of As_4S_6 and As_4O_6. Realgar precipitates when a sodium bicarbonate solution containing arsenious sulfide is strongly heated. As_4S_4 inflames when heated with KNO_3, and because of this property is often used in pyrotechnics.

In the vapor phase at 550° the molecular weight of realgar corresponds to As_4S_4, but at higher temperatures ($\sim 1000°$) As_2S_2 units predominate. Electron-diffraction studies of the vapor species indicate that the structure of As_4S_4 contains the distorted tetrahedral array of arsenic atoms shown in Fig. 12.18, with four sulfur bridges $[d(S-As) = 2.23$ Å$]$ and two As—As

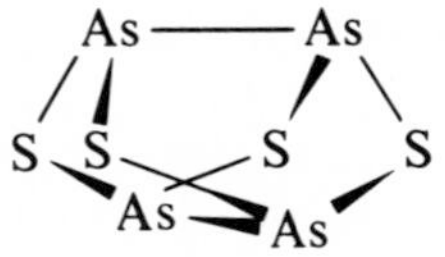

$$d(As-As) = 2.49 \text{ Å}$$
$$d(S-As) = 2.23 \text{ Å}$$
$$\angle (SAsAs) = 100°$$
$$\angle (AsSAs) = 110°$$
$$\angle (SAsS) = 93°$$

FIG. 12.18 Structural parameters for As_4S_4.

bonds $[d(As-As) = 2.49$ Å$]$. The structure in the solid state consists of the same species with virtually the same sulfur–arsenic distances (2.21 Å), but a slightly larger As—As distance (2.54 Å); the latter is longer than the distance in the As_4 molecule (Fig. 12.1).

Arsenic pentasulfide (As_4S_{10}) is an unstable yellow solid which decomposes into arsenious sulfide and sulfur above 95°. It can be prepared by the action of H_2S on a basic solution of Na_3AsO_4.

The Oxygen Derivatives of Antimony   Three oxides of antimony have been reported: Sb_4O_6, Sb_2O_5, and Sb_2O_4. Antimonious oxide (Sb_4O_6) exists in two crystalline forms; the cubic form is stable to 570°. The ortho-rhombic form which is stable above this temperature melts at 650° and boils at 1560°. As in the case of phosphorus and arsenic oxides, the molecules in the vapor phase are dimeric and dissociate to the monomeric form (Sb_2O_3) at very high temperatures. X-Ray studies on the cubic form indicate the presence of a Sb_4O_6 unit which has the same structure as P_4O_6 (Fig. 12.13). These features disappear in the structure of the low-temperature form of anti-monious oxide, which consists of the infinite double chain structure shown in Fig. 12.19.

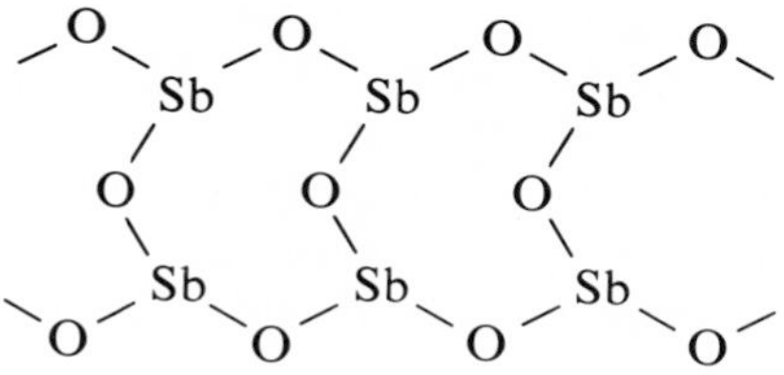

$$d(Sb-O) = 2.00 \text{ Å}$$
$$\angle (SbOSb) = 129°$$
$$\angle (OSbO) = 115°$$

FIG. 12.19   Structural parameters for Sb_2O_3.

Antimonious oxide can be prepared by passing steam over red-hot antimony, by hydrolysis of $SbCl_3$, or by heating Sb_2S_3 in air. If heated in air above 300°, Sb_2O_3 reacts with oxygen to form Sb_2O_4. It reacts with sulfur to give Sb_2S_3 and can be readily reduced to the element by heating with carbon or hydrogen. Antimonious oxide is distinctly amphoteric. $SbCl_3$, $Sb_2(SO_4)_3$, or $Sb(NO_3)_3$ are formed when Sb_2O_3 is dissolved in the cor-responding aqueous acid, while sodium antimonite is formed when Sb_2O_3 is dissolved in NaOH solution.

Antimonic oxide (Sb_2O_5) can be prepared by treating the element with concentrated nitric acid, evaporating the solution, and igniting the residue below about 400°. Above this temperature antimonic oxide looses oxygen to yield Sb_2O_4. Little structural information is available on Sb_2O_5, so it is usually formulated in terms of its empirical composition. Antimonic oxide is an oxidizing agent, being easily reduced to the element by carbon. It is virtually insoluble in water but yields $SbCl_5$ when treated with cold concentrated hydrochloric acid.

Two types of compounds containing anionic antimony exist. These are salts containing the $Sb(OH)_6^-$ ion, which are sometimes formulated as hydrated metaantimonates. For example, the structure of $MSbO_3 \cdot 3H_2O$ (M = Li, Na, Ag) contains the anion $Sb(OH)_6^-$ in which the central atom is surrounded by an octahedral arrangement of hydroxyl groups $[d(Sb—O) = 1.97 \text{ Å}]$. The other broad category of anionic antimony compounds contains SbO_6 units. Compounds with the general formulation $MSbO_3$, $M^{III}SbO_4$, $M^{II}Sb_2O_6$, and $M_2^{II}Sb_2O_7$ fall into this classification. The structures of these compounds involve SbO_6 octahedra joined by corners or edges.

Antimony dioxide (Sb_2O_4) is obtained when the element or any of its oxides or sulfides are heated in air at a temperature of 300–900°. It is an infusible, nonvolatile substance which is also insoluble in water. Crystalline Sb_2O_4 is isomorphous with $SbTaO_4$ which is composed of Sb^{3+} and TaO_4^{3-} ions. Accordingly, it appears that Sb_2O_4 is best formulated as antimony antimonate $[Sb(SbO_4)]$. Many of the reactions of Sb_2O_4 are in accord with this formulation. For example, it dissolves in aqueous HCl in the presence of reducing agents such as iodide ion to give $SbCl_3$. Solution in aqueous base yields a mixture of antimonate and antimonite ions.

The Sulfur Derivatives of Antimony   Antimony forms two sulfides, Sb_2S_3 and Sb_2S_5. Antimonious sulfide (Sb_2S_3 or stibnite) is a gray-black crystalline substance prepared by heating either Sb_2O_3 or the element with sulfur; the substance is also precipitated by H_2S from an aqueous HCl solution of $SbCl_3$. Heating Sb_2S_3 in air gives the corresponding oxide. Antimonious sulfide can be reduced to the element by hydrogen or metallic iron. Like As_2S_3, the substance is insoluble in water, but it dissolves in base to give a mixture of antimonites and thioantimonites and in solutions containing sulfide ion to give thioantimonites. In contrast to As_2S_3, antimonious sulfide is soluble in hot concentrated HCl.

Sb_2S_3 has the unusual structure shown in Fig. 12.20, consisting of interlocking chains with two different types of antimony atoms with respect to the number of neighbors and their relative distances.

Bismuth   Bismuth, the most metallic of the elements in Group VI, also forms two oxides, Bi_2O_5 and Bi_2O_3. Both oxides behave primarily as metal oxides, although anionic derivatives containing bismuth can be prepared by special methods. Bi_2O_3 crystallizes in the cubic structure shown in Fig. 12.21, which is related to the fluorite structure by removing one quarter of the fluoride ions from the latter. The tertrahedrally arranged bismuth atoms in the unit cell can be imagined to have oxygen bridges, so that

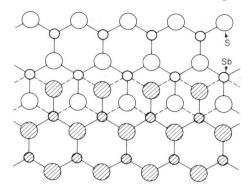

FIG. 12.20 The structure of Sb$_2$S$_3$.

the structure is very nearly that of P$_4$O$_6$ (shown in Fig. 12.13). Bi$_2$O$_3$ is only slightly soluble in water and in concentrated sodium hydroxide solution, but the compound dissolves in aqueous solutions of acids to form the bismuth salts. A white gelatinous precipitate of hydrated bismuth hydroxide is formed when a base is added to an aqueous solution of a bismuth salt.

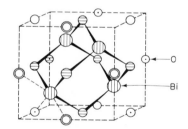

FIG. 12.21 The structure of Bi$_2$O$_3$.

3.5 / Halogen Derivatives of the Group V Elements

Only trivalent halogen-containing compounds of nitrogen (NX$_3$) exist, but the other elements in this family form both trivalent (MX$_3$) and pentavalent derivatives (MX$_5$). In general, the nitrogen compounds are stable to hydrolysis and chemically reactive, whereas the halogen derivatives of the other members of this family behave in the opposite way.

Nitrogen Compounds (*XVI*)   Only two of the nitrogen trihalides have been isolated and characterized as pure substances. Nitrogen trifluoride (*XVIII*) is a colorless gas (mp $-208.5°$, bp $-119°$) prepared by the electrolysis of fused NH$_4$F·HF at 125°. NF$_3$ can also be prepared by the direct fluorination of nitrogen with fluorine. It is the most stable of the nitrogen–

halogen derivatives, being unattacked by water, ammonia, aqueous base, or sulfuric acid. NF_3 reacts with steam to give hydrogen fluoride and oxides of nitrogen; a mixture of NF_3 and hydrogen explodes when ignited:

$$2NF_3 + 3H_2 \rightarrow N_2 + 6HF. \tag{108}$$

The structure of NF_3 corresponds to the expected pyramidal arrangement of atoms with an N—F distance of 1.37 Å and a bond angle of 102°. Nitrogen trifluoride exhibits no donor properties, in contrast to ammonia, although a derivative, F_3NO, formally contains a coordinate-covalent bond. These observations are in agreement with the low dipole moment (0.20 D) of NF_3 compared with that of NH_3 (1.50 D). In both molecules the lone electron pair which occupies an approximately tetrahedral position in an sp^3-type orbital contributes to the molecular moment in the same direction. However, the N—H and N—F individual bond moments are in opposite directions because of the relative electronegativities of hydrogen and fluorine with respect to that of nitrogen (Table 4.8). Thus, the bond moments and the electron-pair moment are in the same direction vectorally for NH_3 but in opposite directions for NF_3.

The electrolytic preparation of NF_3 also yields three other nitrogen–fluorine derivatives in varying amounts: NH_2F (sublimes $\sim 77°$), NHF_2 (mp $-125°$, bp $-70°$), and NF_2 (bp $\sim 125°$) (XIX). Little is known of the reactions and constitution of these products.

The compound F_3NO reacts as a fluorinating agent, yielding FNO as a product, rather than as an oxidizing agent (22). Fluoride-ion acceptors such as BF_3, SbF_3, and AsF_3 form 1:1 adducts with F_3NO, which has been formulated as containing the ion F_2NO^+ in the solid state as well as in solutions of HF, AsF_3, BrF_3, and IF_5 (23).

Nitrogen trichloride, a pale yellow oily liquid (bp 71°), is extremely unstable. It can be prepared by the action of chlorine on an acid solution of NH_4Cl or by the electrolysis of aqueous NH_4Cl. In addition, hypochlorous acid reacts with compounds containing nitrogen in a low valence state, such as ammonia, NH_2OH, N_2H_4, NaN_3 or urea, to give NCl_3:

$$NH_3 + 3HOCl \rightarrow NCl_3 + 3H_2O. \tag{109}$$

The nature of the nitrogen-containing products of the reaction between NH_4Cl and chlorine is strongly dependent upon the acidity of the medium. Below pH 4.5 the major product is NCl_3, at pH 4.5 to 5.0 dichloramine ($NHCl_2$) is formed, and above pH 8.5 monochloramine (NH_2Cl) is produced. Little is known concerning the constitution of the nitrogen chlorides. Both $NHCl_2$ and NCl_3 react with ammonia to give nitrogen and ammonium chloride

$$3NHCl_2 + 7NH_3 \rightarrow 2N_2 + 6NH_4Cl \tag{110}$$

$$NCl_3 + 4NH_3 \rightarrow N_2 + 3NH_4Cl; \tag{111}$$

monochloramine gives hydrazine

$$NH_2Cl + 2NH_3 \rightarrow N_2H_4 + NH_4Cl \tag{112}$$

which can react with excess NH_2Cl to form nitrogen and ammonium chloride

$$2NH_2Cl + N_2H_4 \rightarrow 2NH_4Cl + N_2. \tag{113}$$

Thus, monochloramine ultimately gives the same products in its reaction with ammonia as do NCl_3 and $NHCl_2$, but under suitable conditions the intermediate hydrazine is obtained. This reaction is the basis for the Raschig process for the preparation of hydrazine.

Bromine and iodine derivatives of nitrogen are more poorly characterized than those of chlorine and fluorine. The reaction of bromine with aqueous ammonia under any set of conditions yields only nitrogen

$$2NH_3 + 3NaOBr \rightarrow 3NaBr + N_2 + 3H_2O. \tag{114}$$

However, when the reaction is conducted in ether solution at $-50°$, both NH_2Br and $NHBr_2$ are formed. An explosive purple solid is formed when a mixture of gaseous ammonia and bromine is heated to $100°$ at low pressure and then quenched to $-95°$. The solid has the composition $NBr_3 \cdot 6NH_3$ but little else is known about it.

The reaction of iodine with aqueous ammonia gives a dark precipitate which has been called "nitrogen iodide," but which actually has the composition $NI_3 \cdot NH_3$ (XX). The substance is extremely unstable, giving rise to explosions at the slightest mechanical pressure. Pure nitrogen iodide can be prepared by passing gaseous ammonia over $KIBr_2$

$$3KIBr_2 + 4NH_3 \rightarrow NI_3 + 3NH_4Br + 3KBr. \tag{115}$$

An explosive black powder is obtained when the product is washed with water to remove the soluble ammonium and potassium salts.

Mixed halogen derivatives of nitrogen are prepared by the NaF-catalyzed fluorination of NH_4Cl, which can be made to yield NF_2Cl and $NFCl_2$ (24, 25). The former compound can also be prepared by the action of $SOCl_2$ on N_2F_2 or by the reaction of HNF_2 with organic hypochlorites (26).

Catenated Derivatives. Only two catenated, halogen-containing, nitrogen compounds are known; both are fluorine derivatives. Difluorodiazine, N_2F_2, is a colorless gas which can be prepared by the decomposition of fluorine nitrate (FNO_3) or from difluoroamine

$$2NHF_2 + 2KF \rightarrow 2KHF_2 + N_2F_2. \tag{116}$$

The product is a mixture of *cis* and *trans* isomers, **31** and **32**, which can be separated by gas chromatography. Difluorodiazine is also formed when NF_3

$$\underset{\textbf{31}}{\overset{\displaystyle F}{\underset{\displaystyle F}{\ddot{N}=N}}} \qquad\qquad \underset{\textbf{32}}{\overset{\displaystyle F \qquad F}{\ddot{N}=\ddot{N}}}$$

and mercury vapor are subjected to an electric discharge or when NF_3 is passed over hot metals.

The structural parameters for N_2F_4 which have been obtained from electron-diffraction experiments indicate a N—N bond distance (1.25 Å) which is close to that expected for a double bond (1.20 Å); the N—F distance is normal (1.44 Å). The molecule is planar with an NNF bond angle of $115 \pm 5°$, which suggests that nitrogen sp^2 hybrid orbitals are used in bonding. Pure *trans*-N_2F_2 can be prepared in 45% yield by reacting the mixed isomers with aluminum trichloride

$$4N_2F_2 + Al_2Cl_6 \rightarrow \textit{trans-}N_2F_2 + 3Cl_2 + 2AlF_3 + 3N_2. \quad \textbf{(117)}$$

Tetrafluorohydrazine (*XIX*), N_2F_4, is prepared by the action of NF_3 on hot copper. It is a colorless gas (bp $-73°$) which readily dissociates to radicals

$$N_2F_4 \rightleftarrows 2NF_2 \qquad \Delta H_{298} = 20.3 \text{ kcal mole}^{-1}. \quad \textbf{(118)}$$

The infrared and Raman spectra of N_2F_4 are consistent with the existence of the *trans* form of this molecule as the most stable species in the gaseous, liquid, and solid states (*27, 28*). The difluoroamino radical is a bent species as would be expected, and its presence in all samples of N_2F_4 accounts for the reactivity of the latter substance:

$$N_2F_4 + Cl_2 \rightarrow 2NF_2Cl \quad \textbf{(119)}$$

$$2RI + N_2F_4 \rightarrow 2RNF_2 + I_2 \quad \textbf{(120)}$$

$$2RSH + N_2F_4 \rightarrow 2HNF_2 + RS_2R. \quad \textbf{(121)}$$

The long N—N distance (1.489 Å) observed in N_2F_4 is also consistent with the high reactivity of this substance.

Halogen Derivatives of Phosphorus, Arsenic, Antimony, and Bismuth

Phosphorus Trihalides (XXI, XXII). Phosphorus forms trihalides with all of the halogens; the mixed halogen derivatives PF_2Cl (bp $-47.3°$, mp

$-164.8°$), PF_2Br (bp $-16.1°$, mp $-133.8°$), $PFCl_2$ (bp $-13.85°$, mp $-144.1°$), and $PFBr_2$ (bp $78.4°$, mp $-115°$) are also known. The phosphorus trihalides are all typically low-melting, low-boiling substances. All the trihalides except PF_3 are formed by direct combination of the elements

$$P_4 + 6X_2 \rightarrow 4PX_3. \tag{122}$$

PF_3 is generally best prepared by the fluorination of a phosphorus compound:

$$PCl_3 + AsF_3 \rightarrow PF_3 + AsCl_3 \tag{123}$$

$$2PBr_3 + 3ZnF_2 \rightarrow 2PF_3 + 3ZnBr_2 \tag{124}$$

$$2Cu_3P + 3PbF_2 \rightarrow 2PF_3 + 6Cu + 3Pb. \tag{125}$$

PF_3 is the least reactive of all the phosphorus trihalides. All react with hydroxyl-containing compounds to yield phosphorous acid or its derivatives

$$PX_3 + 3HOR \rightarrow P(OR)_3 + 3HX \tag{126}$$

$$R = H, alkyl.$$

Amides are formed if ammonia or its derivatives react with phosphorus trihalides

$$PX_3 + 3HNR_2 \rightarrow P(NR_2)_3 + 3HX \tag{127}$$

$$R = H, alkyl.$$

Phosphorus trichloride can act as a reducing agent, being oxidized to the pentavalent state in the process:

$$PCl_3 + SO_3 \rightarrow POCl_3 + SO_2 \tag{128}$$

$$3PCl_3 + S_2Cl_2 \rightarrow PCl_5 + 2SPCl_3 \tag{129}$$

$$2PCl_3 + O_2 \rightarrow 2OPCl_3. \tag{130}$$

Although the phosphorus trihalides do not exhibit basic properties in the usual Lewis sense towards simple electron-pair acceptors, complex compounds of the type $Cr(PF_3)_6$ are known.

The available structural data, summarized in Fig. 12.22, indicate that the phosphorus trihalides are pyramidal molecules with the bond angles decreasing steadily from near tetrahedral in PF_3 to $98°$ in PI_3. These data suggest that the σ bond formed between phosphorus and the halogen atom incorporates sp-hybridized phosphorus orbitals. In the case of PF_3, the hybridization is nearly sp^3 but the degree of s character in the hybrid decreases down the series fluorine, chlorine, bromine, iodine. The observed phosphorus-halogen bond distances are essentially those expected on the basis of single-bond radii for PBr_3 and PI_3. However, the P—F bond distance is about 15% shorter than expected; the P—Cl distance is only slightly

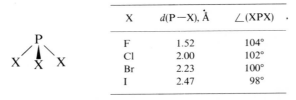

X	$d(P-X)$, Å	$\angle(XPX)$
F	1.52	104°
Cl	2.00	102°
Br	2.23	100°
I	2.47	98°

FIG. 12.22   Structural parameters for phosphorus trihalides.

shorter than expected. Bond shortening in the phosphorus halides has been attributed to the formation of a π bond incorporating the overlap of empty $3d$ orbitals on the phosphorus atom with filled p orbitals on the halogens (**33**). The $d\pi$-$p\pi$ bond formed in this instance is similar to that formed in silicon compounds and discussed in Chapter 11. Presumably the π component of

33

the phosphorus–halogen bond becomes less important for the heavier halogens because of increasingly less effective overlap between the phosphorus $3d$ orbitals and the corresponding np halogen orbitals.

Since the $3d$ orbitals on phosphorus are apparently of sufficiently low energy to become involved in bonding for some phosphorus compounds, an alternative interpretation for the type of hybrid orbitals used in the formation of the σ component is possible. Combinations of d, s, and p orbitals (e.g., d^2sp, dp^3, or d^3p) could be used for the appropriate hybrids.

Catenated Phosphorus–Halogen Derivatives. Several halogen compounds containing P—P bonds are known. Diphosphorus tetrachloride, P_2Cl_4, is a colorless oily liquid (mp $-28°$, bp 180°) which can be prepared by passing PCl_3 vapor through an arc struck between zinc electrodes, or a mixture of PCl_3 and hydrogen through a silent electric discharge. The analogous iodine compound, P_2I_4, is an orange-red solid (mp 124°) and is prepared by the reaction of the elements in carbon disulfide solution.

The structure of P_2I_4, as shown in Fig. 12.23, is essentially that of PI_3, each phosphorus being at the apex of a pyramid. The P—P distance corresponds to twice the single-bond radius; this observation suggests that the

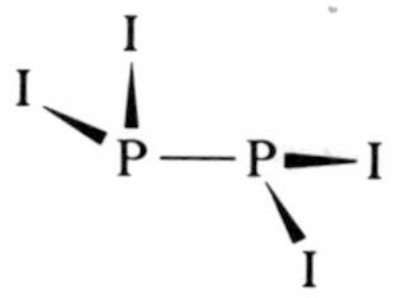

I — P — P — I

$d(\text{P}-\text{I}) = 2.47 \text{ Å}$
$d(\text{P}-\text{P}) = 2.21 \text{ Å}$
$\angle(\text{IPP}) = 94°$
$\angle(\text{IPI}) = 102°$

FIG. 12.23 Structural parameters for P_2I_4.

lone pair on each phosphorus atom is localized and is not involved in $d\pi$-$p\pi$ bonding.

Phosphorus Pentahalides. Phosphorus forms compounds of the type PX_5 with all halogens except iodine. The mixed halogen derivatives PF_3Cl_2, PCl_4F, $PClF_4$, and PF_3Br_2 have also been prepared. Phosphorus penta-fluoride (mp $-83°$, bp $-75°$) is a gaseous substance which can be prepared in several ways:

$$6P_2O_5 + 5CaF_2 \rightarrow 2PF_5 + 5Ca(PO_3)_2 \qquad (131)$$

$$3PCl_5 + 5AsF_3 \rightarrow 3PF_5 + 5AsCl_3. \qquad (132)$$

PCl_5 (mp 160°) and PBr_5 (mp $> 100°$) can be prepared by reaction of the elements. The pentavalent halides are readily attacked by water in two steps. In the first step, phosphoryl halides are formed

$$PX_5 + H_2O \rightarrow OPX_3 + 2HX; \qquad (133)$$

with excess water phosphoric acid is formed

$$PX_5 + 4H_2O \rightarrow H_3PO_4 + 5HX. \qquad (134)$$

The phosphorus pentahalides exhibit an interesting set of structural characteristics. In the vapor state PF_5, PCl_5, and PF_3Cl_2 exist as trigonal bipyramidal molecules, as shown in Fig. 12.24, which is consistent with the electron-pair-repulsion arguments. There is considerable evidence, however, which indicates that a relatively small barrier exists for the interconversion

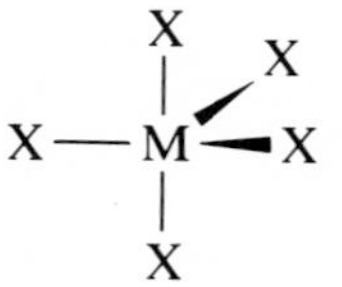

M	X	d(M—X), Å
P	F	1.57
P	Cl	2.04
Sb	Cl	2.31
		2.43

FIG. 12.24 Structural parameters for some Group V element pentahalides.

of an axial position for an equatorial position in five-coordinate phosphorus compounds (XXIII). The most direct process for such an interconversion involves a square-pyramidal intermediate (Fig. 12.25) (29).

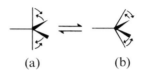

(a)   (b)

FIG. 12.25 The interconversion of axial and equatorial positions in five-coordinate trigonal bipyramidal systems (a) is thought to involve a square-pyramidal intermediate (b).

The chlorine atoms in PF_3Cl_2 are in the axial positions. As in the case of the phosphorus trihalides (Fig. 12.22), the phosphorus–fluorine distances are considerably shorter than the sum of the single-bond radii. Thus, the σ framework for the pentahalides in the vapor state involves dsp^3 orbitals for phosphorus, and the fluorides also possess a $d\pi$-$p\pi$ component.

Experiments on the condensed states of PCl_5 indicate that the molecular species present in the vapor state do not persist in the solid states. X-Ray analysis of the solid compound shows that the crystal consists of tetrahedral PCl_4^+ and octahedral PCl_6^- ions arranged in the cesium-chloride structure. The bond distances in PCl_4^+ (1.98 Å) are only slightly shorter than those in the anion PCl_6^- (2.07 Å), which is essentially that expected from the sum of the single-bond radii (2.09 Å). The σ bonds in the tetrahedral PCl_4^+ species involve sp^3 hybrid phosphorus orbitals, whereas those in the octahedral PCl_6^- ion require d^2sp^3 hybrids. In the solid state PBr_5 is best formulated as $(PBr_4^+)(Br^-)$. Nuclear magnetic resonance spectroscopy (^{31}P) indicates that the PBr_4^- ion is stabilized in the solid state by large cations such as R_4N^+, but PBr_4^- dissociates in nitrobenzene to give PBr_3 and Br^- (30). Presumably bromide ions are too large to form stable PBr_6^- ions. The ability of phosphorus to achieve four-, five-, and six-coordination leads to compounds with very interesting constitutions. For example, two compounds exist with empirical formula PF_3Cl_2. One is a gas at room temperature but the other is a hygroscopic solid which can be sublimed at 135°C; the latter isomer has the constitution $(PCl_4^+)(PF_6^-)$. At elevated temperatures the ionic isomer undergoes disproportionation to PF_5 and PCl_4F; PCl_4F exists as a nonpolar liquid or an ionic solid $[(PCl_4^+)(F^-)]$.

Phosphonitrilic Compounds (XXIV). The reaction of PCl_5 and ammonium chloride in the molten state or in refluxing inert organic solvents yields a series of compounds with the empirical formula $(NPCl_2)_n$. A variety of polymeric compounds can be isolated from the reaction mixture. Those with very large values of n are elastomers which have x-ray patterns typical of fibers, suggesting the presence of linear chains. Well characterized crystalline substances are obtained for $n = 3$ and 4. Both trimer and tetramer react with water, amines, and alcohols to yield phosphorus-substituted derivatives:

$$(NPCl_2)_n + 2nROH \rightarrow [NP(OR)_2]_n + 2nHCl \qquad (135)$$

$$n = 3, 4; R = H, alkyl$$

$$(NPCl_2)_n + 2nHNR_2 \rightarrow [NP(NR_2)_2]_n + 2nHCl \qquad (136)$$

$$n = 3, 4; R = H, alkyl.$$

Using suitable reagents it is possible to prepare a variety of substituted phosphonitrilic derivatives $[(NPX_2)_n, n = 3, 4; X = F, Br, SCN, alkyl]$. Phosphonitrilic halides form addition compounds with Lewis acids such as Al_2X_6 and SbF_5. The compounds $P_3N_3X_6 \cdot AlBr_3$ $(X = Cl, Br)$ and $P_3N_3Br_6 \cdot 2AlBr_3$ have infrared spectra consistent with the presence of an $Al-N$ coordinate-covalent bond (31). On the other hand, complexes such as $(NPF_2)_n \cdot 2SbF_5$ $(n = 3-6)$ have been assigned the fluorine-bridged, nonionic structure **34** (32).

34

X-Ray diffraction data indicate that the trimers and tetramers are cyclic molecules containing alternating phosphorus and nitrogen atoms. The trimer is a planar molecule whereas the tetramer is a puckered ring. The P—N distances in these molecules (Fig. 12.26) are intermediate between that expected for a single bond (1.84 Å) and that expected for a double bond (1.57 Å) between these atoms. The fact that all the P—N distances in a given cyclic structure are the same suggests an electronic problem analogous to the Kekulé structures of benzene. Accordingly, the canonical forms of the types shown in **35** ↔ **36** and **37** ↔ **38** must be important contributing forms in the wave functions for these molecules. Since the phosphorus atom is still pentavalent in these compounds, the π components must involve phosphorus $3d$ orbitals and nitrogen $2p$ orbitals. There is some question concerning the

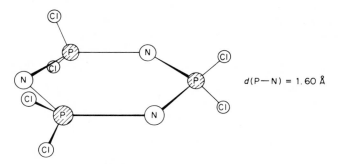

FIG. 12.26   The structure of $P_3N_3Cl_6$.

extent of delocalization in the π system of the trimer; extensive delocalization cannot occur in the tetramer, which exists as a puckered ring.

35   **36**

37   **38**

Arsenic, Antimony, and Bismuth Trihalides. All of the trihalides of these elements (MX_3) exist, and except for BiF_3, which is an infusible white solid, they all have properties characteristic of covalent species. That is, the compounds, which are liquids or low-melting solids, are soluble in organic solvents. The trifluorides are obtained by treating the trioxides with hydrofluoric acid

$$M_2O_3 + 6HF \longrightarrow 2MF_3 + 3H_2O \tag{137}$$

followed by evaporation of the solution. The other trihalides can be prepared by direct reaction of the elements

$$2M + 3X_2 \longrightarrow 2MX_3. \tag{138}$$

These compounds are soluble in aqueous solution, undergoing hydrolysis in the process; the reactions are generally reversible being suppressed by the presence of acid or of an ion such as halide ion which can form complexes. Antimony and bismuth halides yield basic halides upon hydrolysis under certain conditions. Thus, antimonyl halides [(SbO)X] and bismuthyl halides [(BiO)X] have been isolated and characterized. The action of water on any trivalent bismuth compound yields the corresponding bismuthyl derivative

$$BiY_3 + H_2O \rightarrow (BiO)Y + 2HY \qquad (139)$$

$$Y = NO_3{}^-, SO_4{}^{2-}, ClO_4{}^-, \text{halide}.$$

Antimony trichloride has found use as an ionizing solvent for a variety of substances (33). Antimony and bismuth halides form complex compounds with halide ions to give compounds such as K_2SbF_5, $NaSbI_3Br$, and $KBiI_4$.

The trihalides of these elements are all monomeric in the vapor phase or in organic solvents. The structural data for these trihalides, shown in Table 12.7, indicate that the molecules are pyramidal, with smaller bond angles than those of the corresponding phosphorus compounds (Fig. 12.22). These data suggest that, if sp hybrids are used for the σ components of the bonds, the degree of s character is considerably less than in the pure sp^3 hybrid orbital.

TABLE 12.7 Structural parameters of the Group V halides[a]

	$d(M-X)$, Å	$\angle(XMX)$, degrees
AsF_3	1.712	102
$AsCl_3$	2.161	98.4
$AsBr_3$	2.33	100.5
AsI_3	2.55	101.5
SbF_3	2.03	88
$SbCl_3$	2.325	99.5
SbI_3	2.67	99
$BiCl_3$	2.48	100
$BiBr_3$	2.63	100

[a] These molecules are all pyramidal.

Arsenic, Antimony, and Bismuth Pentahalides (XXI). In contrast to phosphorus, only four pentahalides are known for these elements, AsF_5, SbF_5, BiF_5, and $SbCl_5$. Antimony pentachloride can be prepared by the action of excess chlorine on the element (or on $SbCl_3$). Direct reaction between the elements is also the most useful method for the preparation of SbF_5 and BiF_5, the latter compound requiring considerably more drastic conditions

(600°) than usual. Arsenic pentafluoride can be conveniently obtained by treating AsF_3 with a mixture of SbF_5 and bromine

$$2SbF_5 + AsF_3 + Br_2 \rightarrow 2SbF_4Br + AsF_5. \tag{140}$$

Antimony is the only element in this group which forms mixed halides. In addition to SbF_4Br (cf. Eq. 140), two iodine-containing compounds are formed when iodine is dissolved in SbF_5, namely, SbF_4I (mp 80°) and Sb_2F_9I (mp 110–115°).

The antimony pentahalides dissolve in water to form complex compounds, $SbX_5 \cdot H_2O$. Both compounds are good halogenating agents. SbF_5 is often used as a fluorinating agent in organic chemistry; chlorine atoms are replaced by fluorine atoms in such reactions. $SbCl_5$ dissociates readily to chlorine and $SbCl_3$, and this is often the source of the chlorinating species; for example, ethylene forms 1,2-dichloroethane when treated with $SbCl_5$.

Structural studies show that both AsF_5 and $SbCl_5$ have the same structure as PCl_5, which is shown in Fig. 12.24.

Catenated Compounds of Arsenic, Antimony, and Bismuth. No halogen compounds containing metal–metal bonds are known for arsenic or antimony. However, when bismuth is dissolved in molten $BiCl_3$ a black solid with the composition $Bi_{24}Cl_{28}$ is formed (*34*). The structures shown in Fig. 12.27, a metal cluster in the form of the unusual cation $Bi_9{}^{5+}$, and two anions, $BiCl_5{}^{2-}$ and $Bi_2Cl_8{}^{2-}$ (the latter anion containing halogen bridges), combine as $(Bi_9)_2(BiCl_5)_4(Bi_2Cl_8)$, giving the empirical formula $Bi_{24}Cl_{28}$.

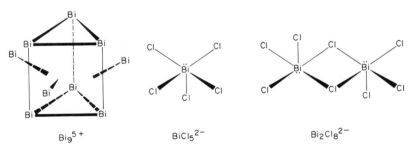

FIG. 12.27   The structures of the species present in $Bi_{24}Cl_{28}$.

3.6 / Hydrogen Compounds of the Group V Elements

The elements of Group V all form monomeric hydrides with the general formula MH_3; some of the properties of these compounds are shown in Table 12.8. The stability of the M—H bond decreases down the family, a fact

TABLE 12.8 The properties of some Group V hydrides

	Mp, °C	Bp, °C	$d(M-H)$, Å	$\angle(HMH)$, degrees	$D(M-H)$, kcal mole^{-1}
NH_3	-77	-33.4	1.008	107.3	93
PH_3	-133	-87.7	1.4206	93.5	78
AsH_3	-116.3	-62.4	1.5192	91.83	71
SbH_3	-88	-18	1.7073	91.30	61

which is reflected in the general properties of these substances. As in the case of the trihalides, the molecules are all pyramidal with bond angles decreasing down the family, suggesting that the degree of s character in the hybrid orbitals used to form the σ framework decreases also. The bond distances correspond well to the sum of the single-bond radii.

Only ammonia can be prepared by a direct combination of the elements

$$N_2 + 3H_2 \rightleftharpoons 2NH_3. \tag{141}$$

Large quantities of ammonia are produced commercially (Haber process) using the reaction shown in Eq. (141), which is carried out in the presence of an iron catalyst at $\sim 500°$ and pressures in the range 200–1000 atm. Small amounts of ammonia can be prepared on a laboratory scale by treating an ammonium salt with a base

$$NH_4Cl + NaOH \rightarrow NH_3 + NaCl + H_2O \tag{142}$$

or by the reduction of nitrogen compounds in a higher oxidation state. For example, nitrates can be quantitatively reduced using active metals such as zinc or aluminum

$$NaNO_3 + 4Zn + 7NaOH \rightarrow NH_3 + 4Na_2ZnO_2 + 2H_2O. \tag{143}$$

The other hydrogen derivatives of elements in Group V can be prepared by treating the corresponding binary compound with water or an acid:

$$2AlP + 3H_2SO_4 \rightarrow Al_2(SO_4)_3 + 2PH_3 \tag{144}$$

$$Na_3As + 3H_2O \rightarrow 3NaOH + AsH_3 \tag{145}$$

$$Mg_3Sb_2 + 6HCl \rightarrow 3MgCl_2 + 2SbH_3 \tag{146}$$

$$Mg_3Bi_2 + 6HCl \rightarrow 3MgCl_2 + 2BiH_3. \tag{147}$$

In addition, phosphine can be prepared by reduction of pentavalent phosphorus compounds (the action of water on a mixture of zinc dust and

$OPCl_3$) or by disproportionation reactions

$$4H_3PO_3 \rightarrow 3H_3PO_4 + PH_3 \tag{148}$$

$$2H_3PO_2 \rightarrow H_3PO_4 + PH_3. \tag{149}$$

Similar possibilities exist for the preparation of AsH_3

$$Na_3AsO_3 + 3HCO_2H \rightarrow 3NaHCO_3 + AsH_3. \tag{150}$$

A very convenient preparation of the hydrides involves the action of an active metal hydride such as $LiAlH_4$ on the corresponding trihalide in an organic solvent.

The hydrides are all good reducing agents. Gaseous ammonia will reduce many metal oxides when the latter are heated

$$3MO + 2NH_3 \rightarrow 3M + N_2 + 3H_2O. \tag{151}$$

Both phosphine and arsine reduce silver nitrate to metallic silver with the intermediate formation of a complex silver salt

$$6AgNO_3 + MH_3 \rightarrow Ag_3M \cdot 3AgNO_3 + 3HNO_3 \tag{152}$$

$$Ag_3M \cdot 3AgNO_3 + 3H_2O \rightarrow 6Ag + 3HNO_3 + H_3MO_3. \tag{153}$$

The hydrides will also react with a variety of oxidizing agents such as the halogens,

$$2NH_3 + 3Cl_2 \rightarrow N_2 + 6HCl \tag{154}$$

$$PH_3 + 4Cl_2 \rightarrow PCl_5 + 3HCl. \tag{155}$$

The hydrides can form two types of compounds. Anionic derivatives containing species such as MH_2^- occur when a proton is removed from the hydride. The alkali and alkaline-earth metals can form amides (NH_2^-), phosphides (PH_2^-), and arsenides (AsH_2^-) when the hydride is allowed to react with a solution of the metal in liquid ammonia (Chapters 8 and 9). It should be recalled that the reaction between the active metals and ammonia is catalyzed by the presence of finely divided metals

$$MH_3 + Na \rightarrow NaMH_2 + \tfrac{1}{2}H_2. \tag{156}$$

The blue metal–ammonia solutions are rapidly decolorized when PH_3 and AsH_3 are added. Although there are no reports of these reactions for stibine and bismuthine, organo-substituted stibides $[MSb(C_6H_5)_2]$ and bismuthides $[MBi(C_6H_5)_2]$ are known. Alkali-metal amides are also formed when dry ammonia gas is passed over the molten metal. The ionic amides are low-melting white crystalline solids which conduct an electric current in the molten state. Electrolysis of such melts gives a mixture of nitrogen and hydrogen at the anode and the metal at the cathode.

The other type of derivative that the hydrides can form involves them in the role of a Lewis base. The simplest of all Lewis acids, the proton, reacts strongly with ammonia and phosphine to form the corresponding tetrahedral "onium" species

$$MH_3 + H^+ \rightarrow MH_4{}^+. \tag{157}$$

A large number of stable ammonium salts are known, but the number of phosphonium compounds is markedly less; neither arsine nor stibine form "onium" salts. The best characterized of the phosphonium salts is the iodide, PH_4I. Phosphonium salts are always thermally less stable than the ammonium salts. The decrease in stability of the phosphonium derivatives appears to correlate with the difference in the basicity of NH_3 and PH_3 in aqueous solution. Ammonia is a weak base in water ($K = 1.8 \times 10^{-5}$), while a solution of phosphine exhibits virtually no basic properties. Both ammonia and phosphine can act as bases towards transition-metal ions; the nature of these interactions is discussed in Chapters 17 and 18.

The ammonium ion behaves like an alkali-metal ion with an ionic radius of 1.48 Å. Ammonium salts are often isomorphous with the corresponding potassium ($r_+ = 1.33$ Å) or rubidium ($r_+ = 1.48$ Å) compounds and have many of the same solubility characteristics as these compounds. In general the $NH_4{}^+$ ion can rotate freely in the crystalline lattice, but in some cases the rotation can be slowed considerably if the lattice contains sites to which hydrogen bonding ($N-H\cdots X$) can occur. For example, ammonium fluoride crystallizes in the open wurtzite structure, in contrast to the other ammonium halides, which have a cubic closest-packed structure. In solid NH_4F, the rotation of the ion is prevented by the formation of four $N-H\cdots F$ bonds from each ammonium ion to the fluoride ions that are in the tetrahedral positions. The polymorphism of many ammonium salts may arise because of differences in hydrogen bonding between the anions and $NH_4{}^+$.

Hydrogen bonding is also evident in the structure of the two known hydrates of ammonia, $NH_3 \cdot H_2O$ (mp 194.15°K) and $2NH_3 \cdot H_2O$ (mp 194.32°K). Neither of these substances contain the $NH_4{}^+$ ion, although the former has the correct composition for "ammonium hydroxide," NH_4OH. Both structures consist of a hydrogen-bonded framework of H_2O and NH_3 molecules, with one strong $N\cdots H-O$ bond [$d(N-O) = 2.84$ Å] and three weaker ones [$d(N-O) = 3.13–3.25$ Å]. These units are crosslinked into a three-dimensional network. The large solubility of ammonia in water suggests that hydrogen bonding also exists in this mixture in the liquid phase. Such solutions are weak electrolytes; for example, $1M$ NH_3 corresponds to a solution which is about 0.5% ionized. Thus, most of the nitrogen-containing species are hydrated (by hydrogen bonding) NH_3 molecules and only a small fraction are in the form of ammonium hydroxide, $NH_4{}^+OH^-$,

$$NH_{3\,(aq)} + H_2O \rightleftharpoons NH_4{}^+ + OH^-. \tag{158}$$

Hydroxylamine   In contrast to the corresponding phosphorus compounds, the nitrogen oxides ONR_3 are not stable, except in certain cases where charge delocalization can occur. For example, pyridine N-oxide **(39)** is a stable compound but alkyl amine oxides are markedly less so : it should be recalled that F_3NO is also an N-oxide. Formally, the N-oxide of ammonia **(40)** exists predominantly in its enol form **(41)** which is known as hydroxylamine

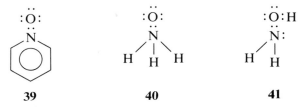

(mp 33°, bp 55–58° at 22 torr). The compound is best prepared in the laboratory by the electrolytic reduction of nitric acid using lead electrodes or by the reduction of nitrate ion with SO_2. Industrially hydroxylamine is prepared by hydrolyzing hydroxamic acid

$$\overset{\displaystyle NOH}{\underset{\displaystyle \parallel}{CH_3C}}OH + H_2O \;\rightarrow\; CH_3CO_2H + HONH_2. \tag{159}$$

The pure liquid is very unstable, liberating N_2, N_2O, and NH_3 continuously; it explodes when heated at normal pressure. Hydroxylamine is a weak acid forming salts such as $Ca(OH)(NH_2O)$ and $Ca(NH_2O)_2$, which are hydrolyzed in aqueous solution. Its more normal behavior is as a base, forming salts such as $[NH_3OH]^+X^-$ with a variety of acids. Hydroxylamine can act as either an oxidizing agent

$$2Fe(OH)_2 + NH_2OH + H_2O \;\rightarrow\; 2Fe(OH)_3 + NH_3 \tag{160}$$

$$NH_2OH + Na_3AsO_3 \;\rightarrow\; Na_3AsO_4 + NH_3 \tag{161}$$

or as a reducing agent

$$NH_2OH + HNO_3 \;\rightarrow\; 2NO + 2H_2O. \tag{162}$$

Although the orientation of the NH_2 group with respect to the OH group is unknown, the structural data available are consistent with an N—O bond distance (1.46 Å) that is very near the single-bond distance (1.48 Å) and bond angles essentially the same as those found in ammonia and water. Similar data are available for the hydroxylammonium ion (NH_3OH^+). Evidence is available that hydroxylamine is coordinated as an amine oxide, H_3NO, in the complex $ZnCl_2 \cdot 2NH_2OH$ (*35*).

Catenated Hydrides   Nitrogen and phosphorus form hydrides with the formulation M_2H_4 containing a bond between atoms of the same kind. Hydrazine (mp 1.8°, bp 113.5°) can be prepared by the careful oxidation of ammonia using NaOCl as an oxidizing agent

$$NH_3 + NaOCl \rightarrow NH_2Cl + NaOH \tag{163}$$

$$NH_2Cl + NH_3 \rightarrow NH_2NH_2 + HCl. \tag{164}$$

These reactions are the basis for the Raschig process for the commercial preparation of hydrazine. Diphosphine (P_2H_4) can be prepared by the reaction of water on crude calcium phosphide which contains Ca_2P_2

$$Ca_2P_2 + 4H_2O \rightarrow 2Ca(OH)_2 + P_2H_4. \tag{165}$$

Phosphine is also liberated in this reaction and can be separated from diphosphine on the basis of the difference in volatility of the two compounds. Diphosphine is usually obtained in any of the reactions used to prepare phosphine. Diphosphine is spontaneously inflammable.

Two lower arsenic hydrides exist, As_2H_2 and As_4H_2, but little is known of the constitution of these compounds or their reactions.

Hydrazine has properties similar to those of ammonia. It is a better reducing agent than ammonia, liberating gold, silver, mercury, and plantinum from their salts in aqueous solution as well as reducing iodates, ferric salts, permanganate ion, H^+, and iodine. Like ammonia, hydrazine is basic; however, two series of salts are obtained upon neutralization. The monoacidic salts, $N_2H_5^+$, are stable in aqueous solutions, but those containing $N_2H_6^{2+}$ ions undergo hydrolysis. Hydrazine forms a very stable monohydrate (mp -51.7, bp 120°); the substance is a fuming liquid which apparently attacks glass and porcelain. It decomposes explosively in the presence of certain metals to give ammonia, hydrogen, and nitrogen among the products.

Hydrazine is a very good ionizing solvent (dielectric constant 53 at 22°), dissolving a variety of substances. Like ammonia it dissolves the alkali metals to give blue solutions, but hydrogen is evolved slowly to give the metal hydrazide

$$N_2H_4 + Na \rightarrow NaN_2H_3 + \tfrac{1}{2}H_2. \tag{166}$$

Liquid hydrazine appears to be highly hydrogen-bonded. The substance in the vapor phase is unassociated and the available structural evidence suggests that free rotation about the N—N bond is hindered, probably because of the lone-pair–lone-pair interaction. The N—N distance is 1.453 Å, which is essentially that obtained from the sum of the single-bond radii

42   **43**   **44**

(1.48 Å). The expected *trans* configuration, **42** and **43**, is not present; rather the best representation is given by rotating the two NH_2 groups to a position 90–95° from the *cis* configuration **44**. The NH_3^+ groups in $N_2H_6^{2+}$ have a staggered configuration.

3.7 / Organic Derivatives of the Group V Elements

In our discussion of the organic derivatives of the Group V elements we shall not consider the carbon compounds of nitrogen because these are usually treated fully in organic chemistry text books. Indeed, the organic chemistry of phosphorus is also so extensive (XXV) that only certain aspects of the chemistry and structure of these compounds which suit the purposes of this text are discussed.

Phosphorus

Many three-coordinate organophosphorus compounds are known, among the most extensively studied being the trialkyl (R_3P) and the mixed halogen (R_nPX_{3-n}) and hydrogen (R_nPH_{3-n}) derivatives (XXVI). Undoubtedly the most convenient method for the preparation of the triorgano derivatives involves the use of the Grignard reagent or other active organometallic compound

$$PX_3 + 3RMgX \rightarrow R_3P + 3MgX_2 \tag{167}$$

$$PX_3 + 3LiR \rightarrow R_3P + 3LiX. \tag{168}$$

The Grignard reaction can be employed with compounds containing P—H bonds, the lower sulfides of phosphorus (P_4S_3), and phosphorus esters $[P(OR)_3]$. In addition, organophosphorus compounds can be prepared by heating phosphorus with alkyl iodides and zinc, alkyl iodides and sodium hydroxide, or alcohols. Alkali-metal phosphides react with alkyl iodides to give the corresponding trialkyl phosphine

$$Na_3P + 3RI \rightarrow 3NaI + R_3P. \tag{169}$$

Mixed halophosphines can be prepared by using the less reactive diorgano-mercury compounds

$$R_2Hg + PCl_3 \rightarrow RPCl_2 + RHgCl \tag{170}$$

$$2R_2Hg + PCl_3 \rightarrow R_2PCl + 2RHgCl; \tag{171}$$

generally, the mixed organomercury halide is a poorer alkylating agent than R_2Hg.

The tricoordinate phosphorus compounds are all pyramidal molecules with bond angles in the range of 93–102°. The trialkyl derivatives are Lewis bases forming phosphonium compounds $[(CH_3)_4PI]$ which are analogous to the ammonium salts. In addition, trialkyl phosphines react with a variety of oxidizing agents to form phosphine oxides (R_3PO); the analogous sulfur compounds (R_3PS) are also known. Oxidation of halophosphines leads to the formation of the corresponding phosphoryl halides $(RPOX_2$ and $R_2POX)$. Trialkylphosphines and the halophosphines react with halogens to form pentacoordinate compounds:

$$R_3P + X_2 \rightarrow R_3PX_2 \tag{172}$$

$$R_2PX + X_2 \rightarrow R_2PX_3 \tag{173}$$

$$RPX_2 + X_2 \rightarrow RPX_4. \tag{174}$$

When these pentavalent halides are heated, trivalent compounds containing a larger proportion of halogen are obtained:

$$R_3PCl_2 \rightarrow R_2PCl + RCl \tag{175}$$

$$R_2PCl_3 \rightarrow RPCl_2 + RCl. \tag{176}$$

Mild oxidation of primary phosphines (RPH_2) yields phosphonous acids, $RP(OH)_2$, whereas the use of stronger oxidizing agents gives phosphonic acids, $RPO(OH)_2$. In the case of the secondary phosphines (R_2PH), the corresponding phosphonous acids $R_2P(OH)$ are more difficult to isolate, oxidation usually leading to the phosphonic acid, $R_2PO(OH)$. Hydrolysis, aminolysis, or alcoholysis of the monohalo- and dihalophosphines gives the expected acids or their derivatives

$$R_2PCl + HY \rightarrow R_2PY + HCl \tag{177}$$

$$RPCl_2 + 2HY \rightarrow RPY_2 + 2HCl \tag{178}$$

$$Y = OH, NR_2, OR.$$

The phosphonyl halides react with these reagents to give phosphonic acids or their derivatives

$$R_2POCl + HY \rightarrow R_2PO(Y) + HCl \tag{179}$$

$$RPOCl_2 + 2HY \rightarrow RPO(Y)_2 + 2HCl. \tag{180}$$

The lower-valent organophosphorus acids exhibit the same keto–enol tautomerism as do the parent acids. Thus, compounds of the type $R_2P(OH)$ are not acidic in aqueous solution, the correct formulation being **45**; esters

$$\begin{array}{cc} \textbf{45} & \textbf{46} \end{array}$$

of this acid can be prepared by an indirect way

$$R_2PCl + R'OH \rightarrow R_2POR' + HCl, \tag{181}$$

and have the structure described by structure **46**. In an analogous way, the phosphonous acids, $RP(OH)_2$, are monobasic in acid solution (**47**), but form diester derivatives (**48**).

$$\begin{array}{cc} \textbf{47} & \textbf{48} \end{array}$$

Catenated Derivatives. Several compounds containing P—P bonds have been described, the simplest being R_2P—PR_2. They may be obtained by the reaction of a monohalophosphine with a secondary phosphine

$$R_2PX + HPR_2 \rightarrow R_2PPR_2 + HX \tag{182}$$

or by a Wurtz-type reaction with a monohalophosphine

$$2R_2PX + 2Na \rightarrow R_2PPR_2 + 2NaX. \tag{183}$$

Little is known of the structure of these compounds; they appear to have many characteristics of the monophosphine R_3P. For example, they form sulfides of the type $R_2P(S)P(S)R_2$. The results of infrared experiments indicate that solid $P_2(CH_3)_4$ is in a *trans* configuration; upon melting the molecule achieves a configuration essentially like that of N_2H_4 (36).

Polymeric compounds with the general formulation $(RP)_n$ are formed when CF_3PI_2 is treated with metallic mercury. A tetramer $(CF_3P)_4$, and pentamer $(CF_3P)_5$, have been isolated. These have been formulated as cyclic compounds **49** and **50**. Recently a trimer, $(PCF_3)_3$, has been reported.

Arsenic (*XXVII*) Three-coordinate organoarsenic compounds can be prepared by the reaction of the Grignard reagent on arsenic halides

$$AsX_3 + 3RMgX \rightarrow AsR_3 + 3MgX_2 \qquad (184)$$

or by the action of alkyl iodides on sodium arsenide

$$Na_3As + 3RI \rightarrow 3NaI + AsR_3. \qquad (185)$$

The latter reaction also can be used to prepare alkyl derivatives of arsine

$$NaAsH_2 + RI \rightarrow NaI + RAsH_2. \qquad (186)$$

Methyl arsine (bp 2°), dimethyl arsine (bp 36°), and trimethyl arsine (bp 50°) are water-clear liquids that inflame in air. In the case of $(CH_3)_3As$, trimethyl arsine oxide has been isolated from the oxidation reaction. Careful oxidation of the other methyl arsines with dry oxygen yields CH_3AsO (mp 95°) and $(CH_3)_2AsO$. The action of hydrogen halides on dimethyl arsine

$$(CH_3)_2AsH + HX \rightarrow (CH_3)_2AsX + H_2 \qquad (187)$$

or on methyl arsine oxide

$$CH_3AsO + 2HX \rightarrow CH_3AsX_2 + H_2O \qquad (188)$$

gives the corresponding methyl arsine halides. When $(CH_3)_2AsCl$ is treated with zinc, a substance containing as As—As bond, historically known as cacodyl, is formed

$$2(CH_3)_2AsCl + Zn \rightarrow (CH_3)_2AsAs(CH_3)_2 + ZnCl_2. \qquad (189)$$

Where known, the organoarsines or their derivatives exhibit pyramidal structures for the molecules, with bond angles generally in the range of 96–102°.

Pentacoordinate organometallic arsenic derivatives are formed when the tricoordinate compounds are treated with the halogens. For example, treatment of trimethyl arsine, dimethyl arsine chloride, or methyl arsine chloride gives the corresponding five-coordinate compound

$$(CH_3)_xAsCl_y + Cl_2 \rightarrow (CH_3)_xAsCl_{y+2} \qquad (190)$$

$$x = 3, 2, 1 \text{ when } y = 0, 1, 2.$$

If the oxidation of methyl arsines is carried out in the presence of water, methyl-substituted arsenic acids are formed

$$CH_3AsH_2 + \tfrac{3}{2}O_2 \xrightarrow{\hspace{2cm}} CH_3AsO(OH)_2 \qquad (191)$$

$$(CH_3)_2AsCl \xrightarrow[(2)\ H_2O]{(1)\ Cl_2} (CH_3)_2AsO(OH). \qquad (192)$$

The diorgano-acids are called arsenic acids whereas the monoorgano-acids are called arsonic acids.

Antimony and Bismuth   Compounds containing Sb—C and Bi—C bonds are best prepared by the reaction of an active organometallic compound such as the Grignard or lithium reagent on the corresponding halides [cf. Eq. (184)]. The reaction of an antimonide or a bismuthide with an alkyl halide [cf. Eq. (185)] is also a useful preparative method. The markedly lower stabilities of SbH_3 and BiH_3 among the hydrides of the elements of Group V are also reflected in the stability of their organo-derivatives. The very unstable compounds $(CH_3)_2SbH$ and CH_3SbH_2 have been prepared and characterized, but the corresponding hydrides of bismuth are unknown. The trivalent organohalogen derivatives of antimony and bismuth are the most common mixed derivatives.

Trialkyl stibines and bismuthines are stable to hydrolysis but they react readily with oxygen; $(CH_3)_3Sb$ forms a stable oxide, $(CH_3)_3SbO$, but at low temperatures $(C_2H_5)_3Bi$ forms an unstable peroxide, $(C_2H_5)_3BiO_2$, which decomposes rapidly to other products. Trimethyl stibine also forms a sulfide, $(CH_3)_3SbS$, and reacts with methyl iodide to form the quaternary salt $(CH_3)_4SbI$. Trialkyl bismuthines are devoid of donor properties. The metal–carbon bonds can be cleaved under certain conditions by halogens

$$R_3M + X_2 \rightarrow R_2MX + RX. \qquad (193)$$

Pentavalent carbon-containing antimony and bismuth compounds can be obtained by the same reactions which lead to the corresponding arsenic compounds. That is, chloride adds readily to the trialkyl

$$R_3M + Cl_2 \rightarrow R_3MCl_2 \tag{194}$$

and alkylhalo derivatives

$$R_2MCl + Cl_2 \rightarrow R_2MCl_3 \tag{195}$$

$$RMCl_2 + Cl_2 \rightarrow RMCl_4. \tag{196}$$

In contrast to compounds of phosphorus and arsenic, antimony and bismuth compounds containing five carbon–antimony bonds are known. SbR_5 ($R = Me$, C_6H_5) and $Bi(C_6H_5)_5$ are surpisingly stable compounds. They are less rapidly oxidized than the corresponding trivalent compounds, but they are more easily hydrolyzed. Oxidation of dialkyl stibines and alkyl stibines followed by hydrolysis of the products form the corresponding stibinic $[R_2SbO(OH)]$ and stibenic $[RSbO(OH)_2]$ acids.

4 / COORDINATION BEHAVIOR

The compounds of the elements of this group can act as either Lewis acids or bases; in several instances both possibilities occur. Trivalent nitrogen compounds, R_3N, act only as Lewis bases forming quaternary ions (R_4N^+) as well as neutral compounds (R_3NAlX_3, R_3NBX_3). Ammonia and the amines are good Lewis bases, but substitution by halogen atoms gives compounds devoid of donor properties. Thus, NF_3 is not known to form coordinate-covalent bonds, presumably because of the electron-withdrawing effect of the fluorine atoms. When electron-releasing alkyl groups are substituted for hydrogen, the resulting amine is generally more basic. The trialkylamines are not usually the most basic of the substituted amines because of steric effects which make coordinate-covalent bond formation difficult. Nitrogen atoms which exhibit oxidation states higher than $3-$ can also act as Lewis bases. For example, NO^+, which is isoelectronic with CO, forms coordinate-covalent bonds to metals to form compounds such as $Co(CO)_3NO$. Such compounds are discussed in detail in Chapter 19.

In contrast to nitrogen, the other elements in Group V form a well known variety of compounds which can behave as both Lewis bases and acids. Many tricoordinate compounds can act as Lewis bases irrespective of the substituents; derivatives bearing electron-withdrawing halogen substituents, as well as electron-releasing alkyl substituents, form four-coordinate ionic or neutral species. Thus, the ionic species MR_4^+ ($R = CH_3$; $M = P$, As, Sb), MCl_4^+ ($M = P$, As, Sb), and neutral compounds such as ZPX_3 ($X = $ halogen, alkyl, OR, OH; $Z = P$, S), X_3AlPXH_3, $Cr(PF_3)_6$, H_3BPF_3, and $Cl_2Pt[P(OR)_3]$ are known to contain a tetrahedral distribution of atoms about the central Group V atom. The trivalent derivatives of these elements can also act as Lewis acids forming compounds of the type $L \cdot MR_3$ where L can be either a neutral (R_2OSbBr_3, $R_3N \cdot BiCl_3$) or an anionic $[M^+BiCl_4^-$, $NaSb(OH)_4$, $MSbCl_4]$ Lewis base. Although little direct evidence exists concerning the structure of such species, the electron-pair-repulsion arguments would predict that they should have a trigonal bipyrimidal arrangement of electron pairs, with one pair unshared. Presumably these species form because of the availability of low-lying d orbitals, which permit the formation of dsp^3 hybrid orbitals. This is not energetically possible in the case of the corresponding nitrogen compounds.

The neutral five-coordinate compounds of the Group V elements which are trigonal bipyrimidal molecules cannot act as bases in the Lewis sense, but they can accept electron pairs from either neutral $[R_2O \cdot SbCl_5$, $R = H$, alkyl; $RCN \cdot SbCl_5]$ or anionic $[PF_6^-$, $Sb(OH)_6^-$, $SbCl_6^-]$ Lewis bases (XXVIII). These species, as would be expected, possess an octahedral arrangement of atoms about the Group V element. The central atom thus uses d^2sp^3 hybrid orbitals to form the σ framework of the molecules or ions.

Arsenic and antimony fluorides are sufficiently acidic that they can form complexes with species such as MF_6^- ($M = $ As and Sb). Thus, the ion $As_2F_{11}^-$ has been isolated in the form of tetraalkylammonium salts and characterized spectroscopically as containing fluorine-bridged species (**51**) (*37, 38*). The infrared spectrum of discrete SbF_5 molecules isolated in an

51

argon or neon matrix suggests that this species has a square pyramidal structure, but in concentrated matrices fluorine-bridged species, $(SbF_5)_n$, occur (*39*). Solutions of CsF or $(C_2H_5)_4N^+SbF_6^-$ in SbF_5 also contain complex fluorine-bridged anions of the type $(Sb_nF_{5n+1})^-$ ($n > 2$) (*40*).

REFERENCES

1. L. H. Bolz, M. E. Boyd, F. A. Mauer, and H. S. Peiser, *Acta Cryst.*, **12**, 247 (1959).
2. Yu. G. Borod'ko and A. E. Krylova, *J. Struct. Chem.*, **8**, 213 (1967).
3. H. Thurn and H. Krebs, *Acta Cryst.*, **B25**, 125 (1969).
4. G. J. Moody and J. D. R. Thomas, *J. Chem. Ed.*, **43**, 205 (1966).
5. R. Bougon, J. Chatelet, J. P. Desmoulens, and P. Plurien, *Compt. Rend.*, **266C**, 1760 (1968).
6. W. A. Guillory and C. E. Hunter, *J. Chem. Phys.*, **50**, 3516 (1969).
7. M. Green and J. W. Linnett, *J. Chem. Soc.*, 4959 (1960).
8. W. G. Fateley, H. A. Bent, and B. Crawford, Jr., *J. Chem. Phys.*, **31**, 204 (1959).
9. H. A. Bent, *Inorg. Chem.*, **2**, 747 (1963).
10. L. O. Anderson and J. Mason, *Chem. Commun.*, 99 (1968).
11. L. Kuhn and E. R. Lippincott, *J. Amer. Chem. Soc.*, **78**, 1820 (1956).
12. B. F. Hoskins, F. D. Whillans, D. H. Dale, and D. C. Hodgkin, *Chem. Commun.*, 69 (1969).
13. L. H. Jones, R. M. Badger, and G. E. Moore, *J. Chem. Phys.*, **19**, 1599 (1951).
14. J. T. Nelson and J. J. Lagowski, *Inorg. Chem.*, **6**, 1292 (1967).
15. J. R. W. Warn and D. Chapman, *Spectrochim. Acta*, **22**, 1371 (1966).
16. R. L. Patton and W. L. Jolly, *Inorg. Chem.*, **8**, 1389, 1392 (1969).
17. R. L. Patton and K. N. Raymond, *Inorg. Chem.*, **8**, 2426 (1969).
18. O. Glemser, A. Müller, D. Böhler, and B. Krebs, *Z. Anorg. Allgem. Chem.*, **357**, 184 (1968).
19. D. Neubauer and J. Weiss, *Z. Anorg. Allgem. Chem.*, **303**, 28 (1960).
20. T. M. Sabine and G. W. Cox, *Acta Cryst.*, **23**, 574 (1967).
21. B. Beagley, D. W. J. Cruickshank, T. G. Hewitt, and K. H. Jost, *Trans. Faraday Soc.*, **65**, 1219 (1969).
22. W. B. Fox, C. A. Wamser, R. Erbeck, D. K. Huggins, J. S. MacKenzie, and R. Jurrik, *Inorg. Chem.*, **8**, 1247 (1969).
23. C. A. Wamser, W. B. Fox, B. Sukornick, J. R. Holmes, B. B. Steward, R. Jurrik, N. Vanderkooi, and D. Gould, *Inorg. Chem.*, **8**, 1249 (1969).
24. A. V. Pankratov and O. M. Sokolov, *Russ. J. Inorg. Chem.*, **13**, 1481 (1968).
25. A. V. Pankratov, O. M. Sokolov, and D. S. Miroshnichenko, *Russ. J. Inorg. Chem.*, **13**, 1618 (1968).
26. K. O. Criste, *Inorg. Chem.*, **8**, 1539 (1969).
27. J. R. Durig and J. W. Clark, *J. Chem. Phys.*, **48**, 3216 (1968).
28. D. F. Koster and F. A. Miller, *Spectrochim. Acta*, **24A**, 1487 (1968).
29. R. S. Berry, *J. Chem. Phys.*, **32**, 933 (1960).
30. K. B. Dillon and T. C. Waddington, *Chem. Commun.*, 1317 (1969).
31. H. E. Coxon and D. B. Sowerby, *J. Chem. Soc.*, A, 3012 (1969).
32. T. Chivers and N. L. Paddock, *J. Chem. Soc.*, A, 1687 (1969).
33. E. C. Baughan, *Electroanal. Chem. Interface Electrochem.*, **29**, 81 (1971), and references therein.
34. A. Hershaft and J. D. Corbett, *Inorg. Chem.*, **2**, 979 (1963).
35. Yu. Ya. Kharitonov and M. A. Sarukhanov, *Russ. J. Inorg. Chem.*, **13**, 186 (1968).
36. J. R. Durig and J. S. DiYorio, *Inorg. Chem.*, **8**, 2796 (1969).
37. S. Brownstein, *Can. J. Chem.*, **47**, 605 (1969).
38. P. A. W. Dean, R. J. Gillespie, and R. Hulme, *Chem. Commun.*, 990 (1969).
39. A. L. K. Aljibury and R. L. Redington, *J. Chem. Phys.*, **52**, 453 (1970).
40. A Commeyras and G. A. Olah, *J. Amer. Chem. Soc.*, **91**, 2929 (1969).

COLLATERAL READINGS

I. H. Schumann, *Angew. Chem. (Intl. Ed.)*, **8**, 937 (1969). The properties of covalent compounds of the Group IV elements containing phosphorous, arsenic, antimony, and bismuth are reviewed. A part of this review deals with the possible participation of *pπ–dπ* components in the bonds formed between the Group IV and Group V elements.

II. G. Henrice-Olivé and S. Olivé, *Angew. Chem. (Intl. Ed.)*, **8**, 650 (1969). A description of several transition-metal complexes which absorb molecular nitrogen from the gas phase, and a discussion of the reaction products.

III. Yu. G. Borod'ko and A. E. Shilov, *Russ. Chem. Revs.*, **38**, 355 (1969). A review which deals with the problems associated with the formation and structure of complexes of molecular nitrogen with transition-metal compounds, metals, and other active species.

IV. R. Juza, *Advan. Inorg. Chem. Radiochem.*, **9**, 81 (1966). A review which deals with the problems associated with the formation and structure of complexes of molecular nitrogen with transition-metal compounds, metals, and other active species.

V. M. N. Hughes, *Quart. Revs.*, **22**, 1 (1968). A review of the preparation, properties, and structure of hyponitrous acid and its derivatives.

VI. J. Goubeau, *Angew. Chem. (Intl. Ed.)*, **5**, 567 (1966). A summary of infrared data pertaining to the force constants and bond orders of nitrogen bonds.

VII. C. C. Addison and N. Logan, *Advan. Inorg. Chem. Radiochem.*, **6**, 71 (1964). A comprehensive survey of the preparation and properties of anhydrous metal nitrates.

VIII. I. R. Beattie, *Prog. Inorg. Chem.*, **5**, 1 (1963). A discussion of the chemistry and physical properties of N_2O_3.

IX. C. Woolf, *Advan. Fluorine Chem.*, **5**, 1 (1965). A review of the preparation, reactions, properties, and structures of nitrogen oxides, nitrosyl derivatives, and nitryl fluoride.

X. F. Seel, *Angew. Chem. (Intl. Ed.)*, **4**, 635 (1965). A discussion of the chemistry of nitrous acid and its derivatives in liquid HF.

XI. W. H. Lee, in *The Chemistry of Nonaqueous Solvents*, Vol. 2 (J. J. Lagowski, ed.), Academic Press, 1967, Chapter 4. A discussion of the solvent properties of anhydrous nitric acid.

XII. W. H. Lee in *The Chemistry of Nonaqueous Solvents*, Vol. 2 (J. J. Lagowski, ed.), Academic Press, 1967, Chapter 3; R. J. Gillespie and E. A. Robinson, in *Nonaqueous Solvent Systems* (T. C. Waddington, ed.), Academic Press, 1965, Chapter 4. The behavior of nitrogen oxides in strongly acidic media is discussed in each of these works.

XIII. M. Becke-Goehring, *Advan. Inorg. Chem. Radiochem.*, **2**, 159 (1960). A survey of the chemistry of the amides and imides of the oxyacids of sulfur. One section deals with S_4N_4.

XIV. J. Goubeau, *Angew. Chem. (Intl. Ed.)*, **8**, 328 (1969). A discussion of spectroscopic investigations on phosphorus–sulphur compounds as related to the character of the P—S bond.

XV. E. Thilo, *Advan. Inorg. Chem. Radiochem.*, **4**, 1 (1962). A detailed discussion of the properties and structures of the condensed phosphates and arsenates. See also, E. Thilo, *Angew. Chem. (Intl. Ed.)*, **4**, 1061 (1965).

XVI. J. W. George, *Prog. Inorg. Chem.*, **2**, 33 (1960). A review of the preparation, properties, and structures of the nitrogen and phosphorus halides and oxyhalides.

XVII. K. A. Becker, K. Plieth, and I. N. Stranski, *Prog. Inorg. Chem.*, **4**, 1 (1962). A survey of the properties and structures of the polymorphic modifications of As_2O_3.

XVIII. C. B. Colburn, *Advan. Fluorine Chem.*, **3**, 92 (1963). A discussion of the chemistry of nitrogen fluorides and their inorganic derivatives, NF_3, N_2F_4, NF_2, N_2F_2, HNF_2, H_2NF, $ClNF_2$, and NF_2NO.

XIX. J. K. Ruff, *Chem. Revs.*, **67**, 665 (1967). A description of the preparation, properties, and reactions of nitrogen fluoride derivatives.

XX. U. Engelhardt and J. Jander, *Fortschr. Chem. Forsch.*, **5**, 663 (1966). A detailed account of the chemistry of NI_3.

XXI. L. Kolditz, *Advan. Inorg. Chem. Radiochem*, **7**, 1 (1965). A survey of the halides of phosphorus, arsenic, antimony, and bismuth in the following oxidation states: $5+$, $3+$, $<3+$.

XXII. R. Schmutzler, *Advan. Fluorine Chem.*, **5**, 31 (1955). A comprehensive survey of inorganic and organic compounds containing a P—F bond.

XXIII. E. L. Muetterties, *Acc. Chem. Res.*, **3**, 266 (1970). A summary of stereochemically nonrigid structures for five-coordinate systems appears as a section of this review.

XXIV. N. L. Paddock and H. T. Searle, *Advan. Inorg. Chem. Radiochem.*, **1**, 347 (1959); C. D. Schmulbach, *Prog. Inorg. Chem.*, **4**, 275 (1962); V. V. Kireev, G. S. Kobesnikov, and I. M. Raigorodski, *Russ. Chem. Revs.*, **38**, 667 (1969). These references contain much detailed information on phosphonitrilic halides and their derivatives.

XXV. R. F. Hudson, *Advan. Inorg. Chem. Radiochem.*, **5**, 347 (1963). A discussion of the structure and reactivity of organophosphorus compounds. The article includes a section on the evidence available for $d\pi$–$p\pi$ bonding. See also L. Maier, *Prog. Inorg. Chem.*, **5**, 27 (1963), for a comprehensive survey of the preparation and properties of primary, secondary, and tertiary phosphines.

XXVI. G. Kamai and G. M. Usacheva, *Russ. Chem. Revs.*, **35**, 601 (1966). A survey of the stereochemistry of trivalent phosphorus and arsenic compounds and their quaternary derivatives, as well as their oxides, sulfides, and organic acid derivatives.

XXVII. W. R. Cullen, *Advan. Organometal. Chem.*, **4**, 145 (1966). A review of organoarsenic chemistry.

XXVIII. M. Webster, *Chem. Revs.*, **66**, 87 (1966). A discussion of the preparation properties, and structures of the addition compounds formed by the Group V pentahalides.

STUDY QUESTIONS

1. Discuss the structure and bonding in the sequence of species PF_3, PF_5, and PF_6^-. Give the most probable canonical forms that contribute to the overall wave function for these species.

2. Give the geometry of the covalently bonded species in the following compounds without recourse to structural data. (a) KH_2AsO_4; (b) $CsNO_3$; (c) P_2H_4; (d) $N_2H_5^+$; (e) HN_3; (f) H_3PO_2; (g) FN_3; (h) H_2NPCl_2; and (i) PCl_4BF_4, a white crystalline solid which does not melt but undergoes decomposition upon heating.

3. Which of the species in Question 2 exhibit multiple bonds incorporating d components?

4. Discuss the idealized hybridization of the atoms in the covalently bonded species in Question 2.

5. Suggest methods to prepare the following compounds incorporating ^{15}N from isotopically enriched molecular nitrogen ($^{15}N_2$): (a) $Na^{15}NH_2$; (b) $Ca_3{}^{15}N_2$; (c) $^{15}N_2O_4$; (d) $K^{15}NO_3$; (e) $^{15}NOBF_4$.

6. When PCl_3 and $P(CH_3)_3$ are heated together in the gas phase, compounds such as $PCl(CH_3)_2$ and $PCl_2(CH_3)$ are formed. Discuss the bonding and structures of the possible transition states in this process.

7. Phosphorus pentachloride (PCl_5) is a white solid which volatilizes at 160° and melts at 148° under pressure. Molten PCl_5 conducts an electric current readily. At 180° PCl_5 vapor has a density which corresponds to $9.3 \ g \ l^{-1}$ corrected to standard conditions. At 250° a sample of PCl_5 exerts a pressure which is two times the value expected. Discuss the structure of the species present in PCl_5 (a) at its melting point, (b) at 180° in the gas phase, and (c) at 250° in the gas phase. In your discussion give a sketch of the geometry of the species indicating the relative positions of all the atoms and the most probable bond angles.

8. Discuss the structure of $N(SiH_3)_3$ in light of the facts that the molecule is devoid of basic character and is a planar molecule.

9. Estimate the standard enthalpy of formation for $NF_{3\,(g)}$.

10. Account for the difference in bond angle between NH_3 (107°) and NF_3 (102°).

11. Although amines are described as possessing a pyramidal structure with bond angles near the tetrahedral angle, compounds of the type $RR'R''N$ have not been resolved into optically active forms. Discuss this apparent anomaly.

12. Discuss possible structures for the following species: (a) $H_2N_2O_2$; (b) $(CH_3)_2O \cdot N_2O_4$; (c) N_3NO; (d) $SPCl_3$.

13. Discuss the Lewis acid–base properties of the following compounds: (a) NH_2OH; (b) $NOCl$; (c) N_2O_3; (d) S_2N_2; (e) N_2F_4; (f) $SPCl_3$; (g) PF_3Cl_2; (h) $BiCl_5{}^{2-}$.

14. Give the most probable canonical structures that contribute to the best wave function for the following molecules: (a) $P_5(CF_3)_5$; (b) S_2N_2; (c) S_3N_2O.

15. Using data that appear in this chapter, calculate the bond moments for PCl_3, $AsCl_3$, and $SbCl_3$; the vapor-phase dipole moments of these compounds are 0.78, 1.59, and 3.9 D, respectively. Compare and comment on these values.

13

GROUP VI
ELEMENTS

1 / INTRODUCTION

The Group VI elements exhibit an ns^2np^4 ground-state configuration (Table 13.1), two of the four p electrons being paired. In practice the stable electronic configuration of the next rare gas can be attained by forming two electron-pair bonds, giving covalent compounds of the type MR_2. As is the case with elements in the previous families, the octet rule is never violated for oxygen, although more than two bonds can be formed to the heavier elements in this family. Thus, for example, SCl_4 and SeF_6 are compounds in which five and six electron pairs, respectively, surround the central atom. The elements of this family also form compounds containing M^{2-} ions, which, of course, also possess the electronic configuration of the next rare gas.

The decreasing electronegativities of the elements in this family with increasing atomic number, as indicated in Table 13.1, are reflected in a decrease

TABLE 13.1 Some properties of the Group VI elements

Property	O	S	Se	Te	Po
Ground-state configuration	$[He] 2s^2 2p^4$	$[Ne] 3s^2 3p^4$	$[Ar] 3d^{10} 4s^2 4p^4$	$[Kr] 4d^{10} 5s^2 5p^4$	$[Xe] 4f^{14} 5d^{10} 6s^2 6p^4$
Ionization potential, kcal					
I	314.0	238.9	225	208	194
II	810.6	540.0	495	429	—
Electronegativity	3.50	2.44	2.48	2.01	1.76
Radii, Å					
ionic (M^{2-})	1.45	1.90	2.02	2.22	2.30
covalent	0.74	1.04	1.17	1.37	—
Mp, °C	−218.4	119^a	217	450	254
Bp, °C	−183	444.6	684.8	1390	962

^{a} Monoclinic.

of the ionic character of the bonds they form. Although sulfur and the heavier elements form compounds, called as a class the chalconides, in which the element can be formulated as the species M^{2-}, only the derivatives of the very electropositive elements are ionic in the operational sense. The change in electronegativity in this family also correlates with a marked decrease in the ability of the elements to form hydrogen bonds ($>M\cdots HX$). There is, of course, the usual increase in metallic properties with increasing atomic number in this family as well as an increase in the tendency to form anionic complex species. The two heaviest elements of this family appear to exhibit some cationic properties in contrast with the other members. Thus, the dioxides PoO_2 and TeO_2 react with hydrogen halides to give the corresponding halides (PoX_4 and TeX_4); polonium also forms typical salts such as $Po(SO_4)_2$. This increase in ability to exhibit cationic behavior parallels that described for the Group V elements, as described in Chapter 12.

2 / THE ELEMENTS

Large quantities of oxygen and sulfur occur in the free state. Oxygen is obtained on a commercial scale from the liquefaction of air, the pure substance being separated from the other components by distillation. Elemental sulfur is mined on a commercial scale, and extensive deposits of sulfur-containing compounds such as gypsum ($CaSO_4$), magnesium sulfate, and various metallic sulfides are known. Selenium and tellurium are markedly less abundant than sulfur and oxygen. These substances are generally found as selenides and tellurides in sulfide ores. Polonium is a product of one of the natural radioactive decay series and is found in uranium- and thorium-containing minerals. Polonium itself is also radioactive, the most abundant isotope, ^{210}Po, decaying by α-emission with a half-life of 138.4 days (*I*).

2.1 / Physical Properties

Oxygen and sulfur are nonmetals in all their properties, but selenium, tellurium, and polonium exhibit some metallic characteristics. For example, gray selenium has a metallic appearance, although its conductivity is not nearly that expected of a metal; however, this allotrope is photoconductive. Gray tellurium is about five orders of magnitude more conducting than selenium, whereas polonium has a conductivity typical of metals.

2.2 / Allotropic Modifications

As shown by the data in Table 13.2, all of the elements in this family except tellurium exist in several allotropic modifications.

TABLE 13.2   Physical properties of the Group VI allotropes

Element	Allotrope	Physical properties	Structure
Oxygen	O_2	mp $-218°$, bp $-183°$, pale blue liquid and solid	linear, $d = 1.21$ Å
	O_3	mp $-249.6°$, bp $-112.3°$, dark indigo-blue liquid and solid	bent, $d = 1.278$ Å, $\angle(OOO) = 116.8°$
	O_4	—	—
Sulfur	orthorhombic	stable below $95.6°$	puckered S_8 ring, $d(S-S) = 2.037$ Å, $\angle(SSS) = 107.8°$, dihedral angle $= 99.3°$
	monoclinic	stable between 95.6 and 119°	—
	rhombohedral	—	molecular-weight data indicate S_6
	amorphous liquid	insoluble in CS_2	—
	$S(\lambda)$	mobile liquid below 160°	S_8 molecule
	$S(\mu)$	viscous liquid above 160°	S_x chains of various lengths
Selenium	vitreous Se	glassy appearance	two nearest neighbors at 2.3 Å
	metallic Se	gray	infinite helical chains, $d = 2.36$ Å, $\angle(SeSeSe) = 103.6°$
	α-Se	red monoclinic	puckered Se_8 ring, $d = 2.34$ Å, $\angle(SeSeSe) = 105°$
	β-Se	red monoclinic	same as for α-Se
	amorphous	red	—
	amorphous	black	—
Tellurium	metallic	gray	infinite helical chains, $d = 2.82$ Å, $\angle(TeTeTe) = 102°$
	amorphous	—	—
Polonium	cubic (α)	—	—
	rhombehedral (β)	—	—

Oxygen   The most common form of oxygen is the O_2 molecule which is paramagnetic. Without the latter information, valence-bond arguments would lead to structure **1** as the predominant canonical form for the wave function since all other forms violate the octet rule. However, the fact that the

$$\ddot{O}::\ddot{O} \qquad\qquad :O\dot{\dot{=}}O:$$

1   **2**

molecule is paramagnetic leads to an additional canonical form (2) incorporating a three-electron bond. The molecular-orbital method perhaps provides a better description of the bonding in the O_2 molecule than does the valence-bond method. The simple molecular-orbital diagram for a homonuclear diatomic molecule incorporating $2s$ and $2p$ orbitals appears in Fig. 13.1. The twelve valence electrons in the molecule are distributed in

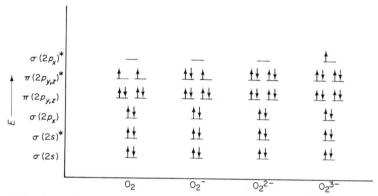

FIG. 13.1 Approximate energy-level diagrams for some binuclear oxygen species.

the available orbitals according to the Pauli principle, giving a paramagnetic species: $[\sigma(2s)]^2 [\sigma^*(2s)]^2 [\sigma(2p_x)]^2 [\pi(2p_z)]^2 [\pi(2p_y)]^2 [\pi^*(2p_z)]^1 [\pi^*(2p_y)]^1$. Inspection of the energy-level diagram shows that the paramagnetism arises from the presence of a pair of degenerate π and π^* orbitals, which leads to two unpaired electrons in the molecule even though there is an even number of valence electrons. According to these arguments the bond order of O_2 is 2.0 (i.e., one full σ bond and two half π bonds) in spite of its paramagnetism. The valence-bond description incorporating canonical forms **1** and **2** suggests a bond order greater than 2. The dissociation energy of oxygen molecules to form the corresponding atoms has been measured as 118.2 kcal mole^{-1}.

Ozone can be prepared by passing oxygen through an electric discharge or exposing it to ultraviolet radiation

$$\tfrac{3}{2}O_2 \rightarrow O_3 \qquad \Delta H = 34.0 \text{ kcal mole}^{-1}. \tag{1}$$

This substance is also produced in certain reactions where molecular oxygen is formed, e.g., the action of fluorine on water or the electrolysis of concentrated H_2SO_4 at high current densities. The fact that the process described by Eq. (1) occurs under the influence of ultraviolet irradiation accounts for the presence of ozone in the upper atmosphere of the earth. The maximum concentration of ozone occurs at about 25 km. Since O_3 has a very intense absorption band with an edge at about 2900 Å, this layer of ozone serves as a filter, protecting the earth's surface from most of the sun's ultraviolet radiation.

Ozone is a symmetrical, but bent, molecule, the parameters being given in Table 13.2. The O—O bond distance (1.278 Å) is greater than that in molecular oxygen (1.21 Å) but shorter than expected for a single bond (1.48 Å). These data suggest that considerable double-bond character exists in the oxygen bonds; canonical forms such as those shown in structures **3–6** must

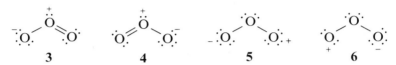

contribute significantly to the wave function of the molecule. The large bond angle, 116.8°, places the terminal oxygen atoms ~ 2.2 Å apart, which is considerably longer than the excited single-bond distance of 1.48 Å. Thus, the

7

valence-bond structure **7** is not a necessary contributor to the wave function to account for the properties of this molecule.

The dissociation energy of ozone

$$O_3 \rightarrow 3O \tag{2}$$

is 143.6 kcal mole^{-1}, which makes the O—O bond in ozone (71.8 kcal mole^{-1}) weaker than that in the oxygen molecule, as expected from the relative O—O bond distances for these two species.

The species O_4 has been detected in liquid oxygen; a sufficient concentration of this species is present in the gaseous phase to be detected spectroscopically (*1–3*). The stoichiometry suggests that O_4 consists of normal oxygen molecules that have dimerized. The fact that O_2 has two unpaired electrons (Fig. 13.1) makes it tempting to formulate O_4 as **8**, but the dimerization energy has been estimated as only 0.13 kcal mole^{-1}, which is far less than would be expected for a normal O—O bond.

O—O
| |
O—O

8

Sulfur   The structural relationships which occur in elemental sulfur are perhaps the most complex among all the elements. Sulfur exhibits the three solid modifications described in Table 13.2. The orthorhombic form is the most stable at room temperature; it consists of the puckered S_8 rings shown in Fig. 13.2, with bond distances (2.04 Å) that are only 2% shorter than that expected for a single bond. Thus, to a very good approximation, the S—S

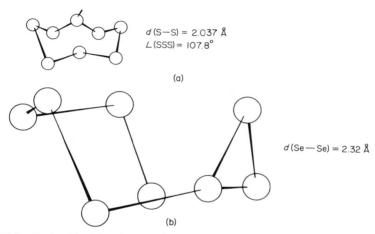

d (S—S) = 2.037 Å
∠(SSS) = 107.8°

(a)

d (Se — Se) = 2.32 Å

(b)

FIG. 13.2   Both sulfur (a) and the $Se_8{}^+$ cation (b) contain the same basic cyclic arrangement of atoms.

bond in this system contains virtually no π character, which could theoretically arise from pπ–dπ interactions. There is evidence that the S_8 ring persists in the monoclinic modification. Sulfur also dissolves in a variety of non-interacting solvents as S_8 molecules. Liquid sulfur exists in at least two discrete forms. Below 160° the liquid is straw yellow and mobile; the available evidence indicates that S_8 molecules predominate at this temperature. Above 160° the liquid darkens and becomes markedly more viscous. It appears that at these temperatures the sulfur rings break and chainlike molecules of indeterminate length are present (4). Sulfur vapor at the boiling point contains S_8 molecules which dissociate into smaller units such as S_6 and S_7 at higher temperatures (5). The S_8 molecules in the vapor phase have virtually the same configuration and structural parameters as in the solid state. Above 800° S_2 predominates and at about 2000° monatomic sulfur is present (II). The bond distance in S_2 molecules is noticeably shorter than that in S_8, suggesting that some double bonding is present in this species. It is unnecessary, however, to invoke dπ–pπ bonding because the molecular-orbital description of S_2 should be the same as for O_2 (shown in Fig. 13.1). The species S_{12}, which has a zig-zag ring structure (6), has been prepared from the reaction between H_2S_x and Cl_2S_y (7); S_{12} has also been detected in sulfur melts (8).

A metastable plastic form of sulfur can be obtained by rapidly quenching the viscous liquid phase. It appears that the plastic modification contains sulfur chains which have not reformed rings, giving a fibrous structure. After a time this plastic form crystallizes to the stable orthorhombic modification. If the vapor at 500°, consisting mainly of S_8 molecules, is cooled rapidly to a film with liquid nitrogen, a purple solid is formed which reverts to a mixture of crystalline and amorphous sulfur upon heating to room temperature. The orthorhombic modification of sulfur is soluble in a variety of solvents; however, it is possible to prepare an amorphous form which is insoluble.

Selenium and Tellurium   The data in Table 13.2 show that selenium possesses many of the characteristics of sulfur. Selenium atoms can bond to each other to form Se_8 molecules which are known in the solid state and in CS_2 solution. Vaporization of selenium and tellurium gives M_8 molecules as well as cyclic polymeric species such as M_5 and M_7 (9). At relatively high temperatures ($\sim 900°$) selenium is diatomic, and becomes monatomic at $\sim 2000°$. Two amorphous and three crystalline forms of selenium are known. If selenium is prepared by reduction of a selenium compound, it precipitates as a red amorphous solid, which turns to a black amorphous form in boiling water. Two of the crystalline forms are remarkable with respect to their properties; both are red and both contain eight-membered rings, but they differ in the rate at which they are converted into the stable gray metallic form. The metallic forms of tellurium and selenium have the same structure.

2.3 / Chemical Properties

Oxygen   Oxygen is known to form compounds with all elements except helium, neon, and argon. Direct combination is difficult to achieve with molecular oxygen in certain instances, even at elevated temperatures, but many oxygen-containing species can be prepared indirectly. For example, oxygen reacts with nearly every metal either at room temperature or at elevated temperatures to form oxides, and with a large number of nonmetals in a similar way. Ozone is a stronger oxidizing agent than is oxygen, reacting under milder conditions. The relative oxidizing ability of O_2 and O_3 can be obtained by comparing the equivalent standard electrode potentials for these substances:

$$O_2 + 4H^+ + 4e^- \rightarrow 2H_2O \qquad E^0 = +1.229 \text{ V} \qquad (3)$$

$$O_3 + 2H^+ + 2e^- \rightarrow O_2 + H_2O \qquad E^0 = +2.07 \text{ V} \qquad (4)$$

$$O_2 + 2H_2O + 4e^- \rightarrow 4OH^- \qquad E^0 = +0.401 \text{ V} \qquad (5)$$

$$O_3 + H_2O + 2e^- \rightarrow O_2 + 2OH^- \qquad E^0 = +1.24 \text{ V}. \qquad (6)$$

The corresponding electrode potentials for the reduction of F_2 ($+2.87$ V) and Cl_2 ($+1.359$ V) indicate that there are few more powerful oxidizing agents than ozone. In many of its reactions ozone is reduced to molecular oxygen; that is, it can be imagined as a carrier for atomic oxygen

$$PbS + 4O_3 \rightarrow PbSO_4 + 4O_2. \tag{7}$$

However, if the reaction can also occur with molecular oxygen, then all three oxygen atoms are consumed

$$3SnCl_2 + 6HCl + O_3 \rightarrow 3SnCl_4 + 3H_2O. \tag{8}$$

Ozone reacts with olefins to form addition compounds

$$R_2C{=}CR_2 + O_3 \rightarrow R_2C \overset{\displaystyle O}{\underset{\displaystyle O-O}{\diagup \diagdown}} CR_2 \tag{9}$$

which can be hydrolyzed to carbonyl derivatives

$$R_2C \overset{\displaystyle O}{\underset{\displaystyle O-O}{\diagup \diagdown}} CR_2 + H_2O \rightarrow R_2C{=}O + O{=}CR_2 + H_2O_2. \tag{10}$$

This sequence of reactions has found extensive use in determining the position of unsaturated bonds in hydrocarbons.

Sulfur   Sulfur is a very reactive element. It combines directly with all elements except the rare gases, nitrogen, tellurium, iodine, gold, platinum, and iridium. Reaction can occur at low temperatures, e.g., with the alkali metals, the alkaline-earth metals, and silver, mercury, lead, and aluminum. With some elements, e.g., nickel, cobalt, and chromium, higher temperatures are required. The element is inert to aqueous HCl but is attacked by oxidizing acids and aqueous base.

$$S + 2H_2SO_4 \rightarrow 3SO_2 + 2H_2O \tag{11}$$

$$S + 6HNO_3 \rightarrow H_2SO_4 + 6NO_2 + 2H_2O \tag{12}$$

$$3S + 6NaOH \rightarrow 2Na_2S + Na_2SO_3 + 3H_2O. \tag{13}$$

Fuming sulfuric acid dissolves sulfur (*III*) to give highly colored paramagnetic solutions containing the species S_4^{2+}, which probably has the same square-planar arrangement of atoms observed in the species Se_4^{2+} and Te_4^{2+}, the structures of which are known (*10*). Powdered sulfur and AsF_5 combine to form a blue solid, $S_8(AsF_6)_2$, or a red solid, S_8AsF_6, depending upon the ratio of reactants taken (*11*). Both substances contain the AsF_6^- ion; the blue compound contains the S_8^+ ion. The temperature-dependent paramagnetism of the red solid suggests that the reversible process

$$S_{16}^{2+} \rightleftharpoons 2S_8^+ \tag{14}$$

occurs. Sulfur will dissolve in solutions containing S^{2-} to form polysulfides

$$S^{2-} + xS \rightarrow S_{(x+1)}^{2-}, \tag{15}$$

and in sulfites to give thiosulfates

$$SO_3^{2-} + S \rightarrow S_2O_3^{2-}. \tag{16}$$

Polysulfides in aqueous solution are involved in equilibria containing S_n^{2-} ($n = 3$–6) species (*12, 13*).

Selenium and Tellurium   Selenium and tellurium have many of the chemical characteristics of sulfur; these are, however, modified by the fact that selenium and tellurium are less readily oxidized than sulfur. Thus when selenium is oxidized by nitric acid, selenous acid is formed

$$Se + 4HNO_3 \rightarrow H_2SeO_3 + 4NO_2 + H_2O, \tag{17}$$

in contrast to the reaction with sulfur [Eq. (12)]. Selenium and tellurium dissolve in strong anhydrous acids such as H_2SO_4 and HSO_3F to yield colored solutions. The species Se_4^{2+} has been identified in the yellow selenium solutions, whereas the green solutions contain Se_8^{2+} (*14, 15*). The green tellurium-containing solutions contain Te_4^{2+}. The cation Se_4^{2+} has a square-planar structure as predicted by molecular-orbital theory (*16*); the cation Se_8^{2+} has a puckered ring structure similar to that of S_8, which is shown in Fig. 13.2. Selenides and tellurides are formed by reaction with the alkali metals, and aqueous solutions of these compounds dissolve the corresponding element to form polyselenides and polytellurides [cf. Eq. (15)]. Selenium and tellurium dissolve in aqueous solutions of sulfides and polysulfides forming mixed chalconide ions of the type SSe_2^{2-} and TeS_3^{2-} (*17*).

3 / COMPOUNDS OF THE ELEMENTS

3.1 / Oxygen

The binary oxygen derivatives of the elements have properties characteristic of ionic and of covalent compounds. Generally, metallic oxides are best formulated as ionic systems, while the nonmetallic oxides are covalent molecules. There are three series of ionic oxides known, the superoxides (O_2^-), the peroxides (O_2^{2-}), and the normal oxides (O^{2-}) (IV, V). Divalent oxygen atoms and the peroxy group are commonly found in covalent species.

The properties of the ionic oxygen-containing species can be understood by recognizing that they are related to O_2 by the successive addition of electrons

$$O_2 \xrightarrow{+e^-} O_2^- \xrightarrow{+e^-} O_2^{2-} \xrightarrow{+2e^-} 2O^{2-}. \qquad (18)$$

If we can assume that the relative order of energies of the molecular orbitals for these diatomic oxygen species is the same, their magnetic properties and bond parameters can be predicted. Thus, from the information given in Fig. 13.1, the electronic configuration of superoxide (O_2^-) should be $[\sigma(2s)]^2$-$[\sigma^*(2s)]^2[\sigma(2p_x)]^2[\pi(2p_y)]^2[\pi(2p_z)]^2[\pi^*(2p_y)]^2[\pi^*(2p_z)]^1$, leading to a para-magnetic ion with a bond order of 1.5. Similarly, the peroxide ion should have the configuration $[\sigma(2s)]^2[\sigma^*(2s)]^2[\sigma(2p_x)]^2[\pi(2p_y)]^2[\pi(2p_z)]^2[\pi^*(2p_y)]^2$-$[\pi^*(2p_z)]^2$, which would predict that this ion is diamagnetic, with a bond order of 1.0. The data in Table 13.3 show that the experimental bond

TABLE 13.3 Bond parameters of some oxygen species

	Theoretical		Experimental	
Species	Bond order	Unpaired electrons	$d(O-O)$, Å	Magnetism[a]
O_2^+	2.5	1	1.12	—
O_2	2	2	1.21	p
O_2^- [b]	1.5	1	1.28	p
O_2^{2-} [c]	1	0	1.49	d

[a] p = paramagnetic, d = diamagnetic.
[b] As determined in KO_2.
[c] As determined in BaO_2 and Na_2O_2.

distances and magnetic properties for these ions are in accord with these predictions. That is, the observed bond distances increase in a regular manner from O_2 to O_2^{2-} and the paramagnetism decreases in this series as expected. The species O_2^{4-} is unstable with respect to two oxide ions $(2O^{2-})$ because its electronic configuration leads to completely filled bonding and anti-bonding orbitals. Under such conditions there is no gain in energy, so the unbonded system $(2O^{2-})$ is more stable than the bonded system (O_2^{4-}). It is interesting to note that the molecular-orbital arguments suggest that the species O_2^{3-} might be sufficiently stable to be detected, but it has not yet been characterized.

A cationic diatomic oxygen species can be obtained from the oxygen molecule by the *loss* of an electron. The dioxygenyl cation, O_2^+, has been detected spectroscopically as a short-lived species. The stable salt $O_2^+PtF_6^-$ forms as an orange solid when molecular oxygen reacts with PtF_6 (*18*). Apparently the ionization potential of O_2 is sufficiently low compared to the electron affinity of PtF_6 that the latter compound oxidizes the former. The dioxygenyl cation is also known in the compounds $O_2^+BF_4^-$ (*19*) and $O_2^+PtF_6^-$ (*20*).

Oxides   Simple oxides in which oxygen is in a $2-$ oxidation state are formed, either directly or indirectly, by virtually all of the elements. Broadly speaking, these compounds can be divided into three classes based upon their structure: (a) molecular oxides, (b) oxides that form giant molecules, and (c) ionic oxides. The divisions between these classes often are not distinct. For example, it is difficult in some cases to decide whether a crystalline oxide is best described as a continuous array of ions or as a giant covalent molecule in which the bonds contain some ionic character. In general, elements which have low electronegativities, such as the active metals, form ionic structures; those with electronegativities comparable to oxygen form covalent molecules. As usual, problems arise with the elements of inter-

TABLE 13.4   **Structural parameters of binary oxides**

Formula	Structure	O^{2-} lattice[a]	Position of counter ion	Example
MO	NaCl	ccp	all octahedral holes occupied	M = Mg, Ca, Sr, Ba
M_2O	antifluorite	ccp	all tetrahedral holes occupied	M = Li, Na, K, Rb
MO_2	fluorite	cubic	$\frac{1}{2}$ cube centers occupied	M = Th, Ce, Hf, Np Pu, Am, Po
	rutile	—	—	M = Ge, Sn, Pb, Ti, V, Nb, Te, Mn
MO	zinc blende	ccp	$\frac{1}{2}$ tetrahedral holes occupied	M = Be
	wurtzite	hcp	$\frac{1}{2}$ tetrahedral holes occupied	M = Zn
M_2O_3	corundum	hcp	$\frac{2}{3}$ octahedral holes occupied	M = Al, Fe, Cr, Ti, V
MO_2	silica	—	—	M = Si, Ge

[a] ccp = cubic closest packing, hcp = hexagonal closest packing (cf. Chapter 2).

mediate electronegativity. More extensive descriptions of the oxygen compounds of the elements occur in various other sections of this book and our discussion here is directed more towards a generalized description of the characteristics of these compounds.

Structural Properties. Molecular oxides are formed by the most nonmetallic elements. For example, carbon (CO_2 and CO), nitrogen (N_2O, NO, N_2O_3, NO_2, and N_2O_5), fluorine (F_2O), sulfur (SO_2 and SO_3), and chlorine (Cl_2O_7) form oxides that are volatile and crystallize in structures containing discrete molecules. One metallic element, osmium, forms an oxide with similar characteristics (OsO_4, mp 41°). The more metallic nonmetals form oxides which are polymerized to some extent (e.g., M_2O_6 and M_2O_5, M = P, As, Sb), yet discrete molecules are still present. Polymerization of the oxides becomes more extensive in the case of the more metallic elements (SiO_2, B_2O_3, BeO), and finally ionic lattices are formed with the most metallic of the elements.

The structures of a large number of crystalline oxides can be derived from a close-packed array of oxide ions with the counter ions occupying either the octahedral or tetrahedral holes; as shown by the data in Table 13.4, coordination numbers of 4, 6, and 8 are common in these structures.

An extensive series of compounds is known which can have the stoichiometry of mixed oxides, e.g., $K_2O \cdot SO_3$, $CaO \cdot CO_2$, $MgO \cdot SiO_2$, $CaO \cdot TiO_2$ (*VI*). Some of these compounds are best formulated as salts of oxyacids in which discrete oxyanions and cations are packed in the most efficient way, as is the case with K_2SO_4 ($K_2O \cdot SO_3$) and $CaCO_3$ ($CaO \cdot CO_2$). In certain cases, e.g., $MgO \cdot SiO_2$, the anions form an infinite structure with cations interspersed within it $[Mg_n(SiO_3)_n]$. However, many structures are known where no discrete anions exist, and the structure is best described as a close-packed arrangement of oxygen ions with the two different cations occupying holes of the appropriate size. This is the case for $CaTiO_3$; there are no discrete "titanate" anions present in this structure.

Chemical Properties. Oxides are conveniently classified as acidic, basic, or amphoteric with respect to their chemical behavior. Although such classification is usually based upon behavior in aqueous systems, similar behavior is observed in the absence of water at high temperatures.

The oxides of the metals are the anhydrides of bases in the water system of compounds. Although discrete oxide ions are present in the solid state, such ions hydrolyze completely to give an equivalent number of hydroxide ions

$$O^{2-} + H_2O \rightarrow 2OH^-, \tag{19}$$

which are the strongest bases that can exist in aqueous systems. The equilibrium constant for Eq. (19) has been established as being greater than

10^{22} (21). Some metallic oxides are very slightly soluble in water, but they will dissolve in acid solutions

$$MO + 2H_3O^+ \rightarrow M^{2+} + 3H_2O \tag{20}$$

$$M = Cu, Mg, Zn, Cd, Hg.$$

The oxides of the nonmetals are the anhydrides of acids, because when they dissolve in water hydroxy compounds are formed which undergo ionization to give hydronium ions

$$SO_3 + H_2O \rightarrow (HO)_2SO_2 \tag{21}$$

$$(HO)_2SO_2 + H_2O \rightarrow H_3O^+ + (HO)SO_3{}^-. \tag{22}$$

Amphoteric oxides give hydroxy compounds in aqueous solutions

$$ZnO + H_2O \rightarrow Zn(OH)_2 \tag{23}$$

that act either as bases

$$Zn(OH)_2 + H_3O^+ \rightarrow Zn(OH)^+ + 2H_2O \tag{24}$$

$$Zn(OH)^+ + H_3O^+ \rightarrow Zn^{2+} + 2H_2O \tag{24a}$$

or acids

$$Zn(OH)_2 + 2OH^- \rightarrow Zn(OH)_4{}^{2-}. \tag{25}$$

The factors which affect the relative acidity of hydroxy compounds are discussed in Chapter 7.

The reaction of aqueous acid solutions with basic solutions leads to neutralization, a process in which a salt is formed "by default." Substances that are salts in the aqueous solvent system can be prepared by direct reaction of the metallic and nonmetallic oxides at elevated temperatures. Thus, calcium oxide reacts with sulfur trioxide to yield calcium sulfate

$$CaO + SO_3 \rightarrow CaSO_4. \tag{26}$$

A large variety of mixed oxides which contain oxyanions can be prepared by this process. Acid–base behavior in fused systems led Lux (22) to suggest that a base is any species which yields oxide ions whereas an acid consumes oxide ions. The reaction shown in Eq. (26) would be interpreted in this view as the base CaO

$$CaO \rightarrow Ca^{2+} + O^{2-} \tag{27}$$

reacting with the acid SO_3

$$SO_3 + O^{2-} \longrightarrow SO_4^{2-}. \tag{28}$$

The salt $CaSO_4$ is again the indirect product of an acid–base reaction.

Hydroxides   Although a large number of metal compounds can be obtained which apparently contain hydroxide ions, this ion is present as a discrete species only in a few cases. All of the alkali-metal hydroxides, MOH, (except lithium), contain OH^- ions which appear to be spherically symmetrical, with an effective radius of 1.3–1.8 Å. Lithium hydroxide can be dehydrated to give the corresponding oxide

$$2LiOH \longrightarrow Li_2O + H_2O, \tag{29}$$

in contrast to the hydroxides of other elements of Group I which dissociate into the elements when heated. In all other (nonalkali) metal hydroxides, the hydroxide group is involved in hydrogen-bonded interactions (**9**) or as a

H
O
M   M

OH···O

9   **10**

bridging group through the oxygen atom (**10**). Thus LiOH crystallizes in the layer structure shown in Fig. 13.3, in which each lithium ion is tetrahedrally

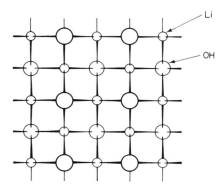

FIG. 13.3 Lithium hydroxide crystallizes in a layer structure, one layer of which is shown here. Each lithium ion is surrounded by four tetrahedrally arranged OH groups. The layers are held together by hydrogen bonds.

surrounded by bridging OH^- ions. Similar structures are observed for divalent hydroxides [$M(OH)_2$; M = Mg, Ca, Mn, Co, Ni, Fe, Cd]. Many hydroxides contain $M(OH)_4$ or $M(OH)_6$ units in which one or more OH groups are shared; hydrogen bonding also is often observed in such structures.

Compounds containing highly charged metal ions often undergo hydrolytic reactions giving covalently bonded species and an acidic solution

$$[M(OH_2)_x]^{n+} + H_2O \rightarrow [M(OH_2)_{x-1}(OH)]^{(n-1)+} + H_3O^+; \quad (30)$$

the metal-containing species are generally highly solvated. Typical examples of such behavior are found among aluminum and ferric salts. Hydroxy-bridged species are often found in partially hydrolyzed solutions of transition-metal ions. For example, at pH 5 a solution of ferric ion yields a precipitate of "ferric hydroxide," which is best described as a hydrated ferric oxide, $Fe_2O_3 \cdot xH_2O$. In very acid solution ($pH < 0$) the iron-containing species is $[Fe(OH_2)_6]^{3+}$, but as the solution becomes more basic ($pH \sim 1$) the co-ordinated water molecules become noticeably acidic, yielding $[Fe(OH_2)_5-(OH)]^{2+}$. In even more basic solutions additional protons are lost, yielding the hydroxy-bridged species (11); the process continues to form higher polymeric units until at $pH \sim 4$ a colloidal suspension is formed.

$$
\begin{array}{c}
\text{H} \\
\text{O} \\
(H_2O)_4Fe \diagup\diagdown Fe(OH_2)_4 \\
\diagdown\diagup \\
\text{O} \\
\text{H}
\end{array}
$$

11

Superoxides (IV)   When the alkali metals potassium, rubidium, or cesium react with excess oxygen, colored crystalline compounds with the formula MO_2 are formed.

$$M + O_2 \rightarrow MO_2. \quad (31)$$

The reaction can be carried out with the pure alkali metal or with the metal–ammonia solutions described in Chapter 8. The lithium derivative has not been prepared in crystalline form, but it is believed to exist in liquid ammonia solutions of lithium metal that have been treated with excess oxygen. Alkaline-earth metal superoxides cannot be prepared by these techniques. However, tetraalkylammonium superoxides were prepared by passing oxygen through the blue solution formed at the cathode during the electrolysis of the corresponding tetraalkylammonium salts (*23*). Tetramethylammonium superoxide has been obtained as a yellow solid (mp 97°) from this reaction. Sodium superoxide has been prepared by heating sodium peroxide with oxygen under pressure

$$Na_2O_2 + O_2 \rightarrow 2NaO_2. \tag{32}$$

As would be expected from their electronic distribution (shown in Fig. 13.1), the superoxides are paramagnetic, strong oxidizing agents, and they react with water to give oxygen and hydroperoxide ion

$$2O_2^- + H_2O \rightarrow O_2 + HO_2^- + OH^-. \tag{33}$$

At ordinary temperatures, KO_2, RbO_2, and CsO_2 are isostructural with CaC_2 (Fig. 2.5), crystallizing in a distorted structure because the O_2^- ions can rotate in the crystal lattice.

Peroxides *(IV, V)*   Ionic peroxides are known only for the alkali metals (M_2O_2) and certain divalent ions (MO_2; $M = Ca, Ba, Sr, Zn, Cd$). Compounds such as PbO_2 and MnO_2 are normal oxides of tetravalent ions rather than peroxides of divalent ions. Care must be exercised in interpreting the reactions of oxygen compounds of certain cations, such as those of polyvalent metals, because both the cation and the peroxide ion are good oxidizing agents.

The ionic peroxides can be formed by heating the active metal in air. For example, heating sodium in air gives the normal oxide as the first product, which yields the peroxide Na_2O_2 on further heating. Peroxides dissolve in water to yield elemental oxygen

$$O_2^{2-} + H_2O \rightarrow HO_2^- + OH^- \tag{34}$$

$$2HO_2^- \rightarrow 2OH^- + O_2 \tag{35}$$

Acidifying cold aqueous solutions of metal peroxides gives a solution of hydrogen peroxide

$$O_2^{2-} + 2H^+ \rightarrow H_2O_2. \tag{36}$$

Solid peroxyhydrates of the alkaline-earth metals ($MO_2 \cdot 8H_2O$) can be prepared by treating salts of these elements in aqueous solution with hydrogen peroxide or with Na_2O_2.

The peroxides are good oxidizing agents towards a variety of substances. For example, fused sodium peroxide is used to decompose insoluble silicates and oxides, to convert carbon-containing compounds to carbonates, and to oxidize metals. Towards suitable oxidizing agents such as permanganate ion and oxygen, the peroxide ion can act as a reducing agent.

Compounds containing covalently bound peroxy groups are discussed in a subsequent section.

Ozonides *(IV)*   When dry potassium hydroxide is treated with a stream of dry ozone, reddish-brown crystals containing $\sim 90\%$ KO_3 are formed. The mixture can be purified by extraction with liquid ammonia in which KO_3 is soluble. The pure substance is unstable at room temperature, decomposing to KO_2 and elemental oxygen. Ammonium ozonide has also been reported (*24*). KO_3 exhibits a paramagnetism very near that expected for one electron. The ion appears to be bent ($\sim 100°$) with oxygen–oxygen distances of about 1.2 Å (*25*).

3.2 / Sulfur, Selenium, and Tellurium

The remaining elements in Group VI, often called the chalcogens, are less electronegative than oxygen, as shown by the data in Table 13.1, but they form compounds with nearly as many elements as does oxygen. Such binary compounds can be classified in the same general terms as the oxides. That is, the binary compounds form ionic lattices with very electropositive metals and giant molecules with the more electronegative metals. Since the chalcogens are less electronegative than oxygen, there is less ionic character in any bond formed by these elements than in the corresponding oxide. The nonmetals yield distinct molecular species which may be monomeric (CM_2; M = S, Se, Te) or polymeric (P_4S_{10}). The metal chalconides do not, in general, exhibit the formulas or properties of the corresponding oxides. For example, iron forms three oxides (FeO, Fe_2O_3, and Fe_3O_4) and two sulfides (FeS and FeS_2), only one of the sulfides having the same formula as an oxide.

Ionic Derivatives   The metal chalconides can be prepared by direct reaction with the metal. In general, reaction occurs readily when the mixture is heated, although some reactions, e.g., that between mercury and sulfur, occur at room temperature. The known metal chalconides are isostructural with the corresponding oxides, as indicated by the data in Table 13.5. All of the alkaline-earth chalconides except those of beryllium, and magnesium telluride, crystallize in the NaCl-type structure (Table 13.5).

The ionic derivatives are soluble in water, giving alkaline solutions which arise from hydrolytic reactions. For example, sulfide ions are extensively hydrolyzed

$$S^{2-} + H_2O \rightarrow SH^- + OH^- \tag{37}$$

because H_2S is a very weak acid. We should also expect the other alkali-metal chalconides to react in this way, since the corresponding hydrides are weaker acids than is H_2S.

TABLE 13.5   Crystal structures[a] of the Group II oxides and chalconides

Group element	O	S	Se	Te
Be	W	Z	Z	Z
Mg	NaCl	NaCl	NaCl	W
Ca	NaCl	NaCl	NaCl	NaCl
Sr	NaCl	NaCl	NaCl	NaCl
Ba	NaCl	NaCl	NaCl	NaCl

[a] W = wurtzite; Z = zinc blende.

Polysulfides can be prepared by dissolving sulfur in aqueous solutions of alkali or alkaline-earth metal sulfides, and by direct reaction of the elements or by reaction of metal–ammonia solutions with sulfur. The latter method has also been used to prepare polyselenides and polytellurides of the alkali metals. Alkali-metal polychalconides containing anions with the composition M_x^{2-} (M = S, Se, Te; $x = 2, 3, 4, 5, 6, 7$) have been prepared, as well as similar compounds of unspecified composition (*VII*). Although structural relationships have not been extensively developed in these species, the structures shown in Fig. 13.4 have been reported. Discrete S_2 units are known to exist in the two forms of FeS_2, pyrites and marcasite. The pyrites structure is closely related to the NaCl structure with the centers of the S_2 groups lying at the chloride positions in this structure. The structure of FeS_2 is not fully ionic, the S—S bond distance in FeS_2 being longer than expected (2.08 Å) for a single bond. A similar bond distance is observed in BaS_3 which contains a bent S_3^{2-} ion. Higher polysulfides also contain chains of sulfur atoms, and the S—S distances in these are more nearly that expected for single-bond distances. In the case of the S_6^{2-} ion one of the distances is markedly shorter than the expected single-bond distance, suggesting that multiple bonding might be a factor in the structure.

Continuous Structures   As shown by the data in Table 13.5, many elements form compounds with sulfur that are best described as continuous lattices composed of atoms bound by more or less covalent bonds. Some of the sulfides resemble alloys in their physical properties in the sense that they are electrical conductors and exhibit a metallic luster and a high reflectivity. Their composition often does not correspond to the usual valence of the metals, e.g., Co_9S_8 and Co_9S_4. The selenides and tellurides of some transition elements resemble alloys even more than the sulfides. Some

FeS_2 S—S $d(S—S) = 2.14$ Å

BaS_3

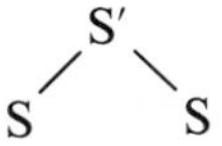

$$d(S'—S) = 2.15 \text{ Å}$$
$$\angle(SS'S) = 103°$$

$BaS_4 \cdot H_2O$

$$d(S^1—S) = 2.03 \text{ Å}$$
$$d(S—S^2) = 2.03 \text{ Å}$$
$$d(S^1—S^2) = 2.07 \text{ Å}$$
$$\angle(SS^1S^2) = \angle(S^1S^2S) = 104.5°$$
dihedral angle = 75.5°

Ca_2S_6

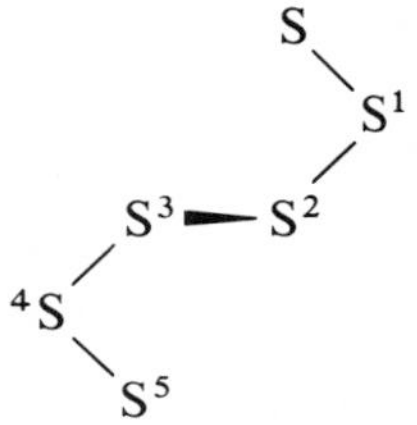

$$d(S—S^1) = 1.99 \text{ Å}$$
$$d(S^2—S^1) = d(S^3—S^4) = 2.11 \text{ Å}$$
$$d(S^2—S^3) = d(S^4—S^5) = 2.03 \text{ Å}$$
$$\angle(SS^1S^2) = \angle(S^2S^3S^4) = \angle(S^3S^4S^5) = 109.6°$$
$$\angle(S^1S^2S^3) = 106.4°$$

FIG. 13.4 Structural parameters for some polysulfides.

transition metals, e.g., chromium and iron, form a range of nonstoichiometric compounds in which random vacancies appear in the basic structural unit. Thus, FeS is rarely obtained in the ratio of 1 : 1, but phases with empirical formulas in the range Fe_6S_7 to $Fe_{11}S_{12}$ have been prepared. Similarly, the chromium–sulfur system contains compounds with formulas CrS, Cr_7S_8, Cr_5S_6, Cr_3S_4, and two forms of Cr_2S_3.

3.3 / Oxygen Derivatives of Sulfur *(VIII)*

Oxygen reacts readily with the other elements in this family to form two principal oxides, MO_2 and MO_3; only the oxide PoO_2 is known for polonium. As might be expected the acidity of the oxides decreases with increasing atomic number, the sulfur oxides being the most acidic. The elements in this family are sufficiently nonmetallic in their characteristics that only polonium shows vestiges of metallic behavior in its oxide.

In addition to the two principal oxides, sulfur forms two other oxides, S_2O and SO_4. Disulfur monoxide, S_2O, can be prepared by the action of an electrical discharge on a low pressure mixture of SO_2 and sulfur or on SO_2 alone. Initially it was believed that this reaction produced sulfur monoxide, SO *(II)*, but subsequent work showed that the reported sulfur monoxide was a mixture of S_2O and SO_2. Disulfur monoxide is reasonably stable in the gaseous state at low pressures, but attempts to condense it lead to a polymeric substance. It reacts with metals to form sulfides, and with halogens to give thionyl halides (X_2SO; $X = Cl$, Br). Disulfur monoxide reacts with water to give sulfurous acid and hydrogen sulfide.

The results of a microwave spectroscopic investigation, shown in Fig. 13.5, indicate that S_2O is a bent molecule. The observed bond distances are very

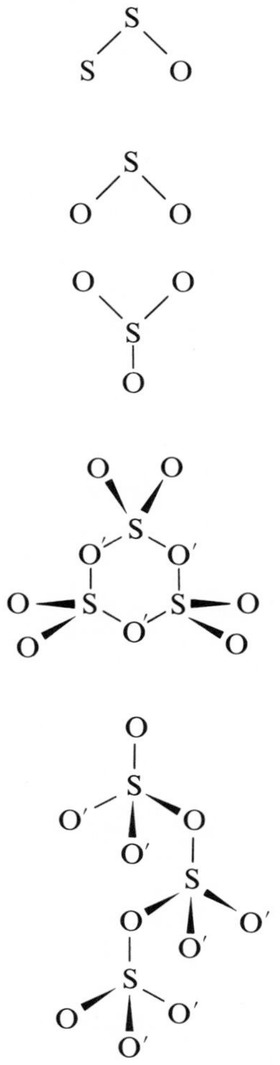

S_2O

$d(S-S) = 1.884$ Å
$d(S-O) = 1.465$ Å
$\angle(SSO) = 118°$

SO_2

$d(S-O) = 1.43$ Å
$\angle(OSO) = 119.5°$

SO_3

$d(S-O) = 1.43$ Å
$\angle(OSO) = 120°$

α-SO_3
(ice-like)

$d(S-O') = 1.60$ Å
$d(S-O) = 1.40$ Å
$\angle(SO'S) = 109°$
$\angle(OSO) = 122°$
$\angle(O'SO') = 100°$

β-SO_3
(asbestos)

$d(O-S) = 1.59$ Å
$d(O'-S) = 1.41$ Å
$\angle(O'SO') = 128°$
$\angle(O'SO) = 107°$
$\angle(SOS) = 121°$

FIG. 13.5 Structural parameters for some sulfur oxides.

close to those expected for S—S (1.88 Å) and S—O (1.49 Å) double bonds. Thus the molecule is best formulated with the central sulfur atom incorporating d orbitals in the bonding scheme (**12**).

12

Sulfur tetroxide, SO_4, can be prepared by subjecting a low-pressure mixture of oxygen and SO_2 to an electric discharge. The product melts at 3° with evolution of oxygen, leaving a liquid residue of composition S_2O_7. Sulfur tetroxide dissolves in concentrated H_2SO_4 in which it exhibits a molecular weight corresponding to SO_4.

An aqueous solution of SO_4 is a strong oxidizing agent. There is little evidence concerning the constitution of SO_4 or of its aqueous solutions.

Oxides   *Sulfur dioxide* is a colorless gas (mp $-75.5°$, bp 10.02°) which is easily condensed to a liquid (critical point, 157.2°). It can be prepared in a variety of ways; the most common include burning sulfur or a metal sulfide in air

$$S + O_2 \rightarrow SO_2 \tag{38}$$

$$CuS + \tfrac{3}{2}O_2 \rightarrow SO_2 + CuO, \tag{39}$$

the action of an acid on a metal sulfite

$$M_2SO_3 + 2H^+ \rightarrow 2M^+ + H_2O + SO_2, \tag{40}$$

or the action of sulfuric acid on a metal below hydrogen in the electromotive series

$$Cu + 2H_2SO_4 \rightarrow CuSO_4 + SO_2 + 2H_2O. \tag{41}$$

Sulfur dioxide in many of its reactions behaves as a reducing agent. Thus, permanganate, dichromate, iodate, the halogens, Fe^{3+}, and Hg_2^{2+} oxidize SO_2 to SO_4^{2-}

$$SO_2 + 2H_2O \rightarrow SO_4^{2-} + 4H^+ + 2e^- \tag{42}$$

in either dilute acid or neutral solution. In certain cases, SO_2 can act as an oxidizing agent. For example, metals such as tin, iron, and magnesium burn in SO_2 to form mixed sulfides and oxides. In very strongly acidic aqueous solution SO_2 can oxidize Fe^{2+}, Sn^{2+}, Cu^+, or Hg_2^{2+}, the reduction product being sulfur or a sulfide

$$4FeCl_2 + SO_2 + 4HCl \rightarrow 4FeCl_3 + 2H_2O + S \tag{43}$$

$$6SnCl_2 + 2SO_2 + 8HCl \rightarrow 5SnCl_4 + 4H_2O + SnS_2. \tag{44}$$

The structure of the SO_2 molecule, shown in Fig. 13.5, has been determined in the crystalline and in the vapor states. Sulfur dioxide is a bent molecule with a bond angle close to that observed for S_2O (Fig. 13.5). The S—O bond distance is shorter than that expected for a single bond (1.78 Å), suggesting that the canonical forms **13** and **14** are the major contributors to the wave function for the molecule. Since sulfur has low-lying $3d$ orbitals available,

13   **14**

it is possible that there is an additional contribution to the multiple bonding which incorporates a $d\pi$–$p\pi$ component. This possibility is particularly tempting since the SO bond distance in SO_2 is shorter than that in S_2O, as indicated by the data in Fig. 13.5. The ultraviolet spectrum of gaseous SO_2 at high temperatures suggests that a second species is present (26) which can either be SOO or OSO with an altered geometry. Attempts to trap this species in a matrix give a product which is normal SO_2 or a dimer, $(SO_2)_2$ (27).

Liquid SO_2 has been systematically studied as a reaction medium (IX). The low dielectric constant for SO_2 (15.4 at 0°C) suggests that it would not be a good solvent for most ionic substances which have high lattice energies. However, its dipole moment (1.62 D) and rather high polarizability (4.33 × 10^{-24} cm^3) indicate that it might be a good solvent for polar covalent molecules which have easily polarizable functional groups. Both predictions are borne out in practice. Liquid SO_2 is a poor solvent for most ionic compounds except iodides or bromides; tetraalkylammonium salts are very soluble. A wide variety of covalent compounds are soluble in SO_2. For example, compounds such as CCl_4, $SiCl_4$, $SnCl_4$, $TiCl_4$, CS_2, $(CH_3)_2O$, C_6H_6, C_6H_5Cl, alcohols, esters, ketones, and aldehydes have been described as either very soluble or soluble in liquid SO_2.

Liquid SO_2 is a very poor conductor of electricity (8 × 10^{-8} Ω^{-1} cm^{-1}); in this respect it is similar to water (6 × 10^{-8} Ω^{-1} cm^{-1}). Early workers attributed the conductivity to autoionization of the solvent

$$2SO_2 \rightleftharpoons SO^{2+} + SO_3^{2-} \tag{45}$$

drawing the parallel to the process responsible for the conductivity of pure water. In this aprotic analogy to the autoionization of water, SO^{2+} and SO_3^{2-} ions were assumed to play the roles of the acidic and basic species, respectively. Substances that increase the concentration of SO^{2+} are acids, while metal sulfites (M_2SO_3) are bases. Indeed, there are many reactions reported which fit this pattern. For example, the reaction between $SOCl_2$ and Cs_2SO_3 can be followed conductometrically in liquid SO_2, a break in the curve occurring at a 1 : 1 mole ratio of reactants. The reaction has been formulated as

$$SOCl_2 + Cs_2SO_3 \rightarrow 2CsCl + 2SO_2 \qquad (46)$$

where it was supposed that $SOCl_2$ would undergo ionization

$$SOCl_2 \rightleftharpoons SO^{2+} + 2Cl^-. \qquad (47)$$

The acidic species formed in this process would react with the base SO_3^{2-}

$$SO^{2+} + SO_3^{2-} \rightarrow 2SO_3 \qquad (48)$$

by transfer of an oxide ion. Experiments with labeled sulfur (*28, 29*) and with labeled oxygen (*30*) in the SO_2–$SOCl_2$ and SO_2–$SOBr_2$ systems indicate that neither the oxygen nor the sulfur atoms exchange between the solute and solvent. Thus, many of the reactions that have been interpreted as acid–base reactions probably would occur in any solvent in which these compounds are soluble. There is apparently no need to invoke solvent involvement in such reactions, except, perhaps, in terms of a general solvation phenomena. Evidence exists that SO_2 does not form solvates, but promotes the ionization of some covalent species such as SO_2X_2 (X = Cl, Br), X_2 (X = Br, Cl), IBr, ICl, and $(C_6H_5)_3CX$ (X = Cl, Br). It appears likely that many of the earlier observations which were interpreted in terms of solvolysis reactions actually arise because of the presence of either molecular oxygen, water, or both substances (*31*). Sulfur dioxide forms numerous solvates in which it acts as an electron-pair acceptor ($R_3N{\cdot}SO_2$), but SO_2 can also act as a Lewis base (e.g., $BF_3{\cdot}SO_2$, $SbF_5{\cdot}SO_2$, $2TiCl_4{\cdot}SO_2$, $2SnBr_4{\cdot}SO_2$, and $Al_2Cl_6{\cdot}SO_2$).

Sulfur trioxide can be prepared in a variety of ways, the easiest being the direct oxidation of sulfur or SO_2. The reaction of SO_2 with atomic oxygen to form SO_3 is rapid, but the reaction with molecular oxygen is exceedingly slow. Numerous catalysts (e.g., V_2O_5, NO, and Pt) can be used to increase the rate of reaction between SO_2 and oxygen. At least three polymorphic forms of SO_3 exist with the structures shown in Fig. 13.5. The α, or ice-like form (mp 16.8°), consists of rings of S_3O_9 molecules made up of SO_4 tetrahedra, each sharing an oxygen atom. The β, or asbestos-like form (mp 32.5°), consists of infinite chains of SO_4 tetrahedra also sharing oxygen atoms. The

γ form of SO_3 (mp 62.6°) has a colloidal appearance, but no structural information is available. Sulfur trioxide in the gaseous phase is monomeric; structural evidence indicates that the molecule is planar with an S—O bond distance of 1.43 Å (Fig. 13.5). The information available on the SO_3 molecule strongly suggests that **15**, **16**, and **17** are the main canonical forms contributing to the wave function, but the very short S—O distance leads to the conclusion that an added $d\pi$–$p\pi$ component is present.

$$\qquad\qquad \mathbf{15} \qquad\qquad\qquad \mathbf{16} \qquad\qquad\qquad \mathbf{17}$$

For all practical purposes the reactions of the various polymorphs of SO_3 are the same, although a few instances exist where one form reacts more readily than the other. For example, the α form is more rapidly hydrated to give sulfuric acid than is the β form. Sulfur trioxide is a good oxidizing agent; above 100° it reacts with sulfur, phosphorus, iron, and zinc. It is also a strong Lewis acid, forming 1 : 1 adducts with a variety of amines.

Sulfurous Acid and its Derivatives   *Sulfurous acid* has not been isolated as a pure substance. Sulfur dioxide is readily soluble in water and the solution behaves as if a dibasic acid is present. For example, aqueous solutions liberate hydrogen from metallic magnesium and can be titrated to two end points ($K_1 = 1.3 \times 10^{-2}$, $K_2 = 1 \times 10^{-7}$). Neutralization of aqueous solutions of SO_2 leads to the normal salts (M_2SO_3) and acid salts ($MHSO_3$) which contain the bisulfite anion. Sulfites can be oxidized by a variety of oxidizing agents to sulfates

$$SO_3{}^{2-} + H_2O \rightarrow SO_4{}^{2-} + 2H^+ + 2e^-. \tag{49}$$

Dry sulfites undergo a disproportionation reaction when heated

$$4M_2SO_3 \rightarrow M_2S + 3M_2SO_4. \tag{50}$$

The sulfite ion is a pyramidal species with S—O distances (1.39 Å) significantly shorter than the bonds in either of the oxides (Fig. 13.5). The pyramidal structure **18** would be predicted on the basis of electron-pair-

$$\mathbf{18}$$

SO_3^{2-}

$$d(S—O) = 1.39 \text{ Å}$$
$$\angle(OSO) = 107.4°$$

$S_2O_3^{2-}$

$$d(S—O) = 1.48 \text{ Å}$$
$$d(S—S) = 1.97 \text{ Å}$$
$$\angle(SSO) = \angle(OSO) = 109.5°$$

$S_2O_4^{2-}$

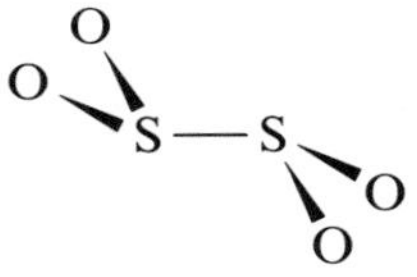

$$d(S—S) = 2.38 \text{ Å}$$
$$d(S—O) = 1.50 \text{ Å}$$
$$\angle(OSO) = 108°$$
$$\angle(SSO) = 98°$$

$S_2O_5^{2-}$

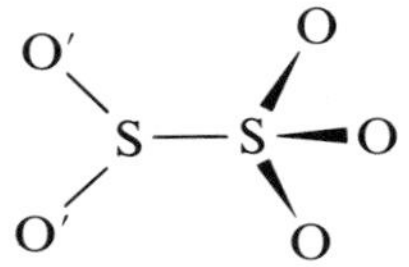

$$d(S—S) = 2.21 \text{ Å}$$
$$d(S—O') = 1.50 \text{ Å}$$
$$d(S—O) = 1.44 \text{ Å}$$

FIG. 13.6 Structural parameters for some sulfur-containing oxyanions.

repulsion arguments, which also lead to three of the canonical forms necessary for the best wave function for the molecule. The very short bond distances in SO_3 indicate that an additional π component can be expected from a $d\pi–p\pi$ interaction.

Sulfites react with various reagents to form anions containing S—S bonds

$$Na_2SO_3 + S \rightarrow Na_2S_2O_3 \tag{51}$$

$$2NaHSO_3 + SO_2 + Zn \rightarrow ZnSO_3 + Na_2S_2O_4 + H_2O \tag{52}$$

$$Na_2SO_3 + SO_2 \rightarrow Na_2S_2O_5; \tag{53}$$

the corresponding acids are not known. The reaction of sulfite ion with elemental sulfur [Eq. (51)] gives the thiosulfate ion ($S_2O_3^{2-}$), which is isostructural with the sulfate ion; the pertinent structural data are shown in Fig. 13.6. The S—O distances in $S_2O_3^{2-}$ are larger than those in SO_3^{2-} or the sulfur oxides but shorter than the expected single-bond distance. In addition, the S—S distance is also shorter than twice the single-bond radius (2.08 Å). Thus, the theoretical description of the $S_2O_3^{2-}$ ion still has many of the factors associated with sulfur species discussed previously.

The reduction of SO_3^{2-} with an active metal [Eq. (52)] leads to the dithionite ion ($S_2O_4^{2-}$), which itself is a very good reducing agent. For example, ions of the less reactive metals such as copper and mercury are readily reduced to the element by $S_2O_4^{2-}$, the oxidation product being SO_3^{2-}. Solutions containing $S_2O_4^{2-}$ ions are used in gas analysis to absorb molecular oxygen. This ion, the structure of which is shown in Fig. 13.6, contains an S—S bond considerably longer than that expected for a single bond, but the S—O distances are near that expected for a double bond formed between these atoms. The geometry of the ion suggests that a lone pair of electrons exists on each sulfur atom. The oxygen atoms in this structure are eclipsed, which is an unusual orientation.

A condensed anion, the pyrosulfate ion ($S_2O_5^{2-}$), is formed when a concentrated solution of NaOH is saturated with SO_2 and the mixture evaporated in an atmosphere of SO_2 [Eq. (53)]. The pyrosulfate ion, the structure of which is shown in Fig. 13.6, contains a long S—S bond and S—O bonds with appreciable amounts of double-bond character.

Derivatives of Sulfurous Acid. Sulfurous acid molecules can be imagined as the precursors of a variety of derivatives in which one or both hydroxyl moieties are replaced by other groups. In certain cases these compounds are chemically interrelated but in other instances the relationship is a formal structural one.

Sulfite esters, $(RO)_2SO$, are prepared by treating thionyl halides with the appropriate alcohol

$$2ROH + Cl_2SO \rightarrow (RO)_2SO + 2HCl. \tag{54}$$

They are useful starting materials for the preparation of sulfate esters, the conversion being effected by oxidizing agents such as $KMnO_4$

$$(RO)_2SO \xrightarrow{[O]} (RO)_2SO_2. \tag{55}$$

This route to sulfate esters is preferred because the direct reaction between sulfuric acid and an alcohol often leads to dehydration products.

All the thionyl halides (X_2SO) except the iodide have been prepared. The mixed halide ClFSO (mp $-139.5°$, bp $12.2°$) has also been prepared. Thionyl fluoride is a colorless gas (mp $-110°$, bp $-44.8°$), whereas the other halides are low-melting liquids (Cl_2SO, mp $-99.5°$, bp $77°$; Br_2SO, mp $-49.5°$, bp $138°$, dec.). All of the other thionyl halides are prepared from Cl_2SO, which itself is readily made by the action of PCl_5 on SO_2

$$PCl_5 + SO_2 \rightarrow SOCl_2 + POCl_3 \tag{56}$$

or the reaction of SO_3 with S_2Cl_2 in the presence of chlorine

$$SO_3 + S_2Cl_2 + 2Cl_2 \rightarrow 3SOCl_2. \tag{57}$$

The latter reaction is carried out in fuming sulfuric acid. Thionyl fluoride is prepared by fluorinating Cl_2SO with SbF_3

$$2SbF_3 + 3SOCl_2 \longrightarrow 3F_2SO + 2SbCl_3. \tag{58}$$

The action of gaseous HBr on thionyl chloride gives thionyl bromide

$$SOCl_2 + 2HBr \longrightarrow SOBr_2 + 2HCl. \tag{59}$$

In general the thionyl halides are sensitive to moisture and react readily with species containing acidic hydrogen atoms to yield hydrogen halides and the corresponding derivatives of sulfurous acid

$$X_2SO + 2HR \longrightarrow R_2SO + 2HX. \tag{60}$$

Thionyl chloride has been used as an aprotic solvent in which the auto-ionization process

$$Cl_2SO \rightleftharpoons ClSO^+ + Cl^- \tag{61}$$

is thought to be established.

Sulfoxides (R_2SO) are prepared by mild oxidation of the corresponding sulfides

$$R_2S \xrightarrow{[O]} R_2SO. \tag{62}$$

Oxidizing agents such as bromine, nitric acid, H_2O_2, or N_2O_4 are useful for this purpose. This method of preparation involves the formation of a sulfonium compound,

$$R_2S + Br_2 \longrightarrow [R_2SBr]^+Br^- \tag{63}$$

which upon hydrolysis yields the sulfoxide

$$[R_2SBr]^+Br^- + H_2O \longrightarrow R_2SO + 2HBr. \tag{64}$$

The sulfoxides are very soluble in water, and the aqueous solutions exhibit weakly basic properties

$$R_2SO + H_2O \rightleftharpoons [R_2SOH]^+OH^-. \tag{65}$$

Although extensive structural data do not exist for the derivatives of sulfurous acid, the available data indicate that the sulfur atom is at the apex of a pyramid with an apical angle of about 105°. This result is consistent with the geometry expected from the electron-pair-repulsion arguments applied to these species.

Sulfuric Acid   At least ten crystalline phases exist in the phase diagram for the system H_2O–SO_3. Of these compounds the most familiar and important are sulfuric acid, $H_2O \cdot SO_3$ or H_2SO_4, and pyrosulfuric acid (sometimes called disulfuric acid), $H_2O \cdot 2SO_3$ or $H_2S_2O_7$. Sulfuric acid is a colorless oily substance (mp 10.36°, bp 320°) which boils with decomposition. It forms a constant-boiling mixture (330°) with water, containing 98.33% H_2SO_4. Sulfuric acid is prepared in vast quantities commercially by oxidizing SO_2 with atmospheric oxygen and dissolving the resultant SO_3 in 98% sulfuric acid; water is added to the $H_2SO_4 \cdot SO_3$ mixture to keep the concentration at 98%.

Sulfuric acid is a strong acid in aqueous solutions, the first dissociation being complete

$$H_2SO_4 + H_2O \longrightarrow H_3O^+ + HSO_4^-; \qquad (66)$$

the second hydrogen ion is less easily removed ($pK_a = 1.9$)

$$HSO_4^- + H_2O \rightleftharpoons H_3O^+ + SO_4^{2-}. \qquad (67)$$

Accordingly, two types of salts can be derived from sulfuric acid, the sulfates and the hydrogen sulfates (or bisulfates). Sulfuric acid is a vigorous dehydrating agent and, when hot, a moderately good oxidizing agent. Many organic compounds which contain the elements of water are dehydrated by sulfuric acid. For example, carbohydrates are charred when treated with concentrated sulfuric acid. Hot concentrated sulfuric acid attacks elements below hydrogen in the electromotive series, such as copper, lead, and mercury, the reduction product being SO_2. In dilute solution it will, of course, react with metals above hydrogen in the electromotive series to liberate hydrogen.

Sulfuric acid crystallizes in the layer structure shown in Fig. 13.7, in which essentially tetrahedral sulfate groups are joined by hydrogen bonds into a continuous sheet. Each molecule of sulfuric acid forms two hydrogen bonds and has two sites to which hydrogen bonds can be formed. The distortion of the sulfate moiety from a true tetrahedral structure is undoubtedly associated with the nature of hydrogen bonds formed in the crystal. The mono- (32) and dihydrates (33) of sulfuric acid are known to have a structure in which hydronium ions are hydrogen bonded to HSO_4^- and SO_4^{2-} ions, respectively. The structures of a variety of ionic sulfates show that the SO_4^{2-} ion is perfectly tetrahedral with S—O bond distances of 1.44 Å, which is shorter than the expected double-bond distance (1.49 Å), suggesting that an additional $d\pi$–$p\pi$ component is present in the bond.

Anhydrous sulfuric acid has been extensively studied as a solvent (X). It is a good solvent for many classes of compounds because of its high dielectric constant (100 at 25°), high polarity, and ability to form hydrogen bonds.

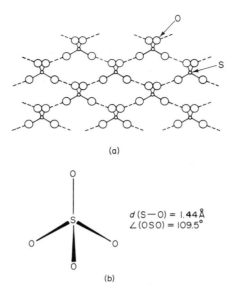

(a)

$d\,(S-O) = 1.44\,\text{Å}$
$\angle\,(OSO) = 109.5°$

(b)

FIG. 13.7   Sulfuric acid crystallizes in a puckered layer structure. Each tetrahedral SO_4 group is joined to four others by hydrogen bonds. The positions of the hydrogen atoms are not shown. The hydrogen bonds are indicated by dashed lines.

Acids and bases in the sulfuric-acid solvent system are defined, according to the usual solvent-system concepts

$$2H_2SO_4 \rightleftharpoons H_3SO_4^+ + HSO_4^-, \tag{68}$$

as substances that increase the concentration of the solvated proton or of the bisulfate ion, respectively.

$$HA + H_2SO_4 \rightleftharpoons H_3SO_4^+ + A^- \tag{69}$$

$$B + H_2SO_4 \rightarrow HB^+ + HSO_4^-. \tag{70}$$

Since sulfuric acid is inherently a highly acidic medium (*34*) the largest class of electrolytes in this solvent act as bases. Just as water exerts a leveling effect on the strengths of acids, sulfuric acid is a leveling solvent for bases. Of course, metal hydrogen sulfates (e.g., $KHSO_4$, $NaHSO_4$) act as bases in sulfuric acid, but a large number of organic compounds also exhibit basic behavior. Sulfuric acid will protonate alcohols, anhydrides, ketones, carboxylic acids, esters, phosphines, nitro-containing compounds, and nitriles as well as the more basic organic derivatives such as amines or amides. Water, nitric acid, and phosphoric acid are examples of inorganic compounds that are protonated in sulfuric acid and act as bases. Disulfuric acid ($H_2S_2O_7$), the higher polysulfuric acids, fluorosulfuric acid (HSO_3F), and complex acids of the type $H[B(HSO_4)_4]$ and $H_2[M(HSO_4)_6]$ ($M = Sn, Pb$) are sufficiently strong to act as acids in pure sulfuric acid [Eq. (69)]. The complex acids can be prepared by allowing the corresponding hydroxy compound to react with disulfuric acid

$$(HO)_3B + 3H_2S_2O_7 \rightarrow H_3SO_4^+ + B(HSO_4)_4^- + H_2SO_4. \quad (71)$$

In addition to the ions arising from the autoprotolysis reaction [Eq. (68)], pure sulfuric acid contains other species. Sulfuric acid undergoes slight dissociation into its components

$$H_2SO_4 \rightleftharpoons H_2O + SO_3 \quad (72)$$

which interact with more sulfuric acid to form disulfuric acid

$$SO_3 + H_2SO_4 \rightarrow H_2S_2O_7 \quad (73)$$

and products arising from the protonation of water

$$H_2SO_4 + H_2O \rightarrow H_3O^+ + HSO_4^-. \quad (74)$$

Disulfuric acid acts as an acid in anhydrous sulfuric acid

$$H_2S_2O_7 + H_2SO_4 \rightleftharpoons HS_2O_7^- + H_3SO_4^+. \quad (75)$$

It has been the custom to describe the species present in anhydrous sulfuric acid solution in terms of the autoprotolysis equilibrium [Eq. (68)], equilibria (74) and (75), and the self-dehydration reaction

$$2H_2SO_4 \rightleftharpoons H_3O^+ + HS_2O_7^-. \quad (76)$$

A detailed cryoscopic study of sulfuric acid solutions containing metal hydrogen sulfate, water, or disulfuric acid yielded the values of the equilibrium constants for these processes shown in Table 13.6.

TABLE 13.6   Values of the equilibrium constants for equilibria established in anhydrous sulfuric acid

Reaction	Constant	Value at 10°
Eq. (68)	$[H_3SO_4^+][HSO_4^-]$	1.7×10^{-4}
Eq. (76)	$[H_3O^+][HS_2O_7^-]$	3.5×10^{-5}
Eq. (75)	$\dfrac{[H_3SO_4^+][HS_2O_7^-]}{[H_2S_2O_7]}$	1.4×10^{-2}
Eq. (74)	$\dfrac{[H_3O^+][HSO_4^-]}{[H_2O]}$	1

A number of polysulfuric acids or their salts are known. In addition to H_2SO_4 and $H_2S_2O_7$, the phase diagram for the H_2O–SO_3 system indicates the existence of $H_2S_3O_{10}$ and $H_2S_4O_{13}$. Spectroscopic evidence is also available which supports the existence of such species. Solutions containing SO_3 dissolved in H_2SO_4 are called oleum, the equimolar mixture being predominantly $H_2S_2O_7$. Salts of disulfuric acid can be prepared by heating metal bisulfates

$$2MHSO_4 \rightarrow M_2S_2O_7 + H_2O. \tag{77}$$

Although little is known about the structure of these condensed acids, their anions have been isolated as suitable salts and investigated by x-ray analysis. The structural data given in Fig. 13.8 show that these anions consist of almost tetrahedral SO_4 units with bridging oxygen atoms (*35*).

$S_3O_{10}^{2-}$

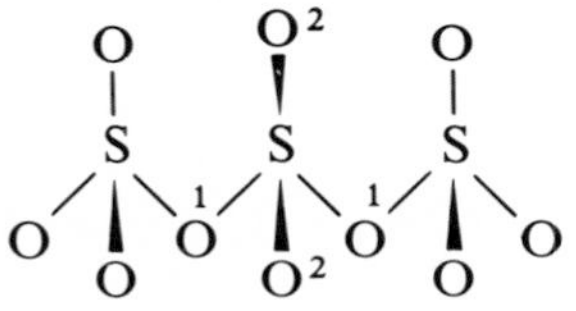

$d(S—O^1) = 1.55$ Å
$d(S—O^2) = 1.44$ Å
$d(S—O) = 1.50$ Å
$\angle(OSO) = 132°$
$\angle(SO^1S) = 122°$
$\angle(O^1SO^2) = 113°$

$HS_2O_7^-$

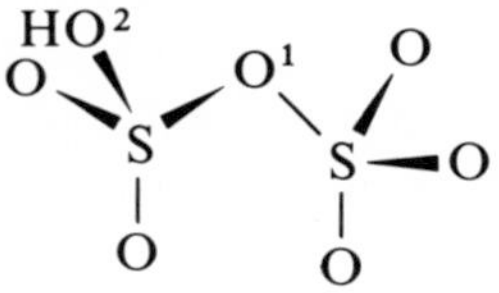

$d(S—O^2) = 1.64$ Å
$d(S—O) = 1.47$ Å
$d(S—O^1) = 1.60$ Å
$\angle(SO^1S) = 114°$
$\angle(O^1SO) = 110°$
$\angle(OSO) = 118°$

FIG. 13.8 The structural parameters for some condensed sulfate ions.

Peroxy Acids. Two derivatives of sulfuric acid are known which contain peroxy bonds (—O—O—). Peroxydisulfuric acid ($H_2S_2O_8$) is a white hygroscopic crystalline compound (mp 65°) which can be prepared by the electrolysis of cold aqueous sulfuric acid using lead electrodes and a very high current density. The action of anhydrous H_2O_2 on chlorosulfuric acid also produces peroxydisulfuric acid. X-Ray analysis of the ammonium or cesium salts of this acid shows that the anion consists of two tetrahedral SO_4 units joined through an oxygen–oxygen bond (as shown in Fig. 13.9) which is shorter than that found in hydrogen peroxide or ionic peroxides.

Peroxydisulfuric acid and its salts are strong oxidizing agents, as might be expected. Silver nitrate acts as a catalyst in promoting many oxidation reactions of peroxydisulfuric acid, such as the conversion of Cr^{3+} to $Cr_2O_7^{2-}$

$S_2O_8{}^{2-}$

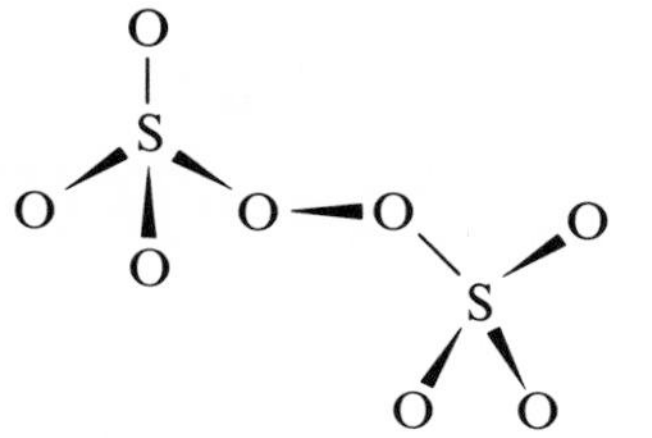

$d(O-O) = 1.31 \text{ Å}$
$d(S-O) = 1.50 \text{ Å}$
$\angle(SOO) = 128°$

FIG. 13.9 The structural parameters for the peroxydisulfate ion.

or Mn^{2+} to $MnO_4{}^-$. It has been suggested that the mechanism of such reactions involves the formation of Ag^{3+} which acts as the oxidant.

Peroxydisulfuric acid undergoes stepwise hydrolysis to form peroxysulfuric acid and hydrogen peroxide

$$HO_3SOOSO_3H + H_2O \rightarrow HO_3SOOH + H_2SO_4 \tag{78}$$

$$HO_3SOOH + H_2O \rightarrow H_2SO_4 + H_2O_2. \tag{79}$$

Peroxysulfuric acid (sometimes called Caro's acid) can be isolated as white hygroscopic crystals (mp 45°). It is less stable than peroxydisulfuric acid and is a good oxidizing agent. In contrast to the situation with peroxydisulfuric acid, no salts are known of peroxysulfuric acid.

Derivatives of Sulfuric Acid. Several classes of compounds exist, as shown in Fig. 13.10, which could be considered as derivatives of sulfuric acid in which one or two hydroxyl groups have been replaced by other groups. Perhaps the simplest derivatives of sulfuric acid are organic diesters of the type $(RO)_2SO_2$. Sulfonic acids (RSO_3H) are prepared by treating aromatic hydrocarbons with oleum

$$ArH + H_2SO_4 \rightarrow ArSO_3H + H_2O \tag{80}$$

or by oxidizing mercaptans

$$RSH \xrightarrow{[O]} RSO_3H. \tag{81}$$

These substances are monobasic acids and form metal salts $(RSO_3{}^-M^+)$ or esters (RSO_2OR'). Sulfonic acids can be converted into the corresponding acid chlorides by the action of PCl_5.

Acid halides of sulfuric acid (X_2SO_2) can be prepared by the direct interaction of SO_2 and the halogen

$$X_2 + SO_2 \rightarrow X_2SO_2 \tag{82}$$

$$X = F, Cl.$$

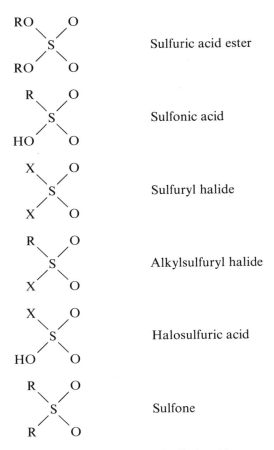

FIG. 13.10   Derivatives of sulfuric acid.

Sulfuryl fluoride is a remarkably inert colorless gas. It is sparingly soluble in water and is not hydrolyzed at 150°. The compound is, however, attacked by alcoholic solutions of KOH. Sulfuryl fluoride can be heated with oxygen, sulfur, or chlorine at red heat with no change. Sulfuryl chloride, a colorless liquid, is generally more reactive than the fluoride. It is a good solvent, dissolving SO_2, I_2, As, and a variety of metal halides. A mixed sulfuryl chlorofluoride exists as a gaseous substance.

Sulfones have the general formulation R_2SO_2. They are generally stable colorless solids which can be made by the careful oxidation of dialkyl sulfides

$$R_2S \xrightarrow{[O]} R_2SO_2 \qquad (83)$$

or by heating sulfuryl chloride with aromatic hydrocarbons under Friedel–

Crafts conditions

$$2ArH + Cl_2SO_2 \xrightarrow{AlCl_3} Ar_2SO_2 + 2HCl. \tag{84}$$

Halosulfonic acids, HSO_3X ($X = Cl, F$), are colorless liquids which can be prepared from the action of HF (in the form of KHF_2) or HCl on SO_3 dissolved in H_2SO_4. They are strong acids in their own right but hydrolyze readily.

$$HSO_3X + H_2O \rightarrow HX + H_2SO_4. \tag{85}$$

Fluorosulfonate salts are stable, in contrast to the corresponding chlorosulfonates. Alkali-metal fluorosulfonates can be prepared by the action of the alkali-metal fluoride on SO_3

$$MF + SO_3 \rightarrow MSO_3F. \tag{86}$$

The fluorosulfonates are only slowly hydrolyzed by water, so they can be crystallized from this solvent.

Thiosulfuric Acids (VII). A variety of sulfuric acids exist in which an oxygen atom has been replaced by sulfur. The simplest of these compounds is thiosulfuric acid, $(HO)_2S_2O$, which has never been isolated. The anion of this acid, the structure of which is given in Fig. 13.6, is well characterized and does indeed contain an S—S bond. Thiosulfates are readily oxidized by oxygen or iodine. With oxygen free sulfur is formed; reaction with iodine yields a polysulfide.

A series of polythionic acids can be formed which have the general formulation $(HO)_2S_nO_4$ where $n = 2, 3, 4, 5, 6$ (Fig. 13.11). The first member of the series, dithionic acid, is discussed in a previous section. Trithionic acid, $(HO)SO_2SSO_2(OH)$, is formally analogous to disulfuric acid. The other three members of the family contain successively four-, five, and six-membered sulfur chains. Formalistically, tetrathionic acid appears to be the sulfur analog of peroxydisulfuric acid.

Little is known of the free thionic acids, but numerous salts have been synthesized and characterized. Dithionates are obtained by treating acidic SO_2 solutions with mild oxidizing agents such as MnO_2

$$2SO_3{}^{2-} \rightarrow [O_3SSO_3]^{2-} + 2e^-. \tag{87}$$

Solutions of the higher polythionic acids can be prepared from thiosulfates by the action of the appropriate oxidizing agents:

$$2Na_2S_2O_3 + 3SO_2 \rightarrow 2Na_2S_3O_6 + S \tag{88}$$

$$2Na_2S_2O_3 + 4H_2O_2 \rightarrow Na_2S_3O_6 + Na_2SO_4 + 4H_2O \tag{89}$$

Dithionic acid
$(HO)SO_2SO_2(OH)$

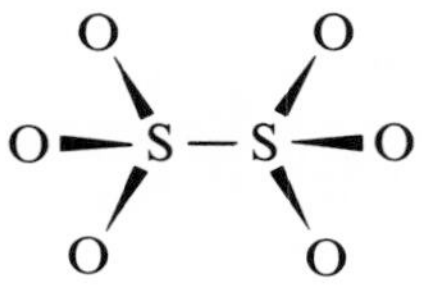

$d(S—O) = 1.50$ Å
$d(S—S) = 2.08$ Å

Trithionic acid
$(HO)SO_2SSO_2(OH)$

—

—

Tetrathionic acid
$(HO)SO_2S_2SO_2(OH)$

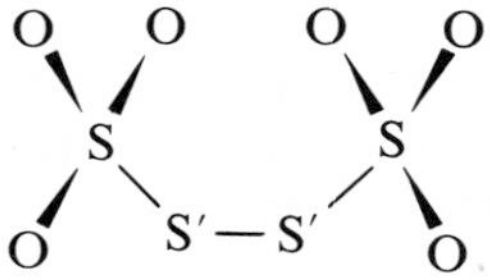

$d(S'—S') = 2.02$ Å
$d(S—O) = 1.41$ Å
$d(S'—S) = 2.13$ Å
$\angle(OSS') = 108°$
$\angle(SS'S') = 103°$
$\angle(OSO) = 111°$

Pentathionic acid
$(HO)SO_2S_3SO_2(OH)$

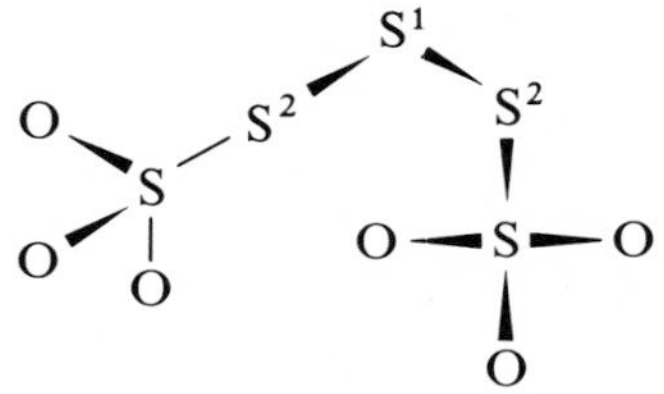

$d(S^1—S^2) = 2.04$ Å
$d(S^2—S) = 2.14$ Å
$d(S—O) = 1.43$ Å
$\angle(OSS^2) = 105°$
$\angle(SS^2S^1) = 103°$
$\angle(S^2S^1S^2) = 106°$

Hexathionic acid
$(HO)SO_2S_4SO_2(OH)$

—

—

FIG. 13.11   Structural parameters of some salts of thionic acids.

$$2Na_2S_2O_3 + I_2 \rightarrow Na_2S_4O_6 + 2NaI. \tag{90}$$

The potassium salt of pentathionic acid can be obtained by passing H_2S into aqueous SO_2 until all of the latter substance has reacted. Addition of KOH to the mixture, followed by slow evaporation of the solvent, yields crystals of $K_2S_5O_6$. Alkali-metal hexathionates are prepared by adding concentrated HCl to a solution containing a mixture of the corresponding alkali-metal thiosulfate and nitrite. The polythionates or their acid solutions slowly deposit sulfur.

Although nothing is known of the constitution of the polythionic acids, the structures of some of the ions as they exist in crystalline salts have been determined; the relevant parameters are summarized in Fig. 13.11. The general features show the existence of S—S chains which have internuclear distances close to that expected for single bonds. In this respect, the structures of polythionates are similar to those of the polysulfides (shown in Fig. 13.4). The bond angles are near the tetrahedral value, as would be expected, and the S—O distances are generally shorter than the expected double-bond values.

3.4 / Oxygen Derivatives of Selenium and Tellurium *(XI)*

Oxides   Like sulfur, selenium and tellurium form two principal oxides. The crystalline colorless dioxides, MO_2 (M = Se, Te), are formed when the corresponding metal is burned in air. Both compounds readily sublime at about 300° and can be melted under pressure. Liquid SeO_2 is yellow while TeO_2 in the liquid phase has a markedly darker yellow color. Selenium dioxide is a mild oxidizing agent, being easily reduced to the element by SO_2, HI, sulfur, and a variety of organic substances; it finds use as an oxidizing agent in organic chemistry. Both dioxides readily absorb the hydrogen halides to form compounds such as $SeO_2 \cdot 5HF$, $SeO_2 \cdot 4HCl$, $TeO_2 \cdot 3HCl$, and $MO_2 \cdot 2HCl$ (M = Se, Te). Presently, the constitution of these compounds is unknown. Selenium dioxide is very soluble in water, but TeO_2 is markedly less soluble. Both compounds dissolve in solutions of strong bases to give the corresponding selenites (SeO_3^{2-}) and tellurites (TeO_3^{2-}).

In contrast to the existence of SO_2 as a simple molecular species, SeO_2 and TeO_2 crystallize in polymeric structures. Crystalline SeO_2 consists of infinite chains of SeO_2 units with bridging oxygen atoms, as shown in Fig. 13.12; adjacent SeO_3 units are not coplanar. The bridging SeO distances are

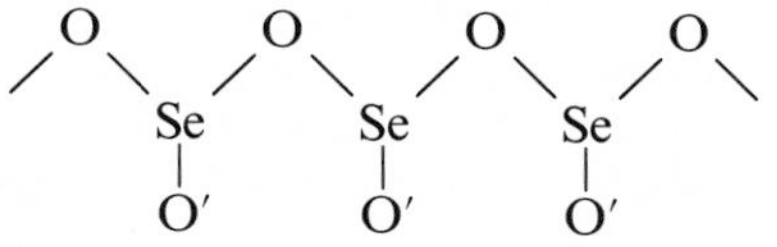

$$\angle(\text{O}'\text{SeO}) = 90°$$
$$\angle(\text{SeOSe}) = 125°$$
$$\angle(\text{OSeO}) = 98°$$

FIG. 13.12 Structural parameters for SeO_2.

1.78 Å whereas the nonbridged distances are slightly shorter, 1.73 Å. The vapor density of selenium dioxide corresponds to the molecular species SeO_2; however, mass spectroscopic experiments indicate that the vapor over selenium dioxide also contains $(SeO_2)_2$ in addition to the monomer (*36*). The dimer has also been identified in the solid state when the vapor is trapped in a low-temperature matrix. Electron diffraction results indicate that the SeO distance in the vapor phase is 1.61 Å, suggesting that a considerable π component is introduced when depolymerization occurs. Tellurium dioxide crystallizes in the rutile structure (Fig. 2.4). Mass spectral studies on subliming tellurium dioxide reveal the presence of $(TeO)_n$ and $(TeO_2)_n$ ($n = 1$–4) as well as Te_2 and O_2 (37).

Both selenium and tellurium also form trioxides, MO_3. Selenium trioxide, a white deliquescent solid (mp 118°), can be prepared by oxidizing selenium with atomic oxygen or treating K_2SeO_4 with SO_3

$$K_2SeO_4 + 3SO_3 \rightarrow K_2S_3O_{10} + SeO_3. \tag{91}$$

Attempts to dehydrate selenic acid (H_2SeO_4) lead to a mixture of SeO_2 and oxygen. On the other hand, dehydration of orthotelluric acid (H_6TeO_6) is the best method for the preparation of TeO_3. Both trioxides dissolve in water to form the corresponding oxyacids.

Selenium trioxide, like SO_3, appears to exist in two crystalline forms. The cubic form can be converted into a needle-like asbestos form by heating. When the yellow-orange TeO_3 is heated for an extended period of time it changes to a more dense gray form which is less reactive than the original modification. Little is known of the detailed structure of any of the forms of selenium or tellurium trioxide.

Oxyacids   Selenous acid (H_2SeO_3) has been isolated as a colorless, hygroscopic, crystalline substance from an aqueous solution of SeO_2. The substance is easily dehydrated to give SeO_2. Selenous acid is dibasic ($K_1 = 4 \times 10^{-3}$; $K_2 = 10^{-8}$) forming two series of salts, the selenites (M_2SeO_3) and the hydrogen selenites ($MHSeO_3$). In acid solution selenous acid is readily oxidized to selenic acid or reduced to the element. The crystal structure of selenous acid shows a puckered layer of SeO_3 units held together by hydrogen bonds. The SeO_3^{2-} ion is nonplanar, as expected from electron-pair-repulsion arguments, with an OSeO bond angle of 100° and three equal Se—O bond distances at 1.74 Å.

Tellurous acid (*XI*) has not been isolated, but alkali-metal tellurites (M_2TeO_3) and hydrogen tellurites ($MHTeO_3$) can be prepared by dissolving TeO_2 in the appropriate amount of the aqueous alkali-metal hydroxide. Aqueous alkaline solutions of the tellurites are easily oxidized to tellurates by air. In acid solution tellurites can be readily reduced to the element by SO_2 and certain metals (Zn, Sn, Cu, Hg). Although polytellurous acids have not been characterized, salts of these substances are known (e.g., $K_2Te_2O_5$ and $K_2Te_4O_{13}$).

Selenic acid, H_2SeO_4, exists as colorless crystals (mp 57°) which melt to a thick oily liquid. It is made by the oxidation of selenium, selenium dioxide, or selenous acid with strong oxidizing agents such as chlorine, bromine, chloric acid, or $KMnO_4$. A crystalline phase corresponding to diselenic acid, $H_2Se_2O_7$, exists in the phase diagram of the system SeO_3–H_2O.

Selenic acid has many of the characteristics of sulfuric acid; it is hygroscopic, dissolves in water with the evolution of heat, and chars organic compounds containing the elements of water. The proton-donating ability of H_2SeO_4 is about 92 % that of H_2SO_4 in the concentration range 50–75 % in aqueous solutions (38). Selenic acid is, however, a stronger oxidizing agent than is sulfuric acid. For example, it attacks HCl to form chlorine. Thus, a hot concentrated mixture of HCl and H_2SeO_4 behaves like aqua regia, attacking gold and platinum. Selenic acid can be reduced to the element using SO_2 or zinc dust. As is the case with sulfuric acid, selenic acid forms two series of salts, M_2SeO_4 and $MHSeO_4$. The selenates are isostructural with sulfates (shown in Fig. 13.7); selenic acid crystallizes in the same puckered layer structure as does sulfuric acid.

A simple telluric acid (*XI*) analogous to sulfuric acid and selenic acid is unknown. When elemental tellurium is oxidized with a mixture of chromic acid and nitric acid, or with H_2O_2, a white crystalline substance, orthotelluric acid, with the formula $Te(OH)_6$ is formed. It might be tempting to formulate this substance as $H_2TeO_4 \cdot 2H_2O$ which would be analogous to the substances that exist in the phase diagram of the system MO_3–H_2O (M = S, Se), but x-ray diffraction results indicate that crystalline orthotelluric acid consists of octahedral $Te(OH)_6$ groups held together by hydrogen bonds.

Orthotelluric acid is a weak electrolyte; since the first two dissociation constants are small ($K_1 = 1.53 \times 10^{-8}$, $K_2 = 4.7 \times 10^{-11}$) it acts for all practical purposes as a dibasic acid in aqueous solution. The apparent acid strength can be enhanced by adding polyhydroxy-organic compounds, such as glycerol, to an aqueous solution of the substance. Orthotelluric acid loses water when heated to form TeO_3, which upon further heating yields TeO_2. The acid is easily reduced to the metal by SO_2, Zn, Fe^{2+}, or N_2H_4. Alkali-metal orthotellurates of the type $M_2H_4TeO_6$ can be prepared by neutralization of orthotelluric acid in aqueous solution. Salts in which all of the hydrogen atoms have been replaced are not easy to prepare, although the salt Ag_6TeO_6 has been reported. Tellurates analogous to the sulfates and selenates also can be prepared by oxidizing tellurites in alkaline solution with chlorine

$$K_2TeO_3 + 2KOH + Cl_2 \rightarrow K_2TeO_4 + 2KCl + H_2O. \qquad (92)$$

Potassium tellurate is isostructural with the corresponding sulfates and selenates, but the compound forms a hydrate ($K_2TeO_4 \cdot 5H_2O$) which might be a derivative of potassium orthotellurate ($K_2H_4TeO_6 \cdot 3H_2O$).

Polymeric telluric acids have been reported (*XI*). Heating an aqueous solution of orthotelluric acid yields a colloidal solution. Pure orthotelluric acid melts at about 140° and loses water to form a syrup which contains polymerized telluric acids. Salts of condensed telluric acids, such as $K_2Te_2O_7\cdot$ $4H_2O$ and $K_2Te_4O_{13}\cdot4H_2O$, have been isolated. Unfortunately, little is known of the structure of these substances.

Derivatives of Selenium and Tellurium Oxyacids   Analogues of the sulfurous and sulfuric acid derivatives are known for selenium and tellurium, but not all of these derivatives have been described for each of the elements. Selenoxides (R_2SeO) and telluroxides (R_2TeO) can be prepared by oxidizing the corresponding selenides or tellurides

$$R_2M \xrightarrow{[O]} R_2MO \tag{93}$$

$$M = Se, Te$$

using $KMnO_4$, $K_2Cr_2O_7$, or HNO_3 in the case of selenium, or atmospheric oxygen for the tellurium derivatives. An additional route to these compounds is available through the dialkyl dihalides

$$R_2SeCl_2 + 2KOH \rightarrow R_2SeO + 2KCl + H_2O \tag{94}$$

$$R_2TeCl_2 + Ag_2O \rightarrow R_2TeO + 2AgCl. \tag{95}$$

Selenium oxyhalides, X_2SeO (X = F, Cl, Br), have been prepared, but the corresponding tellurium compounds have not been reported. The selenium compounds incorporating halogen atoms react with compounds containing potentially acidic hydrogen atoms [cf. Eq. (60)]. For example, the reaction of selenium oxychloride with alcohols yields the esters of selenous acid $[(RO)_2SeO]$ [cf. Eq. (54)]. The corresponding tellurous acid esters $[(RO)_2TeO]$ are not known. Phenylseleninic and phenyltellurinic acids $[C_6H_5MO(OH)]$ can be prepared by oxidizing diphenyl selenide or telluride with nitric acid. Phenylseleninic acid forms a hydrate which has been formulated as a base, $[C_6H_5Se(OH)_2]OH$; the existence of the silver salt $C_6H_5SeO(OAg)$ indicates the amphoteric nature of this hydroxy compound. Phenyltellurinic acid is also amphoteric, being very soluble in aqueous acid as well as in base.

The known derivatives of selenic and telluric acid are less well characterized than those of the lower-valent acids. Selenones and tellurones (R_2MO_2) have been prepared by reactions similar to those used to prepare sulfones [Eq. (83)]. A few esters of selenic acid have been prepared, and selenonic acids ($RSeO_3H$) have been prepared by the oxidation of the corresponding selenic acids. No analogous tellurium derivatives have been reported.

Polyselenonic acids, $(HO)_2Se_xO_3$, appear to be considerably less stable than the corresponding sulfur compounds, but derivatives of selenothionic acids have been prepared. For example, selenium dissolves in potassium sulfate solution to form $K_2SeS_2O_6$. The barium salt of selenopentathionic acid $BaSeS_4O_6 \cdot 2H_2O$, has also been isolated; the structure of the anion in this compound is essentially that of the pentathionate, shown in Fig. 13.11, with the selenium atom occupying the middle position of the sulfur chain.

3.5 / Halogen Compounds

The halogen derivatives of sulfur, selenium, and tellurium are discussed in this section. The chemistry and properties of the oxygen halides are more appropriately discussed in Chapter 14.

Sulfur (XII), selenium, and tellurium form compounds of the types MX_2, MX_4, and MX_6. Compounds containing two atoms of a Group VI element are known for each element (M_2X_2, M = S, Se; M_2X_{10}, M = S, Te). The binary iodides are unknown for sulfur and selenium, probably because the oxidizing power of iodine is not sufficiently large to form stable derivatives of these elements. As might be expected, fluorine forms compounds in which the Group VI element occurs in the highest oxidation state.

Monohalides (M_2X_2)   The direct reaction of either chlorine or bromine with sulfur or selenium produces the corresponding monohalides

$$2M + X_2 \rightarrow M_2X_2. \qquad (96)$$

Sulfur monofluoride is prepared by reacting AgF with molten sulfur

$$2AgF + 3S \rightarrow Ag_2S + S_2F_2. \qquad (97)$$

The compounds are generally colored and have physical properties characteristic of covalent substances, as shown in Table 13.7. Molecular-weight measurements in the vapor phase and in a variety of solvents verify that the correct formulation of these compounds is M_2X_2.

TABLE 13.7   Some properties of sulfur and selenium monohalides

Compound	Mp, °C	Bp, °C	Color at room temperature
S_2F_2	105.5	−99	colorless
S_2Cl_2	−80	137	golden yellow
S_2Br_2	−46	57[a]	garnet red
Se_2Cl_2	—	127	yellow-brown
Se_2Br_2	—	—	red

[a] At 0.22 torr.

The monohalides react with water, the reaction being the most vigorous for the compound with the lightest halogen or the lightest Group VI element. Depending upon the conditions, a mixture of products arises from the hydrolysis reaction. For example, S_2Cl_2 yields HCl, sulfur dioxide, and sulfur upon hydrolysis. The last two of these substances can, of course, interact to form polythionic acids if excess water is present. Similar products are obtained when the selenium monohalides react with water. Sulfur monochloride dissolves sulfur, iodine, some metallic halides, and certain organic compounds. It is soluble in benzene and ether.

Of the monohalides only the structures of S_2Cl_2 and S_2Br_2, shown in Fig. 13.13, are known. These molecules contain an S—S bond, each sulfur atom

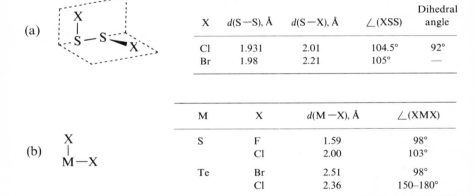

(a)

X	$d(S-S)$, Å	$d(S-X)$, Å	$\angle(XSS)$	Dihedral angle
Cl	1.931	2.01	104.5°	92°
Br	1.98	2.21	105°	—

(b)

M	X	$d(M-X)$, Å	$\angle(XMX)$
S	F	1.59	98°
	Cl	2.00	103°
Te	Br	2.51	98°
	Cl	2.36	150–180°

FIG. 13.13   Structural parameters for (a) S_2X_2 and (b) SX_2 (X = Cl, Br).

carrying a halogen atom; the general geometry is similar to that of H_2O_2 with a dihedral angle of 92°. The experimental bond distances are virtually those expected for single bonds formed between the respective atoms. Presumably, the other monohalides have similar structures, although there is the

$$\begin{array}{c} X \\ \diagdown \\ M-M \\ \diagup \\ X \end{array}$$

19

possibility that the compounds could be formulated as **19**. Interestingly enough, a compound with the molecular formula S_2F_2 but with the constitution shown in **19** has been isolated. This is an isomer of the compound which carries fluorine atoms on both sulfur atoms (*39–41*).

Binary halogen compounds containing chains of sulfur atoms greater than two have been prepared by passing a mixture of S_2Cl_2 and hydrogen through a hot tube or by reacting S_2Cl_2 with hydrogen polysulfides (*42*)

$$2S_2Cl_2 + H_2S_n \rightarrow Cl_2S_{n+4} + 2HCl. \tag{98}$$

Using these two methods it has been possible to prepare polysulfur dichlorides, S_nCl_2, with chains eight atoms long. The chlorides can be converted into the corresponding bromides by treatment with HBr

$$Cl_2S_n + 2HBr \rightarrow Br_2S_n + 2HCl. \tag{99}$$

Dihalides (MX_2)   The dichlorides of sulfur, selenium, and tellurium have been prepared, and only the dibromides of the two heaviest elements in this family have been characterized. The compound SF_2 has been reported to form when SCl_2 at low pressure is passed over HgF_2; the product disproportionates rapidly according to the following reactions (*43*):

$$2SF_2 \rightarrow S + SF_4 \tag{100}$$

$$SF_2 + S \rightarrow SSF_2. \tag{101}$$

Sulfur difluoride also has been prepared from the decomposition of SF_6 in a radiofrequency discharge (*44*). Sulfur dichloride (mp $-70°$, bp $59°$) can be prepared by the clorination of sulfur monochloride at room temperature

$$S_2Cl_2 + Cl_2 \rightleftharpoons 2SCl_2. \tag{102}$$

The reaction is reversible even at room temperature. The selenium dihalides can only exist in the vapor state at elevated temperatures. When either $SeCl_4$ or Se_2Cl_2 is maintained between $190°$ and $600°$, $SeCl_2$ is one of the components in equilibrium. A mixture of selenium and bromine in the temperature range $250–500°$ contains $SeBr_2$. The trend in stability of the dihalides established for sulfur and selenium is reversed in the case of tellurium. Both $TeCl_2$ (mp $175°$, bp $324°$) and $TeBr_2$ (mp $210°$, bp $339°$) exist as deeply colored solids.

The structural data shown in Fig. 13.13 for the dihalides indicate that they are bent molecules, as expected from the electron–pair repulsion arguments.

Tetrahalides (MX_4)   The pattern of stability of the tetrahalides nicely illustrates the interplay of the oxidizing power of the halogen and the resistance to oxidation of the Group VI elements. Sulfur, the most difficult of these elements to oxidize, forms tetrahalides (SF_4 and SCl_4) with the two halogens which are the most powerful oxidizing agents of that family of compounds. Sulfur tetrachloride is a marginally stable compound (Table 13.8). Selenium forms a tetrabromide in addition to SeF_4 and $SeCl_4$, and all of the tellurium tetrahalides are known. As shown in Table 13.8, these

TABLE 13.8   Properties of the Group VI tetrahalides

Compound	Mp, °C	Bp, °C	Color
SF_4	−121	−40.4	colorless gas
$SCl_4{}^a$	−31	—	yellow solid melting to red liquid
SeF_4	9.5	106	colorless liquid
$SeCl_4$	305^b	196^c	colorless crystals
$SeBr_4$	—	sublimes^d	yellow crystals
TeF_4	—	—	white needles
$TeCl_4$	225	390	white crystals, orange vapor
$TeBr_4$	380	∼420	dark yellow or red crystals
TeI_4	259	—	black crystals

^{a} Stable only in the presence of chlorine under pressure.
^{b} Under pressure.
^{c} At 1 atm, sublimes
^{d} Dissociates at 70–80°

compounds have typically covalent properties, and derivatives containing the heavier halogens are colored. Sulfur tetrafluoride (XIII) is prepared by the action of NaF on sulfur dichloride in acetonitrile

$$3SCl_2 + 4NaF \rightarrow SF_4 + S_2Cl_2 + 4NaCl. \tag{103}$$

It is a useful fluorinating agent, converting C=O and P=O groups into CF_2 and PF_2 moieties, and carboxylic acids and P(O)OH units into CF_3 and PF_3, respectively. The other tetrahalides can be made by direct halogenation of the elements or of the lower halides

$$M + X_2 \rightarrow MX_4 \tag{104}$$

$$M = Te, X = Cl, Br, I$$
$$M = Se, X = F, Cl$$

$$S_2Cl_2 + 3Cl_2 \rightarrow 2SCl_4. \tag{105}$$

Structural information is available for SF_4, SeF_4, and $TeCl_4$. These molecules are best described as distorted trigonal-bipyramidal structures in which one of the equatorial positions is occupied by a lone pair of electrons

(20). The crystal structure of TeF_4, on the other hand, indicates the presence of five-coordinate, square-planar, tellurium atoms linked together by fluorine

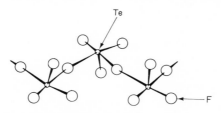

20

bridges (Fig. 13.14) (45). A polymeric structure also apparently is present in benzene or toluene solutions of tellurium halides, where molecular weight measurements indicate the presence of $(TeX_4)_3$ (X = Cl, Br, I) (46).

FIG. 13.14   Structure of TeF_4.

Sulfur tetrafluoride forms adducts with typical Lewis acids such as BF_3 and the Group V pentafluorides. The results of infrared spectral studies indicate that fluorine-bridged ionic structures such as $(SF_3)^+(BF_4)^-$ and $(SF_3)^+(MF_6)^-$ (M = P, As, Sb) are present in these adducts (47). In anhydrous HF the adducts are completely dissociated; SF_4 itself is weakly ionized in this solvent.

Sulfur tetrachloride is stable only in the presence of chlorine under pressure and decomposes at its melting point. It forms addition compounds with the chlorides of polyvalent elements, $SCl_4 \cdot MCl_3$ (M = Al, Tl, Fe), $2SCl_4 \cdot SnCl_4$, and $SCl_4 \cdot SbCl_5$; it would be tempting to formulate these as the chlorosulfonium salts of complex ions, $[SCl_3^+][MCl_4^-]$, $[SCl_3^+]_2[SnCl_6^{2-}]$, and $[SCl_3^+][SbCl_6^-]$. Unfortunately there is little evidence to support such a formulation other than the existence of stable sulfonium cations in other systems. In fact, some of the adducts, e.g., $SCl_4 \cdot SbCl_5$, do not have the solubility characteristics associated with an ionic formulation; for example, $SCl_4 \cdot SbCl_5$ has a low melting point and is soluble in organic solvents. In addition, some adducts such as $SCl_4 \cdot TiCl_4$ and $SCl_4 \cdot 2AsF_3$ do not have the stoichiometry consistent with that expected if complex ions of the type indicated were formed. Selenium and tellurium tetrahalides also form adducts with Lewis acids ($TeCl_4 \cdot SO_3$, $TeCl_4 \cdot 2AlCl_3$, $SeCl_4 \cdot AuCl_3$, $SeCl_4 \cdot SbCl_5$) as well as with Lewis bases $[SeF_4 \cdot C_5H_5N \cdot (C_2H_5)_2O$ and $TeCl_4 \cdot 2(C_2H_5)_2O]$.

Organic Derivatives of the Tetrahalides   Mixed organohalogen derivatives of selenium (XIV) and tellurium such as RMX_3, R_2MX_2, and R_3MX can be prepared in a number of ways. For example, treatment of the tetrahalide with one mole of Grignard reagent gives the organotrihalo compound

$$RMgX + TeX_4 \rightarrow RTeX_3 + MgX_2. \tag{106}$$

Dihalides can be prepared by treating a diorgano compound with halogen

$$R_2M + X_2 \rightarrow R_2MX_2 \tag{107}$$

$$M = Te, Se$$

and the monohalo derivatives can be formed from the reaction of a diorgano derivative with an alkyl halide

$$R_2M + R'I \rightarrow R_2R'MI. \tag{108}$$

Indirect evidence suggests that these compounds are best formulated as selenonium or telluronium derivatives, i.e., $[RMX_2{}^+]X^-$, $[R_2MX^+]X^-$, and $[R_3M^+]X^-$. As indicated earlier, there is direct structural evidence that some tetravalent derivatives of the Group VI elements exhibit a molecular structure related to a distorted trigonal bipyramid (**20**). The ionic formulation of a tetravalent compound of this family presumably involves pyramidal cations of the types $RMX_2{}^+$, R_2MX^+, and R_3M^+ (**21–23**).

Alkylsulfonium compounds are colorless crystalline salts that can be prepared by the reaction of an alkyl sulfide with HI

$$2R_2S + HI \rightarrow [R_3S^+]I^- + HSR \tag{109}$$

or with an alkyl iodide

$$R_2S + RI \rightarrow [R_3S^+]I^-. \tag{110}$$

The anion in these compounds can be replaced by other anions using standard metathetic reactions; the sulfonium cation is unchanged during such reactions. Sulfonium ions with three different groups attached to the sulfur, e.g., $[S(CH_3)(C_2H_5)(CH_2CO_2H)^+]$, have been resolved into optical isomers (**24** and **25**) indicating that the ion is nonplanar, which is in accordance with the results of electron-pair-repulsion arguments. Spectroscopic data also suggest

that the ion $S(CH_3)_3{}^+$ is pyramidal, with the sulfur atom at the apex of the pyramid.

Hexafluorides (MF_6)   All of the Group VI elements form hexafluorides that are very volatile, as shown by the data in Table 13.9; no other hexahalogen derivatives are known. These compounds are prepared by exhaustive fluorination of the corresponding elements. The hexafluorides are remarkably

TABLE 13.9   Some properties of the Group VI hexafluorides

Compound	Mp, °C	Bp, °C
SF_6	− 50.7	− 63.8 (sub.)
S_2F_{10}	− 92	29
SeF_6	− 35	− 47 (sub.)
TeF_6	− 38	− 39 (sub.)
Te_2F_{10}	− 34	53

unreactive substances, the reactivity with a given reagent increasing with increasing molecular weight. For example, SF_6 is not attacked by water, acids, or bases at room temperature; it is resistant to the action of carbon, copper, or magnesium at red heat, and will not react with sodium below its boiling point. It reacts with sulfur vapor or hydrogen at 400°. Selenium hexafluoride is nearly as resistant to attack by common reagents, whereas tellurium hexafluoride is slowly hydrolyzed by water to give orthotelluric acid.

Sulfur and tellurium form binary fluorides containing two Group VI elements bound to each other. Disulfur decafluoride, S_2F_{10}, can be obtained from the photochemical reaction of SF_5Cl with H_2

$$2SF_5Cl + H_2 \rightarrow S_2F_{10} + 2HCl. \tag{111}$$

Curiously, this substance is both very poisonous and very unreactive. The general reactivity of S_2F_{10} is similar to that of SF_6. At high temperatures it is a powerful oxidizing agent often yielding fluorinated derivatives when it reacts. Ditellurium decafluoride is formed in low yield when tellurium is fluorinated.

The structure of the hexafluoride molecules shown in Fig. 13.15 is that of a perfect octahedron, as expected. The bond distances in each compound are significantly shorter than expected on the basis of the usual single-bond radii,

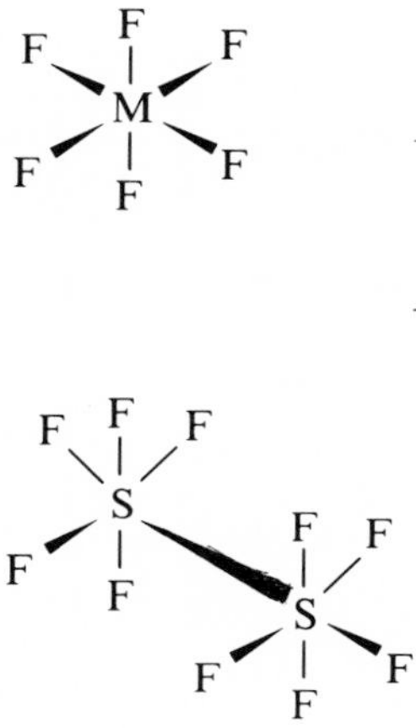

	$d(M-F)$, Å	
M	Exp.	Theor.
S	1.58	1.76
Se	1.68	1.89
Te	1.82	2.09

$d(S-F) = 1.56$ Å
$d(S-S) = 2.21$ Å

FIG. 13.15   Structural parameters for MF_6 and S_2F_{10}.

suggesting that multiple bonds of $d\pi$-$p\pi$ type might be present in the molecule (*48, 49*). The structure of disulfur decafluoride corresponds to two octahedral SF_5 groups connected at one corner through S—S bonds. The S—F bond distances are essentially those found in SF_6, but there is a surprisingly long S—S bond (2.21 Å). It has been suggested that repulsion of the fluorine atoms in the SF_5 groups is the cause of the increased S—S bond distance.

3.6   /   Hydrides

All of the elements in Group VI form binary hydrides, with sulfur forming the most extensive series of catenated derivatives; the properties of some of these compounds are given in Table 13.10. This trend is an extension of the tendency of sulfur to form compounds containing chains of atoms found in the halogen derivatives, the chalconides, and the oxyanions. The stability of the hydrides of these elements decreases with increasing atomic number, the extreme in instability being reached in PoH_2, which decomposes at $-180°$.

TABLE 13.10   **Some properties of the Group VI hydrides**

Compound	Mp, °C	Bp, °C	M—H bond energy, kcal mole^{-1}
H_2O	0	100	111
H_2O_2	-0.89	152	—
H_2S	-85.5	-60.5	88
H_2S_2	-89	-71	—
H_2S_3	-52	—	—
H_2S_4	—	—	—
H_2Se	-41.5	-66	73
H_2Te	-20	-51.2	63

Oxygen   The most important of all the hydrides is water. The details of the chemistry of this substance are not discussed here; rather they appear as is necessary throughout this volume (e.g., Chapter 6). A material known as "polywater" or "anomalous water" is formed when water vapor condenses in pyrex or quartz capillaries (50). This substance exhibits markedly different properties than those of water; for example, it has a higher density ($\leqslant 1.4$ g cm^{-3}), lower freezing point ($-40°$), higher viscosity, lower vapor pressure, and a different vibrational spectrum than water. Although there have been suggestions that a new form of water is present as either tetrameric clusters, H_8O_4, held together by strong multicentered bonds (52, 53), or a polydispersed polymer held together by O—H—O units (54), its properties are also consistent with highly concentrated solutions or gels formed by materials leached from the capillary walls (50, 51). The evidence available is persuasive against the existence of a new form of water.

Hydrogen peroxide, H_2O_2, is a pale blue, syrupy liquid which is miscible with water and soluble in ether. Aqueous solutions of hydrogen peroxide can be prepared from the hydrolysis of peroxydisulfuric acid

$$H_2S_2O_8 + 2H_2O \rightarrow 2H_2SO_4 + H_2O_2, \qquad (112)$$

which in turn is prepared by the electrolysis of H_2SO_4 or $(NH_4)_2SO_4$–H_2SO_4 solutions at a high current density. Acidification of peroxides

$$BaO_2 + H_2SO_4 \rightarrow BaSO_4 + H_2O_2 \qquad (113)$$

also yields solutions of hydrogen peroxide. The pure substance can be obtained from its aqueous solutions by evaporation and distillation under reduced pressure. Crystalline H_2O_2 can be obtained by cooling a concentrated aqueous solution. Hydrogen peroxide is also produced on a large scale by the autoxidation of anthraquinols.

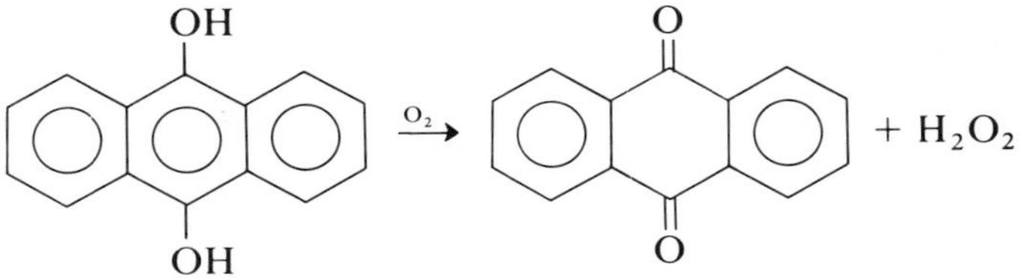

$$\text{(114)}$$

26   **27**

The process can be made cyclic by reducing the quinone (**27**) to the quinol (**26**) with hydrogen over platinum. The H_2O_2 formed is removed from the organic phase by countercurrent extraction with water. The entire process is relatively inexpensive if large quantities of hydrogen are available because the oxidation step [Eq. (114)] can be carried out with atmospheric oxygen. Thus, the process effectively corresponds to the combination of oxygen and hydrogen using anthraquinones as intermediates.

Pure hydrogen peroxide and its aqueous solutions are unstable with respect to oxygen evolution

$$H_2O_{2\,(l)} \longrightarrow H_2O_{(l)} + \tfrac{1}{2}O_{2\,(g)} \qquad \Delta H = 23.45 \text{ kcal.} \qquad \text{(115)}$$

The pure substance decomposes explosively at $144°$; decomposition of the aqueous solutions is promoted by rough surfaces or finely divided metals. Aqueous solutions can be stabilized by acids or various organic compounds such as glycine and acetanilide. Hydrogen peroxide resembles water in many of its physical properties. For example, it has a dielectric constant of 93 at $25°C$. The relatively high boiling point shown in Table 13.10 suggests that the liquid is extensively hydrogen bonded. In aqueous solution, hydrogen peroxide is a weak acid ($K_1 = 1.5 \times 10^{-12}$). Concentrated aqueous solutions of H_2O_2 liberate CO_2 from Na_2CO_3 and react with $Ba(OH)_2$. Hydrogen peroxide is about 10^6 times less basic than water, but the species $H_3O_2^+$ has been suggested as being present in the very strongly oxidizing solutions formed when a mixture of H_2O_2 and HBF_4 is dissolved in sulfolane (55). The H_2O_2 molecule is nonplanar (Fig. 13.16), a characteristic structure observed for the compounds of these elements with the general constitution

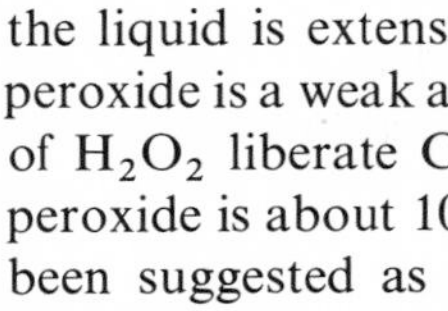

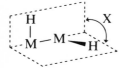

M	$d(M-M)$, Å	$d(M-H)$, Å	$\angle(MMH)$	X
O	1.49	0.97	97°	94°
S	2.55	1.33	95°	90°

FIG. 13.16   Structural parameters for H_2O_2 and H_2S_2.

XMMX. The bond distances are normal, but the HOO bond angle is smaller than the HOH bond angle in water, suggesting a change in the nature of the oxygen hybrid orbitals used in bonding.

Hydrogen peroxide can act either as an oxidizing agent

$$H_2O_2 + 2e^- + 2H^+ \rightarrow 2H_2O \qquad E^0 = 1.77 \text{ V} \qquad (116)$$

or a reducing agent

$$H_2O_2 \rightarrow O_2 + 2H^+ + 2e^- \qquad E^0 = -0.68 \text{ V}. \qquad (117)$$

It is among the most powerful oxidizing agents available; only very strong oxidizing agents such as $HCrO_4^-$, MnO_4^- and Ce^{4+} will oxidize H_2O_2. Experiments using labeled oxygen show that when H_2O_2 acts as a reducing agent [Eq. (117)] both oxygen atoms in the O_2 produced arise from the original H_2O_2 molecule, suggesting that the oxidizing agents remove electrons from this substance. Many reactions of H_2O_2 involve radical species such as HO_2 and HO (*56*), so that it is difficult to formulate a general mechanism for reactions involving H_2O_2 at this time.

TABLE 13.11   Peroxyacids

Name	Structure
Organic peroxyacids	R—C(=O)OOH
Peroxysulfuric acid (Caro's acid)	(HO)S(=O)(=O)OOH
Peroxydisulfuric acid	(HO)S(=O)(=O)—OO—S(=O)(=O)(OH)
Peroxycarbonic acid	HOC(=O)OOH
Peroxydicarbonic acid	HOC(=O)OOC(=O)H
Peroxynitrous acid	O=NOOH
Peroxynitric acid	HON(=O)OOH
Peroxydiphosphoric acid	$(HO)_2PO(=O)OP(=O)(OH)_2$

A variety of formal derivatives of H_2O_2 are known. Among the most extensively investigated types are the peroxyacids, some of which are summarized in Table 13.11, in which an —O— moiety has been formally replaced by an —O_2— group. Thus, the peroxy derivative can carry a terminal OOH or a bridging —O_2— group.

The products formed during the electrical dissociation of water vapor contain H_2O_3 and H_2O_4. An infrared spectral study of the products trapped in a matrix at low temperatures indicates that these substances contain a zig-zag oxygen-chain backbone (57).

Sulfur   Binary hydrides of sulfur of the general formula H_2S_n are known for $n = 1$–6. Hydrogen sulfide (H_2S) can be prepared in a variety of ways, the most common being the reaction of a metal sulfide with concentrated acid. In contrast to water, hydrogen sulfide is a gas under ordinary conditions, indicating a very low degree of hydrogen bonding in the liquid phase; structural data on solid H_2S indicate the presence of hydrogen bonds (58). Although it has a low dielectric constant (8.3 at $-78.6°$), indicating that it is a poor solvent for ionic substances, H_2S has been investigated as a nonaqueous solvent (XV). Hydrogen sulfide burns in oxygen, yielding elemental sulfur or SO_2 depending upon the temperature and the amount of oxygen available. It also reacts with the halogens to liberate sulfur

$$H_2S + X_2 \rightarrow 2HX + S. \tag{118}$$

Hydrogen sulfide is soluble in alcohol and in water, the aqueous solution being weakly acidic ($K_1 = 1.9 \times 10^{-8}$, $K_2 = 1.2 \times 10^{-15}$). Two types of metal salts are known, the normal sulfides (M_2S) and the hydrogen sulfides (MHS). The chemistry of H_2S in aqueous solutions is dominated by the fact that many cations form extremely insoluble sulfides, and the precipitation of these substances can be governed by the pH of the solution. Thus, although solutions containing S^{2-} are extensively hydrolyzed to give basic solutions, the solubility of many sulfides is markedly less than that of the corresponding hydroxides.

Like water, H_2S is a bent molecule with a bond angle of 92.2°; the S—H bond distance (1.34 Å) is normal.

Hydrogen polysulfides (VII), sometimes called sulfanes, can be prepared in several ways including acidification of aqueous solutions of alkali-metal polysulfides

$$Na_2S_n + 2HCl \rightarrow 2NaCl + H_2S_n \tag{119}$$

and the reaction of sulfur chlorides with H_2S (59).

$$S_nCl_2 + 2H_2S \rightarrow 2HCl + H_2S_{n+2} \tag{120}$$

The oily products obtained from reaction (119) can be fractionated to give the relatively pure hydrogen polysulfides, H_2S_n, $n = 1–5$. The reaction shown in Eq. (120) has been used to prepare specific members of the series starting with pure materials. The hydrogen polysulfides in general are soluble in organic solvents. They readily decompose to give H_2S and sulfur, but this reaction is not reversible. Only the structure of H_2S_2 is known; the molecule is nonplanar and has the same geometric features as H_2O_2 (shown in Fig. 13.16). Quantum mechanical calculations indicate that the barrier for rotation about the S—S bond in H_2S_2 can best be accounted for on the basis of $d\pi$-$p\pi$ bonding, rather than repulsions of the unpaired electrons on the sulfur atoms (60, 61).

Several types of derivatives of the sulfanes are known. The metal polysulfides can be imagined as the salts of the corresponding hydrogen polysulfides, although they are not prepared by neutralization reactions. It should be recalled from Fig. 13.4 that the metal polysulfides also contain chains of sulfur atoms. Thus, although the molecular structures of the higher hydrogen polysulfides are not known, there is no reason not to believe that these molecules also contain S—S chains. This suggestion is supported by the fact that covalently bonded organic derivatives such as $(IC_2H_4)_2S_3$ and $(CF_3)_2S_3$ consist of molecules containing S_3 chains (62, 63).

Selenium and Tellurium   Only the simple hydrides of selenium and tellurium are known (Table 13.10). These are gaseous substances at room temperature, which more or less rapidly decompose to the elements. The substances can be prepared by the action of acids on the corresponding chalconide

$$Al_2M_3 + 6HCl \rightarrow Al_2Cl_6 + 3H_2M \qquad (121)$$

$$M = Te, Se.$$

No hydrogen polyselenides or polytellurides are known, although some derivatives such as R_2Se_2 and $(p\text{-}ClC_6H_4)_2Te_2$ have been prepared. Hydrogen selenide and telluride are soluble in water and are more acidic than H_2S in this solvent (H_2Se, $K_1 = 1.3 \times 10^{-4}$; H_2Te, $K_1 = 2.3 \times 10^{-3}$). The aqueous solutions are unstable, depositing the elements over a period of time.

As expected, the H_2Se molecule is bent ($\angle\,(HSeH) = 91°$) with a normal H—Se bond distance (1.46 Å).

4 / COORDINATION BEHAVIOR

The elements in this family can be divided into the usual two groups on the basis of the number of covalent bonds which can be formed. The data in

Table 13.12 indicate that the first member of the family, oxygen, never exceeds a coordination number of four, whereas the heavier members of the family exhibit six-coordinate structures as a maximum.

TABLE 13.12 **The coordination behavior of the Group VI elements**

	Coordination number				
Element	2	3	4	5	6
Oxygen	R_2O	$[H_3O^+]X^-$ $(ClHg)_3O^+$	$H_2O_{(s)}$ $R_2O \cdot 2InR_3$ $[Be_4O(CH_3CO_2)_6]$	— — —	— — —
Sulfur	H_2S	$[R_3S^+]X^-$ $R_2S \cdot HgCl_2$	SF_4	—	SF_6 S_2F_{10}
Selenium	R_2Se	$[R_3Se^+]X^-$	SeF_4 R_2SeX_2	$M^+[SeF_5{}^-]$	SeF_6
Tellurium	X_2Te	$[R_3Te^+]X^-$	$TeCl_4$	$K[CH_3TeI_4]$	TeF_6 Te_2F_{10} $Te(OH)_6$

Normally oxygen forms two bonds in covalent compounds of the type R_2O, which are essentially tetrahedral structures with lone pairs occupying two positions. However, three-coordinate oxygen atoms are found in the hydronium ion (H_3O^+), which is observed in the structures of the monohydrates of several strong acids. The hydronium ion is pyramidal with one lone electron pair. Thus, with respect to electron pairs, the hydronium ion is tetrahedral. An interesting variation on this type of structure is found in the compound $HgO \cdot 2HgCl_2$. This compound is best formulated as $[(ClHg)_3O^+]Cl^-$, the cation consisting of a planar arrangement of mercury atoms about a three-coordinate oxygen atom, as shown in Fig. 13.17; the $(CH_3Hg)_3O^+$ ion is also planar, but the corresponding sulfur derivative, $(CH_3Hg)_3S^+$; is

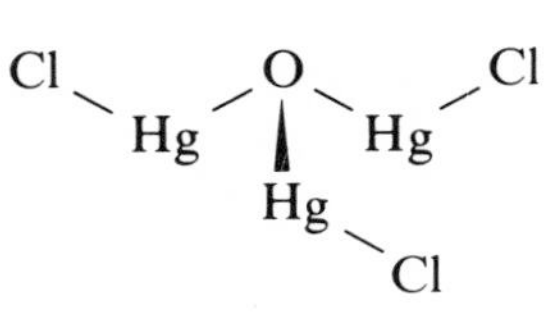

$$d(HgO) = 2.03 \text{ Å}$$
$$d(HgCl) = 2.28 \text{ Å}$$
$$\angle(OHgCl) = 175°$$
$$\angle(HgOHg) = 120°$$

FIG. 13.17 The structural parameters for $(ClHg)_3O^+$.

pyramidal (64). Apparently the lone-pair electrons formally available on the oxygen atom are delocalized in π type bonds into the mercury atoms. Neutral three-coordinate, but pyramidal, species containing oxygen atoms are found in adducts such as R_3BOR_2.

Four-coordinate oxygen atoms are found in basic beryllium salts of certain organic acids. For example, basic beryllium acetate, $Be_4O(CH_3CO_2)_6$, which is an easily volatilized substance, stable to heat, soluble in nondonating solvents, and monomeric in the vapor state, has the structure given in Fig. 13.18, in which a central oxygen atom is tetrahedrally surrounded by four

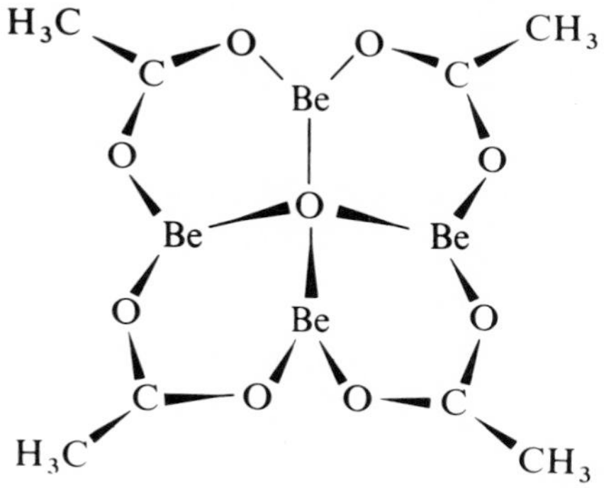

FIG. 13.18 The structure of basic beryllium acetate, $Be_4O(CH_3CO_2)_6$.

beryllium atoms. In addition, four-coordinate oxygen atoms may be present in the adducts $(C_2H_5)_2O\cdot2C_2H_5BCl_2$ and $(C_2H_5)_2O\cdot2In(CH_3)_3$. The oxygen atoms in ice are tetrahedrally surrounded by hydrogen atoms, two of which are formally bonded and two hydrogen bonded.

The elements heavier than oxygen in this family exhibit the same coordination numbers as oxygen, in addition to higher coordination numbers. Two-coordinate bent molecules are found among compounds of the type MH_2, MX_2 (X = halogen), and MR_2. Sulfonium, selenonium, and telluronium compounds with the general formula $R_3M^+X^-$ contain pyramidal cations. However, tetravalent compounds such as SF_4 and $TeCl_4$ basically possess a distorted trigonal-bipyramidal structure in which an electron pair occupies an equatorial position (20). Five-coordinate compounds are known in the form of complex salts such as $MSeF_5$ and $K[CH_3TeI_4]$ which contain the central element in a +4 oxidation state. Although the structures of such species have not been determined, electron-pair-repulsion arguments lead to the conclusion that they might have an octahedral arrangement of electron pairs, five being bound and one unbound. Compounds containing tetravalent, but six-coordinate atoms of this family, such as M_2TeX_6 and M_2SeX_6 (X = halogen), have been prepared. The structure of the anions in these species is not known, but they would be expected to be distorted octahedral.

Six-coordinate perfectly octahedral structures are exhibited by sulfur, selenium, and tellurium in the form of the hexafluorides (MX_6).

REFERENCES

1. R. P. Blickensderfer and G. E. Ewing, *J. Chem. Phys.*, **47**, 331 (1967).
2. T. V. Yagodovskaya, L. I. Nebrasov, and N. P. Klimushina, *Russ. J. Phys. Chem.*, **41**, 474 (1967).
3. B. R. Cairns and G. C. Pimentel, *J. Chem. Phys.*, **43**, 3432 (1965).
4. J. C. Koh and W. Klement, Jr., *J. Phys. Chem.*, **74**, 4280 (1970).
5. G. A. Ozin, *Chem. Commun.*, 1325 (1969).
6. M. Schmidt and E. Wilhelm, *Angew. Chem. (Int. Ed.)*, **5**, 964 (1966).
7. A. Kutoglu and E. Hellner, *Angew. Chem. (Int. Ed.)*, **5**, 965 (1966).
8. M. Schmidt and H. H. Block, *Angew. Chem. (Int. Ed.)*, **6**, 955 (1967).
9. D. J. Meschi and A. W. Searcy, *J. Chem. Phys.*, **51**, 5234 (1969).
10. P. J. Stephens, *Chem. Commun.*, 1496 (1969).
11. R. J. Gillespie and J. Passmore, *Chem. Commun.*, 1333 (1969).
12. A. Teder, *Arkiv Kemi*, **30**, 379 (1969).
13. A. Teder, *Arkiv Kemi*, **31**, 173 (1969).
14. J. Barr, R. J. Gillespie, R. Kapoor, and K. C. Malhotra, *Can. J. Chem.*, **46**, 149 (1968).
15. J. Barr, D. B. Crump, R. J. Gillespie, R. Kapoor, and P. K. Ummat, *Can. J. Chem.*, **46**, 3607 (1969).
16. I. D. Brown, D. B. Crump, R. J. Gillespie, and D. P. Santry, *Chem. Commun.*, 853 (1968).
17. T. N. Griever and I. G. Zartseva, *J. Appl. Chem. (USSR)*, **40**, 1623, 1859 (1967).
18. N. Bartlett, *Proc. Chem. Soc.*, 218 (1962).
19. J. N. Keith, I. J. Solomon, I. Sheft, and H. H. Hyman, *Inorg. Chem.*, **7**, 230 (1968).
20. J. Shamir and J. Binenboym, *Inorg. Chem. Acta*, **2**, 37 (1968).
21. W. L. Jolly, *J. Chem. Ed.*, **44**, 304 (1967).
22. H. Lux, *Z. Electrochem.*, **45**, 303 (1939).
23. A. D. McElroy and J. S. Hashman, *Inorg. Chem.*, **3**, 1798 (1964).
24. I. J. Solomon, K. Hattori, A. J. Kacmarek, G. M. Platz, and M. J. Klein, *J. Amer. Chem. Soc.*, **84**, 34 (1962).
25. L. V. Azároff and I. Cowin, *Proc. Natl. Acad. Sci. U.S.*, **49**, 1 (1963).
26. J. Brown and G. Burns, *Can. J. Chem.*, **47**, 4291 (1969).
27. J. W. Hastie, R. Hauge, and J. L. Margrave, *J. Inorg. Nucl. Chem.*, **31**, 281 (1969).
28. R. E. Johnson, T. H. Norris, and J. L. Huston, *J. Amer. Chem. Soc.*, **73**, 3052 (1951).
29. B. J. Masters and T. H. Norris, *J. Amer. Chem. Soc.*, **77**, 1346 (1955).
30. E. C. M. Grigg and I. Lauder, *Trans. Faraday Soc.*, **46**, 1039 (1950).
31. H. P. Leftin and N. N. Lichten, *J. Amer. Chem. Soc.*, **79**, 2475 (1957).
32. I. Taesler and I. Olovsson, *Acta Cryst.*, **24**, 299 (1968).
33. I. Taesler and I. Olovsson, *J. Chem. Phys.*, **51**, 4213 (1969).
34. R. J. Gillespie and E. A. Robinson, *Can. J. Chem.*, **40**, 644, 658 (1962).
35. R. R. de Vries and F. C. Mijlhoff, *Acta Cryst*, **25B**, 1696 (1969).
36. P. J. Ficalora, J. C. Thompson, and J. L. Margrave, *J. Inorg. Nucl. Chem.*, **31**, 3771 (1969).
37. D. W. Muenow, J. W. Hastie, R. Hauge, R. Bautista, and J. L. Margrave, *Trans. Faraday Soc.*, **65**, 3210 (1969).

38. S. Wasif, *J. Chem. Soc., A*, 142 (1967).
39. R. L. Kuczkowski, *J. Amer. Chem. Soc.*, **86**, 3617 (1964).
40. F. Seel and R. Budentz, *Chem. Ber.*, **98**, 251 (1965).
41. R. D. Brown, R. R. Burden, and G. P. Pez, *Chem. Commun.*, 277 (1965).
42. F. Fehér and M. Baudler, *Z. Anorg. Chem.*, **267**, 293 (1952).
43. F. Seel, E. Heinrich, W. Gombler, and R. Budenz, *Chimia*, **23**, 73 (1969).
44. D. R. Johnson and F. X. Powell, *Science*, **164**, 950 (1969).
45. A. J. Edwards and F. I. Hewaidy, *J. Chem. Soc., A*, 2977 (1968).
46. N. N. Greenwood, B. P. Straughan, and A. E. Wilson, *J. Chem. Soc., A*, 2209 (1968).
47. M. Azeem, M. Browstein, and R. J. Gillespie, *Can. J. Chem.*, **47**, 4159 (1969).
48. I. H. Hillier, *J. Chem. Soc., Ser. A*, 878 (1969).
49. F. Bernardi and C. Zauli, *J. Chem. Soc., A*, 867 (1969).
50. E. Willis, G. R. Rennie, C. Smart, and B. A. Pethica, *Nature*, **222**, 159 (1969).
51. R. E. Davis, D. L. Rousseau, and R. D. Board, *Science*, **171**, 167 (1971).
52. R. W. Bolander, J. L. Kassner, and J. T. Zung, *Nature*, **221**, 1233 (1969).
53. L. J. Bellamy, A. R. Osborn, E. R. Lippincott, and A. R. Bandy, *Chem. Ind. (London)*, 686 (1969).
54. E. R. Lippincott, R. R. Stromberg, W. H. Grant, and G. L. Cessac, *Science*, **164**, 1482 (1969).
55. R. W. Alder and M. C. Whiting, *J. Chem. Soc.*, 4707 (1964).
56. M.-S. Tsao and W. K. Wilmarth, *Advan. Chem. Ser.*, **36**, 113 (1962).
57. P. A. Giguère and K. Herman, *Can. J. Chem.*, **48**, 3474 (1970).
58. E. Sandor and S. O. Ogunade, *Nature*, **224**, 905 (1969).
59. E. Muller and J. B. Hyne, *Can. J. Chem.*, **46**, 2341 (1968).
60. M. E. Schwartz, *J. Chem. Phys.*, **51**, 4182 (1969).
61. A. Veillard and J. Demuynck, *Chem. Phys. Letters*, **4**, 476 (1970).
62. J. Donohue, *J. Amer. Chem. Soc.*, **72**, 2701 (1950).
63. D. P. Stevenson and J. Y. Beach, *J. Amer. Chem. Soc.*, **60**, 2872 (1938).
64. J. H. R. Clarke and L. A. Woodword, *Spectrochim. Acta*, **23A**, 2077 (1967).

COLLATERAL READINGS

I. K. W. Bagnall, *Advan. Inorg. Chem. Radiochem.*, **4**, 197 (1962). A review of the chemistry of polonium.
II. J. Drowart and P. Goldfinger, *Quart. Revs.*, **20**, 545 (1966). A review of the dissociation energies of the Group VI diatomic molecules. Included in the discussion are S_2, Se_2, Te_2, SO, SeO, TeO, TeS, and TeSe.
III. R. J. Gillespie and J. Passmore, *Acc. Chem. Res.*, **4**, 413 (1971). A brief survey of the preparation and properties of the polycations of the Group VI elements.
IV. N.-G. Vannerberg, *Prog. Inorg. Chem.*, **4**, 125 (1962). A comprehensive review of the preparation, properties, and structures of some metallic peroxides, superoxides, and ozonides of the metals.
V. J. A. Connor and E. A. V. Ebsworth, *Advan. Inorg. Chem. Radiochem.*, **6**, 279 (1964). A survey of the peroxycompounds of the transition metals.
VI. R. Ward, *Prog. Inorg. Chem.*, **1**, 465 (1959). A discussion of the structural properties of the mixed metal oxides.
VII. O. Foss, *Advan. Inorg. Chem. Radiochem.*, **2**, 237 (1960). A wide-ranging discussion of the structures of compounds containing chains of sulfur atoms. Included in this discussion are polysulfides, polysulfur hydrides and halides, and polythionates.

VIII. P. W. Schenk and R. Steudel, *Angew. Chem. (Int. Ed.)*, **4**, 402 (1965). A discussion of some of the more recent chemistry of the lower sulfur oxides.

IX. D. F. Burow, in *The Chemistry of Nonaqueous Solvents*, Vol. III (J. J. Lagowski, ed.), Academic Press, 1970, Chapter 2. A survey of the solvent properties of, and solution phenomena in, anhydrous SO_2.

X. W. H. Lee, in *The Chemistry of Nonaqueous Solvents*, Vol. II (J. J. Lagowski, ed.), Academic Press, 1967, Chapter 3; R. J. Gillespie and E. A. Robinson, in *Non-aqueous Solvent Systems* (T. C. Waddington, ed.), Academic Press, 1965, Chapter 4. These papers contain a wealth of information on the use of anhydrous H_2SO_4 as a solvent system.

XI. W. A. Dutton and W. C. Cooper, *Chem. Revs.*, **66**, 657 (1966). A review of the chemistry and properties of the oxides and oxyacids of tellurium.

XII. G. H. Cady, *Advan. Inorg. Chem. Radiochem.*, **2**, 105 (1960); S. M. Williamson, *Prog. Inorg. Chem.*, **7**, 39 (1966). These two reviews contain extensive information on the chemistry and properties of the sulfur fluorides and their derivatives.

XIII. W. C. Smith, *Angew. Chem. (Int. Ed.)*, **1**, 467 (1962). A review of the preparation, reactions, and properties of sulfur tetrafluoride.

XIV. J. Gosselck, *Angew. Chem. (Int. Ed.)*, **2**, 660 (1963). A survey of the chemistry of organoselenium compounds.

XV. F. Fehér in *The Chemistry of Nonaqueous Solvents*, Vol. III (J. J. Lagowski, ed.), Academic Press, 1970, Chapter 4. A survey of the physical and chemical properties of anhydrous H_2S and its ability to act as a solvent.

STUDY QUESTIONS

1. Discuss the observation that SF_6 is markedly more stable toward hydrolysis than are SeF_6 and TeF_6.

2. Discuss the expected order of stability of the alkaline-earth metal peroxides.

3. Give the geometry of the covalently bonded species in the following compounds without recourse to structural data: (a) Cl_2SO; (b) SO_4; (c) H_2S_3; (d) $KTeCl_3$; (e) $[(CH_3)_3Si]_2O$; (f) $ClSO_3H$.

4. Discuss the observation that the temperatures for the dehydration of alkaline-earth-metal hydroxides increase in the following order,

$$Mg(OH)_2 < Ca(OH)_2 < Sr(OH)_3 < Ba(OH)_2,$$

in light of the fact that the magnesium salts are better dehydrating agents than are barium salts.

5. Oxygen forms a very large number of ionic oxides. The process $O_{(g)} + 2e^- \rightarrow O^{2-}_{(g)}$ is endothermic to the extent of 156 kcal mole^{-1}. Discuss these facts.

6. Which of the species in Question 3 exhibit multiple bonds? Discuss the nature of these multiple bonds in terms of the orbitals involved.

7. The experimental standard enthalpies of formation for SiS_2 and $SiSe_2$ are 60 and 27 kcal mole^{-1}, respectively. Calculate the values of these quantities from the electronegativities of the atoms. Comment on the differences between the experimental and expected quantities.

8. Triphenylsilyl iodide and 2,2'-bipyridine form a $1:1$ complex which gives conducting aqueous solutions. What is the most probable structure of cation present in these solutions?

9. The metal–oxygen–metal bond angle in the compound $(Cl_5Ru)_2O$ is $180°$. Sketch the expected structure and give the hybridization for each atom necessary for the valence-bond description of this molecule.

10. Discuss the idealized hybridization for the atoms in the covalently bonded species in Question 3.

11. Discuss the electrode reactions expected when a solution of HNO_3 in anhydrous sulfuric acid is electrolyzed with inert electrodes.

12. Discuss the molecular-orbital description of S_2 assuming that this species is diamagnetic.

13. Describe the bond order in ozone using the internuclear distances in the species O_2^{n-} ($n = 0, 1, 2$).

14. Discuss the main differences in the structures of the ionic chalconides and the corresponding oxygen-containing compounds.

15. Discuss the Lewis acid–base behavior of the following species: (a) $TeCl_3^-$; (b) SO_2; (c) CS_2; (d) $SOCl_2$; (e) SO_3^{2-}; (f) R_2SBr^+; (g) SeO_2; (h) $SeCl_4$.

16. Give the most probable geometrical structures for the ionic species present when the following substances dissolve in anhydrous sulfuric acid to form conducting solutions: (a) CH_3OH; (b) $RC\overset{O}{\overset{\parallel}{-}}O-C\overset{O}{\overset{\parallel}{-}}R$; (c) R_2CO; (d) $RC\overset{O}{\overset{\parallel}{-}}NR_2$; (e) H_3PO_4.

THE HALOGENS

/ **14**

1 / INTRODUCTION

The ground-state configuration of the halogens, ns^2np^5, is one electron short of that of the next rare gas. Accordingly, these elements acquire one electron to form ionic derivatives of the type X^-, or they can form a simple covalent bond, $X-R$. The large electronegativities and electron affinities of these elements, as shown in Table 14.1, are in accord with their tendency to form anionic species, X^-. Covalent bonds formed to the lighter elements in this family are generally very polar; if the molecule formed by the halogens is unsymmetrical, the compounds are among the most polar known. Much of the chemistry of the halogens can be understood in these terms. In addition, the halogens which are heavier than fluorine can expand their coordination number in the same way as the other representative elements in Periods 3–7. Thus, iodine uses all seven outer electrons to form covalent bonds in the compound IF_7 just as sulfur is capable of using all of its outer electrons in

TABLE 14.1 Some properties of the halogens

Property	F	Cl	Br	I
Ground-state configuration	[He] $2s^2 2p^5$	[Ne] $3s^2 3p^5$	[Ar] $3d^{10}4s^2 4p^5$	[Kr] $4d^{10}5s^2 5p^5$
Ionization potential, kcal	402	300	273	241
Electron affinity, kcal	79.2	83.3	77.5	70.6
Electronegativity	4.0	3.0	2.8	2.4
Radii, Å				
ionic (X^-)	1.31	1.81	1.96	2.22
covalent	0.72	0.99	1.14	1.33
Heat of dissociation, X_2, kcal mole^{-1}	37.7	58.2	46.1	36.1
Heat of hydration, X^-, kcal mole^{-1}	121	88	80	70
Molar solubility in H_2O, 25°C	—	9.1×10^{-2}	0.21	1.3×10^{-3}
E^0_{298}, $X^-/\frac{1}{2}X_2$, V	−2.85	−1.36	−1.07	−0.54
Mp, °C	−223	−102	−7.3	114
Bp, °C	−187	−34.6	58.78	183
Color	yellow	pale green	red-brown	violet-blue

forming SF_6. Not only is it possible to achieve high coordination numbers among the heavier halogens, but compounds in which the halogen exhibits a high positive oxidation state, such as in $HClO_4$, are formed. Fluorine, which has the highest electronegativity among all of the elements, shows no tendency to form positively charged species.

As shown by the data in Table 14.1, the first halogen displays properties markedly different from those of the other members of the family. Thus, chlorine, bromine, and iodine possess smoothly varying physical properties, but those of fluorine do not follow the same trends. For example, the ionization potential, electronegativity, heats of dissociation and of hydration, and electrode potentials are all higher for fluorine than expected from the trends established by the other halogens. Also, the ionic and covalent radii, melting and boiling points, and the electron affinity of fluorine are lower than expected. The differences between fluorine and the heavier members of the family appear to be more pronounced than the corresponding differences in other families.

About twenty isotopes of astatine are known but all are relatively short-lived. The most stable isotope has a half-life of 8.3 hours; the chemistry of astatine has been obtained from tracer studies. In general, astatine exhibits behavior expected for the halogen heavier than iodine (*I*).

2 / THE ELEMENTS

Under ordinary conditions the halogens exist as covalent diatomic molecules, X_2, each atom acquiring the electronic configuration of the next rare gas by sharing one pair of electrons. The stability of this structure is reflected in the high dissociation energies for the molecules shown in Table 14.1. Surprisingly, the dissociation energy for fluorine is markedly lower than that of any other halogen. Perhaps the best explanation for this unexpected behavior comes from a consideration of the electronic environment of the X_2 molecule. The internuclear distance for fluorine is the shortest among all the halogens, leading to a greater repulsion of nonbonded electron pairs on the two atoms of the molecule. There have also been suggestions that the heavier halogens may incorporate a $d\pi$-$p\pi$ component in the bond for the molecule X_2, which, of course, is not possible for molecular fluorine. The heavier halogen atoms form multiple bonds to other atoms, using the usual criteria, but little direct evidence for multiple bond formation exists for the diatomic halogen molecules.

2.1 / PREPARATION

The halogens are much too reactive to be found in nature in the elemental state. They occur almost entirely as ionic derivatives of the more active elements. Most of the fluoride-containing minerals, such as CaF_2 (fluorspar or fluorite), Na_3AlF_6 (cryolite), and $3Ca_3(PO_4)_2 \cdot CaFCl$ (fluorapatite) are insoluble. The chlorides and bromides, on the other hand, are more soluble. Large deposits of NaCl are known; brines containing high concentrations of chlorides or bromides occur naturally. Iodine in the form of I^- is found in brines; iodine has also been found naturally in the form of iodates.

Fluorine   The most powerful chemical oxidizing agent, fluorine, is prepared by the electrolytic reduction of F^-. The standard potential for the reduction of F^- is sufficiently high that the process will not occur in aqueous solutions, O_2 being liberated at the anode in preference to F_2 under these conditions. Elemental fluorine is prepared by electrolyzing a mixture of KF dissolved in anhydrous HF. Mixtures containing less than 30% KF are liquid below 0°C, mixtures containing 40–65% KF are liquids in the range 50–100°C, and the equimolar mixture KF·HF is a liquid at 250°. Electrolysis of any of these liquid mixtures using inert electrodes gives H_2 at the cathode and F_2 at the anode. Steel or silver cathodes and amorphous carbon or graphite anodes are usually employed in the electrolysis. The cells are constructed of steel, copper, or monel alloy. The exposed metal parts become coated initially with a thin layer of metal fluoride which is impervious to further reaction.

Chlorine   Chlorine is prepared in large quantities by the electrolysis of brine; hydrogen and aqueous sodium hydroxide also are formed in the process

$$Na^+ + Cl^- + H_2O \rightarrow Na^+ + OH^- + \tfrac{1}{2}Cl_2 + \tfrac{1}{2}H_2. \tag{1}$$

Although chlorine is a powerful oxidizing agent, it can be prepared from Cl^- using chemical oxidants such as MnO_2

$$2Cl^- + 4H^+ + MnO_2 \rightarrow Mn^{2+} + 2H_2O + Cl_2. \tag{2}$$

Generally it is unnecessary to prepare chlorine in the laboratory because it is readily available commercially.

Bromine and Iodine   Bromine is obtained from the oxidation of Br^- found in brines and sea water. Chlorine is used as an oxidizing agent in the process, which is carried out in acid solution

$$2Br^- + Cl_2 \rightarrow Br_2 + 2Cl^-. \tag{3}$$

The liberated bromine is concentrated by sweeping it out of the reaction mixture with a stream of air. Bromine can be prepared conveniently in small quantities by oxidizing Br^- with MnO_2 in acid solution [cf., Eq. (2)].

Iodine is prepared commercially from the oxidation of I^-, as it occurs in brines, using chlorine [cf. Eq. (3)], or from the reduction of iodates; the latter process is used for the commercial preparation of iodine using carbon as a reducing agent. Iodates occur in useful concentrations in certain areas of the Chilean saltpeter beds.

2.2   Chemical Properties

The halogens react primarily as oxidizing agents, reacting with practically all metals

$$2M + nX_2 \rightarrow 2MX_n \tag{4}$$

and many nonmetals to form the corresponding halides. The reactions of the halogens with the representative elements are described in the appropriate chapters of this volume. The more active halogens react also with several classes of compounds. For example, hydrogen is displaced from hydrocarbons

$$X_2 + C_nH_{2n+2} \rightarrow C_nH_{2n+1}X + HX, \tag{5}$$

and oxygen from metal oxides if carbon is also present

$$yX_2 + 2MO_z + 2zC \rightarrow 2zCO + 2MX_y, \tag{6}$$

when these compounds react with halogens. Halogens add readily to unsaturated molecules:

$$C_nH_{2n} + X_2 \rightarrow C_nH_{2n}X_2 \tag{7}$$

$$X = Cl, Br$$

$$CO + X_2 \rightarrow X_2CO \tag{8}$$

$$X = Cl, Br$$

$$SO_2 + X_2 \rightarrow SO_2X_2 \tag{9}$$

Two important reactions occur when halogens are passed into water: water can be oxidized or the halogen molecule undergoes hydrolysis. A consideration of the standard potentials shown in Table 14.1 for the processes in which the halogen acts as an oxidizing agent

$$X_2 + 2e^- \rightleftharpoons 2X^- \tag{10}$$

and water is oxidized

$$H_2O \rightleftharpoons \tfrac{1}{2}O_2 + 2H^+ + 2e^- \tag{11}$$

shows that fluorine and chlorine are sufficiently strong to oxidize water ($E_0 = 1.23$ V) under standard conditions

$$X_2 + H_2O \rightleftharpoons \tfrac{1}{2}O_2 + 2X^- + 2H^+, \tag{12}$$

but bromine and iodine are not. In fact the partial pressure of oxygen in the air is sufficiently large to effect the oxidation of iodide ion to iodine in aqueous solution. The extent of hydrolysis of halogen molecules

$$X_2 + H_2O \rightleftharpoons H^+ + X^- + HOX \tag{13}$$

decreases with increasing atomic number. Thus, at 25°C the equilibrium constants for Eq. (13) are 4.2×10^{-2}, 7.2×10^{-9}, and 2.0×10^{-13}, respectively, for Cl_2, Br_2, and I_2. We might expect F_2 to be the most extensively hydrolyzed halogen molecule, but the oxidation process [Eq. (11)] is so vigorous that it is the dominant reaction.

Positive Oxidation States. Except for fluorine, the halogens form anions which exhibit positive oxidation states. In addition, systems containing iodine cations are known (II), which is in accord with the general observation that the heaviest member of a family has the most metallic character. The metallic properties of iodine are not fully developed, since it is near the end of a period. On this basis, we might expect astatine to exhibit the most pronounced metallic characteristics of the halogens. However, the practical difficulties associated with studying astatine generally preclude a detailed consideration of the chemistry of this element.

The halogens are too electronegative for simple species such as X^+ to exist. Accordingly, it is not surprising that cationic halogen species can be stabilized by Lewis bases. Compounds with the formulation $[X(py)_2]^+A^-$ (X = halogen, py = pyridine, A = univalent anion) have been prepared by several methods. Two of the most useful methods involve the treatment of a mercurous or silver salt with a mixture of iodine and pyridine in a nonaqueous solvent:

$$AgA + 2py + I_2 \rightarrow [I(py)_2]^+A^- + AgI \tag{14}$$

$$Hg_2A_2 + 8py + 3I_2 \rightarrow 2[I(py)_2]^+A^- + 2Hg(py)_2I_2. \tag{15}$$

Methyl-substituted pyridines can also be used in these reactions. Compounds containing cationic halogens which have been prepared include $X(py)_2NO_3$ (X = Cl, Br, I) and $X(py)_2ClO_4$ (X = Br, I). Such compounds can be imagined formally as salts of the hypothetical base $X(py)_2OH$.

Iodonium cations are the most stable of the halogen derivatives and, consequently, have been the most investigated. The iodonium compounds react with iodide ion in basic solution to give quantitatively free iodine. They slowly hydrolyze and disproportionate to give molecular iodine and iodate ion; treatment with sodium hydroxide yields $[I(py)_2]_2O$, the formal anhydride of the hypothetical base $I(py)_2OH$. As might be expected, cationic halogen compounds are good oxidizing agents. Solutions of the compound $I(py)_2NO_3$ in chloroform react with the active metals as well as with platinum to replace the iodine. The electrolysis of $I(py)_2NO_3$ in chloroform, methanol, or water–pyridine mixtures yields iodine at the cathode, supporting the suggestion that iodine is positively charged in such compounds. Treatment of compounds containing $I(py)_2^+$ with KCl or KBr gives the mixed halides with only one coordinated pyridine molecule, $I(py)Cl$ or $I(py)Br$. Other derivatives containing one pyridine molecule are known, e.g., $I(py)NO_3$, $I(py)A$ (A = acetate, benzoate, succinate, phthalate). Such compounds do not appear to be ionic because thay have markedly lower conductivities in acetone than do the compounds containing two pyridine molecules. Thus, these compounds are best formulated as $[I(py)A]$ in which the iodine is still two-coordinate as it is in the ionic species $[I(py)_2]^+A^-$.

Although no direct evidence is available concerning the structure of the cation $I(py)_3^+$, the structure of $I(py)Cl$ has been determined to be that shown

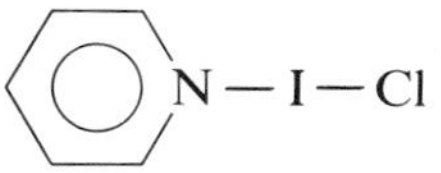

$$d(N-I) = 2.26 \text{ Å}$$
$$d(I-Cl) = 2.51 \text{ Å}$$
$$\angle(NICl) = 180°$$

FIG. 14.1   The structure of the pyridine–ICl adduct.

in Fig. 14.1. The molecule exhibits a linear $N-I-Cl$ arrangement of atoms (**1**), as expected from electron-pair-repulsion arguments, with an $I-Cl$ distance slightly larger than the expected single-bond distance (2.42 Å). The

$$C_5H_5N-\ddot{\underset{..}{I}}-Cl$$

1

$N-I$ distance is essentially the same as that observed in the iodine–trimethyl-amine adduct. The linear coordination of the I^+ species shown in Fig. 14.1 is evident also in the cation $Br(py)_2{}^+$ (*1*), as well as in the thiourea complex $[(H_2N)_2CS]_2I^+I^-$ (*2*). Both Br_2 and Cl_2 have been reported to form $X_2{}^+$ (*3, 4*) although there is some question concerning the interpretation of the data for the chlorine species (*5*).

The nmr spectra of alkyl halides dissolved in the mixed solvent system SbF_5-SO_3 indicate the presence of dialkylhalonium ions, $RR'X^+$ (X = Cl, Br, I; R, R' = CH_3, C_2H_5, C_3H_7) (*6–8*). Such species have been suggested as intermediates in the chlorine exchange reaction between $GaCl_3$ and CH_3Cl.

Solutions of iodine in very acidic media exhibit properties attributed to the presence of cationic iodine-containing species. Iodine dissolved in anhydrous sulfuric acid gives solutions with an absorption spectrum similar to that of iodine vapor, indicating that the major species present is I_2. On the other hand, solutions of iodine in 30 % oleum or in fluorosulfuric acid containing $S_2O_6F_2$ (*9, 10*) are brown and contain $I_2{}^+$, $I_3{}^+$, or $I_5{}^+$ cations depending upon the concentration. At low temperatures, in fluorosulfuric acid, the species $I_2{}^+$ dimerizes to $I_4{}^{2+}$ (*11*). The brown solutions turn blue when an oxidizing agent such as KIO_3 is added, the blue color being attributed to I^+

$$2I_3{}^+ + 9H_2SO_4 + IO_3{}^- \rightarrow 7I^+ + 3H_3O^+ + 9HSO_4{}^-. \quad (16)$$

Iodine and $S_2O_6F_2$ dissolve in fluorosulfuric acid (HSO_3F) to give solutions which exhibit spectra characteristic of I^+ and $I_3{}^+$. Solutions of iodine in IF_5 are brown, the spectra being similar to those of pyridine–iodine solutions; the addition of water gives dark blue paramagnetic solutions similar in color to the sulfuric acid solutions. The blue species in IF_5 solutions migrates to the cathode in electrolysis experiments. Such properties are to be expected for species containing I^+. The abnormally high conductivity of molten iodine has been attributed to the formation of cationic iodine species

$$3I_2 \rightleftharpoons I_3^+ + I_3^-. \tag{17}$$

2.3 / Physical Properties

The halogens form diatomic molecules; under normal conditions fluorine and chlorine are gases, while bromine and iodine form volatile condensed phases. The diatomic molecules which occur in the vapor phase persist in the solid state. Spectroscopic measurements on the vapor phase (*12–15*) and on solutions (*16*) suggest that the equilibrium

$$2X_2 \rightleftharpoons X_4 \tag{18}$$

$$X = I, Br$$

occurs. Evidence for the existence of I_4 adsorbed on silica gel also is available (*17*).

The molecules in crystalline Cl_2, Br_2, and I_2 are arranged in layers with interatomic distances essentially the same as those observed in the vapor state; the relevant data are collected in Table 14.2. The packing of molecules

TABLE 14.2   Structural parameters for the halogen molecules

	Interatomic distances, Å		Intermolecular distance, Å		
	Vapor	Solid	Within layer	Between layers	$2r,^a$ Å
F_2	1.42	—	—	—	—
Cl_2	1.99	2.02	3.34	3.69	3.6
Br_2	2.28	2.27	—	3.78	3.9
I_2	2.67	2.68	3.54	4.06	4.3

$^a r$ = van der Waals radius.

is such, however, that the distance between adjacent atoms of two different molecules within a layer is less than twice the van der Waals radius. These data suggest that some degree of intermolecular bonding occurs in the condensed states of the halogens. The Raman spectra of X_2 crystals at low temperatures also indicate strong intermolecular forces, those in Br_2 being relatively stronger than the forces between Cl_2 molecules (*18–20*). A study of the quadrupole coupling in crystalline I_2 indicates that the interlayer contacts correspond to about 9% covalent bonding (*21*). Similar interactions are observed in other systems containing halogen molecules.

The halogens, except for fluorine, which is very reactive, are slightly soluble in water and readily soluble in a variety of organic solvents. The species which exist in aqueous solution are described in the previous section, but it is also possible to obtain crystalline hydrates of chlorine ($Cl_2 \cdot 7.3H_2O$) and bromine ($Br_2 \cdot 8.5H_2O$). These compounds are formed when water is crystallized in the presence of halogens. The halogen hydrates are not compounds in the chemical sense; rather they exist because the halogen molecule is incorporated into a hydrogen-bonded open arrangement of water molecules which acts as a trap for the halogen molecule. Systems in which a small molecule is trapped in a lattice formed by other molecules are called clathrate compounds. Normally water crystallizes in an open structure in which each molecule is hydrogen-bonded to four other molecules. Although

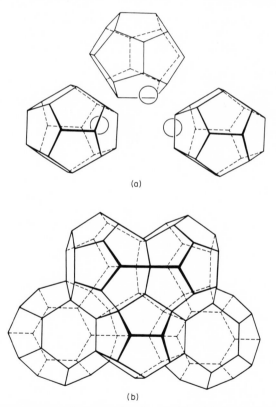

(a)

(b)

FIG. 14.2   The arrangement of water molecules in (a) $Cl_2 \cdot 7.3H_2O$ and (b) $Cl_2 \cdot 8.5H_2O$ crystals. Hydrogen bonds are formed along the edges of the polyhedra indicated. Hydrogen bonding occurs between dodecahedra in (a) as well as between dodecahedra and interstitial water molecules indicated by the open circles. The dodecahedra in the higher hydrate (b) are grouped with tetrakaidecahedra.

the structure has been described as an open structure, it is, relatively speaking, compact compared to other possible three-dimensional frameworks in which the tetrahedral angle is maintained. These more open structures are unstable unless the voids within them are occupied. One such open structure for water, shown in Fig. 14.2, involves 46 water molecules per unit cell arranged to form two pentagonal dodecahedra and six polyhedra with two hexagonal faces and eight pentagonal faces. The chlorine molecules occupy the latter polyhedra, giving an empirical formula of $Cl_2 \cdot 7.3H_2O$. Oxygen atoms occupy the corners of the polyhedra shown in Fig. 14.2, and hydrogen bonds of the type $O\!-\!H\cdots O$ are directed along the edges. The $O\!-\!O$ distance in this structure is 2.78 Å, which is very close to that observed for normal ice (2.76 Å).

The halogens also form crystalline adducts with Lewis bases such as ethers ($C_4H_8O_2 \cdot Br_2$), organic sulfides ($C_4H_8S_2 \cdot I_2$), ketones [$(CH_3)_2CO \cdot Br_2$], and amines [$(CH_3)_3N \cdot I_2$]. The adducts all have a basic chain structure containing linear halogen–halogen units, as shown in Fig. 14.3; the $X\!-\!X$ bond distance

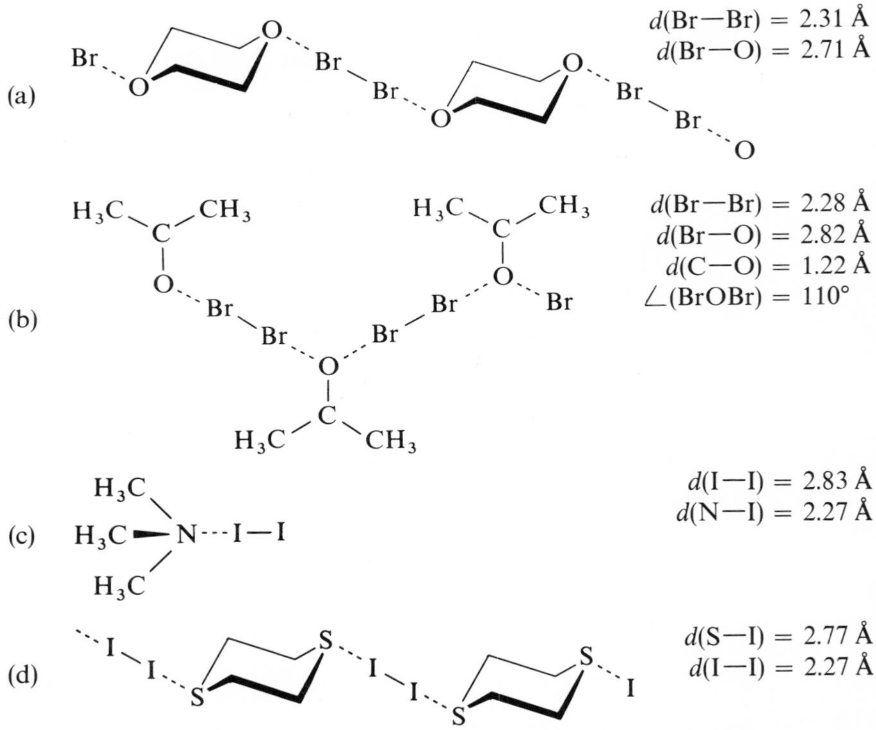

$$d(Br\!-\!Br) = 2.31 \text{ Å}$$
$$d(Br\!-\!O) = 2.71 \text{ Å}$$

$$d(Br\!-\!Br) = 2.28 \text{ Å}$$
$$d(Br\!-\!O) = 2.82 \text{ Å}$$
$$d(C\!-\!O) = 1.22 \text{ Å}$$
$$\angle(BrOBr) = 110°$$

$$d(I\!-\!I) = 2.83 \text{ Å}$$
$$d(N\!-\!I) = 2.27 \text{ Å}$$

$$d(S\!-\!I) = 2.77 \text{ Å}$$
$$d(I\!-\!I) = 2.27 \text{ Å}$$

FIG. 14.3 The structural parameters of some halogen adducts with typical Lewis bases.

in these adducts is essentially that found in the free halogen molecule (Table 14.2). The stability of these adducts is attributed to the contribution of two major canonical forms (**2** and **3**) to the wave function; structure **2** predominates in the resonance hybrid (*III*).

$$X—X:B \qquad\qquad X—X:^- \ B^+$$

$$\textbf{2} \qquad\qquad\qquad \textbf{3}$$

In addition to the adducts of the halogens which can be isolated as crystalline compounds, weaker complexes can be detected in solution (*IV*). The most thoroughly studied examples involve iodine, but chlorine and bromine show similar behavior. Such complexes have been extensively studied using spectroscopic techniques. For example, iodine dissolves in a variety of organic solvents, forming solutions that are colored purple, brown, and every possible shade in between. The purple solutions formed by CCl_4 and saturated hydrocarbons have the same visible absorption spectrum as does iodine in the vapor phase and they presumably contain molecular iodine. However, solutions formed in any solvent such as ether or alcohol containing a basic center exhibit a new feature in their spectra. This is usually a single very intense band in the region 4500–3500 Å. This band is attributed to a charge-transfer transition in which the wave function of the excited state contains a greater proportion of structure **3**, hence the name charge-transfer complex.

3 / COMPOUNDS OF THE ELEMENTS

3.1 / Binary Halogen Compounds

The halogens form compounds with all of the elements except the three lightest rare gases. The binary halides represent one of the most important classes of compounds from theoretical, synthetic, and practically useful standpoints. Because the electronegativities exhibited by the halogens vary widely (Table 14.1), it is not surprising that the binary halides of the elements possess a wide range of properties. Indeed, there is nearly a continuous gradation in the nature of the bonds formed by the halogens from essentially completely ionic to nearly completely covalent. Such a gradation is, of course, reflected in the chemical and physical characteristics of the compounds in question. In spite of these facts, there is a general propensity to classify the halides into ionic and covalent halides. We shall maintain this historical classification in our discussion for the sake of convenience, although the artificial character of such a division will soon be apparent. Much of the detailed chemistry and structural considerations of the halides is included in the discussions of other families of elements. We deal here only with general considerations.

Ionic Halides The ionic halides are sometimes called the salt-like halides. These compounds exist as collections of discrete halogen anions associated with counter ions. The energies available under ambient conditions are not sufficient to overcome the electrostatic interactions between these species, hence the collection of ions exists as a regular array of charged species. Thus, the lattice energy and the relative sizes of the ions in question govern the chemical and physical properties of the ionic halides, as discussed in Chapter 2. It is likely that only the alkali metals and some of the alkaline-earth metals form halides that are truly ionic; the other metal halides exhibit some degree of covalent character. Even in the case of the Group I and II halides, there is evidence that compounds containing highly polarizing cations (Li^+, Be^{2+}, Mg^{2+}) and easily polarizable anions (I^- and to some extent Br^-) exhibit a considerable degree of covalent character. The balance between the polarizability of the halide and the polarizing effect of the cation can be used to understand the trends in a series of halides. The effect of increasing the charge density of a metal ion, and thus its polarizing ability, on the melting point of a halide is seen in the following series of compounds:

	KCl	$CaCl_2$	$SiCl_4$	$TiCl_4$
mp, °C	790	772	-70	-30

Assuming that the melting point is a reasonable measure of the forces acting between the species which comprise a solid substance, a decrease in melting point reflects an increasing degree of covalent character in the M—Cl bond. At the one extreme the lattice of solid KCl consists of individual ions held together by electrostatic forces, whereas that of $TiCl_4$ consists of molecules which are held in the crystal lattice by markedly weaker van der Waals forces. The calculated lattice energies for KCl and $CaCl_2$ are 168 and 531 kcal mole^{-1}, respectively, which is opposite the trend suggested by their melting points. However, the corresponding fluorides exhibit the predicted trends in these properties:

	KF	CaF_2
mp, °C	880	1330
Lattice energy, kcal mole^{-1}	194	623

It would, of course, be expected that of the halides the fluoride ion is the least polarizable and accordingly would form the most perfectly ionic derivatives. The influence of the polarizability of the halogen on the degree of covalent character of a halide can be illustrated by the melting points of the lithium and calcium halides:

	F	Cl	Br	I
mp of LiX, °C	870	617	547	446
Lattice energy, kcal mole^{-1}	247	202	182	169
mp of CaX$_2$, °C	1330	772	760	718
Lattice energy, kcal mole^{-1}	623	537	509	487

The decrease in melting points from the fluoride to the iodide in these two series can also be taken as a reflection of the increase in covalent character of the compounds, although the argument is somewhat masked by the fact that the lattice energies also decrease in the same order.

If an element can form compounds in two or more oxidation states, the halide with the higher oxidation state is usually the more covalent. For example, $SnCl_2$ melts at 246.8°C, while the melting point of $SnCl_4$ is -30.2°C. This trend in properties can be understood by extending the previous arguments on the relative polarizing ability of ions as related to their charge density. Thus, except for the most electropositive elements, the halides of any series of metals can display a wide range of properties, but such variations can be qualitatively interpreted in terms of the factors described here.

The structure of the ionic halides generally reflects the trends in increasing covalency as indicated by the physical properties. As a matter of convenience the structures of the predominantly ionic halides can be considered in one of three classes: (1) three-dimensional structures, (2) layer structures, or (3) chain structures.

In the case of the most electropositive elements, regular three-dimensional arrays are observed for all the halides, as would be expected for completely ionic substances. However, most metals exhibit different structural types for their halogen derivatives. Three-dimensional structures such as the fluorite structure shown in Fig. 14.4a are invariably formed by metal fluorides, whereas the chlorides, bromides, and iodides of the same element exhibit layers as in Fig. 14.4b, or chain-like structures. For example, FeF_3 and BiF_3 form three-dimensional structures, but the corresponding iodides exhibit a layer structure. Both layer and chain structures are exhibited by binary halides which have a high degree of covalent character. Preference for one or the other of these structures stems basically from the valence of the element and the maximum coordination number it can exhibit. Thus, layer structures containing MX_6 octahedra sharing edges occur among the di- and trihalides of the elements in Periods 3–7, e.g., MI_2 (M = Cd, Ca, Mg, Zn, Pb, Mn, Ni, Fe, Co), MBr_2 (M = Mg, Fe, Zn, Co, Cd, Ni), MCl_2 (M = V, Ti, Cd, Fe, Co, Ni, Mg, Mn), MCl_3 (M = Cr, Fe, Ba, Sc, Ti, V), MBr_3 (M = Fe, Ti), and

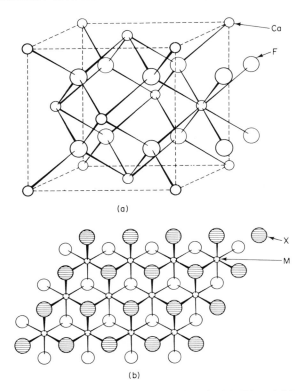

(a)

(b)

FIG. 14.4   Two structures for MX_2 species: (a) fluorite (CaF_2) and (b) a close-packed structure of X atoms with a layer of metal atoms in the octahedral holes.

MI_3 (M = Bi, Sb, As). From a crystallographic point of view, these compounds can be imagined as consisting of closest-packed halide ions with the metal ions in the octahedral holes. It should be recalled from Chapter 2 that there are as many octahedral holes as there are closest-packed ions. The structures of the divalent halides arise from filling half the octahedral holes with cations corresponding to the composition MX_2. If the metal atoms are placed between alternate closest-packed layers of halide ions, the layer structure shown in Fig. 14.4b occurs. If one-third of the octahedral holes are filled with metal atoms, i.e., placing metal atoms in two-thirds of the holes in between alternate close-packed layers, as shown in Fig. 14.5, the compound has the composition MX_3. Both MX_2 and MX_3 structures can be considered as giant molecules within the layers, with only relatively weak van der Waals forces holding the layers together. To complete the crystallographic argument, when all the octahedral holes are filled in the closest-packed structure of halide ions, the compounds have the composition MX with no unique layer in the structure. This, of course, is the sodium-chloride type structure observed for many compounds which appear to be virtually completely ionic.

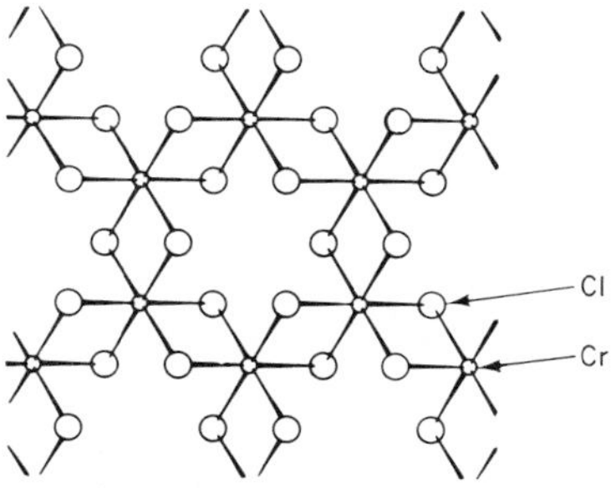

FIG. 14.5 The structure of $CrCl_3$. Metal atoms are octahedrally surrounded by anions with octahedral edges shared.

Halides with some degree of covalent character which crystallize in chain structures are generally formed by elements which exhibit a coordination number of 4 for electronic reasons. For example, both $BeCl_2$, as described in Chapter 9, and $PdCl_2$ crystallize in structures consisting of infinite chains containing halogen bridges. In $BeCl_2$, the basic unit is a $BeCl_4$ tetrahedron which shares opposite edges. The basic unit in $PdCl_2$ is a planar $PdCl_4$ group sharing opposite sides.

Complex Halides   A very large number of chemical systems are known with an empirical composition corresponding to a mixture of two binary halides, $aAX_n \cdot bBX_m$, e.g., $CsI \cdot AgI$, $3NaF \cdot AlF_3$, $3NH_4F \cdot ZrF_4$, $3NaF \cdot TaF_5$, and $3CsCl \cdot 2TlCl_3$. As a matter of convenience, such mixed halides or double salts are written as $A_aB_bX_{n+m}$ where A represents the more electropositive element in the mixture. Understanding the nature of such mixed halides usually requires a considerable experimental effort because of the variety of structural arrangements possible. Little can be done with only general empirical formulations; for example, $KMgF_3$ and $CsAuCl_3$ have distinctly different structures although they have the same formulation. The latter compound is actually a triple salt containing gold in two different oxidation states (Au^+ and Au^{3+}) and coordination numbers; the correct formulation is $Cs_2[AuCl_2][AuCl_4]$. Halides formulated as $A_aB_bX_{n+m}$ do not always contain complex B_bX_{n+m} anions which can become associated with different cations by suitable metathetical reactions. In many instances the "anions" B_bX_{n+m} can be identified crystallographically as coherent units but they have little significance from a chemical viewpoint. Thus, $CaPbF_6$ has a three-dimensional structure containing no discernable complex anion; however, $SrPbF_6$

contains linear arrangements of PbF_6 octahedra sharing opposite corners, and the remaining fluorine is present as F^- in the crystal lattice. The compound $BaPbF_6$ contains finite octahedral PbF_6^{2-} ions and Ba^{2+} ions arranged in a distorted CsCl structure. The PbF_6^{2-} ion is known to exist in solution, but the chains of PbF_6 octahedra which would correspond to the anion $(PbF_5)_n^{n-}$ observed in crystalline $SrPbF_6$ do not exist in solution. Other possible arrangements of the halide ions in the mixed metal halides complicate the issue. The formulation B_bX_{n+m} could correspond to discrete polynuclear complex anions in the solid state or in solution; for example, the species $Tl_2Cl_9^{3-}$ is present in the mixed halide $3CsCl \cdot 2TlCl_3$, indicating the correct formulation to be $Cs_3Tl_2Cl_9$. The formulation B_bX_{n+m} could also correspond to a mixture of a stable complex haloanion and a halide ion. For example, Cs_3CoCl_5 does not contain five-coordinate cobalt, but is rather expressed as $Cs_3(CoCl_4)Cl$. The role of the halides as coordinating agents toward transition-metal ions is described in Chapter 17.

Covalent Halides   The covalently bound halides are formed by the nonmetals. They consist of molecular units held together by relatively strong covalent bonds, but the molecules are held in a crystalline lattice by weak intermolecular forces. Thus it takes relatively little energy to separate the molecular units from each other.

The halogen atoms in covalent halides generally crystallize in a close-packed structure of halogen atoms, with the counter atom occupying a hole in this array. For example, the structures of crystalline GeI_4 and of SnI_4, shown in Fig. 14.6, consist of tetrahedral molecules arranged so that the

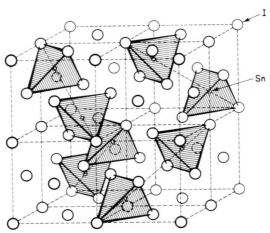

FIG. 14.6 The structure of crystalline SnI_4. The iodine atoms are at the points of a cubic face-centered lattice with one-eighth of the tetrahedral holes occupied by tin atoms, which leads to discrete SnI_4 molecules.

iodine atoms are close packed with germanium or tin atoms at one-eighth of the tetrahedral sites. This gives the correct composition, MX_4, since there are twice as many tetrahedral holes as close-packed halogen atoms. The interactions between the metal and the tetrahedrally disposed halogen atoms are large compared with the interactions between tetrahedra. It is apparent from Fig. 14.6 that the presence of tin or germanium atoms in the remaining tetrahedral sites in the structure would add rigidity to the overall structure, since the halogens would then be shared. In fact the layer structure of CaI_2 contains the same arrangement of iodide ions as the close-packed structure of SnI_4 shown in Fig. 14.6. In the former case the octahedral holes are occupied, in the latter one-eighth of the tetrahedral holes are filled. In the case of $AlBr_3$ the halogen atoms are still close-packed, but the aluminum atoms occupy pairs of adjacent tetrahedral holes. In contrast to the structure of GeI_4 shown in Fig. 14.6, this arrangement leads to two $AlBr_4$ tetrahedra sharing edges. Since the forces acting between tetrahedra are weak, this substance is relatively easily vaporized, but the molecules are dimeric, Al_2Br_6. It should now be apparent that descriptions of the crystal structures of many of the covalent halides are basically a continuation of those of the ionic halides. Usually the most interesting aspects of covalent compounds to chemists are the arrangements of the atoms which make up the molecule and not the structure of the crystalline molecular species. The latter arrangement is governed primarily by the shapes of the molecules and is not of general interest.

The covalent halides are generally characterized by their relatively high volatility, solubility in nonpolar solvents, and a lack of conductivity in the molten state. They react with solvents containing active hydrogen atoms to give the corresponding hydrogen halide. In the case of water, the covalent halides react to give a hydroxy compound which is acidic

$$M-X + H_2O \rightarrow M-OH + HX. \tag{19}$$

Thus, the covalent halides are often called acid halides. Solvolysis reactions also occur with alcohols, ammonia, and substituted amines (RNH_2, R_2NH). A more detailed description of the chemical properties of covalent halides appears in Chapters 10–12.

3.2 / Hydrogen Derivatives

One of the most important classes of covalent halogen compounds is that formed with hydrogen. These compounds can be prepared by direct reaction of the elements

$$H_2 + X_2 \rightarrow 2HX, \tag{20}$$

by heating an anhydrous metal halide with a strong nonvolatile acid

$$MX + HY \rightarrow MY + HX_{(g)}, \qquad (21)$$

by hydrolysis of a nonmetal halide such as PX_3

$$PX_3 + 3H_2O \rightarrow P(OH)_3 + 3HX_{(g)}, \qquad (22)$$

by reduction of a noble-metal halide with hydrogen

$$2AgX + H_{2\,(g)} \rightarrow 2Ag + 2HX_{(g)}, \qquad (23)$$

or by reaction of the halogen with a covalent hydride

$$2X_2 + H_2S \rightarrow SX_2 + 2HX \qquad (24)$$

$$4X_2 + CH_4 \rightarrow CX_4 + 4HX. \qquad (25)$$

Because of the ease of oxidation of Br^- and I^- (Table 14.1) it is important that a nonoxidizing acid such as H_3PO_4 be used when HBr and HI are prepared according to the process shown in Eq. (21); most strong acids that are also oxidizing agents, such as H_2SO_4 and HNO_3, will produce the corresponding halogen from these halides.

The pertinent physical properties of the hydrogen halides are collected in Table 14.3; they are volatile substances, as would be expected, and very soluble in water. Except for HF, which is a relatively weak acid ($K = 7.2 \times 10^{-4}$), the hydrogen halides are strong electrolytes in aqueous solution.

TABLE 14.3 Some physical properties of the hydrogen halides

Property	HF	HCl	HBr	HI
Mp, °C	−83.0	−114.6	−88.5	−50.9
Bp, °C	19.5	−84.1	−67.0	−35.0
Trouton's constant, cal mole^{-1}	24.7	20.4	20.4	19.8
ΔH, kcal mole^{-1} at 20°C	−64.4	−21.9	−7.3	−1.32
Percent dissociation at 1000°C	—	1.4×10^{-2}	5×10^{-1}	33
Molal solubility at 1 atm and 20°C	1.77	1.18	0.61	0.45
Apparent percent dissociation, 1N aqueous solution at 18°C	10	93	93	95
Dielectric constant	175 (−75°)	93 (−95°)	7.0 (−85°)	3.4 (−50°)
Specific conductivity × 10^7, Ω^{-1} cm^{-1}	0.1 (−80°)	0.0035 (−85°)	0.0014 (−84°)	0.00085 (−45°)
d(H—X), Å	0.917	1.275	1.410	1.609
Bond energy, kcal mole^{-1}	136	103	87	71 •
Dipole moment, D	1.98	1.03	0.79	0.38

Considerable evidence exists for hydrogen bonding in anhydrous HF. For example, the value of Trouton's constant, the entropy of vaporization of a substance at its normal boiling point, is high (Table 14.3) suggesting that the liquid is associated; Trouton's constant for the other hydrogen halides is that expected for unassociated liquids. Since each fluorine atom in a molecule of hydrogen fluoride carries only one hydrogen atom and three unshared electron pairs, the association cannot lead to a three-dimensional structure. Above 80° at one atmosphere pressure HF vapor is monomeric; however, at lower temperatures and higher pressures vapor-density data indicate varying degrees of association. It was initially suggested that the data obtained in the temperature range $-39°$ to $+16°$ could be explained in terms of a monomer–hexamer equilibrium

$$6HF \rightleftharpoons (HF)_6, \tag{26}$$

the hexamer being assigned a cyclic hydrogen-bonded structure on the basis of valence-bond arguments. However additional pressure–volume measurements, data from dielectric polarization experiments, and infrared data led to the conclusion that the species present in the gas phase are short-chain polymers of mixed composition with tetramers and hexamers predominating. Electron diffraction patterns, on the other hand, are best interpreted in terms of the presence of only a monomer and a puckered cyclic hexamer (22). The mean FFF angle is 104° and the ring is very flexible, passing readily through the possible conformations expected for a six-membered ring. The zig-zag chain structure shown in Fig. 14.7 was reported

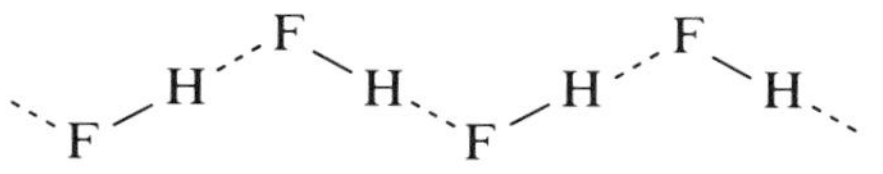

$$d(H—F) = 0.92 \text{ Å}$$
$$d(H\cdots F) = 1.57 \text{ Å}$$
$$\angle(FFF) = 120.1°$$

FIG. 14.7   The structural parameters of solid HF.

for solid HF based upon x-ray diffraction experiments, the bond angle being larger than that reported for the gas phase (23). Hydrogen-bonded structures involving HF are also known in solvated metal fluorides. The best characterized are the so-called bifluorides of the alkali metals, MHF_2, which contain discrete linear F—H—F anions; the F—F distance in the potassium salt (2.26 Å) is markedly less than that in solid HF (2.49 Å), which is probably due to the fact that HF is hydrogen-bonded to an anion in this case. Solvates containing a larger number of HF molecules exist ($KF\cdot2HF$, $KF\cdot3HF$, and $KF\cdot4HF$), but little is known of their structures.

Rather less evidence exists for hydrogen bonding in the other hydrogen halides in the liquid or vapor state. The data in Table 14.4 permit a comparison of the proton chemical shifts of a series of hydrides in the gas phase with similar data obtained on the liquid phase, which shows that the proton signal moves to lower fields in compounds which are known to form strong hydrogen bonds in the liquid phase (24). The association shifts shown in Table 14.4 cannot be accounted for on the basis of bulk susceptibility corrections, and it is argued that these data indicate the existence of hydrogen

TABLE 14.4   Association proton
nmr shifts of liquid hydrides

Compound	Liquid association shift, ppm^a	Temperature °C
CH_4	0	-98
NH_3	1.05	-77
H_2O	4.58	0
H_2S	1.50	-61
HF	>6.65	-60
HCl	2.05	-86
HBr	1.78	-67
HI	2.25	-5
HCN	1.65	-18

^{a} The proton nmr shifts are at lower field relative to the corresponding signals in the gas phase.

bonding in the liquid state. Using such criteria, all of the hydrogen halides are hydrogen-bonded liquids; in contrast, only the value of Trouton's constant for HF, given in Table 14.3, is abnormal. Infrared data on solid HCl and HBr suggest that the molecules in these substances exist as zig-zag chains rather than as oriented dipoles (25, 26). Evidence for the association of HX molecules (X = Cl, Br, I) has also been obtained from infrared measurements on these substances isolated in low-temperature matrices (27).

The anhydrous hydrogen halides have drawn interest as solvents, HF being the most extensively investigated of these compounds, although interest in the other compounds as solvents has been high in recent years (V). A consideration of the HF solvent system is presented in Chapter 7 (VI). The low conductivities of pure HCl, HBr, and HI, shown in Table 14.3, have been interpreted in terms of a self-ionization process

$$3HX \rightleftharpoons H_2X^+ + HX_2^- \tag{27}$$

in which solvation of both the anion and cation occurs. A great deal of evidence is available for the existence of the solvated proton in HF, but the proof for the existence of similar species in the other hydrogen halides is tenuous. The species H_2Cl^+ and H_2Br^+ have been detected in the mass spectra of the pure compounds, and the proton affinity of HCl has been estimated as > 120 kcal mole^{-1} from such data (28). The existence of the bihalide anions HX_2^- is on a much firmer basis (29). A variety of compounds containing bihalide anions have been prepared and characterized, including those containing the same halogen atoms (HBr_2^-, HI_2^-) as well as those with different halogen atoms ($BrHCl^-$, $IHCl^-$) (VII). The stability of such species is markedly enhanced if large cations, such as R_4N^+, are present as counter ions.

A surprisingly large number of covalent species are soluble in liquid HCl, HBr, and HI. Ionic compounds are not, in general, very soluble because of the low dielectric constants of the hydrogen halides (Table 14.3). It has been suggested that the radius of the cation in an ionic compound must be greater than 2.2 Å before an appreciable solubility is exhibited in these liquids. Acid–base phenomena in these solutions can be interpreted in one of two ways. That is, acids can be defined as proton donors or halide ion acceptors, depending upon the emphasis to be stressed in the process under consideration. Thus, compounds of boron

$$BX_3 + 2HX \longrightarrow H_2X^+ + BX_4^- \tag{28}$$

and the elements in Group IV, Group V, and Group VII

$$SnX_4 + 4HX \longrightarrow 2H_2X^+ + SnX_6^{2-} \tag{29}$$

$$2R_4NCl + 3PF_5 \longrightarrow 2R_4N^+ + 2PF_6^- + PF_3Cl_2 \tag{30}$$

$$ICl + 2HCl \longrightarrow H_2Cl^+ + ICl_2^- \tag{31}$$

behave as acids in the liquid hydrogen halides. Two types of solutes can act as bases in liquid HCl, HBr, and HI. Ionic compounds which contain halide ion and dissolve in these substances are bases, as are covalent compounds containing electron pairs which can be protonated. Tetraalkylammonium halides are sufficiently soluble in the hydrogen halides to give reasonably high concentrations of halide ions. In addition, some covalent halogen compounds can also ionize to give halide ions. For example, the phosphorus pentahalides act as halide ion donors rather than as acceptors when dissolved in hydrogen halides:

$$PX_5 + HX \longrightarrow PX_4^+ + HX_2^-. \tag{32}$$

$$X = Cl, Br.$$

Triphenylmethyl chloride also acts as a base in liquid HCl in the sense of increasing the halide ion concentration; in this case ionization occurs presumably because of the stability of the triphenylmethyl carbonium ion. Many types of compounds can be protonated in liquid hydrogen halides to give basic solutions by displacement of equilibrium (27) to the right. Thus, amines, PH_3, $(C_6H_5)_3P$, $(C_6H_5)_3As$, oxygen-containing compounds such as ROH, ROR, carbonyl compounds, and olefins such as 1,1-diphenylethylene act as bases in liquid hydrogen halides. Many of the reactions of acids and bases in these solvents have been investigated by conductometric methods, and the interpretations have been supported in some instances by isolation of the corresponding salts.

3.3 / Oxygen Derivatives

Oxides (*VIII*)   All of the halogens form binary oxygen derivatives. In a chemical sense the oxygen compounds formed with all of the halogens except fluorine are properly called oxides because oxygen is the most electronegative of these elements. However, the fluorine compounds are named fluorides on the basis of the same argument. The chlorine and iodine oxides have been the most extensively investigated of this series of compounds. As usual, there is a difference in the characteristics and behavior of the fluorine derivatives when they are compared to analogous compounds of the heavier elements in this family.

Fluorine forms several oxygen compounds with the general formula O_nF_2 ($n = 1$–6) by reaction of the elements in a glow discharge at low temperatures (*IX*). The compounds with $n = 3$–6 are unstable at room temperature and decompose to OF_2 and oxygen. The synthesis of polyoxygen difluorides apparently proceeds through the formation of the O_2F radical, which is relatively longlived (*30, 31*). Indeed, the radical has been shown to exist at room temperature in equilibrium with BF_3 (*32*) ⁻

$$O_2BF_{4(s)} \rightleftharpoons O_2F + BF_3 \qquad (33)$$

Oxygen difluoride is a colorless gas (bp $-144.8°C$) formed when fluorine reacts with 2% aqueous sodium hydroxide.

$$2F_2 + 2NaOH \rightarrow 2NaF + OF_2 + H_2O \qquad (34)$$

Since OF_2 reacts with water

$$OF_2 + H_2O \rightarrow O_2 + 2HF \qquad (35)$$

it is important that the reagents be in contact for a minimum time. Traces of OF_2 have been observed in the fluorine produced when moist KHF_2 is electrolyzed.

In contrast to its chlorine analogue, OF_2 is a comparatively stable compound, decomposing slowly to the elements only at 250°C. It is a powerful oxidizing agent, as might be expected, reacting with metals, sulfur, phosphorus, the halogens, and hydrogen to give a mixture of the expected oxides and fluorides.

Dioxygen difluoride, O_2F_2 is a red liquid prepared by the action of a glow discharge on a low-temperature mixture of fluorine and oxygen at low pressure. The substance freezes at $-160°C$ to an orange solid; however, above $-95°$ it undergoes sufficient thermal decomposition to prevent an accurate determination of its boiling point. The vapor density at $-100°C$ corresponds to the formula O_2F_2.

Of the oxygen fluorides, only the structures of OF_2 and O_2F_2 are known. As shown in Fig. 14.8, oxygen difluoride is a bent molecule with an O—F distance only slightly shorter than that expected for a single bond (1.46 Å).

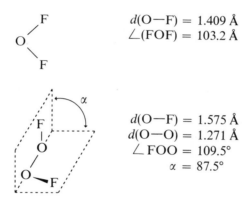

$$d(O-F) = 1.409 \text{ Å}$$
$$\angle(FOF) = 103.2 \text{ Å}$$

$$d(O-F) = 1.575 \text{ Å}$$
$$d(O-O) = 1.271 \text{ Å}$$
$$\angle FOO = 109.5°$$
$$\alpha = 87.5°$$

FIG. 14.8 The structural parameters of OF_2 and O_2F_2.

O_2F_2 has the same general geometry as does H_2O_2, but the O—O distance is shorter than that in H_2O_2 (cf. Fig. 13.16). Thus, the ionic canonical forms **4** and **5**, together with the covalent form **6**, are significant contributors to the wave function:

$$\overset{-}{F} \ O{=}\overset{+}{O}{-}F \qquad F{-}\overset{+}{O}{=}O \ \overset{-}{F} \qquad F{-}O{-}O{-}F$$

4   **5**   **6**

A molecular-orbital description involving three-centered bonds has also been suggested (*33*).

Chlorine forms five oxides: Cl_2O, Cl_2O_3, ClO_2, Cl_2O_6, and Cl_2O_7. Some properties of these substances appear in Table 14.5. Chlorine monoxide,

TABLE 14.5   Some properties of the halogen oxides

	Mp, °C	Bp, °C	Color
Cl_2O	−116	2	red-brown liquid, brown gas
Cl_2O_3	−45^a	—	dark-brown solid
ClO_2	−59	11.0	orange-red solid, red liquid, orange-red gas
Cl_2O_6	3.5	(203)^b	orange solid, red liquid
Cl_2O_7	−91.5	83	colorless liquid and solid
Br_2O	−17.5	—^{a}	brown liquid and solid
BrO_2	—^{c}	—	yellow solid
Br_2O_6	—^{d}	—	white solid
I_2O_4	130^a	—	yellow crystals
I_4O_9	75^a	—	pale yellow solid
I_2O_5	275^a	—	white solid

^{a} Decomposes.
^{b} Estimated.
^{c} Stable only below −40°C.
^{d} Stable only below −80°C.

Cl_2O, is most readily prepared by passing chlorine over freshly precipitated mercuric oxide

$$2Cl_2 + 2HgO \rightarrow HgO \cdot HgCl_2 + Cl_2O. \qquad (36)$$

It is formally the anhydride of hypochlorous acid and can be prepared by distilling an aqueous solution of this acid at reduced pressure. Chlorine dioxide, ClO_2, decomposes explosively above 45° after an induction period, during which time a build-up of the intermediate oxide Cl_2O_3 occurs (34, 35). The latter compound is an unstable dark-brown solid which decomposes at −45° (36). Chlorine dioxide is a powerful oxidizing agent attacking organic matter and converting most metals into a mixture of oxides and chlorides.

Chlorine monoxide, Cl_2O, is a bent molecule, as shown in Fig. 14.9, with a larger bond angle than that found in OF_2. This increase in bond angle of the heavier members of a family compared to the corresponding compound of the Period 2 element has been observed for other families. The Cl—O bond distance (1.70 Å) is slightly shorter than that calculated for the expected single bond formed between these elements (1.73 Å).

Chlorine dioxide, ClO_2, is a very reactive and unstable gas; it has been known to explode violently in the pure state as well as when mixed with air if its partial pressure exceeds 70 torr. Generally this compound is prepared

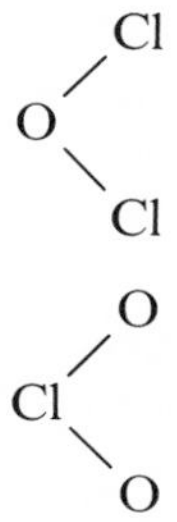

$$d(O-Cl) = 1.70 \text{ Å}$$
$$\angle(ClOCl) = 110.8 \text{ Å}$$

$$d(Cl-O) = 1.49 \text{ Å}$$
$$\angle(OClO) = 116.5°$$

FIG. 14.9 The structural parameters of Cl_2O and ClO_2.

by reducing metal chlorates. The most satisfactory process uses oxalic acid as a reducing agent

$$2KClO_3 + 2H_2C_2O_4 \rightarrow 2ClO_2 + 2CO_2 + K_2C_2O_4 + 2H_2O \quad (37)$$

because the oxidation product is CO_2 which serves as a diluent for the extremely explosive ClO_2. Commercially, chlorine dioxide is prepared by using SO_2 as a reducing agent

$$2NaClO_3 + SO_2 + H_2SO_4 \rightarrow 2ClO_2 + 2NaHSO_4. \quad (38)$$

In spite of its instability, ClO_2 is produced in large quantities as a bleaching agent for cellulose, for water purification, and for the production of chlorites. It always is prepared as needed because of its explosive nature in the gaseous or condensed states; however, ClO_2 persists in aqueous solutions for considerable periods of time.

Chlorine dioxide is a powerful oxidizing agent

$$ClO_2 + 4H^+ + 5e^- \rightarrow Cl^- + 2H_2O \qquad E^0_{298} = 1.50 \text{ V} \quad (39)$$

reacting with hydrocarbons and metals to give chlorites. With aqueous metal hydroxides it gives mixtures of chlorites and chlorates

$$2ClO_2 + 2KOH \rightarrow KClO_2 + KClO_3 + H_2O. \quad (40)$$

Chlorine dioxide is very soluble in water, yielding a yellow crystalline hydrate, $ClO_2 \cdot 6H_2O$, which is stable up to 18°. The aqueous solutions are stable in the dark but decompose when exposed to light yielding a mixture of Cl_2, O_2, Cl_2O_6, and Cl_2O_7.

Chlorine dioxide contains an odd number of electrons, but it exhibits little tendency to dimerize. Its vapor density is normal and there appears to be no association in the liquid state, in aqueous solution, or in CCl_4 solutions. The results of an electron diffraction study, given in Fig. 14.9, show that the molecule is bent, with an O—Cl bond that is considerably shorter than that expected for a single-bonded structure (1.73 Å). The fact that ClO_2 does not exhibit properties associated with odd-electron molecules suggests that the odd electron is delocalized over the entire molecule. In valence-bond terms this would require the use of canonical structures incorporating three-electron bonds.

Dichlorine hexoxide, Cl_2O_6, a red liquid, is formed when ozone reacts with ClO_2 or when ClO_2 is irradiated with ultraviolet light. It decomposes, even at 0°C, to ClO_2 and oxygen, and eventually to Cl_2 and O_2. An explosion occurs when the substance is added to liquid water, but the liquid mixture formed when the gaseous compound and water vapor are condensed contains chloric acid and perchloric acid

$$Cl_2O_6 + H_2O \rightarrow HClO_3 + HClO_4. \tag{41}$$

As expected, Cl_2O_6 is a powerful oxidizing agent.

Dichlorine hexoxide appears to dissociate to the monomer

$$Cl_2O_6 \rightleftharpoons 2ClO_3 \tag{42}$$

under certain conditions. The pure liquid is diamagnetic, suggesting that most of the species present are Cl_2O_6 molecules. The vapor, however, is paramagnetic, indicating the formation of the monomer ClO_3. The molecular weight in CCl_4 solution corresponds essentially to the formula Cl_2O_6, but aqueous solutions of the substance are paramagnetic, indicating that an appreciable amount of monomer is present. Unfortunately there are no reports concerning the structure of this compound. A reasonable suggestion for the structure of the dimer involves the formation of a Cl—Cl bond (7) which could be cleaved homolytically to give a paramagnetic monomer (8).

Structures **7** and **8** parallel the relationship between the structures of NO_2 and its dimer described in Chapter 12.

Dichlorine heptoxide, Cl_2O_7, is the anhydride of perchloric acid and can be prepared by dehydrating this acid with P_2O_5. The compound is the most stable of the chlorine oxides, but it detonates when subjected to shock. It is the poorest oxidizing agent of the chlorine oxides. There are no reports of a direct structure determination of this compound, but the infrared spectrum is consistent with the formulation O_3Cl—O—ClO_3 in which the central Cl—O—Cl bond angle would be 128°.

Bromine. In contrast to chlorine and iodine, the oxides of bromine have not been extensively studied. The data in Table 14.5 indicate that these compounds are thermally less stable than the chlorine oxides; nothing is known of their structures.

Bromine monoxide, Br_2O, is a dark brown liquid which can be prepared by passing bromine over mercuric oxide [cf. Eq. (36)]. The substance decomposes at temperatures above $-40°C$ to bromine and oxygen; hence it is poorly characterized in terms of its physical properties. Bromine monoxide dissolves in CCl_4 to give a stable green solution in which it exhibits a molecular weight corresponding to Br_2O. This solution reacts with aqueous base to form hypobromite ion, but when pure Br_2O reacts with aqueous base a mixture of hypobromite and bromate ions is formed. Bromine monoxide is an oxidizing agent capable of converting I^- to iodine.

Bromine dioxide, BrO_2, is formed as a yellow solid when a low-pressure mixture of oxygen and bromine is subjected to an electrical discharge at liquid-air temperatures. The substance is stable below $-40°$ but decomposes into the elements at higher temperatures. It can be pyrolyzed in a vacuum to Br_2O and a white solid which may be Br_2O_7 (the latter point is not certain). Bromine dioxide, like the analogous ClO_2, dissolves in water to give bromine-containing anions in different oxidation states, namely, Br^-, OBr^-, BrO_2^-, and BrO_3^-.

When bromine is treated with ozone at about $0°C$ a white solid is formed which has been formulated as Br_3O_8 or BrO_3. The product is stable at $-80°$; a transition between two modifications has been reported at $-35°C$. It is soluble in water, giving a colorless solution which is acidic and oxidizing.

Iodine (X). Of the three oxides reported for iodine, only I_2O_7 appears to be related to the analogous compounds of the other halogens. The compound I_2O_4, which appears to be the dimer of the dioxide IO_2, has been formulated as iodyl iodate $[(IO)(IO_3)]$. It has been suggested that the oxide I_4O_9 is best formulated as $I(IO_3)_3$, that is as the iodate of trivalent iodine. Both of these descriptions emphasize the basic nature of iodine.

The compound I_2O_4 is formed by warming iodic acid (HIO_3) with concentrated sulfuric acid for several days. The product is insoluble in cold water but dissolves in hot water to yield iodic acid and iodine

$$5I_2O_4 + 4H_2O \rightarrow 8HIO_3 + I_2; \tag{43}$$

in aqueous base it yields a mixture of iodate and iodide ions. It is an oxidizing agent, giving a mixture of chlorine and ICl when treated with hydrochloric acid. Above $85°$ I_2O_4 decomposes into I_2O_5 and iodine. Spectroscopic data suggest that I_2O_4 consists of a series of I—O chains crosslinked by IO_3 units (*37*). The formulation of I_2O_4 as iodyl iodate is supported by the fact that iodyl sulfate $[(IO)_2(SO_4)_2]$, formed as a yellow solid when H_2SO_4 acts upon HIO_3 (or I_2O_7), can be hydrolyzed to I_2O_4.

The compound I_4O_9 is a very hygroscopic, pale yellow solid prepared by treating iodine with a mixture of ozone and oxygen at room temperature or by warming iodic acid with orthophosphoric acid. The compound decomposes above 75° to give I_2O_5, I_2, and O_2. Water reacts with I_4O_9 to give iodine and iodic acid.

Iodine pentoxide, which is the anhydride of iodic acid, can be prepared as a white infusible solid by pyrolyzing I_2O_4 or I_4O_9, oxidizing I_2 with nitric acid, or dehydrating HIO_3 at about 200°C. It behaves as an oxidizing agent, the reduction product being I_2. Iodine pentoxide has found use as a reagent for the quantitative determination of carbon monoxide. The I_2 which is formed when carbon monoxide reacts with anhydrous I_2O_5 can be determined using conventional iodometric techniques.

Oxyacids   All of the halogens form oxyacids in which the halogen has acquired a formal positive oxidation state; some of the properties of these compounds are summarized in Table 14.6. The hypohalous acids (HXO),

TABLE 14.6   Some properties of the oxyacids of chlorine, bromine, and iodine

Oxyacid formula	X		
	Cl	Br	I
HXO	—[a] $K_a = 3 \times 10^{-8}$	—[a] $K_a = 2 \times 10^{-9}$	—[a] $K_a = 4.5 \times 10^{-13}$ $K_b = 3 \times 10^{-10}$
HXO_2	—[a] $K_a = \sim 10^{-2}$	neither acid nor salt known	neither acid nor salt known
HXO_3	—[a]	—[a]	decomposes at 110°
HXO_4	mp -112°C bp 19°C[c] strong acid	—[b]	HIO_4, sublimes at 110° H_3IO_5, only salts known H_5IO_6, mp 112°

[a] Acid known only in solution; salts characterized.
[b] Acid identified by its mass spectrum.
[c] Under 111 torr pressure.

except hypofluorous acid, are known only in solution and are weak acids. Iodous acid (HIO_2) is unknown, but its salts have been well characterized. Of the halic acids (HXO_3), only iodic acid has been isolated in the pure state; the other halic acids exist only in solution. The perhalic acids are known for chlorine, bromine, and iodine, as are several hydrates of periodic acid. The known oxyanions of the halogens are all oxidizing agents and are chemically related through disproportionation reactions. The standard potentials, shown in Fig. 14.10, reflect some important trends in the oxidizing power of the halogens and the oxyacids. There are some interesting and important

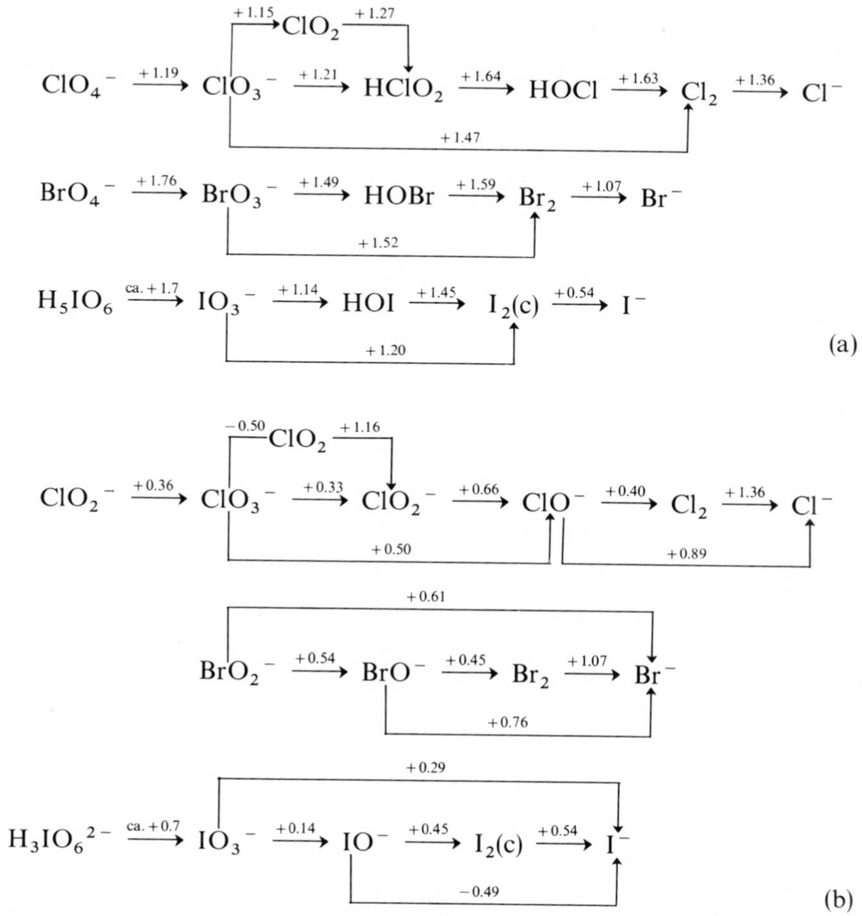

(a)

(b)

FIG. 14.10 Standard half-cell potentials for halogen-containing species: (a) acid solution; (b) basic solution.

reversals in these trends. For example, the relative oxidizing powers of BrO_3^- and IO_3^- in acid and in basic solutions are reversed; a similar reversal occurs for the perhalate ions in acid solution. Such data are important in understanding why, for example, chlorate ion in acid solution will oxidize iodine to iodate but not bromine to bromate and why bromate will oxidize chlorine to chlorate and iodine to iodate. Much useful synthetic work can be accomplished by using the redox relationships described in Fig. 14.10.

Hypohalous Acids. When fluorine is passed into liquid water, the reaction products include HF, O_2, and OF_2. However, if fluorine gas at low pressure is passed over water at $0°C$, the gaseous products also include hypofluorous acid, HOF (*38*). This substance, the only hypohalous acid isolated and characterized at this point in time, is a white solid melting at $-117°$. It appears to be marginally stable at room temperature, decomposing to oxygen and HF

$$2HOF \rightarrow 2HF + O_2. \tag{44}$$

Hypofluorous acid reacts with water, forming H_2O_2, not oxygen

$$HOF + H_2O \rightarrow HF + H_2O_2. \tag{45}$$

Chlorine, bromine, and iodine are slightly soluble in water (Table 14.1), but not all of the halogen exists in the molecular form, X_2, in solution. A disproportionation reaction occurs for a portion of the dissolved halogen molecules

$$X_2 + H_2O \rightleftharpoons H^+ + X^- + HOX; \tag{46}$$

the extent of this reaction decreases with increasing atomic number. The fraction of X_2 present in a saturated solution which undergoes disproportionation is Cl_2, 0.33; Br_2, 0.007; and I_2, 0.005. The hypohalous acids are all very weak (Table 14.6), the acidity decreasing with increasing atomic number. This trend parallels the increasing metallic character of the halogens in the same series. Indeed, it has been estimated that hypoiodous acid has a basic ionization constant

$$HOI \rightleftharpoons HO^- + I^+ \tag{47}$$

several orders of magnitude larger than its ionization constant as an acid (Table 14.6).

The dissolution of halogens in aqueous basic solution leads to a more complex series of species. First a very rapid reaction occurs to form hypohalite ion

$$X_2 + 2OH^- \rightarrow X^- + OX^- + H_2O, \tag{48}$$

but the hypohalite ion undergoes further disproportionation, giving halate ion

$$3OX^- \rightleftharpoons 2X^- + XO_3^-. \tag{49}$$

The rate of the latter reaction increases with increasing atomic number of the halogen. Thus, ClO^- disproportionates very slowly at room temperature but rapidly at temperatures above 75°; hypobromite disproportionates rapidly at room temperature, but the reaction is sufficiently slow at 0°C that large quantities of BrO^- can be kept under these conditions. The rate of disproportionation of IO^- is sufficiently rapid at all temperatures that solution of I_2 in aqueous base gives quantitative yields of iodate, IO_3^-

$$3I_2 + 6OH^- \longrightarrow 5I^- + IO_3^- + 3H_2O. \tag{50}$$

Thus, it is possible to obtain aqueous solutions containing XO^- or XO_3^- by dissolving the corresponding halogen in aqueous base under different conditions.

The most suitable method for the preparation of hypohalite anions involves the reaction of halogen with an aqueous suspension of HgO

$$2X_2 + 2HgO + H_2O \rightleftharpoons HgO \cdot HgX_2 + 2HOX. \tag{51}$$

The halide reduction product is removed from solution and HgO is sufficiently insoluble to keep the solution from becoming too basic, which would encourage the disproportionation reaction shown in Eq. (49) to occur. The hypohalous acids and the corresponding hypohalites are oxidizing agents, as shown by the data in Fig. 14.10. Little is known of the constitution of the free acids because attempts to prepare pure samples of HOX (X = Cl, Br, I) lead to decomposition products. For example, if a 25% aqueous solution of HClO is distilled at low pressure, pure Cl_2O is evolved. The only hypohalite salts that have been isolated are those of the alkali metals and the alkaline-earth metals. On the basis of the electron-pair-repulsion argument we might expect the hypohalous acids to be bent with an XOH bond angle near 109° (9).

$$H \colon \overset{\cdot\cdot}{\underset{\cdot\cdot}{O}} \colon \overset{\cdot\cdot}{\underset{\cdot\cdot}{X}} \colon$$

9

Halous Acids. Of all the halous acids (HXO_2), only chlorous acid and its salts are known; although indirect evidence for the existence of bromous and iodous acid is available, it is not compelling. Chlorous acid has no true anhydride; ClO_2 does not react with water, but undergoes disporportination with aqueous base to give a mixture of chlorite and chlorate ions [Eq. (40)]. An aqueous solution of chlorous acid can be prepared by allowing a suspension of $BaClO_2$ to react with sulfuric acid and removing the less soluble $BaSO_4$ by filtration. Aqueous solutions of chlorous acid are very unstable, decomposing rapidly to ClO_2, ClO_3^-, and Cl^-. Alkali and alkaline-earth metal chlorites can be prepared by the reaction of ClO_2 with the corresponding peroxides:

$$Na_2O_2 + 2ClO_2 \rightarrow 2NaClO_2 + O_2 \qquad (52)$$

$$BaO_2 + 2ClO_2 \rightarrow Ba(ClO_2)_2 + O_2. \qquad (53)$$

Aqueous solutions of $HClO_2$ are colorless, but they soon turn yellow because the acid disproportionates, yielding ClO_2. A variety of metal chlorites have been prepared from the corresponding metal peroxides [Eqs. (52) and (53)] or the direct reaction of ClO_2 with a metal. Chlorites have found use as bleaching or oxidizing agents.

The structure of the chlorite ion in crystalline NH_4ClO_2 has been established. In agreement with electron-pair-repulsion arguments, the chlorite ion is bent (10); the experimental OClO angle is 110.5° and the O—Cl bond

$$:\!\overset{..}{\underset{..}{O}}\!:\overset{..}{\underset{..}{Cl}}\!:\overset{..}{\underset{..}{O}}\!:{}^{-}$$

10

distance is 1.64 Å. The latter is considerably shorter than the value expected (1.73 Å) for a single bond formed between these elements, suggesting that $d\pi$-$p\pi$ bonding is an important consideration in this species.

Halic Acids. Although chloric and bromic acid are known only in aqueous solution, iodic acid has been characterized as a white crystalline substance. Iodic acid is best prepared by dissolving I_2O_5 in a minimum of hot water; upon cooling, crystals of HIO_3 separate. Iodic acid decomposes when heated; HI_3O_8 is formed at 110°, and a mixture of I_2O_5 and water is produced at 195°. Aqueous solutions of chloric and bromic acid can be prepared by treating the barium salt with sulfuric acid and removing the precipitated $BaSO_4$ by filtration. Attempts to concentrate these aqueous solutions beyond the point where they contain less than about seven moles of water per mole of acid lead to decomposition of the acid.

$$3HClO_3 \rightarrow HClO_4 + 2O_2 + Cl_2 + H_2O \qquad (54)$$

$$4HBrO_3 \rightarrow 2Br_2 + 5O_2 + 2H_2O. \qquad (55)$$

It should be recalled [Eq. (40)] that solutions of metal halates can be formed by reacting the appropriate halogen with a hot solution of the corresponding metal hydroxide. Commercially alkali-metal chlorates are prepared by electrolyzing hot aqueous halide solutions and allowing the halogen liberated at the anode to react with the basic solution formed at the cathode.

All of the halate ions are pyramidal, as shown by the data in Fig. 14.11; these ions have abnormally short X—O bond distances compared with the expected values for single-bonded species. Thus, there are two major canonical

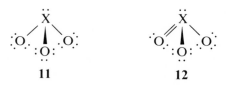

			$d(XO)$, Å	
	X	∠(OXO)	Expt.	Theor.
	Cl	106.7°	1.57	1.73
	Br	111.8°	1.68	1.88
	I	99°	1.82	2.07

FIG. 14.11   The structural parameters of halate ions.

forms which contribute to the wave function for these species, **11** and three forms of type **12**. The structure of iodic acid has basically the same features as that of the IO_3^- ion, shown in Fig. 14.11, except that the I—O distance

11   **12**

involving the OH oxygen atom is slightly longer (1.89 Å) than the other two I—O distances (1.81 Å).

Perhalic Acids. Perchloric (*XI*) and periodic acids (*X*) are well characterized compounds, as shown by the data in Table 14.6; perbrombic acid and its salts have been isolated recently. Anhydrous perchloric acid can be prepared by distilling a mixture of potassium perchlorate and concentrated sulfuric acid under reduced pressure. Anhydrous perchloric acid is hygroscopic; it dissolves in water to form a series of hydrates, $HClO_4 \cdot nH_2O$ ($n = 1, 2, 2.5, 3, 3.5$). The anhydrous acid is a vigorous oxidizing agent toward organic compounds, the reaction often occurring with explosive violence. It is a strong acid in aqueous media and even acts as a proton donor in such acidic media as H_2SO_4, HNO_3, and acetic acid. Perchlorate ion is an oxidizing agent (Fig. 14.10), but only towards very strong reducing agents. Perchlorate can be prepared by the electrolytic oxidation of chlorates, by the solution of metals more active than hydrogen in aqueous perchloric acid, or by the neutralization of metal oxides or hydroxides with perchloric acid.

The data in Fig. 14.12 indicate that the perchlorate ion is tetrahedral with bond distances shorter than those in the chlorate ion (Fig. 14.11), suggesting even more double-bond character in the Cl—O bonds. There is no information on the structure of anhydrous perchloric acid. The monohydrate,

$HClO_4 \cdot H_2O$, appears to be better formulated as hydronium perchlorate, $H_3O^+ClO_4^-$. This substance melts at 50° without decomposition to form a very viscous liquid which itself can be heated to 110° without decomposition. The Cl—O bond distance in the monohydrate has been reported as 1.42 Å, which is slightly shorter than that in the metal perchlorates given in Fig. 14.11.

Although numerous earlier attempts to synthesize perbromates were unsuccessful (39) and reasons for their non-existence discussed (40), aqueous solutions containing this species were prepared by electrochemically oxidizing solutions containing BrO_3^- (41). Chemical oxidation of BrO_3^- can be effected in aqueous solution with XeF_2. Perbromic acid has been identified by its mass spectrum (42).

Three iodine-containing acids or their salts are known in which iodine exhibits an oxidation number of 7: HIO_4, H_3IO_5, and H_5IO_6. These acids can be imagined as hydrates of the hypothetical oxide I_2O_7. In these terms the acids HIO_4, H_3IO_5, and H_5IO_6 are the 1-, 3-, and 5-hydrates, respectively. The best known of these acids is H_5IO_6, called paraperiodic acid; it is prepared commercially by the electrolytic oxidation of iodic acid. A convenient laboratory preparation of this substance involves treatment of the barium salt with concentrated HNO_3. Paraperiodic acid melts with decomposition at 140° to give iodic acid, water, and oxygen. Under vacuum at 80° it loses water to form HIO_4, which is known as metaperiodic acid. Salts of mesoperiodic acid, H_3IO_5, are also known, although the pure acid has never been reported.

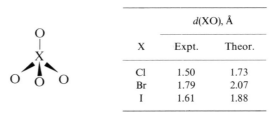

| | | $d(XO)$, Å | |
X	Expt.	Theor.
Cl	1.50	1.73
Br	1.79	2.07
I	1.61	1.88

FIG. 14.12 The bond distances of the tetrahedral perhalate ions.

Salts of the periodic acids can be formed by several indirect methods. For example, oxidation of $NaIO_3$ in aqueous alkali with chlorine or peroxydisulfate ion gives $Na_3H_2IO_6$, which is the acid salt of H_5IO_6. This salt can be converted into the meta salt ($NaIO_4$) by treatment with concentrated nitric acid.

The periodic acids are strong acids and are stronger oxidizing agents than iodic acid. They are often used as reagents for analytical purposes because they generally react smoothly and rapidly. Ozone can sometimes be detected when periodic acids react, but peroxides are never formed.

The metaperiodate ion, IO_4^-, is isostructural with the perchlorate ion (Fig. 14.12) as would be expected. The results of a variety of physical measurements on aqueous periodate solutions indicate that the species IO_4^-, $H_3IO_6^{2-}$, $H_2IO_6^{3-}$, and $H_2I_2O_{10}^{4-}$ are present (43, 44). Paraperiodic acid usually acts as a dibasic acid ($K_1 = 2.3 \times 10^{-2}$, $K_2 = 10^{-6}$, $K_3 = 2.5 \times 10^{-13}$) commonly forming acid salts with the stoichiometry $M_2H_3IO_6$. The structure of the ammonium salt shows the presence of IO_6^{5-} octahedra linked by hydrogen bonds. Thus it would appear that paraperiodic acid is best formulated as $(HO)_5IO$.

3.4 / Interhalogen Compounds (XII, XIII)

The halogens form a series of mixed binary derivatives called the interhalogen compounds; some properties of these compounds appear in Table 14.7. Compounds with the stoichiometry XX' for all possible combinations of the halogens are known. Some compounds of the type XX_3', XX_5', and

TABLE 14.7 **Some physical properties of interhalogen compounds**

Type	Compound	Mp, °C	Bp, °C	Trouton's constant, cal mole^{-1}	Appearance
XX'	ClF	−156	−100	28.0	colorless gas
	BrF	−33	−20	20.5	pale brown gas
	IF	—	—	—	detected spectroscopically
	BrCl	~ −66	~5	—	—
	ICl	27.2 (α)	97	—	ruby-red crystals
		13.9 (β)	—	—	brown-red tablets
	IBr	36	116	—	black solid
XX_3'	ClF_3	−76	12	23.1	colorless gas
	BrF_3	8.8	126	25.7	yellow-green liquid
	IF_3	−28 (dec.)	—	—	yellow powder
	ICl_3	101/16 atm.	64 (sub.)	—	orange solid
XX_5'	ClF_5	> −196	—	—	—
	BrF_5	−61.3	41	23.7	colorless liquid
	IF_5	9.6	98	—	colorless liquid
XX_7'	IF_7	45	5	26.4	colorless gas

XX_7' have also been isolated and characterized. In the case of the interhalogens with higher coordination numbers the following observations can be made. (a) Fluorine is never found as the central atom in the structure of such derivatives, which is in accord with the general observation that only the elements in periods higher than 2 can expand their coordination number and that fluorine seldom achieves a formal positive oxidation state. (b) The halogen with the largest atomic number is the central atom in the interhalogen compounds. (c) The greater the electronegativity difference between the halogens, the greater will be the number of bonds formed to the central atom of the structure. For example, the electronegativity differences in the halogen fluorides, as shown in Table 14.8, decrease in the order $FI > FBr > FCl$; iodine forms all four types of fluorides; bromine only three (BrF, BrF_3, BrF_5), and chlorine two (ClF, ClF_3). Chlorine pentafluoride has been reported as an unstable white solid at $-196°$. Only two iodine chlorides (ICl and ICl_3) and one iodine bromide (IBr) are known.

TABLE 14.8   Electronegativity differences for the halogens and the stoichiometry of the known interhalogens

System	$X_x - X'^a_x$	Compounds formed
F—I	1.6	IF, IF_3, IF_5, IF_7
F—Br	1.2	BrF, BrF_3, BrF_5
F—Cl	1.0	ClF, ClF_3, ClF_5
Cl—I	0.8	ICl, ICl_3
Br—I	0.4	IBr
Cl—Br	0.3	$BrCl$
X—X^b	0.0	XX^b

[a] Electronegativities taken from Table 14.1.
[b] XX represents the halogens F_2, Cl_2, Br_2, and I_2.

The trends described in (c) can also be formulated in terms of the size of the central halogen atom, i.e., the larger the central halogen the greater the number of bonds it can form to another halogen atom. This, of course, would be expected because the electronegativity and the size of the halogens vary in a regular manner with increasing atomic number. It should be emphasized that the factors which are discussed do not necessarily represent the fundamental reasons for the observed stability of the interhalogen compounds. Since all of the interhalogens contain an even number of halogen atoms, we would expect them to be diamagnetic, which is in agreement with the experimental observations.

The diatomic interhalogen compounds XX′ have properties which are generally intermediate between those of the pure substances X_2 and X'_2; see for example the comparison of the melting points of these substances in Fig. 14.13. As the coordination number (and hence the formal oxidation state) of the central halogen increases, the compounds become less intensely colored.

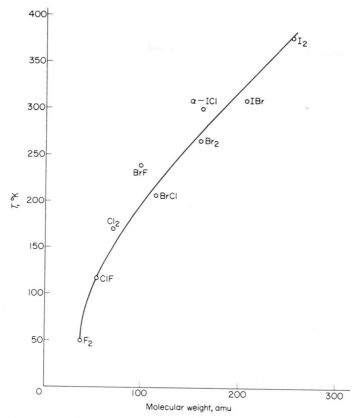

FIG. 14.13 Melting points of the halogens and of the diatomic interhalogen compounds.

Compounds of the Type XX′   The diatomic interhalogens are generally prepared by direct combination of the corresponding halogens

$$X_2 + X_2' \rightarrow 2XX'. \tag{56}$$

Chemically these interhalogens are similar in their properties to the halogens. They act as oxidizing agents towards metals and nonmetals, producing mixed halide products. These interhalogens undergo hydrolysis

$$XX' + H_2O \rightleftharpoons H^+ + X'^- + HOX \tag{57}$$

in much the same way as do the halogens [Eq. (46)]. The more electropositive halogen forms the hypohalous acid, as would be expected. The diatomic interhalogens exhibit varying stabilities, ClF being the most stable. Bromine fluoride disproportionates into bromine and a mixture of the higher bromine fluorides. Attempted preparation of IF under conditions that are successful for the preparation of the other halogen monofluorides leads only to minute amounts of a substance that has been spectroscopically identified as IF. Phase studies of the bromine–chlorine system do not indicate the formation of any bromine chlorides; however, spectroscopic investigations on equimolar mixtures of Br_2 and Cl_2 in the gas phase or in CCl_4 solution indicate that about 20 % BrCl is present at 20° in these systems. Like ClF, ICl and IBr are stable compounds, but some dissociation occurs at elevated temperatures; at 100°C ICl and IBr are about 2 % and 15 % dissociated, respectively. Of the diatomic interhalogen compounds, ICl and IBr have been the most extensively studied.

Iodine monochloride can be prepared by the direct interaction of the elements [Eq. (56)] or by the action of aqueous HCl on a mixture of potassium iodate and potassium iodide

$$6HCl + KIO_3 + 2KI \rightarrow 3KCl + 3ICl + 3H_2O. \tag{58}$$

The hydrolysis of ICl [Eq. (57)] is suppressed by the presence of excess chloride ions, because of the formation of the complex ion ICl_2^-

$$ICl + Cl^- \rightleftharpoons ICl_2^-. \tag{59}$$

Iodine monochloride can be extracted from aqueous solution by organic solvents, but if the aqueous phase contains a high concentration of Cl^- extraction will not occur. This observation lends support to the contention that ICl_2^- exists in such systems.

Iodine monochloride is soluble in a variety of solvents, forming brown solutions in noninteracting solvents such as CS_2 and CCl_4, and yellow solutions in basic solvents such as ether and ethyl alcohol. The reactions of ICl in these solutions suggests that the iodine atom in the molecule bears the positive charge. Indeed, ICl forms stable charge-transfer-type complexes (cf. Fig. 14.3) with pyridine and dioxane in which the base molecule is associated with the iodine atom in this complex, as shown in Fig. 14.14. The conductance of liquid ICl ($4.6 \times 10^{-3} \Omega^{-1} cm^{-1}$ at 35°) is about two orders of magnitude greater than that of liquid iodine, and it has been suggested that the following dissociation process occurs:

$$2ICl \rightleftharpoons I^+ + ICl_2^-. \tag{60}$$

The chlorides of the larger alkali metals are soluble in liquid ICl, giving solutions that are more conducting than the pure solvent, a fact which indicates that the concentration of ICl_2^- is increased in these systems. Salts of type $MICl_2$ [M = K, Rb, Cs, $(CH_3)_4N$] are known. The usual Lewis

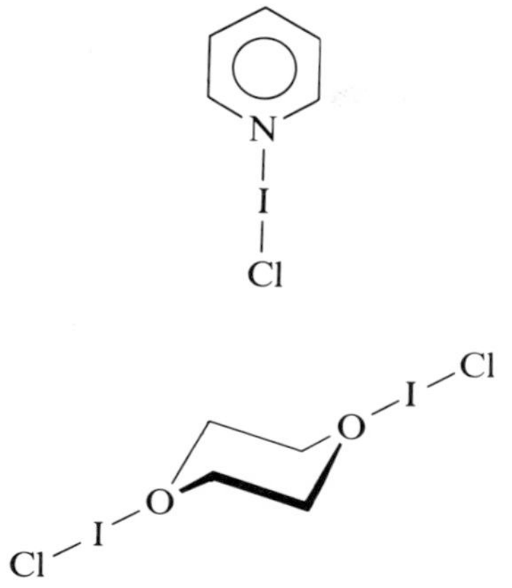

$d(N-I) = 2.26$ Å
$d(I-Cl) = 2.51$ Å
$\angle(NICl) = 180°$

$d(O-I) = 2.60$ Å
$d(I-Cl) = 2.30$ Å
$\angle(OICl) = 180°$

FIG. 14.14 The structural parameters of some interhalogen complex compounds.

formulation of ICl_2^- (13) indicates that this species should be linear, a fact which is borne out by experimental evidence.

13

Interhalogens of the type XX' are diatomic in the gas phase, but there appear to be rather strong interactions between ICl molecules in the solid state. The interatomic distances shown in Table 14.9 suggest that BrCl and ICl are essentially bonded by single bonds; the fluorides apparently contain

TABLE 14.9 Structural parameters for the interhalogens, XX'

| | $d(X-X')$, Å | |
Molecule	Experimental	Expected[a]
Cl—F	1.63	1.71
Br—F	1.76	1.86
Br—Cl	2.14	2.13
I—Cl	2.32	2.32
I—Br	2.52	2.47

[a] Single-bond distance.

a significant contribution of a double-bonded canonical form (**14**) to the

$$^{-}X{=}F^{+}$$

14

overall wave function. Structural data for the crystalline state, summarized in Fig. 14.15, reveal an I—I distance between two ICl molecules (3.08 Å) is significantly shorter than either the intermolecular distance in solid I_2 (3.54 Å) or twice the van der Waals radii for iodine (4.3 Å). Similar weak interactions have been observed for iodine in polyiodide ions.

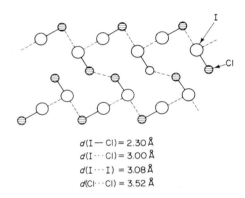

$d(\text{I—Cl}) = 2.30\ \text{Å}$
$d(\text{I}\cdots\text{Cl}) = 3.00\ \text{Å}$
$d(\text{I}\cdots\text{I}) = 3.08\ \text{Å}$
$d(\text{Cl}\cdots\text{Cl}) = 3.52\ \text{Å}$

FIG. 14.15   The molecular arrangement in crystalline ICl.

Compounds of the Type XX$'_3$ The four interhalogen compounds of this type, IF_3, BrF_3, ClF_3, and ICl_3, are prepared by direct reaction of the elements under appropriate conditions

$$X_2 + 3X'_2 \rightarrow 2XX'_3 \tag{61}$$

or by the reaction of a lower interhalogen with a halogen

$$ClF + F_2 \rightarrow ClF_3. \tag{62}$$

In the cases where XX'_3 represents the highest attainable coordination number for the central halogen (ClF_3), the reactions can be carried out with an excess of the latter. However in instances where a higher oxidation state can be formed, the reaction must be controlled either by diluting the more active halogen (e.g., in the preparation of BrF_3) or by using less vigorous conditions (e.g., conducting the reaction between I_2 and F_2 at $-78°$).

Chlorine trifluoride, like fluorine, is a powerful oxidizing agent. It reacts with almost all of the elements except the rare gases, nitrogen, and a few metals which give fluorides that form impervious films. The reaction of ClF_3 with metals appears to be catalyzed by moisture. Chlorine trifluoride reacts vigorously with most organic compounds, often with combustion. Chlorine trifluoride is a planar T-shaped molecule, as shown in Fig. 14.16;

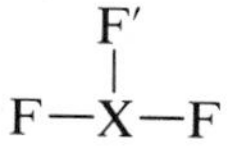

| | | | $d(X-F)$, Å | | |
| | | | Expt. | | |
	X	$\angle(FXF')$	X—F	X—F'	Theor.
	Cl	87.5°	1.698	1.598	1.71
	Br	86.2°	1.810	1.721	1.86

F'
|
F—X—F

FIG. 14.16 The structural parameters of ClF_3 and BrF_3.

its structure follows from the electron-pair-repulsion arguments which minimize nonbonded electron pairs by placing them in the trigonal plane of the bipyramid (**15**). The nonbonded–bonded pair repulsions lead to a

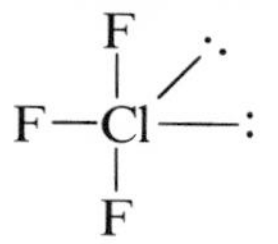

F
| ⋰ ∙∙
F—Cl——:
|
F

15

decrease in the FClF angle from its ideal value of 90°. As observed in other trigonal bipyramidal arrangements of electron pairs, the axial bonds are longer than those in the equatorial positions. In the case of BrF_3, only one of the equatorial positions is occupied by a fluorine atom. It appears from the bond distances that, as in the case of most fluorides, a significant amount of double bonding is present in the molecule. In the solid state ClF_3 exhibits the same T-shaped structure.

Bromine trifluoride is slightly less reactive than ClF_3 and exhibits many of the same chemical properties as that compound. It fluorinates and brominates CCl_4 and CI_4. Many metal and nonmetal oxides are fluorinated by BrF_3

$$4BrF_3 + 3SiO_2 \rightarrow 3SiF_4 + 2Br_2 + 3O_2 \qquad (63)$$

$$2WO_3 + 4BrF_3 \rightarrow 2WF_6 + 2Br_2 + 3O_2 \qquad (64)$$

as are salts of oxyacids. In the latter reactions the corresponding salts of fluoroacids are formed.

Liquid BrF_3 has a remarkably high specific conductance $(8 \times 10^{-3}$ Ω^{-1} cm^{-1} at 20°) which leads to the suggestion that the compound undergoes ionization in the liquid state. There is considerable evidence that the anion formed in the self-ionization is the solvated fluoride ion, in which case the correct process is given by

$$2BrF_3 \rightleftharpoons BrF_2{}^+ + BrF_4{}^-. \tag{65}$$

In this aprotic solvent system, species which increase the concentration of $BrF_2{}^+$ are acids, while bases increase the concentration of $BrF_4{}^-$ (XIV). Compounds containing the cation of the BrF_3 solvent system can be prepared by reacting fluorides which form anionic complexes with BrF_3

$$MF_4 + 2BrF_3 \rightarrow (BrF_2)_2MF_6 \tag{66}$$

$$M = Sn, Ti$$

$$MF_5 + BrF_3 \rightarrow (BrF_2)MF_6 \tag{67}$$

$$M = Sb, Bi, Nb, Ta.$$

A variety of such compounds have been characterized. Compounds containing $BrF_2{}^+$ ions can also be prepared from the corresponding metal oxides using excess BrF_3. Many metals, including silver, gold, niobium, and tantalum, dissolve in liquid BrF_3, forming compounds which contain the $BrF_2{}^+$ cation. Bases in the BrF_3 solvent system can be prepared by dissolving metal fluorides in this substance. Upon evaporation of the solvent, metal tetrafluorobrominates are formed

$$M + BrF_3 \rightarrow MBrF_4 \tag{68}$$

$$M = Ag, K, \tfrac{1}{2}Ba.$$

Both soluble acids $[(BrF_2)MF_6]$ and bases $[M(BrF_4)]$ give solutions in BrF_3 that are very good conductors. The reactions of acids with bases in this solvent system can be followed using conductometric titration methods, a sharp minimum in the conductivity occurring at the equivalence point

$$AgBrF_4 + (BrF_2)SbF_6 \rightarrow AgSbF_6 + 2BrF_3 \tag{69}$$

$$2KBrF_4 + (BrF_2)_2SnF_6 \rightarrow K_2SnF_6 + 4BrF_3. \tag{70}$$

The validity of this interpretation is supported by the isolation of the salt formed in these reactions.

The specific conductance of ClF_3 is markedly lower $(5 \times 10^{-9}\,\Omega^{-1}$ cm$^{-1})$ than that reported for BrF_3; little information is available on possible acid–base behavior in ClF_3. Chlorine trifluoride dissolves SbF_5 and AsF_6 to give monosolvates which have been formulated as $[ClF_2{}^+][MF_6{}^-]$ (M = Sb, As). The anion in the ClF_3 solvent system, $ClF_4{}^-$, has been obtained as a rubidium salt when RbCl is fluorinated in a stream of elemental fluorine (45). The reaction of ClF_3 with BrF_3 in the gaseous state yields a colorless solid, $ClF_3 \cdot BrF_3$, which appears to be ionic, suggesting the formulation $[ClF_2][BrF_4]$.

The BrF_3 molecule has the same general structural features as ClF_3, which leads to a similar interpretation for the bonding in this compound. The $BrF_2{}^+$ ion should be bent on the basis of the number of electron pairs on the central atom (16). The structure of solid $[BrF_2{}^+][SbF_6{}^-]$ contains angular

$$\overset{\displaystyle \cdot\cdot\atop \cdot Br \cdot}{\underset{F \qquad F}{\diagup \quad \diagdown}} \quad +$$

16

BrF_2 units linked by *cis* fluorine bridges to octahedral $SbF_6{}^-$ ions, giving the bromine atom a distorted square-planar environment of fluorine atoms (46).

The structure of the basic species in BrF_3 is well established. The results of x-ray diffraction experiments on crystalline $KBrF_4$ indicate the presence of square-planar $BrF_4{}^-$ ions containing four equal Br—F bonds (1.88 Å) (46). This configuration also would be predicted from the Lewis formulation of the ion (17). The large value of Trouton's constant for BrF_3 (Table 14.7)

$$\underset{F \qquad F}{\overset{F \qquad F}{\diagdown \; | \; \diagup}} \; Br \; \diagup \; | \; \diagdown$$

17

indicates that considerable association must be present in the liquid state. The most probable cause for this association is the presence of planar bridged species (18), a suggestion that is consistent with the known structure

$$\underset{F \qquad F \qquad F}{\overset{F \qquad F \qquad F}{Br \qquad Br}}$$

18

of ICl_3 in the solid state, shown in Fig. 14.17.

Although IF_3 is not stable above $-28°$, it forms stable derivatives of types MIF_4 (M = alkali metal, NO^+) and $IF_3 \cdot NF_5$ (M = As, Sb); attempts to form nitrogen-base adducts led to disproportion reactions (47). Alkyl

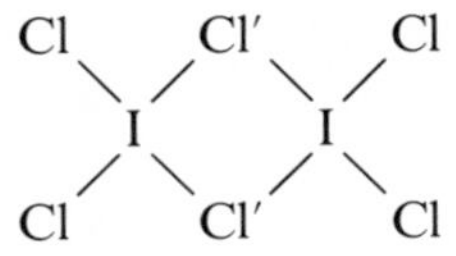

Cl   Cl′   Cl $d(I—Cl) = 2.39$ Å
                       $d(I—Cl′) = 2.70$ Å
    I   I $\angle(ICl′I) = 96°$
                       $\angle(Cl′ICl′) = 84°$
Cl   Cl′   Cl $\angle(ClICl) = 94°$

FIG. 14.17 Iodine trichloride exists as a dimeric species in the solid state.

derivatives of IF_3, such as R_fIF_2 ($R_f = C_2F_5$, C_4F_9) are more stable than the parent compound (48); little is known of the structures of these derivatives.

Iodine trichloride is considerably less reactive and thermally less stable than BrF_3 or ClF_3. Vapor-density measurements indicate that at 77° ICl_3 is completely transformed into an equimolar mixture of ICl and Cl_2. Spectroscopic evidence suggests that ICl_3 also exists as a mixture of ICl and Cl_2 in carbon tetrachloride solution. Iodine trichloride is hydrolyzed by water

$$2ICl_3 + 3H_2O \rightarrow ICl + HIO_3 + 5HCl, \tag{71}$$

but it can be isolated from concentrated aqueous solutions by the addition of concentrated H_2SO_4. The direct combination of Al_2Cl_6 or $SbCl_5$ with ICl_3 gives adducts ($ICl_3 \cdot AlCl_3$ and $ICl_3 \cdot SbCl_5$) which might be formulated as salts containing the ICl_2^+ ion; however, direct structural evidence does not completely support this suggestion.

The structure of ICl_3 cannot be obtained in the vapor state because of its thermal instability. However, x-ray diffraction data for crystalline iodine trichloride, summarized in Fig. 14.17, indicate that it exists as a dimer with all of the atoms of the molecule in one plane. This structure is in marked contrast with that observed for solid ClF_3, which consists of discrete T-shaped molecules. Thus the properties and structures of ClF_3, BrF_3, and ICl_3 in the gaseous, liquid, and solid states appear to form a consistent pattern. There is no interaction between chlorine trifluoride molecules in any of the phases; the normal value of Trouton's constant (Table 14.7) supports this contention. Bromine trifluoride is monomeric in the gas state, and the liquid is associated; although no structural data are available on the solid it is highly probable that its structure would be similar to that of ICl_3. Iodine trichloride is dimeric in the solid state; unfortunately it is not sufficiently stable to obtain high concentrations of the molecules in either of the other phases.

The structure of $ICl_3 \cdot 2SbCl_5$, shown in Fig. 14.18, indicates that it does not contain the species $[ICl_2^+]$, $[ICl_4^-]$, or $[SbCl_4^+]$. Chains of $SbCl_6$ octahedra linked through chlorine bridges by angular ICl_2 groups are present. The iodine atom has essentially the same planar environment as it exhibits in the ICl_3 dimer and the antimony atom exhibits its normal coordination

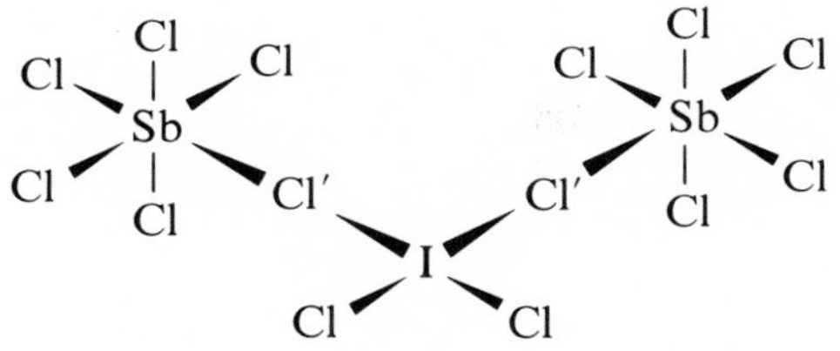

$$d(\text{I—Cl}') = 2.90 \text{ Å}$$
$$d(\text{I—Cl}) = 2.31 \text{ Å}$$
$$d(\text{Sb—Cl}') = 2.43 \text{ Å}$$
$$d(\text{Sb—Cl}) = 2.31 \text{ Å}$$
$$\angle \text{ClICl} = 91.5°$$

FIG. 14.18 The complex compound $(\text{SbCl}_5)_2 \cdot \text{ICl}_3$ contains a bridging ICl_3 unit.

behavior. As usual, the bridging chlorine distances are longer than the non-bridged distances.

Compounds of the Type XX'_5 Three interhalogens with this formulation are known, ClF_5, BrF_5, and IF_5. Chlorine pentafluoride has been prepared from the elements using an excess of fluorine; the reaction occurs in one hour at 250 atm and 350°. Relatively little is known about the chemistry of this compound. Bromine pentafluoride is prepared by heating a mixture of BrF_3 and F_2 to 200°. The compound is thermally stable; it is a very powerful fluorinating agent, comparable in its action to ClF_3. It reacts with most of the elements (except the rare gases, oxygen, and nitrogen), a variety of metallic and nonmetallic binary compounds, and organic compounds; fluorides are the predominant reaction products. The reactions are often explosive and find little preparative use.

Crystalline BrF_5 consists of distorted pyramidal molecules, as would be expected from its Lewis formulation (**19**). The presence of the lone electron

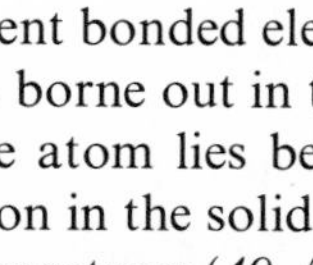

19

pair should repel the four adjacent bonded electron pairs to give bond angles less than 90°. This prediction is borne out in the observed structure, given in Fig. 14.19; the central bromine atom lies below the plane containing four fluorine atoms. The configuration in the solid state apparently persists in the liquid because the fluorine nmr spectrum (49–51) shows two signals. Quantum mechanical calculations indicate that the d orbitals on the central bromine atom play only a minor role in the bonding in BrF_5 (52). Iodine pentafluoride can be prepared by passing fluorine diluted with nitrogen over iodine or I_2O_5. A less hazardous method involves the use of AgF:

$$3\text{I}_2 + 5\text{AgF} \rightarrow 5\text{AgI} + \text{IF}_5. \tag{72}$$

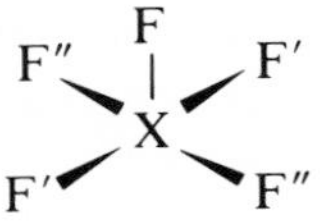

X	$d(X-F)$, Å	$\angle$(FXF)
Br	1.68 (F)	86.5° (FF')
	1.81 (F')	80.5° (FF')
	1.75 (F")	85.4° (FF")
Cl	1.72 (F)	
	1.62 (F', F")	

FIG. 14.19   The structural parameters of BrF_5 and ClF_5.

Like BrF_5, IF_5 is unusually stable to heat, undergoing no decomposition up to 400°. It has a specific conductance ($2 \times 10^{-6}\ \Omega^{-1}\ cm^{-1}$ at 20°) sufficiently high to suggest that the liquid might be ionized

$$2IF_5 \rightleftharpoons IF_4^+ + IF_6^-. \tag{73}$$

Although there is little direct evidence to support the existence of an acid–base system based on the ions in Eq. (73), some indirect observations are helpful. The high value of Trouton's constant given in Table 14.7 suggests that the liquid is associated, probably through fluorine-bridged species. This conclusion is supported by comparison of the Raman spectra of this substance in the gaseous and liquid phases (53).

The adducts $IF_5 \cdot SbF_5$, $KF \cdot IF_5$, and $CsF \cdot IF_5$ can be isolated from a solution of the corresponding substances in IF_5. The salt $KSbF_6$ is isolated when IF_5 solutions of the first named substances are mixed and the solvent evaporated. Compounds containing the hexafluoroiodate anion $[M(IF_6)]$ can be prepared from the reaction of $AgIF_6$ with the appropriate chloride in acetonitrile solution (54). Spectroscopic evidence (55, 56) suggests that the IF_6^- ion is not octahedral, in agreement with the Lewis formulation (20).

20

However, spectroscopic evidence also indicates that $CsBrF_6$ contains an octahedral anion (57). The crystal structure of $IF_5 \cdot SbF_5$ clearly indicates the presence of IF_4^+ ions structurally similar to SF_4, with which it is isoelectronic (58).

Iodine pentafluoride is less reactive than BrF_5. It reacts with many substances but the products have not been well characterized. Metals react slowly, but the heavier nonmetals and metalloids react with incandescence.

Metal and nonmetal oxides form oxyfluorides

$$M_2O_5 + 3IF_5 \rightarrow 2MOF_3 + 3IOF_3; \tag{74}$$

$$M = P, V$$

in some cases solvates are formed

$$WO_3 + 2IF_5 \rightarrow WO_3 \cdot 2IF_5 \tag{75}$$

$$KX + IF_5 \rightarrow KX \cdot IF_5 \tag{76}$$

$$X = IO_3^-, F^-.$$

Substituted derivatives of IF_5 such as RIF_4 ($R = C_6H_5, p\text{-}CH_3C_6H_4$) have been reported (48).

The structure of IF_5 is unknown, but the results of spectroscopic experiments indicate that this molecule is geometrically equivalent to BrF_5.

Compounds of the Type XX_7' Only one compound of this type is known. Iodine heptafluoride is prepared by fluorinating iodine or IF_5. Unless care is taken to exclude moisture, the product is contaminated with IF_5O, which can also arise from the reaction of IF_7 with SiO_2. Iodine heptafluoride appears to be similar to BrF_5 and ClF_3 in its chemical reactivity.

The Lewis formulation for IF_7 indicates that the most probable structures are either a pentagonal bipyramid (21) or a distorted octahedron (22). The

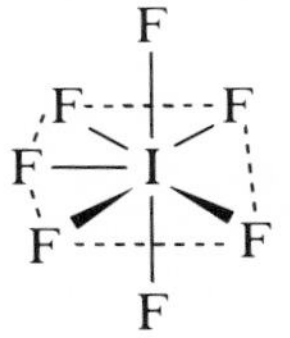

21    22

experimental evidence for the structure of IF_7 appears to be contradictory. The results of infrared and Raman spectroscopy favor the pentagonal bipyramid structure (21) (59, 60) as do electron diffraction data (61, 62). However, crystallographic results indicate that the structure of IF_7 in the solid state is closer to 22 (63, 64).

No single molecular geometry is capable of accounting for the structural data pertaining to IF_7. It appears that the best model for this system involves an intramolecular rearrangement between several geometries, the most probable being the conversion of the trigonal bipyramidal arrangement into the distorted octahedral structure shown in Fig. 14.20 (65). The rapid exchange of fluorine atoms among possible positions in the structure is supported by nmr spectroscopy, since only one resonance peak is found in the spectrum (*66*).

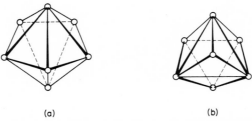

(a)   (b)

FIG. 14.20   Seven-coordinate species such as IF_7 can exist in either of two structures: (a) a pentagonal bipyramid or (b) a distorted octahedral structure in which the seventh ligand occupies a position in one of the octahedral faces. In some cases these structures are easily interconvertible.

3.5 / Polyhalide Ions (*XIII, XV*)

The propensity of the halogens to form interhalogen compounds extends to the formation of similar ionic species. Several such ions are mentioned in the discussion of the chemistry of the interhalogens, e.g., ICl_2^-, BrF_2^+, BrF_4^-, IF_4^+, and IF_6^-. The anions in this list are members of a large class of ions called the polyhalide anions which are formally the products of the reaction between a halogen or interhalogen molecule and a halide ion:

$$X_2' + X^- \rightarrow X_2'X^- \tag{77}$$

$$X'X_3'' + X^- \rightarrow XX'X_3''^- \tag{78}$$

$$X'X_5'' + X^- \rightarrow XX'X_5''^-. \tag{79}$$

The halogens involved may or may not be of the same kind. A list of the principle types of polyhalide species which have been characterized appears in Table 14.10. For the most part, these species are obtained by crystallizing a halide in the presence of the halogen or interhalogen compound. The most stable polyhalide salts, which are commonly colored, appear to be those containing large counter ions such as Rb^+, Cs^+, and the tetraalkylammonium ions (RN_4^+). The compounds are generally thermally unstable, decomposing to the metal halide and the neutral halogen; the metal halide formed in the decomposition contains the most electronegative of the halogens

$$MICl_4 \rightarrow MCl + ICl_3 \tag{80}$$

$$MICl_2 \rightarrow MCl + ICl. \tag{81}$$

TABLE 14.10 The stoichiometry of several types of polyhalide ions

		Number of halogens in the anion					
3			5		7		9
Br_3^-	ClF_2^-	$IBrF^-$	ClF_4^-	$IFCl_3^-$	BrF_6^-	I_7^-	I_9^-
I_3^-	IBr_2^-	$IClBr^-$	BrF_4^-		IF_6^-	BrI_6^-	
	ICl_2^-		IF_4^-			$ClBr_6^-$	
	$BrCl_2^-$		ICl_4^-				
	BrI_2^-						
	$ClBr_2^-$						

The thermal stability for a given type of anion increases with increasing size of the cation. The metal polyhalides can often be converted into compounds containing other polyhalides by the action of halogens (67, 68):

$$IBr_2^- + Cl_2 \rightarrow ICl_2^- + Br_2 \tag{82}$$

$$ClIBr^- + ICl \rightarrow ICl_2^- + IBr \tag{83}$$

$$ICl_2^- + Cl_2 \rightarrow ICl_4^- \tag{84}$$

$$Cl^- + 2F_2 \rightarrow ClF_4^-. \tag{85}$$

The metal polyhalides are unstable in water, leading to a mixture of products which appear to arise from hydrolysis of the interhalogen compound released

$$5ICl_4^- + 9H_2O \rightleftharpoons I_2 + 3IO_3^- + 20Cl^- + 18H^+. \tag{86}$$

The species I_3^- is known to be stable in aqueous solution

$$I_2 + I^- \rightleftharpoons I_3^- \tag{87}$$

the value for the equilibrium constant for Eq. (87) being 1.4×10^{-3} at 25°. No anhydrous acid containing a polyhalide ion is known, but the orange compound $HICl_4 \cdot 4H_2O$ can be isolated by cooling a mixture of concentrated HCl and ICl_3.

All of the known structures of the polyhalides containing three atoms are linear; most species are symmetrical but some, like IBr_2^-, are unsymmetrical (69, 70). Polyhalide ions containing five halogens are all square-planar

23

except IF_4^-, which appears to have a *cis* configuration for the two unpaired electrons (**23**) (*57*).

Iodine forms a remarkable group of polyhalides, MI_n, where $n = 3, 5, 7, 9$, and M is a large counter ion. The triiodide ion is analogous to the other polyhalides containing three halogen atoms. However, the structure of the pentadiodide ion in $(CH_3)_4NI_5$ bears no relationship to the structure of the other anions containing five halogens. This anion is planar and V-shaped, as shown in Fig. 14.21, with I—I distances suggesting that it consists of I^- weakly bonded to two polarized I_2 molecules. The central I—I—I angle is 95°

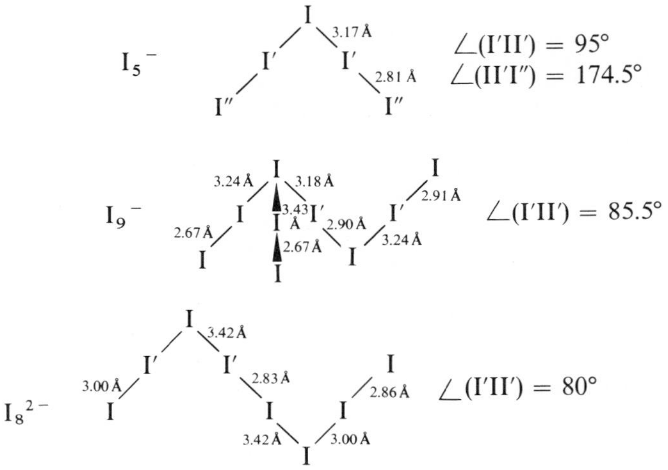

FIG. 14.21   Structural parameters for the polyiodide species I_5^-, I_9^-, and I_8^{2-}.

but the three atoms that form the arms of the V are not collinear. The long I—I distance from the center iodine is well within the van der Waals radii and the intermolecular distance observed in crystalline I_2 and given in Table 14.2. The terminal I—I distance is about 0.13 Å longer than the distance observed in the I_2 molecule. The I_7^- anion consists of an array of a linear I_3^- ion and two I_2 molecules in the proportion $I_3^- : 2I_2$. The arrangement of iodine atoms in the compound $(CH_3)_4NI_9$, given in Fig. 14.21, is even more complex.

The compound Cs_2I_8 contains the I_8^{2-} ion, which consists of an association of two I_3^- units and an I_2 molecule, as shown in Fig. 14.21. Two different I—I distances exist in the I_3^- units (2.86 Å and 3.00 Å), which are slightly shorter and longer, respectively, than that found in the symmetrical I_3^- ion (2.91 Å). Again the I—I distance in the I_2 unit is longer than that observed in free I_2, and the intermolecular distances reflect rather weak bonding.

4 / PSEUDOHALOGENS (*XVI*)

A number of complex radicals, called halogenides or pseudohalogens, are known which behave like halogens. Table 14.11 contains a list of the better characterized pseudohalogens. The azide ion, N_3^-, is generally considered a pseudohalogen ion even though the free pseudohalogen $(N_3)_2$ has not been

TABLE 14.11   **Properties of some pseudohalogens**

Formula	Name	mp °C	bp °C
$(CN)_2$	cyanogen	-27.9	21.17
$(SCN)_2$	thiocyanogen	-2	$-^a$
$(SeCN)_2$	selenocyanogen	—	—
$(SCSN_3)_2$	azidocarbondisulfide	—	—
$(OCN)_2$	oxycyanogen	—	—

^{a} Polymerizes irreversibly at room temperature.

isolated. The pseudohalogens in general exhibit physical and chemical properties characteristic of the halogens. For example, the free pseudohalogens are easily volatile compounds. More important is the similarity of chemical properties of the pseudohalogens and the halogens, which may be summarized as follows.

The pseudohalogens are oxidizing agents. They combine with many metals to form salts that have characteristics similar to those formed by the halogens. For example, the silver, mercurous, and plumbous salts of pseudohalogens are insoluble in water. The pseudohalogens react with aqueous base

$$(CN)_2 + 2OH^- \rightarrow CN^- + OCN^- + H_2O, \tag{88}$$

add to unsaturated molecules

$$(SCN)_2 + H_2C{=}CH_2 \rightarrow \underset{\underset{\text{NCS}}{|}}{H_2C}{-}\underset{\underset{\text{SCN}}{|}}{CH_2}, \tag{89}$$

and will displace less powerful oxidizing agents from their binary salts

$$2SCN^- + Br_2 \rightarrow (SCN)_2 + 2Br^- \tag{90}$$

$$2I^- + (SCN)_2 \rightarrow I_2 + 2SCN^- \tag{91}$$

in much the same way as do the halogens. In some instances, pseudohalide ions can be oxidized to the free pseudohalogen

$$MnO_2 + 4H^+ + 2SCN^- \rightarrow Mn^{2+} + 2H_2O + (SCN)_2. \qquad (92)$$

The pseudohalogens form hydrogen derivatives that are similar to the hydrogen halides. Generally the compounds are strong acids in aqueous solution, analogous to HCl, HBr, and HI, but in some instances (e.g., HCN) they are weak acids like HF. The pseudohalogens form covalent derivatives of the nonmetals which have properties very similar to those of the corresponding halogen derivatives. The most extensively investigated compounds are the isocyanates, cyanates, isothiocyanates, and thiocyanates of silicon (71–77), germanium (78), phosphorus (79, 80), arsenic (81), antimony (82), and sulfur (78). Solutions of the compounds are prepared by reacting the appropriate nonmetal halide with a silver pseudohalide in benzene and removing the precipitated silver halide

$$M-Cl + AgX \rightarrow MX + AgCl. \qquad (93)$$

The pseudohalogens form compounds similar to the interhalogens by combining with each other (e.g., CNN_3, CSSCN, CNSeCN, and $CNSCSN_3$) and with the halogens. The latter type of compound can correspond either to diatomic interhalogens [XCN, XSCN, XN_3 (*XVII*)] or to the higher interhalogens [$SCNCl_3$, $SCNBr_3$, $I(SCN)_3$, $I(OCN)_3$, $(SCSN_3)Br_3$].

The pseudohalide ions form compounds such as $NH_4(SCN)_3$ and CsI_2CN which contain ions that appear to be analogous to the polyhalide ions. In addition, unstable ions of types $(SeCN)I_2^-$ and $(SeCN)_2I^-$, which are analogous to the I_3^- ion, have been detected in aqueous solution. The pseudohalide ions can often act as good bases forming coordination compounds with transition-metal ions. Other reactions of the pseudohalogens appear in more appropriate places in this volume.

5 / STEREOCHEMISTRY AND BONDING

The compounds of the halogens exhibit the greatest display of structural types of any of the families of elements as shown in Table 14.12. At the one extreme, simple diatomic molecules are stable species, as in the hydrogen halides (HX), the halogen molecules (X_2), and the interhalogens (XX'). At the other extreme is a compound, IF_7, in which all seven valence electrons are involved in the formation of single bonds. Between these extremes are species in which halogen aoms form two (Cl_2O, ICl_2^-), three (IO_3^-, ClF_3), four (ClO_4^-, BrF_4^-), five (BrF_5), and six (IF_6^-) bonds with a wide variety of geometrical arrangements.

Basically two hybridization schemes can be used to describe the bonding in these species, viz., hybrids involving s and p orbitals and those involving s, p, and d orbitals. Except for the halogen dioxides (XO_2), the configurations

TABLE 14.12   The configuration and hybridization for some halogen-containing species

Number of bonds	Example	Configuration	Hybridization of central halogen atom
2	$ClO_2{}^a$	bent, 116.5°	sp^2
	$ClO_2{}^-$	bent, 110.5°	sp^3
	$ICl_2{}^-$	linear	dsp^3
3	$XO_3{}^-$	pyramidal	sp^3
	ClF_3	T-shaped	dsp^3
4	$XO_4{}^-$	tetrahedral	sp^3
	BrF_4	planar	d^2sp^3
5	BrF_5	tetragonal pyramid	d^2sp^3
6	$IF_6{}^-$	distorted octahedral or distorted pentagonal bipyramid	d^3sp^3
7	IF_7	distorted octahedral or pentagonal bipyramid	d^3sp^3

^{a} Paramagnetic species.

employed are based on tetrahedral, trigonal-bipyramidal, or octahedral distributions of electron pairs. The ability of the halogens with atomic number greater than that of fluorine to form compounds with higher oxidation states parallels the behavior observed in other families. The interatomic distances in halogen-containing compounds reflect multiple-bond contributions to the bonding in these species using the usual criteria (40). In all instances the d orbitals on the central halogen atoms must contribute to the π component in these bonds except in the case of the halogen dioxide (e.g., ClO_2) where the bond angle (116.5°) suggests that (nearly) sp^2 hybrid orbitals of the halogen atom contribute to the σ framework. If this interpretation is correct, the unhybridized p orbital on the halogen atom can contribute to the π system.

REFERENCES

1. I. Haque and J. L. Wood, *J. Mol. Struc.*, **2**, 217 (1968).
2. H. Hope and G. H.-Y. Lin, *Chem. Commun.*, 169 (1970).
3. G. A. Olah and M. B. Comisarow, *J. Amer. Chem. Soc.*, **91**, 2172 (1969).

4. A. J. Edwards, G. R. Jones, and R. J. C. Sills, *Chem. Commun.*, 1527 (1968).
5. R. S. Eachus, T. P. Sleight, and M. C. R. Symons, *Nature*, **222**, 769 (1969).
6. R. J. Gillespie and K. C. Malhotra, *Inorg. Chem.*, **8**, 1751 (1969).
7. R. J. Gillespie and M. J. Morton, *J. Mol. Spectrosc.*, **30**, 178 (1969).
8. G. A. Olah and J. R. DeMember, *J. Amer. Chem. Soc.*, **91**, 2113 (1969).
9. R. J. Gillespie and B. J. Milne, *Inorg. Chem.*, **5**, 1577 (1966).
10. R. J. Gillespie and B. J. Milne, *Inorg. Chem.*, **5**, 1236 (1966).
11. R. J. Gillespie, J. B. Milne, and M. J. Morton, *Inorg. Chem.*, **7**, 2221 (1968).
12. A. A. Passchier and N. W. Gregory, *J. Phys. Chem.*, **72**, 2697 (1968).
13. M. Tameres, W. K. Duerksen, and J. M. Goodenow, *J. Phys. Chem.*, **72**, 966 (1968).
14. A. A. Passchier, J. D. Christian, and N. W. Gregory, *J. Phys. Chem.*, **71**, 937 (1967).
15. E. A. Ogryzlo and B. C. Sanctuary, *J. Phys. Chem.*, **69**, 4422 (1965).
16. D. D. Eley, F. L. Isack, and C. H. Rochester, *J. Chem. Soc., A*, 1651 (1968).
17. G. Kortüm and H. Koffer, *Ber. Bunsenges. Phys. Chem.*, **67**, 67 (1963).
18. M. Suzuki, T. Yokoyama, and M. Ito, *J. Chem. Phys.*, **50**, 3392 (1969).
19. M. Suzuki, T. Yokoyama, and M. Ito, *J. Chem. Phys.*, **51**, 1929 (1969).
20. J. E. Cahill and G. E. Leroi, *J. Chem. Phys.*, **51**, 4514 (1969).
21. C. H. Townes and B. P. Dailey, *J. Chem. Phys.*, **20**, 35 (1952).
22. J. Janzen and L. S. Bartell, *J. Chem. Phys.*, **50**, 3611 (1969).
23. M. Atoji and W. N. Lipscomb, *Acta Cryst.*, **7**, 173 (1954).
24. W. G. Schneider, H. J. Bernstein, and J. A. Pople, *J. Chem. Phys.*, **28**, 601 (1958).
25. D. F. Hornig and W. E. Osberg, *J. Chem. Phys.*, **23**, 662 (1955).
26. D. F. Hornig and G. L. Hiebert, *J. Chem. Phys.*, **27**, 753 (1957).
27. A. J. Barnes, H. E. Hallam, and G. F. Scrimshaw, *Trans. Faraday Soc.*, **65**, 3150, 3159, 3172 (1969).
28. F. H. Field and F. W. Lampe, *J. Amer. Chem. Soc.*, **80**, 5583 (1958).
29. F. Klanberg and H. W. Kohlschütter, *Z. Naturforsch.*, **16B**, 69 (1961).
30. F. Neumayr and N. Vanderkooi, Jr., *Inorg. Chem.*, **4**, 1234 (1965).
31. R. W. Fessenden and R. H. Schuler, *J. Chem. Phys.*, **44**, 434 (1966).
32. J. N. Keith, I. J. Soloman, I. Sheft, and H. H. Hyman, *Inorg. Chem.*, **7**, 230 (1968).
33. R. H. Jackson, *J. Chem. Soc.*, 4585 (1962).
34. E. T. McHale and G. von Elbe, *J. Phys. Chem.*, **72**, 1849 (1968).
35. H.-J. Schumacher and G. Steiger, *Z. Phys. Chem.*, **B7**, 363 (1930).
36. E. T. McHale and G. von Elbe, *J. Amer. Chem. Soc.*, **89**, 2795 (1967).
37. W. E. Dasent and T. C. Waddington, *J. Inorg. Nucl. Chem.*, **25**, 132 (1963).
38. M. H. Studier and E. H. Appelman, *J. Amer. Chem. Soc.*, **93**, 2349 (1971).
39. G. M. Bancroft and H. D. Gesser, *J. Inorg. Nucl. Chem.*, **27**, 1545 (1965).
40. D. S. Urch, *J. Inorg. Nucl. Chem.*, **25**, 771 (1963).
41. E. H. Appelman, *J. Amer. Chem. Soc.*, **90**, 1900 (1968).
42. M. H. Studier, *J. Amer. Chem. Soc.*, **90**, 1901 (1968).
43. J. Aveston, *J. Chem. Soc., A*, 273 (1969).
44. G. J. Buist, W. C. P. Hipperson, and J. D. Lewis, *J. Chem. Soc., A*, 307 (1969).
45. J. Shamir and N. Parchi, *Spectrosc. Lett.*, **4**, 57 (1971).
46. A. J. Edwards and G. R. Jones, *J. Chem. Soc., A*, 1467 (1969).
47. M. Schmeisser, W. Ludovici, D. Naumann, P. Sartori, and E. Sharf, *Chem. Ber.*, **101**, 4214 (1968).
48. C. S. Rondesvedt, Jr., *J. Amer. Chem. Soc.*, **91**, 3054 (1969).
49. H. S. Gutowsky, D. W. McCall, and C. P. Slichter, *J. Chem. Phys.*, **21**, 279 (1953).
50. E. L. Muetterties and W. D. Phillips, *J. Amer. Chem. Soc.*, **79**, 322 (1957).
51. E. W. Muetterties and W. D. Phillips, *J. Amer. Chem. Soc.*, **81**, 1084 (1959).
52. R. S. Berry, M. Tameres, C. J. Ballhausen, and H. Johansen, *Acta Chem. Scand.*, **22**, 231 (1968).

53. H. Selig and H. Holzman, *Israel J. Chem.*, **7**, 417 (1969).
54. H. Meinert and H. Klamm, *Z. Chem.*, **8**, 195 (1968).
55. H. Klamm, H. Meinert, P. Reich, and K. Witke, *Z. Chem.*, **8**, 469 (1968).
56. S. Bukshpan, J. Soriano, and J. Shamir, *Chem. Phys. Letters*, **4**, 241 (1969).
57. J. Shamir and I. Yaroslavsky, *Israel J. Chem.*, **7**, 495 (1969).
58. H. W. Baird and H. F. Giles, *Acta Cryst.*, **A25**, S115 (1969).
59. R. Lord, M. A. Lynch, Jr., W. C. Schumb, and E. J. Slowinski, Jr., *J. Amer. Chem. Soc.*, **72**, 522 (1950).
60. H. H. Claasen, E. L. Gasner, and H. Selig, *J. Chem. Phys.*, **49**, 1803 (1968).
61. R. E. LaVilla and S. H. Bauer, *J. Chem. Phys.*, **33**, 182 (1960).
62. H. B. Thompson, Jr., and L. S. Bartell, *Trans. Amer. Cryst. Assoc.*, **2**, 190 (1966).
63. R. D. Burbank and F. N. Bensen, Jr., *J. Chem. Phys.*, **27**, 981 (1957).
64. J. Donohue, *Acta Cryst.*, **18**, 1018 (1965).
65. R. D. Burbank and N. Bartlett, *Chem. Commun.*, 645 (1968).
66. N. Bartlett, S. Beaton, L. W. Reeves, and E. J. Wells, *Can. J. Chem.*, **42**, 2531 (1964).
67. H. W. Cremer and D. R. Duncan, *J. Chem. Soc.*, 2243 (1931).
68. H. W. Cremer and D. R. Duncan, *J. Chem. Soc.*, 2031 (1932).
69. G. L. Breneman and R. D. Willett, *Acta Cryst.*, **B25**, 1073 (1969).
70. J. E. Davies and E. K. Nunn, *Chem. Commun.*, 1374 (1969).
71. G. S. Forbes and H. H. Anderson, *J. Amer. Chem. Soc.*, **67**, 1911 (1945).
72. G. S. Forbes and H. H. Anderson, *J. Amer. Chem. Soc.*, **66**, 934 (1944).
73. G. S. Forbes and H. H. Anderson, *J. Amer. Chem. Soc.*, **69**, 1241 (1947).
74. G. S. Forbes and H. H. Anderson, *J. Amer. Chem. Soc.*, **70**, 1043 (1948).
75. H. H. Anderson, *J. Amer. Chem. Soc.*, **69**, 3049 (1947).
76. H. H. Anderson, *J. Amer. Chem. Soc.*, **71**, 1801 (1949).
77. H. H. Anderson, *J. Amer. Chem. Soc.*, **72**, 196 (1950).
78. G. S. Forbes and H. H. Anderson, *J. Amer. Chem. Soc.*, **65**, 2271 (1943).
79. H. H. Anderson, *J. Amer. Chem. Soc.*, **67**, 223 (1945).
80. H. H. Anderson, *J. Amer. Chem. Soc.*, **67**, 2176 (1945).
81. H. H. Anderson, *J. Amer. Chem. Soc.*, **69**, 2495 (1947).
82. H. H. Anderson, *J. Amer. Chem. Soc.*, **64**, 1757 (1942).

COLLATERAL READINGS

I. A. H. W. Aten, Jr., *Advan. Inorg. Chem. Radiochem.*, **6**, 207 (1964); V. D. Nefedov, Ju. V. Norseev, M. A. Toropova, and V. A. Khalkin, *Russ. Chem. Revs.*, **37**, 87 (1968). These two articles contain a summary of the chemistry of astatine.
II. J. Arotsky and M. C. R. Symons, *Quart. Revs.*, **16**, 282 (1962). A review of the preparation and properties of cationic chlorine, bromine, and iodine species.
III. H. A. Bent, *Chem. Revs.*, **68**, 587 (1968); C. K. Prout and J. D. Wright, *Angew. Chem. (Intl. Ed.)*, **7**, 659 (1968). These two reviews contain discussions of structural implications of electron donor–acceptor molecular complexes as well as theoretical arguments for their stability.
IV. L. J. Andrews and R. M. Keefer, *Advan. Inorg. Chem. Radiochem.*, **3**, 91 (1961). A discussion of the methods used to detect molecular complexes of the halogens in solution and the stability of such complexes.

V. M. E. Peach and T. C. Waddington, in *Nonaqueous Solvent Systems* (T. C. Waddington, ed), Academic Press, New York, 1965, Chapter 3; F. Klanberg, in *The Chemistry of Nonaqueous Solvents*, Vol. 2 (J. J. Lagowski, ed.), Academic Press, New York, 1967, Chapter 1. These articles contain a wealth of information on solution phenomena in anhydrous hydrogen halides.

VI. H. H. Hyman and J. J. Katz, in *Nonaqueous Solvent Systems* (T. C. Waddington, ed.), Academic Press, 1965, Chapter 2; M. Kilpatrick and J. G. Jones, in *The Chemistry of Nonaqueous Solvents*, Vol. 2 (J. J. Lagowski, ed.), Academic Press, 1967, Chapter 2. The references contain much information on the properties of anhydrous HF as well as the solutions it forms with a variety of solutes. Experimental techniques for handling HF are also discussed.

VII. D. G. Tuck, *Prog. Inorg. Chem.*, **9**, 161 (1968). A detailed discussion of the preparation, properties, and theoretical aspects of hydrogen bonding in HX_2^- and HXY^- anions.

VIII. M. Schmeisser and K. Brändle, *Advan. Inorg. Chem. Radiochem.*, **5**, 41 (1963). A survey of the chemistry of the oxides of the halogens. Included also is a discussion of the halogen oxyfluorides.

IX. I. V. Nikitin and V. Ya. Rosolovskii, *Russ. Chem. Revs.*, **39**, 545 (1970). A review of the reactions of fluorine and nonmetal fluorides in an electric discharge. Among the reactions discussed is the synthesis of oxygen fluorides and chlorine pentafluoride.

X. M. Dratovsky and L. Pacesova, *Russ. Chem. Revs.*, **37**, 243 (1968). A discussion of the chemistry of oxygen-containing compounds of iodine.

XI. G. S. Pearson, *Advan. Inorg. Chem. Radiochem.*, **8**, 177 (1966). A comprehensive survey of the chemical and physical properties of anhydrous perchloric acid.

XII. L. Stein in *Halogen Chemistry*, Vol. 1 (V. Gutmann ed.), Academic Press, New York, 1967, Chapter 3; W. K. R. Musgrave, *Advan. Fluorine Chem.*, **1**, 1 (1960). Both articles are reviews discussing the preparation and properties of the halogen fluorides. The second review also contains a discussion of the uses of these substances in organic chemistry.

XIII. E. H. Wiebenga, E. E. Havinga, and K. H. Boswijk, *Advan. Inorg. Chem. Radiochem.*, **3**, 133 (1961). A survey of the structure and properties of the interhalogen compounds and the polyhalide complexes.

XIV. A. G. Sharpe in *Nonaqueous Solvent Systems* (T. C. Waddington, ed.), Academic Press, New York, 1965, Chapter 7. A summary of the halogens and interhalogens as nonaqueous solvent systems.

XV. A. I. Popov in *Halogen Chemistry*, Vol. 1 (V. Gutmann, ed.), Academic Press, New York, 1967, Chapter 4. An extensive survey of the preparation, properties, stability, and structure of polyhalogen complex ions.

XVI. J. S. Thayer and R. West, *Advan. Organomet. Chem.*, **5**, 169 (1967). A review of the preparation, structural and physical properties, chemical reactions, and bonding of organometallic pseudohalides.

XVII. K. Dehnicke, *Angew. Chem. (Intl. Ed.)*, **6**, 240 (1967). A survey of the reactions of halogen azides.

STUDY QUESTIONS

1. Describe the physical and chemical properties of hydrogen astatide.

2. Explain the basis for the reversal in melting points in the following pairs of compounds: (a) NaCl, 800°C and NaBr, 755°C: (b) CH_3Cl, −97°C and CH_3Br, −93°C.

3. Discuss the factors which account for the observation that rubidium polyhalides are less stable toward thermal decomposition than are the corresponding cesium compounds.

4. Describe the chemical and physical properties expected for astatine.

5. Neither liquid BrF_5 nor AsF_5 are good conductors of electricity; however, a mixture of these substances is a markedly better conductor than either of the pure substances. (a) Discuss the geometry of all the species present in this mixture. (b) Discuss the bonding present in these species.

6. Describe the geometry of the covalently bonded species in the following compounds without recourse to structural data: (a) $KICl_2$; (b) BrF_3; (c) $NaClO_2$; (d) KIF_6; (e) $NaIO_6$; (f) $INO_3 \cdot py$; (g) Br_2O.

7. Pure liquid IF_5 is a surprisingly good conductor of an electric current. Suggest the geometry of all species present in this system and the hybridization of the central atom in each.

8. Describe methods for the preparation of $(BrF_2)_2TeF_6$ and $AgSbF_6$.

9. Give the idealized hybridization of the atoms in the covalently bonded species in Question 6.

10. The dipole moment and internuclear distance for IBr are 1.2 D and 2.47 Å, respectively. Estimate the per cent ionic character of this compound.

11. Using the information in Question 10, calculate the electronegativity difference for IBr. Compare and discuss this value with that obtained for IBr from standard electronegativity tables.

12. Discuss the similarities and differences in the structures of the Period 2 element chlorides.

13. Discuss the observations that HBF_4 is a strong acid in water yet phase studies on the $HF-BF_3$ system do not reveal the presence of a 1 : 1 compound.

14. Give the most probable geometrical structures for the ionic species present when the following substances are dissolved in anhydrous hydrogen halides: (a) $SnCl_4$; (b) ICl; (c) BCl_3; (d) ROH, (e) $CH_3\overset{\displaystyle O}{\overset{\|}{C}}-OCH_3$; (f) $(CH_3)_2CO$.

15. Discuss the Lewis acid–base character of the following species: (a) ClO_3^-; (b) ClF_3; (c) ICN; (d) BrO_2; (e) I_2: (f) SCN^-; (g) IF_4.

/ **15**

THE RARE GASES

1 / INTRODUCTION

The electronic configuration of rare-gas atoms consists of filled ns and np orbitals, giving each atom a closed-shell structure. Classically such electronic configurations have been considered the epitome of an inert electronic structure; for example, many of the Lewis arguments concerning the nature of ionic and covalent compounds, as well as derivations of this theory, are based on the supposed inertness of such configurations. Since 1962 a considerable number of compounds containing chemically bonded rare gas atoms have been reported. As might be expected from their electronic configurations, shown in Table 15.1, the rare gases are monatomic, a conclusion supported by the ratio of their heat capacities (C_p/C_v), which is the value predicted (1.667) for a monatomic gas. The physical properties of the rare gases given in Table 15.1 show trends which are essentially ideal because no permanent dipolar or electrostatic interactions occur among the species

TABLE 15.1 Some physical properties of the rare gases

	He	Ne	Ar	Kr	Xe	Rn
Ground-state configuration	$1s^2$	$[\text{He}]2s^22p^6$	$[\text{Ne}]3s^23p^6$	$[\text{Ar}]3d^{10}4s^24p^6$	$[\text{Kr}]4d^{10}5s^25p^6$	$[\text{Xe}]4f^{14}5d^{10}6s^26p^6$
Ionization potential, eV	24.58	21.56	15.76	14.00	12.13	10.75
Atomic radius, Å	—	1.60	1.91	2.00	2.2	—
Mp, °K	(0.9)	24.43	83.9	104	133	202
Bp, °K	4.22	27.2	87.4	121.3	103.9	211.3
$\Delta H_{(\text{vap})}$, kcal mole^{-1}	0.025	0.405	1.600	2.240	3.100	3.600
C_p/C_v	1.65	1.64	1.65	1.69	1.67	—
Polarizability, $\times 10^{-24}$ cm^3	0.202	0.392	1.629	2.460	4.000	5.419

which constitute the sample. Thus, the boiling and melting points vary regularly with increasing atomic number, as do many other physical properties.

The fact that rare-gas atoms can be condensed to give liquid and solid phases at low temperatures suggests that weak interactive forces are operative in the spherically symmetrical, closed-shell electronic systems. It was suggested by London that electronic motions in a symmetrical molecule or atom give rise to fluctuating dipoles which time average to zero (*1*). These London forces cause the rare gases to form condensed phases at temperatures which are sufficiently low that the kinetic energy is not an important factor. The London, or dispersion, forces theoretically are directly proportional to the polarizability of the atoms, which is a measure of how loosely the electrons in the system are held, and inversely proportional to their ionization potential; a more detailed discussion of dispersion forces appears in Chapter 7. This prediction correlates well with the data in Table 15.1 which show that the polarizabilities of the rare gases increase with increasing atomic weight but the ionization potential decreases in the same order. Thus, we would expect the London forces to increase with increasing atomic number, which should also be the order of increasing melting or boiling points.

With the exception of helium, the rare gases crystallize in the cubic closest-packed structure. Helium apparently is hexagonal close-packed in the solid state.

2 / OCCURRENCE AND GENERAL PROPERTIES

In 1785 Cavendish reported that all of the nitrogen in air could not be removed by sparking with oxygen; a residue of unreacted gas always remained. Nearly a hundred years later, Rayleigh showed that the density of nitrogen obtained from a chemical reaction was less than that obtained from air after the removal of all components known at the time. Ramsay and his co-workers almost immediately after that observation isolated samples of the noble gases from the atmosphere. Except for helium and radon, the atmosphere is still the only known source of the rare gases. The data in Table 15.2 indicate that these gases exist in low concentrations in the atmosphere; they are conveniently obtained as by-products in the preparation of oxygen and nitrogen from air. The process involves liquefaction of dry, CO_2-free air; distillation of this liquid gives a volatile fraction containing nitrogen, neon, and helium, while oxygen, argon, krypton, and xenon remain in the liquid state. The remaining liquid is fractionated to yield argon, contaminated with oxygen; the latter substance is removed by passing the gaseous mixture, with hydrogen added, over a hot copper–copper-oxide mixture. Further separation of the rare gases is effected by selective adsorption and desorption with charcoal at temperatures ranging from $-190°$ to $-100°$. Helium is a common component of natural gas; concentrations of about 7% have been reported in some natural-gas wells. The element is also found trapped in certain radioactive minerals, where it is undoubtedly formed when α particles arising from radioactive decay processes are electrically neutralized. Radon,

TABLE 15.2   Concentration of the rare gases in the atmosphere

	Per cent composition	
Rare gas	Volume	Weight
He	5×10^{-4}	6.9×10^{-5}
Ne	1.5×10^{-3}	1.2×10^{-3}
Ar	0.94	1.3
Kr	1.1×10^{-4}	2.9×10^{-4}
Xe	9×10^{-6}	4.1×10^{-5}

all isotopes of which are radioactive, is a product of the three natural radioactive decay series. The longest-lived isotope of radon (^{222}Rn), formed in the decay of radium (^{226}Ra, α; $t_{1/2} = 1.62 \times 10^{-3}$ years), decays by α emission with a half-life of 3.825 days. Accordingly, discussions of the properties of the rare gases usually exclude radon.

The properties of helium in the condensed state are unique among those of the elements. Naturally occurring helium contains a small amount of a lighter isotope (^{3}He), which appears to behave as a normal liquid. ^{4}He condenses at 4.18°K and 1 atm to give a substance which also appears to behave normally, but at 2.178°K (the λ point) some of the properties of this liquid, e.g., specific heat, viscosity, thermal conductivity, compressibility, and surface tension, change abruptly. In addition, a change in the temperature coefficients of density, vapor pressure, dielectric constant, and refractive index occurs at the λ point. Above the λ point the substance is called He I, while below this temperature it is called He II. Liquid He II exhibits some extraordinary properties; it has a low viscosity ($\sim 10^{-3}$ times that of H_2 gas), a very high thermal conductivity (~ 800 times that of helium at room temperature), and flow properties unknown for any other liquid. Initially unequal levels of liquid He II in concentric containers become equalized by the flow of a thin film of the liquid up the walls of the container which has the higher level over the top and into the container which has the lower level. In all its properties liquid He II appears to be best described as a superfluid; a convincing theoretical interpretation of the basis of this behavior is not presently available.

3 / CHEMICAL PROPERTIES

The rare gases exhibit two types of compound formation. The rare-gas clathrates are actually inclusion compounds, that is compounds containing physically trapped rare-gas atoms. Compounds containing chemically bound rare-gas atoms have been only recently prepared and characterized.

3.1 / Clathrates

Rare-gas hydrates and deuterohydrates are formed when H_2O or D_2O are crystallized in an atmosphere of the rare gas. The reported compositions for these hydrates are very near $R \cdot 6H_2O$ and $R \cdot 6D_2O$, where R is a rare-gas atom. The stability of these hydrates decreases with decreasing atomic number of the rare gas. Thus, xenon hydrate is very easily formed at 0°C under a pressure of slightly more than one atmosphere of xenon; the dissociation pressure of this substance is 1 atm at 0°C. The dissociation pressures for krypton (14.5 atm) and argon (98.5 atm) hydrates are progressively higher at 0°C; neon and helium do not form hydrates even at very high pressures.

The structure of rare-gas hydrates is essentially that described for the halogen hydrates in Chapter 14. It will be recalled that several stable arrangements of the hydrogen-bonded water framework are possible in addition to the structure observed for ice. Such hydrogen-bonded molecules have relatively large cavities in which molecules of an appropriate size may be trapped, as shown in Fig. 14.2. The relative instability of the rare-gas hydrates is a reflection of the fact that their structures are maintained only by hydrogen bonds and the polarizing effect of the host molecules on the rare-gas atom. The influence of the latter factor is apparent in the increasing stability of the rare-gas hydrates with increasing atomic number of the rare gas, which is also in increasing order of polarizability; the pertinent data appear in Table 15.1.

A more stable lattice of hydrogen-bonded molecules which contains relatively large cavities is exhibited by certain hydroxy compounds, such as p-dihydroxybenzene (2, 3). These substances form two infinite interpenetrating arrays of hydrogen-bonded molecules containing sufficiently large cavities to accommodate the rare gases argon, krypton, and xenon. Indeed, even molecules which have larger dimensions, such as CO_2, H_2C_2, HCO_2H, and CH_3OH, can be incorporated in these structures. The structure of crystalline p-dihydroxybenzene contains one cavity for every three molecules, in agreement with the composition of these clathrates, which approaches $3C_6H_4(OH)_2 \cdot R$ (R = Xe, Kr, Ar). Similar clathrates in which phenol forms the host lattice have been reported (4–6).

3.2 / Compounds of the Elements (*I*)

The extremely stable electronic configurations of the rare gases, shown in Table 15.1, suggest that only highly electronegative atoms or groups can compete successfully with these atoms for their electrons to form either ionic or covalent species. Before the successful preparation of the first rare-gas compound, several suggestions had been made concerning the possibility of chemically combining rare-gas atoms. An unsuccessful attempt was made to combine xenon with fluorine under the influence of an electric discharge (*7*), but 30 years later similar experiments proved successful (*8*). In addition, stable rare-gas–halogen compounds were predicted on theoretical grounds before they were prepared (*9*).

Although we might expect the heavier rare gases to accommodate more than four electron pairs in their valence shells, it is highly unlikely that the neutral atoms are sufficiently electronegative to form a bond to a Lewis base. Thus, using the electronegativities of the elements and the ionization potentials of the rare gases as crude indicators of possible compound formation, we might expect that the most probable candidates for such compounds are the largest rare gases, radon and xenon, and the most electronegative of the elements, fluorine (4.0), oxygen (3.5), nitrogen (3.0), and chlorine (3.0). From a practical standpoint radon is excluded from these considerations because all of its isotopes are radioactive and have relatively short half-lives. In fact, the majority of the known rare-gas compounds contain xenon bonded to fluorine or oxygen. Krypton forms fewer and less stable compounds; evidence exists for unstable rare-gas compounds containing chlorine, but no nitrogen-containing compounds have been characterized. Thus far stable compounds containing the rare-gas atom in $1+$, $2+$, $4+$, and $6+$ formal oxidation states have been prepared.

Halogen Derivatives

Ionic Derivatives. The observation that certain metal hexafluorides, e.g., PtF_6, are sufficiently strong oxidizing agents to form dioxygenyl compounds

$$O_2 + PtF_6 \rightarrow O_2^+PtF_6^- \tag{1}$$

(see also Chapter 13) leads to the suggestion that xenon might be similarly oxidized, since the ionization potentials of O_2 (12.2 eV) and xenon (12.13 eV) are virtually the same (*10*). Xenon reacts with PtF_6 to yield an orange-red solid which is insoluble in CCl_4 and has been formulated as an ionic compound

$$Xe + PtF_6 \rightarrow Xe^+PtF_6^-. \tag{2}$$

The lattice energy of $Xe^+PtF_6^+$ has been calculated (*10*) as 110 kcal mole^{-1}, which is slightly less than that of the corresponding oxygenyl compound

(*11*). The substance hydrolyzes rapidly in water and decomposes at 160°C to give xenon and $Xe(PtF_6)_2$, a red solid. Other xenon hexafluorometallates such as $XeRuF_6$, $XeRhF_6$, and $XePuF_6$ have been reported, although the analogous krypton compounds did not form (*12, 13*). Heating xenon with certain metal hexafluorides leads to a lower-valent metal fluoride and a mixture of xenon fluorides.

Radon has been oxidized in anhydrous HF using the chlorine fluorides, bromine fluorides, IF_7, and NiF_6^{2-} ion (*14*). Electromigration studies of the stable solutions suggest that the radon is present as either Rn^{2+} or RnF^+.

Covalent Compounds. The xenon fluorides can be prepared by direct fluorination of xenon under a variety of conditions

$$Xe + F_2 \longrightarrow XeF_2 \tag{3}$$

$$XeF_2 + F_2 \longrightarrow XeF_4 \tag{4}$$

$$XeF_4 + F_2 \longrightarrow XeF_6 \tag{5}$$

including heat, ultraviolet irradiation, and electric discharge; some properties of the known xenon fluorides are given in Table 15.3. Xenon difluoride

TABLE 15.3 **Some properties of the rare-gas fluorides**

Oxidation state	Compound	Mp, °C	Vapor pressure, torr at 25°C	Color	
				Solid	Gas
2+	XeF_2	129	3.8	white	colorless
	KrF_2	<0 (subl.)	—	white	—
4+	XeF_4	~114	3	white	colorless
6+	XeF_6	49.5	28	white (<42°C) yellow (>42°C)	yellow
	$XeOF_4$	−46.2	28	white	colorless

is readily fluorinated [Eq. (4)], hence it must be removed from the reaction zone if a relatively pure sample is required. In practice, a mixture of xenon and fluorine is circulated through a closed loop containing a section heated to ~ 400° and one cooled to −50°. Relatively pure XeF_2 condenses in the latter section. Xenon difluoride can be also prepared by using O_2F_2 at −118° as a fluorinating agent (*15*). Xenon difluoride has been reported as the product formed in the high-voltage discharge (6000 V) of a mixture of CF_4 and xenon. Other fluorine-containing compounds, such as SiF_4, CF_3OF, and FSO_3F have been suggested or reported to act as fluorine sources in the preparation of xenon fluorides.

Xenon tetrafluoride forms readily as the major product when fluorine and xenon are allowed to react and XeF_2 is not removed. Indeed, it is possible to form this compound by allowing sunlight to fall on a mixture of the elements sealed in a quartz vessel. The preparation of xenon hexafluoride requires more severe conditions, i.e., a high proportion of fluorine ($F_2 : Xe = 20 : 1$), high pressures (50–500 atm), elevated temperatures (200–700°C), and extended periods of heating.

Krypton is less reactive with PtF_6 or RhF_6 and does not react with fluorine under ultraviolet irradiation. However KrF_2 is formed when a mixture of the elements is subjected to an electric discharge at 84°K or irradiated with an electron beam at 123°K. Krypton difluoride is markedly less stable than XeF_2, decomposing rapidly at 0°C. Krypton tetrafluoride has been prepared as an unstable colorless solid by subjecting the elements to an electric discharge at low temperatures.

The preparation of radon fluorides has been established using radiochemical techniques. Neither the stoichiometry nor the constitution of these fluorine-containing compounds has been established because of the radiochemical instability of radon.

The reaction between xenon and chlorine under the influence of a microwave discharge has been studied. The infrared spectra of the products, condensed at 20°K, contain evidence for a linear, symmetrical compound, $XeCl_2$ (*16*). A similar experiment revealed the existence of $KrCl_2$ in a low temperature matrix (*17*).

Xenon difluoride is a white solid which has a sufficiently high vapor pressure to permit its sublimation under mild conditions. The compound exhibits characteristic bands in the mass spectrum ($XeF_2{}^+$, $XeF_2{}^{++}$, and Xe^+) and in the infrared region (549 cm^{-1}) which can be used to monitor the formation or loss of this species.

Xenon difluoride is decomposed by water, aqueous KOH, and sulfuric acid. A 0.15 M solution of XeF_2 can be prepared in water, in which it has a half-life of about seven hours at 0°C. In alkaline solution elemental xenon is rapidly formed

$$XeF_2 + 2OH^- \rightarrow Xe + \tfrac{1}{2}O_2 + 2F^- + H_2O. \tag{6}$$

It reacts with iodide ion

$$XeF_2 + 2I^- \rightarrow Xe + I_2 + 2F^-, \tag{7}$$

hydrogen

$$XeF_2 + H_2 \rightarrow Xe + 2HF, \tag{8}$$

and ammonia

$$\tfrac{3}{2}XeF_2 + 4NH_3 \rightarrow \tfrac{3}{2}Xe + 3NH_4F + \tfrac{1}{2}N_2 \tag{9}$$

to give elemental xenon. The reaction with hydrogen at 200° [Eq. (8)] has been used to establish the composition of XeF_2. Xenon difluoride can fluorinate nitrogen oxides

$$2NO + XeF_2 \rightarrow 2FNO + Xe \tag{10}$$

$$2NO_2 + XeF_2 \rightarrow 2O_2NF + Xe. \tag{11}$$

Adducts of XeF_2 with the Lewis acids MF_5 (M = P, As, Sb, Pt, Ir, Os, Ru, Rh, Ta, Nb, I) have been studied conductometrically (*18, 19*). Several types of adducts are formed, viz., $XeF_2 \cdot MF_5$, $XeF_2 \cdot 2MF_5$, and $2XeF_2 \cdot MF_5$. X-Ray data (*20, 21*) indicate that the 1 : 1 adducts are best formulated as $XeF^+MF_6^-$, the 1 : 2 adducts as $Xe_2F_3^+MF_6^-$, and the 2 : 1 adducts as $XeF^+M_2F_{11}^-$. The $Xe_2F_3^+$ ion consists of two linear XeF_2 groups sharing a fluorine bridge. In the case of the 2 : 1 adducts the XeF^+ unit is also coordinated to a fluorine atom on the M_2F_{11} group.

Electron-pair-repulsion arguments for xenon difluoride, which is isoelectronic with ICl_2^-, suggest that the molecule should be linear (**1**).

$$F\overset{\cdot\cdot}{\underset{\cdot\cdot}{-Xe}}\overset{\cdot\cdot}{-}F$$

1

Diffraction experiments on the compound in the crystalline state show that linear XeF_2 molecules are present; the relevant structural data appear in Table 15.4.

Xenon tetrafluoride is a volatile white solid which apparently can exist in two modifications. This substance can be distinguished from XeF_2 by spectroscopic means since its mass spectrum contains the species XeF_3^+ and its infrared spectrum exhibits a band at $590\,cm^{-1}$. Xenon tetrafluoride is reduced by hydrogen

$$XeF_4 + 2H_2 \rightarrow Xe + 4HF, \tag{12}$$

mercury

$$XeF_4 + 4Hg \rightarrow Xe + 2Hg_2F_2, \tag{13}$$

xenon

$$XeF_4 + Xe \rightarrow 2XeF_2, \tag{14}$$

TABLE 15.4   The structural parameters of the rare-gas fluorides

Compound	Molecular shape	M—F bond distances, Å
XeF$_2$ (solid)	linear	2.00
(gas)	linear	1.977
KrF$_2$ (gas)	linear	1.889
XeF$_4$ (solid)	planar	1.95
(gas)	planar	1.94
KrF$_4$ (gas)	planar	1.89
XeF$_6$ (gas)	distorted octahedral	1.91
XeOF$_4$^{a}	tetragonal pyramid	1.900

^{a} $d(Xe—O) = 1.703$ Å.

and water

$$XeF_4 + 2H_2O \rightarrow Xe + 4HF + O_2. \tag{15}$$

The reaction with water is pH-dependent; in acid or neutral solution 50% of the oxygen and 70% of the xenon expected on the basis of Eq. (15) are obtained. However, in 6N NaOH the gaseous products are formed in accordance with Eq. (15). Xenon trioxide has been isolated from the hydrolysis of XeF$_4$, suggesting a rather complex disproportionation process. Aqueous iodide ion is oxidized by XeF$_4$,

$$XeF_4 + 4I^- \rightarrow Xe + 2I_2 + 4F^- \tag{16}$$

but the oxidation of water [Eq. (15)] is a competing reaction; about 10% of the XeF$_4$ reacts with water when aqueous iodide ion is oxidized. Xenon tetrafluoride reacts with the oxides of nitrogen [cf. Eqs. (10) and (11)] and fluorinates unsaturated organic compounds

$$2CF_3CF{=}CF_2 + XeF_4 \rightarrow 2CF_3CF_2CF_3 + Xe. \tag{17}$$

Attempts to dissolve XeF$_4$ in ethanol, tetrahydrofuran, dioxane, or cyclopentadiene lead to a very vigorous reaction, although the compound apparently dissolves smoothly in CS$_2$, CCl$_4$, dimethylformamide, cyclohexane, or diethyl ether; the ether solution exhibits strong oxidizing characteristics. Solutions of XeF$_4$ in trifluoroacetic anhydride, after treatment with trifluoroacetic acid, yield a yellow solid reported to be (CF$_3$CO$_2$)$_4$Xe.

Neither alkali-metal fluorides nor SbF$_5$ form complexes with XeF$_4$. When XeF$_4$ is dissolved in SbF$_5$ a gas is evolved and a yellow complex (mp 63°C) with the stoichiometry $XeF_2 \cdot 2SbF_5$ is formed. A similar straw-colored tantalum complex ($XeF_2 \cdot 2TaF_5$, mp 81°) forms when XeF_4 is added to TaF_5.

Electron-pair-repulsion arguments suggest xenon tetrafluoride should be a planar molecule (**2**); the data in Table 15.4 verify this prediction. A mixed

$$\overset{\displaystyle F}{} \underset{\displaystyle F}{\overset{\displaystyle \cdots}{\underset{\displaystyle \cdots}{Xe}}} \overset{\displaystyle F}{\underset{\displaystyle F}{}}$$

2

$XeF_2 \cdot XeF_4$ lattice complex is known which contains both linear and square-planar molecules.

Xenon hexafluoride, a pale-yellow solid, is the least volatile and least stable of the xenon fluorides. A violent reaction occurs when the compound is added to water, but under controlled conditions, e.g., moist air or low temperatures, hydrolysis proceeds smoothly. Xenon trioxide can be obtained by evaporating the excess water in such experiments. Incomplete hydrolysis of XeF_6 yields $XeOF_4$ (mp 46.2°). Aqueous XeF_6 solutions react with I^- to give iodine [cf. Eq. (7)].

Xenon hexafluoride or its derivatives form more complexes than do the other xenon fluorides. The reaction of xenon, fluorine, and SbF_5 yields three compounds, $XeF_6 \cdot 2SbF_5$ (mp 108°), $XeF_6 \cdot SbF_5$ (mp 205°), and $2XeF_6 \cdot SbF_5$ (mp 100°), which have different x-ray powder patterns. Xenon hexafluoride reacts with Lewis acids producing compounds such as $XeF_6 \cdot BF_3$, $XeF_6 \cdot AsF_5$, and $2XeF_6 \cdot VF_5$; it reacts with alkali-metal fluorides, giving species formulated as $MXeF_7$. The latter are thermally unstable and yield octafluoroxenates

$$2MXeF_7 \longrightarrow M_2XeF_8 + XeF_6. \tag{18}$$

Rubidium and cesium octafluoroxenates are stable to 400°C. They do, however, hydrolyze in aqueous solutions to give xenon-containing oxyanions. Xenon oxytetrafluorides also yield adducts with some alkali-metal fluorides, e.g., $MF \cdot XeOF_4$ (M = K and Cs) and $3RbF \cdot 2XeOF_4$, as well as with Lewis acids, $2SbF_5 \cdot XeOF_4$.

Incomplete electron-diffraction data suggest that the structure of XeF_6 appears to be a distorted octahedron, although some spectroscopic data have been interpreted in terms of a symmetrical octahedral arrangement. The electron-pair-repulsion model favors the latter structure, which is isoelectronic with IF_7; a process similar to that suggested for the interconversion of the two possible structures for IF_7, shown in Fig. 14.20, has been suggested to explain the available structural information on XeF_6 (*22*). Xenon oxytetrafluoride, on the other hand, exhibits a square-pyramidal arrangement of atoms, in agreement with its Lewis formulation. Unfortunately, little is known of the structure of the XeF_6 or $XeOF_4$ adducts. In one instance, $BF_3 \cdot XeF_6$, there is infrared evidence indicating that a fluoride-ion transfer has occurred, hence a possible formulation for this substance involves the $XeF_5{}^+$ ion, i.e., $XeF_5{}^+ BF_4{}^-$.

Oxygen Derivatives

Derivatives of Xe(II). Although xenon fluorides react more or less vigorously with water, the controlled hydrolysis of these compounds leads to reasonably stable solutions containing xenon–oxygen species. Xenon difluoride dissolves in water to give a transient yellow color; some have suggested this arises from the formation of a very unstable xenon oxide

$$XeF_2 + H_2O \rightarrow XeO + 2HF \tag{19}$$

which eventually yields xenon

$$XeO \rightarrow Xe + \tfrac{1}{2}O_2. \tag{20}$$

The unidentified species present in these solutions appears to be reasonably stable in that it can be extracted into CCl_4; XeF_2 can be obtained unchanged from these solutions by vacuum distillation. Aqueous base leads to rapid reduction of XeF_2 [Eq. (6)]. Aqueous solutions of XeF_2 are good oxidizing agents and can oxidize Ag^+ to Ag^{2+}, Cl^- to Cl_2, $IO_3{}^-$ to $IO_4{}^-$, $BrO_2{}^-$ to $BrO_4{}^-$, Co^{2+} to Co^{3+}, and Ce^{3+} to Ce^{4+}. In acetonitrile XeF_2 does not act as an oxidizing agent unless traces of HF are present; under these conditions it will oxidize iodine to IF_7. The active species in this reaction is thought to be XeF^+ or $Xe_2F_3{}^+$ (23).

Xenon difluoride reacts with very strong oxyacids, HOR [R = ClO_3 (24), SO_2F (24), TeF_5 (25–27), and $-CO-CF_3$ (27)], to give derivatives that are formally related to $Xe(OH)_2$

$$XeF_2 + HOR \rightarrow FXeOR + HF \tag{21}$$

$$XeF_2 + 2HOR \rightarrow Xe(OR)_2 + 2HF. \tag{22}$$

$Xe(OTeF_5)_2$ (mp 35–37°) is thermally stable to 130°, above which temperature it decomposes

$$Xe(OTeF_5)_2 \rightarrow Xe + \tfrac{1}{2}O_2 + O(TeF_5)_2. \tag{23}$$

The trifluoroacetate $[Xe(O_2CCF_3)_2]$, a dark yellow substance, is markedly less stable; it decomposes by thermal or mechanical shock at $-20°C$, forming xenon, CO_2, and hexafluoroethane quantitatively. The reaction of xenon difluoride with $Xe(OTeF_5)_2$ gives a mixed derivative in quantitative yield

$$XeF_2 + Xe(OTeF_5)_2 \longrightarrow 2FXeOTeF_5. \tag{24}$$

$Xe(OTeF_5)_2$ forms a yellow $1:1$ adduct (mp, 160°C) with arsenic penta-fluoride, which has been formulated as $[XeOTeF_5]^+[AsF_6]^-$ on the basis of its Raman spectrum. Xenon difluoride will, however, displace $FXeOTeF_5$ from this compound, suggesting that the relative order of F^- donor strengths is $XeF_2 > FXeOTeF_5$ (28).

The crystal structure of $Xe(OTeF_5)_2$ shows a planar arrangement of groups about the linear $O—Xe—O$ unit with *trans* TeF_5 moieties (25, 26). The linear arrangement of the $O—Xe—O$ unit is consistent with the linear structure observed for XeF_2.

Xenon Trioxide and Its Derivatives (II). Xenon tetrafluoride undergoes disproportionation when it is hydrolyzed

$$3XeF_4 + 6H_2O \longrightarrow XeO_3 + 2Xe + \tfrac{3}{2}O_2 + 12HF. \tag{25}$$

The product, XeO_3, can be isolated as a dangerously explosive white crystal-line compound by careful evaporation of the solvent. Xenon trioxide can also be obtained from the hydrolysis of $XeOF_4$. The Raman spectra of concentrated aqueous solutions of XeO_3 are consistent with the presence of a pyramidal molecule, but other species are also present. Xenon trioxide is a weak base, the equilibrium

$$XeO_3 + OH^- \rightleftharpoons HXeO_4^- \tag{26}$$

being important at pH 11. Crystalline acid salts of xenic acid ($MHXeO_4$, $M = Na, K, Rb, Cs$) have been obtained by removal of solvent from a mixture containing equivalent amounts of the alkali hydroxide and XeO_3.

Xenic acid is also a powerful oxidizing agent. It can oxidize the same species as does an aqueous XeF_2 solution, and is sufficiently powerful to oxidize Br^- to BrO_3^-, Mn^{2+} to MnO_4^-, and Pu^{3+} to Pu^{4+}. The advantage of using xenic acid as an oxidizing agent lies in the fact that the reduction product is a gas.

In spite of its extreme instability, the structure of XeO_3 has been established. The molecule is pyramidal $[\angle(OXeO) = 103°, d(Xe—O) = 1.76 \text{ Å}]$ in the crystalline state, in agreement with the Raman data on solutions of this compound. Thus, the structure of XeO_3 is very similar to that of the IO_3^- ion, with which it is isoelectronic (3).

3

Perxenic Acid and Its Derivatives. Aqueous solutions of perxenic acid, H_4XeO_6, arise from the disproportionation of xenic acid in basic solution

$$2HXeO_4^- + 2OH^- \rightarrow XeO_6^{4-} + Xe + O_2 + 2H_2O. \qquad (27)$$

Thus, unless care is taken, attempts to work with basic solutions of xenic acid lead to ambiguous results. Indeed, early attempts to isolate salts of xenic acid using conventional techniques gave salts of perxenic acid. Aqueous solutions of perxenic acid can also be prepared by passing ozone through a basic aqueous solution containing Xe(VI) species. Spectroscopic evidence suggests that several perxenic acid species are present in such solutions; $HXeO_6^{3-}$ appears to be the predominant species at pH 11, but as the acidity increases $H_2XeO_6^{2-}$ is formed. The hydrated perxenate salts shown in Table 15.5 have been isolated.

TABLE 15.5   Structural parameters of some salts of perxenic acid

Salt	$\angle$(OXeO), degrees	d(Xe—O), Å
$K_4XeO_6 \cdot 9H_2O$	88.8–91.2	1.86
$Na_4XeO_6 \cdot 6H_2O$	87	1.84
$Na_4XeO_6 \cdot 8H_2O$	87.4–92.6	1.86

Aqueous solutions of perxenic acid are very strong oxidizing agents. Such solutions will oxidize Cr^{2+} to $Cr_2O_7^{2-}$, ClO_3^- to ClO_4^-, Np(IV) to Np(VI), and Am(III) to Am(VI), as well as the species mentioned earlier that are oxidized by xenic acid and solutions of XeF_2. The aqueous redox potentials for xenon species are summarized in Fig. 15.1.

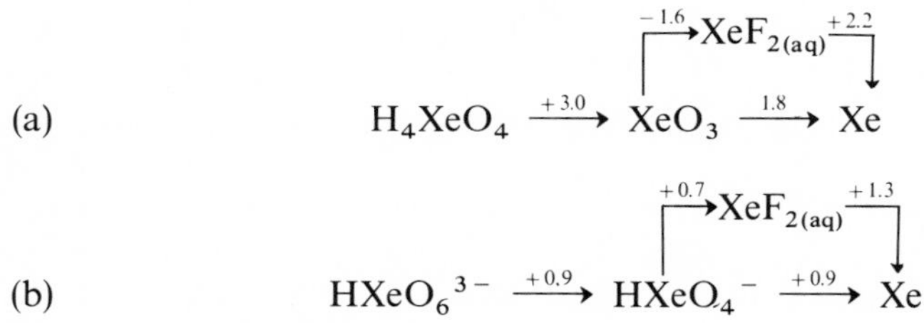

(a)

$$H_4XeO_4 \xrightarrow{+3.0} XeO_3 \xrightarrow{1.8} Xe$$

with $\xrightarrow{-1.6} XeF_{2(aq)} \xrightarrow{+2.2}$ branch

(b)

$$HXeO_6^{3-} \xrightarrow{+0.9} HXeO_4^- \xrightarrow{+0.9} Xe$$

with $\xrightarrow{+0.7} XeF_{2(aq)} \xrightarrow{+1.3}$ branch

FIG. 15.1   The potential diagrams for the redox behavior of xenon-containing species in (a) acidic and (b) basic solutions.

Although the structures of perxenic acid or XeO_4, which has been reported, are not known, crystallographic data on the hydrated alkali-metal perxenates indicate that the anion exhibits approximately octahedral symmetry, in agreement with the structure of the isoelectronic periodate ion IO_6^{5-} and the electron-pair-repulsion model (**4**).

$$\begin{bmatrix} & & O & & \\ O & & | & O & \\ & \diagdown & Xe & \diagup & \\ O & \diagup & | & \diagdown & O \\ & & O & & \end{bmatrix}^{4-}$$

4

3.3 / Stereochemistry and Bonding

With the limited information available, which is summarized in Table 15.6, it appears that the stereochemistry of xenon-containing species follows that of the isoelectronic iodine compounds (cf. Table 14.12). All the species except

TABLE 15.6   Stereochemistry of some rare-gas compounds

Number of bonds	Number of unshared electron pairs	Example	Configuration	Hybridization
2	3	XeF_2	linear	dsp^3
		$Xe(OTeF_5)_2$	linear	dsp^3
3	1	XeO_3	pyramidal	sp^3
4	2	XeF_4	planar	d^2sp^3
5	1	XeF_5^+	—[a]	d^2sp^3
6	1	XeF_6	distorted octahedral	d^3sp^3
	0	XeO_6^{4-}	octahedral	d^2sp^3

[a] Structure unknown.

XeO_3 require d orbitals to form the conventional hybrid orbitals on the central atom for bond formation. The energy requirements for using $5d$ orbitals in the hybridization scheme appear to be too high for stable bond formation, although this may not be the case for $4d$ orbitals (29–31).

REFERENCES

1. F. London, *Trans. Faraday Soc.*, **33**, 8 (1937).
2. H. M. Powell and P. Reisz, *Nature*, **161**, 52 (1948).
3. H. M. Powell, *J. Chem. Soc.*, 298, 300, 468 (1950).
4. P. H. Lahr, H. L. Williams, *J. Phys. Chem.*, **63**, 1432 (1959).

5. B. A. Nikitin, *Izv. Akad. Nauk. SSSR*, **24**, 571 (1940).
6. M. von Stakelberg, *Rec. Trav. Chim.*, **75**, 902 (1956).
7. D. M. Yost and A. L. Kaye, *J. Amer. Chem. Soc.*, **55**, 3890 (1933).
8. A. D. Kirshenbaum, L. V. Streng, A. G. Streng, and A. V. Grosse, *J. Amer. Chem. Soc.*, **85**, 360 (1963).
9. G. C. Pimentel, *J. Chem. Phys.*, **19**, 446 (1951).
10. N. Bartlett, *Proc. Chem. Soc. (London)*, 218 (1962).
11. N. Bartlett and D. G. Lohman, *J. Chem. Soc.*, 5253 (1962).
12. N. Bartlett, N. K. Jha, P. R. Rao, and M. Booth, *Chem. Eng. News*, 4 Feb., 1963.
13. C. L. Chernick, H. H. Claasen, P. R. Fields, H. H. Hyman, J. G. Malm, W. M. Manning, M. S. Matheson, L. A. Quarterman, F. Schreiner, H. H. Selig, I. Sheft, S. Siegel, E. N. Sloth, L. Stein, M. H. Studier, J. L. Weeds, and M. H. Zirin, *Science*, **138**, 136 (1962).
14. L. Stein, *Science*, **168**, 362 (1970).
15. S. I. Morrow and A. R. Young, II, *Inorg. Chem.*, **3**, 759 (1964).
16. L. Y. Nelson and B. C. Pimentel, *Inorg. Chem.*, **6**, 1758 (1967).
17. D. Boal and G. A. Ozin, *Spectrosc. Lett.*, **4**, 43 (1971).
18. F. O. Sladky and N. Bartlett, *J. Chem. Soc.*, A 2188 (1969).
19. J. H. Holloway and J. G. Knowles, *J. Chem. Soc.*, *A*, 756 (1969); D. Martin, *Compt. Rend.*, **265C**, 919 (1967).
20. F. O. Sladky, P. A. Bulliner, N. Bartlett, B. G. DeBoer, and A. Zalkin, *Chem. Commun.*, 1048 (1968).
21. V. M. McRae, R. D. Peacock, and D. R. Russel, *Chem. Commun.*, 62 (1969).
22. R. D. Burbank and N. Bartlett, *Chem. Commun.*, 645 (1968).
23. N. Bartlett and F. O. Sladky, *Chem. Commun.*, 1046 (1968).
24. N. Bartlett, M. Wechsberg, F. O. Sladky, P. A. Bulliner, G. R. Jones, and R. D. Burbank, *Chem. Commun.*, 703 (1969).
25. F. O. Sladky, *Angew. Chem. (Intl. Ed.)*, **8**, 373, 523 (1969).
26. F. O. Sladky, *Monatsh. Chem.*, **101**, 1559 (1970).
27. F. O. Sladky, *Monatsh. Chem.*, **101**, 1571 (1970).
28. F. O. Sladky, *Monatsh. Chem.*, **101**, 1578 (1970).
29. J. G. Malm, H. Selig, J. Jortner, and S. A. Rice, *Chem. Rev.*, **65**, 199 (1965).
30. C. A. Coulson, *J. Chem. Soc.*, 1442 (1964).
31. R. C. Catton and K. A. R. Mitchell, *Can. J. Chem.*, **48**, 2695 (1970).

COLLATERAL READINGS

I. R. Hoppe, *Angew. Chem. (Intl. Ed.)*, **3**, 538 (1964); G. J. Moody and J. D. R. Thomas, *Rev. Pure Appl. Chem.*, **16**, 1 (1966); J. H. Holloway, *Prog. Inorg. Chem.*, **6**, 241 (1964). These reviews contain numerous references to the chemistry of the rare gases, the structures of the compounds formed, and their properties.
II. B. Jaselskis, *Rec. Chem. Prog.*, **31**, 103 (1970). A detailed survey of the behavior of XeO_3 in aqueous and nonaqueous solutions.

STUDY QUESTIONS

1. Several binary covalent compounds of xenon have been prepared; however, no such ionic compounds are known. Predict the stability of xenon monofluoride (XeF) assuming that it would form a crystal lattice of the same structure and lattice energy as CsF.

2. Solutions of XeF_6 in anhydrous HF conduct an electric current. Suggest a possible mechanism for the ionization process and give the structure of these ions.

3. Discuss the possible Lewis acid–base chemistry of XeF_4O.

4. When xenon tetrafluoride undergoes alkaline hydrolysis the following species have been isolated or postulated to exist in the reaction sequence.

$$XeF_4 \rightarrow XeF_4O \rightarrow XeO_5{}^{4-} \rightarrow XeO_6{}^{6-}$$

Give the geometrical arrangement of the atoms in each xenon-containing species.

5. Discuss the basis for the relationship among the first ionization potentials of the rare gases.

6. Discuss the structural relationships between the iodine fluorides and the xenon fluorides.

7. Using simple molecular-orbital theory discuss the bonding in the species XeF^+ which is postulated to exist in compounds` of the type $XeF_2 \cdot AsF_5$.

8. Without recourse to structural data, predict the geometry of the following species: (a) $XeF_3{}^+$; (b) F_3XeO; (c) $XeF_8{}^{2-}$; (d) $XeF_5{}^+$; (e) $XeOTeF_5{}^+$; (f) $Xe(OTeF_5)_2$.

THE TRANSITION
ELEMENTS

1 / INTRODUCTION

The transition elements appear in the fourth, fifth, and sixth periods of the periodic table (see Fig. 1.27) and exhibit electronic configurations in which d and f electrons play an important role. These elements can be divided conveniently into two groups, the d-block and the f-block elements. The d-block elements have partially filled d orbitals in their neutral atoms, as shown in Table 16.1, or in the atoms as they exist in compounds. It is important to recognize that limiting the definition of a transition element to the electron distribution of the neutral atom excludes copper, silver, and gold, which have ground-state configurations $ns^1(n-1)d^{10}$, as well as zinc, cadmium, and mercury $[ns^2(n-1)d^{10}]$ from the d-block elements (cf. Table 16.1). On the other hand, if the existence of d electrons in the chemically combined atoms is the only criterion, the species Sc^{3+}, Y^{3+}, La^{3+}, and Ac^{3+} would be excluded, but the species Cu^{2+}, Ag^{2+}, and Au^{3+} would be included. Thus,

593

some species of the *d*-block elements do not exhibit the properties associated with the presence of *d* electrons. It is apparent that classifications based upon electronic configurations of certain species become somewhat arbitrary.

The *f*-block elements, sometimes called the inner transition elements, have partially filled *f* orbitals as shown in Table 16.2, either in their neutral atoms or in some of their oxidation states. The series of elements in the sixth period which incorporate 4*f* electrons in their configuration are called the lanthanides. The actinides are the elements in the seventh period which possess 5*f* electrons.

2 / PHYSICAL PROPERTIES OF THE ELEMENTS

2.1 / The *d*-Block Elements

These elements are all characteristically metallic and generally exhibit high melting and boiling points, as shown by the data in Table 16.1; the last members in the family melt and boil at markedly lower temperatures compared to the other members in the family. Mercury, the last member of the last transition series is the notable exception among the metals, being a liquid under normal conditions.

The electronic configurations of the *d*-block elements represent a break in the pattern established for the representative elements as described in Chapter 1, in the sense that the orbitals with increasing value of the quantum number *l* are not filled successively within a given principal quantum shell. The relative order of energies of atomic orbitals is shown in Fig. 16.1. For hydrogen, the subshells associated with a given principal quantum number are degenerate, but this degeneracy is removed in more complex atoms. In general, the energy of a given atomic orbital decreases with increasing atomic number of the element, but the decrease is not smooth in all cases. Because electrons with different secondary quantum numbers penetrate to different extents to the nucleus in a multielectron atom (see Fig. 1.19), they vary in their effectiveness in screening outer electrons from the nucleus. The energy of any electron in an atom is given by

$$E = -\frac{2\pi^2 \mu e^4 (Z^*)^2}{n^2 h^2} \tag{1}$$

where Z^* is the effective nuclear charge felt by that electron and all other symbols have their usual meaning. Values for Z^* can be estimated from the screening constants assigned to different types of electrons, as described in Chapter 2.

Principal quantum
number

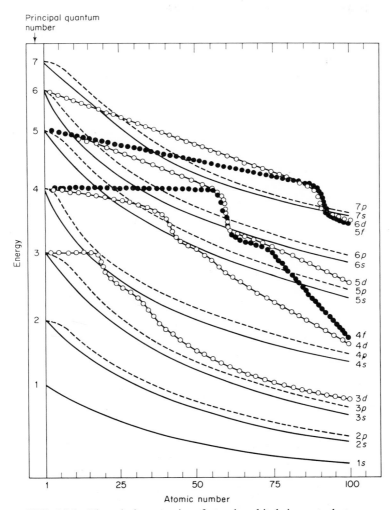

FIG. 16.1 The relative energies of atomic orbitals in neutral atoms.

Because of the differences in the penetration of orbitals with different values of l within a given quantum shell, the energies of the subshells do not decrease in a parallel fashion; s and p orbitals, which are the most penetrating, show a more nearly parallel decrease in energies than do d and f orbitals, as shown in Fig. 16.1. Thus, for the first 18 elements a regular order of filling is observed for the sequence of orbitals $1s$, $2s$, $2p$, $3s$, $3p$. It is interesting to note from Fig. 16.1 that the energy of the $3d$ orbital is essentially unchanged as the atomic number increases, an effect which can be attributed to the fact that the $3d$ electrons are less penetrating than any of the s or p orbitals of

TABLE 16.1 Some properties of the d-block elements

	Sc $(3d^1 4s^2)$	Ti $(3d^2 4s^2)$	V $(3d^3 4s^2)$	Cr $(3d^5 4s^1)$	Mn $(3d^5 4s^2)$	Fe $(3d^6 4s^2)$	Co $(3d^7 4s^2)$	Ni $(3d^8 4s^2)$	Cu $(3d^{10} 4s^1)$	Zn $(3d^{10} 4s^2)$
Mp, °C	1539	1680	1920	1900	1250	1539	1492	1453	1083	419
Bp, °C	2727	3300	3400	2600	2100	2500	2900	2820	2580	919
Atomic radius, Å[a]	1.64	1.47	1.35	1.29	1.27	1.26	1.25	1.25	1.28	1.37
Ionic radius, M^{2+}, Å	—	0.90	0.88	0.84	0.80	0.76	0.74	0.72	—	0.62
First ionization potential, eV	6.54	6.83	6.74	6.76	7.43	7.90	7.86	7.83	7.72	9.39
Oxidation states[b]	3	-1,0,2, 3,4	-1,0,2, 3,4,5	-2,-1,0, 2,3,4,5, 6	-1,0,1, 2,3,4,5, 6,7	0,2,3,4, 5,6	0,2,3,4	0,2,3,(4)	1,2,3	2

	Y $(4d^1 5s^2)$	Zr $(4d^2 5s^2)$	Nb $(4d^4 5s^1)$	Mo $(4d^5 5s^1)$	Tc $(4d^6 5s^1)$	Ru $(4d^7 5s^1)$	Rh $(4d^8 5s^1)$	Pd $(4d^{10} 5s^0)$	Ag $(4d^{10} 5s^1)$	Cd $(4d^{10} 5s^2)$
Mp, °C	1509	1850	2420	2620	2140	2400	1960	1552	961	321
Bp, °C	2927	4400	5100	4600		3900	3900	3200	2180	778
Atomic radius, Å[a]	1.82	1.60	1.47	1.40	1.35	1.34	1.34	1.37	1.44	1.52
First ionization potential, eV	6.6	6.95	6.77	7.18	7.45	7.5	7.7	8.33	7.57	8.99
Oxidation states[b]	3	2,3,4	2,3,4,5	0,2,3,4, 5,6	0,4,5,6, 7	0,3,4,5, 6,7,8	0,(1),2, 3,4,6	0,(1),2, 3,4	1,2,(3)	2

	La $(5d^1 6s^2)$	Hf $(5d^2 6s^2)$	Ta $(5d^3 6s^2)$	W $(5d^4 6s^2)$	Re $(5d^5 6s^2)$	Os $(5d^6 6s^2)$	Ir $(5d^7 6s^2)$	Pt $(5d^9 6s^1)$	Au $(5d^{10} 6s^1)$	Hg $(5d^{10} 6s^2)$
Mp, °C	920	2000	3000	3390	3170	2700	2443	1769	1063	-39
Bp, °C	3469	5100	6000	5700		4600		3800	2660	357
Atomic radius, Å[a]	1.877	1.59	1.47	1.41	1.37	1.35	1.36	1.39	1.44	1.55
First ionization potential, eV	5.61	5.5	5	7.98	7.87	8.7	9.2	8.96	9.22	10.43
Oxidation states[b]	3	2,3,4	2,3,4,5	0,2,3,4, 5,6	0,2,3,4, 5,6,7	0,(1),2, 3,4,5,6, 7,8	0,(2),3, 4,5,6	0,2,4,5, 6	1,3	1,2

[a] For 12 coordination.
[b] Excluding the element itself. Oxidation states in brackets are poorly characterized. Oxidation states listed in italics are the most stable.

lower energy; the radial distribution curves supporting this observation are shown in Fig. 16.2. Accordingly, electrons in these orbitals are maximally effective in screening the $3d$ orbital, and the energy of this orbital consequently changes little with increasing atomic charge. As the first 18 elements

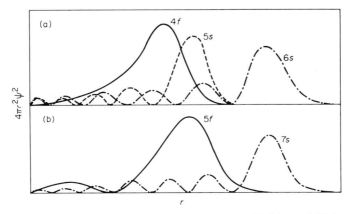

FIG. 16.2   The radial distribution functions for (a) the lanthanides and (b) the actinides.

are built up in the Aufbau process, the energies of all higher orbitals are also affected by the increase in nuclear charge. Most important, as shown in Fig. 16.1, at atomic number 18 the energy of the $4s$ orbital has decreased to the point where it lies below that of the $3d$ orbital, so that the next electron enters the $4s$ orbital. However, the radial distribution curves indicate that the $4s$ electron density lies outside of that of the $3d$ orbitals; the $4s$ electrons cannot effectively screen the $3d$ orbitals from the increased nuclear charge which causes a decrease in the energy of the $3d$ orbital below that of the $4p$ orbital. Accordingly, the next electron is added to the $3d$ orbital which, of course, is the start of the first transition series; since the $3d$ orbital can accommodate a maximum of ten electrons, the first transition series contains ten elements [scandium (21) to zinc (30)]. The electronic configurations of the elements in the first transition series, as shown in Table 16.1, generally reflect the sequential addition of electrons to the $3d$ level. Figure 16.1 shows that the energy of the $3d$ orbital closely parallels that of the $4p$ level; apparently these energies are sufficiently close that in certain cases the electronic configurations might be different from those predicted from the relative energies shown in Fig. 16.1. Such differences are observed at the center of the first transition series, where chromium has the configuration $[Ar]3d^5 4s^1$ rather than the expected $[Ar]3d^4 4s^2$ configuration, and at the end where copper has the configuration $[Ar]3d^{10}4s^1$ rather than $[Ar]3d^9 4s^2$. These changes in the expected order of filling have been attributed to added stability of a half-filled orbital. The

TABLE 16.2 Some properties of the f-block elements

	La $(5d^16s^2)$	Ce $(4f^26s^2)$	Pr $(4f^36s^2)$	Nd $(4f^46s^2)$	Pm $(4f^56s^2)$	Sm $(4f^66s^2)$	Eu $(4f^76s^2)$	Gd $(4f^75d^16s^2)$	Tb $(4f^96s^2)$	Dy $(4f^{10}6s^2)$
Mp, °C	920	795	935	1024	—	1072	826	1312	1356	1407
Bp, °C	4200	3468	3127	3027	—	1900	1439	3000	2800	2600
Atomic radius, Å	1.877	1.82	1.828	1.821	—	1.802	2.042	1.802	1.782	1.773
Ionic radius, M^{3+}, Å	1.061	1.034	1.013	0.995	0.979	0.964	0.950	0.938	0.923	0.908
First ionization potential, eV	5.61	5.60	5.48	5.5	—	5.6	5.67	6.16	5.98	—
$-E^0$, M/M^{3+}	2.52	2.48	2.47	2.44	2.42	2.41	2.41	2.40	2.39	2.35
Oxidation states	3	3, 4	3, 4	2, 3, 4	2, 3	2, 3	2, 3	3	3, 4	3, 4

	Ho $(4f^{11}6s^2)$	Er $(4f^{12}6s^2)$	Tm $(4f^{13}6s^2)$	Yb $(4f^{14}6s^2)$	Lu $(4f^{14}5d^16s^2)$	Ac $(6d^17s^2)$	Th $(6d^27s^2)$	Pa $(5f^26d^17s^2)^a(5f^36d^17s^2)$	U $(5f^36d^17s^2)$	Np $(5f^57s^2)$
Mp, °C	1461	1497	1545	824	1652	1050	1750	—	1132	640
Bp, °C	2600	2900	1727	1427	3327	—	3500–4200	—	3818	—
Atomic radius, Å	1.766	1.757	1.746	1.940	1.734	—	—	—	—	—
Ionic radius, M^{3+}, Å^d	0.894	0.881	0.869	0.859	0.848	1.11	(1.08)	(1.08)	1.03	1.01
First ionization potential, eV	—	6.08	5.81	6.22	6.15	—	6.95	—	6.08	—
$-E^0$, M/M^{3+}	2.32	2.30	2.28	2.27	2.25	2.6	—	—	1.80	1.83
Oxidation states	3, 4	2, 3	2, 3	2, 3	3	3	3, 4	3, 4, 5	3, 4, 5, 6	3, 4, 5, 6, 7

	Pu	Am	Cm	Bk	Cf	Es	Fm	Md	No[a]	Lw
	$(5f^6 7s^2)$	$(5f^7 7s^2)$	$(5f^7 6d^1 7s^2)$	$(5f^8 6d^1 7s^2)^b$	$(5f^{10} 7s^2)$	$(5f^{11} 7s^2)$	$(5f^{12} 7s^2)$	$(5f^{13} 7s^2)$	$(5f^{14} 7s^2)$	$(5f^{14} 6d^1 7s^2)$
Mp, °C	639.5	1100								
Bp, °C	3235	—								
Atomic radius, Å										
Ionic radius, M^{3+}, Å	1.00	0.99	0.98							
First ionization potential, eV	5.8	6.0								
$-E^0$, M/M^{3+}	2.03	2.32								
Oxidation states	3, 4, 5, 6, 7	3, 4, 5, 6	3, 4	3, 4	3	3	3	2, 3	3	3

[a] Or $5f^1 6d^2 7s^2$.
[b] Or $5f^9 7s^2$.
[c] The name nobelium has been suggested for element 102, but this is not officially recognized by IUPAC.
[d] Estimated values in parentheses.

variation in first ionization potentials, shown in Fig. 16.3, of the *d*-block elements reflect the relative stabilities of their electronic configurations. There is generally an increase in ionization potential from the first member of the series to the last, although fluctuations occur in several places which may be correlated with electronic configurations. For example, in Fig. 16.3 two

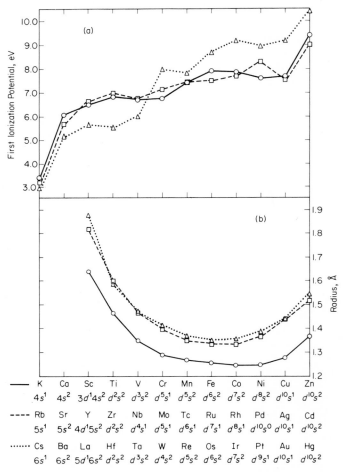

	K	Ca	Sc	Ti	V	Cr	Mn	Fe	Co	Ni	Cu	Zn
——	$4s^1$	$4s^2$	$3d^14s^2$	d^2s^2	d^3s^2	d^5s^1	d^5s^2	d^6s^2	d^7s^2	d^8s^2	$d^{10}s^1$	$d^{10}s^2$
	Rb	Sr	Y	Zr	Nb	Mo	Tc	Ru	Rh	Pd	Ag	Cd
- - - -	$5s^1$	$5s^2$	$4d^15s^2$	d^2s^2	d^4s^1	d^5s^1	d^6s^1	d^7s^1	d^8s^1	$d^{10}s^0$	$d^{10}s^1$	$d^{10}s^2$
	Cs	Ba	La	Hf	Ta	W	Re	Os	Ir	Pt	Au	Hg
······	$6s^1$	$6s^2$	$5d^16s^2$	d^2s^2	d^3s^2	d^4s^2	d^5s^2	d^6s^2	d^7s^2	d^9s^1	$d^{10}s^1$	$d^{10}s^2$

FIG. 16.3 The first ionization potentials (a) and atomic radii (b) for the *d*-block elements.

grossly similar trends are apparent within the first *d* series elements; the ionization potentials of the first four elements, scandium (21) through chromium (24), appear to form a related group distinctly different from that formed by the remainder of the elements. It is interesting to note that minima in these curves, shown in Fig. 16.3, occur in the regions where the elements exhibit an electron configuration different than that expected from the sequential filling of the d orbitals. Both the first and second series of d-elements show generally similar values of the ionization potentials, but the third d-series exhibits a markedly different pattern which probably results from the fact that the $4f$ series of elements must be generated before the $5d$ series begins.

Inspection of Fig. 16.1 shows that a similar decrease in energy of the $4d$ and $5d$ orbitals occurs relative to the corresponding $5p$ and $6p$ orbitals. Thus, there are two more transition series in the periodic table. As shown in Fig. 16.1, the energy of the $4d$ orbital decreases to a value just below that of the $5s$ orbital so that the configuration of most of these elements, as they appear in Table 16.1, shows the presence of a partially filled $5s$ shell. Indeed, only the first two elements [yttrium (39) and zirconium (40)] and the last element [cadmium (48)] in this series exhibit a filled $5s$ orbital. In the case of the first two elements, the data in Fig. 16.1 indicate that the $4d$ level is above the $6s$ level, but at higher atomic numbers the order is reversed. The start of the third transition series is interrupted by the appearance of a stable $4f$ level. However, the order of filling of the $5d$ level for the elements lanthanum (57) through mercury (80) is more regular than for the previous two transition series because the relative energies of the $5d$ and $6s$ levels diverge steadily with increasing atomic number; the electronic configurations of the elements in the third transition series appear in Table 16.1.

The electronic configurations also reflect other characteristics of the d-block elements. Thus, the sequential addition of an electron into a d-orbital is reflected in a smooth variation of atomic dimensions. d-Orbitals are less penetrating than the next-lowest-energy s-orbital, as shown by the radial distribution curves in Fig. 16.2, but they have their main electron density within that of the $(n + 1)s$-orbital. The magnitude of this variation is not proportionately as large as that observed for the representative elements. The data shown in Fig. 16.3 pertaining to the atomic radii of the d-block elements contain an unusual feature. In a given family of representative elements, we are accustomed to an increase in the size of the atoms in a given family with increasing atomic number. The data in Fig. 16.3 show an increase between the first two d series of elements, but the radii of the elements in the third d series are virtually the same as those of the corresponding members of the second d series. This phenomenon is associated with the intervention of the $4f$ orbital, which must be filled before the $5d$ series of elements is begun. The data in Fig. 16.4 indicate that the f-series of elements exhibit a regular decrease in atomic radii, called the lanthanide contraction, which essentially compensates for the expected increase in the atomic size with increasing atomic number. The net result of the lanthanide contraction is that the second and third d series elements exhibit similar radii and have very similar chemical and physical properties, such as lattice energies, ionization potentials, and solvation energies; the properties of these elements are much more similar than would be expected on the basis of the usual familial relationships. Thus, the chemistry of a family of d-block elements is roughly divided into that of

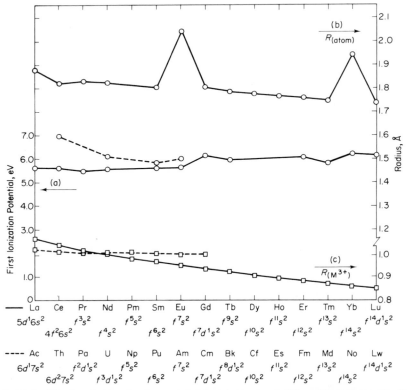

FIG. 16.4   The first ionization potentials (a) and the atomic (b) and ionic (c) radii of the f-block elements.

the first member of the family and that associated with the next two elements which have virtually identical properties. A detailed discussion of the lanthanide contraction appears in the next section.

2.2 / The f-Block Elements (l)

Like the d-block elements, all of these elements have metallic properties; that is, they exhibit a metallic luster, are good conductors of electricity and heat, and have typically high melting and boiling points, as shown by the data in Table 16.2.

The electronic structures of the early elements of the f-block are not particularly regular because the $(n - 1)d$ and $(n - 2)f$ orbital energies are very similar; such irregularities are more prevalent in the actinides $(5f)$ than in the lanthanides $(4f)$. The data in Fig. 16.1 show that the energies of the $4f$ and $5d$ orbitals are virtually the same beginning near atomic number 57; similar behavior is observed for the $6d$ and $5f$ orbitals at atomic number 89. The ground-state configurations of several members of the actinide series, given in Table 16.2, incorporate electrons in the $6d$ level, illustrating the closeness of the energies of the $6d$ and $5f$ orbitals.

The radial distribution curve for $4f$ orbitals, shown in Fig. 16.2, indicates that the mean electron density associated with these orbitals lies within that of the $6s$ orbital, although the $6s$ orbital is more penetrating than the $4f$ orbital. This relationship is manifested in the relative sizes of the atoms of the f-block elements and by the trends in the first ionization potentials given in Fig. 16.4. Except for europium (63) and ytterbium (70), the atomic radii for the lanthanide elements are remarkably constant; a small decrease in size occurs with increasing atomic number. The exceptions to this trend occur at the points where the $5d$ orbital is half-filled and completely filled. The M^{3+} ions in the lanthanide series have a $4f^n$ $(n = 0–14)$ configuration in which the f orbital acquires one electron for each element in the series. Accordingly, these ions show a regular decrease in size with increasing atomic number, as illustrated in Fig. 16.4. There are no exceptions in this sequence, since Eu^{3+} and Yb^{3+} exhibit no unusual electronic features with respect to their neighbors. The lanthanide contraction has an indirect but profound effect on the properties of the second and third d-series elements, as described in the previous section. Presently available data on M^{3+} cations of the actinide elements suggest that a similar behavior exists in this series, but the variation in size within these ions is even less than that observed in the lanthanides. The ionization potentials of the lanthanides, depicted in Fig. 16.4, also show relatively little variation, in contrast to those of the d-block elements given in Fig. 16.3.

At this point it should be apparent why in discussions of the chemistry of the transition metals the classification of elements strictly into a d and f block is unsatisfactory in many cases. The electronic configurations of the first and last members of the d and f series of elements and some of the common ions they form are listed in Table 16.3. Consider first the d-block elements. The first members of each d series exhibit the electronic configuration $(n - 1)d^1ns^2$. The most common oxidation state of these elements is M^{3+}, in which the ion has a rare-gas configuration. Lutetium (71), in addition, has a filled $4f$ orbital. The first elements of the f series have the configuration

$(n - 1)d^1ns^2$ rather than $(n - 2)f^1ns^2$ because of the effect of inner-electron shielding on the relative energies of the corresponding $(n - 1)d$ and $(n - 2)f$ orbitals, as shown in Fig. 16.1. Hence, strictly speaking, these elements should not be considered as part of the f block on the basis of the electronic structure of the *neutral atom*. The common oxidation state for the lanthanides and

TABLE 16.3   The electronic configurations of the first and last members of the d- and f-block elements

	First member	Last member
d-Block	Sc [Ar]$3d^14s^2$	Zn [Ar]$3d^{10}4s^2$
	Sc [Ar]$^{3+}$	Zn [Ar]$^{2+}3d^{10}$
	Y [Kr]$4d^15s^2$	Cd [Kr]$4d^{10}5s^2$
	Y [Kr]$^{3+}$	Cd [Kr]$^{2+}4d^{10}$
	Lu [Xe]$4f^{14}5d^16s^2$	Hg [Xe]$4f^{14}5d^{10}6s^2$
	Lu [Xe]$^{3+}4f^{14}$	Hg [Xe]$^{2+}4f^{14}5d^{10}$
f-Block	La [Xe]$5d^16s^2$	Yb [Xe]$4f^{14}6s^2$
	La [Xe]$^{3+}$	Yb [Xe]$^{2+}4f^{14}$
		Yb [Xe]$^{3+}4f^{13}$
	Ac [Rn]$6d^17s^2$	No [Rn]$5f^{14}7s^2$
	Ac [Rn]$^{3+}$	No [Rn]$^{3+}5f^{13}$

actinides is M^{3+}, in which state these ions possess the corresponding rare-gas configuration. Thus, each of the first members of the d and f series forms ions with the same charge. The atomic and ionic radii of yttrium (Y, 1.801 Å; Y^{3+}, 0.88 Å) are similar to the corresponding values of the lanthanides as a result of the lanthanide contraction. Accordingly, the chemistry of yttrium is much better understood when it is related to that of the lanthanides rather than when it is included in a discussion of the d-block elements. Although scandium forms smaller atomic and ionic species (Sc, 1.641 Å; Sc^{3+}, 0.68 Å), its chemistry is sufficiently similar to that of the lanthanides that it too is discussed with the lanthanide elements rather than with the d-series elements. Lutetium, the first element which follows the lanthanides, starts the last d-series, but it forms a M^{3+} cation with a filled $4f$ orbital. Thus, lutetium on the basis of its atomic and ionic radii also appears to be a member of the lanthanide series and its chemistry reflects this suggestion.

The last members of the d-block elements, i.e., zinc, cadmium, and mercury, are generally considered as a group apart from other members of their respective series because they form ions in which all the orbitals are filled.

This configuration leads to generally different properties than those associated with the other members of the d series.

3 / CHEMICAL PROPERTIES

3.1 / The d-Block Elements

In general the d-block elements are relatively unreactive with oxygen, the halogens, sulfur, nitrogen, hydrogen, and water vapor under ordinary conditions, but at elevated temperatures reaction occurs more or less readily with these reagents. Often when massive pieces of these elements do react they form impervious product films which prevent further reaction. However, in a finely divided state some of the elements such as iron and nickel are highly reactive, bursting into flame when exposed to air. The transition elements form interstitial compounds with nitrogen (Chapter 12), hydrogen (Chapter 6), and carbon (Chapter 11).

3.2 / The f-Block Elements

The lanthanide metals are soft, ductile, and more reactive with the common reagents than are the d-block elements; thus, the lanthanides react slowly with the halogens and oxygen at room temperature

$$2M + 3X_2 \rightarrow 2MX_3 \tag{2}$$

$$4M + 3O_2 \rightarrow 2M_2O_3 \tag{3}$$

but burn readily in these reagents above 200°C. They react with sulfur at its boiling point

$$2M + 3S \rightarrow M_2S_3 \tag{4}$$

and with nitrogen above 1000°C

$$2M + N_2 \rightarrow 2MN. \tag{5}$$

Above 300° the lanthanides react rapidly with hydrogen to form two classes of hydrides, MH_2 and MH_3; reaction with boron and carbon at high temperatures gives borides (MB_4, MB_6) and carbides (MC_2, M_2C_3, MC, M_2C, M_3C, M_4C), respectively. The metals react slowly with water under normal conditions, but rapidly at elevated temperatures, to form hydrogen and the corresponding hydrous oxides.

4 / OXIDATION STATES

4.1 / The d-Block Elements (I)

d-Block elements show a wide range of oxidation states, as shown by the data collected in Table 16.1. The first members of any d series of elements exhibit relatively few oxidation states; the number increases more or less regularly to the center of the series and then decreases for the last members. Apparently the elements in the first half of a given series use all the s and d electrons in achieving the maximum number of oxidation states; the number of oxidation states for the last half of the series decreases because the number of unpaired d electrons decreases steadily. Species that exhibit low oxidation states can exist as discretely charged ions either in solution or in a crystalline lattice. For example, Cr^{2+}, Fe^{2+}, Ag^+, and Cu^{2+} are well known ionic species. However, the elements in higher oxidation states are very polarizing so that these species could exist only in an environment of highly electronegative atoms. Thus, only oxyanions or binary compounds of oxygen and fluorine are known for the higher oxidation states of the d-block elements, as shown by the data in Table 16.4 (II–IV). Stable higher oxidation states appear to be favored by the heavier elements in a given family, a trend which is also present in the representative elements. For example, the 6+ oxidation state of chromium is found in CrO_4^{2-}, which is a good oxidizing agent in aqueous solution, but the corresponding species containing molybdenum and tungsten (MoO_4^{2-} and WO_4^{2-}) are markedly more difficult to reduce. Care must be exercised, of course, in discussing the stability of a substance because two kinds of stability exist. The thermodynamic stability is directly related to the relative energy difference between the products and reactants, whereas kinetic stability refers to the rate at which a compound decomposes. Thus, it is possible to have a compound which is thermodynamically unstable but for which the decomposition process is exceedingly slow. It should be apparent that the existence of a substance under a set of conditions is not necessarily related to the relative thermodynamic stability of the substance, since kinetic factors may be involved.

The factors associated with the relative ease of oxidation of any element in aqueous solutions are best described in terms of the energy cycle shown in Fig. 16.5 in which the oxidizing agent chosen for reference is $H_{(aq)}^+$. The standard free-energy change for the process

$$M_{(s)} + nH_{(aq)}^+ \longrightarrow M_{(aq)}^{n+} + \frac{n}{2}H_2 \qquad (6)$$

can be expressed in terms of the heats of sublimation (ΔH_s), the summation of the appropriate ionization potentials (ΣI), the heats of hydration of the

TABLE 16.4 Some typical compounds of the d-block elements[a]

Oxidation state	Ti	V	Cr	Mn	Fe	Co	Ni	Cu
2+	TiO TiX$_2$	VO VX$_2$	CrO CrF$_2$(CrX$_2$)	MnO MnF$_2$, MnX$_2$	FeO FeF$_2$, FeX$_2$	CoO CoF$_2$, CoX$_2$	NiO NiF$_2$, NiX$_2$	CuO CuF$_2$
3+	Ti$_2$O$_3$ TiF$_3$	V$_2$O$_3$ VF$_3$, VX$_3$	Cr$_2$O$_3$ CrF$_3$, CrX$_3$	Mn$_2$O$_3$ MnF$_3$	Fe$_2$O$_3$ FeF$_3$, FeX$_3$	[Rh$_2$O$_3$] CoF$_3$	[PdF$_3$]	[Au$_2$O$_3$] [AuF$_3$]
4+	TiO$_2$ TiF$_4$	VO$_2$ VF$_4$, VX$_4$	CrO$_2$ CrF$_4$	MnO$_2$ MnF$_4$, TcCl$_4$	[RuO$_2$] [RuF$_4$]	[RhO$_2$] [RhF$_4$]	[PtO$_2$] [PdF$_4$]	
5+		V$_2$O$_5$, VO$_4$$^{3-}$ VF$_5$	CrF$_5$	[TcF$_5$]	[RuF$_5$]	[RhF$_5$]	[PtF$_5$]	
6+			CrO$_3$, (Cr$_2$O$_7$$^{2-}$) [MoF$_6$]	ReO$_3$ [ReF$_6$]	[RuF$_6$]	[RhF$_6$]	[PtF$_6$]	
7+				Mn$_2$O$_7$, (MnO$_4$$^-$) [ReF$_7$]			[PtF$_7$]	
8+					[RuO$_4$]			

[a] X = Cl, Br, I.

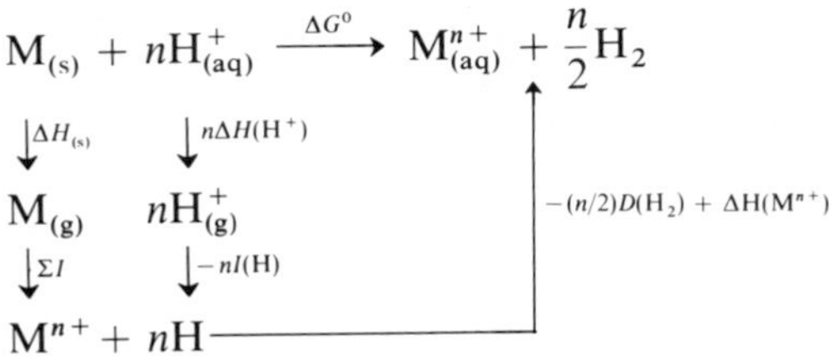

$$M_{(s)} + nH^+_{(aq)} \xrightarrow{\Delta G^\circ} M^{n+}_{(aq)} + \frac{n}{2}H_2$$

$$\downarrow \Delta H_{(s)} \qquad \downarrow n\Delta H(H^+)$$

$$M_{(g)} \qquad nH^+_{(g)} \qquad \qquad \uparrow -(n/2)D(H_2) + \Delta H(M^{n+})$$

$$\downarrow \Sigma I \qquad \downarrow -nI(H)$$

$$M^{n+} + nH$$

FIG. 16.5   An energy cycle for the oxidation of a metal by hydrogen ion.

ionic species [$\Delta H(H^+)$ and $\Delta H(M^{n+})$], the ionization potential of hydrogen [$I(H)$], and the dissociation energy of $H_2 [D(H_2)]$

$$\Delta G^\circ = \Delta H_s + \Sigma I + n\Delta H(H^+) - nI(H) + \Delta H(M^{n+}) - \frac{n}{2}D(H_2). \quad (7)$$

If all the free-energy changes, expressed in electron volts, are made with reference to the same oxidizing agent, Eq. (7) can be rewritten as

$$\Delta G^\circ = \Delta H_s + \Sigma I + \Delta H(M^{(n-1)+}) - 4.66n \quad (8)$$

by substituting the appropriate values for the ionization potential of hydrogen (13.60 eV), the dissociation energy for H_2 (4.48 eV), and the hydration energy (2–4) of H^+ (11.18 eV). Since the hydration energy, $\Delta H(M^{n+})$ for an ion of radius r is given by

$$\Delta H(M^{n+}) = -\frac{7.25(n)^2}{r + 0.85} \text{ eV} \quad (9)$$

it should be possible to estimate ΔG° for Eq. (6) from available data. Inspection of Eq. (8) shows that the magnitude and the sign of the free-energy change depend upon the way in which ΔH_s and ΣI, which are positive, balance against the sum of $\Delta H(M^{n+})$ and 4.66n, which is negative. Pertinent data, related to the data shown in Fig. 16.5, for M^{2+} ions of the first d-block elements are presented graphically in Fig. 16.6. The quantity $\Delta H(M^{2+}) -$ (2)(4.66) decreases regularly because of the steady decrease in $\Delta H(M^{2+})$, which is to be expected from Eq. (9) for a series of ions with smoothly varying radii (cf. Table 16.1). An interesting interplay of factors occurring in the two terms which constitute the sum $\Delta H_s + \Sigma I$ appears in Fig. 16.6. Starting with vanadium, the decrease in the heat of sublimation (ΔH_s) across the series is almost immediately interrupted by a marked drop in the value for manganese. On the other hand, there are sharp upward fluctuations in the steadily increasing value of ΣI at chromium and at copper; both are associated with abnormally large values of the second ionization potential compared with that of the neighboring elements, which stems from the presence of a filled or half-filled d orbital in the ionic species. The increase in ΣI for chromium virtually compensates for the decrease in ΔH_s, so that the trend in ΔG^0 is maintained. However, in the case of manganese, ΣI is normal and the sharp decrease in ΔH_s is not compensated, which leads to a decrease in ΔG^0. The decrease in ΔG^0 for zinc can also be traced to an abnormally low value for ΔH_s.

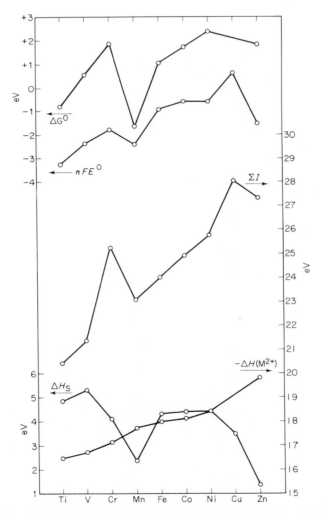

FIG. 16.6 The heats of sublimation (ΔH_s), total ionization potentials (ΣI), heat of hydration [$\Delta H(M^{2+})$], and the calculated free-energy change for the process described in Eq. (6) for the first d-block elements. Shown also are the experimental values for the oxidation of the metals by H^+ ($-nFE^0$).

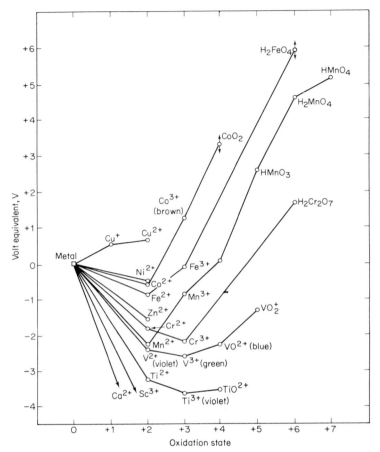

FIG 16.7 Oxidation state diagram for the first series of d-block elements; the potential of the H^+/H_2 half cell $[a(H^+) = 1; E^0 = 0]$ is taken as the reference.

It should be pointed out that the model described by the energy cycle in Fig. 16.5 cannot be wholly correct. The standard free energy for Eq. (6) can also be estimated from the standard potential (E^0) for the corresponding electrochemical cell, using the following equation:

$$\Delta G^0 = -nFE^0 \tag{10}$$

However, the product $-nFE^0$ for the first d-series elements, which is also plotted in Fig. 16.6, exhibits the same general trends as the ΔG^0 values calculated from Eq. (8), indicating that the arguments presented in its derivation have a qualitative validity.

The oxidation-state diagram for the first transition series (Fig. 16.7) shows several interesting features, the most obvious being that only one element, copper, forms a stable $1+$ oxidation state. With increasing atomic number, the metals become progressively more difficult to oxidize to the common M^{2+} state; the lower oxidation states are also generally more difficult to oxidize in the same order. The most stable oxidation states in

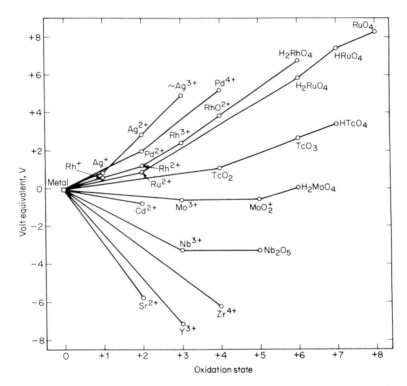

FIG. 16.8 Oxidation-state diagram for the metals of the second series of d-block elements; the potential of the H^+/H_2 half cell $[a(H^+) = 1; E^0 = 0]$ is taken as reference.

the series are the $2+$ and $3+$ states; hence it is not surprising that these are among the products formed when species in higher and lower oxidation states react. The higher states of cobalt (Co^{3+}), iron (FeO_4^{2-}), and manganese (MnO_4^-) are sufficiently strong oxidizing agents in aqueous solutions to liberate oxygen. However, the kinetics of this reaction for MnO_4^- are sufficiently slow under normal conditions that it finds wide use as an analytical titrant. This is a good example of the important difference between kinetic and thermodynamic stability.

All the metals of the first transition series except copper behave as reducing agents with respect to hydrogen evolution. Normally copper will not dissolve in aqueous acid unless the latter contains an oxidizing anion such as NO_3^-. However, copper will dissolve in acids containing species which can form a strong complex with Cu^{1+}. For example, copper will not ordinarily dissolve in aqueous HCl; however, when urea, which forms strong complexes with copper ions, is added hydrogen is rapidly evolved from solution. The oxidation-state diagram for the first series of d-block elements, shown in Fig. 16.7, indicates that Ti^{2+}, V^{2+}, and Cr^{2+} are sufficiently strong reducing agents to liberate hydrogen gas from acid solution. Again the reactions are sufficiently slow to make the ions useful reagents in aqueous solutions. The lower oxidation states of the metals are basic, the higher states acidic, and the $4+$-state is usually amphoteric.

The oxidation-state diagrams for the second and third d-series elements which appear in Figs. 16.8 and 16.9 show very similar features and are distinctly different from the diagram in Fig. 16.7 for the first d-series elements.

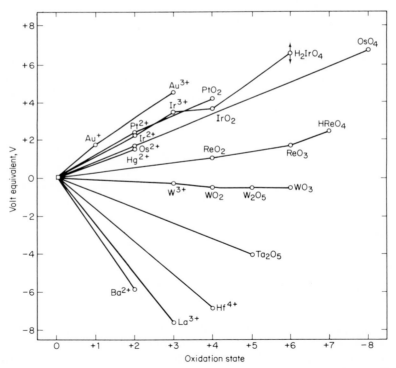

FIG. 16.9 Oxidation-state diagram for the metals of the third series of d-block elements; the potential of the H^+/H_2 half cell $[a(H^+) = 1; E^0 = 0]$ is taken as reference.

TABLE 16.5 Electronic configurations
and some properties of d-block ions

Configuration	Example	Unpaired electrons	Color
$3d^0$	Ti^{4+}	0	colorless
	Sc^{3+}	0	colorless
$3d^1$	Ti^{3+}	1	purple
	V^{4+}	1	blue
$3d^2$	V^{3+}	2	green
$3d^3$	Cr^{3+}	3	violet
	V^{2+}	3	violet
$3d^4$	Mn^{3+}	4	violet
	Cr^{2+}	4	blue
$3d^5$	Mn^{2+}	5	pink
	Fe^{3+}	5	yellow
$3d^6$	Fe^{2+}	4	green
$3d^7$	Co^{2+}	3	pink
$3d^8$	Ni^{2+}	2	green
$3d^9$	Cu^{2+}	1	blue
$3d^{10}$	Zn^{2+}	0	colorless

The similarity of these oxidation-state diagrams can also be traced to the effect of the lanthanide contraction on the properties of the elements of the third d series, as discussed previously. The metals in the second and third series are less easily oxidized than the corresponding elements in the first d series. In addition, there is a decrease in the ease of oxidation of the elements along these series; that is, the elements become less electropositive with increasing atomic number.

In contrast to the representative elements, most of the compounds of the d-block elements are colored in the solid state and in aqueous solution, as shown by the data in Table 16.5. The spectra of these compounds generally contain broad absorption bands of the type given in Fig. 16.10. Many of the compounds of the transition elements are also paramagnetic. The data in Table 16.5 suggest that both the paramagnetism and the spectra appear to be correlated with the number of d electrons in the ion. Thus, ions with no unpaired d electrons at either the beginning (Sc^{3+}, Ti^{4+}) or the end (Zn^{2+}) of the series are colorless, much like most of the ions of the representative elements. However, ions containing unpaired electrons exhibit a color, i.e., they absorb electromagnetic radiation in the visible region. The characteristic color of these ions in aqueous solutions can be changed by the addition of other substances. For example, copper sulfate dissolves in water to give a blue solution; the addition of ammonia to this solution produces a rich purple color whereas the addition of hydrochloric acid yields a bright yellow solution. This phenomenon, which is attributed to the formation of complex ions, is widespread in transition-metal chemistry and is discussed in detail in Chapters 17 and 18.

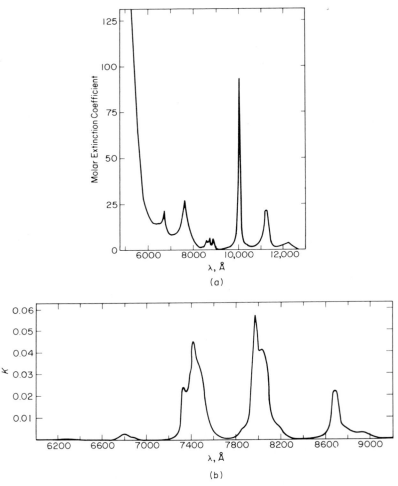

FIG. 16.10   The absorption spectra of typical *f*-block elements in aqueous solution: (a) Am^{4+} in 0.1 M $Na_4P_2O_7$, (b) Nd^{3+}. The absorbances in the neodymium spectrum are reported on the basis of grams per liter instead of the usual molar units.

4.2 / The *f*-Block Elements (*V*)

The lanthanides exhibit a markedly narrower range of oxidation states than do the *d*-block elements, as indicated by the data collected in Table 16.2;

what little is known of the chemistry of the actinides indicates that they achieve a greater variety of oxidation states than do the lanthanides. Most of the compounds of the lanthanides involve the 3+ oxidation state, e.g., MF_3 and M_2O_3.

Although sufficient information is not available to quantitatively analyze the factors associated with oxidation of the lanthanides to trivalent cations

$$M_{(s)} + 3H^+_{(aq)} \rightarrow M^{3+}_{(aq)} + \tfrac{3}{2}H_2, \tag{11}$$

using the energy cycle shown in Fig. 16.5 some qualitative observations can be made by recalling that the standard free-energy change for a redox couple [cf. Eq. (8)] is measured by the standard electrode potential [Eq. (10)]. The steady decrease in E^0 from -2.48 to -2.25 V across the lanthanide series, as indicated in Table 16.2, suggests that the sum of the energy terms in Eq. (8) expressed for the lanthanides contains no marked fluctuations. In addition, if Eq. (9) reasonably accurately reflects the variation in hydration energy for the 3+ ions in this series, it is apparent that there is a relatively smooth, but small, change in the heat of hydration $[\Delta H(M^{3+})]$ as a result of the variation in the ionic radius shown in Table 16.2. Thus it appears that the sum of the remaining terms in Eq. (8), ΣI and ΔH_s, contain no marked fluctuations.

TABLE 16.6   **The ionic radii and standard electrode potentials of several lanthanides**

	Sm^{2+}	Eu^{2+}	Yb^{2+}
Radius, Å	1.13	1.12	1.13
E^0 for M^{3+}/M^{2+}, V	-1.55	-0.43	-1.15

In contrast to the other lanthanides, samarium, europium, and ytterbium form stable 2+ oxidation states which are sufficiently strong reducing agents, as indicated by the data in Table 16.6, to liberate hydrogen from water under normal conditions; however, the kinetics for the reaction of Eu^{2+} with water are slow. The crystal radii of the 2+ states lie between those of Ca^{2+} (0.94 Å) and Sr^{2+} (1.29 Å); accordingly, these ions behave in much the same way as do the heavier alkaline-earth metal ions. The similarity of europium and ytterbium to the alkaline-earth metals is reinforced by the observation that these lanthanides dissolve in liquid ammonia (5–8) forming solutions of solvated electrons (Chapter 9).

A stable 4+ oxidation state is observed for cerium. Although the oxidation potential for Ce^{4+} is sufficiently large ($Ce^{4+}/Ce^{3+} = +1.74$ V) to oxidize water the reaction rate is slow, making aqueous solutions of Ce^{4+} good analytical oxidizing agents. Several other lanthanides also show 4+ oxidation states, but these are stable only in a crystalline lattice, e.g., MO_2 (M = Pr, Nd, Tb, Dy).

TABLE 16.7   The electronic configurations and some properties of the lanthanide ions

Configura-tion	Example	Unpaired electrons	Color
$4f^0$	La^{3+}	0	colorless
$4f^1$	Ce^{3+}	1	colorless
$4f^2$	Pr^{3+}	2	green
$4f^3$	Nd^{3+}	3	red
$4f^4$	Pm^{3+}	4	pink
$4f^5$	Sm^{3+}	5	yellow
$4f^6$	Eu^{3+}	6	pink
$4f^7$	Gd^{3+}	7	colorless
$4f^8$	Tb^{3+}	6	pink
$4f^9$	Dy^{3+}	5	yellow
$4f^{10}$	Ho^{3+}	4	yellow
$4f^{11}$	Er^{3+}	3	red
$4f^{12}$	Tm^{3+}	2	green
$4f^{13}$	Yb^{3+}	1	colorless
$4f^{14}$	Lu^{3+}	0	colorless

The trivalent lanthanide cations also exhibit striking colors in their crystalline salts and in aqueous solution; the pertinent data are collected in Table 16.7 (9). The trivalent ions are increasingly paramagnetic with increasing atomic number until the middle of the series (Gd^{3+}), after which point a steady decrease occurs. The species with no unpaired f electrons (La^{3+} and Lu^{3+}) or one f electron (Ce^{3+} and Yb^{3+}, and Gd^{3+}, which has a

TABLE 16.8   Color of actinide ions in aqueous solution

Element	M^{3+}	M^{4+}	MO_2^+	MO_2^{2+}
Ac	colorless	—	—	—
Th	—	colorless	—	—
Pa	—	colorless	—	—
U	reddish	green	—	—
Np	violet	yellow-green	—	yellow
Pu	blue	tan	—	yellow-orange
Am	pink	—	yellow	brown
Cm	colorless	—	—	—

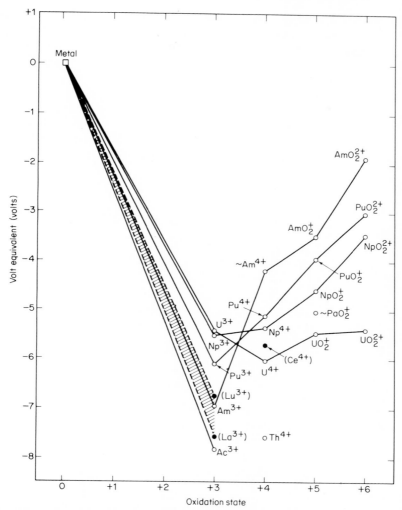

FIG. 16.11 Oxidation-state diagram for the actinide elements; the potential of the H^+/H_2 half cell $[a(H^+) = 1; E^0 = 0]$ is taken as reference. The potentials for the rare-earth elements fall within a very narrow range, as indicated by the shaded portion.

half-filled f shell), are colorless. The spectra of the lanthanides generally contain more and sharper absorption bands than those of the d-block elements.

The actinide elements show a common $3+$ oxidation state; the data in Table 16.2 indicate that elements in the first half of the series preferentially exhibit higher oxidation states, the extreme occurring at neptunium, in the compound Li_5NpO_6 (*10, VI*). The oxidation-state diagram for the actinide elements appears in Fig. 16.11. The higher oxidation states are stabilized as simple ions in crystalline lattices, e.g., MO_2 (M = Th, Pa, U, Np, Ru, Am), Pa_2O_5, and UO_3. Binary systems containing mixed oxidation states are also possible, e.g., $PaO_{2.3}$, U_4O_9, and U_3O_8. In aqueous solution the higher oxidation states exist as complex oxy-ions (MO_2^+, MO_2^{2+}, MO_5^{3+}; M = U, Pa, Np, Am) which in some instances can be isolated in salts (*11*).

Just as in the case of the lanthanides, some of the oxidation states of the actinides are characteristically colored (Table 16.8). The absorption spectra of aqueous solutions of these substances exhibit sharp bands. The actinides show a variation in magnetic properties similar to that of the lanthanides.

REFERENCES

1. T. Moeller, *J. Chem. Ed.*, **47**, 417 (1970).
2. L. Benjamin and V. Gold, *Trans. Faraday Soc.*, **50**, 797 (1954).
3. W. M. Latimer, K. S. Pitzer, and C. M. Slansky, *J. Chem. Phys.*, **7**, 108 (1939).
4. E. J. W. Verwey, *Rec. Trav. Chim.*, **61**, 127 (1942).
5. R. Juza and C. Handenfeldt, *Naturwiss.*, **55**, 229 (1968).
6. L. L. Pytlewski and J. K. Howell, *Chem. Commun.*, 1280 (1967).
7. J. S. Anderson, N. J. Clark, and I. J. McColm, *J. Inorg. Nucl. Chem.*, **30**, 105 (1968).
8. S. Salot and J. C. Warf, *J. Amer. Chem. Soc.*, **90**, 1932 (1968).
9. D. G. Karraker, *J. Chem. Ed.*, **47**, 424 (1970).
10. C. Keller and H. Seiffert, *Inorg. Nucl. Chem. Letters*, **5**, 51 (1969).
11. V. I. Spitsyn, A. D. Gelman, N. N. Krot, M. P. Mefodiyeva, F. A. Zakharova, Yu. A. Komkov, V. P. Shilov, and I. V. Smirnova, *J. Inorg. Nucl. Chem.*, **31**, 2733 (1961).

COLLATERAL READINGS

I. R. S. Nyholm and M. L. Tobe, *Advan. Inorg. Chem. Radiochem.*, **5**, 1 (1963). A review of the factors involved in stabilizing oxidation states of the transition metals.
II. A. Carrington and M. C. R. Symons, *Chem. Revs.*, **63**, 443 (1963). A survey of the structure and reactivity of the oxyanions of the transition metals.
III. J. Selbin, *Angew. Chem. (Int. Ed.)*, **5**, 712 (1966). A discussion of the range of stability of transition-metal oxycations in aqueous solution.
IV. R. D. Peacock, *Prog. Inorg. Chem.*, **2**, 193 (1960). A description of the chemistry of some fluorine compounds of the transition metals. Included in this discussion are the simple fluorides, oxyfluorides, and complex oxyfluorides.
V. L. B. Asprey and B. B. Cunningham, *Prog. Inorg. Chem.*, **2**, 267 (1960). A review concerned with unusual oxidation states of some actinide and lanthanide elements.
VI. N. Hodge, *Advan. Fluorine Chem.*, **2**, 138 (1961). A review in which the chemistry of the fluorides of the actinide elements is discussed.

STUDY QUESTIONS

1. Sketch the radial distribution functions for $4s$, $4p$, $4d$, and $4f$ electrons on the same relative scale.

2. Without reference to the periodic table, which atom in each of the following pairs with the indicated atomic numbers would be expected to have the higher ionization potential: 21 and 22; 46 and 47; 28 and 29; 11 and 29; 48 and 80; 30 and 48.

3. Without reference to the periodic table, give the expected ground-state configuration of the atoms with the following atomic numbers: 24, 21, 29, 45, 60, 80, and 96.

4. Describe the general chemical properties that might be expected for the unknown series of elements in which the differentiating electrons occur in $5g$ orbitals.

5. Explain the observation that europium compounds are often found in high concentrations among calcium-containing minerals.

6. Discuss the factors leading to the observation that the actinides exhibit higher oxidation states than do the lanthanides.

7. Discuss the observation that all the divalent oxides of the elements in the first transition series except CrO and CuO crystallize in the $NaCl$ structure.

8. Compare the relative ease of hydrolysis of Fe^{2+} and Fe^{3+}.

9. Estimate the effective nuclear charge for the following species: Sc, V^{2+}, Co^{3+}, Zn^{2+}, Pm^{3+}.

10. Discuss the position of elemental lutetium in the periodic chart using its electronic configuration and physical properties to support your argument. Consider the chemical properties of lutetium, including the chemistry of the most prevalent oxidation state (Lu^{3+}) in your discussion.

11. Without reference to a periodic chart, discuss the expected oxidation states for the elements with atomic numbers 61 through 64.

12. Describe the lanthanide contraction and its consequences.

13. What factors are responsible for the resistance of copper to oxidation by hydrogen ion when the other elements in the first d block form the M^{2+} state under the same conditions?

14. Discuss the general differences in the chemistry of the first d-block series of elements with respect to the other two d-block series.

17

PROPERTIES OF
TRANSITION-METAL
COORDINATION
COMPOUNDS

1 / INTRODUCTION

The formation of species containing coordinate-covalent bonds, i.e., the product of the reaction between a Lewis acid and a Lewis base, is mentioned in the discussions of the chemistry of the representative elements. Virtually all of the representative elements in Periods 3 and higher form compounds which can behave as Lewis acids, e.g., ICl_3, $SiCl_4$, PF_5, SiF_4. In addition, the elements occurring before carbon in Period 2 also give compounds which behave as Lewis acids, e.g., $BeCl_2$, BF_3. Lewis bases, on the other hand, are species which carry pairs of electrons that are not initially involved in bonding; such species are usually anions, such as the halide ions, or simple neutral molecules such as amines and ethers or their derivatives. Table 17.1 contains several examples of Lewis acid–base interactions involving the representative elements.

Transition-metal ions can also act as Lewis acids, forming compounds which can be imagined as containing coordinate-covalent bonds. For

621

TABLE 17.1   Examples of Lewis acid–base interactions among species of the representative elements

Group	Period	Lewis acid	Lewis base	Coordinately-bound species
7	5	ICl_3	Cl^-	ICl_4^- $(KICl_4)$
6	4	$SeCl_4$	Cl^-	$SeCl_6^{2-}$ (K_2SeCl_6)
5	3	PF_5	F^-	PF_6^- (KPF_6)
4	3	SiF_4	F^-	SiF_6^{2-} (K_2SiF_6)
3	3	$AlCl_3$	NH_3	$AlCl_3 \cdot NH_3$
			Cl^-	$AlCl_4^-$ $(KAlCl_4)$
3	2	BF_3	F^-	BF_4^- (KBF_4)
			R_2O	$BF_3 \cdot OR_2$
2	2	$BeCl_2$	Cl^-	$BeCl_4^{2-}$ (K_2BeCl_4)
			R_3N	$BeCl_2 \cdot 2R_3N$
2	3	$MgCl_2$	Cl^-	$MgCl_4^{2-}$ (K_2MgCl_4)
			NH_3	$Mg(NH_3)_6^{2+}$
				$[Mg(NH_3)_6(ClO_4)_2]$

example, anhydrous copper sulfate, $CuSO_4$, is a colorless substance, but dissolution in water gives a sky-blue solution, while in liquid ammonia a purple solution forms. If potassium chloride is added to the aqueous solution a bright green solution forms. It is possible to isolate compounds with formulations such as $CuSO_4 \cdot 4H_2O$,* $CuSO_4 \cdot 4NH_3$, and $CuCl_2 \cdot 2KCl$ from these solutions. The first two compounds appear to be solvates and the third a double salt. However, a better description of these compounds, based on a variety of data, involves species in which the copper ion acts as a Lewis acid towards H_2O, NH_3, and Cl^- as Lewis bases, viz., $[Cu(H_2O)_4]$-SO_4, $[Cu(NH_3)_4]SO_4$, $K_2[CuCl_4]$. The chemistry of transition-metal ions is dominated by their ability to act as electron-pair acceptors towards Lewis bases, which in this context are often called ligands. As might be expected, the number of ligands associated with various transition-metal ions and their arrangement in space is characteristic of the metal ion. In this chapter we consider the coordination behavior and general properties of transition-metal ions as elucidated from a variety of experimental techniques. A discussion of the theories of bonding in transition-metal coordination compounds appears in Chapter 18.

In the previous paragraph it was stated without evidence that the ligands are associated with the metal ions in coordination compounds. The analysis of the products formed from simple chemical reactions is sufficient to show that certain groupings of ligands and transition-metal ions persist in these processes. As an example, consider the complexes of trivalent cobalt which

---

* The common form of "hydrated" copper sulfate is $CuSO_4 \cdot 5H_2O$ in which one of the water molecules is hydrogen-bonded in the lattice.

undergo substitution reactions in aqueous solutions only very slowly. The air oxidation of an ammoniacal aqueous solution containing $CoCl_2$ and NH_4Cl leads, among other compounds, to a red crystalline substance with the empirical formulation $CoCl_3 \cdot 5NH_3 \cdot H_2O$. Treatment of this substance with silver salts yields compounds in which the halogen has been replaced by another anion

$$CoCl_3 \cdot 5NH_3 \cdot H_2O + 3AgNO_3 \xrightarrow{\text{(aq)}} Co(NO_3)_3 \cdot 5NH_3 \cdot H_2O + 3AgCl \quad (1)$$

$$2CoCl_3 \cdot 5NH_3 \cdot H_2O + 3Ag_2SO_4 \xrightarrow{\text{(aq)}} Co_2(SO_4)_3 \cdot 10NH_3 \cdot 2H_2O + 6AgCl.$$
$$(2)$$

All the cobalt-containing compounds in these reactions exhibit $Co : NH_3 : H_2O$ ratios of $1 : 5 : 1$. Thus, $[Co(NH_3)_5H_2O]^{3+}$ appears to persist as a unit unchanged in a series of reactions, much the same way as does the ammonium ion, NH_4^+. Historically, many metal-containing complex ions incorporating Co^{3+}, Cr^{3+}, Pt^{4+}, and Pt^{2+} were shown to exist by such simple experiments. The formulation of complex ions containing metal ions can be traced to Alfred Werner, who suggested that all metals possess two types of valence, primary and secondary.* The primary valence is the normal ionic valence associated with the metal, and can be satisfied only by anions. The secondary valence can be satisfied by either anions or neutral species. Just as each metal has a characteristic primary valence, it also possesses a characteristic secondary valence; thus in the example cited above cobalt has a primary valence of $3+$ and a secondary valence of 6. The suggestion that anions can satisfy either primary or secondary valences arose from the reactions of complex compounds such as $CoCl_3 \cdot 5NH_3$

$$CoCl_3 \cdot 5NH_3 + AgNO_3 \xrightarrow{\text{(aq)}} Co(NO_3)_2Cl \cdot 5NH_3 \quad (3)$$

$$CoCl_3 \cdot 5NH_3 + H_2SO_4 \xrightarrow{\text{(aq)}} Co(SO_4)Cl \cdot 5NH_3. \quad (4)$$

It is significant that the cobalt-containing products in Eqs. (3) and (4) still retain one chlorine atom. The products are characterized by a $Co : Cl : NH_3$ ratio of $1 : 1 : 5$, which suggested to Werner that the complex ion $[Co(NH_3)Cl]^{2+}$ is present in all these compounds. Thus cobalt exhibits a primary valence of $3+$ (although the charge on the complex ion is $2+$) and a secondary valence of 6. Werner's term secondary valence has given way to the modern term coordination number, which is used in much the same sense as it is in crystal chemistry for describing the number of nearest neighbors. The validity of formulating the compounds $CoCl_3 \cdot 5NH_3$ and

---

* Such species are sometimes called Werner complexes.

$CoCl_3 \cdot 5NH_3 \cdot H_2O$ as $[Co(NH_3)_5Cl]Cl_2$ and $[Co(NH_3)_5H_2O]Cl_3$, respectively, is supported by conductivity measurements. Inspection of the molar conductances, determined at the same dilution, of a variety of ionic compounds, given in Table 17.2, shows that all compounds with the same number

TABLE 17.2 Molar conductivities of some typical electrolytes

	Concentration, M			
Salt	0^a	0.003	0.03	0.05
NaCl	126.45	123.74	118.51	111.06
KCl	149.86	146.95	141.27	133.37
NaI	126.94	124.25	119.24	112.79
KNO_3	144.96	141.84	132.82	136.31
$MgCl_2$	258.80	248.22	219.10	206.16
$CuCl_2$	276.68	260.72	240.72	216.94
Na_2SO_4	259.8	248.30	224.88	195.50
$LaCl_3$	437.4	371.0	365.4	318.6
$K_3Fe(CN)_6$	523.5	489.0	—	—
$K_4Fe(CN)_6$	738.0	668.96	538.52	430.80

^{a} Infinite dilution values obtained by extrapolation.

of ions fall within a relatively narrow range of conductivity. Thus, salts such as MX_2 or M_2X which give three ions exhibit a conductivity distinctly different than MX which gives only two ions in solution. Experimentally the conductivities of $[Co(NH_3)_5Cl]Cl_2$ and $[Co(NH_3)_5H_2O]Cl_3$ fall in the range expected for substances containing three and four ions, respectively. Conductivity measurements in nonaqueous solutions have also been useful in elucidating the number of ions present in a coordination compound, especially if the compound is unstable in aqueous solutions. Conductivity data obtained with nonaqueous solvents require more care in interpretation since extensive conductivity data similar to those shown in Table 17.2 for aqueous solutions may not be available for comparison; an additional difficulty in using nonaqueous solvents arises from the necessity of employing solvents with a sufficiently high dielectric constant to assure that ion-association is not important (Chapter 7).

The ligands which satisfy the characteristic coordination number of a transition-metal ion are said to occupy the coordination sphere of the ion, which is usually depicted by square brackets in the line formula of the compound. A list of some typical transition-metal coordination compounds is given in Table 17.3. Coordination numbers of 4 and 6 are the most prevalent, but species with lower and higher coordination numbers have been identified.

TABLE 17.3 Examples of known coordination numbers for transition metals and the representative elements

Coordination number	Geometry	Examples	
		Transition metals	Representative elements
2	linear	$Ag(NH_3)_2^+$; $Hg(NH_3)_2^{2+}$; $M(CN)_2^-$, $M = Cu^+$, Ag^+, Au^+; $AuCl_2^-$	$BeCl_{2(g)}$
3	trigonal-planar	$Ag(R_3P)_3^+$, HgI_3^-, $Ag(R_2S)_3^+$, $Cu(R_3P)_3^+$, $Au(AsR_3)_3^+$	BCl_3, ICl_3
4	tetrahedral	$ZnCl_4^{2-}$, $CdCl_4^{2-}$, HgI_4^{2-}, $FeBr_4^-$	BF_4^-
	planar	$Pt(NH_3)_2Cl_2$, $Ni(CN)_4^{2-}$	ICl_4^-
5	trigonal-bipyramidal	$Fe(CO)_5$, $Cu(terpy)Cl_2$,[a] $Ni(R_3P)_2Br_3$	PF_5
	square-pyramidal	$NiBr_2 \cdot triArs$[b], $Ni(PR_3)_2Br_3$	BrF_5
6	octahedral	$Cr(CN)_6^{3-}$, $Co(NH_3)_6^{3+}$, $Cr(CO)_6$	PF_6^-, SiF_6^{2-}
7	pentagonal-bipyramidal	ZrF_7^{3-}, HfF_7^{3-}, UF_7^{3-}	IF_7
	distorted trigonal prism	NbF_7^{2-}, TaF_7^{2-}	—
	distorted octahedron	$NbOF_6^{3-}$	—
8	dodecahedron (bis-diphenoid)	$Mo(CN)_8^{4-}$	—
	square antiprism	TaF_8^{3-}, $Zr(acac)_4$[c]	—
9	distorted trigonal prism	$Nd(H_2O)_9^{3+}$, UCl_9^{6-}, $La(OH)_9^{6-}$, ReH_9^{2-}	—

[a] terpy = 2,2′,2″-terpyridine.
[b] triArs = $(CH_3)_2AsC_2H_4As(CH_3)C_2H_4As(CH_3)_2$.
[c] acac = acetylacetonate.

Werner was unable to suggest a satisfactory theoretical basis for the existence of the secondary valence. Today we correlate the ability of transition-metal ions to form complex compounds with their ability to act as electron-pair acceptors (Lewis acids) forming coordinate-covalent bonds with the ligands.

2 / STEREOCHEMISTRY

The observation that a given number of ligands can occupy the coordination sphere of a transition-metal ion naturally leads to a consideration of the configuration of complex ions. Originally, Werner deduced the configuration of six- and four-coordinate species on the basis of the number of isomers which could be isolated for certain compounds, an argument reminiscent of the method used to establish the structure of benzene. For example, $PtCl_2$ forms two different complexes with ammonia, both with the empirical formula $PtCl_2 \cdot 2NH_3$. This immediately rules out a tetrahedral arrangement of ligands about platinum because all tetrahedral positions are equivalent; there is only one possible arrangement of two pairs of two groups occupying tetrahedral positions. However, if the ligands occupied the corners of a square about the central metal ion, the two isomers shown in Fig. 17.1

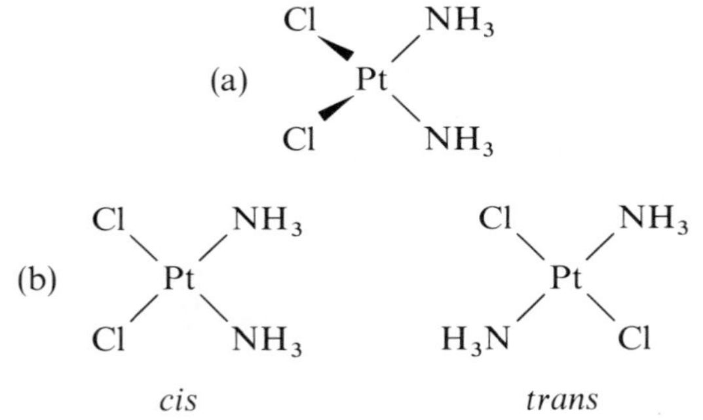

FIG. 17.1 Four-coordinate complexes of the type $[PtCl_2(NH_3)_2]$ exist in two isomeric forms. A tetrahedral arrangement of ligands (a) should yield only one form, but a planar arrangement (b) should give two isomers.

would be possible for $PtCl_2 \cdot 2NH_3$. In one isomer, identical ligands are opposite each other (or *trans*); in the other the same groups are adjacent to each other (or *cis*). Werner also established the stereochemistry of six-coordinate cobalt using arguments based on the abundance of isomers isolated for compounds containing complex ions of the type $[Ma_4b_2]^{n+}$, e.g., $[Co(NH_3)_4Cl_2]^{1+}$. If the ligands were hexagonally arranged in a plane about the cobalt, the three isomers shown in Fig. 17.2 should result. Similar considerations for a prismatic arrangement also lead to three isomers for this species. However, if the ligands occupied the corners of an octahedron, only two isomers would result. Werner could never isolate more than two isomers for six-coordinate complexes of the type $[Ma_2b_2]^{n+}$, which led him to postulate the octahedral geometry for these complex ions.

Modern methods of direct structure determination, such as x-ray diffraction, support the validity of the arguments based on isomer abundance. In fact, a number of different physical methods are now available which are useful in establishing the stereochemistry of coordination compounds. These

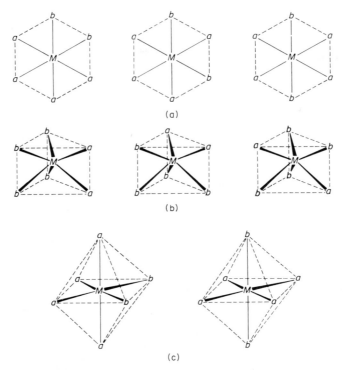

FIG. 17.2   The expected number of isomers for three different geometrical arrangements of ligands in compounds of the type Ma_4b_2. Only two isomers have been isolated in complexes with this stoichiometry, supporting the octahedral arrangement (c) of ligands over the planar (a) or prismatic (b) distribution.

are discussed in a subsequent section. Although a variety of complex compounds have been isolated and characterized, the limited number of structural types shown in Fig. 17.3 have been observed. Many of the arrangements of ligands about the central metal ion (i.e., 2-, 3-, 4-, 5-, 6- and 7-coordination) reflect molecular geometries observed for the species formed by the representative elements, but transition-metal ions, as a class, can achieve higher coordination numbers and different geometries for a given coordination number. For example, five-coordinate trigonal-bipyramidal complexes are known, but square-pyramidal structures have also been observed for this coordination. Some seven-coordinate species appear to be structurally similar to IF_7, but others exhibit the distorted prismatic arrangement shown in Fig. 17.3. The ligands in eight-coordinate complex ions are arranged at the corners of either a square antiprism or a dodecahedron (*I*). All the nine-coordinate species of known structure possess the ligands in a distorted prismatic arrangement. A survey of the coordination

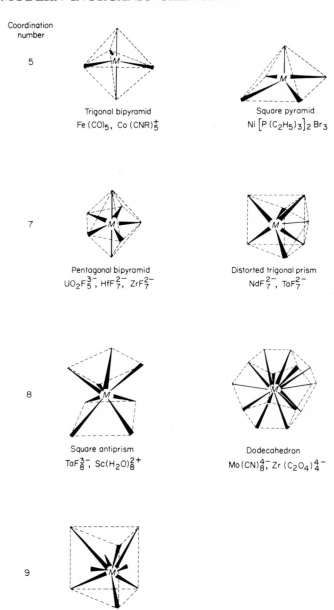

Coordination number

5

Trigonal bipyramid
$Fe(CO)_5$, $Co(CNR)_5^+$

Square pyramid
$Ni[P(C_2H_5)_3]_2 Br_3$

7

Pentagonal bipyramid
$UO_2F_5^{3-}$, HfF_7^{2-}, ZrF_7^{2-}

Distorted trigonal prism
NdF_7^{2-}, TaF_7^{2-}

8

Square antiprism
TaF_8^{3-}, $Sc(H_2O)_8^{2+}$

Dodecahedron
$Mo(CN)_8^{4-}$, $Zr(C_2O_4)_4^{4-}$

9

Face–centered trigonal prism
ReH_9^{2-}, $Nd(H_2O)_9^{3+}$

FIG. 17.3   Some examples of the geometries attained by five-, seven-, eight-, and nine-coordinate transition-metal species.

numbers and the corresponding geometrical arrangements of the ligands which have been observed for transition-metal species is presented in Table 17.4. The data collected in this table lead to several general conclusions. First, 6 and 4 are the most common coordination numbers exhibited by the transition metals. The six-coordinate species are octahedral, some such as CuX_6^{n+} being distorted; the four-coordinate species exist either as tetrahedral or square-planar complexes. In some instances tetrahedral complexes can also be distorted. Generally the metals of the d-block elements in their

TABLE 17.4   The coordination numbers observed for various oxidation states of the d-block elements[a]

Oxidation state	Ti	V	Cr	Mn	Fe	Co	Ni	Cu	Zn
−1	6	6	6	4s, 5, 6		4	4		
0	6	6	6	6	5, 6	4	4		
1		6, 7	6	6	6	4, 5, 6	4	2, 3, 4	
2		6	6, 7	4, 4s, 5, 6, 7	4, 5, 6	4, 4s, 5, 6	4, 4s, 5, 6	4d, 4s, 5, 6d	4, 5, 6
3	6	4, 5, 6	4d, 6	5, 6	4, 6, 7	4, 6	5, 6	4, 6	
4	5, 6, 7, 8	5, 6, 8	4, 6	6	6	6	6		
5		6	4, 6, 8	4	4				
6			6	4	4				
7				3, 4					

Oxidation state	Zr	Nb	Mo	Tc	Ru	Rh	Pd	Ag	Cd
−1		6				4, 5			
0	6		6	6	5		4		
1						4s, 5		2, 3, 4, 6	
2			5, 6, 7, 9	5, 6	5, 6	4s, 5, 6	4s, 5, 6	4s	4
3			6, 8	6	6	6		4s, 6	
4	4, 6, 7, 8	6, 8	6, 8	4, 6	6, 8	6	6		
5		5, 6, 7, 8	5, 6, 8	5	5, 6		5, 6		
6			4, 5, 6, 8		4, 6	6	6		
7				4	4				

Table 17.4 (continued)

Oxidation state	Hf	Ta	W	Re	Os	Ir	Pt	Au	Hg
−1		6		5					
0			6	6	5		4		
1				6		4s, 5		2, 4	
2			7, 9	6	6		4s, 5, 6		4, 5, 6
3			7	5, 6	6	5, 6		4s, 5, 6	
4	4, 6, 7, 8	6	6, 8	4, 6, 7	6	6	6		
5		5, 6, 7, 8	6, 8	5, 6, 7, 8	6	6	5		
6			4, 5, 6, 8	6, 7, 8	4, 5, 6	6			
7				4, 6, 7, 8, 9	6				
8					5, 6				

[a] In the cases where the geometries are known, the following symbols are used: 4 = tetrahedral, 4s = square-planar, 4d = distorted tetrahedral, 6 = octahedral, 6d = distorted octahedral. The other coordination numbers generally correspond to the geometries shown in Fig. 17.3.

common oxidation states exhibit high coordination numbers at the beginning of a series, with the lower coordination numbers being attained at the end of the series. The heavier metals in any family can attain higher coordination numbers than the lighter members. It is not unusual to find a metal in the same oxidation state exhibiting different coordination numbers. Finally, the lanthanides and actinides show a remarkably consistent coordination behavior which parallels their constancy of valence. Unusually low or high oxidation states can often be stabilized by complex-ion formation (*II*). Complexes with odd coordination numbers are not usually achieved except in unusual circumstances. Coordination number 5 is the most common of the odd coordination numbers (*III*).

3 / LIGANDS

One of the best methods for classifying complex compounds is on the basis of the nature of the ligands involved. Generally speaking, the ligands supply the electron density for the covalent bond formed to the transition-metal atom. Thus, it is not surprising that the ligand atoms directly attached to the transition-metal atom are those found at the extreme right of the representative elements, as shown in Table 17.5, where the number of electrons available exceeds the number of normal covalent bonds the ligand

TABLE 17.5   **Some typical ligands**

	C	N	O	F
Neutral species	$CO, C_2H_4,$ $C_6H_6,$ CNR	$N_2, NH_3, NR_3,$ $NH_2CH_2CH_2NH_2,$ $NO, N_2H_4, RCN,$ bipyridyl	$R_2O, ROH, R_3PO,$ $R_3AsO, R_2SO,$ $CH_3C(O)CH_2C(O)CH_3$	—
Ions	CN^-	$NO_2{}^-, N_3{}^-, NCS^-,$ $NH_2{}^-$	$OH^-, O_2{}^-, O_2{}^{2-},$ $CO_3{}^{2-}, SO_4{}^{2-},$ $S_2O_3{}^{2-}, RCO_2{}^-,$ $VO_2{}^-, C_2O_4{}^{2-}$	F^-

	P	S	Cl
Neutral species	$PH_3, PR_3, PX_3,$ $R_2PCH_2CH_2PR_2$	$R_2S, RSCH_2CH_2SR$	—
Ions	$PR_2{}^-$	$SCN^-, S_2O_3{}^{2-}$	Cl^-

	As	Se	Br
Neutral species	$R_3As, AsCl_3,$ $R_2AsCH_2CH_2AsR_2$	R_2Se	—
Ions	$AsR_2{}^-$	$SeSCN^-$	Br^-

	Sb	Te	I
Neutral species	R_3Sb	R_2Te	—
Ions	$SbR_2{}^-$	—	I^-

atom can form. The ligand atoms can be associated with neutral molecules such as the trivalent Group V derivatives, MR_3 (IV), where the substituents can be organic groups, hydrogen, or halogens (V), as well as similar divalent derivatives of the Group VI elements, R_2M. In each of these cases the ligand atom has at least one pair of unshared electrons (**1** and **2**). Other neutral ligands include oxidized derivatives of the Group V elements (R_3MO), binary oxides (CO and NO), and organic derivatives containing the CN moiety (RCN and RNC). Even organic compounds such as C_2H_4 and C_6H_6

containing unsaturated systems can act as ligands toward transition-metal atoms. The chemistry and properties of the organometallic derivatives of

1   2

the transition metals are properly considered as a distinct part of transition-metal coordination chemistry and are treated in detail in Chapter 19.

It should not be surprising that virtually all of the anions can act as ligands since these species carry a negative charge, a fact which suggests that an unshared pair of electrons is available. However, a few anions such as the tetraalkyl borates (BR_4^-) must be excluded from this generalization. Inspection of Table 17.5 indicates that many anions are known to act as ligands. Theoretically some anions have two or more possible sites for coordination. For example, the Lewis formulations for CN^- (3) (VI), SCN^- (4), and NO_2^- (5) show two possible coordination sites. In each

$$:C\equiv N:$$

3

$$:\ddot{S}C\equiv N:$$

4

5

instance complex compounds in which the metal is coordinated to one or the other site have been isolated. The point is considered in more detail in the next section.

Although the data in Table 17.4 show predominantly the coordination behavior of the lower oxidation states of the transition metals, it should be recalled from Chapter 16 that some of the transition metals can exhibit valence states as high as $8+$. In some ways the oxyanions, in which the transition metals exhibit their highest oxidation states, can be considered as being composed of a highly charged cation surrounded by oxide, O^{2-}, ligands. In this view the oxidation states M^{7+}, M^{6+}, and M^{5+} give rise to the tetrahedral complex oxyanions MO_4^- (MnO_4^-, ReO_4^-), MO_4^{2-} (CrO_4^{2-}, MnO_4^{2-}, ReO_4^{2-}, MoO_4^{2-}, WO_4^{2-}), and MO_4^{3-} (VO_4^{3-}). Of course, it is not correct to describe these species as collections of the corresponding ions because the high polarizing power of atoms in high oxidation states leads to significant covalent bonding, but this point of view is often useful in certain discussions.

3.1 / Polydentate Ligands

Many species have only one point of attachment when they form complex ions with metal atoms; such ligands are called monodentate ligands.* Some ligands could have several lone electron pairs and thus form several co-ordinate-covalent bonds. Such ligands are called polydentate ligands. If, because of geometrical considerations, the lone electron pairs are sufficiently well separated to form bonds to the same metal atom, the ligand is called a chelate.† For example, triethylenediamine (**6**) is bidentate but the two amine groups are constrained to be opposite each other and hence can form coordinate bonds to two different metals. Ethylenediamine (**7**), however,

$$
\begin{array}{c}
\quad\; CH_2 - CH_2 \\
:N \!-\! CH_2 - CH_2 \!-\! N: \\
\quad\; CH_2 - CH_2
\end{array}
\qquad\qquad
H_2\ddot{N} \underset{CH_2 - CH_2}{\diagdown \diagup} \ddot{N}H_2
$$

6   **7**

can also form two metal–nitrogen coordinate bonds, but because of the geometry of the molecule, the bonds are formed to the same atom, giving a five-membered ring. As in the chemistry of carbon compounds, five- and six-membered rings are unstrained and are easily formed by chelating ligands. Although bidentate chelates are the most widely investigated ligands, species with three, four, five, and six coordination sites are also known; examples of such ligands appear in Table 17.6.

Polydentate chelates usually incorporate a two- or three-carbon chain between the ligand atoms, which provides sufficient separation of the latter for formation of five- or six-membered heterocyclic systems. In some instances, such as with carboxylate ions (*VII*), four-membered rings can be formed. In addition, *N,N*-diethylthiocarbamate ion (**8**), CO_3^{2-} (**9**), and

$$
\left[(C_2H_5)_2NC \underset{\ddot{S}:}{\overset{\ddot{S}..}{\diagup\!\!\!\diagdown}} \right]^{-}
\qquad
\left[O\!=\!C \underset{\ddot{O}:}{\overset{\ddot{O}:}{\diagup\diagdown}} \right]^{2-}
\qquad
\left[O\!=\!N \underset{\ddot{O}:}{\overset{\ddot{O}:}{\diagup\diagdown}} \right]^{-}
$$

8   **9**   **10**

NO_3^{-} (**10**) (*VIII*) are known to act as bidentate ligands. As indicated in Table 17.6, chelate ligands may be neutral species or charged species.

---

* Dentate: "having a tooth-like projection," i.e., an electron pair for bond formation.
† From the Greek for claw.

TABLE 17.6 Some typical polydentate ligands

Name	Structure
Bidentate ligands	
ethylenediamine (en)	H_2N ... NH_2 / CH_2—CH_2
bipyridyl (bipy)	N N (bipyridyl ring structure)
o-phenylenebisdimethylarsine (diars)	H_3C, CH_3 — As ; As — H_3C, CH_3
8-quinolinolate ion	$\left[\ \text{N} \quad \text{O} \ \right]^{1-}$
N,N-diethylthiocarbamate ion	$\left[(C_2H_5)_2N{-}C \begin{array}{c} S \\ S \end{array} \right]^{1-}$
acetylacetonate ion (acac$^-$)	$\left[H_3C{-}C{=}O \ \ O{=}C{-}CH_3 \ ; \ \overset{\mid}{C}{-}H \right]^{1-}$
Tridentate ligands	
diethylenetriamine (dien)	H_2N ... $\overset{\mid}{N}$ ... NH_2 / CH_2—CH_2 H CH_2CH_2
iminodiacetate ion	$\left[\begin{array}{c} O \\ O \end{array}{=}C ... \overset{\mid}{N} ... C{=}\begin{array}{c} O \\ O \end{array} \ ; \ CH_2 \ H \ CH_2 \right]^{2-}$

Table 17.6 (continued)

Name	Structure

Tetradentate ligands

triethylenetetramine (trien)

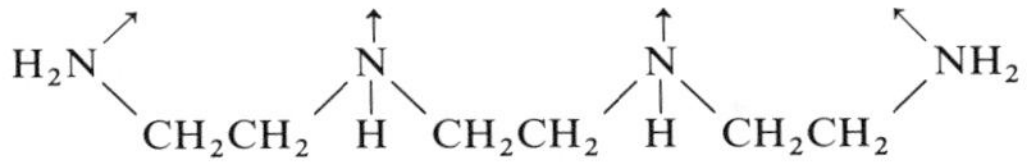

$$H_2N-CH_2CH_2-\overset{H}{\underset{|}{N}}-CH_2CH_2-\overset{H}{\underset{|}{N}}-CH_2CH_2-NH_2$$

Pentadentate ligands

ethylenediaminetriacetate ion

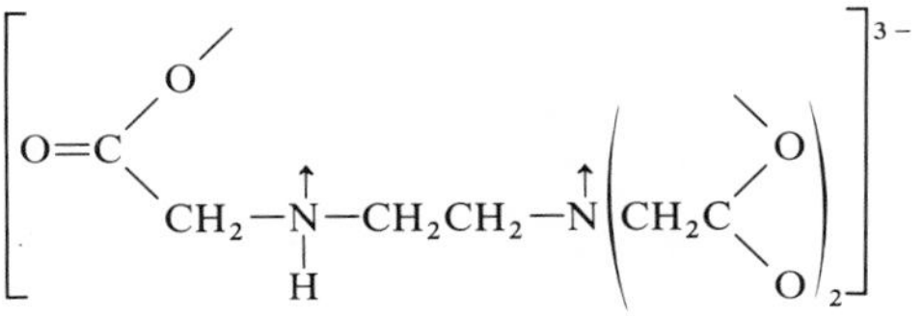

$$\left[O{=}C\overset{O}{\diagdown}-CH_2-\overset{H}{\underset{|}{N}}-CH_2CH_2-N\left(CH_2C\overset{O}{\diagup}_{O}\right)_2 \right]^{3-}$$

Hexadentate ligands

1,8-bis(salicylideneamino)-3,6-dithiooctane

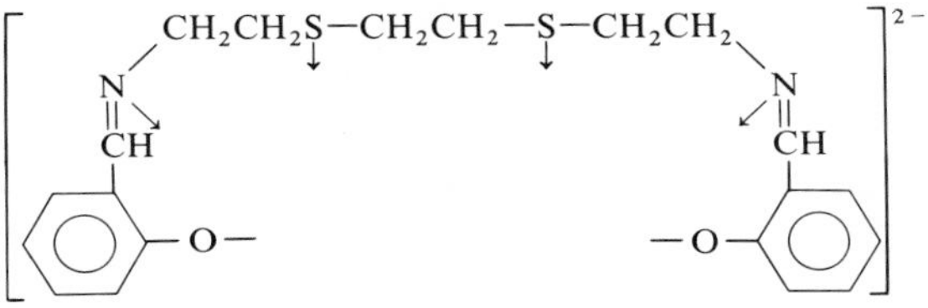

$$\left[\begin{array}{c} CH_2CH_2S-CH_2CH_2-S-CH_2CH_2 \\ N{=}CH \qquad\qquad\qquad CH{=}N \\ -O- \qquad -O- \end{array} \right]^{2-}$$

ethylenediaminetetraacetate ion (EDTA)

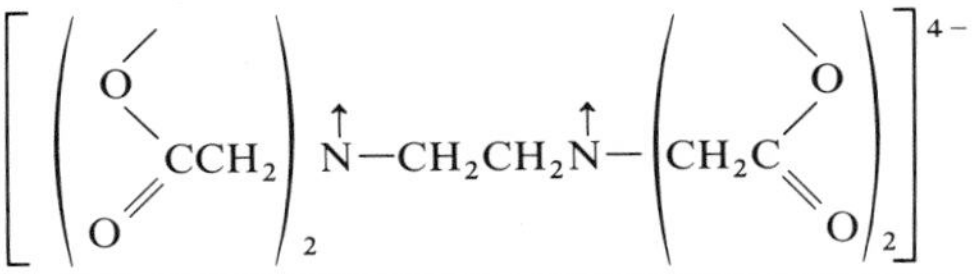

$$\left[\left(\overset{O}{\underset{O}{\diagup\diagdown}}CCH_2\right)_2 N-CH_2CH_2N-\left(CH_2C\overset{O}{\underset{O}{\diagdown\diagup}}\right)_2 \right]^{4-}$$

Charged chelates may satisfy one primary and one secondary valence [e.g., acetylacetonate **(11)** (*IX*) and 8-quinolinolate **(12)**], or two secondary and two primary valences (e.g., **8**, **9**, and **13**).

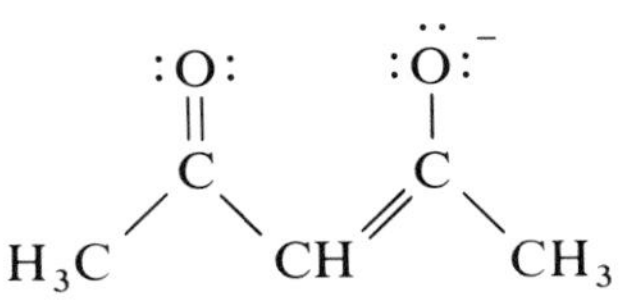

| **11** | **12** | **13** |

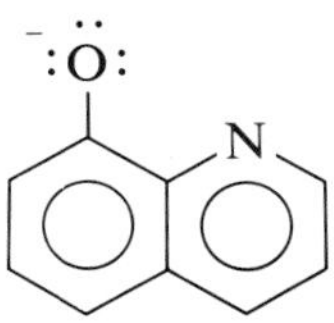

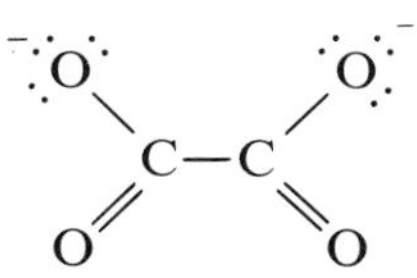

3.2 / Bridging Ligands

Under certain conditions, ligands with more than one lone pair can coordinate to two or more metal atoms to form polynuclear complex compounds. Among the bridging ligands are monatomic ions such as the halides in species like $PtCl_2 \cdot NH_3$ (14) or simple ions in which a single atom is still the bridging group, for example anions of the Group V (15) (M = N, P)

14   15   16

and Group VI (16) (M = O, S) elements. Oxygen (OH^-) and nitrogen (NH_2^-) serve as bridging atoms in polynuclear complexes formed by certain transition-metal ions in aqueous (17) and in liquid-ammonia solutions (18). A single atom in a neutral ligand can also act at a bridging group. For example,

17   18

certain divalent derivatives of the Group VI elements, such as R_2S (19), and carbon in CO (20) are bridging ligands of this type.

19   20

Ligands with two coordination positions on two different atoms can form polynuclear complexes. Neutral species such as 6 behave in this way as well as do many ionic ligands. Thus, the cyanide ion in $Pd(CN)_2$ forms a continuous two-dimensional structure (21) containing both Pd—C and Pd—N bonds, and the peroxide ion acts as a bridge in 22. Most neutral polydentate ligands do not form polynuclear complexes except under unusual

$$-Pd-C\equiv N-Pd-$$

(structure 21 with Pd, N, C triple bonds)

21

$$\left[(NH_3)_4Co \underset{O-O}{\overset{NH_2}{\diagdown}} CoCl_2(NH_3)_2 \right]^{2+}$$

22

circumstances. Thus, the tridentate phosphine*

$$(H_5C_6)_2PH_2C-H_2C \underset{}{\overset{C_6H_5}{\underset{P}{\diagup}}} CH_2CH_2P(C_6H_5)_2$$

23

forms dinuclear (**24–26**) as well as trinuclear (**27**) complexes (*1*).

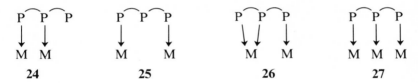

   24   **25**   **26**   **27**

3.3 / Metal Clusters (X)

Polynuclear complexes which contain a metal–metal bond as opposed to a bridging ligand are generally discussed as a separate class. Such compounds are found among elements which exhibit stable low valence states. In a formal way, metal-containing species can also act as ligands towards other metal atoms. For example, both Cl^- (cf. **29**) and the $(CO)_5Mn$ moiety (cf. **30**) replace CO in $Mn(CO)_6$ (**28**). The compound $Mn_2(CO)_{10}$ (**30**) contains

(structure 28: Mn with six CO ligands)

28

(structure 29: Mn with five CO ligands and one Cl)

29

(structure 30: two Mn atoms each bearing CO ligands, Mn–Mn bond)

30

---

* This ligand is symbolized as P‿P‿P.

a metal–metal bond with an octahedral orientation of the ligands about each metal atom.

Binuclear metal clusters can contain direct metal–metal bonds with no bridging groups, as in $Mo_2(CO)_{10}$ (**30**) and $(Re_2Cl_8)^{2-}$ (**31**); bridging groups

$[Cl_4ReReCl_4]^{2-}$

31

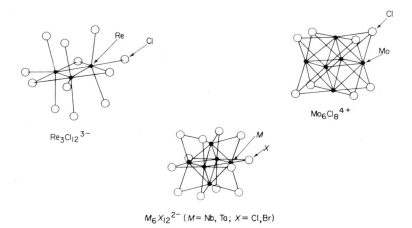

32

are present in compounds such as $Co_2(CO)_8$ (**20**) or $Fe_2(CO)_9$ (**32**).

The two types of trinuclear species containing metal–metal bonds are known; both possess a triangular arrangement of metal atoms. In one (e.g., Nb_3Cl_8) there are two halogen bridges, whereas the other (e.g., Re_3X_9)

FIG. 17.4 The structures of several metal clusters.

contains a single halogen bridge. A similar relationship is observed in the two hexanuclear metal clusters (Fig. 17.4) which possess an octahedral arrangement of metal atoms. In one case (e.g., $Mo_6X_8^{4+}$) a single ligand bridges three metal atoms; in the other ($M_6Cl_{12}^{2+}$, M = Ta, Nb) the ligands bridge two metal atoms.

4 / ISOMERISM

The variety of coordination geometries and the extensive number of ligands with different modes of attachment lead to an interesting display of isomerism among the coordination compounds. Isomerism can be discussed under two general categories: stereoisomers which arise because of the arrangement of ligands about the transition-metal atom, and isomers which occur because ligands can be attached to the metal in different ways.

4.1 / Stereoisomerism

Complex species which exhibit coordination numbers 2 and 3 cannot exhibit stereoisomerism. Four-coordinate species which have a tetrahedral arrangement of ligands yield stereoisomers only when all four ligands are different, which, of course, leads to an enantiomorphic pair. Optical isomers are discussed in a subsequent section.

Four-coordinate planar complex ions exhibit geometrical isomerism if they possess equal numbers of two kinds of ligands, i.e., Ma_2b_2. The ligands of the same kind can be opposite each other (**33**) or adjacent to each other (**34**), such arrangements being designated *trans* and *cis*, respectively, after the

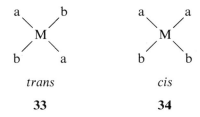

| *trans* | *cis* |
| **33** | **34** |

spatial analogy found in the disubstituted ethylenes ($CHX=CXH$). The stereochemistry of neutral planar complexes such as $PtCl_2 \cdot 2NH_3$ can be elucidated by dipole-moment measurements, the *cis* form being nonzero within experimental error. As a subsidiary point, the fact that the compound $PtCl_2 \cdot 2NH_3$ gives one isomer with a zero dipole moment is evidence that this isomer, at least, cannot be tetrahedral. The correctness of the *cis–trans* assignment in these platinum compounds is verified by several simple experiments. For example, *cis*-$PtCl_2(NH_3)_2$ reacts readily with the bidentate bivalent oxalate ion (**13**) to give $Pt(NH_3)_2C_2O_4$, which is logically the *cis* isomer since it is generally assumed that bidentate ligands can only bridge adjacent coordination positions. In addition, complex ions containing Pt^{2+} bearing four different ligands [Ptabcd] can be obtained in three isomeric forms (**35–37**). If the four-coordinate Pt^{2+} complexes were tetrahedral, only two isomers, which would be enantiomorphs, should be possible. *Cis* and *trans* isomers are also possible in planar binuclear complex species. For

example, platinum forms a complex species with the empirical formula $PtCl_2 \cdot PR_3$ which is dimeric in benzene solution. The latter fact leads to a chlorine-bridged formulation which has three possible arrangements of the nonbridged ligands (**38, 39** and **40**). The first two isomers have a *cis–trans*

relationship of the ligands, whereas the last is analogous to the structure of 1,1-dichloroethylene, sometimes called *gem*-dichloroethylene.

Bidentate ligands carrying substituents on one or more of the heterocyclic ring atoms can form geometrical isomers. For example, two molecules of a methyl-substituted ethylenediamine can be complexed in a planar arrangement in two different ways which correspond to a *cis* orientation (**41**) of the methyl groups with respect to the median plane of the ring atoms; the other possible orientation is *trans* (**42**). There are, in fact, two *cis* and two *trans* isomers with respect to this orientation. It is apparent that very complex arrangements of geometrical isomers can arise if substitution on the ligands is considered.

4.2 / Octahedral Stereoisomerism

The majority of six-coordinate complexes have an octahedral arrangement of ligands. Since all the corners of a regular octahedron are equivalent, there are no stereoisomers of complexes of the type $[Ma_5b]$. However, when two ligands of the same kind are present $[Ma_4b_2]$, the two possible arrangements of ligands shown in Fig. 17.5 occur, *cis* and *trans*. If the coordination compound contains equal numbers of two types of ligands, Ma_3b_3, two isomers are again possible as shown in Fig. 17.5.

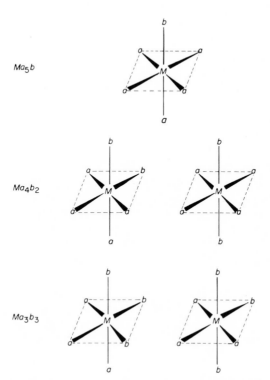

FIG. 17.5   The possible isomers for an octahedral arrangement of only two ligands.

4.3 / Optical Isomers

The absence of a plane of symmetry in a complex ion is symptomatic of the existence of two isomers of that species that differ only in the direction they rotate the plane of polarization of a beam of polarized light. In other words, if a species does not have a plane of symmetry, its mirror image is nonsuperimposable and the two are enantiomorphs. The most obvious way a complex species can exhibit optical isomerism is if the ligand has an asymmetric center. However, many enantiomorphic pairs of complex species are known in which the metal atom is the asymmetric center. Thus, tetrahedral and octahedral complexes which carry all different ligands have no plane of symmetry and must exist in enantiomorphic forms. As might be imagined, such species are difficult to prepare; however, the introduction of bidentate ligands into the coordination sphere often leads to optically active structures. For example, optically active tetrahedral complex ions can arise if they incorporate bidentate ligands containing two different ligand atoms such as 8-quinolinol (**43**). Such ligands can form two tetrahedral complexes, **44** and **45**, that are nonsuperimposable mirror images and hence

43   44   45

are optical isomers. Planar complexes are rarely optically active, but the unsymmetrical chelate ligands can be used to make such structures asymmetric. An interesting example of this situation occurs in the Pt^{2+} complex containing two different substituted ethylenediamine molecules (**46**). The planar structure (**46**) has no plane of symmetry and should exist in optically active forms. However, a plane of symmetry would exist if the complex were tetrahedral (**47**). The fact that a complex with this formulation can be resolved

46   47

into optically active forms is evidence that the geometrical arrangement must be planar (2).

The introduction of two bidentate ligands (C—C) into an octahedral structure [Ma₂(C—C)₂] also leads to *cis* and *trans* isomers, as shown in Fig. 17.6. The *trans* form contains a plane of symmetry, but the *cis* form does not, leading to the existence of an enantiomorphic pair of compounds. When

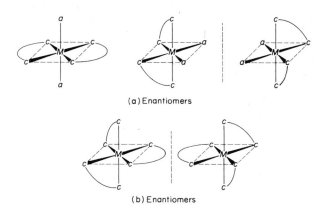

(a) Enantiomers

(b) Enantiomers

FIG. 17.6   Octahedral complexes can be optically active if (a) two or (b) three bidentate ligands are present in the coordination sphere. In the former instance the *cis* isomer carries the asymmetric center.

three bidentate ligands are attached to a metal atom giving an octahedral complex [M(C—C)₃], the product has no plane of symmetry and it must exist in enantiomorphic forms.

Optical isomerism is not limited to mononuclear complexes. Polynuclear ions containing bridging groups can also yield enantiomorphic structures. For example, the binuclear Co^{3+} ion **48** has been obtained in three forms;

$$\left[(en)_2Co \underset{NO_2}{\overset{NH_2}{<>}} Co(en)_2 \right]^{4+}$$

48

the two shown in Fig. 17.7 are optically active. The third is optically inactive, being an internally optically compensated, or *meso*, form.

Mixtures of optically active species can be resolved using several experimental techniques. Ionic species are often resolved by treating a solution of the enantiomorphs with an optically active counter ion to give a mixture of diastereoisomers* which can be fractionally crystallized to give the pure diastereomers. The pure complex optical isomer can then be obtained from the appropriate diastereomer. Optically active cationic complexes such as

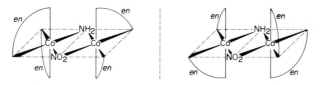

FIG. 17.7   Enantiomers of $[(en)_2Co(NH_2)(NO_2)Co(en)_2]^{4+}$.

$[Cr(en)_3]^{3+}$ and *cis*-$[Co(en)_2Cl_2]^+$ have been resolved with organic anions such as D-tartrate or D-α-bromocamphor sulfonate. A number of optically active nitrogen bases such as strychnine, brucine, and morphine are available for the resolution of optically active anionic complexes such as $[M(C_2O_4)_3]^{3-}$, where $M = Cr^{3+}$, Co^{3+}, Rh^{3+}, and Ir^{3+}.

Partial resolution of optically active complexes has been obtained by selective adsorption on optically active quartz (*3, 4*), or by taking advantage of the difference in the reaction rates of diastereomers (*5*). The resolution of optically active isomers depends upon the kinetic stability of the complexes, and it is not always possible to effect a resolution of a mixture of isomers. For example, the complex ions $Al(C_2O_4)_6^{3+}$, $Fe(C_2O_4)_6^{3+}$, $Ni(en)_3^{2+}$, and $Zn(en)_3^{2+}$, which should have octahedral structures and hence exist in optically active forms as shown in Fig. 17.6, have never been resolved, in contrast to the ions $Cr(C_2O_4)_3^{3-}$, $Co(C_2O_4)_3^{3-}$, and $Co(en)_3^{3+}$. Ions of the former group exchange ligands rapidly, as determined by tracer experiments, leading to racemization of enantiomers as rapidly as they are formed. The pure optically active isomers of the latter group of complexes are kinetically stable to racemization on a relatively long time base.

5 / OTHER TYPES OF ISOMERISM

In addition to the isomers which arise from purely geometrical considerations, isomerism in complexes can be traced to the way in which ligands satisfy the coordination positions and/or primary valences of transition-metal atoms.

---

* Diastereoisomers (or diastereomers) are optically active isomers that are not mirror images. Diastereoisomers have small but measurable differences in physical properties such as solubility which can be used as a basis for their separation.

5.1 / Ionization Isomerism

Ionization isomers are compounds which have the same composition but which give different ions in solution. Such situations arise because anions can act as ligands, satisfying coordination positions, and as counter ions, maintaining electrical neutrality. Among the numerous examples of ionization isomerism are the following:

$$[Co(NH_3)_5Br]C_2O_4, \qquad [Co(NH_3)_5C_2O_4]Br$$

$$[Co(NH_3)_5NO_3]SO_4, \qquad [Co(NH_3)_5SO_4]NO_3$$

$$[Pt(NH_3)_4Cl_2]Br_2, \qquad [Pt(NH_3)_4Br_2]Cl_2$$

$$[Co(en)_2(Cl)(NO_2)]SCN, \qquad [Co(en)_2(SCN)(NO_2)]Cl,$$

$$Co[(en)_2(SCN)(Cl)]NO_2.$$

In each case, simple chemical tests show that the anion is different. It is interesting to note that some divalent anions which can act as bidentate ligands, such as $C_2O_4^{2-}$ (13), also occupy one coordination position in other complex species, as in the first pair of ionization isomers listed above.

The fact that water can act as a ligand or can occupy lattice positions in a crystal structure independent of the metal atoms leads to a variation of ionization isomerism which is sometimes called hydrate isomerism. The classic case of hydrate isomerism involves three of the compounds with the empirical formulation $CrCl_3 \cdot 6H_2O$. The characteristics of these compounds are summarized in Table 17.7. The difference in color clearly indicates that

TABLE 17.7   Properties of the isomers of $CrCl_3 \cdot 6H_2O$

Color	Total number of ions in solution[a]	Moles AgCl precipitated formula weight	Formulation
violet	4	3	$[Cr(H_2O)_6]Cl_3$
green	3	2	$[Cr(H_2O)_5Cl]Cl_2 \cdot H_2O$
green	2	1	$[Cr(H_2O)_4Cl_2]Cl \cdot 2H_2O$

[a] Determined by conductivity measurements.

there are at least two isomers, and the difference in the number of ions formed in solution confirms the presence of three isomers, as does the number of Cl^- ions detected outside the coordination sphere. The latter information suggests the formulation of the complex ions in terms of a replacement of Cl^- for H_2O ligands, assuming that the coordination number remains 6 for all species. This leads to the conclusion that the two green isomers must contain some water of hydration, which is verified by the fact that these isomers, shown in Table 17.7, lose one and two moles of water, respectively, when stored over concentrated sulfuric acid. Obviously, the products formed by dehydration are no longer isomeric. Numerous examples of hydrate isomerism also are available, and perhaps one of the more interesting involves the compounds $[Co(NH_3)_3(H_2O)_2Cl]Br_2$ and $[Co(NH_3)_3(H_2O)(Cl)(Br)]$-$Br \cdot H_2O$ which involve both hydrate and ionization isomerism. Presumably another isomer could exist with this composition, i.e., $[Co(NH_3)_3(H_2O)Br_2]$-$Cl \cdot H_2O$. Hydrate isomerism itself should be a special case of solvate isomerism in which solvent molecules that can act as ligands can also occupy noncoordination lattice sites. Although general solvate isomerism has not been extensively studied, there have been reports of the formation of transition-metal ammoniates which lose NH_3 sufficiently easily to suggest that some of the ammonia present in the original formulation is not bound to the metal atom.

5.2 / Linkage Isomerism

Ligands which can form coordinate-covalent bonds at two or more non-equivalent sites lead to linkage isomers. Historically, the nitrite ion (5) was the first to be investigated in this respect. Cobalt forms several pairs of compounds containing a NO_2^- ion within the coordination sphere which exhibit different colors: $[Co(NH_3)_5NO_2]Cl_2$, $[Co(en)_2NO_2]Cl$, and $[Co(NH_3)_2(py)_2NO_2]Cl$. Two isomers exist for the first compound. The red isomer contains an O-bonded NO_2^- ligand (49), whereas in the yellow-brown

$$[(NH_3)_5Co(ONO)]Cl_2 \qquad\qquad [(NH_3)_5Co(NO_2)]Cl_2$$

49 **50**

isomer this ligand is N-bonded (50). The red isomer slowly converts into the yellow-brown compound in solution and in the solid state. Several other ligands such as SCN^- [cf. (51)], $S_2O_3^{2-}$ [cf. (52)], and SO_3^{2-} are known

$$(CO)_5MnSCN \text{ and } (CO)_5MnNCS$$

51

$$(H_3N)_5Co-OSO_2S^+ \text{ and } (H_3N)_5Co-SSO_3^+$$

52

to form linkage isomers. There are many ligands which theoretically can exhibit linkage isomerism for which only one isomer has been isolated.

5.3 / Coordination Isomerism

Coordination isomerism occurs when both the anion and cation are complex species with a different distribution of ligands. The same metal ion in the same oxidation state may be present in both ions

$$[Cr(NH_3)_6][Cr(SCN)_6] \text{ and } [Cr(NH_3)_4(SCN)_2][Cr(NH_3)_2(SCN)_4]$$

$$[Pt(NH_3)_6][PtCl_4] \text{ and } [Pt(NH_3)_3Cl][Pt(NH_3)Cl_3],$$

or the metal may be present in two different oxidation states

$$[Pt(NH_3)_4][PtCl_6] \text{ and } [Pt(NH_3)_4Cl_2][PtCl_4].$$

Coordination isomers also may involve two different metals

$$[Co(en)_3][Cr(CN)_6] \text{ and } [Cr(en)_3][Co(CN)_6]$$

$$[Cu(NH_3)_4][PtCl_4] \text{ and } [Pt(NH_3)_4][CuCl_4].$$

In fact, it is possible in some instances to isolate isomers in which the distribution of ligands in the cation and anion vary in a regular manner, as in the series $[Co(en)_3][Cr(C_2O_4)_3]$, $[Co(en)_2C_2O_4][Cr(en)(C_2O_4)_2]$, $[Cr(en)_2C_2O_4][Co(en)(C_2O_4)_2]$ and $[Cr(en)_3][Co(C_2O_4)_3]$.

A variation of coordination isomerism occurs in polynuclear complexes where the ligands can be arranged in different ways with respect to the metal ions. For example, there are two possible distributions (**53** and **54**) of chlorine

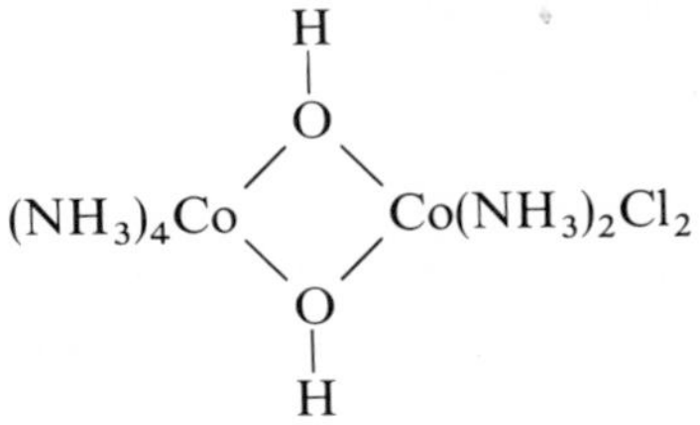

53

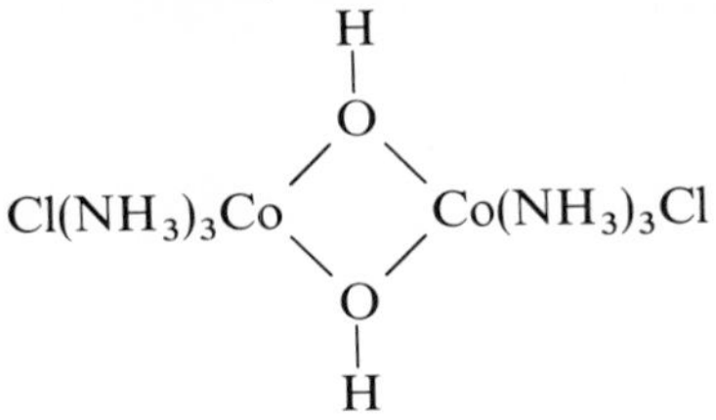

54

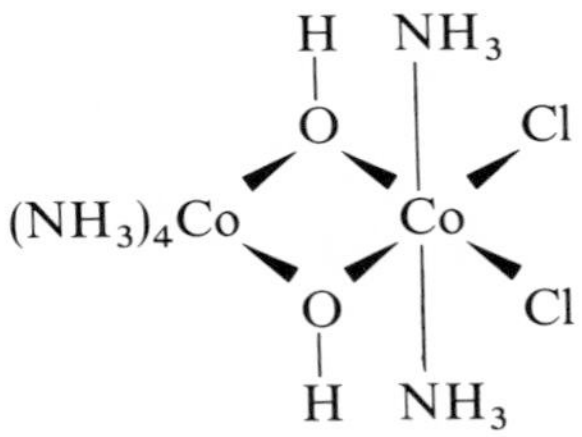

55

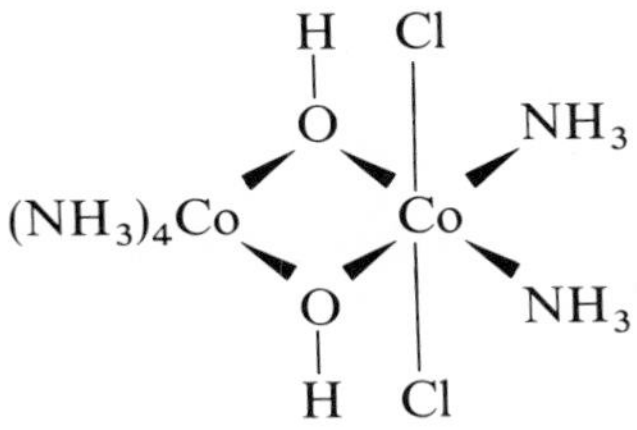

56

atoms in the binuclear species with the empirical formulation $[Co_2(NH_3)_6$-$(OH)_2Cl_2]^{2+}$. Of course three stereoisomers (55-57) are possible for this formulation; 57 is optically active.

57

5.4 / Polymerization Isomers

Polymerization isomers have the same empirical formulas but differ in their molecular weights. The name represents a formalism because there is no polymerization of smaller units into larger units in the usual sense of the term. The differences in the isomers still represent differences in arrangements of ligands and not in numbers of monomers. A relatively simple example of polymerization isomerism occurs in the chemistry of platinum where three different compounds exist with the empirical formula $Pt(NH_3)_2$-Cl_2; their molecular weights correspond to $Pt(NH_3)_2Cl_2$, $[Pt(NH_3)_2Cl_2]_2$, and $[Pt(NH_3)_2Cl_2]_3$. Experimental results show that these compounds should be formulated as $Pt(NH_3)_2Cl_2$, $[Pt(NH_3)_4][PtCl_4]$, and $[Pt(NH_3)_4]$-$[Pt(NH_3)Cl_3]_2$, respectively. In fact, the coordination isomer of the last-named compound, $[Pt(NH_3)_3Cl]_2[PtCl_4]$, also has been identified. All these platinum-containing species are planar complexes, which also leads to the possibility of *cis–trans* isomers for the first member of the family. Table 17.8 lists some coordination polymers isolated for octahedral cobalt complexes.

TABLE 17.8   Octahedral coordination polymers

Molecular formula	Formulation
$Co(NH_3)_3(NO_2)_3$	$[Co(NH_3)_3(NO_2)_3]$
$Co_2(NH_3)_6(NO_2)_6$	$[Co(NH_3)_6][Co(NO_2)_6]$
	$[Co(NH_3)_4(NO_2)_2][Co(NH_3)_2(NO_2)_4]$
$Co_3(NH_3)_9(NO_2)_9$	$[Co(NH_3)_5NO_2][Co(NH_3)_2(NO_2)_4]_2$
$Co_4(NH_3)_{12}(NO_2)_{12}$	$[Co(NH_3)_6][Co(NH_3)_2(NO_2)_4]_3$
	$[Co(NH_3)_4(NO_2)_2]_3[Co(NO_2)_6]$
$Co_5(NH_3)_{15}(NO_2)_{15}$	$[Co(NH_3)_5NO_2]_3[Co(NO_2)_6]_2$

Inspection of the results listed in Table 17.8 indicates that all of the possible isomeric species have not been characterized for a given molecular formula.

6 / STABILITY OF COMPLEX SPECIES

Stability is, unfortunately, a term which has acquired several loosely defined meanings. In our discussion of complex species we consider two aspects of stability—kinetic stability and thermodynamic stability. Kinetic stability refers to the rate and mechanism of a particular reaction of complex ions, e.g., substitution, isomerization, and racemization. It is apparent, for example, that kinetic stability is necessary if a mixture of optical isomers is to be resolved. Thermodynamic stability, on the other hand, is directly related to metal–ligand bond energies. Thus, a complex ion such as $Co(NH_3)_6^{3+}$ might be thermodynamically unstable with respect to another such as $Co(H_2O)_6^{3+}$, but because of kinetic factors the transformation of one into the other could be an exceedingly slow process. At the other extreme, a thermodynamically stable complex ion could still exchange ligands rapidly with the surrounding medium; such exchange processes, of course, cannot lead to thermodynamically unstable species.

6.1 / Thermodynamic Properties

Ideally, the energetics of coordinate-covalent bond formation between a metal M of unspecified charge and the ligand L could be obtained by measuring the enthalpy for the process

$$M_{(g)} + nL_{(g)} \rightleftharpoons ML_{n(g)}. \qquad \Delta H \qquad (5)$$

It also might be of interest to establish the enthalpy for the process corresponding to sequential formation of the coordinate bonds in a complex

$$M_{(g)} + L_{(g)} \rightleftharpoons ML_{(g)} \qquad \Delta H_1 \qquad (6)$$

$$ML_{(g)} + L_{(g)} \rightleftharpoons ML_{2(g)} \qquad \Delta H_2 \qquad (7)$$

$$ML_{2(g)} + L_{(g)} \rightleftharpoons ML_{3(g)} \qquad \Delta H_3 \qquad (8)$$

$$\vdots \qquad \vdots \qquad \vdots \qquad \vdots$$

$$ML_{(n-1)(g)} + L_{(g)} \rightleftharpoons ML_{n(g)} \qquad \Delta H_n. \qquad (9)$$

Unfortunately the necessary data for establishing these quantities, directly or indirectly, are not generally available, so that the results of other, less ideal experiments must be used. The enthalpies of reactions **6–9** have been determined in solution, the solvent most often used being water

$$[M(H_2O)_n]_{(aq)} + L_{(aq)} \rightleftharpoons [M(H_2O)_{(n-1)}L]_{aq} + H_2O \qquad \Delta H_1 \quad (10)$$

$$[M(H_2O)_{(n-1)}L]_{(aq)} + L_{(aq)} \rightleftharpoons [M(H_2O)_{(n-2)}L_2]_{(aq)} + H_2O \qquad \Delta H_2 \quad (11)$$

$$[M(H_2O)_{(n-2)}L_2]_{(aq)} + L_{(aq)} \rightleftharpoons [M(H_2O)_{(n-3)}L_3]_{(aq)} + H_2O \qquad \Delta H_3 \quad (12)$$

$$\vdots \qquad\qquad \vdots \qquad\qquad \vdots \qquad\qquad\qquad \vdots \qquad \vdots$$

$$[M(H_2O)L_{(n-1)}]_{(aq)} + L_{(aq)} \rightleftharpoons [ML_n]_{(aq)} \qquad\qquad + H_2O \qquad \Delta H_4. \quad (13)$$

Thus, the enthalpies for reactions **10–13** correspond to the differences in the bond energies of the $M-OH_2$ and $M-L$ bonds; they also include the enthalpies of solvation of the complex species and of the ligands, which in some instances can be estimated.

In practical terms, the enthalpies of the processes involving the stepwise addition of ligands to a transition-metal ion [Eqs. (10)–(13)] are determined by measuring the equilibrium constants for these processes, which are written in the conventional way:

$$K_1 = \frac{[M(H_2O)_{(n-1)}L]}{[M(H_2O)_n][L]} \qquad (14)$$

$$K_2 = \frac{[M(H_2O)_{(n-2)}L_2]}{[M(H_2O)_{(n-1)}L][L]} \qquad (15)$$

$$K_3 = \frac{[M(H_2O)_{(n-3)}L_3]}{[M(H_2O)_{(n-2)}L_2][L]} \qquad (16)$$

$$\vdots \qquad\qquad \vdots$$

$$K_4 = \frac{[ML_n]}{[M(H_2O)L_{(n-1)}][L]} \qquad (17)$$

The individual equilibrium are called the stepwise formation (or stability) constants. The overall formation constant, β_n, corresponds to the process

$$M(H_2O)_n + nL \rightleftharpoons ML_n + nH_2O, \qquad (18)$$

and is simply the product of the stepwise formation constants

$$\beta_n = K_1 K_2 K_3 \cdots K_n = \prod_{i=1}^{i=n} K_i. \qquad (19)$$

The overall equilibrium constant for any sequence of consecutive equilibria is related to the total standard free-energy change for these steps by the usual relationship

$$\Delta G^0 = -2.303\, RT \log \beta_n. \tag{20}$$

Since the standard free-energy change is related to the standard enthalpy (ΔH^0) and entropy (ΔS^0) changes for the process by

$$\Delta G^0 = \Delta H^0 - T\Delta S^0, \tag{21}$$

all the important thermodynamic quantities can be obtained from the measurement of β_n at several temperatures using

$$\Delta S^0 - \Delta H^0 \left[\frac{1}{T}\right] = 2.303\, R \log \beta. \tag{22}$$

In practice the concentration of the complex or the free ligand is determined in mixtures containing known total analytical concentrations of the metal ion and the ligand. The analytical methods for determining the concentrations of free ligands or the metal in any of its complexed forms depends upon the system. In some instances it is possible to use spectroscopic techniques if the system exhibits a spectrum containing analytically useful absorption bands. Electrometric methods, i.e., polarography or potentiometry, have been used in other cases. Distribution methods are based upon studying the change in apparent distribution coefficient between water and an immiscible phase of one of the species present in the system (usually the ligand) as the composition of the system is changed. In all of these measurements care must be taken that the activity coefficients of the species in question remain constant. Since there is usually a relatively large change in concentration of ionic species during the course of these experiments, the measurements are made at constant ionic strength using appropriate salts containing noncomplexing or weakly complexing anions. Alkali-metal perchlorates are often used for this purpose. A detailed description of the experimental and calculational methods used for studying equilibria involving complex ions in solution is available (XIII) as is a summary of the results of many such experiments (XII).

A graphical example of the results of such experiments is given in Fig. 17.8 for the copper–ammonia system. As the concentration of the ligand increases the proportion of copper in the various complex ions increases to a maximum and then decreases. That is, certain complexes predominate within a given ligand concentration range, but the ranges for each complex overlap. Only in the case of very low or very high ligand concentrations is a single species present.

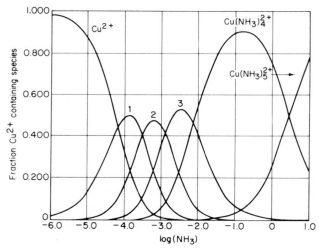

FIG. 17.8 The complex ions present in aqueous solutions containing various proportions of Cu^{2+} and NH_3. The number of NH_3 molecules in the complex ion is indicated on the three center curves.

The interpretation of the nature of complex ions present in solutions is generally a difficult task. For example, it is important to know the coordination numbers of the complex ions being studied so that the stoichiometry of the overall reaction, and hence the number of stepwise constants to be determined, can be established. As is the case with the chemistry of the representative elements, a knowledge of the structure of a complex compound in the solid state is no guarantee that it will have the same structure in solution. Indeed, there are established cases, e.g., Co^{2+}, where the coordination number of the metal ion changes with the nature of the ligand or with the temperature of the system. An additional difficulty arises in the study of complex ions in solution because the hydrated species are also acidic

$$M(H_2O)_n + H_2O \rightarrow H_3O^+ + M(H_2O)_{(n-1)}(OH)^-. \qquad (23)$$

As shown by the data in Table 17.9, the acidities of such ions vary widely. In some cases the concentration of deprotonated species is sufficiently small that it is unimportant in the analysis of the system; in other instances it cannot be ignored.

Available data on the stability of complex ions in aqueous solution reveal several trends. The magnitude of stability constants for divalent ions of the first transition series with oxygen or nitrogen ligands decreases in the order

$$Mn^{2+} < Fe^{2+} < Co^{2+} < Ni^{2+} < Cu^{2+} < Zn^{2+}$$

| r^{2+}, Å | 0.91 | 0.83 | 0.82 | 0.78 | 0.69 | 0.74 |

TABLE 17.9 Acid dissociation
constants of some complexes

Complex	pK_a
$Co(NH_3)_6{}^{3+}$	> 14
$Pt(en)_2Cl_2{}^{2+}$	11.0
$Pt(NH_3)_5Cl^{3+}$	8.4
$Pt(NH_3)_6{}^{4+}$	7.2
$[Co(NH_3)_5H_2O]^{3+}$	5.7
$[Al(OH_2)_6]^{3+}$	5.0
$[Cr(OH_2)_6]^{3+}$	3.9
$[Fe(OH_2)_6]^{3+}$	2.2
$[Pt(NH_3)_4(H_2O)_2]^{4+}$	2.0

which parallels the trends in the radii of the $2+$ ions (6). This relationship is also observed in the lanthanide series, where the stability constants for a given ligand, as shown in Table 17.10, are virtually invariant (XII). In complexes of the same ligand with a metal in two different oxidation states,

TABLE 17.10 Stability constants (25°) for
complexes of the lanthanides with various ligands

M^{3+}	r^{3+}, Å	Log $\beta_3{}^a$			
		NTA	HEDTA	EDTA	DCTA
Y	0.88	11.18	14.65	18.09	19.41
La	1.061	10.36	13.46	15.50	16.35
Ce	1.034	10.83	14.11	15.98	—
Pr	1.013	11.07	14.61	16.40	17.23
Nd	0.995	11.26	11.86	16.61	17.69
Pm	0.979	—	—	—	—
Sm	0.964	11.53	15.28	17.14	18.63
Eu	0.950	11.52	15.35	17.35	18.77
Gd	0.938	11.54	15.22	17.37	18.80
Tb	0.923	11.59	15.32	17.93	19.30
Dy	0.908	11.74	15.30	18.30	19.69
Ho	0.894	11.90	15.32	—	18.89
Er	0.881	12.03	15.42	18.85	20.20
Tm	0.869	12.22	15.59	19.32	20.46
Yb	0.859	12.40	15.88	19.51	20.80
Lu	0.848	12.49	15.88	19.83	20.91

^{a} NTA = nitrilotriacetic acid; HEDTA = N-hydroxyethylenediaminetriacetic acid; EDTA = ethylenediaminetetraacetic acid; DCTA = 1,2-diaminocyclohexanetetraacetic acid.

those formed with the higher oxidation state are more stable; an example of this effect is shown by the data given in Table 17.11. This observation is, of course, predictable on the basis of simple electrostatic considerations.

TABLE 17.11   Stability constants
for cobalt in two oxidation states

	CN^{-a}	$NH_3{}^a$	en^b
Co^{2+}	19.09	4.90	13.82
Co^{3+}	64	33.66	48.69

$^a \beta_6$.
$^b \beta_3$.

Chelating ligands such as ethylenediamine exhibit a markedly greater stability than ligands such as ammonia containing the same ligand atom. This observation can be readily understood in terms of the thermodynamic factors which determine the value of the equilibrium constant [Eq. (22)]. Large values of β are favored by large negative enthalpy and positive entropy changes. Although we might expect an enthalpy difference in the N—M bond strength between NH_3 and ethylenediamine, the magnitude should not be so large as to give the experimentally observed difference in $\log \beta$ for these two ligands (Table 17.12). An analysis of such systems shows that the increased stability arises predominantly from an increase in the entropy during the

TABLE 17.12   Stability constants for some ammine complexes

Metal complex^a	Mn^{2+}	Fe^{2+}	Co^{2+}	Ni^{2+}	Cu^{2+}	Zn^{2+}	Cd^{2+}
			$\log \beta$				
$M(NH_3)^b$	—	ca 3.7	5.3	7.8	12.6	9.1	6.9
$M(en)_2{}^b$	4.9	7.7	10.9	14.5	20.2	11.2	10.3
$M(trien)^b$	4.9	7.8	11.0	14.1	20.5	12.1	10.0
$M(tren)^b$	5.8	8.8	12.8	14.0	18.8	14.6	12.3
$M(dien)_2{}^c$	7.0	10.4	14.1	18.9	21.3	14.4	13.8
$M(penten)^c$	9.4	11.2	15.8	19.3	22.4	16.2	16.8

^{a} en = $(NH_2)_2C_2H_4$; trien = $NH_2(C_2H_4)NH(C_2H_4)NH(C_2H_4)NH_2$; tren = $(NH_2C_2H_4)_3N$; dien = $(NH_2C_2H_4)_2NH$; penten = $(NH_2C_2H_4)_2N(C_2H_4)N(C_2H_4NH_2)_2$.
^{b} Four-coordinate.
^{c} Six-coordinate.

reaction. Qualitatively this result can be understood in terms of the usual equivalence between entropy and the degree of disorder in a system. In the case of the formation of a complex species containing monodentate ligands in aqueous solution, as in the reaction

$$Ni(H_2O)_6^{2+} + 6NH_3 \rightleftharpoons Ni(NH_3)_6^{2+} + 6H_2O, \tag{24}$$

an equivalent number of water molecules are displaced by the incoming ligands. Since there is no net increase in the total number of molecules in the system, we would expect only a small entropy change during this process. The results show that generally the entropy change for reactions involving monodentate ligands is of the order of 2 eu. However, in chelation reactions

$$Ni(H_2O)_6^{2+} + 3en \rightleftharpoons Ni(en)_3^{2+} + 6H_2O \tag{25}$$

each ligand displaces two water molecules, leading to a net increase in the number of molecules in the system, which suggests a considerably larger entropy change. Reactions involving the formation of complexes with polydentate ligands have entropy changes between 25 and 60 eu. The general validity of this argument is supported by the data in Table 17.12, which show an increase in the stability of transition-metal complexes with an increase in the number of complexing atoms in the chelate.

In general there is a difference in the relative stability of the complexes formed by ligands containing elements in the second period (N, O, F) and those of the corresponding elements in the same family. Two extreme cases can be recognized (XIV). The class a metals are those forming their most stable complexes with ligands containing atoms which are the first members in their family; for these metals the general order of stability of analogous complexes is N > P, O > S, and F > Cl. The class b metals are those which form more stable complexes with ligands in Period 3 or in subsequent periods than with ligands in the second period. The class b elements are those predominantly found in a triangular area near the center of the periodic table:

$$Cu^{1+}$$

$$Rh^{3+} \quad Pd^{2+} \quad Ag^{1+} \quad Cd^{2+}$$

$$Ir^{3+} \quad Pt^{2+} \quad Au^{1+} \quad Hg^{2+} \quad Tl^{1+}, Tl^{3+} \quad Pb^{2+}.$$

The class a metals include most of the ions with empty or completely filled $(n-1)d$ orbitals; thus, the cations of the representative elements and transition-metal ions with $(n-1)d^{1-3}$ configurations are in class a. The basis of classification for the metal ions into these two groups is related to the distinction made between hard and soft acids and bases, as described in

Chapter 10. In this context, class *a* metals are hard acids and class *b* metals soft acids.

6.2 / Kinetic Stability

Although the rates of ligand replacement in complex ions vary widely, complex ions fall into two broad classifications. Inert complexes are those for which reactions occur relatively slowly so that the rate can be studied by conventional methods. Labile complexes undergo reactions much more rapidly. In this respect the distinction between inertness and thermodynamic stability must be clearly made. Thermodynamic stability implies inertness, but the converse is not necessarily true. For example, $Co(NH_3)_6^{3+}$ is thermodynamically unstable with respect to $Co(NH_3)_5(H_2O)^{3+}$, yet the former ion persists in aqueous solution for extended periods of time because the reaction

$$Co(NH_3)_6^{3+} + H_2O \rightarrow Co(NH_3)_5(H_2O)^{3+} + NH_3 \qquad (26)$$

is extremely slow; that is, there is a high kinetic barrier for this process.

6.3 / Experimental Methods (*XV*)

The mechanisms of chemical reactions are elucidated by experimentally determining the rate law for the reactions in question. For example, the reaction between two compounds

$$aA + bB \rightarrow \text{products} \qquad (27)$$

might represent an experimentally determined rate law of the form

$$\text{Rate} = k[A]^m[B]^n, \qquad (28)$$

where the square brackets represent the molar concentration of the indicated species, k is the specific rate constant, and the numerical values of m and n are experimentally determined. The values of m and n need not necessarily be integral and may bear no relationship to the factors a and b in Eq. (27). The overall order of the reaction is the sum $m + n$; the orders with respect to A and B are m and n, respectively. It is important to note that the order of the reaction is not predictable from the balanced chemical equation corresponding to the reaction. In general the rate law need not be as simple as described in Eq. (28); it may contain a series of additive terms

$$\text{Rate} = k_1[A]^l + k_2[A]^m[B]^n. \qquad (29)$$

Experimentally, it is found that specific rate constants vary with temperature in the same general way as do equilibrium constants. Thus, plots of $\log k$ vs $1/T$ are often linear relationships. A formal analogy exists between the relationship of the equilibrium constant to the thermodynamic functions of state and the specific rate constant to the energy (ΔE^{+}), enthalpy (ΔH^{+}), and entropy (ΔS^{+}) of activation as determined from the temperature dependence of the specific rate constant. However, it must be stressed that ΔE^{+}, ΔS^{+}, and ΔH^{+} are not thermodynamic functions.

Experimentally the form of the rate law for a given reaction is determined by following the change in concentration of reactants (or in some cases, the products) as a function of time. Generally these methods fall into three broad categories. The so-called "static methods" involve conventional analytical techniques and are applicable to reactions with a half-life of more than a minute. For example, the rate of replacement of Cl^{-} by water in the complex ion $[Co(NH_3)_5Cl]^{2+}$

$$[Co(NH_3)_5Cl]^{2+} + H_2O \rightarrow [Co(NH_3)_5(H_2O)]^{3+} + Cl^{-} \tag{30}$$

is sufficiently slow ($k = 1 \times 10^{-5}$ at 25°) that it can be followed by analytically determining Cl^{-} in aliquots taken of the reaction mixture using Ag^{+} to precipitate AgCl. Instrumental methods of analysis are also useful in some instances. Thus, since $[Co(NH_3)_5Cl]^{2+}$ has a different visible spectrum from that of the product in Eq. (30), the rate of this process can be followed using spectroscopic measurements. Conductivity methods would also prove useful for following the rate of the process in Eq. (30).

Stop-flow methods (7) involving rapid mixing techniques can be used to follow reactions which have a half-life greater than 10^{-3} seconds. Such techniques necessarily require that a change in some physical property (light absorption, conductivity, etc.) be monitored; ordinary chemical analytical methods are too slow to be useful. Finally, relaxation methods (8) have been developed which will measure rates of reaction to the limit set by diffusion processes ($k = \sim 10^{10}$ sec^{-1}). Relaxation methods involve following the rate of return of a system, which is initially at equilibrium, after it has been disturbed momentarily by creating a pressure or temperature jump. In some cases, nmr methods can be used to determine the rates of certain reactions.

6.4 / Mechanisms of Reactions

The mechanism of the reaction is the series of stepwise reactions which are consistent with the experimentally observed rate law. That is, the steps in the postulated mechanism must lead to a rate expression which is mathematically the same as the experimental rate law. Usually it is not possible to extract an unambiguous mechanism from the experimental data. As an example of the difficulty in elucidating a mechanism for a reaction, let us consider one possible interpretation for the reaction

$$A + B \rightarrow C + D \tag{31}$$

which appears to be similar to the substitution reaction given in Eq. (30). We might consider that the mechanism involves an intermediate species X

$$A + B \xrightarrow{k_1} X \tag{32}$$

$$X \xrightarrow{k_2} A + B \tag{33}$$

$$X \xrightarrow{k_3} C + D \tag{34}$$

sometimes called the activated complex, which can then react in either of two ways. One path leads back to the formation of the original reactants [Eq. (33)], whereas the other gives the final products [Eq. (34)]. Each step proceeds at a characteristic rate, defined by an appropriate specific rate constant. A detailed analysis of this system of equations leads to the overall rate expression

$$\text{Rate} = \frac{k_1 k_3 [A] [B]}{k_2 + k_3}. \tag{35}$$

However, since the specific rate constants are unknown and are to be extracted from the experimental data, Eq. (35) becomes

$$\text{Rate} = k'[A][B], \tag{36}$$

where $k' = k_1 k_3/(k_2 + k_3)$. In this form it is apparent that Eq. (36) also represents a mechanism which assumes that reaction (31) occurs in one step. Moreover, if one of the reactants, e.g., B, is present in large excess, so that its concentration does not change appreciably during the course of the reaction, Eq. (36) becomes

$$\text{Rate} = k''[A], \tag{37}$$

which is effectively a first-order reaction. Such situations often occur for complex ions when solvent molecules displace coordinated ligands, as in Eq. (30).

A variety of factors may influence the rates at which complex ions react. The nature of the central metal ion and the charge on the complex are important considerations which derive from the electronic configuration of the transition metal. In addition, the nature of the ligands leaving, entering, or not participating—but within the coordination sphere—affects the rate of the reaction. Finally, environmental effects such as the nature of the solvent and the presence of other ligands or metal ions in the solvent affect reaction rates.

In spite of the general difficulties associated with the interpretation of kinetic data, progress has been made in interpreting the processes by which complex ions react in solution. These processes can be generally classified as substitution, racemization, and electron-transfer reactions. In addition, the reactions of the ligands when they are coordinated are of interest.

6.5 / Substitution Reactions (*XVI*)

Two limiting mechanisms have been suggested for ligand substitution reactions

$$[ML_n] + X \rightarrow [ML_{(n-1)}X] + L \tag{38}$$

which parallel those commonly accepted for the reactions of organic compounds. In one case the complex species undergoes dissociation

$$[ML_n] \underset{k_2}{\overset{k_1}{\rightleftharpoons}} [ML_{(n-1)}] + L \tag{39}$$
$$\textbf{58}$$

to form an intermediate with a lower coordination number (**58**) which then reacts with the substituting ligand to form the product

$$[ML_{(n-1)}] + X \overset{k_3}{\longrightarrow} [ML_{(n-1)}X]. \tag{40}$$
$$\textbf{58}$$

This mechanism is designated S_N1 (substitution, nucleophilic, first order). The rate law for an S_N1 mechanism is given by

$$\text{Rate} = \frac{k_1 k_3 [ML_n][X]}{k_2[L] + k_3[X]}, \tag{41}$$

which, if the rate of the second step is much greater than that of the first $(k_3 \gg k_2)$, reduces to

$$\text{Rate} = k_1[ML_n]. \tag{42}$$

The other general mechanism involves a slow bimolecular process

$$[ML_n] + X \overset{k_1}{\longrightarrow} [ML_nX] \tag{43}$$

to form an intermediate with a higher coordination number than the original complex, which then rapidly yields the product

$$[ML_nX] \overset{k_2}{\longrightarrow} [ML_{(n-1)}X] + L. \tag{44}$$

This is described as an S_N2 (substitution, nucleophilic, second order) mechanism and satisfies the rate law

$$\text{Rate} = k_1[ML_n][X] \tag{45}$$

if $k_2 \gg k_1$. Relatively few substitution processes involving complex ions occur by purely S_N1 or S_N2 mechanisms, but it is usual to discuss the results in these terms as a matter of convenience.

Substitution reactions may be conveniently classified in terms of the nature of the displacing ligand, which is most often a solvent molecule; such processes are important in aqueous solutions, in which case they are called hydrolysis reactions. Ionic ligands may displace solvent molecules from the solvation sphere; such reactions are called anation processes. Finally, it might be possible for anions to displace other anions in complex species. Such processes do not occur directly but usually involve a hydrolysis reaction followed by anation.

By far the most kinetic data available have been collected for octahedral and planar complexes in aqueous solution.

Octahedral Complexes (*XVII*)   For hydrolysis reactions [Eq. (38), $X = H_2O$], the limiting rate laws for an S_N1 mechanism [Eq. (42)] and an S_N2 mechanism [Eq. (45)] assume the same form

$$\text{Rate} \approx k[ML_n], \tag{46}$$

since the concentration of the incoming ligand is essentially constant in the latter case. At the other extreme, the rate law for the S_N1 mechanism would experimentally appear to be second order if $k_2[L] \gg k_3[X]$

$$\text{Rate} \approx \frac{k_1 k_3}{k_2[L]}[ML_n][X] \approx k'[ML_n][X]. \tag{47}$$

This situation could occur by compensation of not too dissimilar rates for the first and second steps by the relatively large and essentially constant concentration of water as the incoming ligand. Thus, it might not be easily possible to distinguish the mechanism of this reaction on the basis of the experimental rate law.

Table 17.13 contains a selection of experimental results obtained for the hydrolysis of some octahedral complexes. Although the experimental rate laws do not give an unambiguous assignment to the mechanism of these reactions, ancillary information supports the suggestion that these hydrolysis reactions generally follow the S_N1 mechanism. For example, the rates of acid hydrolysis reactions generally follow the order of thermodynamic stability of the complex ions.

TABLE 17.13   Acid hydrolysis rate constants (k, sec^{-1}) for the complex $M(NH_3)_5X$ (M = Co and Cr)

M	NO$_3^-$	I$^-$	Br$^-$	Cl$^-$
		X		
Cr	—	$2.4 \times 10^{-4\,a}$	5×10^{-5}	1.0×10^{-5}
Co	2.7×10^{-5}	10^{-5}	6.3×10^{-6}	1.7×10^{-6}

M	NH$_3$	NSC$^-$	H$_2$O	SO$_4^{2-}$
		X		
Cr	2.5×10^{-5}	1×10^{-7}	$10^{-6\,b}$	—
Co	very slow	3×10^{-9}	7×10^{-6}	10^{-6}

^{a} 0°C.
^{b} 40°C.

The rates at which hydrolysis of metal complexes occur in basic solutions provide some insight into the mechanism of these reactions. The experimental rate law for ligands containing protons, e.g., NH_3, is of the form

$$\text{Rate} = k[M(NH_3)_5L][OH^-] \tag{48}$$

and the data in Table 17.14 show that the rates for base hydrolysis are

TABLE 17.14   Acid and base hydrolysis rate constants (25°C) for several complex ions

Complex	k	
	Acid, sec^{-1}	Base, l mole^{-1} sec^{-1}
cis-[Cr(en)$_2$Cl$_2$]$^+$	3.3×10^{-4}	2.8×10^{-2}
trans-[Cr(en)$_2$Cl$_2$]$^+$	3.8×10^{-5}	3.7×10^{-2}
[Fe(phen)$_3$]$^{2+}$	7.2×10^{-5}	1.0×10^{-2}
trans-[Co(en)$_2$F$_2$]$^+$	10^{-5}	63
cis-[Co(en)$_2$Cl$_2$]$^+$	3.5×10^{-4}	10^3
trans-[Co(en)$_2$Cl$_2$]$^+$	3.9×10^{-5}	3×10^3

markedly higher than for acid hydrolysis (Table 17.14). The suggested mechanism for this rate law involves rapid establishment of an equilibrium between the metal complex and its conjugate acid

$$[M(NH_3)_5X]^{n+} + OH^- \xrightleftharpoons{\text{fast}} [M(NH_3)_4(NH_2)X]^{(n-1)+} + H_2O, \quad (49)$$

followed by a slow dissociation step

$$[M(NH_3)_4(NH_2)X]^{(n-1)+} \xrightarrow{\text{slow}} [M(NH_3)_4(NH_2)]^{n+} + X^-, \quad (50)$$

and a rapid addition of a solvent molecule

$$[M(NH_3)_4(NH_2)]^{n+} + H_2O \xrightarrow{\text{fast}} [M(NH_3)_5(OH)]^{n+}. \quad (51)$$

The last two steps are equivalent to those involved in a normal S_N1 mechanism; accordingly the basic hydrolysis process is called S_N1CB (substitution nucleophilic, first order, conjugate base). The experimental rate law is the same as would be expected for an S_N2 reaction involving the hydroxide ion; accordingly, on the basis of this information alone it is impossible to distinguish between the two mechanisms. There are other pieces of evidence which favor the S_N1CB mechanism. As expected, the rate of base hydrolysis of complex ions which contain no easily deprotonated ligands is essentially the same as that for acid hydrolysis. The difference in the rates of acid and base hydrolysis is now interpreted in terms of a greater ability of the leaving ligand in the deprotonated species. A study of the stereochemical changes which occur during base hydrolysis strongly indicates that the general mechanism outlined in Eqs. (49)–(51) is essentially correct.

When water is the ligand substituted in aqueous solutions the process is called ligand exchange (*XVIII*). That is, water molecules in the coordination sphere change places with water that exists in the bulk solvent. Such processes are very fast and must be studied by relaxation methods. In general the rate of exchange increases with the size of the metal ion which is also in the order of the $M-OH_2$ bond strength. Thus, it appears that the primary factor involved in the exchange process is the breaking of the $M-OH_2$ bond which suggests that the S_N1 mechanism is the best description of the process.

Anation reactions essentially follow the same process as described for water exchange. The formation of a five-coordinate intermediate by loss of a water molecule

$$ML_5(H_2O) \rightarrow ML_5 + H_2O \quad (52)$$

appears to be the first step in the anation reaction. The intermediate can either recombine with another water molecule

$$ML_5 + H_2O \rightarrow ML_5(H_2O), \quad (53)$$

which corresponds to the slow step in the mechanism of hydrolysis, or with a ligand

$$ML_5 + X \rightarrow ML_5X. \tag{54}$$

A number of difficulties arise in the study of anation reactions. Such reactions are generally slower than water exchange, which is a competing process. The establishment of an ion-pair equilibrium between the free anions and complex cations which affects the rate law also confuses the interpretation of the data for anation processes. The latter problem has been eliminated by studying the anation of $[Co(CN)_5(H_2O)]^{2-}$ with various anions (9); in this case the mechanism is consistent with Eqs. (52) and (53).

The mechanism for the substitution of one ligand for another

$$ML_5X + Y \rightarrow ML_5Y + X \tag{55}$$

involves the formation of an aquo complex

$$ML_5X + H_2O \rightarrow ML_5(H_2O) + X \tag{56}$$

which occurs by the general hydrolysis process [Eqs. (49)–(51)]. The aquo complex formed then reacts with the incoming ligand by way of the same intermediate, as is suggested for the water-exchange process. Several observations support this two-step process. The rates of reaction of a given ion ML_5X with different ligands show very little dependence upon the nature of the ligand. This observation suggests that all these reactions go through the same intermediate, that is, the aquo complex [cf. Eq. (56)]. In addition the rates for all ions are essentially the same and are nearly equal to the rate observed for water exchange.

Planar Complexes

The largest amount of information available for this class of species exists for Pt^{2+} complexes, although some data also exist for Ni^{2+}, Pd^{2+}, Au^{3+}, Rh^{1+}, and Ir^{1+}. In the case of the reactions of square-planar platinum complexes there is a retention of configuration when substitution reactions occur. That is, *cis* isomers lead to *cis* products, and *trans* isomers to *trans* products. This observation suggests that a dissociative process is not involved in the formation of the transition state. The substitution of square-planar complexes

$$ML_4 + X \rightarrow ML_3X + L \tag{57}$$

follows a modified bimolecular rate law

$$\text{Rate} = k_1[ML_4] + k_2[ML_4][X]. \tag{58}$$

Substitution reactions of square-planar complexes have been interpreted in terms of a two-path process, both paths involving a trigonal-bipyramidal intermediate and an S_N2 process. The existence of a higher-coordinate transition state is suggested by evidence that five-coordinate complexes containing platinum [e.g., $M_2[Pt(SnCl_3)_5]$] and other elements that are normally square-planar have been identified. In one path the incoming ligand is a water molecule

$$ML_4 + H_2O \xrightarrow{k_1} ML_4(H_2O) \tag{59}$$

whereas the substituting ligand is incoming in the second path

$$ML_4 + X \xrightarrow{k_2} ML_4X. \tag{60}$$

Experimentally the constant k_1 is relatively insensitive to changes in the net charge on the complex as shown by the data in Table 17.15.

TABLE 17.15   Effect of charge on k_1 at 20°C

Complex	$k_1 \times 10^{+5}, \sec^{-1}$ [a]
$PtCl_4{}^{2-}$	3.9
$Pt(NH_3)Cl_3{}^-$	0.62[b]
$Pt(NH_3)Cl_3{}^-$	5.6[c]
cis-$Pt(NH_3)_2Cl_2$	2.5
trans-$Pt(NH_3)_2Cl_2$	9.8
$Pt(NH_3)_3Cl^+$	2.6

[a] For the reaction $Pt(NH_3)_nCl_{(4-n)} + H_2O \rightarrow Pt(NH_3)_n(H_2O)Cl_{(3-n)} + Cl^-$.
[b] For replacement of Cl^- trans to NH_3.
[c] For replacement of Cl^- cis to NH_3.

An interesting phenomenon associated with the reactivity of square-planar complexes called the trans-effect has been described (XIX). The trans-effect can be defined in several ways, but for our purposes we shall be concerned with the effect of a coordinated ligand upon the rate of substitution of ligands trans to it in a metal complex. Extensive qualitative observations on the products observed during the synthesis of planar complexes which provide information on the trans-labilizing ability of ligands are supported by a variety of quantitative kinetic data. The order of decreasing trans-effect for some common ligands is CO, CN^-, $C_2H > PR_3$, $H^- > CH_3{}^-$ $SC(NH_2)_2 > C_6H_5$, $NO_2{}^-$, I^-, $SCN^- > Br^-$, $Cl^- > py$, NH_3, OH^-, H_2O. The kinetic range given in this series is about 10^6. That is, ligands with the largest trans-effects (CO, CN^-) labilize groups trans to them to the extent that the latter are replaced

$$[PtCl_3L]^{n+} + Y \rightarrow trans\text{-}[PtLCl_2Y]^{(n+1)} + Cl^- \tag{61}$$

$\sim 10^6$ times as fast as is the case with ligands with the smallest *trans*-effects (NH_3, H_2O).

6.6 / Racemization of Optical Isomers (*XX*)

It will be recalled that octahedral complexes of the type *cis*-$M(\widehat{aa})_2X_2$ and $M(\widehat{aa})_3$ can exist in optically active forms. In the case of the racemization of *cis*-$M(\widehat{aa})_2X_2$ two processes can occur: hydrolysis of the monodentate ligands, which probably occurs by an S_N1 mechanism, and loss of optical activity. The data in Table 17.16 show that *cis* isomers react with the formation of *cis*-aquo products, whereas the *trans* isomers give a mixture of *trans* and *cis* products with the proportions dependent upon the nature of the complex. Similar results are observed for anation reactions on the *cis* isomer. Such reactions presumably occur by mechanisms described previously for ordinary substitution reactions. For the most part, the aquation reactions of cobalt complexes occur with retention of configuration. That is, the $(+)$-$Co(en)_2Cl_2^+$ enantiomer gives rise to a hydrolysis product with the same optical characteristics, $(+)$-$Co(en)_2(H_2O)Cl^+$. On the other hand, one of the enantiomers of *cis*-$Co(en)_2(H_2O)_2^{3+}$ undergoes isomerism to the *trans* form which establishes an equilibrium with the other enantiomer

$$(+)\text{-}cis\text{-}Co(en)_2(H_2O)_2^{3+} \rightleftharpoons trans\text{-}Co(en)_2(H_2O)_2^{3+} \tag{62}$$

$$\rightleftharpoons (-)\text{-}cis\text{-}Co(en)_2(H_2O)_2^{3+}.$$

TABLE 17.16 Products formed in substitution reactions of some geometrical complexes of the type $M(en)_2X_2$

Reactant	Reagent	Product	Per cent *cis* isomer
cis-$Co(en)_2Cl_2^+$	H_2O	$Co(en)_2(Cl)(H_2O)^{2+}$	100
trans-$Co(en)_2Cl_2^+$	H_2O	$Co(en)_2(Cl)(H_2O)^{2+}$	35
cis-$Co(en)_2ClBr^+$	H_2O	$Co(en)_2(Br)(H_2O)^{2+}$	100
trans-$Co(en)_2ClBr^+$	H_2O	$Co(en)_2(Br)(H_2O)^{2+}$	50
cis-$Cr(en)_2(NCS)Cl^+$	H_2O	$Co(en)_2(Cl)(H_2)^{2+}$	98
trans-$Pt(en)_2Cl_2^{2+}$	Br^-	$Pt(en)_2(Br)_2^{2+}$	0
trans-$Pt(en)_2BrCl^{2+}$	py	$Pt(en)_2(Br)(py)^{3+}$	0
cis-$Rh(en)_2Cl_2^+$	OH^-	$Rh(en)_2(OH)_2^+$	100
trans-$Rh(en)_2Cl_2^+$	OH^-	$Rh(en)_2(OH)_2^+$	0

The rate of isomerization is, within experimental error, the same as the rate of racemization. A similar relationship is observed between the rates of isomerization and racemization of $cis\text{-}Cr(C_2O_4)_2(H_2O)_2{}^-$ in the pH range 3–7. The hydrolysis of cis complexes very often leads to observations that are difficult to unravel experimentally. For example, the rate of loss of optical activity of a methanol solution of $cis\text{-}Co(en)_2Cl_2{}^+$ is equal to the rate of exchange of radioactive chlorine with one chloro group $(10, 11)$ suggesting that the complex dissociates to a symmetrical five-coordinate intermediate which leads directly to loss of optical activity. Methanol is apparently a poor competitor for coordination compared with the chloride ion; when the chloride ion reenters the coordination sphere it does so to form 70% $trans$ isomer and 30% cis isomer, which is racemic. The rate of loss of optical activity is the same in the presence of other ligands forming $Co(en)_2ClX^+$ $(X = Br^-,\ NCS^-,\ NO_3{}^-)$ which suggests that the five-coordinate intermediate is a common precursor. However, competing processes occur in the presence of water because it is a better ligand than methanol. In the hydrolysis reaction of $cis\text{-}Co(en)_2Cl_2{}^+$ the product is predominantly $cis\text{-}Co(en)_2(H_2O)Cl^{2+}$ with a retention of configuration, as shown in Table 17.16, but this product can now undergo isomerization to the $trans$ form

$$(+)\text{-}cis\text{-}Co(en)_2Cl_2{}^+ \rightarrow (+)\text{-}cis\text{-}(en)_2(H_2O)Cl^{2+}$$

$$\rightarrow trans\text{-}Co(en)_2(H_2O)Cl^{2+}. \tag{63}$$

Racemization of $(+)\text{-}cis\text{-}Co(en)_2(H_2O)Cl^{2+}$ is about 25 times faster than replacement of the remaining chloride ion; it is suggested that the process involves loss of coordinated water forming a five-coordinate intermediate, $Co(en)_2Cl^{2+}$, which can yield a $trans$ aquo complex [Eq. (63)], the original cis configuration, or the enantiomorph. Similar results are obtained for aqueous solutions of $cis\text{-}Co(en)_2XCl^{n+}$ $(X = NH_3,\ NO_2{}^-,\ NCS^-),\ (+)\text{-}cis\text{-}Co(en)_2F_2{}^+$, and $(+)\text{-}cis\text{-}Cr(en)_2Cl_2{}^+$.

The racemization of octahedral complexes containing three chelate ligands has also been widely studied. Two types of mechanisms appear to be operative in these cases. An intermolecular mechanism is consistent with the data available for the racemization of $Ni(\widehat{aa})_3{}^{2+}$ ($\widehat{aa}$ = o-phenylenediamine or bipyridyl). The rate of exchange of bound ligand with free ligand in the solvent is the same as the rate of racemization, indicating that the basic mechanism involves the steps

$$(+)\text{-}Ni(\widehat{aa})_3{}^{2+} \underset{+\,\widehat{aa}}{\overset{-\,\widehat{aa}}{\rightleftharpoons}} Ni(\widehat{aa})_2{}^{2+}$$

$$\underset{-\,\widehat{aa}}{\overset{+\,\widehat{aa}}{\rightleftharpoons}} (-)\text{-}Ni(\widehat{aa})_3{}^{2+}. \tag{64}$$

The other possible mechanism for racemization involves an intramolecular process in which one end of a bidentate ligand becomes detached leading to either a five-coordinate intermediate (**59**) or a six-coordinate intermediate containing a solvent molecule (**60**). In such processes the rates of racemization

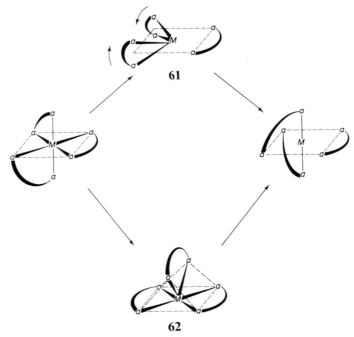

59   **60**

are greater than the rates of exchange of the chelate ligand. The complex ions $M(C_2O_4)_3{}^{3-}$ (M = Co^{3+}, Cr^{3+}) and $M(acac)_3{}^{3+}$ (M = Co^{3+}, Cr^{3+}) racemize by this mechanism. The two other possible processes which could lead to intramolecular racemization (given in Fig. 17.9) involve a twisting of the original structure. In one case (Fig. 17.9, **61**) there is a concerted exchange of two pairs of ligand positions by a rhombic twist. The equivalent effect can

61

62

FIG. 17.9 Two possible routes for the racemization of an octahedral complex containing three bidentate ligands.

be obtained by twisting the structure to form a trigonal prismatic inter-mediate (Fig. 17.9, **62**). Although there has been much interest in the race-mization of octahedral complexes, the nature of the intermediate in intra-molecular processes is still unknown.

6.7 / Reactions of Coordinated Ligands (*XXI*)

Very often ligands which are coordinated to transition-metal ions undergo reactions which are not observed for the free ligand. That is, the coordination process leads to an enhancement of certain characteristics exhibited by ligand species. It is convenient to discuss such reactions in terms of conven-tional types of reactions.

Protonation Reactions   One of the most common reactions of co-ordinated ligands involves proton transfer. For example, coordinated water molecules are more acidic than the free ligands

$$M(H_2O)_n^{m+} + H_2O \rightarrow M(H_2O)_{(n-1)}(OH)^{(m-1)+} + H_3O^+. \quad (65)$$

The data in Table 17.17 indicate that the aquo complexes of many transition and nontransition metal ions are often extensively hydrolyzed in aqueous

**TABLE 17.17   Hydrolysis constants
of some aquo complexes at 25°C**

Ion[a]	pK_a
Ca^{2+}	12.6
Sc^{3+}	5.1
Ti^{3+}	1.1
V^{3+}	2.8
Cr^{3+}	3.8
Mn^{2+}	10.6
Fe^{2+}	9.5
Fe^{3+}	2.2
Co^{2+}	8.9
Co^{3+}	0.7
Ni^{2+}	10.6
Cu^{2+}	6.8
Zn^{2+}	8.8
$Co(NH_3)_5(H_2O)^{3+}$	6.6
$cis\text{-}Co(NH_3)_4(H_2O)_2^{3+}$	6.0
$Ru(NH_3)_5(H_2O)^{3+}$	4.2
$Pt(NH_3)_5(H_2O)^{4+}$	strong acid

[a] Where no other ligands are present the data for the aquo complexes are given.

solution. As might be expected, complexes of the highest oxidation state of a given element are most extensively hydrolyzed. Where data are available, it appears that the first member of a family is a weaker acid than the corresponding complexes of the heavier members of the family. In certain species hydrolysis reactions are the first steps in olation processes, i.e., reactions in which hydroxo complexes lose water to form polynuclear oxo-bridged species

$$2M(H_2O)_{(n-1)}(OH)^{(m-1)+} \longrightarrow$$

$$[(H_2O)_{(n-1)}M{-}O{-}M(H_2O)_{(n-1)}]^{2(m-1)+} + H_2O. \quad (66)$$

The latter may act as a base, becoming protonated to form a hydroxy-bridged complex

$$[(H_2O)_{(n-1)}M{-}O{-}M(H_2O)_{(n-1)}]^{2(m-1)} + H_3O^+ \longrightarrow$$

$$\overset{\displaystyle H}{\underset{\displaystyle |}{}}$$
$$[(H_2O)_{(n-1)}M{-}O{-}M(H_2O)_{(n-1)}]^{2m-1} + H_2O. \quad (67)$$

Thus, the evaluation of the extent of hydrolysis reactions [Eq. (65)] from experimental data often becomes difficult because of the existence of competing acid–base reactions [Eqs. (66) and (67)] (*12–14*). Ammine complexes can also act as acids under suitable conditions

$$M(NH_2R)_n^{m+} \longrightarrow [M(NHR)(NH_2R)_{(n-1)}]^{(m-1)+} + H^+. \quad (68)$$

The ammine complexes of the first series of *d*-block elements are generally very weak acids in aqueous solution; however, their acidity is more noticeable in strongly basic solvents such as liquid ammonia, as shown by the data collected in Table 17.18. Generally the ammine complexes are weaker than the corresponding aquo complexes. Extensive deprotonation of osmium, iridium, rhodium, and platinum ethylenediamine complexes has been

TABLE 17.18 Ionization of some ammine complexes at 25°C

Ion	pK_a
$Co(NH_3)_6^{3+}$	14
$Co(en)_3^{3+}$	12.5 (7[a])
$Rh(NH_3)_6^{3+}$	14
$Pt(NH_3)_6^{4+}$	7.2
$Pt(en)_3^{4+}$	5.5
$Os(en)_3^{4+}$	strong acid
$Au(en)_2^{3+}$	6.5

[a] Determined in liquid ammonia $-75°C$.

achieved in liquid ammonia using KNH_2 (*15*), e.g.,

$$Pt(en)_2I_2 + KNH_2 \rightarrow [Pt(en - H)(en)]I* + NH_3 + KI \quad (69)$$

$$[Pt(en - H)(en)]I + KNH_2 \rightarrow Pt(en - H)_2 + NH_3 + KI \quad (70)$$

$$Pt(en - H)_2 + KNH_2 \rightarrow K[Pt(en - H)(en - 2H)] + NH_3. \quad (71)$$

Many of the intermediate species have been isolated and characterized, and some of these reactions are reversible upon the addition of NH_4^+.

Substitution Reactions   In some instances coordinated ligands have been shown to undergo substitution reactions. For example, the cyclopentadienyl rings in ferrocene, $(C_5H_5)_2Fe$, one of the more widely studied organometallic derivatives, can be substituted in much the same way as any conventional aromatic system. Metal acetylacetonates, which incorporate the enol form of acetylacetone, appear to contain an aromatic-type heterocyclic ring (**63**). This suggestion is supported by the fact that under certain

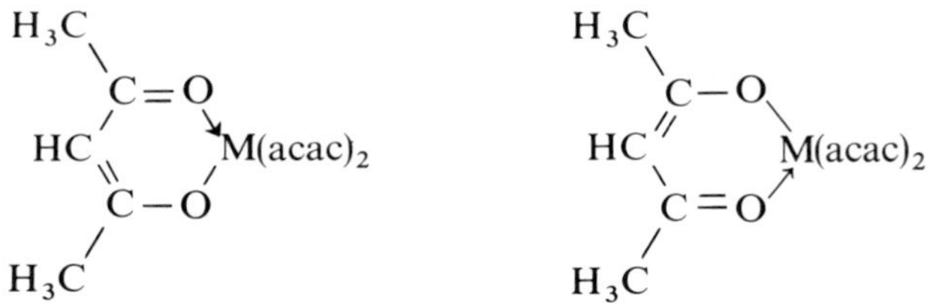

63

conditions the unique C—H bond in the hetero ring can be easily substituted

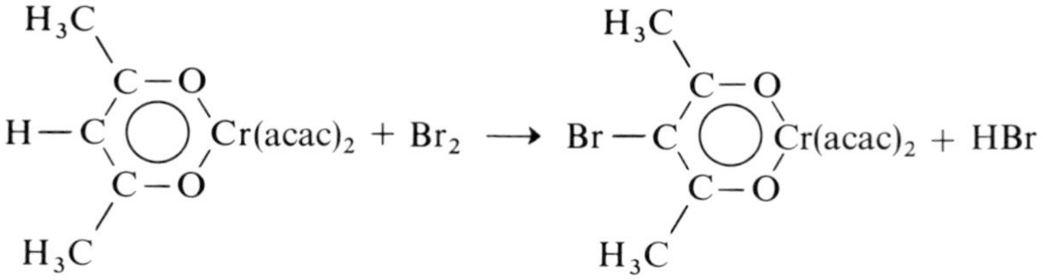

$$(72)$$

An indirect substitution of the N—H bond in some complexes has been achieved using the deprotonated intermediates [cf. Eq. (68)] which when treated with alkyl iodides yield the corresponding *N*-alkyl derivatives

$$[Pt(en - H)(en)]I + CH_3I \rightarrow [Pt(NH_2-C_2H_4-NHCH_3)(en)]I_2. \quad (73)$$

---

* The symbol (en − H) represents a molecule of ethylenediamine which has lost a proton.

Addition Reactions   Addition reactions of complexed ligands can occur if the ligand contains a point of unsaturation. For example, sulfur can readily expand its coordination number, and ligands containing this atom can undergo addition reactions which stem from this fact

$$L_nMSR + BrR \rightarrow [L_nMSR_2]Br. \tag{74}$$

An interesting example of such reactions involves the stepwise formation of macrocyclic ligands (16, 17)

$$\tag{75}$$

A type of addition reaction involves the insertion of a saturated molecule into a transition-metal–ligand bond

$$L_nM-X + Z \rightleftharpoons L_nM-ZX. \tag{76}$$

Reactions such as this have been often observed with metal–carbon bonds

$$L_2M(CO)_4{}^+ + OR^- \underset{HX}{\overset{KOH}{\rightleftharpoons}} L_2M(CO)_3-\overset{\displaystyle O}{\overset{\displaystyle \|}{C}}-OR \tag{77}$$

$$M = Mn, Re; L = P(C_6H_5)_3$$

$$(CO)_5MnCH_3 + CO \rightarrow (CO)_5Mn-\overset{\displaystyle O}{\overset{\displaystyle \|}{C}}-CH_3 \tag{78}$$

$$(C_5H_5)Fe(CO)_2R + SO_2 \rightarrow (C_5H_5)Fe(CO)_2SO_2R \tag{79}$$

$$HCo(CO)_4 + C_2H_4 \rightarrow (C_2H_5)Co(CO)_4; \tag{80}$$

under certain conditions these reactions are reversible [Eq. (77)].

Insertion reactions are believed to be critical steps in some catalytic reactions (XXII) that are commercially important, such as the oxo process ($18, 19$)

$$R_2C{=}CR_2 + CO + H_2 \xrightarrow{Co^{2+}} R_2CHCR_2{-}\overset{\overset{\displaystyle O}{\|}}{C}{-}H, \tag{81}$$

the Wacker process ($20, 21$)

$$C_2H_4 + \tfrac{1}{2}O_2 \xrightarrow[Cu^{2+}]{PdCl_4{}^{2-}} CH_3{-}CHO, \tag{82}$$

the hydration of acetylenes (22)

$$RC{\equiv}CR' + H_2O \xrightarrow{Hg^{2+}} R{-}\overset{\overset{\displaystyle O}{\|}}{C}{-}CH_2R' \tag{83}$$

and the polymerization of olefins using the Ziegler–Natta-type catalysts ($23–25$)

$$nH_2C{=}CH_2 \xrightarrow[Al_2(C_2H_5)_6]{TiCl_4} {-}({-}CH_2CH_2{-})_n{-}. \tag{84}$$

Oxidation–Reduction Reactions (XXIII)   Oxidation–reduction reactions are common processes observed for transition-metal species. Since it is known that most of these ions are strongly hydrated in solution to form complex ions, an interesting question concerning the nature of the redox mechanism arises. We know, for example, that it is possible to oxidize Fe^{2+} in aqueous solution with appropriate oxidizing agents, the process usually being described by

$$Fe^{2+} \rightarrow Fe^{3+} + e^-, \tag{85}$$

which is only a general description of what happens, not how it happens. Solvated species must be involved in such processes

$$M(H_2O)_n{}^{m+} \rightarrow M(H_2O)_{n'}{}^{(m+1)} + e^-, \tag{86}$$

the change in oxidation state occurring by electron transfer to the oxidizing agent through the coordination sphere of the complex ion which is oxidized. However, the equivalent process should occur by other paths. For example, the transfer of a hydrogen *atom* from coordinated water to the oxidizing agent

$$[M(H_2O)_n]^{m+} + ox \rightarrow [M(H_2O)_{(n-1)}(OH)]^{m+} + H{\cdot}ox \tag{87}$$

also corresponds to oxidation since the remaining hydroxyl radical can be converted into a coordinated hydroxide ion by electron transfer from the metal. Thus, the problem of studying redox mechanisms of coordinated metal ions can become quite involved.

Electron-transfer reactions can be conveniently classified into two groups depending upon their mechanisms. In the outer-sphere mechanism the reactants maintain their coordination spheres, and electron transfer occurs through the ligand shells by a tunneling process. The inner-sphere mechanism involves an activated complex with a ligand bridge between the metal atoms.

The inner-sphere mechanism has been demonstrated for many reactions, most involving $Co^{3+}(NH_3)_5L$ as the oxidizing agent, where L is a ligand capable of acting as a bridge and $M^{2+}_{(aq)}$ (M = Cr, V, Ru, Fe) is the reducing agent

$$M^{2+}_{(aq)} + Co^{3+}(NH_3)_5L + 5H^+ + 5H_2O$$

$$\rightarrow M^{3+}(H_2O)_5L + Co^{2+}_{(aq)} + 5NH_4{}^+. \quad (88)$$

The products always contain the ligand bound to the oxidized form of M. The second-order rate constants for several inner-sphere reactions involving the Cr^{2+}/Co^{3+} couple for which an inner-sphere mechanism is reasonably unambiguous are listed in Table 17.19. Both Cr^{3+} and Co^{2+} form inert

TABLE 17.19 Second-order rate constants at 25°C for some inner-sphere redox reactions involving the $Cr^{2+}_{(aq)}/Co^{3+}(NH_3)_5L$ couple

L	k, $l\,mole^{-1}\,sec^{-1}$
H_2O	0.5
OH^-	1.5×10^6
F^-	2.5×10^5
Cl^-	6×10^5
Br^-	1.4×10^6
I^-	3×10^6
$N_3{}^-$	3×10^5
NCS^-	19
$SO_4{}^{2-}$	18
$PO_4{}^{3-}$	4.8×10^9

complexes but $Co^{2+}_{(aq)}$ and $Cr^{2+}_{(aq)}$ are labile. These results suggest that the activated complex for the process in Eq. (88) involves the bridged species

$$[(NH_3)_5Co\cdots L\cdots Cr(H_2O)_5]^{4+},$$

64

reduction occurring because the neutral ligand is transferred from Co^{3+} to Cr^{2+}. Since Co^{2+} species are labile the cobalt-containing product formed from **64** rapidly exchanges with the solvent to form $Co^{2+}_{(aq)}$. The other product, $Cr^{3+}(H_2O)_5L$, is inert. Some interesting correlations of electron-transfer rates for reactions involving bridging organic ligands

$$Co(NH_3)_5L + Cr^{2+}_{(aq)} + 5H^+ \overset{H_2O}{\rightarrow} Co^{2+}_{(aq)} + Cr^{3+}(H_2O)_5L + 5NH_4^+ \quad (89)$$

have been reported; the data are collected in Table 17.20. Complexes containing the aliphatic carboxylates generally undergo electron transfer at a

TABLE 17.20   Second-order rate constants
at 25°C for some inner-sphere redox reactions
involving the $Cr^{2+}_{(aq)}/Co^{3+}(NH_3)_5L$ couple

L	k, l mole^{-1} sec^{-1}
acetate, $CH_3CO_2^-$	0.15
butyrate, $C_3H_7CO_2^-$	0.08
H-succinate, $^-O_2C-C_2H_4-CO_2H$	0.27
H-maleate, $^-O_2C-CH=CH-CO_2H$	180
H-oxalate, $^-O_2C-CO_2H$	400

markedly lower rate than those incorporating maleate or oxalate ligands (**65** and **66**). The latter are highly conjugated systems and probably serve as more efficient conductors of electrons than aliphatic ligands.

65   **66**

The electron exchange which occurs between inert Cr^{3+} and labile $Cr^{2+}_{(aq)}$

$$Cr(H_2O)_5L^{2+} + Cr^{*2+}_{(aq)} \rightarrow Cr^*(H_2O)_5L^{2+} + Cr^{2+}_{(aq)} \quad (90)$$

$$L = F, Cl, Br$$

also involves a bridged complex. Reaction (90) apparently involves no net change, in contrast to reaction (88), however, the process can be followed using radioactive chromium as a tracer.

Electron-exchange reactions which occur by way of the outer-sphere mechanism also have been observed in systems in which no apparent net chemical change occurs

$$Fe^*(CN)_6^{3-} + Fe(CN)_6^{4-} \rightleftharpoons Fe^*(CN)_6^{4-} + Fe(CN)_6^{3-}, \qquad (91)$$

but kinetic data for reactions involving two different metals are also available, and are summarized in Table 17.21. Generally the outer-sphere mechanism is operative when both reactant molecules exchange their ligands slowly,

**TABLE 17.21   Second-order rate constants at 25°C
for some redox reactions which occur by
the outer-sphere mechanism**

Reaction^a	k, l mole^{-1} sec^{-1}
$MnO_4^- + MnO_4^{2-}$	3.6×10^3
$Fe(CN)_6^{4-} + Fe(CN)_6^{3-}$	7.4×10^2
$Fe(CN)_6^{4-} + Fe(phen)_3^{3+}$	10^8
$Fe(phen)_3^{2+} + Fe(phen)_3^{3+}$	10^5
$Fe(bipy)_3^{2+} + Fe(phen)_3^{3+}$	10^8
$Os(bipy)_3^{2+} + Mo(CN)_8^{3-}$	2×10^9
$Os(bipy)_3^{2+} + Os(bipy)_3^{3+}$	5×10^4
$Ru(phen)_3^{2+} + RhCl_6^{2-}$	2.5×10^9
$Fe(CN)_6^{4-} + IrCl_6^{2-}$	3.8×10^5

^{a} phen = o-phenanthroline; bipy = 2,2'-bipyridyl.

but the observed rates of electron transfer are relatively large. Although the details are not known, the tunneling process is related to the fact that the wave function which describes an atomic orbital has a finite value at large distances from the nucleus, so that electron transfer between two nuclei can occur at distances that are considerably larger than those corresponding to the collision of the species. In solution, the energy barrier through which the electron must tunnel is considerably more altered than would be expected in the gas phase because of the intervention of solvent molecules. However, ligands which contain structural features such as unsaturated centers or single-atom ligands that facilitate the movement of electrons to the surface of the complex should give redox systems for which electron transfer is very rapid. In this respect it is interesting to compare the data in Table 17.21 on such complexes with the rate of electron exchange for the couples $Co(en)_3^{2+}/$ $Co(en)_3^{3+}$ ($k_2 = 5 \times 10^{-5}$ liter mole^{-1} sec^{-1}) and $Co(NH_3)_6^{2+}/Co(NH_3)_6^{3+}$ ($k_2 < 10^{-8}$ liter mole^{-1} sec^{-1}).

An interesting system involving reduction of coordinated ligands which has received attention recently involves molecular nitrogen (XXIV). Treatment of $RuCl_3$ or $K_2[RuCl_5(H_2O)]$ with hydrazine leads to the complex ion $[Ru(NH_3)_5N_2]^{2+}$ containing nitrogen in two oxidation states (26, 27). A variety of complexes containing molecular nitrogen and transition metals such as iron, osmium, cobalt, rhodium, iridium, and nickel have been prepared. In some instances, these complexes can be prepared directly from molecular nitrogen

$$[Ru(NH_3)_5(H_2O)]^{2+} + N_2 \rightarrow [Ru(NH_3)_5N_2]^{2+} + H_2O. \qquad (92)$$

Complexes containing molecular nitrogen are of interest because they serve as models for biological systems which fix atmospheric nitrogen. Although little success has been reported in the reduction of complexed N_2 to ammonia (28), atmospheric nitrogen has been reduced in the presence of transition-metal derivatives (29–32).

REFERENCES

1. R. B. King, P. N. Kapoor, and R. N. Kapoor, *Inorg. Chem.*, **10**, 1841 (1971).
2. W. H. Mills and T. H. H. Quibell, *J. Chem. Soc.*, 839 (1935).
3. R. Tsuchida, M. Kobayashi and A. Nakamura, *J. Chem. Soc.* (*Japan*), **56**, 1339 (1935).
4. R. Tsuchida, M. Kobayashi, and A. Nakamura, *Bull. Chem. Soc. Japan*, **11**, 38 (1936).
5. H. B. Jonassen, J. C. Bailar, Jr., and E. H. Huffman, *J. Amer. Chem. Soc.*, **70**, 756 (1948).
6. H. Irving and R. J. P. Williams, *J. Chem. Soc.*, 3192 (1953).
7. Q. H. Gibson, *Disc. Faraday Soc.*, **17**, 137 (1954).
8. M. Eigen, *Pure Appl. Chem.*, **6**, 97 (1963).
9. A. Haim and W. K. Wilmarth, *Inorg. Chem.*, **1**, 573 (1962).
10. D. D. Brown and C. K. Ingold, *J. Chem. Soc.*, 2680 (1953).
11. D. D. Brown and R. S. Nyholm, *J. Chem. Soc.*, 2696 (1953).
12. L. G. Sillén, *Acta Chem. Scand.*, **8**, 299, 318 (1954).
13. L. G. Sillén, *Acta Chem. Scand.*, **16**, 159 (1962).
14. N. Ingri and L. G. Sillén, *Acta Chem. Scand.*, **16**, 173 (1962).
15. G. W. Watt, P. W. Alexander, and B. S. Manhas, *J. Amer. Chem. Soc.*, **89**, 1483 (1967), and references therein.
16. D. H. Busch, D. C. Jicha, M. C. Thompson, J. W. Wrathall, and E. Blinn, *J. Amer. Chem. Soc.*, **86**, 3642 (1964).
17. M. C. Thompson and D. H. Busch, *J. Amer. Chem. Soc.*, **86**, 3651 (1964).
18. M. Johnson, *J. Chem. Soc.*, 4859 (1963).
19. R. W. Goetz and M. Orchin, *J. Amer. Chem. Soc.*, **85**, 1549 (1963).
20. J. Smidt, W. Hafner, J. Jura, R. Sieber, J. Sedlmeier, and A. Sabel, *Angew. Chem.* (*Intl. Ed.*), **1**, 80 (1962).
21. A. Aguilo, *Advan. Organometal. Chem.*, **5**, 321 (1967).
22. M. Miocque, N. M. Hung, and V. Q. Yen, *Ann. Chim.*, **8**, 157 (1963).
23. J. K. Stille, *Chem. Revs.*, **58**, 541 (1958).
24. A. D. Ketley and F. X. Worber, *Science*, **145**, 667 (1954).
25. F. Dawans and P. Teyssie, *Bull. Soc. Chim. France*, **10**, 2376 (1963).
26. D. E. Harrison and H. Taube, *J. Amer. Chem. Soc.*, **89**, 5706 (1967).
27. D. E. Harrison, E. Weissberger, H. Taube, *Science*, **159**, 320 (1968).

28. Yu. G. Borod'ko, A. K. Shilova, and A. E. Shilov, *Russ. J. Phys. Chem.*, **44**, 248 (1970).
29. M. E. Vol'pin and V. B. Shur, *Dokl. Akad. Nauk SSSR*, **156**, 1102 (1964).
30. G. Henrici-Olive and S. Olive, *Angew. Chem. (Intl. Ed.)*, **6**, 73 (1967).
31. H. Brintzinger, *J. Amer. Chem. Soc.*, **88**, 4305, 4307 (1966).
32. E. E. van Tamelen, G. Boche, S. W. Ela, and R. B. Fechter, *J. Amer. Chem. Soc.*, **89**, 5707 (1967).

COLLATERAL READINGS

I. S. J. Lippard, *Prog. Inorg. Chem.*, **8**, 109 (1967). A review concerned with eight-coordinate transition-metal complexes.

II. R. S. Nyholm and M. L. Tobe, *Advan. Inorg. Chem. Radiochem.*, **5**, 1 (1963). A discussion of the effect of environment upon the stability of oxidation states of transition metals.

III. R. S. Nyholm and M. L. Tobe in *Essays on Coordination Chemistry* (W. Schneider, ed.), Birkhauser Verlag, Basel, 1964; E. L. Meutterties and R. A. Shunn, *Quart. Revs.*, **20**, 245 (1966); L. Sacconi, *Pure Applied Chem.*, **17**, 95 (1968). These reviews contain considerable information on 5-coordinate complexes of transition metals.

IV. G. Booth, *Advan. Inorg. Chem. Radiochem.*, **6**, 1 (1964). An extensive discussion of the complexes of the transition metals with phosphines, arsines, and stibines.

V. Th. Kruck, *Angew. Chem. (Intl. Ed.)*, **6**, 53 (1967). A survey of the transition-metal complexes formed by PF_3.

VI. B. M. Chadwick and A. G. Sharpe, *Advan. Inorg. Chem. Radiochem.*, **8**, 84 (1966); L. Malatesta, *Prog. Inorg. Chem.*, **1**, 283 (1959). These two reviews contain extensive information on transition-metal complexes with CN^-. The first deals with C-bonded complexes whereas the second deals with N-bonded systems.

VII. C. Oldham, *Prog. Inorg. Chem.*, **10**, 223 (1968). A portion of this review is a discussion of the transition-metal complexes formed by simple carboxylic acids.

VIII. C. C. Addison and D. Sutton, *Prog. Inorg. Chem.*, **8**, 195 (1967). A survey of transition-metal complexes containing the nitrate ion.

IX. J. P. Fackler, Jr., *Prog. Inorg. Chem.*, **7**, 361 (1966). A portion of this review deals with transition-metal complexes formed by β-diketones.

X. F. A. Cotton, *Quart. Revs.*, **20**, 389 (1966); M. C. Baird, *Prog. Inorg. Chem.*, **9**, 1 (1968). These two reviews contain information on transition-metal compounds containing clusters of metal atoms.

XI. T. Moeller, D. F. Martin, L. C. Thomson, R. Ferrus, G. R. Feistal and W. J. Randall, *Chem. Revs.*, **65**, 1 (1965). A review of the coordination chemistry of yttrium and the rare-earth-metal ions.

XII. L. G. Sillén and A. E. Martell, *Stability Constants of Metal–Ion Complexes*, The Chemical Society, London 1964; L. G. Sillén and A. E. Martell, *Stability Constants of Metal–Ion Complexes, Supplement I*, The Chemical Society, London, 1971. These two volumes contain comprehensive listings of stability constant data.

XIII. F. J. C. Rossotti and H. Rossotti, *The Determination of Stability Constants*, McGraw-Hill, New York, 1961. A comprehensive treatise on the methods used to establish stability constants for complexes in solution.

XIV. S. Ahrland, J. Chatt, and N. Davis, *Quart. Revs.*, **12**, 265 (1958). A discussion of the relative affinities of ligand atoms for acceptor molecules and ions. The classification of atoms into class *a* and *b* acceptors is developed here.

XV. E. Caldin, *Fast Reactions in Solution*, Wiley, New York, 1964; A. A. Frost and R. G. Peason, *Kinetics and Mechanisms*, Wiley, New York, 1961; I. Amdur and G. G. Hammes, *Chemical Kinetics*, McGraw-Hill, New York, 1966. Each of these volumes contains extensive information on the theory and techniques used to investigate solution kinetics.

XVI. F. Basolo and R. G. Pearson, *Advan. Inorg. Chem. Radiochem.*, **3**, 1 (1965); F. Basolo, *Pure Appl. Chem.*, **17**, 37 (1968). These reviews contain surveys of the mechanisms of substitution reactions of metal complexes.

XVII. M. L. Tobe, *Acc. Chem. Res.*, **3**, 377 (1970). A survey of the possible mechanisms by which base hydrolysis of octahedral complexes may occur.

XVIII. R. G. Pearson and M. M. Anderson, *Angew. Chem. (Intl. Ed.)*, **4**, 281 (1965). A review on the exchange rates of ligands in complex ions.

XIX. F. Basolo and R. G. Pearson, *Prog. Inorg. Chem.*, **4**, 381 (1962); J. M. Pratt and R. G. Thorp, *Advan. Inorg. Chem. Radiochem.*, **12**, 375 (1969). These reviews contain surveys of the *trans*-effect in a variety of systems.

XX. J. C. Bailar, Jr., *Pure Appl. Chem.*, **16**, 91 (1966). A discussion of optical inversions which occur in the reactions of cobalt complexes.

XXI. Q. Fernando, *Advan. Inorg. Chem. Radiochem.*, **7**, 185 (1965); Yu. N. Kukushkin, *Russ. Chem. Revs.*, **39**, 169 (1970). These references contain extensive information on the reactions undergone by coordinated ligands.

XXII. G. N. Schrauzer, *Angew. Chem. (Intl. Ed.)*, **3**, 185 (1964). A survey of the role of coordination chemistry in catalytic processes.

XXIII. H. Taube, *Advan. Inorg. Chem. Radiochem.*, **1**, 1 (1959); R. T. M. Fraser, *Pure Appl. Chem.*, **11**, 64 (1961); J. Halpern, *Quart. Rev.*, **15**, 207 (1961); A. G. Sykes, *Essays Chem.*, **1**, 25 (1970). These reviews contain detailed information on the mechanisms of electron-transfer reactions.

XXIV. A. D. Allen and F. Bottomley, *Acc. Chem. Res.*, **1**, 360 (1968); Yu. G. Borod'ko and A. E. Shilov, *Russ. Chem. Revs.*, **38**, 355 (1969). Surveys of the coordination chemistry of molecular nitrogen.

STUDY QUESTIONS

1. Give the structural formulas for all the possible isomers for the species with the following empirical formulas indicating those isomers that are optically active: (a) $Co(en)_2ClCO_3$; (b) $[Cr(en)(NH_3)_2BrCl]Cl$; (c) $PtCl_2(H_2O)_2$; (d) $[Cr(H_2O)_3(OH)Cl_2](NO_3)_2$; (e) $[Co(en)_2(H_2O)Br][CoCl_4]$; (f) $Ir(en)_3Cl_3(H_2O)$; (g) $Pd(en)_2CO_3Cl_2$.

2. A complex of chromium has the *empirical* formula $[Cr(SO_4)(OH)(NH_3)_2]$. It is found to exist in six isomeric forms, four of which can be resolved into *d, l* pairs. What are the configurations of all of the isomers? Show both the *d* and *l* forms of the optically active isomers.

3. Sketch the structure of ferric acetylacetonate. What is the expected paramagnetism of this compound?

4. How many isomers are possible for an octahedral complex with the formulation $M(abc_2d_2)$ containing monodentate ligands? Which of these isomers are optically active?

5. The protons present in the complex $[Co(NH_3)_6]Cl_3$ are found to exchange slowly with deuterium when this compound is dissolved in D_2O. Suggest a mechanism for this process.

6. Give the possible structures for the following compounds: $PtCl_4 \cdot 6NH_3$; $PtCl_4 \cdot 5NH_3$; $PtCl_4 \cdot 4NH_3$; $PtCl_4 \cdot 3NH_3$; $PtCl_4 \cdot 2NH_3$. The last compound is a nonelectrolyte.

7. Nickel acetate tetrahydrate $[Ni(C_2H_3O_2)_2 \cdot 4H_2O]$ is a monomeric species which can lose two moles of water to form another monomeric species, and this species in turn can lose another two moles of water to form a third monomeric species. Give the most probable structure(s) for these three nickel acetates indicating any isomers which may be formed.

8. Give examples of and sketch the following species: (a) the *trans* isomer of an octahedral complex; (b) an optically active tetrahedral complex; (c) an optically active octahedral complex containing at least one bidentate ligand; (d) an optically active planar complex.

9. A solid compound with the formula $Co(en)_2Cl_3$ dissolves in water to give a conducting solution. (a) What is the constitution of the species that are possible in a sample of the solid compound? Which, if any, of these species would be optically active? (b) What is the constitution of the species that are possible in solution? Which, if any, of these species would be optically active?

10. Draw the possible isomers for the planar complex $[Pt(NH_3)Cl_2]_2$. Suggest a possible method for distinguishing among these possible species.

11. Explain the observation that the compound $PdCl_2 \cdot 2PF_3$ is much more stable than the compound $PdCl_2 \cdot 2NH_3$ even though the compound $BF_3 \cdot NH_3$ is much more stable than the compound $BF_3 \cdot PF_3$.

12. The compound $Ni(en)_2(CN)_2$ can be formulated as either $[Ni(en)_2(CN)_2]$ or $[Ni(en)_2](CN)_2$. Give the simple crystal-field interpretation of both formulations.

13. Predict the most probable number of unpaired electrons in the following complex compounds: (a) $Cr(CO)_6$; (b) $Zn(NH_3)_4^{2+}$; (c) $Ti(H_2O)_6^+$; (d) $Mn(Cl)_6^{4-}$.

14. Which of the compounds in Question 13 would be expected to be labile and which would be inert?

15. A dark green compound with the empirical formula $Co(NH_3)_3$-$(H_2O)Br_3$ dissolves in water to give a deep green solution which becomes red upon standing. The molar conductivity of the solution increases during this time:

	Molar conductivity
Freshly prepared solution (green)	151
5 min after first measurement	288.0
10 min after first measurement	325.0
15 min after first measurement	340.7
20 min after first measurement	347.8
40 min after first measurement (red)	363.5

It was also observed that one mole of bromine per mole of the green compound, $Co(NH_3)_3(H_2O)Br_3$, could be precipitated as silver bromide per formula weight of the compound. On the basis of the above information answer the following questions. (a) What is the most probable structure of the green compound? Give the structures of all possible isomers (indicating any optical isomers) and their expected magnetic properties. (b) What is the most probable structure of the compound which gives the red solution? Give the structure of all possible isomers (indicating any optical isomers).

16. Although positively charged species are not normally expected to act as good ligands, discuss three positively charged species which might be possible ligands.

17. Discuss the optical activity of the species $Co(en')Cl_3$ where en is N-methylethylenediamine.

18. Suggest a reason why complexes of Fe^{2+} are less stable than the corresponding complexes of both Zn^{2+} and Fe^{3+}.

19. The successive stability constants for the species formed in aqueous solutions when ethylenediamine reacts with Ni^{2+} are as follows: $\log K_1 = 7.5$, $\log K_2 = 6.4$, and $\log K_3 = 4.4$. Calculate β_3. What species are present in the equilibrium described by K_3? Give the geometry of the nickel-containing species involved in the equilibrium described by K_3.

20. Suggest a possible mechanism to account for the fact that isotopic oxygen (O*) does not enter the coordination sphere in the reaction

$$[Co(NH_3)_5CO_3]^+ + 2H_3O^{*+} \rightarrow [Co(NH_3)_5(H_2O)]^{3+} + 2H_2O^* + CO_2$$

21. Which of the following complex ions might be expected to be the most poisonous: $Cr(CN)_6^{3-}$, $Fe(CN)_6^{3-}$, or $Fe(CN)_6^{4-}$?

/ **18**

TRANSITION-METAL COMPLEXES

THEORETICAL ASPECTS

1 / INTRODUCTION

Ever since it was recognized that some transition-metal compounds exhibit a capacity to react with ligands above and beyond that expected on the basis of the normal valence rules, there has been an active interest in the theoretical explanation of these effects. Werner's recognition of the existence of two types of valences—primary and secondary—was an important organizational aspect of coordination chemistry. The consequence of this suggestion was far reaching in terms of stereochemical and structural considerations, even though no adequate theory existed for predicting such valences or the nature of the bonding between ligands and transition-metal atoms. From a modern point of view several theoretical arguments are available which address themselves to the structure of and bonding in transition-metal coordination compounds. A single theory does not exist which encompasses all aspects of this problem; rather several theories, successful in different aspects of coordination chemistry, are available. We describe these theories here in a semihistorical manner because such an exposition contributes to an understanding of the experimental basis upon which they were modified and leads to an appreciation of why certain terms of concepts are still found in the modern literature. Again, we must remember that a theory need not be entirely correct from a fundamental point of view to be useful in correlating the available data.

2 / THE SIDGWICK THEORY

The Lewis interpretation of the nature of the normal covalent bond and his adaptation to coordinate-covalent bond formation in species such as $BF_3 \cdot NH_3$ (**1**) was extended to transition-metal complexes by Sidgwick (*1–3*)

$$
\begin{array}{ccc}
F & & H \\
 \diagdown & & \diagup \\
F \blacktriangleleft B &:& N \blacktriangleleft H \\
 \diagup & & \diagdown \\
F & & H
\end{array}
$$

1

and Lowry (*4*). In this view, the primary valences of Werner were the ordinary electrovalences associated with ionic compounds and, hence, could only be satisfied by an appropriate number of counter ions. Secondary valences, on the other hand, were satisfied by the formation of coordinate-covalent bonds between the transition-metal atom and the ligand which may or may not be charged. Thus, anions can satisfy both primary and secondary valences whereas neutral ligands can satisfy only secondary valence. Coordinate-covalent bonds are sometimes called dative or semipolar bonds.

According to Sidgwick, transition-metal atoms acquire sufficient electrons in forming coordinate-covalent bonds to achieve the next rare-gas configuration. In this respect the Sidgwick model followed one of the postulates of the Lewis theory of covalent bond formation for the representative elements as described in Chapter 1. The effective atomic number (EAN) of a metal atom in a complex ion represents the total number of electrons associated with that ion. In the Sidgwick theory, transition-metal ions form complexes in such a way as to achieve an effective atomic number near that of the next rare gas. The following example illustrates the way in which the effective atomic number of a complex ion, viz., $[Co(NH_3)_6]^{3+}$, is determined.

Species	Number of electrons
Co^{3+}	$(Z - 3) = 27 - 3 = 24$
$6NH_3$ groups	$6 \times 2 = 12$
EAN of cobalt in the complex	36

The effective atomic number of cobalt in this complex is 36, which is the same as the atomic number of the next rare gas, krypton. Thus cobalt, which cannot attain a rare-gas configuration with its primary valence, can achieve this structure by forming six coordinate-covalent bonds. Using the predecessor of the electron-pair-repulsion arguments (5), Sidgwick suggested that the six ligands in complexes of this kind are disposed in an octahedral arrangement about the central metal ion. The effective atomic numbers of the metals in the first transition series of several common complexes appear in Table 18.1. It is apparent from these data that the effective atomic number

TABLE 18.1   Some typical complex compounds of the first *d*-block elements and their effective atomic numbers

$Sc(NH_3)^{3+}$ (30)	$Ti(H_2O)_6^{4+}$ (30)	$V(CO)_6$ (35)	$Cr(CO)_6$ (36)
	$Ti(H_2O)_6^{3+}$ (31)	$V(NH_3)_6^{3+}$ (32)	$Cr(NH_3)_6^{3+}$ (33)
		VF_6^- (30)	
$Mn(CO)_5Cl$ (36)	$Fe(CO)_5$ (36)	$Co(CN)_4^{4-}$ (35)	$Ni(CO)_4$ (36)
$Mn(H_2O)_6^{2+}$ (35)	$FeCl_4^{2-}$ (32)	$CoCl_4^{2-}$ (33)	$Ni(CN)_4^{2-}$ (34)
$MnCl_6^{2-}$ (33)	$Fe(CN)_6^{4-}$ (36)	$Co(NH_3)_6^{2+}$ (37)	$Ni(NCS)_6^{4-}$ (38)
	$Fe(CN)_6^{3-}$ (35)	CoF_6^{3-} (36)	$Ni(diars)_2Cl_2^+$ (37)
$Cu(CN)_4^{3-}$ (36)	$ZnCl_4^{2-}$ (36)		
$CuCl_4^{2-}$ (35)			

of a transition-metal atom in a complex generally is near that of the next rare gas, although there are numerous examples where complex species have markedly lower $[Ti(H_2O)_6^{4+}]$ or higher $[Ni(NCS)_6^{4+}]$ effective atomic numbers. The EAN correlates reasonably well with the fact that the elements at the end of the series possess low coordination numbers for their normal valence states; those at the beginning of the series exhibit the higher coordination numbers required to achieve a rare-gas configuration. In addition, the observation that the higher valence state of an element generally exhibits the higher coordination number is reflected in the EAN of the metal in these complexes. Although at this point in time the EAN concept is of little interest in coordination chemistry, it has found some use in correlating certain aspects of the chemistry of organometallic transition-metal derivatives, as described in Chapter 19.

3 / VALENCE-BOND THEORY

Although the valence-bond theory was not developed specifically for coordination compounds, it has had some success when applied to such systems. It should be recalled from Chapter 4 that the valence-bond method for describing the electron density in a covalent bond involves the use of canonical forms which contribute to the overall wave function of the molecule. Each canonical form corresponds to a certain electron distribution arising from the overlap of atomic or hybrid orbitals. Thus, for the normal covalent bond in the diatomic molecule AB, canonical forms (2–4) might contribute

$$A:B \qquad\qquad A^-:B^+ \qquad\qquad A^+:B^-$$

$$\textbf{2} \qquad\qquad\qquad \textbf{3} \qquad\qquad\qquad \textbf{4}$$

to the wave function for the molecule

$$\Psi_{AB} = C_2\Psi_2 + C_3\Psi_3 + C_4\Psi_4. \tag{1}$$

The valence-bond description of a coordinate-covalent bond is developed in much the same way. For example, the coordinate-covalent bond formed between a Lewis acid and a ligand ($M \leftarrow L$) is described by similar canonical structures, one of which involves the overlap of an empty orbital on the acid with a filled orbital on the ligand. Valence-bond theory was originally applied to the bonding in complex ions in an attempt to describe the nature of the bonds formed, the stereochemistry of the complexes and their magnetic properties.

As is the case with the valence-bond description of the representative elements, atomic orbitals are generally not the most efficient for bonding and hybrid orbitals must be used to provide maximum overlap. The relative bond strengths formed by the common atomic and hybrid orbitals are given in Table 4.3. In Chapter 17 we saw that two coordination numbers—four and six—are commonly observed for complex ions. Moreover, three structural types (square-planar, tetrahedral, octahedral) are commonly found among these species. Although three hybridization schemes are necessary to account for the structures of most of the complex species known, the hybrid orbitals shown in Table 18.2 can be constructed for virtually all of the geometrical arrangements of ligands about the transition-metal atom. It should be recognized that structures with coordination numbers greater than 4, as well as the square-planar structure, require hybrid orbitals incorporating a d component. Fortunately, transition-metal atoms have d orbitals as a part of their valence shell. Thus, the valence-bond description of complex ions involves using empty hybrid orbitals with the appropriate geometry on the metal atom overlapping with filled orbitals on the ligand. In general the

TABLE 18.2   **Hybrid orbitals and their configurations for transition-metal ions with various coordination numbers**

Coordination number	Configuration	
	Orbital	Spatial
2	sp	linear
3	sp^2	trigonal
4	sp^3	tetrahedral
4	$(d_{x^2-y^2})sp^2$	planar
5	$(d_{z^2})sp^3$	trigonal bipyramid
5	$(d_{x^2-y^2})sp^3$	square pyramid
6	d^2sp^3	octahedral

electron distribution for the metal atom in its ground state is not compatible with this requirement and an excited state configuration must be used, just as in the case of most of the compounds of the representative elements (Chapter 4). An outline of the application of valence-bond theory to complex ions is best illustrated using an example, viz., $[Co(H_2O)_6]^{3+}$. The ion Co^{3+} has six $3d$ electrons, and it might be expected to exhibit the configuration shown in **5** in the gaseous state if Hund's rule is obeyed. The ion $[Co(H_2O)_6]^{3+}$

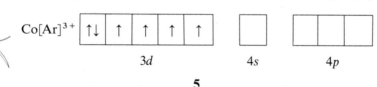

$$Co[Ar]^{3+}$$

$$3d \qquad 4s \qquad 4p$$

5

contains an octahedral arrangement of ligands which require vacant metal hybrid orbitals of the type d^2sp^3, but configuration **5** shows electrons present in the lowest lying d orbitals available, i.e., $3d$. Thus, to create a vacant set of hybrid orbitals with the appropriate symmetry two electrons must be promoted in energy. In this case the electrons can be either paired in the $3d$ shell or placed in an orbital of higher energy than $4p$. Of these two possibilities the former requires less energy, giving the excited configuration shown in **6**. The empty $3d$, $4s$, and $4p$ orbitals are now available to form a d^2sp^3 hybrid

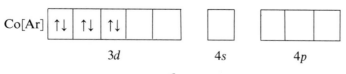

$$Co[Ar]$$

$$3d \qquad 4s \qquad 4p$$

6

which overlaps with the filled ligand orbitals to form six metal-ligand bonds (**7**). Thus, the valence-bond method leads to the conclusion that

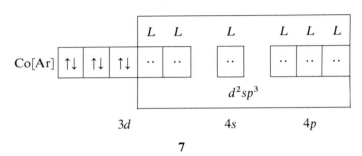

7

$[Co(H_2O)_3]^{3+}$ should be diamagnetic, as it is experimentally observed to be. Other examples of the application of valence-bond arguments to bonding in different complexes appear in Table 18.3.

At this point it might appear that only the nature of the transition-metal ion governs the details of how the valence-bond theory is applied to a given complex species. However it is observed that the same transition metal forms complex ions with different ligands which experimentally possess different magnetic moments. For example, Fe^{3+} forms an octahedral complex with F^- which shows a magnetic moment equivalent to five unpaired electrons, but the octahedral complex of Fe^{3+} formed with CN^- exhibits a magnetic moment equivalent to one unpaired electron. Other examples of this situation appear in Table 18.4. Clearly the ligand must play an important role in determining the nature of the bond in such cases. In the valence-bond theory the difference in magnetic moments in such cases is attributed to different hybrids used in the bonding scheme. Thus, the description of $[Fe(CN)_6]^{3-}$ involves the use of $3d$ orbitals in the d^2sp^3 hybrid (**8**) leading to a distribution of the five metal electrons among three degenerate orbitals; the resulting

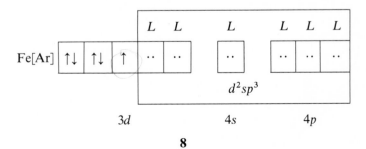

8

electronic configuration corresponds to one electron unpaired. On the other hand, the bonding in $(FeF_6)^{3-}$ is supposed to involve $4d$ orbitals in the octahedral hybrid (**9**) which leaves all of the lower lying $3d$ orbitals available

TABLE 18.3 Examples of valence-bond descriptions of some common coordination species

Species	Geometry	Valence-bond electron description	Magnetic moment (Bohr magnetons) Calculated[a]	Experimental

Species	Geometry	Valence-bond electron description ($3d$ — $4s$ — $4p$)	Calculated[a]	Experimental
$Cr(H_2O)_6{}^{3+}$	octahedral	$3d$: ↑ ↑ ↑ · · · · $4s$: · · $4p$: · · · · · · — d^2sp^3	3.87	3.7
$Fe(CN)_6{}^{4-}$	octahedral	$3d$: ↑↓ ↑↓ ↑↓ · · · · $4s$: · · $4p$: · · · · · · — d^2sp^3	0	0
$Ni(CN)_4{}^{2-}$	square planar	$3d$: ↑↓ ↑↓ ↑↓ ↑↓ · · $4s$: · · $4p$: · · · · — dsp^2	0	0
$Cu(NH_3)_4{}^{2+}$	square planar	$3d$: ↑↓ ↑↓ ↑↓ ↑↓ · · $4s$: · · $4p$: · · · · ↑ — dsp^2	1.73	1.82
$Ni(NH_3)_4{}^{2+}$	tetrahedral	$3d$: ↑↓ ↑↓ ↑↓ ↑ ↑ $4s$: · · $4p$: · · · · · · — sp^3	2.83	2.63
$MnCl_4{}^{2-}$	tetrahedral	$3d$: ↑ ↑ ↑ ↑ ↑ $4s$: · · $4p$: · · · · · · — sp^3	5.92	5.87
$ZnCl_4{}^{2-}$	tetrahedral	$3d$: ↑↓ ↑↓ ↑↓ ↑↓ ↑↓ $4s$: · · $4p$: · · · · · · — sp^3	0	0

[a] Calculated from the spin-only formula $\mu = \sqrt{n(n + 2)}$ where n is the number of unpaired electrons.

TABLE 18.4   Observed magnetic moments in some transition-metal complexes

Metal ion	Number of d electrons	Ligand	Structure	Magnetic moments	
				Calculated[a]	Observed
Fe^{2+}	4	CN^-	octahedral	4.90	5.25
Fe^{2+}	4	H_2O	octahedral	4.90	0
Co^{3+}	6	F^-	octahedral	0	4.96
Co^{3+}	6	NH_3	octahedral	0	0

[a] $\mu = \sqrt{n(n + 2)}$ where n is the number of unpaired electrons.

to be filled by the metal electrons. In the case of $[FeF_6]^{3-}$ the five degenerate $3d$ orbitals are half-filled, leading to a complex with five unpaired electrons.

Fe[Ar]

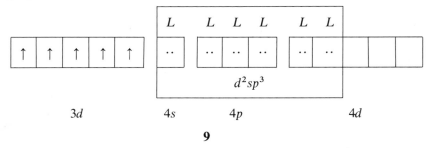

9

Originally complexes in which outer d orbitals are used in forming the hybrids (as in **9**) were called ionic complexes; those using inner orbitals for the hybrids, as in **8**, were termed covalent complexes. These are unfortunate choices for names because of the implications arising from the use of this terminology with the compounds of the representative elements. For example, the compound iron acetylacetonate, $Fe(acac)_3$, has a magnetic moment equivalent to five unpaired electrons suggesting that the $4d$ (**9**) rather than the $3d$ orbitals (**8**) are used in bonding. Calling this species an ionic complex gives an impression which is contrary to the observation that $Fe(acac)_3$ is soluble in nonpolar solvents. To overcome this problem the inner orbital complexes have been called hyperligated complexes; coordination compounds using outer orbitals are called hypoligated complexes. Perhaps the most useful description of such complexes involves terms that reflect experimentally observed properties rather than those which focus attention on theoretically derived quantities. Accordingly, such complexes are very often called high-spin and low-spin, or spin-free and spin-paired complexes.

As in the case of the representative elements, the application of valence-bond theory to coordination compounds can accommodate the formation of multiple bonds between the ligand and the metal atom. Generally, multiple-bond formation can be inferred from the same criteria used for the representative elements (Chapter 4), such as bond shortening. However, it is often difficult to establish the value of the expected single-bond radius for the transition-metal atom in the oxidation state as it occurs in the complex. For example, the radius of nickel (1.39 Å) in the tetrahedral complex $Ni(CO)_4$ has been obtained by extrapolation from the covalent tetrahedral radii of copper (1.35 Å) and zinc (1.31 Å). Using this value and the covalent radius of triply bound carbon (0.63 Å), the expected Ni—C distance in $Ni(CO)_4$ is 2.02 Å; the experimentally observed distance in this molecule is 1.82 $\pm$ 0.02 Å, suggesting that canonical forms of the type **10** as well as the expected

$$(CO)_3Ni{=}C{=}O \qquad\qquad (CO)_3Ni{-}C{\equiv}O$$

10 **11**

single-bonded structure (**11**) make important contributions to the wave-function for this molecule. Although earlier attempts at establishing the theoretical value of the single bonds between metal atoms and ligands may have been based on questionable assumptions, considerable progress has been made recently to alleviate such problems (*6*). Other physical methods have been used to investigate the extent of ligand–metal multiple bonding in complex species. For example, with certain ligands such as carbon monoxide vibrational spectroscopy gives useful information on the nature of the metal–carbon bond. The C—O stretching frequency for a ligand singly bonded to a metal atom (**11**) occurs at higher energy than when the metal–carbon bond has some multiple-bond character (**10**) (*6, 7*). Finally, the presence of metal–ligand multiple bonds has been inferred from unusual stability relationships in similar compounds. For example, although amines are more basic than the corresponding phosphines, some metal–amine complexes are less stable than the corresponding phosphine complexes. It will be recalled that this is one of the criteria used to separate the elements into the *a* and *b* classifications discussed in Chapter 17.

The canonical forms in the valence-bond theory which lead to a shortening of the metal–ligand bond can involve at least three different types of orbital overlap, as shown in Table 18.5. Ligands with empty *p* or *d* orbitals can accept electron density from the appropriate filled *d* orbitals of the metal atom. Unsaturated species containing a Period 2 element, such as carbon monoxide or the isoelectronic ion CN^-, appear in the former class of ligands, whereas trivalent derivatives of phosphorus and the heavier Group V elements are typical examples of the latter class of metal. Since the metal atoms in these complexes donate the electronic density to the π component, it is not surprising that such bonding is found primarily in the complexes of the d-block elements at the end of the series in lower oxidation states. Such metal species would be expected to contain a large concentration of d-electron density. A few cases are known where the ligand supplies the π

TABLE 18.5   Three possible metal–ligand π interactions

Canonical form	Type	Orbital representation	Typical ligands	Typical metals
M $\rightleftarrows$ L	$d\pi$-$p\pi$		CO, CN⁻ CNR, NO	Later members of the transition elements in a low oxidation state
M $\rightleftarrows$ L	$d\pi$-$d\pi$		PF₃, PR₃ AsR₃, SR₃	Later members of the transition elements in a low oxidation state
M $\rightleftharpoons$ L	$p\pi$-$d\pi$		O²⁻, F⁻ NH₂⁻	Early members of the transition elements in high oxidation states

density, in which case the transition-metal ion must have a vacant d orbital available of the appropriate symmetry. Metals at the beginning of the series of the d-block elements in their higher oxidation states are the best suited for this type of π-bond formation.

If the bonding in complex molecules is described as simply the formation of a coordinate-covalent bond, problems arising from unfavorable charge distribution arise. Consider as an example the ion $[Co(NH_3)_6]^{3+}$, which requires the formation of six metal–ligand coordinate-covalent bonds. The free metal ion carries a $3+$ formal charge, but upon bond formation the metal acquires a share of half of twelve electrons which gives it a formal charge of $3-$. Intuitively from past experience, this is clearly an undesirable situation. However, if the metal–ligand bond possesses some ionic character in the sense of having ionic canonical forms contributing to the molecular wave function, it is possible to distribute the positive charge of the original cobalt atom over all of the atoms in the system and not require that the metal atom acquire a large negative charge. This mechanism for dissipating the charge of the metal ion gains some support from the fact that complexed ammonia or water molecules are often more acidic than the corresponding free ligands (Chapter 17). The ability of some ligands to form π bonds by using electron density from the appropriate metal orbits, as shown in Table 18.5, represents a more efficient way to remove electric charge from the metal atom.

Certain aspects of the valence-bond theory can be used to correlate the lability of complex ions with their electronic configurations. Experimental observations on the lability and magnetic properties of some complex species are shown in Table 18.6. As before, the magnetic data give an indication

TABLE 18.6   **The lability, magnetic properties, and metal hybridization for some complex ions**

Compound	Class	Number of unpaired electrons	d^2sp^3 hybridzation			
			$3d$	$4s$ $4p$		$4d$
$V(NH_3)_6^{3+}$	labile	2	↑ ↑			
$Cr(CN)_6^{3-}$	inert	3	↑ ↑ ↑			
$MnCl_6^{3-}$	labile	4	↑ ↑ ↑ ↑			
$Co(CN)_6^{3-}$	inert	0	↑↓ ↑↓ ↑↓			

of the type of hybrid orbital used by the metal atom for bonding. From hybrid designations in Table 18.6 it is apparent that the labile complexes possess vacant, low-lying, unhybridized metal orbitals which can act as a point of attack by an incoming ligand. The inert complexes have electron density in the unhybridized metal orbitals, suggesting that they are not as readily available for bond formation to a ligand. This argument is essentially the same as that used to explain the difference in the relative ease of hydrolysis of CCl_4 and $SiCl_4$.

Presently the valence-bond theory is not often used in discussing the nature of coordination compounds for several important reasons. Although the magnetic criteria of bond type is a useful device it is usually used a posteriori in valence-bond arguments and does not lead to an understanding of detailed magnetic properties. Generally the valence-bond theory provides only qualitative agreement with experimental observations and does not account for the difference in the observed stabilities of different stereochemical arrangements. Perhaps its most serious drawback is its inability to treat the spectroscopic properties of transition-metal complexes in any detail.

4 / ELECTROSTATIC MODELS

An attempt was made to calculate the energy of formation of complex ions using a simple electrostatic model incorporating conventional potential-energy equations $(8–11)$. As might be expected the theory leads to a prediction that linear, tetrahedral, and octahedral structures are the most stable for two-, four-, and six-coordinate species. Unfortunately the existence of stable four-coordinate planar complexes is not predicted by the ionic model.

Basically the ionic model involves a calculation of the total energy, E, in terms of (a) the attraction of the metal ion and the ligand, E_A; (b) the mutual electrostatic repulsion of the ligands, E_R; (c) the interaction between induced dipoles, E_I; and (d) the van der Waals repulsion between the ligands and the metal atom, E_r:

$$E = E_A + E_R + E_I + E_r. \tag{2}$$

The treatment is reminiscent of the method used to calculate lattice energies of ionic compounds, as given in Chapter 2. Monatomic ionic ligands can be treated as charged spheres with radii equal to the experimentally observed crystallographic values. In the case of uncharged ligands, dipole moments and a detailed knowledge of the structure of the ligand are the important quantities for the calculation. Once the general equation for a complex of a given geometry is available, the calculation of the energies of all systems of that geometry can be made if the experimental parameters are known. The results of such calculations, which appear in Table 18.7, expressed in terms of the bond energies of the complex

$$M^{m+}_{(g)} + nL_{(g)} \longrightarrow ML^{m+}_{n(g)}, \tag{3}$$

are surprisingly good. However, even in these favorable cases the results are misleading, since other properties such as the relative stability of complex species, magnetic properties, and spectral data are not easily correlated. Indeed, other than the suggestion that the available electrons are placed into

TABLE 18.7   **Bond energies for complex species**

	Bond energy, kcal	
Complex	Experimental[a]	Calculated
AlF_6^{3-}	233	212
$Co(NH_3)_6^{3+}$	134	117
$Cr(H_2O)_6^{3+}$	122	111
$Fe(H_2O)_6^{3+}$	116	109
$Zn(NH_3)_4^{2+}$	89	86
$Fe(H_2O)_6^{2+}$	58	50

[a] Determined from a thermochemical cycle corresponding to the process described in Eq. (3).

unperturbed d orbitals according to Hund's rule, there is no simple way to predict the gross magnetic properties of the complex using the electrostatic model. This procedure, of course, gives the wrong results and it does not suggest why, for example, $Fe(CN)_6^{3-}$ should have a different magnetic moment than FeF_6^{3-}.

5 / CRYSTAL-FIELD THEORY

Another electrostatic approach to an understanding of the properties of complex species, crystal-field theory, was suggested by Bethe in 1929, but it did not find use by chemists until about 25 years later. Rather than attempting to calculate the metal–ligand bond energies as described above, crystal-field theory considers the effect of the electrostatic fields of the ligands on the energy of the d orbitals of the metal (12–14).

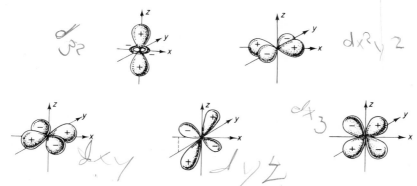

FIG. 18.1 d-Orbital representation for a free gaseous transition-metal ion.

To illustrate the principles of the crystal-field theory, consider an octa-hedral complex species, ML_6^{n+}. The central metal ion in the gas phase contains a degenerate set of five d orbitals which can be conveniently de-scribed with reference to the normal Cartesian coordinate system. The boundary surfaces of the d orbitals and their arrangement in space is shown in Fig. 18.1. We can imagine the influence of the six ligands on the d orbitals by

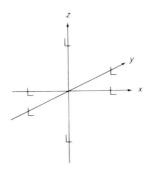

FIG. 18.2 Ligand orientation with respect to the metal–orbital axes used in the crystal-field theory.

placing the ligands on the Cartesian axes at infinity and moving them toward the metal atom until the experimentally observed metal–ligand internuclear distance is attained; this is shown schematically in Fig. 18.2. Recalling that ligands are either anions or polar species, the negative end of which is co-ordinated to the metal ion, we see that the metal comes under the influence of a negative electric field in this process. If the electric field arising from the ligands were spherically symmetrical we would expect that the energy of all the d orbitals would be raised to the same extent, i.e., it would require a higher energy to maintain an electron in an orbital in the presence of a negative field than without this field, as shown in Fig. 18.3. However, the six ligands generate an octahedral field which interacts differently with the d orbitals than does a spherical field. The orbitals that lie along the Cartesian axes, $d_{x^2-y^2}$ and d_{z^2}, are perturbed more than the orbitals which lie between these axes, d_{xy}, d_{xz}, and d_{yz}. Thus, in an octahedral field, the five degenerate d orbitals split into two groups. This result, achieved in an intuitive manner, can be more rigorously obtained by applying group theory to the symmetry of the d orbitals (1). The two sets of d orbitals present in an octahedral field are identified several different ways in the literature. The higher-energy set has been called e_g, $d\gamma$, or γ_3; the corresponding labels for the set at lower energy T_{2g}, $d\varepsilon$, or γ_5. The energy difference between the two sets of orbitals which arise from an octahedral field is measured in terms of the parameter Δ_0 or $10Dq$. Under the influence of a purely electrostatic field the average energy

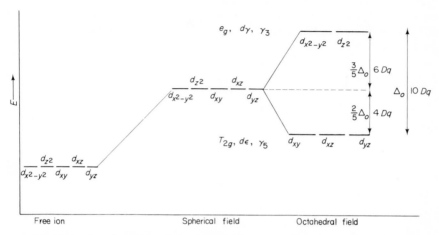

FIG. 18.3   Generalized splitting pattern for d-orbitals in an octahedral field.

of the perturbed orbitals is zero. Thus, the higher e_g levels must be $6Dq$ or $\frac{3}{5}\Delta_0$ greater in energy than the average, while the lower T_{2g} levels are $4Dq$ or $\frac{2}{5}\Delta_0$ below the average value. In other words, if the original set of d orbitals in the gaseous ion contained five electrons, an octahedral perturbation of the levels would increase the energy of the electrons in the two e_g orbitals by $2 \times 6Dq$ (or $2 \times \frac{3}{5}\Delta_0$). The gain in energy when electrons occupy the T_{2g} orbitals compared to the energy they would have if they occupied orbitals corresponding to the average energy is called the crystal-field stabilization energy (CFSE). The crystal-field stabilization energy is equal in magnitude to Dq (or Δ_0) per electron.

The discussion of crystal-field effects up to this point has been in terms of a symmetrical field; that is, all ligands of the same kind at the same distance. A change in either, or both, of these factors will produce an unsymmetrical field at the site of the metal atom. For example, moving two *trans* ligands away from the plane of the remaining ligands leads to a tetragonal arrangement of ligands, a commonly observed distorted octahedral structure; the relative order of energy levels for such distortions is given in Fig. 18.4. Under a tetragonal field the intensity of interaction at the *trans* positions decreases with respect to the other positions, leading to a splitting of both the e_g levels and the T_{2g} levels. The relative magnitude of the change in the energies of these two levels can be deduced by considering the symmetry of the individual orbitals involved. For example, if we take the tetragonal distortion to occur along the Z axis, the energy of the d_{z^2} orbital will decrease with respect to that in the octahedral field. In a similar manner the energies of the d_{yz} and d_{xz} orbitals decrease with respect to that of the d_{xy} orbital and a part of the degeneracy in the T_{2g} orbitals is removed. A tetragonal distortion which occurs by moving the *trans* ligands together gives rise to the same type of splitting pattern, but the relative energies of the split levels are reversed as shown in Fig. 18.4.

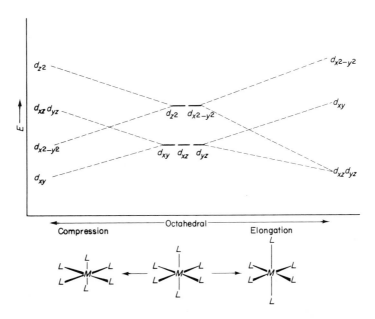

FIG. 18.4 A symmetrical octahedral ligand field can be tetragonally distorted by moving *trans* ligands toward or away from the central metal atom. A similar distortion of the field occurs in complexes of the type *trans*-Ma_2b_4. Tetragonal distortions give rise to additional splittings of degenerate orbitals.

The extreme case of tetragonal distortion occurs when the *trans* ligands of an octahedral field are sufficiently far removed so that they have no effect on the d orbitals of the transition metal; this arrangement is equivalent to a square-planar complex. In this case, the energy of the d_{z^2} levels is virtually the same as that of the degenerate d_{xy} and d_{yz} orbitals (Fig. 18.4.).

Crystal-field effects on the d orbitals of transition metals surrounded by any number of ligands also can be deduced using similar arguments. Figure 18.5 shows the relative energies of the d orbitals for some of the more commonly observed coordination numbers in complex species. In the case of tetrahedral complexes, the pattern for the splitting of the d orbitals is the reverse of that found in octahedral complexes, as shown in Fig. 18.6. The relative magnitudes of the tetrahedral crystal field splitting (Δ_t) and the octahedral

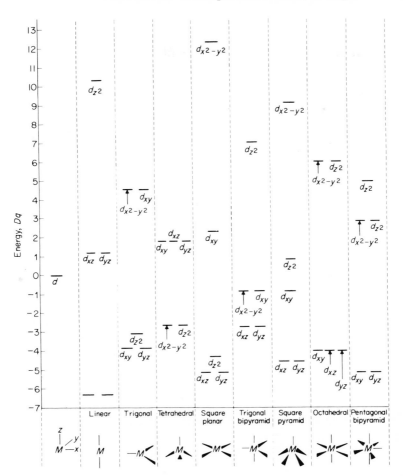

FIG. 18.5   Splittings of d-orbitals under the influence of different crystal fields. The energy is given in units of Dq.

splitting (Δ_0) are given by

$$\Delta_t = \tfrac{4}{9}\Delta_0 \tag{4}$$

for the case where all factors, i.e., the kind of ligand and metal and the metal–ligand distance, are the same. Of course it is difficult to conceive of a series of complex ions with different coordination numbers where these factors are exactly the same, but Eq. (4) indicates that, in general, crystal-field effects in the tetrahedral case should be about half those observed for octahedral coordination.

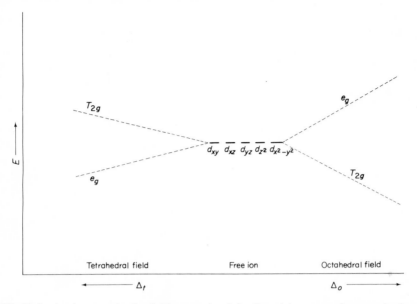

FIG. 18.6   An increase in the field strength of the ligand increases the magnitude of the splitting of the e_g and T_{2g} orbitals in octahedral and tetrahedral complexes.

6 / THE PROPERTIES OF TRANSITION-METAL IONS

Let us now consider the details of applying crystal-field arguments to the properties of transition-metal ions, concentrating predominantly on the two most commonly observed coordination geometries—octahedral and tetrahedral. Intuitively it might be expected that the crystal-field splitting parameter Δ would be a function of the ligand, since only by accident would two ligands be capable of generating the same field (Fig. 18.6). Thus, ligands should be classified in terms of the relative strengths of the fields they generate. It is possible to obtain the relative splitting of the d-orbital energies from quantum mechanical perturbation calculations on the free metal ion using hydrogen-like wave functions and idealizing the ligands as a set of point charges or dipoles (*15, 16*). Such calculations yield a new set of metal wave functions that are linear combinations of the original functions. The difference in energies of the two set of new levels ($10Dq$ in Fig. 18.3) is related to the electron repulsion terms of the free ion

$$10Dq \approx \frac{5eq\overline{a^4}}{3r^5} \approx \frac{5e\mu\overline{a^4}}{r^6};$$ (5)

e is the electronic charge; q or μ is the charge or dipole moment if the ligand; r is the metal–ligand distance; and $\overline{a^4}$ is the average value of the fourth power of the distance of the electron from the nucleus, a quantity which is related to the (usually unknown) radial dependence of the wave function. In general, the quantities necessary to complete detailed calculations of this type for specific complex ions are not available and the crystal-field splitting parameters are usually derived empirically from spectroscopic data. The electronic spectrum of an ion with a d^1 configuration is the simplest case from which the crystal-field splitting parameter can be obtained. The ground-state configuration of a d^1 system in an octahedral field should have an electron in one of the T_{2g} orbitals. The spectrum of $Ti(H_2O)_6^{3+}$, shown in Fig. 18.7, consists of a weak band in the visible at about 5000 Å and a very intense band in the ultraviolet region. Since the band in the visible region changes position when other ligands are substituted for water, it is reasonable to assume that this transition involves transfer of the one d electron in the

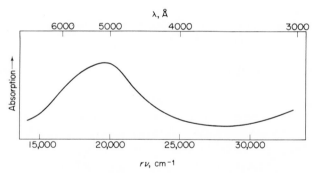

FIG. 18.7   The electronic spectrum of $Ti(H_2O)_6^{3+}$.

system from the lower T_{2g} orbital to the e_g orbital; the energy of this process also corresponds to $10Dq$ or Δ_0 (Fig. 18.3). Thus, for water $10Dq$ or Δ_0 is about 20,400 cm^{-1}, and accordingly $4Dq$, which is the crystal-field stabiliza-tion energy, is 8160 cm^{-1} ($\sim$ 1 eV or 23 kcal). Detailed analysis of the spectra of a variety of complex species shows that certain generalizations obtain with respect to the value of the crystal-field splitting parameters for octahedral complexes. In the first series of d-block elements, Δ_0 for the divalent aquo complexes is $\sim$ 28 kcal, and it increases 40% to 80% for the corresponding trivalent aquo species. The crystal-field splitting parameter for the second series of d-block elements is 50% larger than that of the first series; the third d-block series is about twice as large as the first

$$0.75(\Delta_0)_{4d} \approx (\Delta_0)_{3d} \approx 0.5(\Delta_0)_{5d}. \tag{6}$$

The order of increasing crystal-field splitting of some metals in their common oxidation states is given by the series: $Mn^{2+} < Co^{2+} \approx Ni^{2+} < V^{2+} < Fe^{3+} < Cr^{3+} < Co^{3+} < Mn^{4+} < Mo^{3+} < Rh^{3+} < Ir^{3+} < Re^{4+} < Pt^{4+}$. As indicated previously [Eq. (4)] the crystal field splitting for tetrahedral complexes is about 40–50% that of octahedral complexes.

The common ligands have been arranged in order of increasing crystal-field splitting from a study of their effects on the spectra of transition-metal ions, the usual order being given by (17–19) $I^- < Br^- < Cl^- \approx SCN^- \approx N_3 < F^- < OC(NH_2)_2 < OH^- < C_2O_4{}^2 \approx H_2O < NCS^- < NC^- < NH_2CH_2CO_2{}^- < NH_3 \approx C_5H_5N < en \approx SO_3{}^{2-} < NH_2OH < NO_2{}^- < $ phenylenediamine $< CN^-$. This order is usually called the spectrochemical series or the Fajans–Tsuchida series.

6.1 / Magnetic Properties (II)

The fact that the crystal-field splitting is dependent upon the nature of the ligand suggests that the magnetic properties of a complex can be related to its spectroscopic properties through the spectral chemical series. It should be recalled that Hund's rule insures maximum multiplicity for electrons placed in degenerate orbitals. Another way of looking at Hund's rule is that electron pairing is an energetically unfavorable process. Consider the situation in which two electrons are present in two degenerate orbitals [(a) in Fig. 18.8]. If the energy difference between the orbitals (ΔE) were to increase, a point would be reached [(b) in Fig. 18.8] where ΔE would be less than the energy required for electron pairing (E_p) and the two electrons would still be unpaired. Beyond this point [(c) in Fig. 18.8] the pairing energy would

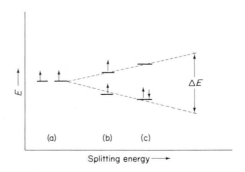

FIG. 18.8   Under the influence of small ligand fields the pairing energy for two electrons is large compared to the splitting of the orbitals and the electrons are unpaired (b) as in the free ion (a). However, when the splitting energy is larger than the pairing energy the system is diamagnetic (c).

be less than the splitting, and the lowest state for the system corresponds to two paired electrons. This argument can be applied to the splitting of the metal d orbitals into two sets by the ligand field in octahedral and tetrahedral complexes (cf. Fig. 18.6). On the basis of magnetic behavior, complexes can be divided into two groups—low-spin and high-spin. In the crystal-field interpretation the high-spin complexes occur when the pairing energy is greater than the crystal-field splitting, which corresponds to the presence of weak-field ligands. If the pairing energy is less than the crystal-field splitting low-spin complexes result. Table 18.8 gives the electronic distribution expected for weak-field and strong-field octahedral and tetrahedral complexes.

TABLE 18.8 Electron distributions for weak- and strong-field octahedral and tetrahedral complexes

| Number of d electrons | Octahedral complexes | | | | Tetrahedral complexes | | | |
| | Weak field, $P > \Delta_0$ | | Strong field, $P < \Delta_0$ | | Weak field, $P > \Delta_t$ | | Strong field, $P < \Delta_t$ | |
	T_{2g}	e_g	T_{2g}	e_g	e_g	T_{2g}	e_g	T_{2g}
1	↑		↑		↑		↑	
2	↑↑		↑↑		↑↑		↑↑	
3	↑↑↑		↑↑↑		↑↑	↑	↑↓↑	
4	↑↑↑	↑	↑↓↑↑		↑↑	↑↑	↑↓↑↓	
5	↑↑↑	↑↑	↑↓↑↓↑		↑↑	↑↑↑	↑↓↑↓	↑
6	↑↓↑↑	↑↑	↑↓↑↓↑↓		↑↓↑	↑↑↑	↑↓↑↓	↑↑
7	↑↓↑↓↑	↑↑	↑↓↑↓↑↓	↑	↑↓↑↓	↑↑↑	↑↓↑↓	↑↑↑
8	↑↓↑↓↑↓	↑↑	↑↓↑↓↑↓	↑↑	↑↓↑↓	↑↓↑↑	↑↓↑↓	↑↓↑↑
9	↑↓↑↓↑↓	↑↓↑	↑↓↑↓↑↓	↑↓↑	↑↓↑↓	↑↓↑↓↑	↑↓↑↓	↑↓↑↓↑
10	↑↓↑↓↑↓	↑↓↑↓	↑↓↑↓↑↓	↑↓↑↓	↑↓↑↓	↑↓↑↓↑↓	↑↓↑↓	↑↓↑↓↑↓

It is apparent from Table 18.8 that there is no difference in the d-electron configuration for d^1, d^2, d^3, d^8, d^9, and d^{10} systems in the octahedral case. However, d^4, d^5, d^6, and d^7 systems exhibit different electronic distributions depending upon whether the ligands generate strong or weak crystal fields. In other words, ligands high in the spectrochemical series should form low-spin complexes whereas ligands at the other end of the series should yield high-spin complexes. Thus, it is apparent why the complex ion $Fe(CN)_6^{3-}$ has a spin corresponding to one unpaired electron but $Fe(H_2O)_6^{3+}$ has five unpaired electrons; the electron distributions for these species are given in Fig. 18.9.

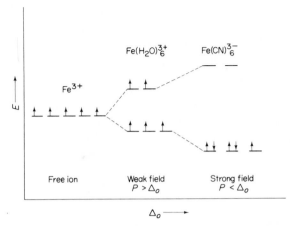

FIG. 18.9 The energy-level diagram for weak- and strong-field Fe^{3+} complexes. The latter are high-spin but the former exhibit a decrease in paramagnetism.

Similar results are obtained for tetrahedral complexes from the information in Table 18.8. That is, d^1, d^2, d^7, d^8, d^9, and d^{10} systems have the same magnetic properties irrespective of the strength of the crystal field. The magnetic properties of d^3, d^4, d^5, and d^6 ions, however, depend upon the nature of the ligands. Presently there is little evidence for the existence of low-spin tetrahedral complexes, presumably because the tetrahedral crystal-field splitting factor is about half that of the octahedral field [Eq. (4)]. Thus, even with the very-strong-field ligands the tetrahedral field is not sufficiently large to overcome the pairing energy.

Theoretical estimates have been made of the pairing energy from spectroscopic data on gaseous ions. This information together with the theoretical calculations of crystal-field splittings from the electron-repulsion parameter [Eq. (5)] serve as the basis for a verification of the crystal-field arguments presented here. The results are summarized in Table 18.9. There is some reason to believe that the pairing energies as estimated for the free ions are about 20% lower in the complexes; however, this would not affect the conclusions drawn from Table 18.9, namely that the predicted spin states correspond to those experimentally observed.

The magnetic behavior of distorted octahedral structures can also be interpreted using crystal-field arguments. It should be recalled that such distortions arise by placing a pair of *trans* ligands further from the metal ion than the four ligands; the extreme of this type of distortion occurs when the metal has achieved planar coordination. Consider the distortion of an octahedral complex for a d^8 system, which is illustrated in Fig. 18.10. The regular octahedral structure contains two unpaired electrons. A weak tetragonal

TABLE 18.9 Average electron-pairing energies and
crystal-field splittings for some octahedral complex ions

Electron configuration	Example	Energy, kcal		Predicted spin state
		P	Δ_0	
d^4	$Cr(H_2O)_6{}^{2+}$	67.0	39.8	high
	$Mn(H_2O)_6{}^{3+}$	79.8	59.9	high
d^5	$Mn(H_2O)_6{}^{2+}$	72.7	22.3	high
	$Fe(H_2O)_6{}^{3+}$	85.5	39.1	high
d^6	$Fe(H_2O)_6{}^{2+}$	50.2	29.7	high
	$Fe(CN)_6{}^{4-}$	50.2	94.1	low
	$CoF_6{}^{3-}$	59.9	37.1	high
	$Co(NH_3)_6{}^{3+}$	59.9	65.6	low
d^7	$Co(H_2O)_6{}^{2+}$	64.2	26.6	high

distortion splits both the original e_g and T_{2g} levels; under these conditions the assignment of a single crystal-field splitting parameter is impossible because of the presence of four different levels, one of which is degenerate. In the case of a d^8 ion, however, the important quantity is the energy difference (E_t) between the originally degenerate e_g orbitals, which arises because of the tetragonal distortion. Thus, in a weakly distorted octahedral field a d^8 system gives a high-spin complex because the pairing energy is greater than E_t. As more distortion occurs, E_t becomes greater than the pairing energy

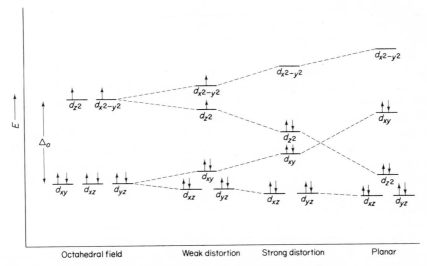

FIG. 18.10 The energy-level diagram for an octahedral field d^8 system as it undergoes distortion to a planar arrangement of ligands.

and the complex becomes low-spin. If the structure is distorted to the extreme, i.e., to give a planar complex, its spin remains low.

It should be noted that the basic arguments presented here on the effect of distorting a regular structure are independent of the arguments for why distortions should occur in certain systems.

6.2 / Electronic Spectra of Complex Ions (*III*)

We have seen earlier in this chapter how the spectra of d^1 ions can be used to establish the order of ligands in the spectrochemical series. Before proceeding to the spectra of more complex ions two factors, other than the position of the absorption band attributed to a d-d transition which appears in the spectrum of $Ti(H_2O)_6^{3+}$ (Fig. 18.7.), should be discussed. The relatively low intensity of the band is characteristic of d-d transitions which occur in complex ions. Normally such transitions would be "forbidden" in the free, uncomplexed metal ion by the Laporte selection rule ($\Delta l = \pm 1$). The fact that such transitions occur in complexes suggests that the interaction between the ligands and ions is not entirely coulombic and that the metal orbitals involved are not pure d orbitals but are mixed to some extent with p and/or f orbitals. In other words, the principal assumption of the crystal-field theory is not entirely correct. Another feature of interest in the spectrum of $Ti(H_2O)_6^{3+}$ is the large bandwidth compared with the widths observed in the spectra of gaseous ions. This suggests that there are a series of closely lying excited and/or ground states giving rise to a relatively broad span of energies rather than the expected sharp span. Degeneracies in the excited and/or ground states can be removed by a series of mechanisms, the most common of which is a coupling of molecular vibrations.

The spectra of d^9 ions such as Cu^{2+} are remarkably similar to those of the d^1 species in the sense that they both exhibit only one weak d-d transition. Formalistically, a d^9 ion can be imagined as a d^{10} ion to which has been added a "positive hole" (or a positron). The crystal-field arguments developed for the d^1 system are now valid for the d^9 ion except that the pattern of the energy-level diagram shown in Fig. 18.3 will be inverted because a positive hole is the most stable where an electron is the least stable. Using this formalism, the ground state of the octahedral d^9 complex ion can be described as a positron occupying an e_g orbital; the spectrum of the ion arises from the transition of the positron to a higher T_{2g} orbital, the energy difference Δ_0 being given by the position of the absorption band. The hole formalism is a useful device in discussing the spectra of complex ions containing more than one electron, since we can now relate the arguments for a d^n ion to those of an ion with a d^{10-n} configuration.

To describe the spectra of metal complexes containing other than d^1 or d^9 ions requires a detailed discussion of the interaction of the Russell–Saunders states of a d^n system with the electrostatic fields of the ligands. The details of the construction of such energy-level diagrams are beyond the scope of this text; the interested reader is directed to Refs. (20)–(22).

6.3 / Thermodynamic Properties (*IV*)

The stabilizing effect of the crystal field (Fig. 18.3) is reflected in the measured energies of interaction between the metal ion and ligand. For example, the hydration energies of divalent ions of the first d-block elements

$$M^{2+} + (6 + n)H_2O \rightarrow [M(H_2O)_6]^{2+}_{(aq)}, \tag{7}$$

as determined from thermochemical cycles, exhibit the periodic variation shown in Fig. 18.11. The ions Ca^{2+} (d^0), Mn^{2+} (d^5) and Zn (d^{10}) have a spherical distribution of charge, in the sense that all of the d orbitals are either empty, singly occupied, or doubly occupied. The data for these ions lie essentially on a straight line, as would be expected for a group of ions with regularly decreasing radii. No intermediate ions are apparently more strongly hydrated than those with a spherical charge distribution; accordingly, the difference between the actual hydration energy of an ion and the line joining ions with a spherical charge distribution corresponds to the crystal-

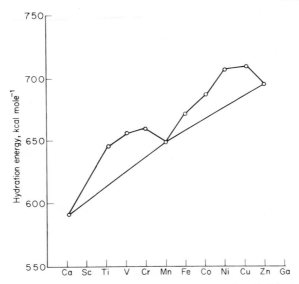

FIG. 18.11 The hydration energy of the 2+ ions of Period 3 elements.

field stabilization which arises because of the splitting of degenerate d levels in the free ion. A correction of the heats of hydration for the crystal-field stabilization energies as determined from spectral studies gives hydration energies which are very close to those predicted from the ions with spherical charge distributions. Similar results are obtained for the lattice energies of divalent halides shown in Fig. 18.12. Thus, estimates of crystal-field effects can be obtained from either spectral or thermodynamic data.

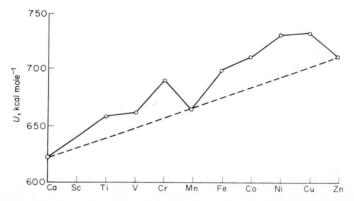

FIG. 18.12   The lattice energies of the divalent halides of Period 3 elements.

Crystal-field effects are also observed in the variation of the radii of complex ions. Figure 18.13 shows the octahedral radii of the divalent ions of the first d-block ions. The radii of Ca^{2+}, Mn^{2+}, and Zn^{2+} decrease smoothly

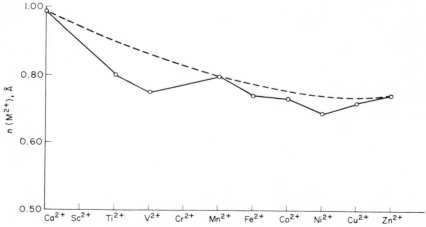

FIG. 18.13   The octahedral radii of the 2+ ions of Period 3 elements.

as would be expected for ions screened by a spherical charge distribution with increasing nuclear charge. The d^1, d^2, and d^3 ions have smaller radii than expected because the electrons occupy e_g orbitals which are concentrated in the region between the ligands in these species. The nuclei of these ions are less shielded than if they had a spherical distribution of electrons, which leads to a decrease in the metal–ligand distance or a smaller effective metal atom radius. The d^4 and d^5 ions in a weak-field complex have electrons added successively to the T_{2g} orbitals. The first T_{2g} electron causes an increase in the relative radius of the d^4 ion compared to that of the d^3 ion because the T_{2g} orbitals are pointing directly at the ligands. This leads to a repulsion of the ligands, or an increase in the effective octahedral radius. The d^5 ion has a spherically symmetrical electron density and its radius is proportionally smaller than that of the d^0 system, but larger than that of the d^4 system. The addition of electrons beyond the d^5 structure leads to a repetition of the previous pattern, except that the d levels now become doubly filled.

6.4 / Stereochemistry (V)

Crystal-field theory also provides the basis for understanding the stereochemical consequences of forming complex ions. For example, although many of the six-coordinate complexes are known to be regular octahedra, some exist as distorted octahedra. The Jahn–Teller theorem describes the characteristics associated with the energy of a system which necessarily lead to a distortion of its molecular geometry. Briefly, the Jahn–Teller theorem states that a nonlinear molecule in a degenerate electronic state will undergo a distortion of its nuclear framework to attain a lower symmetry and energy, thereby removing the original degeneracy. The theorem does not address itself to the mechanism by which the degeneracy is removed. The source of the driving energy for the Jahn–Teller distortion can be easily understood using Cu^{2+}, which is a d^9 system, as an example. In an octahedral field the copper d orbitals are split by the crystal field of the ligands in the usual way, shown in Fig. 18.14. Without distortion of the octahedral structure the electron distribution is given by $(T_{2g})^6(e_g)^3$. If we assume that the electrons in the e_g orbitals are distributed so that two are paired in the d_{z^2} orbital, both the direction of the Jahn–Teller distortion and the stabilizing effect become apparent. Both the d_{z^2} and $d_{x^2-y^2}$ orbitals are pointing directly toward the ligands. The ligands along the Z axis are thus more shielded from the nuclear charge of copper than are those in the $X-Y$ plane, the net effect being a movement of the ligands along the Z axis away from the metal relative to those in the $X-Y$ plane, which move towards the metal. Thus, the resulting structure is an elongated octahedron. If the arrangement of electrons in the original regular octahedral structure were $(d_{z^2})^1(d_{x^2-y^2})^2$, similar arguments would lead to a distorted octahedron which has two short bonds relative to four long bonds, that is, the octahedron would be flattened.

The energy relationships arising from Jahn–Teller distortions can be obtained from Fig. 18.14. Let us continue to assume that the electrons are paired in the d_{z^2} orbital. Under these conditions both the e_g and T_{2g} levels

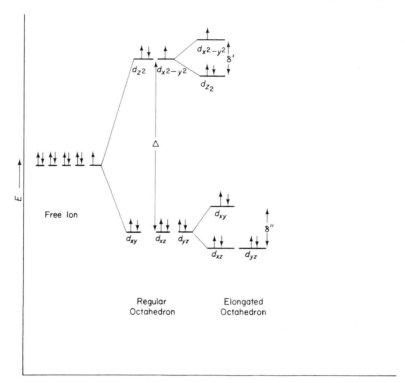

FIG. 18.14   The Jahn–Teller tetragonal distortion of octahedral Cu^{2+} gives rise to an additional splitting of the e_g and T_{2g} orbitals. The ligand-field splitting is considerably larger than the Jahn–Teller splittings, i.e., $\Delta \gg \delta' > \delta''$.

are split, the splitting obeying the center of gravity rule. For the d^9 system there is no gain in energy because of the splitting of the T_{2g} orbitals; the latter are completely filled and the net gain in energy from the lower orbitals is compensated by a loss in energy for the higher orbital

$$4(-\tfrac{1}{3}\delta_2) + 2(+\tfrac{2}{3}\delta_2) = 0. \tag{8}$$

However, Jahn–Teller splitting of the e_g level leads to a net gain in energy

$$2(-\tfrac{1}{2}\delta_1) + 1(+\tfrac{1}{2}\delta_1) = -\tfrac{1}{2}\delta_1. \tag{9}$$

The Jahn–Teller stabilization energy is small compared with crystal-field effects, however it is the driving energy for the distortion of the regular octahedral structure. Jahn–Teller distortions have been detected for copper in its crystalline compounds, as summarized by the data in Table 18.10; in some instances the distortion is sufficiently large to suggest that the compounds are essentially planar.

TABLE 18.10   Jahn–Teller distortions in solid copper compounds

Compound	Metal–ligand distances, Å	
	Planar	Axial
CuF_2	1.93	2.27
$CuCl_2$	2.30	2.95
$CuBr_2$	2.40	3.18
$CuCl_2 \cdot 4H_2O$	2.01 ($2H_2O$)	2.98
	2.31 (2Cl)	
$CuCl_2 \cdot 2py$	2.02 (2py)	3.05
	2.28 (2Cl)	

Jahn–Teller distortions of octahedral complexes would be expected whenever the e_g orbitals are occupied by an odd number of electrons. Thus, a Jahn–Teller effect should be observed for d^4 high-spin systems which arise with weak-crystal-field ligands and low-spin d^7 systems in addition to the d^9 systems. Evidence for Jahn–Teller distortion is sometimes observed spectroscopically. For example, the single d-d band for Ti $(H_2O)_6^{3+}$, shown in Fig. 18.7 is not symmetrical, as might be expected for a single transition. The asymmetry of the band has been attributed to a Jahn–Teller distortion which splits the energy of the excited states, giving rise to two possible transitions; presumably the bands arising from the d-d transitions to the two closely lying Jahn–Teller states lead to the observed unsymmetrical band envelope.

Crystal-field theory also gives some insight into the basis of stereochemical changes from metal to metal. Table 18.11 contains the one-electron crystal-field stabilization energies calculated for octahedral, square-planar, and tetrahedral complexes with d^n configurations. In this calculation electron interactions have been neglected and it is assumed that the orbitals exhibit integral electron occupancy. The energies indicated may not be entirely valid if these assumptions are subjected to detailed scrutiny, but the general results are useful in discussing stereochemistry.

Before discussing stereochemistry using the data in Table 18.11, it should be recalled that the total energy associated with complex formation involves

TABLE 18.11   Ligand-field stabilization energies, Dq

System	Octahedral		Square-planar		Tetrahedral	
	Weak-field	Strong-field	Weak-field	Strong-field	Weak-field	Strong-field
d^0	0	0	0	0	0	0
d^1	4	4	5.14	5.14	2.67	2.67
d^2	8	8	10.28	10.28	5.34	5.34
d^3	12	12	14.56	14.56	3.56	8.01
d^4	6	16	12.28	19.70	1.78	10.68
d^5	0	20	0	24.84	0	8.90
d^6	4	24	5.14	29.12	2.67	7.12
d^7	8	18	10.28	26.84	5.34	5.34
d^8	12	12	14.56	24.56	3.56	3.56
d^9	6	6	12.28	12.28	1.78	1.78
d^{10}	0	0	0	0	0	0

several factors, i.e., metal–ligand attraction, ligand–ligand repulsion, and electron-pairing energy. These factors, on an absolute scale, are one or two orders of magnitude larger than the ligand-field stabilization energy, and they usually vary incrementally in a more or less regular manner along a series of transition-metal ions. Thus, ligand-field effects are often the deciding factors in stability arguments. Considering first the relative one-electron ligand-field stabilization energies for octahedral and square-planar complexes shown in Table 18.11, we note that for weak-field ligands only the d^4 (e.g., Cr^{2+}) and d^9 (e.g., Cu^{2+}) systems are markedly different in their crystal-field stabilization energies. It will be recalled that for both of these octahedral structures Jahn–Teller distortion is expected. In the case of copper the distortion corresponds to an elongation of the regular octahedral array of ligands. The data in Table 18.11 also suggest planar coordination in all other cases except d^0, d^5, and d^{10}. At least part of this prediction is in good agreement with observation, since planar complexes have been reported for Ni^{2+}, Pd^{2+}, Pt^{2+}, and Au^{3+} (d^8) and for Co^{2+} and Rh^{2+} (d^7). The ligand-field stabilization energies for octahedral and tetrahedral complexes shown in Table 18.11 also are in agreement with experimental observation. Tetrahedral complexes containing strong-field ligands are not common; the crystal-field splitting for a tetrahedral arrangement is about half that for the same ligand in an octahedral arrangement [Eq. (4)]. In the cases where some tetrahedral complexes have been observed, e.g., the d^7 system CoX_4^{2+} (X = halogen), the loss in crystal-field energy is relatively small and is probably compensated by other factors. The basis for the choice of a tetrahedral arrangement over an octahedral arrangement in the d^0, d^5, and d^{10} systems where neither structure possesses crystal-field stabilization energy must be found in other factors, such as lower ligand–ligand repulsion energies in the tetrahedral structures compared with the same ligands in octahedral arrangements. Finally, the preferred arrangement of four ligands in a planar structure compared with a tetrahedral structure is also reflected in Table 18.11.

6.5 / Critique of Crystal-Field Theory (VI)

There is ample evidence that the initial postulate of the crystal-field theory, i.e., that the effect of the ligand on the metal is purely electrostatic, cannot be correct. For example, the experimentally observed spectrochemical series cannot be interpreted solely on the basis of charge density on the ligand. Inspection of the order of ligands shows that OH^- has a smaller splitting than H_2O, F^- smaller than CN^-, and 1,10-phenanthroline a greater splitting than many ionic ligands. Direct evidence for some degree of covalent bonding in the metal–ligand interactions comes from nmr (23, 24) and esr (25–27) experiments. The following simple example will suffice to illustrate how such conclusions can be reached from nmr data. The nmr spectrum of the paramagnetic compound $V(acac)_3$ (12) shows that the resonance signal for the proton attached to the heterocyclic ring is considerably shifted from that observed for isostructural, but the diamagnetic, aluminum compound. The

$$\left[M \begin{array}{c} O-C \overset{CH_3}{\diagdown} \\ \diagup \qquad \diagdown \\ \qquad\qquad CH \\ \diagdown \qquad \diagup \\ O-C \diagdown \\ \qquad\quad CH_3 \end{array} \right]_3$$

12

shift in the position of the ring-proton resonance can be accounted for by delocalization of the paramagnetic electron density into the acetylacetonate ring. Using simple crystal-field theory the electrons are restricted to the T_{2g} metal orbitals; however, the nmr results indicate that the ring proton feels this electron density, suggesting that the interaction between the ligand and the metal cannot be simply electrostatic. Electron delocalization has been detected also in complexes containing ligands which might be expected to correspond most nearly to the electrostatic assumptions of the crystal-field theory. Thus, the results of fluorine nmr experiments on octahedral complexes of the type MF_6^{2-} (M = Mn, Fe, Co) have been interpreted in terms of a 2–5% delocalization of electron density on the ligand. The esr spectrum of $IrCl_6^{2-}$ suggests that the single unpaired electron is only 70% localized on the metal atom. The results of many other experimental methods have been used to provide evidence for the delocalization of electron density from the metal atom to the ligand; these include nuclear quadrupole spectroscopy (*28*), Mössbauer spectroscopy (*29, 30*), and magnetic susceptibility measurements (*31*).

In spite of the difficulties which arise from the basic assumptions of crystal-field theory, it has had enormous success in correlating the observed properties of coordination compounds. The crystal-field formalism gives the basis for many predictions even if it is not physically correct.

7 / MOLECULAR-ORBITAL THEORY

Conceptually, the most accurate description of the bonding in complex compounds is given by molecular-orbital theory, the application of which yields polynuclear molecular orbitals encompassing the entire set of atoms. It might be appropriate at this point to recall some of the principles associated with molecular-orbital theory as described in Chapter 4. Generally, molecular orbitals are constructed by combining the appropriate ligand orbitals with the atomic orbitals of the metal, the *d* orbitals playing an important part in the combination. As always, the combinations must be consistent with the inherent symmetry of the orbitals involved; in the general case, identification of the symmetrically acceptable combinations can be systematically made by applying the principles of group theory to the problem. This technique is beyond the scope of this book, but the interested student can find the details in a recent work (*1*). Our approach here will be qualitative, intuitive, and descriptive (*32*).

Consider an octahedral complex in which the metal atom is placed at the origin of the usual Cartesian coordinate system and the ligands are placed on the axes as indicated in Fig. 18.2. The metal orbitals available for bond formation are of the type ns, $(n - 1)d$, and np; the ligands have an appropriate orbital such as a p or hybrid orbital directed towards the metal atom. The nine metal orbitals, shown in Table 18.12 are members of one of four symmetry classes which are designated according to their origin in group theory. Briefly, A_{1g} orbitals are nondegenerate and have the full symmetry of the molecule, e_g are doubly degenerate orbitals which differ in spatial orientation, and T_{2g} and T_{1u} are triply degenerate orbitals which differ in spatial orientation. The subscripts g (German *gerade* for even) and u (German *ungerade* for odd) indicate whether the wave function undergoes a change in sign upon inversion through the origin. Orbitals which are centrosymmetric are called *gerade* and carry the subscript g; noncentrosymmetric orbitals are *ungerade* (*u*).

TABLE 18.12   Symmetry classification for octahedral complexes with no π bonding

Symmetry class	Metal orbital	Composite ligand orbitals
A_{1g}	ϕ_{ns}	$\Sigma = \dfrac{1}{\sqrt{6}}(\sigma_x + \sigma_{-x} + \sigma_y + \sigma_{-y} + \sigma_z + \sigma_{-z})$
e_g	$\phi_{(n-1)d_{z^2}}$	$\Sigma_{z^2} = \dfrac{1}{2\sqrt{3}}(2\sigma_z + 2\sigma_{-z} - \sigma_x - \sigma_{-x} - \sigma_y - \sigma_{-y})$
	$\phi_{(n-1)d_{x^2-y^2}}$	$\Sigma_{x^2-y^2} = \dfrac{1}{\sqrt{2}}(\sigma_x + \sigma_{-x} - \sigma_y - \sigma_{-y})$
T_{1u}	ϕ_{np_z}	$\Sigma_z = \dfrac{1}{\sqrt{2}}(\sigma_z - \sigma_{-z})$
	ϕ_{np_x}	$\Sigma_x = \dfrac{1}{\sqrt{2}}(\sigma_x - \sigma_{-x})$
	ϕ_{np_y}	$\Sigma_y = \dfrac{1}{\sqrt{2}}(\sigma_y - \sigma_{-y})$
T_{2g}	$\phi_{(n-1)d_{xy}}$	

TABLE 18.12—continued

Symmetry class	Metal orbital	Composite ligand orbitals
T_{2g}	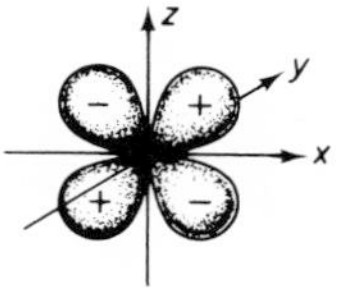 $\phi_{(n-1)d_{xz}}$	—
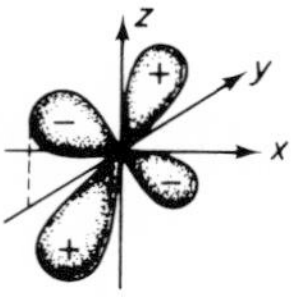	$\phi_{(n-1)d_{yz}}$	—

The six ligand orbitals which go into the formation of the σ component of the metal–ligand bond are generalized in this discussion as ϕ. The individual ϕ orbitals are combined into six composite orbitals which are constructed to give effective overlap with the metal orbital with the same symmetry characteristics; this overlap leads to the σ component of the metal–ligand bond. The ligand orbitals also may be classed according to their symmetry; their distribution is shown in Table 18.12. It should be noted that there are no ligand orbitals in the symmetry class T_{2g} which are capable of contributing to the σ component of the metal–ligand bond. We shall return to π bonding shortly.

Each metal orbital can now be combined with the corresponding ligand orbital using the LCAO method described in Chapter 4 to yield the usual bonding and antibonding pair of molecular orbitals (cf. Fig. 4.9). It is generally assumed that the bonding and antibonding molecular orbitals are symmetrically placed above and below the mean energy of the metal and ligand orbitals from which they were derived. The energy-level diagram shown in Fig. 18.15 results when all the metal orbitals are combined with the ligand orbitals of the same symmetry class. The molecular orbitals in this diagram are designated by the symmetry of the orbitals from which they are derived. When such energy-level diagrams are used to interpret the nature of the bonding in a complex species it is assumed that a given molecular orbital has the character associated with the atomic orbital which is nearest in energy. Thus, the E_g, A_{1g}, and T_{1u} molecular orbitals, as shown in Fig. 18.15, are primarily ligand orbitals, and the electrons in these orbitals are primarily on the ligands. The remaining molecular orbitals are primarily metal-like. It

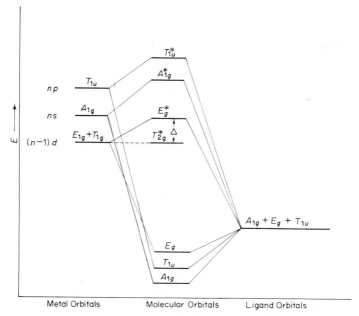

FIG. 18.15 A generalized molecular-orbital diagram for an octahedral complex in which only σ bonding is important.

should be noted that the T_{2g} orbitals which do not enter into σ bonding (Fig. 18.15) are unchanged in energy and become nonbonding in the complex.

Octahedral complexes, of course, contain six ligands which contribute twelve electrons to the bonding scheme. Thus, the total number of electrons to be placed in the available molecular orbitals is $12 + d^n$, where d^n represents the number of d electrons associated with the metal. Since the molecular orbitals of symmetry E_g and T_{1u} are doubly and triply degenerate, respectively, the maximum number of bonding electrons is given by the configuration $(T_{1u})^6(A_{1g})^2(E_g)^4$. Thus, there will always be d^n electrons available to be placed in the molecular orbitals which are higher in energy than the bonding orbitals. The next available molecular orbitals are T_{2g} (nonbonding) and $E_g{}^*$, which is exactly the set that arises in the crystal-field splitting of the metal d orbitals shown in Fig. 18.3. In other words, the distribution of the d^n electrons in the crystal field theory and the molecular orbital theory must be made among the same orbitals using the same arguments concerning the relative energies required to pair an electron or to place it in a slightly-higher-energy orbital. However, the molecular-orbital theory provides a means to accommodate the observations previously mentioned which indicate the covalent nature of the metal–ligand interaction. For example,

Fig. 18.15 indicates that all octahedral complexes should have, in addition to the d–d transitions, a band which corresponds to a transfer of charge from predominantly ligand orbitals to predominantly metal orbitals. The intense bands in the ultraviolet region (cf. Fig. 18.7) observed for many complexes correspond to transitions of σ electrons to empty T_{2g} or $E_g{}^+$ orbitals. Moreover, it is now possible to understand how the electron density on the ligand can be affected by unpaired metal electrons.

Multiple metal–ligand bonding can be described by molecular–orbital theory also. The metal atom in a complex contains two types of orbitals (i.e., T_{1u} and T_{2g}) which are of the correct symmetry for π bonding. In a complex species the metal orbitals in the symmetry class $T_{1u}(np_x, np_y, np_z)$ are involved in the formation of σ molecular orbitals and are presumably unavailable for π bonding. This leaves only orbitals of the T_{2g} symmetry class (d_{xy}, d_{xz}, d_{yz}) for the formation of a π component.

The symmetry classes of ligand orbitals which can form π bonds can be deduced from group theory (I). In an octahedral complex the six ligands which have a pair of mutually perpendicular orbitals capable of forming π bonds give rise to twelve orbitals that belong to four symmetry classes, T_{1g}, T_{2g}, T_{1u}, and T_{2u}. Since there are no metal orbitals of the symmetry classes T_{1g} and T_{2u}, these ligand orbitals become nonbonding in the molecule. The ligand T_{1u} orbitals will not combine with the metal T_{1u} orbitals because these orbitals are already involved in σ bonding. Thus, the ligand T_{2g} and the metal orbitals in the same symmetry class are the only remaining candidates to form the metal–ligand π molecular orbital. It will be recalled from Fig. 18.15 that the metal T_{2g} orbital is nonbonding in octahedral complexes with ligands that engage only in σ bonding. The interaction of ligand orbitals of the same symmetry class, however, leads to a T_{2g} molecular orbital which is of different energy. Two important cases exist, as shown in Fig. 18.16. If the T_{2g} ligand orbital is empty and of higher energy than the metal orbital, the resulting molecular T_{2g} orbital is stabilized relative to a pure metal T_{2g} orbital; the latter is nonbonding in complexes which contain no π-bonding ligands. Thus, the energy difference, Δ, between the molecular T_{2g} and $E_g{}^*$ (Fig. 18.16b) levels increases relative to the difference where metal–ligand π bonding does not exist (Fig. 18.16a). The other important case involves ligand T_{2g} orbitals (Fig. 18.17c). In this case the T_{2g} molecular orbital is also stabilized relative to the original metal orbital, but since the original ligand T_{2g} orbitals contained electrons the molecular T_{2g} orbital is filled. Thus, it is the difference between the $T_{2g}{}^*$ and $E_g{}^*$ orbitals which determines the value of Δ (Fig. 18.16c), which is less than that in the non-π-bonded case (Fig. 18.16a).

The inclusion of π orbitals in the molecular-orbital description of complex species provides the basis for a more complete understanding of the properties of such substances. Thus, we might expect effects due to π bonding with ligand atoms which have empty T_{2g}-type orbitals, such as the trivalent

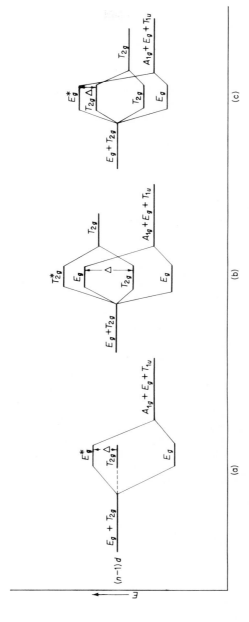

FIG. 18.16 The relative order of molecular orbitals for an octahedral complex involving π bonding [(b) and (c)] compared with the same complex where π bonding is not important (a).

derivatives of Group V elements (PR_3, AsR_3, etc.) and the divalent derivatives of Group VI elements (SR_2, TeR_2, etc.) in Periods 3 and higher. Metal–
ligand π bonding is also possible with ligands such as the halides and derivatives of the Period 6 elements that have both higher-energy empty and lower-
energy filled T_{2g} orbitals, such as O^{2-}, F^-, OH^-, and NH_2^-, can also form π
bonds to transition-metal ions. It is believed that π bonding also occurs when
saturated carbon-containing ligands are bound to metal ions. In such cases
anti-bonding π ligand orbitals interact with the T_{2g} metal orbitals. For example,
in the case of cyanide ion or the isoelectronic CO molecule all of the bonding
σ and π orbitals are filled (see Fig. 4.16). These species have anti-bonding π
orbitals, shown in Fig. 18.17, which have the correct symmetry to interact

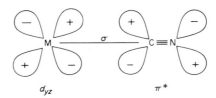

FIG. 18.17 An orbital representation of the interaction of metal orbitals with antibonding π orbitals in a metal–cyanide complex which leads to a decrease in the C—N
bond order.

with the ligand T_{2g} orbitals. If the resulting π molecular orbital contains
electrons, the CN or CO bond order must decrease because electron density
is now present in an antibonding component of this bond; this effect is
observed as a decrease in the CO or CN infrared stretching frequency of
these ligands upon complexation. Moreover, the increase in stability of the
metal–CN bond accounts for the observation that this ligand is at the high-
energy end of the spectrochemical series, a position which is not intuitively
expected on the basis of the simple crystal-field theory.

The molecular-orbital theory is equally successful when applied to tetrahedral complexes and complexes exhibiting other geometries. At this point
in time the quantitative application of molecular-orbital theory to complex
species involves a considerable calculational effort, during the course of
which rather arbitrary decisions must be made. There is generally little
agreement concerning the basis for choosing values for certain parameters.
The interested reader is directed to Refs. (33)–(38), which are helpful in
understanding the nature of the problems involved.

REFERENCES

1. N. V. Sidgwick, *J. Chem. Soc.*, **123**, 725 (1923).
2. N. V. Sidgwick, *Trans. Faraday Soc.*, **19**, 469 (1923).

3. N. V. Sidgwick, *J. Soc. Chem. Ind.*, **42**, 901, 1203 (1923).

4. T. Lowry, *J. Soc. Chem. Ind.*, **42**, 316 (1923).

5. N. V. Sidgwick and H. M. Powell, *Proc. Royal Soc. A*, **176**, 164 (1940).

6. F. A. Cotton and R. M. Wing, *Inorg. Chem.*, **4**, 314 (1965).

7. F. A. Cotton, *Inorg. Chem.*, **3**, 702 (1964).

8. F. J. Garrick, *Phil. Mag.*, **9**, 131 (1930).

9. F. J. Garrick, *Phil. Mag.*, **10**, 71, 76 (1930).

10. F. J. Garrick, *Phil. Mag.*, **11**, 741 (1931).

11. F. J. Garrick, *Phil. Mag.*, **14**, 914 (1932).

12. S. F. A. Kettle, *J. Chem. Ed.*, **46**, 339 (1969).

13. R. Krishnamurthy and W. B. Schaap, *J. Chem. Ed.*, **46**, 799 (1969).

14. R. Krishnamurthy and W. B. Schaap, *J. Chem. Ed.*, **47**, 433 (1970).

15. C. J. Ballhausen and C. K. Jorgensen, *Kgl. Dansk. Videnskab. Selskab. Mat. Fys. Medd.*, **29**, No. 14 (1955).

16. C. J. Ballhausen, *Kgl. Dansk. Videnskab. Selskab. Mat. Fys. Medd.*, **29**, No. 4 (1954).

17. C. K. Jorgensen, *Absorption Spectra and Chemical Bonding in Complexes*, Pergamon Press, London, 1961.

18. C. K. Jorgensen, *Advan. Chem. Phys.*, **5**, 33 (1962).

19. D. F. Shriver, S. A. Shriver, and S. E. Anderson, *Inorg. Chem.*, **4**, 725 (1965).

20. C. J. Ballhausen, *Introduction to Ligand Field Theory*, McGraw-Hill, New York, 1962.

21. B. N. Figgis, *Introduction to Ligand Fields*, Wiley, London, 1966.

22. Y. Tanabe and S. Sugano, *J. Phys. Soc. Japan*, **9**, 753, 766 (1954).

23. R. G. Shulman and V. Jaccarino, *Phys. Rev.*, **108**, 1219 (1957).

24. R. G. Shulman and K. Knox, *Phys. Rev.*, **119**, 94 (1960).

25. J. Owen and K. W. H. Stevens, *Nature*, **171**, 836 (1953).

26. J. H. E. Griffiths and J. Owen, *Proc. Roy. Soc. A*, **226**, 96 (1954).

27. J. S. van Wieringen, *Disc. Faraday Soc.*, **19**, 118 (1955).

28. C. H. Townes and B. P. Dailey, *J. Chem. Phys.*, **17**, 782 (1949).

29. N. N. Greenwood, *Chem. Britain*, **3**, 56 (1967).

30. R. H. Herber, R. B. King, and G. K. Wertheim, *Inorg. Chem.*, **3**, 101 (1964).

31. J. Owen, *Disc. Faraday Soc.*, **19**, 127 (1955).

32. S. F. A. Kettle, *J. Chem. Ed.*, **43**, 21, 652 (1966).

33. H. Basch, A. Veste, and H. B. Gray, *J. Chem. Phys.*, **44**, 10 (1966).

34. R. F. Fenske, K. G. Caulton, D. D. Radtke, and C. C. Sweeney, *Inorg. Chem.*, **5**, 951, 960 (1966).

35. F. A. Cotton and C. B. Harris, *Inorg. Chem.*, **6**, 369, 376 (1967).

36. H. Basch and H. B. Gray, *Inorg. Chem.*, **6**, 365 (1967).

37. R. F. Fenske and C. C. Sweeney, *Inorg. Chem.*, **3**, 1105 (1964).

38. R. F. Fenske, *Inorg. Chem.*, **4**, 33 (1965).

COLLATERAL READINGS

I. F. A. Cotton, *Chemical Applications of Group Theory*, Interscience, New York, 1963. A very useful description of group-theoretical arguments applied to various chemically interesting areas.

II. B. N. Figgis and J. Lewis, *Prog. Inorg. Chem.*, **6**, 37 (1964). A discussion of the magnetic properties of transition-metal complexes and their interpretation in terms of ligand-field theory.

III. C. J. Ballhausen, *Prog. Inorg. Chem.*, **2**, 251 (1960); J. Ferguson, *Pure Appl. Chem.*, **14**, 1 (1964); J. Ferguson, *Prog. Inorg. Chem.*, **12**, 159 (1970); A. B. P. Lever, *J. Chem. Ed.*, **45**, 711 (1968); N. S. Hush and R. J. M. Hobbs, *Prog. Inorg. Chem.*, **10**, 259 (1968). These reviews contain various aspects of the interpretation of the electronic spectra of transition-metal complexes.

IV. P. George and D. S. McClure, *Prog. Inorg. Chem.*, **1**, 381 (1959). A discussion of the effect of *d*-orbital splittings on the thermodynamic properties of transition-metal compounds and coordination complexes.

V. J. A. Creighton, *Essays in Chemistry*, **2**, 45 (1971). A short discussion of the Jahn–Teller effect on transition-metal complexes.

VI. F. A. Cotton, *Pure Appl. Chem.*, **16**, 175 (1966). This review on semiempirical calculations of the electronic structures of transition-metal complexes contains a section in which experimental data are presented to estimate covalence.

STUDY QUESTIONS

1. Which of the following complex ions might be expected to be inert: $Mn(CN)_6^{4-}$; $Rh(NH_3)_6^{3+}$; RuF_6^{2-}; $V(H_2O)_6^{3+}$?

2. The tetrahedral complex ion $CoCl_4^{2-}$ is a high-spin complex. Describe the bonding in this species in terms of valence-bond arguments. Describe the bonding in this complex using simple crystal-field arguments.

3. Discuss the observation that Co^{2+} readily forms tetrahedral complexes whereas Ni^{2+} does not.

4. Although $CrCl_3$ is appreciably soluble in water, the rate of solution of the anhydrous material is exceedingly slow. However, the rate can be increased by addition of a trace of Sn^{2+}. Discuss these observations.

5. Nickel is known to form the following compounds: $K_2Ni(CN)_4$, $Ni(en)_2(SCN)_2$, and $Ni(CO)_4$. (a) Using valence-bond arguments, discuss the possible structures for $K_2Ni(CN)_4$. In your discussion include a consideration of the types of bonds formed and the magnetism expected for these structures. (b) Explain how $Ni(CO)_4$ can be both tetrahedral and diamagnetic. (c) Assuming $Ni(en)_2(SCN)_2$ can be formulated as a hexacoordinated species, draw all of the possible isomers.

6. Assuming that Cr^{2+} forms tetrahedral complex ions with a variety of ligands, explain how the expected magnetism for high-field and for low-field ligands would be predicted.

7. Discuss the expected lability of tetrahedral Cr^{2+} complexes.

8. Give an example of a coordination compound in which $d\pi$-$p\pi$ bonding might occur between the metal and ligand.

9. Determine the effective atomic number of the metal atom in each of the following species: (a) $AlCl_3 \cdot (CH_3)_2O$; (b) $Cr(CO)_6$; (c) $TiCl_6^{2-}$; (d) $Cu(H_2O)_4^{2+}$; (e) $V(CN)_6^{4-}$; (f) $PdCl_2(NH_3)_2$; (g) $Ni(PF_3)_4$; (h) $[PdCl_2-P(C_6H_5)_3]_2$; (i) PtF_6^-.

10. Using ligand-field arguments discuss the magnetism of $Mn(H_2O)_6^{3+}$.

11. Discuss the magnetism of the tetrahedral species $FeCl_4^-$ using (a) valence-bond arguments, and (b) simple crystal-field arguments. (c) What is the effective atomic number of iron in this complex?

12. Using valence-bond arguments, how could you determine which orbitals are used in the formation of the complex ion $MnCl_6^{4-}$ from magnetic measurements alone?

13. Discuss the spectrum of $[Ti(H_2O)_6]^{3+}$ shown in Fig. 18.7 in terms of the Jahn–Teller effect.

14. Discuss the fact that $Co(NCS)_4^-$ exhibits a magnetic moment of 4.4 Bohr magnetons in terms of the spectrochemical series.

15. Consider the compound $Ni(en)_2(SCN)_2$ in answering the following questions. Assuming that this compound is an octahedral complex, draw all possible isomers, indicating those that are optically active. Describe how to determine, if it *is* possible to determine, whether this compound is an octahedral complex, a tetrahedral complex, or a square-planar complex, using magnetic data only.

16. Assume that $Ni(en)_2(CN)_2$ can be formulated as either an octahedral or a square-planar complex. Give the simple crystal-field description of this compound for both formulations.

17. The reduction of $K_2[Ni(CN)_4]$ with sodium amalgam yields a red diamagnetic compound with the empirical formula $K_2[Ni(CN)_3]$. (a) Suggest a possible structure for the red compound. (b) Give a theoretical description for the constitution of the red compound.

18. Using simple crystal-field theory explain why $Mn(H_2O)_6^{2+}$ is colorless.

19. Explain how the complex ion $[Co(CN)_6]^{3-}$ can be inert to substitution reactions while the complex ion $[Co(Br)_6]^{3-}$ is labile with respect to substitution of Br^- by other ligands.

20. Using crystal-field arguments, predict the expected coordination number of Mo^{5+}.

21. Estimate the spin-only magnetic moment for a d^6 ion in octahedral and tetrahedral fields generated by weak- and strong-field ligands.

22. Estimate the difference in relative energies of an octahedral field and a tetrahedral field for d^4 high-spin complexes in terms of Dq.

23. Discuss the observation that chromous acetate monohydrate is dimeric and paramagnetic.

24. Consider the compounds described in Question 7 of Chapter 17. (a) Give the expected electronic distribution for the nickel atoms in each compound using valence-bond arguments. (b) Describe the structure of each compound using crystal-field theory.

25. Nickel tetracarbonyl $[Ni(CO)_4]$ is known to be diamagnetic. Using valence-bond arguments, predict the structure of this compound. Explain the fact that the experimentally determined length of the Ni—C bond in nickel tetracarbonyl is about 20% shorter than would be expected from theory.

ORGANIC DERIVATIVES OF THE TRANSITION METALS

1 / INTRODUCTION

Since the early 1950's the chemistry of compounds containing transition metals bonded to carbon-containing moieties has attracted enormous interest. These compounds are of practical and theoretical interest. Some of these compounds possess catalytic activity in important processes. The structures of certain organic derivatives of transition metals provide the basis for rigorous tests of bonding theories.

The chemistry of the organic derivatives of transition metals is best discussed for our purposes in the familiar terms of coordination chemistry. That is, the organic moiety is considered as a ligand which coordinates to the transition-metal atom. Previously, the ligand–metal σ bond was described conceptually as being formed by the donation of a pair of electrons from the ligand atom to the metal. These ideas are continued in discussing the organic derivatives of the transition metals, except that in some cases the ligands can formally donate more than two electrons to the metal atom to form a bond. Table 19.1 contains examples of the types of carbon-containing ligands which fall within this classification scheme. As in most classification attempts,

TABLE 19.1   Classification of organic ligands

Number of electrons	Ligand	Example Ligand	Example Complex
1	alkyl	$\cdot R$	$[(CH_3)_3PtCl]_4,$ $R_3PAu(CH_3)_3$
	aryl	$\cdot C_6H_5$	$(C_6H_5)AuCl_3,$ $(C_6H_5)_3Cr[O(C_2H_5)_2]_3$
	acyl	$\cdot \overset{\displaystyle O}{\overset{\|}{C}}-R$	$R\overset{\displaystyle O}{\overset{\|}{C}}-Co(CO)_2L_2$
2	carbonyl	$:CO$	$Cr(CO)_6$
	alkene	$R_2C{=}CR_2$	$Pt(C_2H_4)_2Cl_2$
3	allyl	$H_2C\overset{CH}{\diagup\;\diagdown}CH_2$	$Ni(C_3H_5)_2$
	cyclopropenyl	$HC{=}{=}CH$ with CH bridge	$[Ni(CO)(C_6H_5)_3(C_3H_3)Br]_2$
4	butadiene	$H_2C\overset{CH-CH}{\diagup\quad\diagdown}CH_2$	$Fe(CO)_3(C_4H_6)$
	cyclobutadiene	$\begin{matrix}HC-CH\\\|\quad\|\\HC-CH\end{matrix}$	$Fe(CO)_3(C_4H_4)$
5	cyclopentadienyl	(cyclopentadienyl structure)	$Fe(C_5H_5)_2$
6	benzene	(benzene structure)	$Cr(C_6H_6)_2$
	cycloheptatriene	(cycloheptatriene structure)	$Mo(C_7H_8)(CO)_3$
7	cycloheptatrienyl	(cycloheptatrienyl structure)	$[Mo(C_7H_7)(CO)_3]BF_4$

some rather arbitrary assignments, which eventually lead to ambiguities, arise. For example, the compounds listed in Table 19.1 as one-electron ligands correspond to organometallic derivatives σ-bonded to a metal atom (**1**), but they could also be interpreted as containing a metal ion in a positive

$$M:R \qquad\qquad M^+:R^-$$

$$\mathbf{1} \qquad\qquad\quad \mathbf{2}$$

oxidation state forming a coordinate-covalent bond with a carbanion ligand (**2**). Thus, the methyl groups in $R_3PAu(CH_3)_3$ might be classified as one-electron ligands attached to gold atoms or methide ions (CH_3^-) coordinated to Au^{3+}. The object of this discussion is not to describe the constitution of the one-electron ligands, but to recognize that the classification in Table 19.1 is arbitrary in some cases.

Inspection of the ligands listed in Table 19.1 shows that most of the organic ligands which donate an even number of electrons to the transition-metal ion contain unsaturated groups. For example, the two ethylene molecules in the complex $Pt(C_2H_4)_2Cl_2$ apparently occupy coordination positions in much the same way as do the ammonia molecules in the complex $Pt(NH_3)_2Cl_2$. Conceptually, the π component in the carbon–carbon double bond of the ethylene molecule is equivalent to the unshared electron pair in ammonia. In neither case is the electron pair necessary to maintain the integrity of the atoms that make up the free ligand molecules, and, in this sense, both molecules are unsaturated. Thus, unsaturation in carbon compounds, whether conjugated or unconjugated, appears to act as the source of electron density in forming complexes. Except for the one-electron ligands, carbon-containing ligands which donate odd numbers of electrons to the transition-metal atom are formally organic radicals which also have unsaturated sites.

Broadly speaking, two types of bonding have been observed for organic ligands. In one case, a single carbon atom of the ligand is attached directly to a metal atom through a σ bond (**1**); one-electron ligands and the two-electron ligand CO fall into this classification. The one-electron ligands are formally substitution products of organic molecules. The other broad class of compounds involves electron donation from the entire organic moiety; in such cases there is no apparent preferential bond formed between the metal atom and any single carbon atom. Ligands bonded in this way are said to be π-bonded. This broad distinction between σ- and π-bonded ligands is important in the case of, for example, the allyl ligand which can be either σ-bonded (**3**) or π-bonded (**4**). In the first instance (**3**), the allyl group acts as a

3   **4**

one-electron ligand, while in the second case (**4**) it is a three-electron ligand. Examples of both types of behavior for the allyl moiety are known. Similar possibilities occur for the five- and seven-electron ligands. Generally, an attempt is made to reflect the mode of bonding in complexes containing such ligands, if it is known, in the formulations of these compounds, e.g., π-$C_5H_5Fe(CO)_2$-σ-C_5H_5 and π-$C_5H_5Fe(CO)_2$-σ-$CH_2CH{=}CHCH_3$.

2 / ELECTRONIC CONSIDERATIONS (*l*)

The use of the effective atomic number concept in rationalizing the stoichiometry of coordination compounds is discussed in Chapter 18. It will be recalled that, in general, complex ions formed from elements near the beginning of the transition series do not obey the EAN concept (see Table 18.1), which is basically an extension of the Lewis ideas concerning stable electronic configurations to include atoms which possess low-energy d orbitals. A large number of organometallic derivatives of the transition elements, however, do obey the EAN rule described in Chapter 18, which makes it useful in understanding the unusual structures observed in these systems. The modern expression of the ideas embodied in Sidgwick's effective atomic number is the 18-electron rule, which recognized the fact that ns, $(n-1)d$, and np orbitals of transition-metal atoms are generally used for bonding; a stable electronic configuration should arise when these orbitals are filled, i.e., $2 + 10 + 6 = 18$ electrons. Application of the 18-electron rule requires a bookkeeping scheme to account for the electrons possessed by the metal atom (the number of which depends upon its formal oxidation state) and those donated by the ligands; the latter number is the basis for the classification of ligands shown in Table 19.1. For example, iron forms a complex with butadiene, $Fe(CO)_3C_4H_6$, which has been formulated as **5**. The ligands are neutral, which means that iron is formally zero-valent and, hence, possesses eight electrons $(4s^2 3d^6)$. Three carbon monoxide ligands

5

6

donate two electrons each for a total of six electrons; the butadiene moiety is a four-electron ligand. Thus, the iron atom has a total of $8 + 6 + 4 = 18$

electrons associated with it. A second example should suffice to illustrate the 18-electron rule. Rhodium forms a mixed π complex, $Rh(\pi\text{-}C_5H_5)(\pi\text{-}C_2H_4)_2$, which has been formulated as (**6**). Since, according to the information in Table 19.1, both C_5H_5 and C_2H_4 are considered neutral ligands, the metal atom is formally zero-valent and contains nine electrons ($5s^2 4d^7$). Cyclopentadiene is a five-electron ligand, and each ethylene molecule contributes two electrons, leading to $9 + 5 + 2(2) = 18$ electrons associated with the transition-metal atom. The 18-electron rule is often useful in predicting the structure of organometallic derivatives of the transition metals; however, as in the case of the representative elements, there are occasional "violations" of the rule. In these instances, the compounds have unusual properties which reflect such "violations."

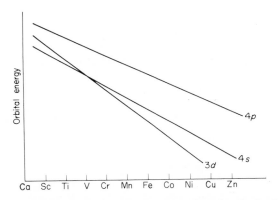

FIG. 19.1   The relative energies of valence-shell orbitals for the Period 3 elements.

The basis for the 18-electron rule lies in the fact that the ns, np, and $(n - 1)d$ orbitals of transition metals are of sufficiently similar energies to constitute the valence orbital. Figure 19.1 is a schematic representation of the relative energies of these orbitals for the first transition series. Several features in this diagram should be noted. First, the binding energy of the valence orbitals increases with increasing atomic number. The slopes of the lines for $4s$ and $4p$ orbitals are virtually the same and the binding energy of the $3d$ orbitals increases more rapidly than either of these. The 18-electron rule is violated for elements near the beginning or end of the transition series where the $3d$ orbitals are either relatively high in energy or they are so low as to be considered part of the core. This generalization finds support in the fact that there are as yet no known 18-electron complexes of titanium and only a few for copper. The commonly observed 16-electron environment for nickel complexes can also be understood in terms of Fig. 19.1. In this case the $3d$ electrons are marginally important as far as metal–ligand bonding is concerned; a more important consideration is the difference in energy of the $4s$ and $4p$ orbitals.

3 / ONE-ELECTRON LIGANDS

σ-Bonded alkyl, aryl, and acyl groups are among the most common one-electron ligands. Compounds containing these ligands can be prepared by one of several general methods, although some specialized methods are available for specific compounds. Thus, the reaction of metal–halogen derivatives with Grignard reagents or with the more active organoalkyl metals compounds under the usual anaerobic conditions leads to complexes containing one-electron ligands:

$$CrCl_3 + 3C_6H_5MgX \xrightarrow{\text{THF}} Cr(C_6H_5)_3 \cdot 3THF + 3MgXCl \qquad (1)$$

$$\pi\text{-}C_5H_5Fe(CO)_2Cl + RMgX \rightarrow \pi\text{-}C_5H_5Fe(CO)_2R + MgXCl \qquad (2)$$

$$(\pi\text{-}C_5H_5)_2TiCl_2 + 2LiCH_3 \rightarrow (\pi\text{-}C_5H_5)_2Ti(CH_3)_2 + 2LiCl \qquad (3)$$

$$cis\text{-}[(C_2H_5)_3P]_2PtCl_2 + 2LiCH_3 \rightarrow cis\text{-}[(C_2H_5)_3P]_2Pt(CH_3)_2 + 2LiCl \quad (4)$$

$$NiL_2Cl_2 + C_6H_5Li \rightarrow NiL_2(C_6H_5)_2 + 2LiCl. \qquad (5)$$

$$L = R_3P \text{ or bipyridyl}$$

In some instances the more reactive organoalkali-metal alkylating agents form anionic complexes:

$$CrCl_3 + 6LiCH_3 \xrightarrow[-18°]{\text{dioxane}} Li_3[Cr(CH_3)_6] \cdot 3C_4H_8O_2 + 3LiCl \qquad (6)$$

$$CrCl_3 + 6LiC_6H_5 \xrightarrow{(C_2H_5)_2O} Li_3[Cr(C_6H_5)_6] \cdot 4(C_2H_5)_2O + 3LiCl. \qquad (7)$$

Other moderately reactive organometallic alkylating agents have been used to prepare alkyl derivatives of some transition metals

$$CH_3-AlCl_2 + TiCl_4 \rightarrow CH_3TiCl_3 + AlCl_3 \qquad (8)$$

$$(C_2H_5)_4Pb + TiCl_4 \rightarrow C_2H_5TiCl_3 + (C_2H_5)_3PbCl. \qquad (9)$$

A second general method of preparation involves the action of transition-metal-containing anions as nucleophiles on carbon–halogen bonds

$$Na^+[Mn(CO)_5]^- + CH_3I \rightarrow CH_3Mn(CO)_5 + NaI \qquad (10)$$

$$[(CN)_5Co]^{3-} + C_6H_5CH_2Br \rightarrow [(CN)_5CoCH_2C_6H_5]^{2-} + Br^- \tag{11}$$

$$Na^+[Mn(CO)_5]^- + R\!-\!\overset{O}{\overset{\|}{C}}\!-\!Cl \rightarrow R\!-\!\overset{O}{\overset{\|}{C}}\!-\!Mn(CO)_5 + NaCl \tag{12}$$

$$R = CH_3, CF_3, C_6H_5$$

$$Na^+[\pi\text{-}C_5H_5Fe(CO)_2]^- + C_6F_6 \rightarrow \pi\text{-}C_5H_5Fe(CO)_2\text{-}\sigma\text{-}C_6H_5 + NaF \tag{13}$$

$$Na^+[Mn(CO)_5]^- + CF_2\!=\!CFCF_2Cl$$

$$\rightarrow (CO)_5MnCF\!=\!CFCF_2Cl + NaF. \tag{14}$$

The relative nucleophilicity of the carbonyl anions, which are discussed in more detail in a subsequent section, decreases in the following order based on their reactivity with perfluoro-olefins: $[\pi\text{-}C_5H_5Fe(CO)_2]^- > [Re(CO)_5]^- > [Mn(CO)_4P(C_6H_5)_3]^- > [Mn(CO)_5]^- > [Co(CO)_4]^- \approx [Fe(CO)_4]^{2-}$.

One-electron carbon-containing ligands also have been prepared by the addition of a transition-metal–hydrogen bond to an olefin (*II*)

$$trans\text{-}[(C_2H_5)_3P]_2PtClH + C_2H_4 \rightarrow trans\text{-}[(C_2H_5)_3P]_2PtClC_2H_5 \tag{15}$$

$$\pi\text{-}C_5H_5Fe(CO)_2H + CH_2\!=\!CHCN \rightarrow \pi\text{-}C_5H_5Fe(CO)_2CH(CH_3)CN \tag{16}$$

$$Co(CO)_4H + C_2F_4 \rightarrow Co(CO)_4CF_2CF_2H \tag{17}$$

or to an alkyne (Eq. 18)

$$(CO)_5MnH + CF_3C\!\equiv\!CCF_3 \rightarrow (CO)_5Mn(CF_3)C\!=\!CHCF_3. \tag{18}$$

Addition reactions to unsaturated systems have also been observed for other groups bonded to metal atoms:

$$Co_2(CO)_8 + C_2F_4 \rightarrow (CO)_4Co\!-\!CF_2CF_2Co(CO)_4 \tag{19}$$

$$(CH_3)_3Sn\!-\!Mn(CO)_5 + C_2F_4 \rightarrow (CH_3)_3SnCF_2CF_2Mn(CO)_5 \tag{20}$$

$$[(C_2H_5)_3P]_2PtF_2 + C_2F_4 \rightarrow [(C_2H_5)_3P]_2Pt(C_2F_4)F \tag{21}$$

$$Fe(CO)_5 + 2C_2F_4 \rightarrow (CO)_4Fe\!\!\begin{array}{c} CF_2\!-\!CF_2 \\ \diagup \quad \mid \\ \diagdown \quad \mid \\ CF_2\!-\!CF_2 \end{array} + CO. \tag{22}$$

3.1 / Reactions

Transition-metal compounds containing one-electron organic ligands exhibit a wide range of stabilities which depend upon the nature of the other ligands present as well as on the transition metal. Metal–carbon bonds are easily cleaved by conventional oxidizing agents

$$(CH_3)_4Ti + 4I_2 \rightarrow TiI_4 + 4CH_3I \tag{23}$$

$$(C_2H_5)_2AuBr + Br_2 \rightarrow C_2H_5AuBr_2 + C_2H_5Br \tag{24}$$

$$CH_3Mn(CO)_5 + Br_2 \rightarrow BrMn(CO)_5 + CH_3Br, \tag{25}$$

as well as by acids in some cases

$$cis\text{-}[(C_2H_5)_3P]_2Pt(CH_3)_2 + HCl \rightarrow cis\text{-}[(C_2H_5)_3P]_2PtCH_3Cl + CH_4. \tag{26}$$

One-electron carbon ligands can add to olefins

$$\pi\text{-}C_5H_5Fe(CO)_2H + CH_2{=}CHCH{=}CH_2 \rightarrow$$
$$\pi\text{-}C_5H_5Fe(CO)_2CH_2CH{=}CHCH_3 \tag{27}$$

$$Ti{-}R + CH_2{=}CH_2 \xrightarrow{Al_2R_6} Ti{-}CH_2CH_2R. \tag{28}$$

The unsaturated molecules carbon monoxide and sulfur dioxide behave similarly

$$CH_3Mn(CO)_5 + CO \rightarrow CH_3{-}\overset{\displaystyle O}{\overset{\|}{C}}{-}Mn(CO)_5 \tag{29}$$

$$\pi\text{-}C_5H_5Fe(CO)_2CH_3 + SO_2 \rightarrow \pi\text{-}C_5H_5Fe(CO)_2SO_2CH_3. \tag{30}$$

Of the two possible structures for the product in Eq. (30), **7** or **8**, proton magnetic resonance studies support structure **8**. Insertion of carbon monoxide

$$\pi\text{-}C_5H_5{-}\overset{\displaystyle CO}{\underset{\displaystyle CO}{Fe}}{-}\overset{\displaystyle O}{\overset{\|}{S}}{-}OCH_3 \qquad\qquad \pi\text{-}C_5H_5{-}\overset{\displaystyle CO}{\underset{\displaystyle CO}{Fe}}{-}\overset{\displaystyle O}{\underset{\displaystyle O}{\overset{\|}{\underset{\|}{S}}}}{-}CH_3$$

$$\textbf{7} \qquad\qquad\qquad\qquad \textbf{8}$$

in the metal–CH$_3$ bond [Eq. (29)] appears to be an intramolecular process.

Treatment of methylmanganese pentacarbonyl with ^{14}CO produces acetyl-manganese pentacarbonyl with no labeled carbon in the acetyl

$$CH_3Mn(CO)_5 + {}^{14}CO \rightarrow CH_3-\overset{O}{\overset{\|}{C}}-Mn(CO)_4({}^{14}CO). \qquad (31)$$

Acetyl derivatives of manganese are also obtained with other substituting ligands

$$CH_3Mn(CO)_5 + L \xrightarrow{5°} CH_3-\overset{O}{\overset{\|}{C}}-Mn(CO)_4L, \qquad (32)$$
$$L = R_3P, R_3N$$

the product formed being a mixture of *cis* and *trans* isomers, with the latter predominating. The mixture can be converted entirely into the *cis* isomer by heating in hexane. Kinetic studies of the processes shown in Eqs. (31) (*1*) and (32) (*2*) suggest that the rate-controlling step involves breaking the Mn—CH$_3$ bond as the methyl group attacks carbon monoxide while the latter ligand is still attached to the manganese atom

$$\begin{array}{ccc} \overset{CH_3}{\overset{|}{O\equiv C-Mn}} & \rightarrow & \overset{CH_3}{\overset{|}{O=C\equiv Mn}} \end{array} \rightarrow \quad \overset{CH_3}{\overset{|}{\underset{O}{C-Mn.}}} \qquad (33)$$

Similar carbonylation reactions have been observed in other systems

$$C_6H_5Mn(CO)_5 + P(C_6H_5)_3 \rightarrow C_6H_5-\overset{O}{\overset{\|}{C}}-Mn(CO)_4P(C_6H_5)_3 \qquad (34)$$

$$\pi\text{-}C_5H_5Fe(CO)_2CH_3 + P(C_6H_5)_3 \rightarrow$$
$$\pi\text{-}C_5H_5Fe(\overset{O}{\overset{\|}{C}}-CH_3)(CO)P(C_6H_5)_3 \qquad (35)$$

$$trans\text{-}[(C_2H_5)_3P]_2Pt(CH_3)Cl + CO \rightarrow \quad \begin{array}{c} (C_2H_5)_3P \diagdown \quad \diagup \overset{O}{\overset{\|}{C}}-CH_3 \\ Pt \\ \diagup \quad \diagdown \\ Cl \quad\quad P(C_2H_5)_3 \end{array} \qquad (36)$$

Indeed, alkyl derivatives of cobalt carbonyl consist of an equilibrium mixture of the carbonyl and acyl compounds (*III*)

$$RCo(CO)_4 \rightleftharpoons R-\overset{\displaystyle O}{\overset{\|}{C}}-Co(CO)_3. \tag{37}$$

The perfluoroalkyl hydrocarbons do not undergo carbonylation reactions; in this respect, the perfluoroalkyl and -aryl moieties behave more like halogen substituents than like their formal hydrocarbon analogues.

The general order of stability of complexes containing one-electron carbon ligands increases in the following order:

$$\text{alkyl} < C_6H_5 < -\overset{\displaystyle O}{\overset{\|}{C}}-R < CF_3 < C_6F_5.$$

Thus, $(CH_3)_4Ti$ decomposes at room temperature to a mixture of products and $C_2H_5Mn(CO)_5$ decomposes at 90°

$$2C_2H_5Mn(CO)_5 \rightarrow Mn_2(CO)_{10} + C_4H_{10}. \tag{38}$$

Acyl derivatives decarbonylate upon heating

$$R-\overset{\displaystyle O}{\overset{\|}{C}}-Mn(CO)_5 \rightarrow RMn(CO)_5 + CO. \tag{39}$$

$$R = CH_3, C_6H_5, CF_3$$

In fact, reaction (39) is the preferred method for preparing complexes containing one-electron perfluoroalkyl or -aryl derivatives since the starting material is readily prepared from the corresponding acyl halide [Eq. (12)]. Moreover, perfluoroalkyl one-electron ligands are markedly more stable than their hydrogen analogues and perfluoroaryl derivatives are more stable than the perfluoroalkyl compounds.

3.2 / Bonding

The one-electron alkyl ligands, $-CR_3$, can be bound only to the metal atom by a conventional σ bond involving any of a number of possible metal orbitals; the strength of this bond has been discussed in terms of the ionic and covalent contributions to the wave function (*3, 4*). The results suggest that the M—C bond strength should increase with increasing effective electronegativity of the alkyl group and with decreasing atomic number of the metal within any family. The latter effect arises from the fact that the overlap integral should be greater for the lighter elements because of their less diffuse orbitals. Unfortunately at this point there is little information available to test these hypotheses. The greater stability of the perfluoro derivatives compared with the ordinary hydrocarbons may be a reflection of the increased effective electronegativity of the perfluoroalkyl group.

The greater stability of phenyl derivatives compared with the alkyl derivatives has been interpreted as a reflection of multiple bonding which could arise from either donation of electrons from filled $p\pi$ orbitals of the ligand to empty metal orbitals (**9**) or donation of electrons from the appropriate metal orbitals to empty $p\pi^*$ orbitals on the ligand (**10**). Metal–acyl

$$M \xleftarrow[\sigma]{d\pi\text{-}p\pi} L \qquad\qquad M \xrightarrow[\sigma]{d\pi\text{-}p\pi^*} L$$

9   **10**

complexes exhibit a markedly lower infrared frequency than the normal ketonic frequency (5, 6) which has been interpreted in terms of metal–ligand multiple bonding involving the π orbitals of the carbonyl group. Presumably this process involves interaction of the filled $p\pi$ orbital of the ligand with an appropriate empty d orbital of the metal (**11**). Some complexes containing

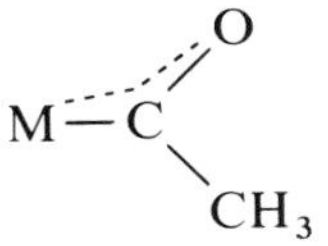

$$M - C \underset{CH_3}{\overset{O}{\big\langle}}$$

11

a one-electron acyl ligand are readily protonated

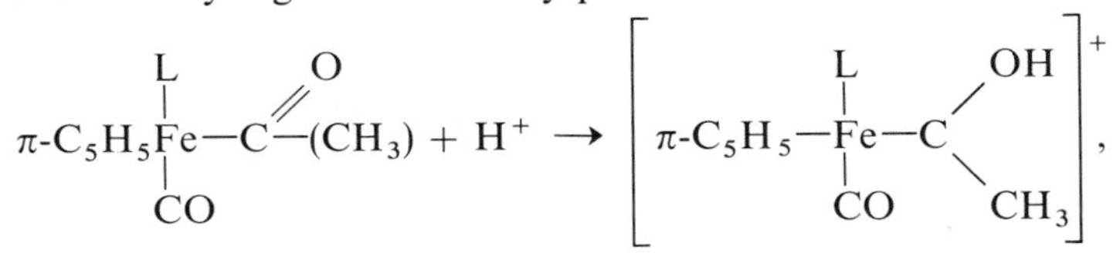

$$\pi\text{-}C_5H_5\underset{CO}{\overset{L}{Fe}}-C{\overset{O}{\parallel}}(CH_3) + H^+ \rightarrow \left[\pi\text{-}C_5H_5-\underset{CO}{\overset{L}{Fe}}-C\underset{CH_3}{\overset{OH}{\big\langle}}\right]^+, \qquad (40)$$

12

which supports this suggestion. Indeed, a metal–"carbene" complex has been isolated with a formulation similar to **12** for which a formal double bond is postulated (**13**) (7).

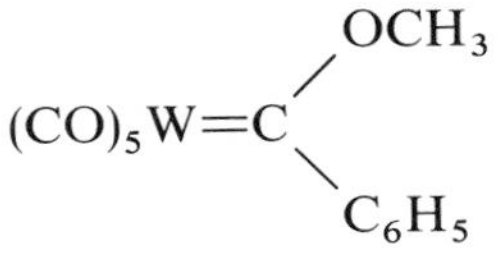

$$(CO)_5W{=}C\underset{C_6H_5}{\overset{OCH_3}{\big\langle}}$$

13

Finally, the noticeable stability of compounds containing metals bonded to perfluorinated carbon moieties and the decrease in the C—F stretching frequency have been attributed (8, 9) to metal–ligand π bonding arising from donation of electron density from metal d orbitals to the σ^* orbitals on the ligand.

4 / TWO-ELECTRON LIGANDS

Presently two types of two-electron carbon ligands are important, viz., σ- and π-bonded. Carbon monoxide, which forms complexes with many transition metals, is the most extensively studied ligand of the first kind. Olefins form the other large class of two-electron ligands. In the case of unsaturated compounds containing several olefinic bonds a distinction between those that are unconjugated, e.g., cyclooacta-1,5-diene (14), or

14   15   16

conjugated, e.g., butadiene (15) and cycloheptatriene (16), must be made. Ligands with two or more unconjugated double bonds are analogous to the more conventional polydentate ligands described in Chapter 17. However, ligands which contain conjugated double bonds are best considered separately because the complexes they form have markedly different characteristics, suggesting that the entire conjugated system of electrons is involved. Accordingly, butadiene (15) and cycloheptatriene are discussed in subsequent sections.

4.1 / Carbon Monoxide

Some of the better characterized binary metal carbonyls are described in Table 19.2. Two types of compounds are known, the simple metal carbonyls and polynuclear species. In general the compounds are easily volatile substances which are also soluble in nonpolar organic solvents. Except for $V(CO)_6$ and the cobalt carbonyls they are air-stable. The metal carbonyls burn readily, and the liquids $Fe(CO)_5$ and $Ni(CO)_4$ should be handled cautiously because they are toxic and form explosive mixtures with air.

TABLE 19.2 The properties of some binary metal carbonyls

	$V(CO)_6$	$Cr(CO)_6$	$Mn_2(CO)_{10}$	$Fe(CO)_5$	$Fe_2(CO)_9$	$Fe_3(CO)_{12}$	$Co_2(CO)_8$	$Co_4(CO)_{12}$	$Ni(CO)_4$
Color	black	colorless	golden	yellow	bronze	dark green	orange	black	colorless
Mp or bp	mp 70° dec.	—[a]	mp 154°[a]	bp 103°	mp 100° dec.	mp 140° dec.[a]	mp 51°[a]	mp 60° dec.	—

	$Mo(CO)_6$	$Tc_2(CO)_{10}$	$Ru(CO)_5$	$Ru_3(CO)_{12}$	$Rh_2(CO)_8$	$Rh_6(CO)_{16}$
Color	colorless	colorless	colorless	green	orange-yellow	black
Mp or bp	mp 180° dec.[a]	—[a]	mp −22°	mp 150° dec.	mp 76°	—

	$W(CO)_6$	$Re_2(CO)_{10}$	$Os(CO)_5$	$Os_3(CO)_{12}$	$Ir_2(CO)_8$	$Pt(CO)_4$
Color	colorless	colorless	colorless	yellow	green-yellow	—
Mp or bp	180° dec.[a]	mp 177°[a]	mp −15°	mp 224°	—	—

[a] Sublimes in vacuo.

Preparation   Of the simple carbonyls, the nickel and iron compounds can be prepared by direct combination of CO with the finely divided elements

$$Ni + 4CO \rightarrow Ni(CO)_4 \tag{41}$$

$$Fe + 5CO \rightarrow Fe(CO)_5. \tag{42}$$

Nickel carbonyl forms readily at room temperature under one atmosphere of carbon monoxide; the reaction is reversible and serves as a method for preparing very pure nickel. In contrast, $Fe(CO)_5$ is formed less readily, temperatures of 200° at a pressure of 100 atm being required to effect formation.

The other simple carbonyls are prepared by reducing compounds of the transition metal in a carbon monoxide atmosphere. A variety of reducing agents have been used in this process, e.g., aluminum alkyls

$$CrCl_3 \xrightarrow[CO]{(C_2H_5)_6Al_2} Cr(CO)_6, \tag{43}$$

the active metals, such as sodium

$$MoCl_5 \xrightarrow{Na + CO} Mo(CO)_6, \tag{44}$$

zinc, and magnesium, and hydrogen

$$2CoCO_3 + 2H_2 + 8CO \xrightarrow[150°]{300 \text{ atm.}} Co_2(CO)_8 + 2CO_2 + 2H_2O. \tag{45}$$

These reactions are generally conducted with suspensions in organic solvents such as ether, tetrahydrofuran, or diglyme, using carbon monoxide pressures in the range of 200–300 atm. In some instances carbon monoxide can act as the reducing agent

$$RuI_3 \xrightarrow[1 \text{ atm}]{CO} Ru(CO)_2I_2$$

$$Ru(CO)_2I_2 \xrightarrow[CO]{Ag} Ru(CO)_5 \tag{46}$$

$$OsO_4 + 9CO \xrightarrow[50 \text{ atm}]{100°} Os(CO)_5 + 4CO_2 \tag{47}$$

$$Re_2S_7 + 17CO \xrightarrow[200 \text{ atm}]{250°} Re_2(CO)_{10} + 7COS \tag{48}$$

Iron pentacarbonyl can be used to form certain other metal carbonyls because its carbon monoxide ligands are labile

$$MCl_5 \xrightarrow[M = Mo, W]{Fe(CO)_5} M(CO)_6. \tag{49}$$

The polynuclear carbonyls are prepared by thermal or photochemical decomposition of the simple carbonyls

$$2Fe(CO)_5 \xrightarrow{hv} Fe_2(CO)_9 + CO \tag{50}$$

or from polynuclear carbonylate anions, which are discussed in a subsequent section.

Structure (*IV*) The transition-metal carbonyls exhibit structures which reflect general adherence to the 18-electron rule. The structural data summarized in Table 19.3 show that the compounds with the stoichiometry $M(CO)_6$ are octahedral; $Fe(CO)_5$ is a trigonal bipyramid, and $Ni(CO)_4$ is

TABLE 19.3   Structural parameters for some mononuclear carbonyls

| Compound | Electrons | | Structure | d, Å | |
	Metal	Ligand		M—C	C—O
$Cr(CO)_6$	6	12	octahedral	1.92 ± 0.04	1.16 ± 0.05
$Mo(CO)_6$	6	12	octahedral	2.08 ± 0.04	1.15 ± 0.05
$W(CO)_6$	6	12	octahedral	2.06 ± 0.04	1.13 ± 0.05
$Fe(CO)_5$	8	10	trigonal bipyramid	1.797 ± 0.015 (axial) 1.842 ± 0.015 (trig)	
$Ni(CO)_4$	10	8	tetrahedral	1.84 ± 0.04	1.15 ± 0.02

tetrahedral. Several electron diffraction studies indicate that the axial Fe—C bond is longer than that in the equatorial position (*10–13*), but x-ray diffraction data (*14*) suggest that the differences are negligible. One exception to the 18-electron rule appears to be $V(CO)_6$ which has only 17 electrons associated with the metal ion; accordingly the compound is paramagnetic. Irrespective of the geometry of the ligands around the vanadium atom, it appears to be sterically difficult to form the dimer $V_2(CO)_{12}$ containing a metal–metal bond which would accommodate to the 18-electron rule. Recent electron spin resonance measurements on $V(CO)_6$ conducted at low temperatures indicate that a monomeric species is present from room temperature down to liquid-nitrogen temperatures. Significant dimerization sets in at about 50°K and persists to liquid-helium temperatures (*15*).

The 18-electron rule accounts for the observation that manganese and cobalt form binuclear carbonyls as the most stable species. A mononuclear manganese carbonyl with the formulation $Mn(CO)_5$ corresponds to a 17-electron system. In contrast to the case of vanadium, the dimeric compound $Mn_2(CO)_{10}$ contains a metal–metal bond, as shown in Fig. 19.2; its structure corresponds to two octahedra sharing a corner. The data show that the carbon monoxide ligands perpendicular to the axis are in a staggered configuration. Although details of the structures of $Tc_2(CO)_{10}$ and $Re_2(CO)_{10}$ are not yet known, it is highly likely that the general structural features for these compounds are the same as those for $Mn_2(CO)_{10}$. Cobalt also forms

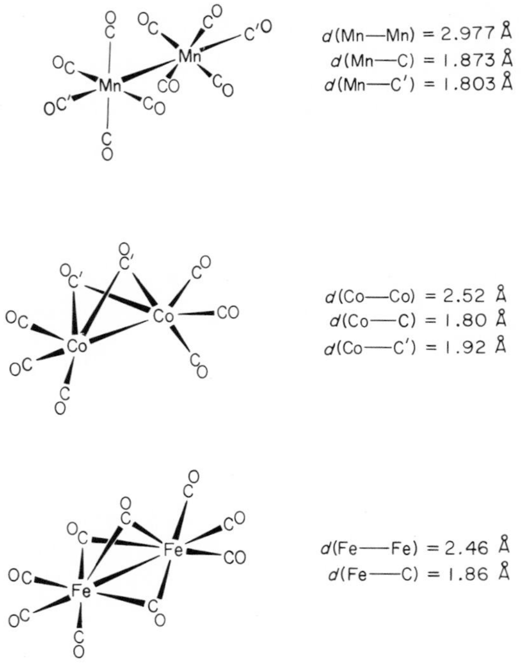

$d(Mn—Mn) = 2.977$ Å
$d(Mn—C) = 1.873$ Å
$d(Mn—C') = 1.803$ Å

$d(Co—Co) = 2.52$ Å
$d(Co—C) = 1.80$ Å
$d(Co—C') = 1.92$ Å

$d(Fe—Fe) = 2.46$ Å
$d(Fe—C) = 1.86$ Å

FIG. 19.2   The structures of $Mn_2(CO)_{10}$, $Co_2(CO)_8$, and $Fe_2(CO)_9$.

a binuclear carbonyl, $Co_2(CO)_8$. The structure of this compound (Fig. 19.2) shows (in addition to a metal–metal bond) bridging carbon monoxide ligands, a feature observed in many polynuclear carbonyls. In terms of the electron formalism, bridging carbonyl groups contribute only one electron to each metal atom and lead to an 18-electron environment for each atom, as indicated by the following calculations:

	Electrons/cobalt atom
neutral cobalt	9
bridging CO	$2 \times 1 = 2$
terminal CO	$3 \times 2 = 6$
$\frac{1}{2}$ of metal–metal bond	1
	18

Without the metal–metal bond both cobalt atoms would be 17-electron systems. It is not immediately apparent why the structure is not simply the dimer $(CO)_4Co—Co(CO)_4$ corresponding to two trigonal bipyramids or square pyramids sharing a corner with no bridging carbonyl groups, since this arrangement would also give an 18-electron system. Infrared evidence for a nonbridged structure for $Co_2(CO)_8$ in equilibrium with a bridged structure in solution is available (16–18). The other binuclear carbonyl, $Fe_2(CO)_9$, contains three carbonyl bridges, as shown in Fig. 19.2. Each iron atom in this compound is associated with 17 electrons, which leads to the suggestion that a metal–metal bond is present.

The structures of two trinuclear carbonyls are known with certainty. $Os_3(CO)_{12}$ contains a cyclic metal cluster with no bridging groups (Fig. 19.3) (19). However, the electronically similar iron derivative $Fe_3(CO)_{12}$ contains a single bridging carbonyl group (Fig. 19.3) (20). The structure of

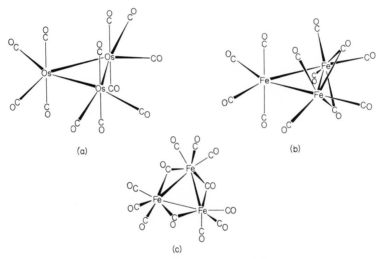

(a)

(b)

(c)

FIG. 19.3   The structures of (a) $Os_3(CO)_{12}$ and (b) $Fe_3(CO)_{12}$. The infrared spectrum of $Fe_3(CO)_{12}$ in solution shows the presence of three bridging CO groups, suggesting that this compound has the structure shown in (c) under these conditions. For structural parameters for (a) and (c) see Ref. (19); for (b) see Ref. (20).

$Fe_3(CO)_{12}$ is related to that of $Fe_2(CO)_9$ by replacement of a bridging carbonyl in the latter structure with a $Fe(CO)_4$ moiety. The infrared spectrum of $Fe_3(CO)_{12}$ in solution suggests that each Fe—Fe bond also has an associated bridging carbonyl group. This structure is supported by an incomplete x-ray analysis, by infrared spectroscopy, and by Mössbauer spectroscopy (21), which indicates the presence of two different iron atoms.

Metal clusters are also present in $M_4(CO)_{12}$ (M = Co, Rh, Ir) and $Rh_6(CO)_{16}$. The cobalt and rhodium compounds $[M_4(CO)_{12}]$ contain a tetrahedral arrangement of metal atoms (Fig. 19.4). Three metal atoms carry terminal carbonyl groups and are bridged by carbonyl groups; the remaining cobalt atom carries three terminal carbonyl groups. Infrared studies of solutions of $Co_4(CO)_{12}$ indicate the presence of four bridging groups. In both structures all the metal atoms obey the 18-electron rule if

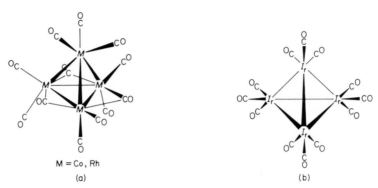

FIG. 19.4   The structures of (a) $Co_4(CO)_{12}$ and $Rh_4(CO)_{12}$, and (b) $Ir_4(CO)_{12}$. For structural parameters, see Ref. (*22*).

metal–metal bonds are assumed. The compound $Ir_4(CO)_{12}$, on the other hand, consists of four $Ir(CO)_3$ clusters without bridging carbonyl groups as shown in Fig. 19.4. The structure of $Rh_6(CO)_{16}$, shown in Fig. 19.5, consists of a hexagonal cluster of metal atoms with carbonyl groups which bridge three of the metal atoms. A polynuclear ruthenium carbonyl originally described as $Ru_6(CO)_{18}$ (*25*) is actually a carbonyl carbide, $Ru_6(CO)_{17}C$, containing an octahedral array of ruthenium atoms surrounding a carbon atom (Fig. 19.5) and one bridging carbonyl group (*24*).

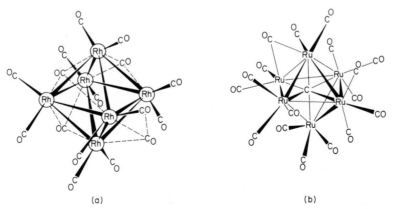

FIG. 19.5   The structures of (a) $Rh_6(CO)_{16}$ and (b) $Ru_6(CO)_{17}C$. For structural parameters, see (a) Ref. (*23*) and (b) Ref. (*24*).

Infrared spectroscopy has been widely used to elucidate the molecular structures of metal carbonyls (V). For the dinuclear carbonyls a C—O stretching frequency is observed in two distinct regions (26). The terminal carbonyl groups absorb near 2100 cm^{-1}, but the bridging groups, which are similar to organic ketones, absorb in the region 1700–1850 cm^{-1}. Thus, the infrared spectrum of $Os_3(CO)_{12}$ shows no bands in the bridging group region, in accord with the structure shown in Fig. 19.3(a), whereas the corresponding iron derivative exhibits absorption bands in both regions, in agreement with the structure shown in Fig. 19.3(c). In addition, information concerning the number of carbonyl groups attached to a metal atom can be obtained from an analysis of the infrared spectrum in favorable cases. In fact, this was the technique used to establish the structure of $Co_4(CO)_{12}$ in solution (27). The method finds extensive use in mixed metal carbonyls where isomers can exist; a common example involves the octahedral disubstituted metal carbonyls, $ML_2(CO)_4$, which can exist as either *cis* or *trans* isomers. A great deal of care is necessary in using infrared data to determine the molecular structure of metal carbonyls because C—O stretching frequencies exhibit marked solvent shifts and, in addition, terminal C—O frequencies move to lower energies if the metal atom also carries ligands that are poor π-electron acceptors.

Bonding   Initially it might appear unusual for carbon monoxide to act as a sufficiently strong ligand to form stable metal complexes; certainly, up to this point there has been little indication that carbon monoxide exhibits significant basic characteristics. The unusual stability of transition-metal-to-carbon bonds in these complexes has been attributed to multiple-bond formation, as shown schematically in Fig. 19.6. The σ component of

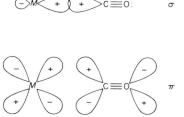

FIG. 19.6   Orbital representation for the σ and π components in a metal carbonyl bond. The π component involves antibonding orbitals of the C—O bond, leading to a decrease in C—O bond order.

this bond involves overlap of the filled *sp* hybrid orbital with an appropriately directed vacant orbital on the metal atom. The latter could be an atomic orbital or an appropriate combination of atomic orbitals which form one of the usual hybrids, e.g, sp^3 in the case of $Ni(CO)_4$, d^2sp^3 for $M(CO)_6$ (M = Cr, Mo, W), and dsp^3 for $Fe(CO)_5$. The π component is believed to arise from the interaction of filled metal d orbitals with the antibonding π orbitals of the carbonyl moiety. The formation of the metal–carbon σ bond results in a net increase of electron density on the metal atom, but the π component leads to a decrease in electron density. The net effect of this process leaves the metal atom little changed, but the electron density donated to the π^* carbonyl orbitals decreases the bond order of the C—O bond. In valence-bond terms, 17 and 18 represent the main contributing canonical

$$M \leftarrow C{\equiv}O{:} \qquad\qquad M \rightleftharpoons C{=}\overset{\cdot}{\underset{\cdot}{O}}{:}$$

$$\textbf{17} \qquad\qquad\qquad \textbf{18}$$

forms in the description of the wave function for the metal–carbon bond. The predicted decrease in the C—O bond order can be seen in the decrease in $v(CO)$ from $2155 \ \mathrm{cm}^{-1}$ for the free ligand to $\sim 2000 \ \mathrm{cm}^{-1}$ for the complexed carbonyl group. This description of the metal–carbon bond is also reflected in $v(CO)$ for an isostructural and isoelectronic series in which the charge on the metal is varied regularly. The data in Table 19.4 for two such

TABLE 19.4   Carbonyl stretching frequencies in some mononuclear species

Complex	$v(CO)$, cm^{-1}
$[Mn(CO)_6]^+$	2096
$Cr(CO)_6$	2000
$[U(CO)_6]^-$	1859
$Ni(CO)_6$	2046
$[Co(CO)_4]^-$	1883
$[Fe(CO)_4]^{2-}$	1788

series of compounds show the predicted decrease in $v(CO)$ as the negative charge on the metal increases. Attempts have been made to establish the existence of metal–carbon π bonding using the conventional bond-shortening arguments given in Chapter 4. Unfortunately there is considerable difficulty in establishing the single-bond radius expected for metal atoms in unconventional oxidation states (VI). However, in some favorable cases it is possible to make a reasonable estimate of the expected single-bond distance between a metal atom and the carbon atom in a carbonyl group (28, 29). Indeed, the most recent attempt (29) indicates that the Mo—C bond in the mixed molybdenum carbonyl π-$C_5H_5Mo(CO)_3(\sigma$-$C_2H_5)$ is about 15% shorter than expected.

Reactions   The carbon monoxide ligand of transition-metal carbonyls can be displaced by a variety of other ligands (*VII*). The products are almost invariably mixed ligand complexes retaining some of the original carbonyl ligands; only in a few instances can all of the carbonyl ligands be replaced. Since the metal atoms in the binary carbonyls are formally zero-valent some ligand displacement reactions correspond to oxidation of the metal

$$Fe(CO)_5 + I_2 \rightarrow Fe(CO)_4I_2 + CO \tag{51}$$

$$Mn_2(CO)_{10} + Br_2 \rightarrow 2Mn(CO)_5Br, \tag{52}$$

whereas other reactions give products in which the metal is still formally zero-valent

$$Fe(CO)_5 + P(C_6H_5)_3 \rightarrow Fe(CO)_4P(C_6H_5)_3 + CO \tag{53}$$

$$Cr(CO)_6 + C_6H_6 \rightarrow Cr(CO)_3C_6H_6 + 3CO. \tag{54}$$

In some oxidation processes [Eq. (51)] a change in the coordination number occurs which can be interpreted as an accommodation to the 18-electron rule. Thus, $Fe(CO)_5$ and the product in Eq. (51) both obey the 18-electron rule even though the coordination number has changed. The metal atom in the compound $Fe(CO)_4I_2$ is formally Fe^{2+} (d^6) with six ligands each donating two electron pairs. In other instances, such as in Eq. (52), the coordination number of the transition-metal atom remains the same even though oxidation occurs. Thus, the coordination number is apparently not the important factor in understanding the constitution of metal carbonyls. Indeed, we shall soon see that the concept of coordination number as described earlier is difficult to apply to many organometallic derivatives of the transition metals.

4.2 / Metal Carbonyl Derivatives

A very large number of mixed metal carbonyls have been formed by the reaction of neutral ligands with binary carbonyl compounds. One of the most useful methods involves the direct displacement of carbon monoxide molecules by the ligands in question

$$M(CO)_6 + L \rightarrow M(CO)_5L + CO. \tag{55}$$

Generally reaction can be effected in a refluxing solution of the metal carbonyl and the ligand in an appropriate solvent, by heating such systems in an autoclave, or by subjecting the mixture to strong ultraviolet radiation. Using the proper conditions it is possible to displace any number of carbonyl groups. For example, when $Mo(CO)_6$ is heated with acetonitrile (CH_3CN), or if the mixture is irradiated under a nitrogen atmosphere, three carbonyl groups are displaced from the metal atom

$$Mo(CO)_6 + 3CH_3CN \rightarrow Mo(CO)_3(NCCH_3)_3 + 3CO. \qquad (56)$$

If the reaction is conducted in hexane under one atmosphere of carbon monoxide only one carbonyl group is displaced,

$$Mo(CO)_6 + CH_3CN \rightarrow Mo(CO)_5(NCCH_3) + CO, \qquad (57)$$

but in a stream of nitrogen under the same conditions an additional carbon monoxide is displaced

$$Mo(CO)_6 + 2CH_3CN \rightarrow Mo(CO)_4(NCCH_3)_2 + 2CO. \qquad (58)$$

Displacement of weakly bound ligands in compounds which also incorporate carbonyl groups has been used with success to form complexes of interest:

$$Mo(CO)_5(NCCH_3) + CH_2{=}CHCN \rightarrow$$

$$Mo(CO)_5(NCCH{=}CH_2) + CH_3CN. \qquad (59)$$

A variety of mixed carbonyl derivatives contain many of the usual monodentate ligands such as amines (R_3N, C_5H_5N), phosphines $(R_3P; X_3P, X = Cl, F)$, arsines $(R_3As, AsCl_3)$, stibines $(R_3Sb, SbCl_3)$, oxygen derivatives (R_2O, R_2CO, ROH, H_2O), and sulfur compounds (R_2S). Mixed carbonyl derivatives containing polydentate ligands such as ethylenediamine, dipyridyl, phenanthroline, and tripyridyl have also been prepared by these methods. In addition, a very large number of carbonyl–arene–metal complexes have been prepared by the direct displacement reaction

$$M(CO)_6 + Ar \rightarrow M(CO)_3Ar + 3CO. \qquad (60)$$

$$M = Cr, Mo, W$$

$$Ar = C_6H_6, \text{ anthracene, phenanthrene}$$

Carbonyl Halides   Halogen derivatives of the metal carbonyls have been prepared by reaction of the transition-metal carbonyl with the appropriate halogens. In some cases direct addition of the halogen molecule occurs to give a compound with a higher coordination number.

$$Fe(CO)_5 + X_2 \rightarrow Fe(CO)_5X_2 \qquad (61)$$

which then decomposes by loss of carbon monoxide

$$Fe(CO)_5X_2 \rightarrow Fe(CO)_4X_2 + CO. \qquad (62)$$

The halogen adducts formed by mixed carbonyl derivatives of the Group VIB metals containing bi- and tridentate ligands are sufficiently stable to be isolated; the properties of some of the adducts are given in Table 19.5. Such adducts can exist as neutral, anionic, or cationic species, but they all

TABLE 19.5   The formulas and colors of some stable, mixed carbonyl derivatives of the Group VIB elements

Compound[a]	Color
$[Mo(CO)_3(dipy)Br_2]$	deep yellow
$[W(CO)_3(dipy)I_2]$	orange
$[Cr(CO)_2(diars)_2Br]Br$	pale yellow
$[Cr(CO)_2(diars)_2I]I_3$	brown
$(C_5H_5NCH_3)[Mo(CO)_4I_3]$	yellow
$(C_5H_5NCH_3)[W(CO)_4I_3]$	yellow

[a] Dipy = 2,2'-dipyridyl;   diars = o-phenylenebisdimethylarsine.

appear to be heptacoordinate. Metal–metal bonds in polynuclear carbonyls are often cleaved when these substances are treated with halogens [cf. Eq. (52)]

$$2BrMn(CO)_5 \xrightarrow{120°} Mn(CO)_4Br_2 + 2CO. \qquad (63)$$

The monohalo derivatives are monomeric and the polyhalo derivatives are often dimeric with halogen bridges, as indicated in Fig. 19.7. Irradiation of metal hexacarbonyls in the presence of halogens yields seven-coordinate dimeric complexes which are presumably halogen bridged (30):

$$2M(CO)_6 + 2X_2 \xrightarrow{uv} [M(CO)_4X_2]_2 + 4CO \qquad (64)$$
$$M = W, Mo; X = Br, I.$$

The halogen bridges can often be broken by treating the dimers with a strong ligand such as pyridine

$$[Mn(CO)_4I_2]_2 + 4py \rightarrow 2Mn(CO)_3(py)I + 2CO \qquad (65)$$

which yields monomeric species. Mixed metal carbonyl halides have also been prepared by the action of carbon monoxide on metal halides

$$nRuI_3 + 2nCO \rightarrow [Ru(CO)_2I_2]_n + \tfrac{n}{2}I_2 \qquad (66)$$

$$2PtCl_2 + 2CO \rightarrow [Pt(CO)Cl_2]_2. \qquad (67)$$

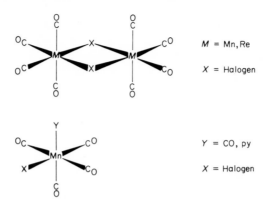

FIG. 19.7   The structures of $[M(CO)_4X]_2$ (M = Mn, Re; X = halogen) and *trans*-$Mn(CO)_4XY$.

Finally, anionic carbonyl halides are obtained when metal carbonyls are treated with ionic halides

$$M(CO)_6 + R_4N^+X^- \rightarrow R_4N^+[M(CO)_5X]^- + CO \qquad \textbf{(68)}$$

$$M = Cr, Mo, W$$

$$Mn_2(CO)_{10} + 2R_4N^+X^- \rightarrow (R_4N^+)_2[Mn_2(CO)_8X_2]^{2-} + 2CO. \quad \textbf{(69)}$$

The metal carbonyl halides are useful reagents for the preparation of compounds containing mixed metal–metal bonds

$$(CO)_5MnI + Na[MR] \rightarrow (CO)_5MnMR + NaI \qquad \textbf{(70)}$$

$$MR = Co(CO)_4{}^-.$$

Carbonylate Anions (*VIII*)   Anionic species containing a metal carbonyl moiety can be prepared in several ways. The metal–metal bonds in polynuclear carbonyls are cleaved by alkali metals, usually in the form of amalgams

$$Co_2(CO)_8 + 2Na \longrightarrow 2Na[Co(CO)_4] \qquad \textbf{(71)}$$

$$Mn_2(CO)_{10} + 2Na \longrightarrow 2Na[Mn(CO)_5] \qquad \textbf{(72)}$$

$$4Ni(CO)_4 + 2Na \xrightarrow{NH_3} Na_2[Ni_4(CO)_9] + 7CO; \qquad \textbf{(73)}$$

ether solvents such as tetrahydrofuran and diglyme are used in these reactions because the alkali-metal carbonylates are usually soluble in these solvents. Treatment of mono- or polynuclear metal carbonyls with strong bases in aqueous solution also yields carbonylate ions

$$Cr(CO)_6 + 3KOH \rightarrow K[HCr(CO)_5] + K_2CO_3 + H_2O \qquad (74)$$

$$2Cr(CO)_6 + 10KOH + H_2O \rightarrow$$
$$K_2[Cr_2(CO)_6(OH)_3H] + 2H_2 + 4HCO_2K + 2K_2CO_3 \quad (75)$$

$$Fe(CO)_5 + 3NaOH \rightarrow Na[HFe(CO)_4] + Na_2CO_3 + H_2O \qquad (76)$$

$$Fe_3(CO)_{12} + 4OH^- \rightarrow [Fe_3(CO)_{11}]^{2-} + CO_3^{2-} + 2H_2O \qquad (77)$$

$$3Ni(CO)_4 + 3\,phen \rightarrow [Ni(phen)_3]^{2+}[Ni_2(CO)_6]^{2-} + 6CO \qquad (78)$$

Carbonylate anions can also be prepared in some instances by the action of carbon monoxide on metal salts under appropriate conditions

$$2Co^{2+} + 11CO + 12OH^- \xrightarrow[\text{KCN}]{H_2O} 2[Co(CO)_4]^- + 3CO_3^{2-} + 6H_2O$$
$$(79)$$

$$TaCl_5 + 6CO + 6Na \xrightarrow[\sim 5000\,\text{psi}]{\text{diglyme}} [Na(\text{diglyme})_2]^+[Ta(CO)_6]^- + 5NaCl.$$
$$(80)$$

As might be expected, the carbonylate anions are easily oxidized by air. They may be isolated from aqueous solutions by precipitation with large cations.

Perhaps the most important use of metal carbonylate anions is their reaction as nucleophiles [cf. Eq. (70)]

$$(CO)_xM^- + RX \rightarrow (CO)_xMR + X^-. \qquad (81)$$

$$R = \text{alkyl, aryl, } SiR_3, PR_2, SR$$

Carbonyl Hydrides (IX)   Acidification of aqueous solutions containing carbonylate anions liberates carbonyl hydrides

$$Na[Co(CO)_4] + H^+ \xrightarrow{aq} HCo(CO)_4 + Na^+ \qquad (82)$$

$$NaHFe(CO)_4 + H^+ \xrightarrow{aq} H_2Fe(CO)_4 + Na^+. \qquad (83)$$

These substances can also be prepared directly from hydrogen

$$Mn_2(CO)_{10} + H_2 \xrightarrow[200°]{200\,\text{atm}} 2HMn(CO)_5 \qquad (84)$$

$$Co + 4CO + \tfrac{1}{2}H_2 \xrightarrow[150°]{50\,\text{atm}} HCo(CO)_4. \qquad (85)$$

The carbonyl hydrides are generally unstable, volatile liquids, as indicated by the data in Table 19.6. These substances are sparingly soluble in water in which they act as acids; in this respect the nomenclature seems to be anomalous.

TABLE 19.6   The properties of some carbonyl hydrides

Compound	Color	Mp, °C	pK_a	$\tau(^1H)$, ppm^a
$HMn(CO)_5$	colorless	-25	~ 7.1	17.5
$H_2Fe(CO)_4$	yellow	-70	~ 4.4	21.1
$H_2Fe_3(CO)_{11}$	red	—	—	25
$HCo(CO)_4$	yellow	-26	strong	20

^{a} $\tau(^1H)$ is the position of the single proton magnetic resonance signal observed for these compounds; reference is $(CH_3)_4Si = 10.00$.

Nitrosyl Derivatives (X)   Nitric oxide can displace carbon monoxide from some metal carbonyls to form mixed derivatives

$$Fe_2(CO)_9 + 4NO \rightarrow 2Fe(CO)_2(NO)_2 + 5CO \qquad (86)$$

$$Co_2(CO)_8 + 2NO \rightarrow 2Co(CO)_3NO + 2CO. \qquad (87)$$

Depending upon the conditions, it is possible to displace all of the ligands in metal carbonyls:

$$Fe(CO)_5 + 4NO \rightarrow Fe(NO)_4 + 5CO. \qquad (88)$$

The characteristic properties of some mixed transition-metal nitrosyl carbonyl derivatives are given in Table 19.7.

The paramagnetic nitric oxide molecule can be considered as a three-electron donor. In a formal way, we can imagine that the nitrosyl group loses an electron to the metal, forming the nitrosonium ion (Chapter 12), which then coordinates to the negatively charged metal atom (**19**). The nitrosonium

$$M \leftarrow : \overset{+}{\underset{}{N}} \equiv O :$$

19

ion is isoelectronic with CO and should exhibit essentially the same distribution of electrons in the available molecular orbitals. Accordingly, electron density can be released from the metal atom into the π^* orbitals of the NO^+ moiety, decreasing the N—O bond order. This general description of the nature of the bonding between NO and a metal atom is supported by the relative position of the NO stretching frequency $[\nu(NO)]$ in the series of

TABLE 19.7 Properties of some metal–nitrosyl derivatives

Compound	Color	Mp, °C
$V(NO)(CO)_5$	red-violet	—
$Mn(NO)_3CO$	dark green	27
$Mn(NO)(CO)_4$	red	−1.5
$Fe(NO)_2(CO)_2$	red	18.4
$Co(NO)(CO)_3$	red	−11
$[Co(NO)_2Br]_2$	black-brown	116
$[Ni(NO)(NO_2)]_n$	blue	25 dec.
$\pi\text{-}C_5H_5NiNO$	red	oil
$(\pi\text{-}C_5H_5)_6Mn_6(NO)_8$	black	220 dec.
$(\pi\text{-}C_5H_5)_3Mn(NO)_3$	black	100 dec.
$\pi\text{-}C_5H_5Cr(NO)_2CH_3$	green	165 dec.
$C_3H_5Fe(NO)(CO)_2$	red	oil

compounds shown in Table 19.8. In nitrosonium salts where the antibonding orbital is empty $v(NO)$ is significantly higher than it is for NO, which contains one electron in this orbital. The NO stretching frequency of the metal nitrosyls varies between 1580 and 1900 cm^{-1}, indicating that in some instances the π^* orbital has received more or less electron density than that found in the free ligand.

TABLE 19.8 The NO stretching
frequencies for several compounds

Compound	$v(NO)$, cm^{-1}
$NO^+BF_4^-$	2250
NO	1878
$Co(CO)_3NO$	1832
$Mn(CO)(NO)_3$	1823, 1734
$\pi\text{-}C_5H_5NiNO$	1820
$Fe(CO)_2(NO)_2$	1810, 1766
$\pi\text{-}C_5H_5Cr(NO)_2CH_3$	1779, 1670
$(\pi\text{-}C_5H_5)_3Mn(NO)_3$	1720. 1497
$V(NO)(CO)_5$	1700
$K_4[Mo(NO)(CN)_5]$	1455

4.3 / Structure and Bonding of Metal Carbonyl Derivatives

Because most of the metal carbonyls incorporating other ligands obey the 18-electron rule, the discussion of the structure and bonding of these derivatives is an extension of that for the binary metal carbonyls. Most of the ligands which can replace carbon monoxide are better Lewis bases than CO; in many cases these ligands are also capable of forming π bonds to the metal atom. The carbonyl stretching frequency in octahedral metal carbonyls has been used to estimate the extent to which a ligand can act as a π acceptor (31). For example, consider the case of a metal atom carrying carbon monoxide and another ligand (20). If the ligand cannot act as a π-acceptor, the accumulated electronic charge on the metal must be dissipated

$$O \equiv C - M - L \qquad O = C \Leftarrow M - L \qquad O = C \Leftarrow M \Rightarrow L$$

$$\mathbf{20} \qquad\qquad\qquad \mathbf{21} \qquad\qquad\qquad \mathbf{22}$$

into the metal–carbon π component formed by overlap of the filled metal d orbitals and the carbon–oxygen antibonding π orbitals (21). In this instance $\nu(CO)$ should decrease relative to the value observed for the corresponding carbonyl compound. On the other hand, if the ligand can also act as a π-acceptor, the electronic charge may also be localized through a π system onto the ligand (22). If the ligand is a better π-acceptor than carbon monoxide, $\nu(CO)$ should increase compared to the unsubstituted carbonyl because electron density would be removed from the π^* carbon monoxide orbital, increasing the bond order. In other words, the relative position of $\nu(CO)$ in a series of metal–carbonyl derivatives with similar geometries should reflect the relative π-acceptor character of the substituents. The following decreasing order of π-acceptor character for various ligands has been determined (32).

$$NO \geqslant CO > \left\{ \begin{matrix} CNR \\ PF_3 \end{matrix} \right\} > \left\{ \begin{matrix} PCl_3 \\ AsCl_3 \\ SbCl_3 \end{matrix} \right\} > PCl_2(OR) > PCl_2R > PCl(OR)_2$$

$$> \left\{ \begin{matrix} PClR_2 \\ P(OR)_3 \end{matrix} \right\} > \left\{ \begin{matrix} PR_3 \\ AsR_3 \\ SbR_3 \\ SR_2 \end{matrix} \right\} > RCN > \left\{ \begin{matrix} NR_3 \\ OR_2 \\ ROH \end{matrix} \right\}.$$

The ligands in this list can form π bonds to the metals in a number of ways. The relationship of nitric oxide to carbon monoxide is discussed in an earlier section; isocyanides that are carbon-bonded to metals, as well as cyanides which are nitrogen-bonded, also possess antibonding π orbitals which can interact with metal d orbitals. Ligands containing atoms in the third and higher periods (P, As, Sb, S) have low-lying empty d orbitals which can accept π electron density, as shown schematically in Fig. 19.8. The change in the degree of metal–carbon π bonding is also reflected in the metal–carbon distance. For example, both the metal–nitrogen and metal–carbon distances

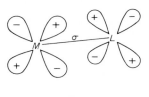

FIG. 19.8   Orbital representation for the $d\pi$–$p\pi$ interaction in metal–ligand bonds.

in the isostructural series of complexes $Mn(CO)_4(NO)$, $Mn(CO)_3(NO)$-$P(C_6H_5)_3$, and $Mn(CO)_2(NO)_2P(C_6H_5)_3$ decrease in the order indicated; substitution of carbon monoxide by the poor π-acceptor $P(C_6H_5)_3$ requires that the remaining metal–carbon π system remove a greater negative charge from the metal atom which corresponds to a stronger $d\pi$-$p\pi$ bond. Finally, the ligands at the end of the series have no low-lying orbitals capable of π bonding.

The structures of the simple binary carbonylate anions follow the general trends described for the metal carbonyls. Thus, mononuclear species such as $:Mn(CO)_5^-$ have the expected octahedral arrangement of electron pairs with one position vacant, i.e., the distribution of ligands about the metal corresponds to a square pyramid. The polynuclear carbonylate anions contain clusters of metal atoms with bridging and nonbridging carbonyl ligands. As in the case of the neutral compounds, there are several possible structures predicted on the basis of the 18-electron rule and it is generally difficult to make an a priori decision concerning the correct structure. Six-, seven-, and 12-metal-atom clusters are known in the anions $[Co_6(CO)_{14}]^{4-}$ (33), $[Co_6(CO)_{15}]^{2-}$ (34), $[Rh_7(CO)_{16}]^{3-}$ (35), and $[Rh_{12}(CO)_{30}]^{2-}$ (36). The structures of these species are shown in Fig. 19.9.

The structures of the carbonyl hydrides apparently follow closely those of the corresponding parent carbonyl with hydrogen atoms acting as either terminal or bridging groups. Thus, the structure of $HMn(CO)_5$ has been described (37) as essentially octahedral with one position occupied by a direct metal–hydrogen bond. It has also been suggested that $H_2Fe(CO)_4$ is an octahedral molecule with *cis* substituents (38). Several polynuclear carbonyl hydride species appear to have hydrogen atoms bridging between two metal atoms. For example, the species $[HFe_3(CO)_{11}]^-$ has the same structure as the carbonyl $Fe_3(CO)_{12}$ (Fig. 19.3), with a bridging hydrogen atom rather than a carbonyl group. The ion $[HCr_2(CO)_{10}]^-$ appears to be unique in that

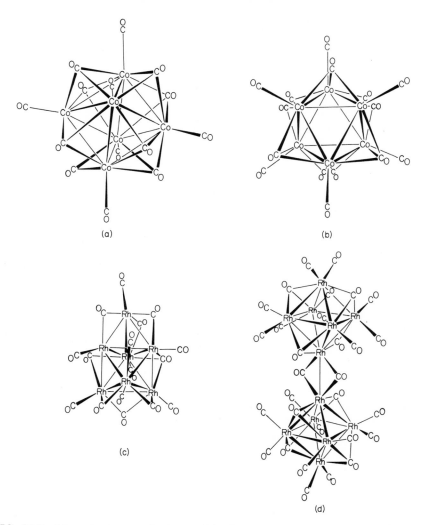

FIG. 19.9 The structures of the carbonylate anions (a) $[Co_6(CO)_{14}]^{4-}$, (b) $[Co_6(CO)_{15}]^{2-}$, (c) $[Rh_7(CO)_{16}]^{3-}$, and (d) $[Rh_{12}(CO)_{30}]^{2-}$. For structural parameters, see (a) Ref. (33), (b) Ref. (34), (c) Ref. (35), and (d) Ref. (36).

no direct metal–metal bond is present; rather the two $Cr(CO)_5$ moieties are connected by a linear $Cr—H—Cr$ arrangement of atoms (39–43). Although the position of the hydrogen atom in these complexes has not been directly observed, inferences obtained from direct structural investigations are supported by the high proton nmr shielding parameters observed for these compounds. It is likely that the best description of the bonding systems in these compounds requires the use of multicentered molecular orbitals as in the case of the boron hydrides described in Chapter 10.

4.4 / Olefins as Two-Electron Ligands (*XI, XII*)

Olefins represent the other large class of two-electron ligands. Complexes containing olefins are prepared most often by displacement of loosely held ligands such as solvent molecules

$$2RhCl_3 + 2H_2O + 6C_2H_4 \rightarrow Rh_2Cl_2(C_2H_4)_4 + 2CH_3CHO + 4HCl,$$

ionic ligands

(89)

$$K_2PtCl_4 + C_2H_4 \xrightarrow{H_2O} K[PtCl_3(C_2H_4)] + KCl, \qquad (90)$$

or neutral ligands

$$[Rh(CO)_2Cl]_2 + 2C_8H_{12} \rightarrow Rh_2Cl_2(C_8H_{12})_2 + 4CO. \qquad (91)$$

They can also be prepared by the direct reaction of a metal salt with the olefin

$$2PdCl_2 + 2C_2R_4 \rightarrow [PdCl_2(C_2R_4)]_2. \qquad (92)$$

Some examples of metal–olefin complexes which have been synthesized appear in Table 19.9.

TABLE 19.9   Some examples of complex compounds containing olefin ligands

Ligand	Mode of attachment	Complex	Reference
Ethylene	monodentate	$Fe(CO)_4C_2H_4$	(44)
		$[(C_2H_4)_2RhCl]_2$	(45)
		$[C_2H_4PtCl_2]_2$	(46)
cis-Cyclooctene	monodentate	$Ag(C_8H_{14})_2NO_3$	(47)
Norbornadiene	monodentate	$[Cu(C_7H_8)Cl]_4$	(48)
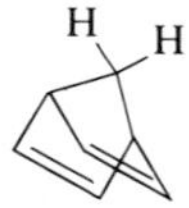	bidentate	$[Ag(C_7H_8)]X$	(49)
	bidentate	$[PdCl_2(C_7H_8)]$	(50)
Cycloocta-1,5-diene (C_8H_{12})	bidentate	$Rh_2Cl_2(C_8H_{12})_2$	(51)
Cyclododeca-1,5,9-triene ($C_{12}H_{18}$)	bidentate	$PdCl_2(C_{12}H_{18})$	(52)
	tridentate	$Ni(C_{12}H_{18})$	(53)

The metal–olefin complexes exhibit a wide range of chemical stability. Complexed olefins are often labile, being easily displaced by ligands such as phosphines. Complexed olefins are also more susceptible to nucleophilic attack than the uncomplexed species.

4.5   /   Structure and Bonding in Olefin Ligands

The stoichiometry of the olefin–metal complexes is that expected if each double bond donated an electron pair to the metal. All of the structures containing metal–olefin bonds have a common feature, namely the ligand appears to occupy coordination positions with the C—C axis perpendicular to the expected normal metal–ligand bond axis (**23**). Examples of several

$$M \leftarrow \begin{matrix} CR_2 \\ || \\ CR_2 \end{matrix}$$

23

such structures appear in Fig. 19.10. It is interesting that benzene appears to complex through only one C—C bond, if internuclear distances are to have their usual meaning in terms of bonding. The important features in the complex $C_6H_6CuAlCl_4$ are (a) the tetrahedral arrangement of ligands about both aluminum and copper with a halogen bridge between the metal atoms, and (b) benzene apparently occupying one coordination position. Curiously, the coordinated C—C distance (1.35 Å) is markedly shorter than the other C—C distances (1.43 Å). On the other hand, silver in the analogous complex $C_6H_6AgAlCl_4$ bonds to an edge of the benzene ring, with the plane at nearly right angles (98°) to the normal bonding direction (60). A similar orientation is observed in the complex $C_6H_6AgClO_4$, shown in Fig. 19.10. The observed C—C distances in the complexes are generally longer than those associated with the free ligand (1.337 Å), but in some cases the increase is marginal. The best theoretical description of the metal–olefin bond (61, 62) still involves the donation of electronic density from the ligand to the metal and back donation from the metal, i.e., a two component interaction. One bonding component in the complex involves the overlap of filled π orbitals on the ligand with an empty orbital of appropriate symmetry on the metal atom, as shown schematically in Fig. 19.11; the back donation of electrons is believed to involve filled metal orbitals overlapping the antibonding π orbitals of the olefin. This description is consistent with some of the physical properties reported for olefin–metal complexes. For example, the decrease in $v(C{=}C)$ of 60–150 cm^{-1} in complexed olefins compared with the free ligand is a consequence of placing electron density in the π^* orbitals of the

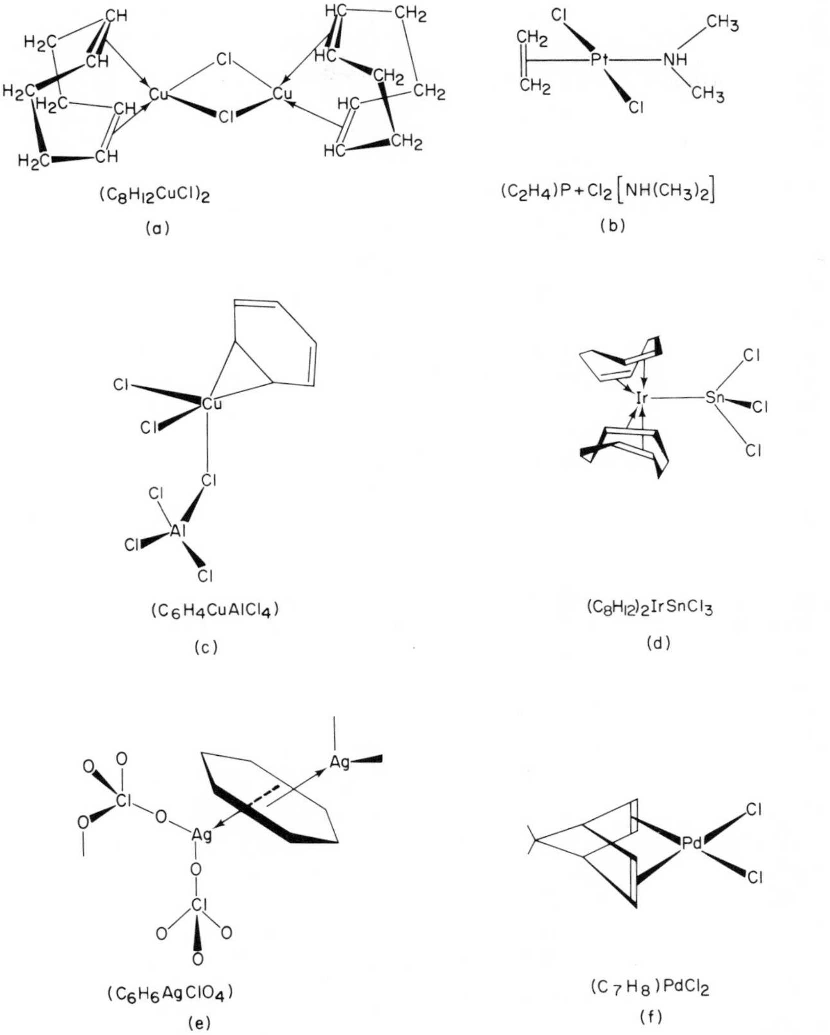

FIG. 19.10   The structures of several metal complexes containing unsaturated ligands. For structural parameters, see (a) Ref. (*54*), (b) Ref. (*55*), (c) Ref. (*56*), (d) Ref. (*57*), (e) Ref. (*58*), and (f) Ref. (*59*).

olefin, which has the effect of decreasing the bond order. X-Ray evidence also indicates that the C=C bond in the complexed olefin is longer, in some cases, than that in the free olefin. In cases where very electronegative groups are attached to the olefin leading to a less effective metal–π interaction but a better $d\pi$-$p\pi^*$ interaction than with an ordinary olefin, it has been suggested

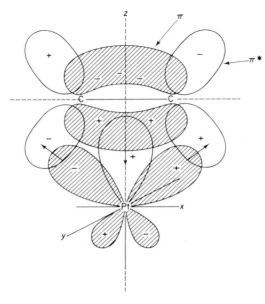

FIG. 19.11 Orbital representation of the metal–olefin bond. Both the bonding and antibonding π orbitals of the ligand are involved in bonding the metal to the olefin.

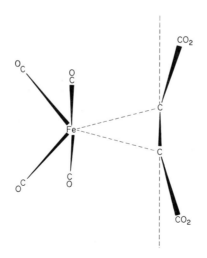

FIG. 19.12 Structure of the fumaric-acid–iron-tetracarbonyl complex.

that these ligands are bound primarily by the latter component. This description is equivalent to changing the hybridization of the bound carbon atoms from sp^2 to sp^3 and the formation of a three-membered ring system (**24**).

24

Evidence for this formulation has been found in ^{19}F nmr spectra of fluoro-olefin complexes (*63, 64*) and from the x-ray structure of the fumaric-acid–iron-tetracarbonyl complex shown in Fig. 19.12. The latter structure exhibits a considerable increase in the carbon–carbon bond distance of the complexed olefin and a geometry in which the carbonyl groups are bent away from the $Fe(CO)_4$ moiety.

5 / THREE-ELECTRON LIGANDS (*XII, XIII*)

The simplest and most widely investigated three-electron ligand is the π-allyl moiety (**25**), although a number of complexes of cyclic π-enyl derivatives (**26**) are also known.

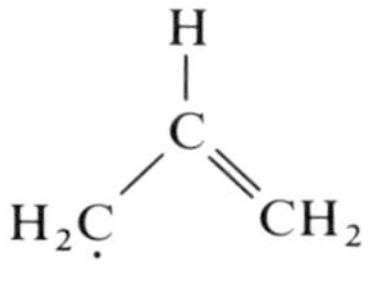

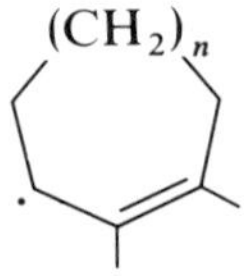

25   **26**

A formal relationship exists between the even- and odd-numbered electron ligands. For example, methyl-substituted monoolefins which are two-electron ligands are related to the three-electron allyl groups by removal of a radical

$$CH_2{=}CHCH_2X \xrightarrow{-X\cdot} CH_2{=}CHCH_2\cdot. \tag{93}$$

Similarly, four- and six-electron ligands are related to the next higher alkenyl ligand

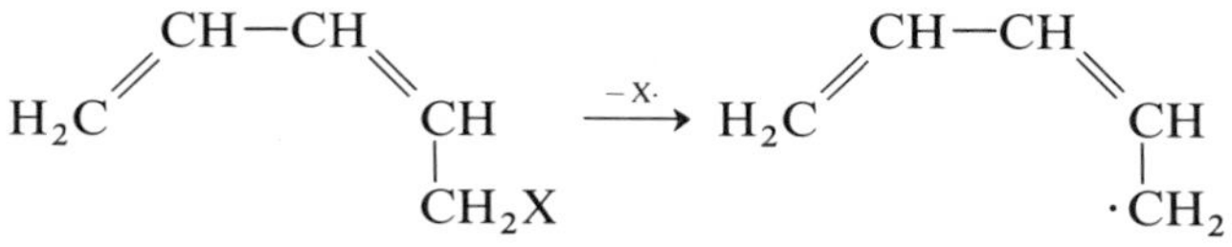

(94)

$$
\begin{array}{ccc}
\underset{HC}{\overset{H}{\diagdown}}\text{C=C}\overset{H}{\diagup}\ X & & \\
& \xrightarrow{-X} & \\
\end{array}
\qquad (95)
$$

It must be stressed that these are formal relationships for the most part, although in some instances the process indicated does represent a method of preparation of the odd-electron ligand.

5.1 / Preparation

Several general routes to π-enyl complexes are known. Allyl Grignard reagents react with some metal salts as well as with other organometallic derivatives to yield compounds containing the π-allyl group

$$2C_3H_5MgBr + NiCl_2 \xrightarrow{-10°} (\pi\text{-}C_3H_5)_2Ni + 2MgBrCl \qquad (96)$$

$$(\pi\text{-}C_5H_5)_2Ni + C_3H_5MgBr \longrightarrow (\pi\text{-}C_5H_5)(\pi\text{-}C_3H_5)Ni + C_5H_5MgBr. \qquad (97)$$

Allyl chloride reacts with carbonylate anions to yield σ-allyl complexes

$$Na^+[Mn(CO)_5]^- + ClC_3H_5 \longrightarrow \sigma\text{-}C_3H_5Mn(CO)_5 + NaCl \qquad (98)$$

which can be converted easily into the corresponding π-allyl derivatives

$$\sigma\text{-}C_3H_5Mn(CO)_5 \xrightarrow{uv} \pi\text{-}C_3H_5Mn(CO)_4 + CO. \qquad (99)$$

Allyl alcohol also is a useful reagent for the formation of π-allyl complexes. π-Enyl complexes have been prepared from olefin complexes which carry an unsaturated ligand that can be converted into a π-allyl group

$$\longrightarrow [\pi\text{-}C_3H_5PdCl_2]_2 + 2HCl \qquad (100)$$

$$+ \ PdCl_2 \longrightarrow [\pi\text{-}(C_6H_9)PdCl]_2 + 2HCl. \qquad (101)$$

Some conjugated olefins undergo addition of an M—X group to yield π-enyl complexes. For example, $PdCl_2$ gives a π-enyl complex directly with butadiene

$$CH_2=CHCH=CH_2 + 2PdCl_2 \rightarrow$$

$$
\begin{array}{c}
\underset{\text{CH}_2\text{Cl}}{\overset{\overset{\displaystyle CH_2}{\overset{|}{HC \cdots}}}{\underset{|}{CH}}} \rightarrow Pd \underset{Cl}{\overset{Cl}{\diamond}} Pd \leftarrow \underset{\underset{\displaystyle CH_2Cl}{\overset{|}{HC}}}{\overset{\overset{\displaystyle H_2C}{\cdots CH}}{}} \quad (102)
\end{array}
$$

but mixed iron carbonyl hydrides give intermediate σ-bonded derivatives

$$\pi\text{-}C_5H_5Fe(CO)_2H + CH_2=CHCH=CH_2 \rightarrow$$

$$\pi\text{-}C_5H_5Fe(CO)_2CH_2CH=CHCH_3 \quad (103)$$

which can be converted into compounds containing a π-enyl system

$$\pi\text{-}C_5H_5Fe(CO)_2CH_2CH=CHCH_3$$

$$\xrightarrow{uv} \pi\text{-}C_5H_5\underset{CO}{\overset{|}{Fe}} \leftarrow \underset{H_2C}{\overset{\overset{\displaystyle CH_3}{\overset{|}{HC}}}{\diagdown CH}} + CO \quad (104)$$

The addition of $Co(CO)_4H$ to butadiene leads directly to the π-enyl complex without the intermediate formation of the σ derivative

$$Co(CO)_4H + CH_2=CHCH=CH_2 \rightarrow (CO)_3Co \leftarrow \underset{H_2C}{\overset{\overset{\displaystyle CH_2}{\overset{|}{HC}}}{\diagdown CH}} + CO \quad (105)$$

5.2 / Properties (*XIV*)

The properties of a few π-enyl complexes are shown in Table 19.10. Pure π-allyl metal complexes are generally less stable than the mixed derivatives shown in Table 19.10. π-Allyl complexes of the metals which appear early in the transition series are highly reactive and exhibit catalytic activity, in

TABLE 19.10   The properties of some complexes containing π-allyl ligands (*XIII*)

Compound	Color	Mp/bp, °C
π-$C_3H_5Mn(CO)_4$	pale yellow	mp 52°
π-$C_3H_5Co(CO)_3$	orange	bp $-33°$ (39 mm)
$[\pi$-$C_3H_5NiBr]_2$	red	mp 93° dec.
$[\pi$-$C_6H_9PdCl]_2$	yellow	mp 85°
π-$C_3H_5Fe(CO)_3Cl$	yellow-brown	mp 85°
π-$C_3H_5Mo(\pi$-$C_5H_5)(CO)_2$	yellow	mp 134° dec.
π-$C_3H_5Co(\pi$-$C_5H_5)I$	red	mp 95.5°

contrast with the complexes of the elements at the ends of the series. Most of our knowledge of the chemistry of the π-allyl complexes comes from the latter group of compounds. As might be expected, the π-allyl group is rather easily displaced by a variety of ligands

$$Ni(\pi\text{-}C_3H_5)_2 + 4CO \rightarrow Ni(CO)_4 + C_6H_{10} \qquad (106)$$

$$Ni(\pi\text{-}C_3H_5)_2 + 4PR_3 \rightarrow Ni(PR_3)_4 + C_6H_{10}. \qquad (107)$$

5.3 / Structure and Bonding (*XV*)

The direct structural information available for π-allyl complexes, as shown in Fig. 19.13, has several interesting features. In some instances the plane of the π-allyl group is essentially perpendicular to the expected metal–ligand vector; however, in other π-allyl derivatives the plane of the π-allyl group is nearly parallel to this line. All the C—C distances in sandwich-type π-allyl compounds are virtually the same (~ 1.36 Å), being longer than that for a simple carbon–carbon double bond and in the same range as that found for complexed olefins. On the other hand, the non-sandwich π-allyl complexes exhibit two C—C distances that are significantly longer than that observed in the sandwich-type complexes. In such compounds two short metal–carbon distances are found. Thus, it appears that two extreme structures are possible for π-allyl groups bound to a metal atom (**27** and **28**).

27

28

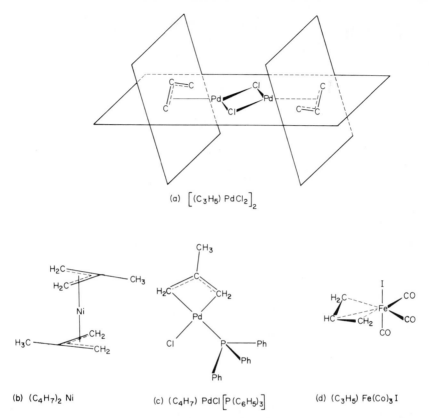

FIG. 19.13   The structures of some metal–allyl complexes. For structural parameters, see (a) Ref. (*65*), (b) Ref. (*66*), (c) Ref. (*67*), and (d) Ref. (*68*).

The bonding of a π-allyl group to a metal atom undoubtedly involves the same general considerations as described for the olefin complexes. It appears that the sandwich-type complexes of the triangular π-allyl groups involve the overlap of filled π orbitals on the ligand with appropriate empty metal orbitals; back donation presumably occurs to the π^* ligand orbitals. In the case of the π-allyl moiety the local symmetry is very low, and the usual symmetry arguments are not sufficient to discern which of the many possible combinations of orbitals are the most suitable for bonding. The non-sandwich π-allyl compounds appear to be similar to the olefin complexes, which are best described in terms of direct metal–ligand bonding through carbon atoms that are nearly sp^3-hybridized.

The variation in the bonding of the allyl group exhibited by the structures shown in Fig. 19.13 is also reflected in the proton nmr spectra of these complexes. Some π-allyl complexes give spectra characteristic of an A_2M_2X system, that is the hydrogen atoms on each carbon atom exhibit characteristic chemical shifts with the expected splitting patterns, as shown in Fig. 19.14.

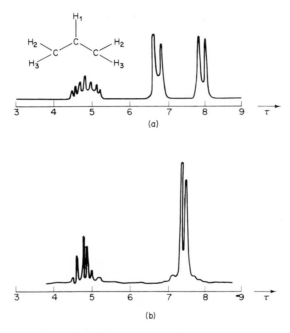

FIG. 19.14   (a) The proton nmr spectrum of a π–allyl ligand; (b) the proton nmr spectrum of $Zr(C_3H_5)_4$ at $-20°$.

However, the nmr spectra of some π-allyl complexes [e.g., $M(C_3H_5)_2$, $M = Ni, Pt, Pd$] (69) indicate that the hydrogen atoms on the terminal carbon atoms are magnetically equivalent at room temperature but not at low temperatures. In other words, the spectra of these compounds change from the pattern shown in Fig. 19.14(a) to that shown in Fig. 19.14(b) as the temperature decreases. The nmr spectra of some allyl complexes suggest that the terminal hydrogen atoms are equivalent at temperatures below $-70°$, whereas other complexes exhibit this characteristic when they are dissolved in basic solvents. Several mechanisms have been invoked to explain the magnetic equivalence of the terminal hydrogen atoms. Where free ligands are available in the system (e.g., solvent molecules), the terminal hydrogen atoms can equilibrate through a rapid exchange process

$$\text{M}-(\pi\text{-allyl}) + \text{L} \rightleftharpoons \text{M}-(\sigma\text{-allyl})\text{L}. \tag{108}$$

For complexes which show equivalent terminal hydrogen atoms under conditions where ligand exchange is not possible, an equilibrium between a π and σ complex has been postulated:

$$
\begin{array}{ccc}
& \text{H} & \\
& | & \\
& \text{C} & \\
\text{H}_2\text{C} \diagdown \overset{|}{\diagup} \text{CH}_2 & \rightleftharpoons & \text{H}_2\text{C} \diagdown \overset{\text{C}=\text{CH}_2}{\diagup} \\
& | & \text{M} \\
& \text{M} &
\end{array}
\qquad (109)
$$

In this interpretation rotation of the plane of the ligand around the metal–ligand axis in the σ complex, followed by formation of a π complex, leads to the observed nmr spectrum.

6 / FOUR-ELECTRON LIGANDS

6.1 / Preparation

Conjugated dienes act as four-electron ligands in which the entire π system apparently acts as the electron donor, in contrast to unconjugated dienes which act as four-electron ligands because they are bidentate. The most widely studied four-electron ligand is butadiene (XVI) and its cyclic derivative cyclobutadiene (XVII). Most of the known complexes containing four-electron ligands are carbonyl derivatives and can be prepared by the direct reaction of the diene with the metal carbonyl. Cyclobutadiene complexes have been prepared by reacting cis-dichlorocyclobutenes

$$
2\ \underset{\text{Cl}}{\overset{\text{Cl}}{\underset{\text{H}}{\overset{\text{H}}{\square}}}} + \text{Fe}_2(\text{CO})_9 \longrightarrow 2\ \overset{\square}{\underset{\substack{\text{C}\\\text{O}}}{\overset{|}{\underset{\text{O}^\text{C}\ \text{Fe}\ ^\text{C}\text{O}}{}}}} + 3\text{CO} + \text{Fe[CO]}_3\text{Cl}_2
$$

$$(110)$$

$$
2\ \overset{\text{H}_3\text{C}}{\underset{\text{H}_3\text{C}}{\square}}\overset{\text{CH}_3\ \text{Cl}}{\underset{\text{CH}_3\ \text{Cl}}{}} + 2\text{Ni(CO)}_4 \longrightarrow \left(\overset{\text{H}_3\text{C}\quad\quad\text{CH}_3}{\underset{\underset{\text{Cl}\quad\text{Cl}}{\text{Ni}}}{\overset{\text{H}_3\text{C}\quad\text{CH}_3}{\square}}} \right)_2 + 8\text{CO} \quad (111)
$$

or acetylenes

$$2C_6H_5C{\equiv}CC_6H_5 + Fe(CO)_5 \rightarrow$$  $$+\ 2CO \qquad (112)$$

$$4C_6H_5C{\equiv}CC_6H_5 + Mo(CO)_6 \rightarrow [(\pi\text{-}C_6H_5)_4C_4]_2Mo(CO)_2 + 4CO \ (113)$$

with a metal carbonyl. In the latter two reactions, the complexed cyclo-butadiene ligands are formed by dimerization of two acetylene groups.

6.2 / Structure and Bonding

The known structures of the four-electron ligands given in Fig. 19.15 indicate that the ligand carbon atoms are essentially planar and equidistant from the metal atom. The substituents on the terminal carbon atoms in the butadiene complexes are bent slightly out of the plane of the ligand and away from the metal atom; a similar relationship occurs in all of the cyclo-butadiene complexes. In the case of the 1,3-diene complexes an increase in the electronegativity of the substituents leads to a more marked deviation from planarity. Thus, it appears that the terminal carbon atoms in 1,3-diene complexes exhibit more or less sp^3 character leading to the formulation (**29**).

29

All of the carbon atoms in the cyclobutadiene complexes also appear to have greater p character than those in the sp^2 hybrids found in the uncom-plexed ligand.

7 / FIVE-ELECTRON LIGANDS (*XVIII, XIX*)

The most common ligand in this class is the cyclopentadienyl moiety (**30**), although other five-electron cyclic ligands are known, i.e., (**31** and **32**).

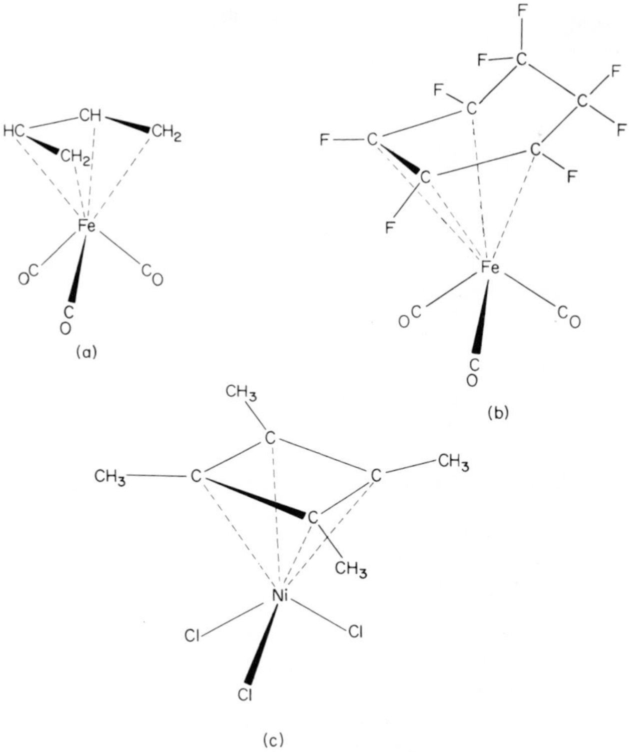

FIG. 19.15 The molecular structures of several metal–diene complexes. For structural parameters, see (a) Ref. (*70*), (b) Ref. (*71*), and (c) Ref. (*72*).

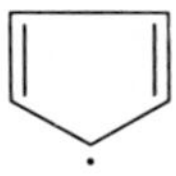

30 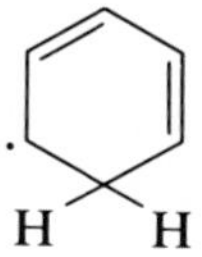   **31** 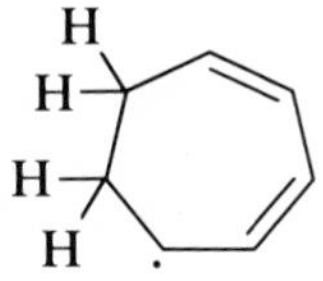   **32**

There are, of course, two ways in which a potential five-electron ligand can be attached to a metal atom. σ-Bonded ligands (**33**) are simple substituted

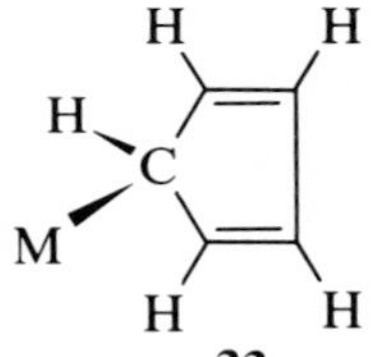

33 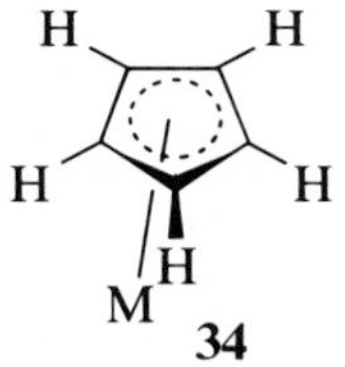   **34**

derivatives of the parent hydrocarbon. When all five electrons are involved (34) a π-bonded system results; ligands bound in this way are known as dienyl ligands.

7.1 / Preparation

Perhaps the most general method for the preparation of cyclopentadienyl complexes involves the reaction of anhydrous metal halides with solutions of alkali-metal cyclopentadienides

$$2NaC_5H_5 + MCl_2 \rightarrow (\pi\text{-}C_5H_5)_2M + 2NaCl. \tag{114}$$

Mixed carbonyl derivatives are obtained when the same reagent is allowed to react with metal carbonyls

$$NaC_5H_5 + W(CO)_6 \rightarrow [\pi\text{-}C_5H_5W(CO)_5]^-Na^+ + CO \tag{115}$$

or with carbonyl halides

$$4NaC_5H_5 + 2Os(CO)_3Cl_2 \rightarrow$$
$$[\pi\text{-}C_5H_5Os(CO)_2]_2 + 2CO + 4NaCl + C_{10}H_{10}. \tag{116}$$

Using reactions such as these, a very large number of pure and mixed transition-metal cyclopentadienyls have been prepared and characterized. A few of these compounds are shown in Table 19.11.

7.2 / Properties

The binary complexes are highly colored, stable to hydrolysis, and many melt without decomposition. The stability of the complexes to oxidation varies widely, but they all react in acid solution with a variety of oxidizing agents to yield cations

$$(\pi\text{-}C_5H_5)_2M \rightarrow (\pi\text{-}C_5H_5)_2M^+ + e^- \tag{117}$$

which can be isolated by precipitation with appropriate counter ions.

The π-cyclopentadienyl carbonyls are also highly colored and soluble in organic solvents; their stoichiometries follow that expected from the 18-electron rule. Because the cyclopentadienyl ring is less labile than a carbonyl group it frequently survives the action of reagents which remove carbonyl groups

$$\pi\text{-}C_5H_5V(CO)_4 \xrightarrow{Cl_2/O_2} \pi\text{-}C_5H_5VO(Cl)_2 + 4CO_2 \tag{118}$$

TABLE 19.11 The properties of some complexes containing a π-cyclopentadienyl ligand[a]

Ti	V	Cr	Mn	Fe	Co	Ni
$[\pi\text{-}(C_5H_5)_2Ti]_2$ green (dec. $>200°$)	$(\pi\text{-}C_5H_5)_2V$ purple ($167°$)	$(\pi\text{-}C_5H_5)_2Cr$ scarlet ($172°$)	$Mn(\pi\text{-}C_5H_5)_2$ amber —	$(\pi\text{-}C_5H_5)_2Fe$ orange ($173°$)	$(\pi\text{-}C_5H_5)_2Co$ purple ($173°$)	$(\pi\text{-}C_5H_5)_2Ni$ green (dec. $173°$)
$(\pi\text{-}C_5H_5)_3Ti$ green (dec. $130°$)						
$(\pi\text{-}C_5H_5)_2Ti(CO)_2$ red-brown (dec. $90°$)	$(\pi\text{-}C_5H_5)V(CO)_4$ orange (dec. $138°$)	$[\pi\text{-}C_5H_5Cr(CO)_3]_2$ green (dec. $163°$)	$\pi\text{-}C_5H_5Mn(CO)_3$ yellow (dec. $76.8°$)	$[\pi\text{-}C_5H_5Fe(CO)_2]_2$ red-purple ($194°$)	$\pi\text{-}C_5H_5Co(CO)_2$ red ($-22°$)	$[\pi\text{-}C_5H_5Ni(CO)]_2$ red (dec. $146°$)
				$\pi\text{-}C_5H_5Fe(CO)_4$	$[\pi\text{-}C_5H_5Co(CO)]_3$	$(\pi\text{-}C_5H_5)_3Ni_3(CO)_2$ green (dec. $200°$)

[a] Melting point given in brackets.

or replace other ligands

$$\pi\text{-}C_5H_5Mo(CO)_3Cl + Cl_2 \rightarrow \pi\text{-}C_5H_5Mo(CO)_2Cl_3 + CO \quad (119)$$

$$\pi\text{-}C_5H_5Mo(CO)_3Cl + NH_3 \rightarrow [\pi\text{-}C_5H_5Mo(CO)_3NH_3]^+Cl^-. \quad (120)$$

The metal–metal bonds in binuclear cyclopentadienyl derivatives can be cleaved by oxidizing agents

$$[\pi\text{-}C_5H_5Fe(CO)_2]_2 + Cl_2 \rightarrow 2\pi\text{-}C_5H_5Fe(CO)_2Cl \quad (121)$$

or reducing agents

$$[\pi\text{-}C_5H_5Fe(CO)_2]_2 + 2Na \rightarrow 2Na^+[\pi\text{-}C_5H_5Fe(CO)_2]^-. \quad (122)$$

in much the same way as those in the binary carbonyls.

Finally, the complexed cyclopentadienyl ring exhibits a wide range of reactions normally associated with aromatic systems (XX). The most extensive work in this respect has been done with $Fe(C_5H_5)_2$, ferrocene, but the ruthenium and osmium analogues have also been investigated, as have some of the mixed carbonyls.

7.3 / Structure and Bonding

The structures of the cyclopentadienyl derivatives contain planar C_5H_5 rings associated with the metal atoms through the flat side of the ring; the derivatives containing two rings exhibit a sandwich-type structure. In the solid state, the carbon rings in ferrocene attain a staggered configuration with respect to each other, but an eclipsed configuration is observed for ruthenocene and osmocene. Curiously, electron-diffraction data on ferrocene in the vapor state show the rings to be eclipsed. All of the C—C bond distances in these compounds are generally equal at ~ 1.42 Å, although in some instances such as $[\pi\text{-}C_5H_5Mo(CO)_3]_2$ there appears to be distortion in the C_5H_5 moiety.

Considerable effort has been expended in describing the electronic structure of the metallocenes using molecular-orbital arguments (73–75). The resulting molecular-orbital diagram for ferrocene, which appears in Fig. 19.16, indicates that the metal $4s$ and $3d_{z^2}$ orbitals interact weakly with the appropriate ring orbitals. The antibonding e_{2g} ligand orbitals interact with the metal $3d_{x^2-y^2}$ and $3d_{xy}$ orbitals, producing slightly bonding molecular orbitals which consist mostly of metal orbitals. This corresponds to a back-donation component in the ligand–metal bond. Ferrocene is an 18-electron system in which the electrons occupy the nine molecular orbitals of lowest energy.

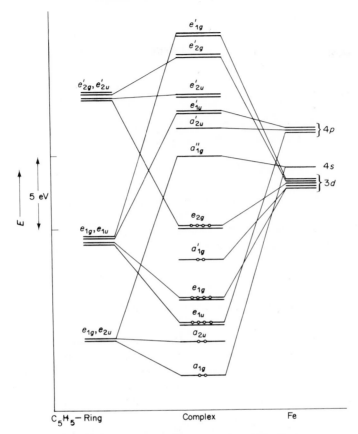

FIG. 19.16 The molecular-orbital diagram for ferrocene.

8 / SIX-ELECTRON LIGANDS (*XVIII*)

The ligands in this class can be divided into two groups: (a) benzene, its derivatives, and other aromatic systems, which are called arene ligands, and (b) other conjugated six-electron ligands such as cycloheptatriene (**35**) and cycloocta-1,3,5-triene (**36**).

      35   **36**

8.1 / Preparation

Bis-arene complexes are prepared most readily using a Friedel–Crafts reaction under reducing conditions

$$3CrCl_3 + 2Al + AlCl_3 + 6C_6H_6 \rightarrow 3[(\pi\text{-}C_6H_6)_2Cr]^+AlCl_4^-. \quad (123)$$

The cationic complexes formed in such reactions can be converted into the neutral compound using a reducing agent

$$Cr(\pi\text{-}C_6H_6)_2^+ + e^- \rightarrow Cr(\pi\text{-}C_6H_6)_2, \quad (124)$$

the most useful being sodium dithionite. This preparative method has been used to form a variety of bis-arene complexes, some of which are described in Table 19.12. Arene carbonyl derivatives can be prepared directly from the metal carbonyl and the arene.

$$Cr(CO)_6 + C_6H_6 \rightarrow (\pi\text{-}C_6H_6)Cr(CO)_3 + 3CO. \quad (125)$$

The reaction occurs readily in a high-boiling ether solvent; ultraviolet irradiation can often be used to advantage.

Neutral bis-arene derivatives of the transition metals are generally highly colored and soluble in organic solvents, sublime readily under vacuum, and are thermally stable to about 300°. In contrast with the metallocenes, they react readily with atmospheric oxygen, being easily oxidized to form cations which undergo disproportionation in aqueous alkaline solution

$$2[(\pi\text{-}C_6H_6)_2Cr]^+ \rightarrow (\pi\text{-}C_6H_6)_2Cr + 2C_6H_6 + Cr^{2+}. \quad (126)$$

In fact, bis-arenechromium compounds are sufficiently good reducing agents to act as electron donors to strong π-acceptor acids such as tetracyano-ethylene (76)

$$(\pi\text{-}C_6H_6)_2Cr + (CN)_4C_2 \rightarrow [(\pi\text{-}C_6H_6)_2Cr]^+[(CN)_4C_2]^-. \quad (127)$$

The neutral arene–metal carbonyls have many of the same physical properties as the bis-arene derivatives. They are, however, much more stable towards oxidation. There are several indications that the $M(CO)_3$ moiety withdraws electrons from the aromatic ring system. For example, $C_6H_5CO_2H$ becomes a weaker acid when it is complexed to a $Cr(CO)_3$ group. Complexed arene rings undergo nucleophilic substitution reactions more readily than do the free ligands (77, 78); correspondingly, the complexed ligands also show reduced reactivity toward electrophiles (79).

TABLE 19.12 The properties of some complexes containing π-benzene ligands[a]

V	Cr	Mn	Fe	Co	Ni
$(\pi\text{-}C_6H_6)_2V$ red-brown (277°)	$(\pi\text{-}C_6H_6)_2Cr$ brown (284°)	$[\pi\text{-}(CH_3)_6C_6]_2Mn^+$ off-white	$[\pi\text{-}(CH_3)_6C_6]_2Fe^+$ deep violet	$[\pi\text{-}(CH_3)_6C_6]_2Co$ red-brown	$[\pi\text{-}(CH_3)_3H_3C_6]Ni$ red-brown
—	$\pi\text{-}C_6H_6Cr(CO)_3$	$\pi\text{-}C_6H_6Mn(CO)_3{}^+$	—	$[\pi\text{-}(C_6H_6)_3Co_3(CO)_2]^+$	—

[a] Melting point given in brackets.

8.2 / Structure and Bonding

There has been considerable controversy concerning the structure of bis-benzenechromium. All investigators are in agreement that the arene rings are eclipsed, but the C—C distances are in dispute. X-Ray data on the solid substance have been reported to indicate that the C—C bonds are (a) all equal within experimental error (*80–83*) and (b) alternately slightly longer and shorter than those in benzene (*84*). Electron-diffraction studies in the vapor phase (*85*) support the former description, but a recent neutron diffraction study indicates that the latter description is correct (*86*).

The stereochemistry of bis-arene metal tricarbonyl depends upon the ring substituents. The available structural information indicates that benzene-

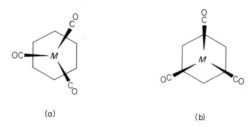

(a)   (b)

FIG. 19.17   The eclipsed (a) and staggered (b) orientations of the tricarbonyl group in metal–arene tricarbonyls.

chromium tricarbonyl and hexamethylbenzenechromium tricarbonyl have staggered configurations, Fig. 19.17, whereas the anisole and *o*-toluidine complexes are eclipsed. There is general agreement that the C—C distances in the ligand are all equivalent (~ 1.40 Å) (*87, 88*).

Complexes formed from conjugated but nonaromatic six-electron ligands show distinctive differences in the C—C distances (Fig. 19.18). In the struc-

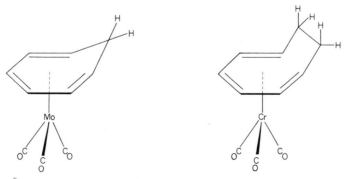

FIG. 19.18   The molecular structures of $C_7H_8Mo(CO)_3$ and $C_8H_{10}Cr(CO)_3$.

tures of both $C_7H_8Mo(CO)_3$ and $C_8H_{10}Cr(CO)_3$ the six carbon atoms of the ligand occupy a plane. The carbon atoms bound to the terminal atoms in the conjugated system are distinctly above this plane, suggesting that the terminal carbon atoms have a degree of p character greater than that expected for sp^2 hybridization.

The molecular-orbital description (73–75) of the bis-benzenechromium derivatives, given in Fig. 19.19, shows some characteristics of that for ferrocene (Fig. 19.16). The differences between the two systems arise because the chromium atomic orbitals are higher than those of iron, but the benzene orbitals are lower than those of the cyclopentadienyl radical. The result is

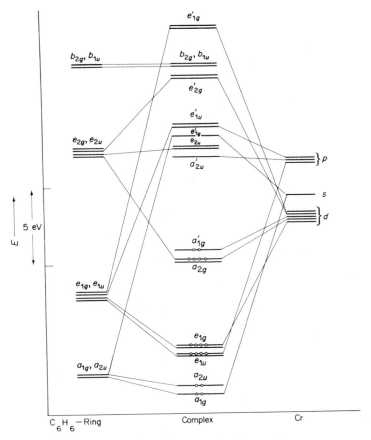

FIG. 19.19 The molecular-orbital diagram for $(C_6H_6)_2Cr$.

that the a_{1g} molecular orbitals contribute less to the metal–ligand bond than those in ferrocene, whereas the e_{1g} molecular orbitals contribute more to the bonding.

9 / SEVEN-ELECTRON LIGANDS (XXI)

The cycloheptatrienyl group (37) has been the most frequently studied seven-electron ligand, but only mixed complexes containing this ligand are

37

known. Complexes containing seven-electron ligands can be prepared by abstracting a hydride ion from cycloheptatriene complexes.

$$(128)$$

38

The neutral cycloheptatriene complex can be obtained again when **38** reacts with strongly basic anions.

The structures of two π-cycloheptatrienyl complexes are known (89, 90). In both instances the C_7 ring is planar with equal C—C distances.

39

Cyclooctatetraene (39) acts as a six-electron ligand when it displaces carbon monoxide from $Mo(CO)_6$

$$Mo(CO)_6 + C_8H_8 \rightarrow \quad \text{[structure]} \quad + 3CO. \qquad (129)$$

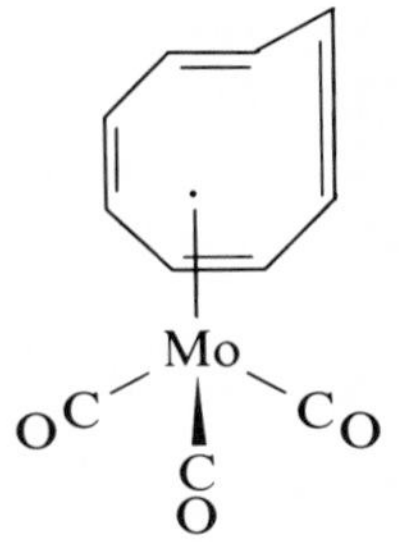

40

The crystal structure of (**40**) shows that one of the carbon–carbon double bonds is nonbonded (*91*). Treatment of this product with acid leads to a complex containing a seven-electron ligand (**41**)

$$\textbf{40} + H^+ \rightarrow \quad \text{[structure]} \qquad (130)$$

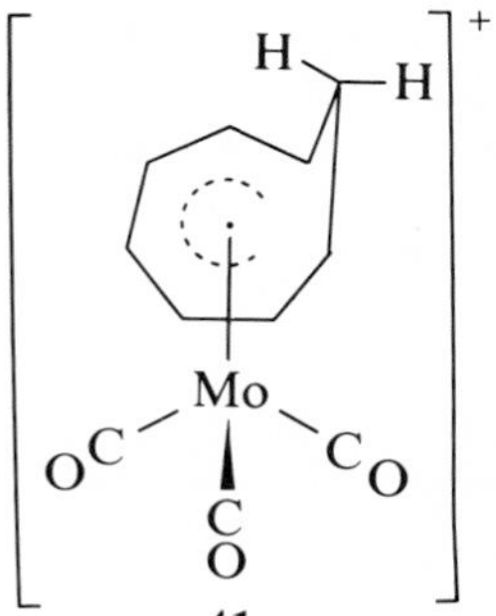

41

10 / FLUXIONAL ORGANOMETALLIC MOLECULES (*XXII*)

Cyclooctatetraene yields complexes with certain transition-metal carbonyls which have some unusual properties. This ligand can act as a 2-, 4-, 6-, or 8-electron donor to transition metals. The 18-electron rule predicts that the complexes of C_8H_8 formed with iron, ruthenium, chromium, molybdenum, and tungsten should have the stoichiometry $C_8H_8M(CO)_3$. In the iron and ruthenium complexes C_8H_8 acts as a four-electron ligand (**42**), whereas it is a 6-electron ligand in the other compounds (**43**). X-Ray studies indicate the validity of this assignment to the respective structures, i.e., the preferred association of the metal carbonyl moieties as indicated schematically by **42** (*92*) and **43** (*93*). However, the proton nmr spectra of these compounds in solution exhibit an interesting temperature dependence. At room temperature the spectra of the iron and ruthenium compounds consist of a single sharp line, contrary to what might be expected from their formulation as **42**. At about $-145°$ the more complex spectrum expected for **40** is observed. Similar phenomena occur for the chromium, molybdenum, and tungsten compounds, but at higher temperature range. That is, the spectrum consistent with the formulation **43** is observed at $-40°$ but consists of a single line in

M = Fe, Ru   M = Cr, Mo, W

42   **43**

the temperature range 50–100°. The available data suggest the existence of a facile exchange process occurring in solution at the higher temperatures, which makes all the protons in the ring equivalent

$$M(CO)_3 \qquad\qquad M(CO)_3 \qquad\qquad\qquad (131)$$

$$M(CO)_3 \qquad\qquad M(CO)_3 \qquad\qquad\qquad (132)$$

In each case detailed analysis of the data suggests that the processes occur by a 1,2-shift mechanism.

Molecules for which several configurations of the constituent atoms are equivalent are called fluxional molecules; since the configurations are equivalent, the molecules of a fluxional system have the same free-energy content. Equations (131) and (132) represent attempts to indicate the nature of the process occurring and should not be interpreted as suggesting that the individual molecular species exist over a long time frame. In this respect fluxional molecules different from tautomers. Molecules with different instantaneous identities are present in a tautomeric equilibrium, in contrast to a fluxional system in which all molecules have the same identities.

Fluxional behavior has been observed in several other organometallic systems (**44**) (*94, 95*) and (**45**); some complexes of the π-allyl moiety also fall into this classification (*102*). In most cases direct structural determination

R

44

$M = \pi\text{-}C_5H_5Fe(CO)_2$ (*96, 98*), $(C_2H_5)_3PCu$ (*97, 100*),

$\sigma\text{-}C_5H_5Hg$ (*97–99*), $\pi\text{-}C_5H_5Cr(NO)_2$ (*101*)

(45)

indicates the existence of the structures expected on the basis of the 18-electron rule, but nmr experiments show the existence of fluxional behavior in solution.

REFERENCES

1. F. Calderazzo and F. A. Cotton, *Inorg. Chem.*, **1**, 30 (1962).

2. R. J. Mawby, F. Basolo, and R. G. Pearson, *J. Amer. Chem. Soc.*, **86**, 3994 (1964).

3. H. H. Jaffé and G. O. Doak, *J. Chem. Phys.*, **21**, 196, 1118 (1953).

4. H. H. Jaffé, *J. Chem. Phys.*, **22**, 1462 (1954).

5. R. B. King, *J. Amer. Chem. Soc.*, **85**, 1918 (1963).

6. D. M. Adams and G. Booth, *J. Chem. Soc.*, 1112 (1962).

7. O. S. Mills and A. D. Redhouse, *Angew. Chem. (Intl. Ed.)*, **4**, 1082 (1965).

8. J. B. Wilford and F. G. A. Stone, *Inorg. Chem.*, **4**, 389 (1965).

9. R. B. King and M. B. Bisnette, *J. Organomet. Chem.*, **2**, 15 (1964).

10. A. Almenningen, A. Haaland, and K. Wahl, *Acta Chem. Scand.*, **23**, 2245 (1969).

11. M. I. Davis and H. P. Hanson, *J. Phys. Chem.*, **69**, 3405 (1965).

12. M. I. Davis and H. P. Hansen, *J. Phys. Chem.*, **71**, 775 (1967).

13. B. Beagley, D. W. J. Cruickshank, P. M. Pinder, A. G. Robiette, and G. M. Sheldrick, *Acta Cryst.*, **B25**, 737 (1969).

14. J. Donohue and A. Caron, *J. Phys. Chem.*, **70**, 603 (1966).

15. H. J. Keller, P. Laubereau, and D. Nöthe, *Z. Naturforsch.* **24B**, 257 (1969).

16. K. Noack, *Helv. Chim. Acta*, **47**, 1555 (1964).

17. K. Noack, *Spectrochim. Acta*, **19**, 1925 (1963).

18. G. Bor, *Spectrochim. Acta*, **19**, 2065 (1963).

19. E. R. Corey and L. F. Dahl, *Inorg. Chem.*, **1**, 521 (1962).

20. C. H. Wei and L. F. Dahl, *J. Amer. Chem. Soc.*, **91**, 1351 (1969).

21. N. E. Erickson and A. W. Fairhall, *Inorg. Chem.*, **4**, 1320 (1965).

22. C. H. Wei, *Inorg. Chem.*, **8**, 2384 (1969).

23. E. R. Corey, L. F. Dahl, and W. Beck, *J. Amer. Chem. Soc.*, **85**, 1202 (1963).

24. A. Sirigu, M. Bianchi, and E. Benedetti, *Chem. Commun.*, 596 (1969).

25. F. Piacenti, M. Bianchi, and E. Benedetti, *Chem. Commun.*, 775 (1967).

26. S. F. A. Kettle and I. Paul, *Advan. Organomet. Chem.*, **10**, 199 (1972).

27. D. L. Smith, *J. Chem. Phys.*, **42**, 1460 (1965).

28. L. Pauling, *Nature of the Chemical Bond*, 3rd ed., Cornell University Press, Ithaca, N.Y., 1960.
29. F. A. Cotton and R. M. Wing, *Inorg. Chem.*, **4**, 314 (1965).
30. R. Colton and C. J. Rix, *Aust. J. Chem.*, **22**, 305 (1968).
31. E. W. Able, M. A. Bennett, and G. Wilkinson, *J. Chem. Soc.*, 2323 (1959).
32. W. D. Horrocks, Jr. and R. C. Taylor, *Inorg. Chem.*, **2**, 723 (1963).
33. V. G. Albano, P. L. Bellon, P. Chini, V. Scatturin, *J. Organomet. Chem.*, **16**, 461 (1969).
34. V. G. Albano, P. Chini, and V. Scatturin, *J. Organomet. Chem.*, **15**, 423 (1968).
35. V. G. Albano, P. L. Bellon, and G. F. Ciani, *Chem. Commun.*, 1024 (1969).
36. V. G. Albano and P. L. Bellon, *J. Organomet. Chem.*, **19**, 405 (1969).
37. S. J. La Placa, W. C. Hamilton, and J. A. Ibers, *Inorg. Chem.*, **3**, 1491 (1964).
38. L. L. Lohr and W. N. Lipscomb, *Inorg. Chem.*, **3**, 22 (1964).
39. R. J. Doedens and L. F. Dahl, *J. Amer. Chem. Soc.*, **87**, 2576 (1965).
40. H. D. Kaez, W. Fellmann, F. R. Wilkes, and L. F. Dahl, *J. Amer. Chem. Soc.*, **87**, 2753 (1965).
41. J. M. Smith, W. Fellmann, and L. H. Jones, *Inorg. Chem.*, **4**, 1361 (1965).
42. U. Anders and W. A. G. Graham, *Chem. Commun.*, 499 (1965).
43. L. B. Handy, P. M. Treichel, L. F. Dahl, and R. G. Hayter, *J. Amer. Chem. Soc.*, **88**, 366 (1965).
44. H. D. Murdoch and E. Weiss, *Helv. Chim. Acta*, **46**, 1588 (1963).
45. R. Cramer, *Inorg. Chem.*, **1**, 722 (1962).
46. J. S. Anderson, *J. Chem. Soc.*, 971 (1934).
47. W. O. Jones, *J. Chem. Soc.*, 1808 (1954).
48. N. C. Baenziger, H. L. Haight, and J. R. Doyle, *Inorg. Chem.*, **3**, 1535 (1964).
49. E. W. Abel, M. A. Bennett, and G. Wilkinson, *J. Chem. Soc.*, 3178 (1959).
50. E. W. Abel, M. A. Bennett, and G. Wilkinson, *J. Chem. Soc.*, 3413 (1959).
51. E. O. Fisher and H. Werner, *Angew. Chem. (Intl. Ed.)*, **2**, 80 (1963); M. A. Bennett, *Chem. Revs.*, **62**, 611 (1962).
52. E. Kuljean and H. Frye, *Z. Naturforsch.*, **19B**, 651 (1964).
53. G. Wilke, *Angew. Chem.*, **72**, 581 (1960).
54. S. E. Manahan, *Inorg. Chem.*, **5**, 2063 (1966).
55. P. R. H. Alderman, P. G. Owston, and M. J. Rowe, *Acta Cryst.*, **13**, 149 (1960).
56. R. W. Turner and E. L. Amma, *J. Amer. Chem. Soc.*, **85**, 4046 (1966).
57. P. Porta, H. M. Powell, R. J. Mawby, and L. M. Venanzi, *J. Chem. Soc.*, A, 455 (1967).
58. R. E. Rundle and J. H. Goring, *J. Amer. Chem. Soc.*, **72**, 5337 (1950).
59. N. C. Baenziger, G. F. Richards, and J. R. Doyle, *Acta Cryst.*, **18**, 924 (1965).
60. R. W. Turner and E. L. Amma, *J. Amer. Chem. Soc.*, **88**, 3243 (1966).
61. M. J. S. Dewar, *Bull. Soc. Chim. France*, **18**, C71 (1951).
62. J. Chatt and L. A. Duncanson, *J. Chem. Soc.*, 2939 (1953).
63. K. F. Watterson and G. Wilkinson, *Chem. Ind.*, 1358 (1960).
64. M. Green, R. B. L. Osborne, A. J. Rest, and F. G. A. Stone, *Chem. Commun.*, 502 (1966).
65. W. E. Oberhansli and L. F. Dahl, *J. Organomet. Chem.*, **3**, 43 (1965).
66. R. Uttech and H. Dietrich, *Z. Krist.*, **122**, 60 (1965).
67. J. Powell, S. D. Robinson, and B. L. Shaw, *Chem. Commun.*, 78 (1965).
68. M. Kh. Minasyan, Yu. T. Struchkov, I. I. Knitskaya, and R. L. Avoyan, *Zh. Strukt. Khim.*, **7**, 903 (1966).
69. J. K. Becconsall, B. E. Job, and S. O'Brien, *J. Chem. Soc.*, A, 423 (1967).
70. O. S. Mills and G. Robinson, *Proc. Chem. Soc. (London)*, 421 (1960).

71. M. R. Churchill and R. Mason, *Proc. Chem. Soc.* (*London*), 226 (1964).
72. J. D. Dunitz, H. C. Mez, O. S. Mills, and H. M. M. Shearer, *Helv. Chim. Acta*, **45**, 647 (1962).
73. E. M. Schustorovich and M. E. Dyatkina, *Doklady Akad. Nauk SSSR*, **128**, 1234 (1959).
74. E. M. Schustorovich and M. E. Dyatkina, *Zh. Strukt. Kim.*, **7**, 139 (1966).
75. F. A. Cotton, *The Chemical Applications of Group Theory*, Interscience, John Wiley and Sons, Inc., New York, 1963.
76. J. W. Fitch, Jr. and J. J. Lagowski, *J. Organomet. Chem.*, **5**, 480 (1966).
77. B. Nichols and M. C. Whiting, *J. Chem. Soc.*, 551 (1959).
78. D. A. Brown and J. R. Raju, *J. Chem. Soc.*, *A*, 40 (1966).
79. R. Riemschneider, O. Becker, and K. Franz, *Monatsh.*, **90**, 571 (1959).
80. J. A. Ibers, *J. Chem. Phys.*, **40**, 3129 (1964).
81. E. Keulen and F. Jellinek, *J. Organomet. Chem.*, **5**, 490 (1966).
82. E. Weiss and E. O. Fischer, *Z. Anorg. Chem.*, **286**, 142 (1956).
83. F. A. Cotton, W. A. Dollase, and J. S. Wood, *J. Amer. Chem. Soc.*, **85**, 1543 (1963).
84. F. Jellinek, *Nature*, **187**, 871 (1960).
85. A. Haaland, *Acta Chem. Scand.*, **19**, 41 (1965).
86. E. F. Förster, G. Albrecht, W. Dürselen, and E. Kurras, *J. Organomet. Chem.*, **19**, 215 (1969).
87. P. Corradinį and G. Allegra, *J. Amer. Chem. Soc.*, **81**, 2271 (1959).
88. M. F. Bailey and L. F. Dahl, *Inorg. Chem.*, **4**, 1314 (1965).
89. G. Engebretson and R. E. Rundle, *J. Amer. Chem. Soc.*, **85**, 481 (1963).
90. G. Allegra and G. Perego, *Ric. Sci*, **1A**, 362 (1961).
91. J. S. McKechnie and I. C. Paul, *J. Amer. Chem. Soc.*, **88**, 5927 (1966).
92. B. Dickens and W. N. Lipscomb, *J. Amer. Chem. Soc.*, **83**, 4862 (1961).
93. B. Dickens and W. N. Lipscomb, *J. Chem. Phys.*, **37**, 2084 (1962).
94. R. B. King and A. Fronzaglia, *J. Amer. Chem. Soc.*, **88**, 709 (1966).
95. F. A. Cotton and M. D. LaPrade, *J. Amer. Chem. Soc.*, **90**, 5418 (1968).
96. F. A. Cotton and T. J. Marks, *J. Amer. Chem. Soc.*, **91**, 7523 (1969).
97. G. Wilkinson and T. S. Piper, *J. Inorg. Nucl. Chem.*, **2**, 32 (1956).
98. M. J. Bennett, Jr., F. A. Cotton, A. Davison, J. W. Faller, S. J. Lippard, and S. M. Morehouse, *J. Amer. Chem. Soc.*, **88**, 4371 (1966).
99. G. Wilkinson and T. S. Piper, *J. Inorg. Nucl. Chem.*, **3**, 104 (1956).
100. G. M. Whitesides and J. S. Flemming, *J. Amer. Chem. Soc.*, **89**, 2855 (1967).
101. F. A. Cotton, A. Musco, and G. Yagupsky, *J. Amer. Chem. Soc.*, **89**, 6136 (1967).
102. J. Powell and B. L. Shaw, *J. Chem. Soc.*, *A*, 1839 (1967).

COLLATERAL READINGS

I. P. R. Mitchell and R. V. Parish, *J. Chem. Ed.*, **46**, 811 (1969). A discussion of applications of the 18-electron rule.
II. M. F. Lappert and B. Prokai, *Advan. Organomet. Chem.*, **5**, 225 (1967); N. S. Vyazankin, G. A. Razuvaev, and O. A. Kruglaya, *Organomet. Chem. Revs.*, **3**, 323 (1968). Both reviews contain information on insertion reactions of metal and metalloid compounds with olefins.
III. R. F. Heck, *Advan. Organomet. Chem.*, **4**, 1 (1966). A review of the synthesis and reactions of alkyl- and arylcobalt tetracarbonyls.
IV. E. W. Abel and F. G. A. Stone, *Quart. Revs.*, **23**, 325 (1969). A survey of the structural considerations of the transition-metal carbonyls.

V. L. M. Haines and M. H. B. Stiddard, *Advan. Inorg. Chem. Radiochem.*, **12**, 53 (1969). A discussion of the vibrational spectra of transition-metal carbonyl complexes and the use of such information to elucidate structures.

VI. B. P. Biryukov and Yu. T. Struchkov, *Russ. Chem. Revs.*, **39**, 789 (1970). A review of metal–metal interactions in π complexes of transition metals. The discussion is based on experimental structural parameters.

VII. G. R. Dobson, I. W. Stolz, and R. K. Sheline, *Advan. Inorg. Chem. Radiochem.*, **8**, 1 (1966). A discussion of the substitution products formed when metal hexacarbonyls react with various ligands.

VIII. R. B. King, *Advan. Organomet. Chem.*, **2**, 157 (1964); M. I. Bruce and F. G. A. Stone, *Angew. Chem. (Intl. Ed.)*, **7**, 747 (1968). Both reviews contain information on the reactions of metal carbonylate anions.

IX. M. L. H. Green and D. J. Jones, *Advan. Inorg. Chem. Radiochem.*, **7**, 115 (1965). This review on the hydride complexes of transition metals contains a section on the carbonyl hydrides.

X. B. F. G. Johnson and J. A. McCleverty, *Prog. Inorg. Chem.*, **7**, 277 (1966). A discussion of the preparation, reactions, properties, and structures of NO complexes of the transition metals.

XI. R. Jones, *Chem. Revs.*, **68**, 785 (1968); H. W. Quinn and J. H. Tsai, *Advan. Inorg. Chem. Radiochem.*, **12**, 217 (1969). Both reviews deal with the olefin complexes of the transition metals.

XII. R. G. Guy and B. L. Shaw, *Advan. Inorg. Chem. Radiochem.*, **4**, 77 (1962). A discussion of olefin, acetylene, and π-allylic complexes of the transition metals.

XIII. M. L. H. Green and P. L. I. Nagy, *Advan. Organomet. Chem.*, **2**, 325 (1964); M. I. Lobach, B. D. Babitskii, and V. A. Kormer, *Russ. Chem. Revs.*, **36**, 476 (1967). Both reviews deal with the properties and structures of π-allyl complexes of the transition metals.

XIV. G. Wilke, B. Bogdanovic, P. Hardt, P. Heimbach, W. Keim, M. Kröner, W. Oberkirch, K. Tanaka, E. Steinrücke, D. Walter, and H. Zimmermann, *Angew. Chem. (Intl. Ed.)*, **5**, 151 (1966). This paper contains a review of the chemistry of metal-allyl complexes.

XV. M. R. Churchill and R. Mason, *Advan. Organomet. Chem.*, **5**, 93 (1967). A survey of structural chemistry of organotransition-metal complexes.

XVI. G. Wilke, *Angew. Chem. (Intl. Ed.)*, **2**, 105 (1963). A survey of olefin–metal π-complexes as intermediates in polymerization processes.

XVII. P. M. Maitlis, *Advan. Organomet. Chem.*, **4**, 95 (1964); R. Pettit, *Pure Appl. Chem.*, **17**, 253 (1968). Both reviews deal with the preparation, structure, and bonding of cyclobutadiene complexes of transition metals.

XVIII. E. O. Fischer and H. P. Fritz, *Advan. Inorg. Chem. Radiochem.*, **1**, 55 (1959); G. Wilkinson and F. A. Cotton, *Prog. Inorg. Chem.*, **1**, 1 (1959). Both reviews contain extensive information on the preparation, properties, and structures of cyclopentadienyl and arene derivatives of transition metals.

XIX. J. Birmingham, *Advan. Organomet. Chem.*, **2**, 365 (1964); W. F. Little, *Survey Prog. Chem.*, **1**, 133 (1963). Both reviews deal exclusively with the synthesis and properties of ferrocenes.

XX. K. Plesske, *Angew. Chem. (Intl. Ed.)*, **1**, 312, 349 (1962). A discussion of the reactions of the cyclopentadienyl ring in ferrocene.

XXI. M. A. Bennett, *Advan. Organomet. Chem.*, **4**, 353 (1966). A survey of the π-complexes formed by seven- and eight-membered ring ligands.

XXII. F. A. Cotton, *Acc. Chem. Res.*, **1**, 257 (1968). A discussion of the fluxional behavior of transition-metal π-complexes.

STUDY QUESTIONS

1. Predict the magnetic behavior of the following compounds: $Fe(CO)_5$, $C_5H_5Mn(CO)_3$, $Fe(CO)_2(NO)_2$, $Cr(CO)_3(PR_3)_3$, $Cr(C_6H_6)_2$.

2. Discuss the factors which make phosphines and arsines more effective than amines in stabilizing low oxidation states of transition metals.

3. Discuss the observation that carbon monoxide can be replaced from $Ni(CO)_4$ by PF_3 or $SbCl_3$, but no reaction occurs with NCl_3, PF_5, or $SbCl_5$.

4. Explain the observation that $HCo(CO)_4$ is a markedly stronger acid than $HCo(CO)_3PR_3$.

5. From the limited experimental data available on transition-metal–carbon compounds it appears that the order of thermal stability of metal–carbon bonds increases in the order: $M-CH_3 < M-C_6H_5 < M-\overset{\displaystyle O}{\overset{\|}{C}}-CH_3 < M-CF_3$. Discuss possible factors which may account for this order.

6. Discuss a possible method of preparation, the structure, and the theoretical bonding aspects for the following substances: (a) the chromium carbonyl derivative of borazine; (b) the iron carbonyl derivative of anthracene; (c) $C_8H_8PtI_2$.

7. The reaction between arenes and $Mo(CO)_6$ in diglyme is first-order with respect to the carbonyl. (a) Suggest two possible mechanisms for this reaction. (b) How would you distinguish experimentally between these possible mechanisms?

8. There are two easily interconvertible isomers with the empirical formula $CH_3Co(CO)_4$. (a) Give the probable structures of the two isomers. (b) Experimentally, how would you distinguish between these isomers?

9. Iridium forms complexes containing molecular oxygen, for example $IrCl_2[P(C_6H_5)_3]_2O_2$ and $Ir(CO)(PR_3)_2ClO_2$, that can exchange molecular oxygen rapidly with their surroundings. (a) Describe the possible constitution of each of the products, as well as the deoxygenated species. (b) Discuss the possible ways in which oxygen may be attached in these complexes. (c) What is the electronic configuration of the deoxygenated species?

10. Discuss the structure and bonding of the products of the following reactions:

(a) $C_4H_6 + PdCl_2 \rightarrow C_8H_{12}Cl_2Pd_2Cl_2$

(b) $(C_6H_5)_3C_3Br + Ni(CO)_4 \rightarrow [(C_6H_5)_3C_3]_2Ni_2Br_2(CO)_2$

(c) $C_6H_8 + Fe_3(CO)_{12} \xrightarrow{CsF} C_6H_8Fe(CO)_3$

(d) $C_5F_6 + Co_2(CO)_8 \rightarrow C_5F_6Co_2(CO)_8$

11. Discuss the common features found in the bonding of three-, five-, and seven-electron ligands to transition-metal atoms.

12. Sketch the π component that is postulated to exist between the metal and the one-electron ligand in compounds containing the $M-CF_3$ moiety.

13. Give examples of transition-metal complexes containing the following organic ligands: (a) a conjugated four-electron ligand; (b) a five-electron ligand; (c) an unconjugated four-electron ligand; (d) a one-electron ligand.

14. When one mole of iron pentacarbonyl reacts with bromine, one mole of carbon monoxide is liberated; only one other product is formed. Discuss the stereochemistry of the iron-containing product.

15. Describe the structures of the following species: (a) $Co(CO)_4^-$; (b) $Mn(CO)_5^-$; (c) $Ni_4(CO)_9^{2-}$; (d) $Cr_2(CO)_6(OH)_3H^{2-}$; (e) $HFe(CO)_4^-$.

INDEX

NOTES

NOTES

NOTES

NOTES

NOTES

NOTES

NOTES

NOTES